Nucleic Acid Testing
for Human Disease

Nucleic Acid Testing for Human Disease

Edited by
Attila Lorincz

CRC Press
Taylor & Francis Group
Boca Raton London New York

CRC Press is an imprint of the
Taylor & Francis Group, an **informa** business

A TAYLOR & FRANCIS BOOK

CRC Press
Taylor & Francis Group
6000 Broken Sound Parkway NW, Suite 300
Boca Raton, FL 33487-2742

© 2006 by Taylor & Francis Group, LLC
CRC Press is an imprint of Taylor & Francis Group, an Informa business

First issued in paperback 2019

No claim to original U.S. Government works

ISBN 13: 978-0-367-45354-1 (pbk)
ISBN 13: 978-1-57444-543-5 (hbk)

Library of Congress Card Number 2005046731

Library of Congress Cataloging-in-Publication Data

Nucleic acid testing for human disease / edited by Attila T. Lorincz.
    p. cm.
  Includes bibliographical references and index.
  ISBN 1-57444-543-X (alk. paper)
  1. Nucleic acids--Diagnostic use. I. Lorincz, Attila T.

RB43.8.N83N83 2006
616.07'56--dc22   2005046731

**Visit the Taylor & Francis Web site at
http://www.taylorandfrancis.com**

**and the CRC Press Web site at
http://www.crcpress.com**

Dedication

I dedicate this book to my wife Joan and my children Eamon and Ilona. Their love and support have sustained me throughout the years and enabled me to focus my energy to undertake this task.

Preface

Nucleic acid tests (NATs) represent a "grab bag" of highly diverse methods that detect and characterize properties of deoxyribonucleic acids (DNAs) and ribonucleic acids (RNAs) in their various forms. Despite the whimsical-sounding name, these NATs are not pesky; they constitute highly sophisticated and remarkably important medical advances that have progressed from relative obscurity in the 1980s to major players in the repertoires of most clinical testing laboratories today.

Tens of millions of routine NATs are now performed for clinical diagnosis, with millions more conducted in research settings. From the perspective of current and expected future benefits to human healthcare, their value is immeasurable. In fact, NATs have the potential in a matter of a few decades to become the predominant molecular methods for testing and monitoring human health.

This book brings together a diverse group of leading researchers from academia, clinical testing laboratories, and industry who describe NAT technologies and related clinical applications. The intent is to give readers sufficient detail of the myriad techniques to allow an understanding of how tests are performed and interpreted, and to present the do's and don'ts of clinical use.

The first part of the book emphasizes aspects of the technology. It includes seven chapters on specific laboratory techniques with supplementary matter on test applications. The techniques covered include target- and signal-amplification-based NAT procedures, microarrays, bead-based multiplex assays, *in situ* hybridization, single nucleotide polymorphism (SNP) analysis, and RNA expression profiling. The final chapter of this section focuses on laboratory issues, including the pressing need for proper clinical validation of tests intended for routine use on humans.

The second part of the book has a stronger focus on specific applications of NATs to diagnose and treat human diseases such as infections, cancers, and inherited disorders. The chapters run the gamut from viral, bacterial, and fungal detection to human genetic exploration by SNP determination, patterns of RNA expression, and the growing relevance of epigenetic changes such as methylation. To round out the book, we include a forward-looking chapter on where the field is seen to be moving from an academic research perspective and an overview of test validation issues.

Good medicine starts with an accurate diagnosis, and it is now abundantly clear that favorable outcomes in medicine are tied to robust NATs. It is hoped that this book will promote understanding of the tremendous and increasing impact of NATs in medicine, and thereby play a role in speeding the evolution of these exciting tools to further improve the human condition.

Attila T. Lorincz, Ph.D.
Gaithersburg, Maryland, USA

Acknowledgments

I am deeply grateful for the assistance, support, and influence of many friends and colleagues who made this book possible. Special thanks to Jennifer Smith, my co-author of Chapter 10, and to the authors of the individual chapters, all of whom took substantial time from their busy lives to write excellent and thought-provoking reviews of their respective fields. Allison Cullen, Irina Nazarenko, Renee Howell, Richard Obiso, Paul Eder, and others on my research and development team helped "hold the fort" while I was engrossed in the myriad distracting details of executing this work.

Mark Schiffman, Eduardo Franco, Ralph Richart, Richard Reid, and other colleagues in my field who appreciate the complexities of the quest for scientific truth have influenced my thinking and research work over the years. Evan Jones and Charles Fleischman provided a conducive environment, and Katherine Mack communicated with all the contributors and the publishing house, and worked tirelessly on the minutiae of the manuscripts to the very end.

Editor

Attila T. Lorincz, Ph.D., has extensive expertise in the genetic probe field and in the pathogenesis and molecular biology of human papillomavirus (HPV). He lectures worldwide and has edited and/or authored numerous reviews, book chapters, and over 125 original research papers in peer-reviewed journals. He is first author of a paper chosen in 2003 as one of 12 classic papers published during the past 50 years in *Obstetrics & Gynecology* (1992; 79: 328–337). This paper discussing the cancer risks of 15 HPV types is the most frequently cited paper ever to appear in the journal, with well over 500 citations to date.

The research Dr. Lorincz has directed in the HPV field since 1984 has led to the cloning and characterization of clinically important virus types, provided a clearer understanding of the natural history of HPV infection, and revealed key evidence of the association of HPV with cervical cancer. Several United States and international patents concerning HPV types 35, 43, 44, and 56 have been awarded for this work. Dr. Lorincz is an inventor of the Hybrid Capture HPV DNA test now recognized as the standard of care for ASC-US triage in the United States and of growing importance for routine cervical cancer prevention worldwide.

Dr. Lorincz earned his doctorate in genetics from Trinity College in Dublin, Ireland. His research career has included two postdoctoral fellowships at the University of California (UC) at San Diego and Santa Barbara. At UC, he made significant contributions to the field of molecular biology and cell cycle control, including cloning and characterization of the first known *cdk* gene described in a paper in *Nature* in 1984. He subsequently moved to industry but has retained many academic affiliations.

He has served as an adjunct associate professor at the Georgetown University Medical School in Washington, D.C., and as a lecturer at The Johns Hopkins University in Baltimore, Maryland. Dr. Lorincz was the recipient of the 1994 American Venereal Disease Association Achievement Award for outstanding contributions to the field of sexually transmitted disease research.

Contributors

Leen J. Blok
Erasmus MC
University Medical Center Rotterdam
Rotterdam, The Netherlands

Adam M. Bressler
Emory University
Atlanta, Georgia, USA

Antoinette A.T.P. Brink
Vrije Universiteit Medical Center
Department of Pathology
Amsterdam, The Netherlands

Thomas B. Broudy
Affymetrix, Inc.
Santa Clara, California, USA

Curt W. Burger
Erasmus MC
University Medical Center Rotterdam
Rotterdam, The Netherlands

Darrell P. Chandler
Argonne National Laboratory
Biochip Technology Center
Argonne, Illinois, USA

Lorraine Comanor
Independent Research Consultant
Palo Alto, California, USA

Manel Esteller
Spanish National Cancer Research Center
Molecular Pathology Program
Cancer Epigenetics Laboratory
Madrid, Spain

Paolo Fortina
Thomas Jefferson University
Center for Translational Medicine
Philadelphia, Pennsylvania, USA

Payman Hanifi-Moghaddam
Erasmus MC
University Medical Center Rotterdam
Rotterdam, The Netherlands

Madhuri Hegde
Baylor College of Medicine
Department of Molecular & Human Genetics
Houston, Texas, USA

Theo J.M. Helmerhorst
Erasmus MC
University Medical Center Rotterdam
Rotterdam, The Netherlands

David A. Hendricks
Independent Consultant
Berkeley, California, USA

W. Mathias Howell
Uppsala University
Department of Genetics & Pathology
Molecular Medicine
Uppsala, Sweden

Mel Krajden
British Columbia Centre for Disease Control
University of British Columbia
Vancouver, British Columbia, Canada

Larry J. Kricka
University of Pennsylvania Medical Center
Department of Pathology & Laboratory
  Medicine
Philadelphia, Pennsylvania, USA

Ulf Landegren
Uppsala University
Department of Genetics & Pathology
Molecular Medicine
Uppsala, Sweden

Chatarina Larsson
Uppsala University
Department of Genetics & Pathology
Molecular Medicine
Uppsala, Sweden

Ricardo V. Lloyd
Mayo Clinic
Department of Laboratory Medicine &
  Pathology
Rochester, Minnesota, USA

Attila T. Lorincz
Digene Corporation
Gaithersburg, Maryland, USA

Chris J.L.M. Meijer
Vrije Universiteit Medical Center
Department of Pathology
Amsterdam, The Netherlands

Christine J. Morrison
Centers for Disease Control & Prevention
National Center for Infectious Diseases
Mycotic Diseases Branch
Atlanta, Georgia, USA

Stephen A. Morse
Centers for Disease Control & Prevention
Bioterrorism Preparedness & Response
  Program
Atlanta, Georgia, USA

Hubert G.M. Niesters
Erasmus MC
University Medical Center Rotterdam
Rotterdam, The Netherlands

Mats Nilsson
Uppsala University
Department of Genetics & Pathology
Molecular Medicine
Uppsala, Sweden

Anne Harwood Peruski
Indiana University School of Medicine
Gary, Indiana, USA

Leonard F. Peruski, Jr.
Centers for Disease Control and Prevention
International Emerging Infections Program
Bangkok, Thailand

Gary W. Procop
Cleveland Clinic
Cleveland, Ohio, USA

Elisabeth Puchhammer-Stöckl
Medical University of Vienna
Institute of Virology
Vienna, Austria

Xiang Qian
Mayo Clinic
Department of Laboratory Medicine &
  Pathology
Rochester, Minnesota, USA

Sue Richards
Oregon Health Sciences University
Department of Molecular & Medical Genetics
Portland, Oregon, USA

Santiago Ropero
Spanish National Cancer Center
Molecular Pathology Program
Madrid, Spain

Julius Schachter
University of California, San Francisco
Chlamydia Research Laboratory
San Francisco, California, USA

Martin Schutten
Erasmus MC
University Medical Center Rotterdam
Rotterdam, The Netherlands

Jennifer S. Smith
University of North Carolina
Department of Epidemiology
Chapel Hill, North Carolina, USA

Peter J.F. Snijders
Vrije Universiteit Medical Center
Department of Pathology
Amsterdam, The Netherlands

Johan Stenberg
Uppsala University
Department of Genetics & Pathology
Molecular Medicine
Uppsala, Sweden

Janet A. Warrington
Affymetrix, Inc.
Santa Clara, California, USA

Belinda Yen-Lieberman
Cleveland Clinic
Cleveland, Ohio, USA

Contents

PART I   Nucleic Acid Diagnostic Technology

PART II Disease Applications of Nucleic Acid Tests

Part I

Nucleic Acid Diagnostic Technology

1 Target Amplification-Based Techniques

Antoinette A.T.P. Brink, Peter J.F. Snijders, and Chris J.L.M. Meijer

CONTENTS

INTRODUCTION

Target amplification-based techniques — in the broadest sense — are molecular biological methods that utilize nucleic acid polymerases, target-specific oligonucleotides (primers), and a mixture of four (deoxy)ribonucleotides to amplify a specific nucleic acid sequence up to a level at which it can be detected, analyzed, or manipulated. The different target amplification-based techniques can be distinguished by the reaction components that are present in addition to the above-mentioned essential three components and by other reaction conditions such as temperature.

Two commercially available target amplification-based techniques that are most widely used in molecular diagnostics are the polymerase chain reaction [PCR, commercialized mainly by Roche (Basel, Switzerland)] and nucleic acid sequence-based amplification [NASBA, commercialized by bioMérieux (Boxtel, The Netherlands)]. Principles, advantages, and disadvantages of these two and several other target amplification-based techniques will be discussed in this chapter. In addition, we will describe several methods for detection of amplification reaction products.

CYCLIC AMPLIFICATION METHODS

The most commonly used type of target amplification-based techniques are those that rely on the property of DNA to become single-stranded (denatured) when heated and double-stranded again (renatured) when cooled down.

POLYMERASE CHAIN REACTION

PCR was the first cyclic DNA target amplification method invented.[1] It is still the most widely used technique (see Figure 1.1). The essential components of the reaction include, besides the target DNA sample, DNA polymerase, magnesium ions that serve as cofactors for the polymerase, a mixture of the deoxyribonucleotides dATP, dCTP, dGTP, and dTTP or equivalent base analogues, a buffering agent such as Tris-HCl, and several picomoles of each of two specific oligonucleotide primers.

The primers usually are 17 to 25 nucleotides long. One of the primers is the so-called forward, sense, upstream, or plus primer that will bind to the antisense strand of the target DNA; the other primer is called the reverse, antisense, downstream, or minus primer and will bind to the sense strand of the target DNA. The orientation of the primers relative to each other is such that if they are extended at their 3' ends by a DNA polymerase, the newly generated product will serve as a target for binding by the other primer.

The size of the region flanked by the two primers can be anywhere between zero bases and several kilobases, depending on the purpose of the PCR, the type of target material, and the capacity of the DNA polymerase used. For example, for sensitive detection of microorganisms in clinical specimens — especially when the DNA quality is suboptimal — the distance between the two primers should be fewer than 200 base pairs and amplification can then be achieved using a standard DNA polymerase. By contrast, long-distance DNA polymerases should be used, for example, in the process of characterization or cloning of unknown translocation partners of immunoglobulin genes.[2]

In the early days of PCR use, regular DNA polymerases became inactive at the denaturation temperature, which meant that fresh DNA polymerase had to be added following each denaturation step. The resulting accumulation of protein in the reaction markedly decreased reaction efficiency. The isolation of a thermostable DNA polymerase from *Thermus aquaticus* bacteria (*Taq* polymerase)[3] was the solution to this problem. In addition, the optimum elongation temperature for this enzyme is 72°C — much higher than the optimum temperature for conventional DNA polymerases. This allows increased primer annealing temperatures, and thus greatly enhances the specificity of the amplification.

As soon as *Taq* polymerase was introduced, use of PCR increased remarkably. Nowadays, thermostable DNA polymerases of other bacteria are also available for this purpose. The best known

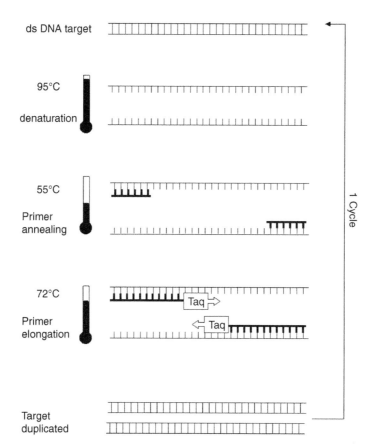

ds DNA target

95°C
denaturation

55°C
Primer
annealing

72°C
Primer
elongation

Taq

Taq

Target
duplicated

1 Cycle

FIGURE 1.1 Principle of PCR. One PCR cycle consists of three events. The target DNA (top) is made single-stranded by heating (usually to a temperature of 94°C or higher). Subsequent cooling to a temperature approaching the Tm of the primers allows the primers to anneal to their specific sequences on both ends of the region to be amplified. The DNA polymerase can then extend the primers using the target DNA strands as templates. In the next cycle, the original target DNA and the newly generated strands become targets for primer annealing and extension; the process becomes exponential (bottom).

of these are *Pfu* DNA polymerase (from *Pyrococcus furiosus*) and Vent$_D^®$ (New England Biolabs, Ipswich, Massachusetts, USA, http://www.neb.com) DNA polymerase (from *Thermococcus litoralis*). *Pfu* polymerase has a very low base incorporation error rate and therefore serves as the polymerase of choice when the PCR product must be free of point mutations. Vent polymerase has a half-life at 95°C that exceeds the half-life of *Taq* polymerase by a factor of 4 and can therefore be used for amplification of target sequences with high numbers of secondary structures. However, DNA polymerases other than *Taq* generally are more difficult to handle and more expensive.

Other cyclic amplification methods such as the ligase chain reaction (LCR) rely on DNA denaturation and renaturation. Some of these methods were developed for specific purposes that could not be covered by a standard PCR.

LIGASE CHAIN REACTION

LCR uses both a thermostable DNA polymerase and a DNA ligase.[4] In addition, the system involves two adjacent oligonucleotides that hybridize to the target DNA and are linked by the ligase only when they are perfectly base-paired at the junction. The ligation product is amplified exponentially by cycling in the presence of a second set of adjacent oligonucleotides complementary to the first. Hence, like PCR, LCR requires a thermocycler and each cycle results in a duplication of the target nucleic acid molecule. It should be noted that a single mismatch of the oligonucleotides with their targets will prevent ligation and amplification — a feature that renders this method particularly useful, for example, for detecting specific point mutations.

AMPLIFICATION OF RNA: REVERSE TRANSCRIPTASE POLYMERASE CHAIN REACTION

PCR can only use DNA as a template. Therefore, if detection of RNA is desired, the target RNA first must be converted to DNA. This is accomplished using an RNA-dependent DNA polymerase called reverse transcriptase (RT) as found in retroviruses. RT will elongate a DNA primer annealed to the RNA sequence that needs to be amplified, using the RNA strand as a template, to generate cDNA that can be used as a target in the PCR. From the bacteria *Thermus thermophilus*, a heat-stable DNA polymerase that can act as reverse transcriptase was isolated,[5] allowing reverse transcription and DNA amplification steps to be performed in the same reaction. Moreover, the ability of such an enzyme to act at high temperatures allows easy reverse transcription of RNAs with high numbers of secondary structures. Unfortunately, the error rate of this polymerase is relatively high. An alternative one-step RT-PCR system containing a mixture of enzymes has been commercialized by Roche (Basel, Switzerland).

ISOTHERMAL AMPLIFICATION METHODS

Unlike the cyclic target amplification-based techniques, isothermal amplification methods are conducted at constant temperatures approaching the optimums for the polymerases involved.

Because of the absence of distinct cycles, the kinetics of isothermal amplification methods is quite different from those of cyclic amplification methods. For example, the kinetics cannot be described by a reaction equation involving a $\log_2$ amplification of target. Although real-time detection methods (see below) can be applied, direct quantification is much more difficult than with real-time PCR methods, for example.

NUCLEIC ACID SEQUENCE-BASED AMPLIFICATION

NASBA,[6] based on the so-called self-sustaining sequence replication (3SR) method (reviewed by Mueller et al.[7]), is the most widely used isothermal amplification method. This type of reaction involves three enzymes: avian myeloblastosis virus reverse transcriptase (AMV-RT), RNAseH, and T7 RNA polymerase. At a reaction temperature of 41°C, these enzymes act together with a pair of specific primers to generate a large amount of reaction product consisting of single-stranded antisense RNA molecules (Figure 1.2). Without additional denaturation steps, NASBA is suitable for specific amplification of RNA even in a background of genomic DNA[6,8] because the highest temperature reached during the process is 65°C to stretch the target RNA and allow primer annealing prior to addition of the enzymes. This temperature is insufficiently high to allow denaturation of DNA.

Because of the nature of the reaction product, this method is mostly used for detection purposes and not for further sequence analysis and/or cloning of the reaction products. NASBA is used in particular for the detection of microorganisms with RNA genomes in non-histological clinical samples, for example, detecting human immunodeficiency virus (HIV) in whole blood[9] and enteroviruses in feces.[10] In studies of microorganisms with DNA genomes, NASBA can be used to specifically detect infections in which microorganisms are transcriptionally active, and this method is probably more relevant clinically than the detection only of their DNA genomes. This technique has been described for detecting *Chlamydia trachomatis* in cervical smears and urine samples and human papillomavirus (HPV) in cervical smears.[11] NASBA can also be used for the detection of unspliced transcripts in a background of genomic DNA.[12]

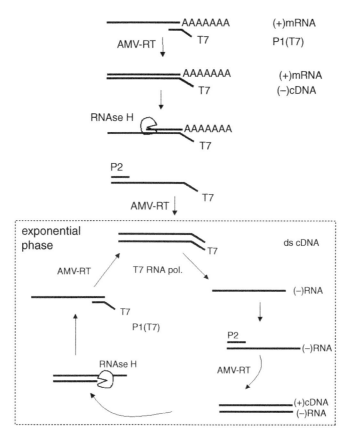

FIGURE 1.2 Principle of NASBA. The reverse primer P1 that carries a T7 RNA polymerase recognition signal anneals to its specific sequence at the 3' end of the target mRNA (top). Avian myeloblastosis virus reverse transcriptase (AMV-RT) elongates primer P1 from its 3' end, using the mRNA as a template to form the first strand of (–)cDNA. RNAseH then removes the original mRNA strand to allow annealing of the forward primer P2 to the cDNA. Primer P2 is elongated by AMV-RT to form the second cDNA strand. The T7 signal is now double-stranded, allowing recognition by T7 RNA polymerase, which starts to transcribe the double-stranded cDNA into single-stranded antisense RNA molecules. These are targets for primer P2 and will serve as templates for AMV-RT to generate (+)cDNA (inset). RNAseH removes the antisense RNA strand to allow annealing of primer P1, which is elongated at its 3' end to generate the second cDNA strand. Simultaneously, the upper cDNA strand is elongated at its 3' end to generate a double-stranded T7 signal. From this point on, amplification is exponential.

STRAND DISPLACEMENT AMPLIFICATION

The low reaction temperature of NASBA[13] and consequently its specificity for RNA are disadvantageous when DNA amplification is desired. DNA amplification would require either additional denaturation steps to precede the actual amplification reaction or the use of polymerases with strand-displacement activity. The latter option is used in the strand-displacement amplification (SDA) method that relies on extensions of the 3' ends at nicks and displacement of the downstream DNA strand by an exonuclease-deficient Klenow DNA polymerase. Exponential amplification results from coupling sense and antisense reactions in which strands displaced from a sense reaction serve as targets for an antisense reaction and vice versa. The reaction is carried out at 37°C and the reaction product consists of double-stranded DNA that can be analyzed as a PCR product.

Loop-Mediated Isothermal Amplification

Loop-mediated isothermal amplification (LAMP)[14] was developed to overcome the relatively low target specificities of NASBA and SDA inherent to their low incubation temperatures. LAMP is carried out at 65°C and employs *Bst* DNA polymerase, which has high strand displacement activity, and a set of four primers (two inner and two outer) that together recognize six distinct sequences on the target DNA. The latter is thought to enhance reaction specificity as compared, for example, to NASBA in which only two primers are used. The LAMP reaction relies on a combination of cDNA synthesis, strand displacement DNA synthesis, and formation of stem–loop DNA by self-primed DNA synthesis. The final reaction product is basically a mixture of stem–loop DNAs with various stem lengths and cauliflower-like structures with multiple loops.

DESIGN OF AMPLIFICATION PRIMERS

General Conditions for Amplification Primers

Amplification primer design can be performed with commercially available software (e.g., Oligo, [Molecular Biology Insights, Cascade, Colorado, USA, http://www.oligo.net], Vector NTI [Invitrogen, http://www.invitrogen.com], Primer Express [Applied Biosystems, Foster City, California, USA, http://www.appliedbiosystems.com]) or shareware on the internet (e.g., Primer3, http://frodo.wi.mit.edu/primer3/primer3_code.html). These programs calculate the optimal primer pair for the sequence to be amplified, taking into account sequence-specific parameters such as GC and repeat contents and user-defined parameters such as desired amplicon lengths. For most amplification methods, primers are approximately 20 bases in length and have GC contents approaching 50%. Furthermore, they should not form internal secondary structures such as hairpin loops and the 3′ ends of the two primers should not be complementary because this would lead to the generation of primer dimers.

While most amplification primer selection can be achieved by computer analysis (*in silico*), some experimental work may be necessary to fully optimize new primers. For example, the standard concentration of magnesium ions in a PCR is 1.5 mM but it may be wise to run a few PCRs with different magnesium concentrations when starting with a new primer pair. Likewise, the annealing temperature and the ramping time (in the case of PCR) may need some optimization.

Application-Specific Conditions for Primers

Most other conditions taken into account during primer design depend on the amplification method, the type and size of sequence to be amplified, and the type of (clinical) material from which the sequence is to be amplified. First, the region flanked by the two primers should neither exceed the capacity of the polymerase used in the reaction nor the fragment lengths of the nucleic acids in the sample. For example, detection of a specific DNA sequence in formalin-fixed, paraffin-embedded material by PCR requires two primers that are preferentially no more than 200 base pairs (bp) apart, because the DNA in this type of material is fragmented to a certain degree.

Moreover, the primers may need modifications depending on the amplification method. For example, the antisense primer in a NASBA reaction must contain a T7 RNA polymerase recognition signal at its 5′ end (which should not form secondary structures with the hybridizing part of the primer), followed by a purine-rich sequence to prevent abortive transcription, and it should preferentially have an adenosine at its 3′ end to increase reverse transcription efficiency.

The specificity of the amplification reaction is dependent on the primer sequence relative to the target. When possible, exactly matching primers should be selected. The exact match with their target sequences — and, importantly, the absence of possible matches with irrelevant sequences — should be confirmed by screening the primers against a sequence database, for example, using the basic local alignment search tool (BLAST, http://www.ncbi.nlm.nih.gov/BLAST). By contrast, if multiple related targets, for example, a broad spectrum of different HPV genotypes, must be amplified, one can design primers that match DNA regions that are highly conserved but not necessarily completely identical among all targets. In that case, subtle base differences among targets can be dealt with by selecting degenerate primers or sets of overlapping primers[15,16] that contain universal base analogues, such as inosine[17,18] and/or primers containing mismatches that are accepted under conditions of primer annealing at reduced stringency.[19] Finally, primers may need modifications such as biotinylation to enable detection of the reaction product (see below).

DETECTION OF AMPLIFICATION PRODUCTS

Probably the easiest way of detecting an amplification product is analysis by agarose gel electrophoresis. This is possible when the expected size of the product is known and the amplification does not give rise to any nonspecific products. For a PCR with well chosen primers and reaction conditions, this is generally not a problem, although an additional step to confirm reaction product specificity remains sensible, especially in diagnostic settings. In other circumstances, such an additional step is mandatory.

For example, NASBA reaction products cannot be judged at the gel level since they usually appear as nondistinct smears because of the rather nonspecific nature of the amplification (due to its low temperature). Methods to confirm the specificity of an amplification product can roughly be divided in three groups: hybridization-based methods, restriction endonuclease-based methods, and direct sequence analysis. All three types can be applied for endpoint detection, whereas hybridization-based methods can also be applied in real-time. This will be discussed below.

ENDPOINT DETECTION

Southern Blotting and Other Forward Hybridization Methods

Agarose gel analysis in fact is a way of endpoint amplification product detection. To confirm product specificity, however, reaction products on agarose gels traditionally were transferred to nitrocellulose or nylon membranes by Southern blotting, followed by hybridization with a radioactively labelled nucleic acid probe specific for the sequence flanked by the amplification primers. Later detection methods became nonradioactive and used chemiluminescence. Southern blotting and hybridization may increase the sensitivity of the detection by a factor of 10.[20] Southern blotting and hybridization can be applied to reaction products of all the amplification techniques mentioned above including NASBA in which the product consists of RNA (although strictly speaking, the process should then be referred to as Northern blotting).

An alternative method of forward hybridization that allows larger throughput is the microtiter plate-based format or enzyme immunoassay (EIA). This involves immobilization of the amplification product in microtiter plates, for example, by incorporating biotin into the amplification product (via biotinylated primers or modified dNTPs) and capturing onto streptavidin-coated microtiter plates. After denaturation, the PCR product is hybridized with the labelled specific probes and the hybrids are detected. For example, probes can be digoxygenin-labelled and the hybrid detected with alkaline phosphatase-conjugated anti-digoxygenin antibodies, followed by a colorimetric reaction. This principle has been described for type-specific[21] and group-specific[22] detection of HPVs.

Reverse Hybridization Methods

Reverse hybridization methods can be defined briefly as hybridizations for which the specific probes but not the amplification product itself are immobilized. This reverse method is particularly suitable if the amplification product has to be hybridized to a wide range of probes, for example, to detect all possible HPV genotypes within a PCR product generated with consensus primers.

Reverse hybridization can be performed in a filter format. the labelled (e.g., biotinylated) amplification product is hybridized to specific oligos immobilized on filters and detected. This detection can be achieved radioactively or nonradioactively, for example via a colorimetric reaction.[23]

Alternative nonradioactive protocols involve the use of chemiluminescence. An example is reverse line blot (RLB) analysis following GP5+/6+[24] consensus PCR used to detect a broad spectrum of HPV genotypes (Figure 1.3). One advantage of chemiluminescence over a colorimetric reaction is the possibility to strip and re-use the membranes containing the oligonucleotide probes.

Reverse hybridization can also be performed in microtiter plate-based formats[25] or even in microarray-based formats.[26] A flow cytometry-based method using Luminex (Austin, Texas, USA) color-coded microsphere beads was developed recently. The different microspheres contain two spectrally distinct fluorochromes present in various concentrations in such a way that a spectral array encompassing 100 different microsphere sets with specific spectral addresses is created. Each oligonucleotide probe is covalently attached to a specific microsphere set with a distinct spectral address. During analysis, individual microspheres are interrogated by two lasers. The first laser excites the fluorochromes within the microsphere and allows identification of the microsphere set and consequently the attached probe. The second laser excites a reporter fluorochrome incorporated

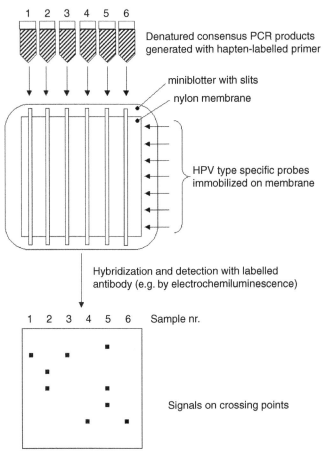

FIGURE 1.3 Principle of reverse line blotting (RLB). The probes (top) are attached to a membrane using a miniblotter with slits. Subsequently, the membrane is rotated 90° in the miniblotter (middle). PCR products are generated using hapten-labelled primers. After denaturation, the single-stranded PCR products are pipetted into the slits of the miniblotter. If the probe sequence is represented in a sample, hybridization will occur on the crossing point of the sample slit with the respective probe line. Subsequently, hybridization is detected using a labelled antibody directed against the hapten and visualized using chemiluminescence (bottom).

in the hybridized PCR product and allows quantification of the PCR product. This system has successfully been used for *Mycobacterium tuberculosis* genotyping.[27]

Restriction-Fragment Length Polymorphism Analysis

Restriction-fragment length polymorphism (RFLP) analysis implies restriction endonuclease digestion of amplification products and comparison of the digestion pattern with known patterns. Not only is this method suitable for confirmation of reaction product specificity, it also allows distinction of genotypes among humans[28] and viruses,[29] for example. One disadvantage is the fact that this technique does not enhance detection sensitivity as hybridization-based methods do. Moreover, the results can be difficult to interpret if the PCR product contains more than one genotype, as is the case for infections with multiple genotypes of a certain virus. Because of these disadvantages and the fact that this method is relatively laborious, it is no longer widely used.

It is obvious that RFLP is suitable only for amplification products that consist of double-stranded DNAs.

Direct Sequence Analysis

Direct sequence analysis of amplification products[30] is the confirmation method of choice when the exact sequence of the region flanked by the amplification primers is not known. Direct analysis is a rather insensitive detection method because a relatively large amount of amplification product is necessary to obtain a positive signal. Moreover, as with RFLP, the presence of multiple genotypes within the amplification product will hamper interpretation of the results, although this can be overcome by cloning of the amplification product and sequencing the separate clones. However, this is a very laborious process and thus used more in research settings than in routine molecular diagnostics.

REAL-TIME MEASUREMENT

Real-time detection of amplification products relies on the increase of a fluorescent signal with each cycle or as a function of reaction time (Figure 1.4), with the exception of a real-time detection method used in combination with LAMP (see below). The fluorescent signal may be derived from specifically hybridizing probes or from intercalating agents.

Because the kinetics of cyclic amplification methods can be described by a logarithmic function, real-time detection can be used for quantification of the original amount of target nucleic acid in the sample: the more cycles it takes for the signal to arise above background, the less target nucleic acid was present in the sample (Figure 1.5a). The amount of target nucleic acid in the sample can usually be extrapolated from a standard curve of known target concentrations. This type of quantification cannot be applied to isothermal amplification methods because of their different kinetics. However, real-time measurement of a NASBA reaction (using molecular beacons; see below), for example, yields an amplification curve resembling that of a cyclic amplification method (Figure 1.5b).

Non-Hybridization Real-Time Detection

The best known method for non-hybridization real-time detection is the measurement of fluorescence obtained with SYBR Green — an intercalating dye that fluoresces only when it is bound to double-stranded DNA. Hence, the fluorescence will increase in time if a double-stranded DNA reaction product is formed.

The method is also suitable to detect the presence of possible nonspecific products after the amplification reaction. Plotting fluorescence as a function of temperature as the thermal cycler heats through the dissociation temperature of the product yields a DNA melting curve. The shape and

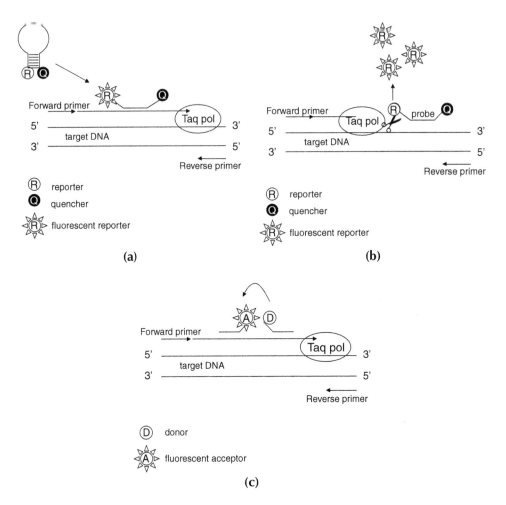

(a)   (b)

(c)

FIGURE 1.4 Principles of hybridization-based real-time detection methods. (a) Molecular beacons are oligos carrying a reporter dye (e.g., FAM or Texas Red) and a quencher [generally the non-fluorogenic molecule 4-[4′dimethylaminophenylazo]benzoic acid (DABCYL)] on their opposite ends. They form a hairpin loop in the absence of their complementary PCR products, which keeps the quencher and the reporter in close proximity and thus prevents the reporter from emitting light. When PCR product is generated, the specific domain of the beacon will hybridize with it during the annealing phase, resulting in opening of the hairpin, separation of the reporter and the quencher, and the emission of light from the reporter. (b) Fluorescent 5′ exonuclease (TaqMan) amplification is carried out in the presence of a probe carrying a reporter dye at the 5′ end and a quencher at the 3′ end. During the annealing phase, the probe will hybridize to the amplification product. When the *Taq* polymerase reaches this hybrid, it will degrade the probe by 5′–3′ exonuclease activity. Once the 5′ reporter is released, it is no longer quenched and will start to emit light. The 5′ reporter molecule in general is a fluoroscein isothiocyanate (FITC) derivative, FAM, VIC or TET, and the quencher is 6-carboxytetramethylrhodamine (TAMRA). (c) Fluorescence resonance energy transfer (FRET, LightCycler). In addition to amplification primers, the amplification reaction contains two probes that can hybridize with the amplification product in the direct vicinity of each other. The donor fluorophore on one probe is excited by a light-emitting diode (LED) source and transfers its energy to the acceptor fluorophore on the other probe only when amplification product is present and both probes are hybridized. The acceptor fluorophore emits light of a longer wavelength, which is detected in specific channels.

position of this DNA melting curve are functions of the GC:AT ratio, length, and sequence and can be used to differentiate amplification products separated by fewer than 2 degrees Celsius in melting temperature. Hence, specific products can be distinguished from nonspecific products.[31]

It is clear that the intercalating dye method can be used only if the reaction product is double-stranded DNA.

For reaction products obtained with the LAMP method, another real-time detection system has been described. It is based on the fact that a large amount of pyrophosphate is produced as a by-product of the LAMP reaction, yielding a white precipitate of magnesium pyrophosphate in the reaction mixture. It is thought that the resulting increase in turbidity of the reaction mixture correlates with the amount of DNA synthesized.[32] However, because of the complicated kinetics of LAMP, this feature cannot be used for quantification. Moreover, because the signal is not correlated with the physical state of the DNA (as is SYBR Green), it cannot be used to confirm specificity of the reaction product by analysis of the melt curve.

HYBRIDIZATION-BASED REAL-TIME DETECTION

To detect amplification products in real time and simultaneously confirm their specificities, hybridization-based real-time detection methods are the best choices. Briefly, these systems use specific probes that can hybridize to part of the region flanked by the amplification primers and that fluoresce upon hybridization. The higher the amount of amplification product, the stronger the fluorescent signal. The way the fluorescence is obtained depends on the type of probe. The best known real-time detection systems are molecular beacons (Figure 1.4a),[33] the fluorescent 5′ exonuclease assay (TaqMan, Applied Biosystems, Figure 1.4b),[34] and fluorescence resonance energy transfer (FRET; LightCycler, Roche, Figure 1.4c).[35]

All three of these methods have specific advantages and disadvantages. In general, molecular beacon and TaqMan assays are performed in a thermocycler that also reads fluorescence. The format is identical to regular PCR methods (e.g., a 96-well plate), whereas LightCycler requires a specialized thermocycler and the reactions occur in capillaries. The cycling times of the LightCycler are very short due to the efficient heat conduction, whereas TaqMan and molecular beacon assays take approximately the same amount of time as regular PCRs. All three methods are in fact equally suitable for detecting PCR products.

Figure 1.5a is a typical amplification plot. Figure 1.5b shows the corresponding calibration line (standard curve) that can be obtained by means of a dilution series of known target concentrations. The multiplicity of the reaction, i.e., the number of different targets that can be amplified and detected simultaneously, depends on the number of different fluorescence filters in the thermocycler. For example, the current generation LightCycler devices have six filter sets that, in theory, allow the detection of up to six different targets. However, optimal use of this capacity is limited simply by the fact that a multitude of amplification primers and probes may lead to loss of sensitivity or specificity of the reaction.

NASBA products consisting of single-stranded RNA can be detected easily with molecular beacons.[36] The amplification plot obtained is similar to that shown in Figure 1.5b, although the x axis represents reaction time instead of cycle number. However, as noted above, the kinetics of NASBA prohibits quantification by means of a standard curve. Instead, an internal control (a so-called calibrator) can be used. The calibrator is amplified by the same primer pair as the target sequence but contains an internally reshuffled sequence of ~20 bp detected by a differentially labelled molecular beacon. The addition of a fixed amount of this calibrator to the reactions allows quantification.[37]

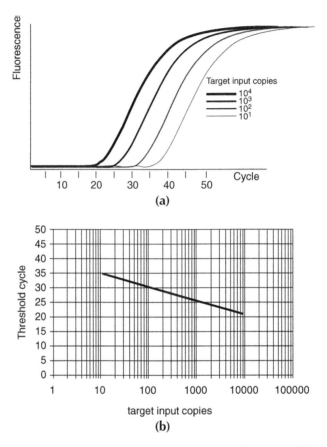

FIGURE 1.5 A real-time cyclic amplification method, in this case PCR, using different amounts of input target material. (a) Real-time amplification curve in which the increase in fluorescence intensity is plotted against the increase in cycle number. The threshold cycle is defined as the cycle in which signal rises above the background. (b) Threshold cycle plotted against the amount of input target material, displayed on a logarithmic scale; a straight line is obtained from which input quantities of unknown samples can be deduced.

VALIDATION OF AMPLIFICATION METHOD

SENSITIVITY

Every amplification assay should include verification that possible negative results are not due to lack of sensitivity of the test. The analytical sensitivity of a test is the minimal amount of target DNA or RNA necessary for reproducible detection, preferably expressed in absolute copy numbers per reaction or per volume unit. For the most accurate estimation, the analytical sensitivity is best determined using dilution series of plasmid clones containing the target DNA sequence. If the target consists of RNA, artificial target molecules can be generated by cloning the corresponding DNA sequence in a plasmid containing RNA polymerase recognition signals (a so-called transcription vector) and subsequently performing *in vitro* transcription.

To mimic as closely as possible the amplification circumstances in clinical material, the artificial target molecules should not be used on their own, but instead should be diluted in a background of irrelevant DNA or irrelevant RNA, depending on the target nucleic acid to be detected. If plasmid clones of the target DNA are not available, cultured cells with known copy numbers of the target gene can be used to determine the sensitivity. However, this is not acceptable for RNA amplification tests because the copy number of an RNA per cell is generally unknown. In general, a test can be considered highly sensitive when it reproducibly detects as few as 10 to 100 target copy molecules per reaction.

If the analytical sensitivity of an assay is not acceptable, amplification conditions can be altered for optimization (e.g., primer annealing temperature, salt concentration, enzyme concentration). If the alteration does not produce the desired result, it is usually necessary to design another primer pair.

SPECIFICITY

An amplification method should not give false positivity due to nonspecific amplification. To make sure that this is not the case, the amplification method should always be tested on irrelevant DNA or RNA prior to use. In general, nonspecific amplification can be avoided by selecting amplification primers carefully by means of a BLAST search (see above). If nonspecific amplification occurs despite careful selection, alteration of amplification conditions can sometimes be helpful.

RISK OF TARGET AMPLIFICATION-BASED TECHNIQUES: CONTAMINATION

Because of the exponential nature of target amplification-based methods,[38] even a small contamination with reaction product generated earlier can be disastrous. It should be noted that the reaction product contains vast amounts of amplifiable molecules. In addition, carry-over from other samples poses a major risk of contamination.

CONTAMINATION-FREE WORKING ENVIRONMENT

Reaction product carry-over can be avoided by using separate laboratories: one for setting up the reaction and one for running the reaction and detecting the reaction product. In addition, many diagnostic laboratories that rely on amplification techniques make it a practice for their staff to not re-enter the reaction set-up room after having worked in the detection room. Obviously, reaction product carry-over is less likely to occur when reaction vials are not opened upon completion of a reaction. This is a great additional advantage of real-time detection assays.

To avoid contamination of reaction components with sample material, an additional separate room for sample work-up can be used. Finally, cross-contamination of samples should be avoided, for example, by using only pipettes fitted with disposable filter tips and never using the same filter tip twice. Examination gloves should be worn, particularly in the presence of risk of contamination from an examiner's hands, for example, when human sequences are amplified or common cutaneous microorganisms such as cutaneous HPVs are assayed.

ALTERNATIVE METHOD TO AVOID CARRY-OVER: USE OF URACIL DNA
GLYCOSYLASE ENZYME

Uracil DNA glycosylase (UDG) cleaves the uracil from the phosphodiester backbone of uracil-containing DNA. The resulting apyrimidinic sites block replication by DNA polymerases and are very labile to acid–base hydrolysis. UDG does not react with free deoxyuridine 5-triphosphate (dUTP) and is inactivated by heat denaturation. These properties can be utilized to prevent reaction product carry-over by incorporating dUTP in all PCR products (by substituting dUTP for dTTP or by incorporating uracil during primer synthesis) and treating all subsequent preassembled reactions with UDG, followed by heat-denaturation of UDG.[39] UDG is commercially available (e.g., AmpErase from Applied Biosystems) and is used in many diagnostic laboratories.

INHIBITION OF AMPLIFICATION

Direct PCR on crude extracts of clinical specimens such as cervical smears and formalin-fixed, paraffin-embedded tissues is usually successful. However, clinical specimens such as whole blood and cytological preparations collected in certain preservation media may contain inhibitors for which many DNA polymerases are sensitive. In such cases, it is important to remove inhibiting substances by proper nucleic acid extraction. In addition, the efficiency of the extraction (and of the amplification itself) can be monitored by the addition of known amounts of an external control prior to the extraction, which is then detected by an additional specific amplification reaction (as described by van Doornum et al.[40]).

CONCLUDING REMARKS

Target amplification-based techniques are very powerful molecular diagnostic and research tools. The choice of amplification technique and the subsequent detection method depends on the type of nucleic acid to be amplified, the desired sensitivity, the amount and type of (clinical) samples to be tested, and the equipment available. For detection of single specific targets, the use of real-time detection systems is highly recommended because of their speed and relatively low contamination danger. For the detection of multiple related sequences, amplification with consensus primers followed by a reverse hybridization technique is very promising. In this respect, microchip technology will allow the detection, for example, of large numbers of (related) pathogens in clinical specimens. In general, we can expect future amplification systems to be faster and more automated.

REFERENCES

1. Saiki RK, Bugawan TL, Horn GT, Mullis KB, Erlich HA. Analysis of enzymatically amplified beta-globin and HLA-DQ alpha DNA with allele-specific oligonucleotide probes. *Nature* 1986; 324 (6093): 163–166.
2. Proffitt J, Fenton J, Pratt G, Yates Z, Morgan G. Isolation and characterisation of recombination events involving immunoglobulin heavy chain switch regions in multiple myeloma using long distance vectorette PCR (LDV-PCR). *Leukemia* 1999; 13 (7): 1100–1107.
3. Saiki RK, Gelfand DH, Stoffel S, Scharf SJ, Higuchi R, Horn GT et al. Primer-directed enzymatic amplification of DNA with a thermostable DNA polymerase. *Science* 1988; 239 (4839): 487–491.
4. Barany F. Genetic disease detection and DNA amplification using cloned thermostable ligase. *Proc Natl Acad Sci USA* 1991; 88 (1): 189–193.
5. Myers TW, Gelfand DH. Reverse transcription and DNA amplification by a *Thermus thermophilus* DNA polymerase. *Biochemistry* 1991; 30 (31): 7661–7666.
6. Compton J. Nucleic acid sequence-based amplification. *Nature* 1991; 350 (6313): 91–92.
7. Mueller JD, Putz B, Hofler H. Self-sustained sequence replication (3SR): an alternative to PCR. *Histochem Cell Biol* 1997; 108 (4–5): 431–437.
8. Morre SA, Sillekens PT, Jacobs MV, de Blok S, Ossewaarde JM, van Aarle P et al. Monitoring of *Chlamydia trachomatis* infections after antibiotic treatment using RNA detection by nucleic acid sequence based amplification. *Mol Pathol* 1998; 51 (3): 149–154.
9. Kievits T, van Gemen B, van Strijp D, Schukkink R, Dircks M, Adriaanse H et al. NASBA isothermal enzymatic *in vitro* nucleic acid amplification optimized for the diagnosis of HIV-1 infection. *J Virol Methods* 1991; 35 (3): 273–286.
10. Landry ML, Garner R, Ferguson D. Rapid enterovirus RNA detection in clinical specimens by using nucleic acid sequence-based amplification. *J Clin Microbiol* 2003; 41 (1): 346–350.
11. Cuschieri KS, Whitley MJ, Cubie HA. Human papillomavirus type-specific DNA and RNA persistence: implications for cervical disease progression and monitoring. *J Med Virol* 2004; 73 (1): 65–70.

12. Brink AA, Vervoort MB, Middeldorp JM, Meijer CJ, van den Brule AJ. Nucleic acid sequence-based amplification: a new method for analysis of spliced and unspliced Epstein–Barr virus latent transcripts and its comparison with reverse transcriptase PCR. *J Clin Microbiol* 1998; 36 (11): 3164–3169. Erratum: *J Clin Microbiol* 1999; 37 (11): 3788.

13. Walker GT, Fraiser MS, Schram JL, Little MC, Nadeau JG, Malinowski DP. Strand displacement amplification: an isothermal, *in vitro* DNA amplification technique. *Nucleic Acids Res* 1992; 20 (7): 1691–1696.

14. Notomi T, Okayama H, Masubuchi H, Yonekawa T, Watanabe K, Amino N et al. Loop-mediated isothermal amplification of DNA. *Nucleic Acids Res* 2000; 28 (12): E63.

15. Gravitt PE, Peyton CL, Apple RJ, Wheeler CM. Genotyping of 27 human papillomavirus types by using L1 consensus PCR products by a single-hybridization, reverse line blot detection method. *J Clin Microbiol* 1998; 36 (10): 3020–3027.

16. Tieben LM, ter Schegget J, Minnaar RP, Bouwes Bavinck JN, Berkhout RJ, Vermeer BJ et al. Detection of cutaneous and genital HPV types in clinical samples by PCR using consensus primers. *J Virol Methods* 1993; 42 (2–3): 265–279.

17. Kleter B, van Doorn LJ, Schrauwen L, Molijn A, Sastrowijoto S, ter Schegget J et al. Development and clinical evaluation of a highly sensitive PCR reverse hybridization line probe assay for detection and identification of anogenital human papillomavirus. *J Clin Microbiol* 1999; 37 (8): 2508–2517.

18. Gregoire L, Arella M, Campione-Piccardo J, Lancaster WD. Amplification of human papillomavirus DNA sequences by using conserved primers. *J Clin Microbiol* 1989; 27 (12): 2660–2665.

19. Roda-Husman AM, Walboomers JM, van den Brule AJ, Meijer CJ, Snijders PJ. The use of general primers GP5 and GP6 elongated at their 3′ ends with adjacent highly conserved sequences improves human papillomavirus detection by PCR. *J Gen Virol* 1995; 76 (Pt 4): 1057–1062.

20. Roda-Husman AM, Walboomers JM, van den Brule AJ, Meijer CJ, Snijders PJ. The use of general primers GP5 and GP6 elongated at their 3′ ends with adjacent highly conserved sequences improves human papillomavirus detection by PCR. *J Gen Virol* 1995; 76 (Pt 4):1057–1062.

21. Jacobs MV, van den Brule AJ, Snijders PJ, Helmerhorst TJ, Meijer CJ, Walboomers JM. A non-radioactive PCR enzyme immunoassay enables a rapid identification of HPV 16 and 18 in cervical scrapes after GP5+/6+ PCR. *J Med Virol* 1996; 49 (3): 223–229.

22. Jacobs MV, Snijders PJ, van den Brule AJ, Helmerhorst TJ, Meijer CJ, Walboomers JM. A general primer GP5+/GP6+-mediated PCR enzyme immunoassay method for rapid detection of 14 high-risk and 6 low-risk human papillomavirus genotypes in cervical scrapings. *J Clin Microbiol* 1997; 35 (3): 791–795.

23. Saiki RK, Walsh PS, Levenson CH, Erlich HA. Genetic analysis of amplified DNA with immobilized sequence-specific oligonucleotide probes. *Proc Natl Acad Sci USA* 1989; 86 (16): 6230–6234.

24. van den Brule AJ, Pol R, Fransen-Daalmeijer N, Schouls LM, Meijer CJ, Snijders PJ. GP5+/6+ PCR followed by reverse line blot analysis enables rapid and high-throughput identification of human papillomavirus genotypes. *J Clin Microbiol* 2002; 40 (3): 779–787.

25. Keller GH, Huang DP, Manak MM. Detection of human immunodeficiency virus type 1 DNA by polymerase chain reaction amplification and capture hybridization in microtiter wells. *J Clin Microbiol* 1991; 29 (3): 638–641.

26. Kim CJ, Jeong JK, Park M, Park TS, Park TC, Namkoong SE et al. HPV oligonucleotide microarray-based detection of HPV genotypes in cervical neoplastic lesions. *Gynecol Oncol* 2003; 89 (2): 210–217.

27. Cowan LS, Diem L, Brake MC, Crawford JT. Transfer of a *Mycobacterium tuberculosis* genotyping method, Spoligotyping, from a reverse line blot hybridization, membrane-based assay to the Luminex multianalyte profiling system. *J Clin Microbiol* 2004; 42 (1): 474–477.

28. Higuchi R, von Beroldingen CH, Sensabaugh GF, Erlich HA. DNA typing from single hairs. *Nature* 1988; 332 (6164): 543–546.

29. Lungu O, Wright TC, Jr., Silverstein S. Typing of human papillomaviruses by polymerase chain reaction amplification with L1 consensus primers and RFLP analysis. *Mol Cell Probes* 1992; 6 (2):145–152.

30. Wong C, Dowling CE, Saiki RK, Higuchi RG, Erlich HA, Kazazian HH, Jr. Characterization of beta-thalassaemia mutations using direct genomic sequencing of amplified single copy DNA. *Nature* 1987; 330 (6146): 384–386.

31. Ririe KM, Rasmussen RP, Wittwer CT. Product differentiation by analysis of DNA melting curves during the polymerase chain reaction. *Anal Biochem* 1997; 245 (2): 154–160.

32. Mori Y, Nagamine K, Tomita N, Notomi T. Detection of loop-mediated isothermal amplification reaction by turbidity derived from magnesium pyrophosphate formation. *Biochem Biophys Res Commun* 2001; 289 (1): 150–154.

33. Tyagi S, Kramer FR. Molecular beacons: probes that fluoresce upon hybridization. *Nat Biotechnol* 1996; 14 (3): 303–308.

34. Heid CA, Stevens J, Livak KJ, Williams PM. Real time quantitative PCR. *Genome Res* 1996; 6 (10): 986–994.

35. Wittwer CT, Ririe KM, Andrew RV, David DA, Gundry RA, Balis UJ. The LightCycler: a microvolume multisample fluorimeter with rapid temperature control. *Biotechniques* 1997; 22 (1): 176–181.

36. Leone G, van Schijndel H, van Gemen B, Kramer FR, Schoen CD. Molecular beacon probes combined with amplification by NASBA enable homogeneous, real-time detection of RNA. *Nucleic Acids Res* 1998; 26 (9): 2150–2155.

37. Weusten JJ, Carpay WM, Oosterlaken TA, van Zuijlen MC, van de Wiel PA. Principles of quantitation of viral loads using nucleic acid sequence-based amplification in combination with homogeneous detection using molecular beacons. *Nucleic Acids Res* 2002; 30 (6): e26.

38. Kwok S, Higuchi R. Avoiding false positives with PCR. *Nature* 1989; 339 (6221): 237–238.

39. Longo MC, Berninger MS, Hartley JL. Use of uracil DNA glycosylase to control carry-over contamination in polymerase chain reactions. *Gene* 1990; 93 (1): 125–128.

40. van Doornum GJ, Guldemeester J, Osterhaus AD, Niesters HG. Diagnosing herpesvirus infections by real-time amplification and rapid culture. *J Clin Microbiol* 2003; 41 (2): 576–580.

2 Signal Amplification-Based Techniques

David A. Hendricks and Lorraine Comanor

CONTENTS

INTRODUCTION

Nucleic acid targets associated with human disease are often present in extremely low copy numbers in biological specimens. Therefore, assays for detection or quantification of signature disease-associated nucleic acids employ amplification technologies to enhance or amplify their output. To date, three different approaches to amplification have been developed (Figure 2.1). With target amplification, assay output is enhanced by generating multiple copies of the nucleic acid target

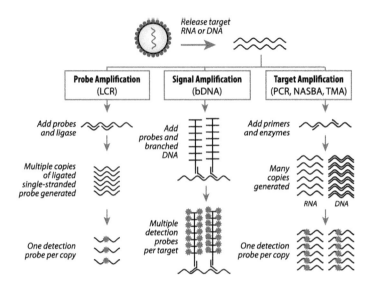

FIGURE 2.1 Three amplification approaches for detection or quantification of disease-associated nucleic acids. Multiple copies of the nucleic acid target sequence are generated to enhance assay output in target amplification technologies, including polymerase chain reaction (PCR), nucleic acid sequence-based amplification (NASBA), and transcription mediated amplification (TMA). Alternatively, multiple copies of a specific probe formed only in the presence of the nucleic acid target sequence are generated to enhance assay output in probe amplification technologies such as ligase chain reaction (LCR). Signal amplification technologies such as branched DNA (bDNA) do not generate multiple copies of the nucleic acid target or a specific probe; they increase the amount of signal generated from each nucleic acid target sequence by development of large signal-generating nucleic acid hybrid complexes.

sequence. Alternatively, probe amplification enhances assay output by increasing the number of copies of a specific configuration of the probe that forms only in the presence of the target sequence. The third approach to output enhancement, signal amplification, relies on increasing the amount of signal on the nucleic acid target from the specimen.

The nature of the clinical information derived from detection, quantification, or characterization of nucleic acid targets defines the performance requirements of assays employing any of the amplification technologies. Assays should detect and/or quantify nucleic acid targets in the range of concentrations found in biological specimens and they should not yield signals unless the specific nucleic acid target is present. Molecular assays must also provide accurate and reproducible results even if inherent genetic variability is present in the nucleic acid target sequence. Conversely, assays should discriminate between closely related genetic variants if identification of a variant has clinical significance. Ideally, assays should be standardized to a common, fixed standard and be expressed as units of measure that have clinical meanings. In many cases, assays should allow multiple results to be derived from one sample in one test.

This chapter focuses on signal amplification technologies. In the first part of the chapter, we review the molecular bases and inherent properties of the signal amplification technologies that support the many commercial kits and laboratory-developed assays available today. We then compare the analytical performances of signal amplification assays and discuss possible ramifications of performance features that could affect clinical decisions. Since the goal of all assays that employ signal amplification is to provide valuable clinical results, in the final part of the chapter we review major human infectious diseases, primarily those with viral etiologies, and summarize the roles of assays that employ signal amplification in better understanding these diseases and in managing care for people afflicted with them.

TABLE 2.1
Overview of Signal Amplification Technologies

Technology	Theoretical Amplification	Commercial Availability*	Representative Applications
Branched DNA (bDNA)	11,564-fold	Chiron; now Bayer HealthCare Diagnostics Division, Tarrytown, New York, USA.	Viral load measurements for HIV, HBV, HCV, CMV, cytokine mRNA; *in situ* detection of HIV-1 and HPV
Cycling Probe (CP)	Several orders of magnitude	ID Biomedical, Vancouver, British Columbia, Canada Excimus Biotech, Inc., Savage, Maryland, USA (CataCleave probe)	Detection of methicillin resistance in *Staphylococcus aureus*; detection of vancomycin resistant *vanA* and *vanB* genes in enterococci; characterization of direct repeats in *Mycobacterium tuberculosis*
Hybrid Capture (HC)	3000-fold	Digene Corporation, Gaithersburg, Maryland, USA	Viral load measurements of HBV and CMV; detection of HPV, *Neisseria gonorrhoeae* and *Chlamydia trachomatis*
Invader	10^6–10^7-fold	Third Wave Technologies, Inc., Madison, Wisconsin, USA	Detection of Factor V Leiden, Factor II, ApoE, Cystic Fibrosis, Tay-Sachs, SNP analysis, analysis of A1298C polymorphism in methylene tetrahydrofolate reductase gene, genotypic analysis of *P450 2D6* gene; analysis of *CYP3A7* gene family
Tyramide Signal Amplification (TSA), also called catalyzed reporter deposition (CARD)	500–1000	Molecular Probes, Inc., Eugene, Oregon, USA	Detection of HPV, HIV, 16S ribosomal RNA, antimicrobial resistance genes, multiple research applications

* Commercial availability dependent on country-specific regulatory status.

SIGNAL AMPLIFICATION TECHNOLOGIES

Five technologies have been identified in the literature as signal amplification methods (Table 2.1). They include the branched DNA (bDNA) technology employed in assays first developed by Chiron Corporation and now produced by Bayer Healthcare's Diagnostics Division; Cycling Probe™ (CP) technology* from ID Biomedical; Hybrid Capture® (HC) technology† employed in assays commercialized by Digene Corporation; Invader® technology‡ in assays from Third Wave Technologies; and Tyramide Signal Amplification (TSA), also called catalyzed reporter deposition (CARD), from Molecular Probes, Inc.

A search of the literature indexed in PubMed®** from 1990 through 2004 identified 1219 citations mentioning any of the five signal amplification technologies in a title or an abstract (Figure

---

* Cycling Probe is a trademark of ID Biomedical, Vancouver, British Columbia, Canada.

† Hybrid Capture is a registered trademark of Digene Corporation, Gaithersburg, Maryland, USA.

‡ Invader is a registered trademark of Third Wave Technologies, Inc., Madison, Wisconsin, USA.

** PubMed is a registered trademark of the National Library of Medicine, Bethesda, Maryland, USA.

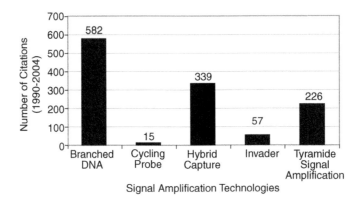

FIGURE 2.2 Number of citations from a search of the literature indexed in PubMed from 1990 through 2004 for the five signal amplification technologies. PubMed was queried for references citing names of the signal amplification technologies in the title or the abstract; full text searches were not conducted.

2.2). The higher numbers of citations for the bDNA (582) and HC (339) technologies may reflect the longer availability of commercial (and often U.S. Food and Drug Administration-cleared or European Union-registered) assays that employ these technologies in the marketplace. Commercial assays employing signal amplification technologies are tools for clinical and research studies; thus, the technology often is not identified in a title or abstract of a citation. Therefore, the numbers of citations noted in the figure probably represent under-estimations of the total studies in which these signal amplification technologies were used. Nonetheless, the large number of citations clearly reflects the widespread adoption of these technologies by the clinical community.

BRANCHED DNA TECHNOLOGY

The bDNA technology[1-3] has evolved through three generations of design and the third generation is used in products today.[4-7] Figure 2.3 is a schematic view of third generation bDNA technology. In brief, a nucleic acid target is released from viral particles or cells or made available in fixed tissues[8] by treatment with proteinase K and detergent that also inactivate RNases and DNases in biological materials. In the microwell format, released target is anchored to the plate by means of capture probes that have sequences complementary to sequences on the nucleic acid target and other sequences that hybridize to oligonucleotides attached to the well surfaces. The same attached oligonucleotides are used in all bDNA-based assays regardless of the target.

Target probes simultaneously hybridize to other sequences on the nucleic acid target. Following removal of excess, non-hybridized probes, a preamplifier molecule with 14 repeat sequences is allowed to hybridize to sequence overhangs on contiguous target probes; simultaneous hybridization to the contiguous target probes is required for stable attachment of the preamplifier to the nucleic acid complex. Then, amplifier molecules with 14 repeat sequences and with overhangs complementary to the repeat sequences on the preamplifier molecules are incubated with the complexes.

After removal of non-hybridized amplifier molecules, the complexes are incubated with excess alkaline phosphate-modified oligomers, called label probes, that hybridize to the repeat sequences on the amplifier. Quantification of nucleic acid targets in specimens is determined using a calibration curve generated for each assay run by testing calibrators with known amounts of target.

The third generation bDNA technology benefits by incorporation of 5-methyl-2′-deoxycytidine (isoMeC) and 2′-deoxyguanosine (isoG) in place of some of the cytosine and guanosine residues in some of the probes.[7] In spite of careful probe design in the earlier generations of bDNA technology, there are possible short stretches of unintended hybridization of capture probes and the other probes of the assay. Such unintended hybridization would produce background noise. Probes containing isoMeC and isoG form very stable complexes with nucleic acids containing

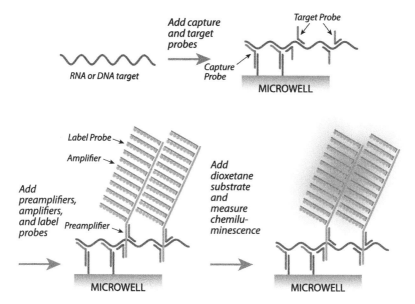

FIGURE 2.3 Third generation branched DNA (bDNA) technology. Nucleic acid target (RNA or DNA) released from viral particles, bacteria, or mammalian cells is anchored to a microwell by capture probes that hybridize with both the target and with oligonucleotides attached to the microwell surface. Target probes simultaneously hybridize to other sequences on the nucleic acid target. A large nucleic acid hybrid complex is developed by sequential additions of preamplifier (binds to target probes), amplifier, and label probes. The oligonucleotide label probes are modified with alkaline phosphatase that produces signal when a dioxetane substrate is added. The chemiluminescent signal is measured and the nucleic acid target is quantified using a calibration curve generated in each run from calibrators with known amounts of nucleic acid target.

complementary isomeric forms but much less stable hybrids with the natural bases.[9,10] Therefore, isomeric forms of cytosine and guanosine were incorporated at some positions selectively in synthesis of all the probes except capture probes.

Multiple capture and target probes, each about 30 nucleotides long, are designed for the bDNA technology to recognize different, conserved sequences in the nucleic acid target. Such long probes form stable hybrids with nucleic acid targets that may contain several base pair mismatches. A software program for design of probes used in the bDNA technology has been described.[11]

The theoretical amplification (possible number of label probes hybridized in the complex) of the third generation bDNA technology could be as great as 11,564 (59 preamplifier molecules × 14 sites for amplifier molecules per preamplifier × 14 sites for label probe per amplifier molecule) for the VERSANT® HIV-1 RNA 3.0 Assay (bDNA).* The theoretical amplifications for the VERSANT HCV RNA 3.0 (bDNA) and VERSANT HBV 3.0 (bDNA) assays are 2352 and 7448, respectively. The lower theoretical amplifications for both of the latter two assays reflect the incorporation of fewer target probes (highly conserved sequences in the nucleic acid target).

CYCLING PROBE TECHNOLOGY

CP technology is an isothermal, linear signal amplification reaction that also may have application for detection of accumulating amplification products in real-time target amplification (e.g., PCR technology) or probe amplification (e.g., rolling circle technology) systems.[12,13] Figure 2.4 depicts CP technology. A single chimeric DNA–RNA–DNA probe is allowed to hybridize to single-stranded DNA with complementary sequences. In the presence of RNase H, the RNA moiety is cleaved and

---

* VERSANT is a registered trademark of Bayer HealthCare Diagnostics Division, Tarrytown, New York, USA.

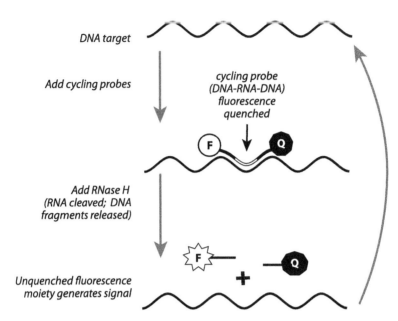

FIGURE 2.4 Cycling Probe (CP) technology. A single-stranded DNA target is allowed to hybridize with a chimeric DNA–RNA–DNA target-specific probe. In some applications of the technology, one of the DNA portions is labeled with a fluorophore (F) and the other portion contains a quencher (Q) that absorbs emissions from the F due to their proximity in the intact probe. RNase H cleaves the RNA portion of the probe when it is part of a target–probe hybrid and the DNA fragments of the probe melt off the nucleic acid target. The resulting DNA fragments can be detected by gel electrophoresis, immunoassays with lateral-flow strip devices or by detection of fluorescence since the emissions of the F no longer are quenched by the Q. The DNA target then is available for subsequent hybridization with another chimeric probe. The CP reaction is linear and isothermal.

the DNA fragments dissociate from the target under hybridization conditions. The target DNA then is available for hybridization with another chimeric probe.

Thus, DNA fragments from multiple probes are generated linearly from each target at a single temperature. The released DNA fragments can be detected by gel analysis, immunoassays that employ lateral-flow strip devices,[14] or detection of fluorescence. Hogrefe and co-workers reported that RNase H catalyzes 7.1 cleavages per second per RNase monomer.[15] The chimeric probes used in the CP technology can be synthesized on a synthesizer by solid phase chemistry techniques. Harvey et al.[13] showed that the ideal length of the RNA portion of the chimeric probe is four purine residues that are perfectly complementary to four bases in the target DNA.

In the CP technology, chimeric probes with fluorescence resonance energy transfer (FRET) couples in the DNA fragments are called catalytically cleavable fluorescence probes, or CataCleave probes.[13] Probes with FRET generally contain both light emission donors and acceptors. In an intact probe, these two moieties are in close proximity and little emission is recorded from the donor as it is quenched by the acceptor. After cleavage and separation of the donor and acceptor, emission from the donor is greatly enhanced.

In the CataCleave probes reported by Harvey et al.,[13] the emission donor and acceptor were fluorescein and rhodamine, respectively. CataCleave probes were employed for detection of accumulating amplification products from real-time PCR reactions. Such reactions employed a thermostable RNase H and a short, isothermal step for detection of amplicons by the CP reactions in each amplification cycle. Likewise, CataCleave probes have been shown to detect amplification products in the isothermal rolling circle technology reported to produce amplification of several orders of magnitude.

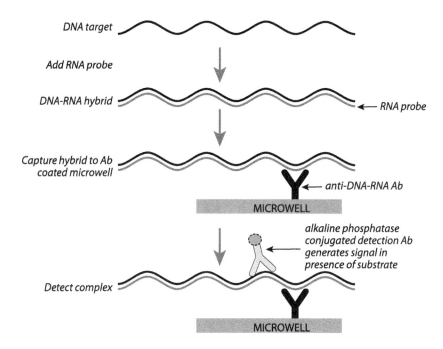

FIGURE 2.5 Hybrid Capture (HC) technology. Genome-length or subgenome-length RNA probes are hybridized to single-stranded DNA targets. The resulting hybrids are attached to a microwell surface via immobilized antibodies (Ab) that recognize and capture RNA–DNA hybrids. Following removal of non-hybridized materials and washing of the microwells, alkaline phosphatase-modified monoclonal antibodies that also recognize RNA–DNA hybrids are added to the wells. Because these antibodies are not sequence-specific, multiple antibodies hybridize to each RNA–DNA hybrid. Chemiluminescence is generated in the presence of alkaline phosphatase and added substrate in proportion to the amount of input single-stranded DNA target.

HYBRID CAPTURE TECHNOLOGY

HC technology has been described in detail for hepatitis B virus (HBV), human papillomavirus (HPV) and human cytomegalovirus (CMV).[16–20] Figure 2.5 illustrates second generation HC technology. To provide single-stranded viral DNA target for HC assays, specimens are treated with a denaturation agent that is expected also to release DNA from matrix. RNA probes of genome length (8 kb for HPV[16] or 3.2 kb for HBV[18]) or subgenome length (4 probes covering 17% or about 40 kb of the unique long region for CMV[21]) are allowed to hybridize in solution in microwells with the single-stranded DNA target. The hybridization mixture then is transferred to another microwell with immobilized antibodies that recognize and capture RNA–DNA hybrids. After capture of the hybrids, microwells are washed and alkaline phosphatase-modified monoclonal antibodies that recognize RNA–DNA hybrids are added to the wells. These antibodies are not sequence-specific; therefore, multiple antibodies are expected to hybridize to each RNA–DNA hybrid. Following removal of excess materials from the wells, substrate is added and chemiluminescence is produced in proportion to the amount of input single-stranded DNA target.

Boeckh and Boivin[20] described the theoretical amplification that can be calculated for the HC technology for quantification of CMV. They assumed that about 1000 antibodies to DNA–RNA hybrids would bind to each 40-kb RNA probe–CMV DNA complex. Assuming that about three alkaline phosphatase molecules are conjugated to each antibody molecule, the total theoretical amplification is 3000-fold.

Calibrators with known amounts of nucleic acid target are tested in separate wells in each run of the assay and the resulting calibration curve (signal output versus target concentration input) is employed for quantification of the DNA target in specimens. Mixtures of RNA probes that recognize different HPV types or HBV genotypes are employed in assays to ensure that genotypic variants are recognized and/or quantified accurately. HPV assays are, at best, semi-quantitative as the number of cervical cells that contain viral nucleic acid is unknown in specimens; the ideal HPV assay also would contain probes that could be used for quantification of housekeeping genes in cervical cells in the specimen.[17]

A more recent modification of the HC technology for detection of specific HPV types has been described.[22] Following formation of hybrids between RNA probes and single-stranded DNA target, hybrids of specific HPV types are captured with biotinylated HPV type-specific oligonucleotides to streptavidin-coated microwells. The captured hybrids are detected with anti-RNA–DNA monoclonal antibodies conjugated with alkaline phosphatase. This modification is especially useful in detecting specific HPV types associated with high risk for cervical carcinoma in single or mixed infections.

INVADER TECHNOLOGY

Invader technology has applications both in detection of single nucleotide changes and in quantification of specific nucleic acids.[23,24] Figure 2.6 represents this technology as applied to DNA targets. A primary oligonucleotide probe is designed with a 3' target-specific region (TSR) that is complementary to a sequence on the single-stranded DNA target and a 5' portion (flap) that does not hybridize to the target. A second oligonucleotide probe, the Invader, hybridizes just upstream of the primary probe and its 3' terminal oligonucleotide overlaps with the 5' terminal oligonucleotide in the TSR of the probe. The 3' terminal base of the Invader probe does not need to bind to the target for the subsequent cleavage to proceed. Cleavase®,* an enzyme from a thermophilic archaebacterium, recognizes the tripartite structure of the Invader probe, primary probe, and target sequence and cleaves the primary probe immediately downstream of the overlap, thus releasing the 5' flap and one nucleotide in the TSR.

The released flap plays a key role in a simultaneous secondary reaction. It acts as an Invader probe and hybridizes to a hairpin FRET cassette. Emission from excited fluorescein is quenched by the nearby quencher in the intact FRET cassette. Cleavage of the Invader structure with the same Cleavase used in the primary reaction results in the separation of the fluorescent dye and quencher; the fluorescein emission, no longer quenched, is enhanced and measured.

Cleavage in both the primary and secondary reactions occurs simultaneously and at one temperature (isothermal). After the probe is cleaved, it melts off the target and is replaced by another probe that is cleaved. It is estimated that 30 to 50 probes per minute are cleaved in the primary reaction, which after 1 hour results in amplification of approximately 10^3-fold. The secondary reaction results in amplification of 10^3- to 10^4-fold. Thus, overall amplification of 10^6- to 10^7-fold is theoretically possible.[23] The sensitivity of this technology has been reported to be at 1,000 to 10,000 copies/ml.[24]

RNA detection and quantification are also possible via the Invader technology along with several small design changes.[23,25,26] The Cleavase used in this application is derived from a eubacterium since the enzyme used with DNA targets cannot recognize tripartite Invader structures that contain RNA. In addition, several other probes, similar in design to those used with DNA targets, are employed. The Invader and primary probes for RNA targets can be designed to span splice junctions in mRNA; thus DNA with intron sequences would not result in cleaved probes.

TYRAMIDE SIGNAL AMPLIFICATION

TSA technology, also called catalyzed reporter deposition (CARD), has application for detection and localization of specific nucleic acids or proteins.[27–29] Figure 2.7 is a schematic of the process.

---

* Cleavase is a registered trademark of Third Wave Technologies, Inc., Madison, Wisconsin, USA.

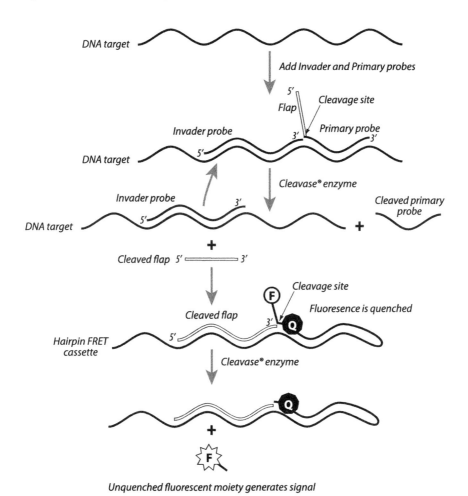

FIGURE 2.6 Schematic of Invader technology. Two different oligonucleotide probes are hybridized to the single-stranded DNA target. The primary oligonucleotide probe has a 3′ target-specific region (TSR) that hybridizes to the DNA target and a 5′ flap that does not hybridize to the DNA target. The Invader probe hybridizes just upstream of the primary probe. The 3′ terminal end of the Invader probe overlaps with the 5′ end of the TSR of the primary probe. The Cleavase enzyme recognizes the tripartite structure of the Invader probe, primary probe, and DNA target sequence and cleaves the primary probe immediately downstream of the overlap, thus releasing the 5′ flap and one nucleotide in the TSR. In a simultaneous secondary reaction, the cleaved flap hybridizes to a hairpin FRET cassette that contains a fluorescent dye (F) and a non-fluorescent quencher (Q). The same Cleavase enzyme used in the primary reaction cleaves the tripartite structure formed in the secondary reaction, thereby separating the fluorescent dye and quencher, resulting in enhancement of fluorescence emission.

TSA may employ nucleic acid probes modified with either labeled tyramide or with moieties that are ligands for labeled tyramide. This technology is dependent on the formation of an activated tyramide, a quinone-like structure, by horseradish peroxidase and hydrogen peroxide. Activated tyramide forms covalent bonds with adjacent phenol groups primarily in tyrosine residues of adjacent proteins. Signal may be produced directly or indirectly from moieties such as biotin, haptens (e.g., trinitrophenol), or fluorochromes conjugated to the tyramide. Because the signal is bound in place, it is intensified and affords a 500- to 1000-fold increase in sensitivity when compared with conventional biotin–avidin complexes in *in situ* hybridization techniques.[30]

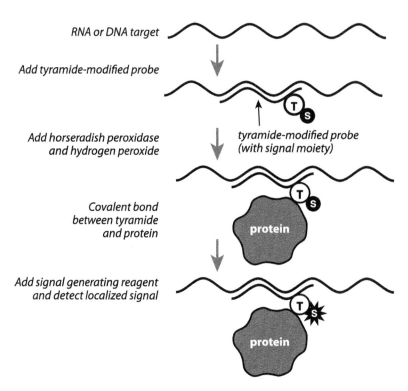

FIGURE 2.7 Tyramide Signal Amplification (TSA), also called catalyzed reporter deposition (CARD) technology. A nucleic acid probe modified with either labeled tyramide or with moieties that are ligands for labeled tyramide is allowed to hybridize to an RNA or DNA target, usually *in situ*. The addition of horseradish peroxidase and hydrogen peroxidase results in the production of activated tyramide that forms covalent bonds with adjacent proteins. Signal then is produced either directly or indirectly from moieties such as biotin, haptens (e.g., trinitrophenol) or fluorochromes conjugated to the tyramide.

The size of the localized signal-generating complex may be increased by using biotin-labeled probes that bind with horseradish peroxidase-modified streptavidin. Deposition of biotin-modified tyramide to the first complex on the target serves as a localized deposit for recognition by another layer of horseradish peroxidase-modified streptavidin. An insoluble brown precipitate may be formed after the addition of a substrate such as 3,3′-diaminobenzidine. Use of TSA may result in detection of single copies of viruses or other microorganisms in cell and tissue preparations.

Other Technologies

In addition to the five signal amplification technologies described above, there are other promising technologies, sometimes called direct detection technologies that, like signal amplification, rely on the accumulation of signal for detection of nucleic acid targets. However, in contrast to signal amplification, direct detection technologies involve the production of signal directly from hybridization of probes; development or accumulation of signal is not required through the use of enzymes such as alkaline phosphatase or other signal-enhancing mechanisms.

One of these technologies relies on resonance light scattering following hybridization of target-specific oligonucleotides modified with gold nanoparticles to signature nucleic acid targets.[31–34] A shift in the color of scattered light occurs when nanoparticles are in close proximity due to the binding of several different probe-modified nanoparticles to neighboring target DNA regions or to complexes formed by nanoparticles bound to multiple targets.[33,34]

Storhoff and co-workers[33] reported that this technology with 50-nm nanoparticles can detect as few as ~20,000 nucleic acid target molecules and that it has been employed to detect the *mecA* gene that confers resistance in methicillin-resistant *Staphylococcus aureus*. Bao and co-workers[32] reported that the deposition of elementary silver on the nanoparticles after they were bound to nucleic acid target resulted in much greater light scattering.

Another promising technology relies on charged transport through DNA duplexes that form monolayers on gold film; this technology perfectly distinguishes Watson–Crick paired hybrids from mismatched hybrids.[35–37] DNA probes derivatized at the 5′ ends with thiol-terminated alkyl chains are allowed to hybridize in solution with target DNA. Hybrids are deposited from the solution onto the gold surface; duplexes are oriented at about 70° to the surface in tightly packed monolayers. Electrons that flow through the DNA duplexes reduce intercalated methylene blue to leucomethylene blue, which in turn reduces solution-borne ferricyanide. Electrons continue to flow in this electro-catalytic cycle as long as there is available ferricyanide. The flow of electrons is diminished if there is a base mismatch (between the probe and target nucleic acid) or damaged DNA. The autocatalytic cycle can be allowed to proceed for longer periods with a resulting increase in signal and, potentially, greater differentiation between perfect and mismatched DNA and probe. Boon and co-workers[37] reported the application of this technology to detection of known single-base changes in the gene that encodes a multifunctional transcription factor for the p53 tumor suppressor gene.

PERFORMANCE FEATURES

The performance features of an assay directly impact its clinical utility. An assay with greater sensitivity may be more suitable to diagnose patients with infections or a signature nucleic acid than an assay with lesser sensitivity. A more sensitive assay also could better assess therapeutic efficacy for microbial or viral suppression and possible eradication. High assay specificity, on the other hand, gives a clinician confidence that a positive result indicates presence of a particular pathogen.

Assay linearity has taken on increasing importance as changes in viral load now have been incorporated into treatment algorithms for certain viral infections. Changes in viral load documented by a quantitative assay should not be influenced by bias due to nonlinearity of the assay. Accuracy also is important because therapeutic decision points in treatment guidelines for viral infections for example, often rely on an absolute number of nucleic acid targets in a specimen. High assay precision is desirable because smaller changes (those not expected from assay and biological variability) can be detected and may have clinical significance. Finally, quantification of nucleic acid targets by assays should not be influenced by genetic variability in the nucleic acid target sequence since absolute target thresholds may be critical for clinical decisions or prognosis.

SENSITIVITY

Historically, ultimate sensitivity has been the driving motivator of many clinical researchers and designers of new assays. Realization of ultimate sensitivity has been elusive because it is difficult to define and all amplification technologies have both theoretical and practical limitations. In the past, sensitivity often was assessed by determining the lowest level of nucleic acid target that could be distinguished from background (e.g., signal-to-noise ratio of 3). Sensitivity, also called assay cut-off or detection cut-off, sometimes was defined as the lowest level of target that produced more signal than that produced by a population of negative specimens. Assessment of sensitivity is becoming more uniformly defined.[38] Today, sensitivity is sometimes expressed as Limit of Detection (LoD) — the lowest level of target that produces signal and/or quantification values above background in 95% of replicates tested. The sensitivities of the signal amplification assays listed in Table 2.2 are cited in terms of both historic measures of sensitivity and LoD (where reported).

TABLE 2.2
Sensitivity, Limits of Detection, and Dynamic Range Values of Representative Assays Employing Different Signal Amplification Technologies

Assay	Technology	Sensitivity [References]	LoD [References]	Dynamic Range [References]
VERSANT HIV-1 RNA 3.0 Assay (bDNA)	bDNA	50/75 HIV-1 RNA copies/ml [4]	75 HIV-1 RNA copies/ml [4]	50/75–500,000 HIV-1 RNA copies/ml [4]
VERSANT HCV RNA 3.0 Assay (bDNA)	bDNA	615/990 HCV RNA IU/ml [5]	990 HCV RNA IU/ml [5]	615–7,700,000 HCV RNA IU/ml [5]
VERSANT HBV DNA 3.0 Assay (bDNA)	bDNA	2,000 HBV DNA copies/ml [6]	3,300 HBV DNA copies/ml [6]	2,000–100,000,000 HBV DNA copies/ml [6]
CMV 2.0 Assay (bDNA)	bDNA	630 CMV DNA copies per 10^6 leukocytes [111]; 900 CMV DNA copies/ml [52]	250 CMV DNA copies per 10^6 leukocytes [111]	630–5,600,000 CMV DNA copies per 10^6 leukocytes [111]
Hybrid Capture II HBV DNA Test*	HC	142,000 HBV DNA copies/ml [112]	Not available	142,000–1,700,000,000 HBV DNA copies/ml [112]
Ultra-sensitive Hybrid Capture II HBV DNA Test	HC	4,700 HBV DNA copies/ml [112]; 8,000 HBV DNA copies/ml [18]	Not available	4,700–57,000,000 HBV DNA copies/ml [112]
Hybrid Capture System CMV DNA Test†	HC	700 CMV DNA copies/ml [113]; 1,400 CMV DNA copies/ml [114]	3,000 CMV-infected target cells [54]	700–277,000 CMV DNA copies/ml [113]; 1,400–560,000 CMV DNA copies/ml [114]; 3,000–100,000 infected target cells [54]
Hybrid Capture II HPV DNA Test‡	HC	5,000 HPV genomes/assay [16]	Not calculated for qualitative assay	Qualitative assay
Hybrid Capture II GC/CT DNA Test	HC	Clinical sensitivity similar to performance from target amplification [115,116]	Not calculated for qualitative assay	Qualitative assay
Hybrid Capture II HSV DNA Test	HC	100,000–200,000 virions [117]	Not calculated for qualitative assay	Qualitative assay
HIV-1 Research Assay Configurations	Invader	<100 HIV-1 RNA copies/ml [26]; 1,000–10,000 HIV-1 RNA copies/reaction in gene expression assays [23]	Not available	Not available
HBV Research Assay Configuration	Invader	0.002 HBV DNA copies/cell and 50 HBV DNA copies/ml serum [24]	Not available	10,000–1,000,000,000 HBV DNA copies/ml serum [24]

* Hybrid Capture II HBV DNA test results are sometimes expressed as picograms/ml, where 1 picogram is equivalent to ca. 283,000 copies HBV DNA [118].

† Hybrid Capture System CMV DNA Test results are sometimes expressed as picograms/ml, where 1 picogram is equivalent to ca. 3,800 CMV DNA copies.

‡ Hybrid Capture II HPV DNA Test results are sometimes expressed as picograms/ml, where 1 picogram is equivalent to ca. 100,000 HPV DNA copies.

Sensitivity values often vary for the same assay, depending on experimental design and the definition of sensitivity chosen by the authors of studies. Studies may be statistically under-powered for assessment of sensitivity, especially if the LoD is being determined. Measures of sensitivity for different assays or technologies often are not assessed with standard volumes of specimen. Thus, sensitivity measures for these assays may not reflect the sensitivity of the underlying technologies. Finally, the sensitivity of signal amplification technologies may vary based on the nature of the specimen tested. Detection of single copies of signature DNA or RNA molecules may be achieved with *in situ* hybridization techniques with the bDNA or TSA technologies.[8,30,39–42]

Analytical sensitivity should not be confused with clinical sensitivity. Clinical sensitivity often is defined as the capacity of an assay to detect a signature nucleic acid in patients known to have the target. Unequivocal identification of such patients can be very difficult. In some clinical settings, specific thresholds of nucleic acid target levels are used to identify people with more active or progressive disease. These thresholds afford varying degrees of sensitivity, specificity, positive predictive value, and negative predictive value. However, a measure of sensitivity of a specific threshold or algorithm rule (to identify a specific disease condition or prognosis) should not be confused with analytical or clinical sensitivity of an assay.

Analytical and clinical sensitivity are not easily predictable since assays often are directed to different regions of the nucleic acid target, methods for assessing sensitivity vary, and possible inhibitors in specimens or collection devices may affect results. The ranges of sensitivity of assays employing signal amplification technologies that are required in the clinical setting are covered in the section on clinical utility later in this chapter.

ACCURACY

Accuracy is a measure of how closely an assay-derived quantitation value approximates the true value. It has not been possible to obtain the true value of a nucleic acid target in biological materials because no established reference quantification methods are available and nucleic acid targets often are heterogeneous. Thus, other quantitative nucleic acid amplification technologies typically are used for comparison of assay quantification values; such evaluations assess correlation of quantification and not accuracy.

In the absence of established and widely accepted methods for determining the true value for the concentration of nucleic acids, the trend is toward standardizing all assays to consistent international standards. Historically, both commercial and research assays yielded widely different results for the same preparations, in part because the methods and standards for calibration of these assays varied.

A means for assessing accuracy of an assay is to employ a well characterized and widely adopted international standard. Recently, the World Health Organization (WHO) developed international standards against which manufacturers could calibrate their commercial kits.[43–46] Alternatively, a universal external panel, such as one developed for assessment of accuracy of quantification of human immunodeficiency virus type 1 (HIV-1),[47] could be considered for universal standardization. A more thorough discussion of the importance of assay standardization is found later in this chapter.

PRECISION

The precision or reproducibility of an assay ultimately describes its capacity to yield similar results regardless of the circumstances (e.g., placement in an assay run, day, instrument, reagent lot, or operator). For qualitative assays, the frequency with which an assay yields a reactive result for true positive specimens with low copy numbers or a non-reactive result from specimens not containing the target nucleic acid is assay precision.

In fact, the concept of LoD is closely related to precision because it is the lowest concentration of target that yields a reactive result in a qualitative assay or a quantification value in a quantitative assay (distinguished from quantification values from specimens that do not contain the analyte) in 95% of the replicates tested. Thus, assays with greater precision will yield reactive results more often than assays with lesser precision for specimens with very low target copy numbers. In quantitative assays, precision defines the smallest change in quantification value that is statistically significant.

Precision may be expressed as percent coefficient of variation (%CV); the %CV is calculated by dividing the standard deviation by the mean value for a set of measurements. Quantification values from replicates of a sample generally are not normally distributed; however, they may be described by parameters of a normal distribution after they are transformed (e.g., $\log_{10}$ transformation). Thus, precision may be expressed as standard deviation of $\log_{10}$ values. Numerical values for both CV and standard deviations of $\log_{10}$ values are independent of the underlying concentration and, thus, can be used to compare precision estimates across the dynamic range of an assay.

Ideal measures of precision should incorporate all sources of variance, including variances within and between runs and across lots, operators, days, instruments, and sites. In the examples shown in Table 2.3, measures of precision (%CV) for the bDNA and HC technologies were as low as 3% and as high as 33%. Except for multi-site studies such as trials for licensing or registration of assays (e.g., FDA approval or clearance; Conformité Européene mark), all sources of analytical variance seldom are included in assessments of precision. In addition, biological variation (actual variation in levels of target in an infected person with stable disease) is seldom considered in overall measures of precision.

In one of the few studies in which all sources of variability were considered, Elbeik and co-workers[5] showed that the smallest change in viral load that could not be explained by inherent total variance (assay plus biologic) in the VERSANT HCV RNA 3.0 assay (bDNA) was 2.6-fold. Similarly, Gleaves and co-workers[4] showed that the VERSANT HIV-1 RNA 3.0 assay (bDNA) could detect changes as low as 2.9-fold. Thus, larger changes indicating either increased or decreased nucleic acid target levels represent true changes in viral load and may therefore indicate a changing clinical picture.

In general, assays that employ signal amplification technologies are considered more precise than assays using target amplification technologies. Target amplification depends on the synthetic and degradative activities of DNA or RNA polymerases required to produce extremely large numbers of amplicons reproducibly, regardless of wide differences in biological compositions of specimens. In addition, reactants in target amplification technologies may become rate-limiting during the geometric accumulation of amplicons, especially with higher target inputs. The real-time target amplification technologies now available are designed to avoid inaccuracy due to rate-limiting reactants.[48]

Standards for acceptable precision evolved in the clinical community based on intuition or experience — but often not on clinical studies. For example, clinicians generally consider that 0.5 $\log_{10}$ changes (three-fold increase or decrease in viral load) are significant or "real" in HIV-1-infected patients.[51] However, it is not clear whether a three-fold change in viral load portends clinical changes. Nonetheless, assays should achieve this prescribed level of precision to support clinicians who are faced with the challenge of managing patients with heterogeneous responses to therapy and disease.

DYNAMIC RANGE

The range of concentrations for which an assay provides acceptably precise and accurate results may be defined as its dynamic range. Some investigators do not include the criterion of precision in the definition of dynamic range. Several different terms with nuanced differences, for example, reporting range, measuring range, and quantitation range, have been described; differentiation of these terms will not be covered in this chapter.

TABLE 2.3
Precision of Representative Quantitative Assays Employing Different Signal Amplification Technologies

Assay	Technology	Sources of Variance Accounted for in %CV Values	Sources of Variance not Accounted for in %CV Values	Percent CV [References]
VERSANT HIV-1 RNA 3.0 Assay (bDNA)	bDNA	Within run and between runs, operators, sites, days, and kit lots	Biological variability	24–39 [4]
VERSANT HCV RNA 3.0 Assay (bDNA)	bDNA	Within runs and between runs, operators, sites, days, and kit lots	Biological variability	17–32 [5]
VERSANT HBV DNA 3.0 Assay (bDNA)	bDNA	Within runs and between runs, operators, sites, days, and kit lots	Biological variability	3–33 within site [18]
Hybrid Capture II HBV DNA Test	HC	Within runs and between runs, sites, and days	Between operators and kit lots, biological variability	3–33 within site [18] 9–29 between site [19]
Hybrid Capture System CMV DNA Test	HC	Within runs and between runs	Between operators, sites, days, and kit lots, biological variability	9–29 within site [119] 16-17 between site [119] 6.3 within site [55]
Hybrid Capture II HPV DNA Test	HC	Within runs	Between runs, operators, sites, days, and kit lots, biological variability	3–4 [120]
HBV Research Configuration Assay	Invader	Within runs	Between runs, operators, sites, days, and kit lots, biological variability	0.04–29 [24]

The performance characteristics of an assay should be rigorously assessed throughout its dynamic range as physicians and patients make clinical decisions based on changes in viral load. For example, if hepatitis C virus (HCV) RNA does not decrease by at least 2 $\log_{10}$ from baseline viral load by the end of the first 12 weeks of therapy in chronically infected patients treated with pegylated interferon and ribavirin, the recommendation is to discontinue therapy.[49] Table 2.2 lists dynamic ranges of assays employing bDNA or HC technologies.

Assay linearity and precision contribute to the establishment of the dynamic range of an assay. Linearity is evaluated by determining how closely the calibration curve or quantification values from a dilution panel approximate a straight line. Stated otherwise, linearity is a measure of how closely an assay's output is proportional to the nucleic acid target concentration.[50] Linearity is sometimes evaluated by the R^2 value derived from a plot of observed versus expected values for a dilution panel.[18] However, the R^2 metric is an insensitive measure of linearity because very significant deviations from linearity result only in small changes in R^2 values.

Today, linearity is assessed by determining whether the bias (difference between expected value and the value obtained from the assay) is within a predefined accepted range. For example, the bias observed throughout the dynamic ranges of the VERSANT HIV-1 RNA 3.0 (bDNA), VERSANT HCV RNA 3.0 (bDNA), and VERSANT HBV DNA 3.0 (bDNA) assays were <0.05 $\log_{10}$, <0.05 $\log_{10}$, and <0.12 $\log_{10}$, respectively.[4-6]

SPECIFICITY

The capacity of an assay to detect or quantify only the nucleic acid target of interest is defined by specificity. Analytical specificity may be assessed by testing specimens known to contain nucleic acids from other organisms that cause pathologic conditions (e.g., testing for other blood-borne viruses, bacteria, or fungi while assessing specificity of an HIV-1-directed assay). Clinical specificity may be assessed by testing specimens from patients who have similar pathologic conditions but are known not to contain the signature nucleic acid. In the cases of qualitative assays, specificity also may be assessed by testing specimens from a "normal" population.

The thoroughness of assessment of specificity varies widely in the literature; in general, such measures are more robust in multi-site studies often required for registration of assays with government health authorities. Specificity measures for the bDNA-based assays in multi-site studies with specimens from non-pathologic populations that are expected not to contain the nucleic acid target are 97.6% (n = 999 specimens), 98.8% (n = 999 specimens), 99.3% (n = 450 specimens), and 97.8% (at disease diagnosis) for the VERSANT HIV-1 RNA 3.0 (bDNA), VERSANT HCV RNA 3.0 (bDNA), VERSANT HBV DNA 3.0 (bDNA), and CMV 2.0 (bDNA) assays, respectively.[4-6,52]

Neisters et al.[18] found a specificity of 98.4% based on testing 15 specimens for a total of 63 results with the ultra-sensitive Hybrid Capture II HBV DNA test; similar results (99.2%; n = 134) were reported by Pawlotsky et al.[53] Caliendo and co-workers[54] reported a specificity of 93.2% for the Hybrid Capture CMV DNA test using uninfected buffy coat cells. Imbert-Marcille et al.[55] tested this assay with 50 specimens from healthy individuals along with a panel of viruses including herpes simplex-2, Epstein–Barr, varicella zoster, and human herpesvirus-6 at concentrations of 10^8 to 10^{10} particles/ml; they found a specificity of 100%. Values for specificity derived from very limited sample sizes and from preparations not encountered in clinical settings (e.g., cell-culture derived samples) should be interpreted with caution.

INCLUSIVITY AND EXCLUSIVITY OF GENETIC VARIANTS

The bDNA and HC signal amplification technologies employ long probes that cover significant portions of target sequences. Because of this design characteristic, they are expected to hybridize robustly to closely related but different members of the same family of nucleic acid targets; thus, they are expected to have very good inclusivity of variant forms of the nucleic acid targets. In comparison, target amplification technologies employ relatively short probes or primer sequences that must recognize all variants of specific targets.

Target amplification often involves elongation of nascent nucleic acid pieces (amplicons) from primers with polymerases — a process that works efficiently only if the primer is base-paired at its 3′ terminal nucleotide with the target. Such a strict requirement produces challenges in primer design for assays based on target amplification technologies. Relatively short probes also are employed by the Invader and CP technologies; thus, the design of assays based on these signal amplification technologies may be under the same constraints as assays based on target amplification technologies.

Many viral nucleic acid targets of clinical importance are known to have genetic variants that can be distinguished by phylogenetic analyses into genotypes or clades. For example, nine to eleven clades of HIV-1,[56–58] six clades of HCV,[59,60] and eight genotypes of HBV[61] were reported. Studies of detection or quantification of genetic variants for HIV-1 and HCV using the bDNA or target amplification technologies have been reported,[62–66] but experimental approaches varied widely. In general, these studies evaluated detection or quantification of nucleic acids by assays using *in vitro*-generated transcripts quantified by analytical methods or viral panels (virus in cell culture fluid or natural specimens diluted in modified serum or plasma) in which the virus was quantified by electron microscopy or consensus of multiple different nucleic acid quantification methods. Rigorous, common methodological approaches will be required by all amplification technologies in the future to establish the true performance of the assay.

While signal amplification technologies must be inclusive for genetically diverse nucleic acid targets, they also are employed in assays that distinguish closely related but genetically and often clinically distinct variants. The HC technology accomplishes discrimination of different HPV types that carry different clinical prognoses using a third generation assay that is currently under development. It involves specific capture of probe RNA-target HPV DNA hybrids to avidin-coated wells with biotin-labeled, HPV type-specific oligonucleotide probes.[67] The captured RNA–DNA hybrids then are detected with alkaline phosphatase-modified monoclonal antibody to RNA–DNA hybrids.

The Invader technology is particularly well suited for discrimination of single base differences because cleavage of the primary probe only occurs if it base-pairs with the target and forms a tripartite structure with the Invader probe recognized by Cleavase. This technology has been employed in research prototype assays to differentiate covalently closed circular DNA from double-stranded linear HBV DNA,[24] mRNA of rat cytochrome P450 2B1 from 2B2,[26] various single nucleotide polymorphisms,[68] and variants of the CYP3A gene family.[23] The CP technology has been applied to detection of specific resistance, but not wild-type, genes in enterococci and *Staphylococcus aureus*.[14,69,70]

STANDARDIZATION

Soon after the introduction of assays developed in-house or available as commercial kits and based on target or signal amplification technologies, it became clear that quantification values for the same specimens were different for different assays. Often, quantification values were expressed in different units such as copies, equivalents, genomes, or picograms per milliliter. Formulas for converting quantification values between these units often were not known or were inaccurate if known. Even when different assays provided results in the same unit, quantification values for the same specimen often were different because the assays were standardized to different calibrators by the manufacturer. Problems with inconsistent standardization may even have hindered the rapid adoption of viral load measures by the clinical community.

Ideally, assays using target or signal amplification technologies would be calibrated to common international standards to permit direct comparisons of quantification values and measures of sensitivity. With universal standardization, findings obtained with one assay can be applied to clinical practice regardless of the assay used. Ultimately, universally standardized assays will contribute to the development of guidelines for patient management. With universal standardization, clinicians and patients will be able to interpret test results without needing to convert the units of measure to those from previous tests.

In the last few years, collaborative studies have been organized to establish international standards and develop calibrated working reagents for HIV-1, HBV, and HCV. In 1999, WHO's Expert Committee on Biological Standardization developed the first international standard for HIV-1 RNA (National Institute for Biological Standards and Controls [NIBSC] code 97/656) for use with nucleic acid tests and assigned it a value of 100,000 international units per milliliter (IU/ml).[43] Also in 1999, WHO developed the HCV standard (NIBSC code 96/790) containing HCV genotype 1a virus and assigned a value of 100,000 IU/ml.[45,71] Subsequently, WHO sponsored the development of panels for HCV genotypes 2 through 6 that could be used to assess the capacities of assays to quantify different HCV genotypes equivalently.[72] The first WHO standard for HBV (NIBSC code 97/746) was assigned a value of 1,000,000 IU/ml.[46] The quantification values of these international standards for HIV-1, HCV, and HBV were derived from values obtained from a number of different laboratory-developed and commercially available quantitative or qualitative assays that employed signal or target amplification technologies.

Some challenges still exist in the implementation of international standards and the IU/ml unit of quantification that accompanies these calibrators. Although the clinical community adopted the measure of IU/ml for HCV RNA in guidelines for management of HCV-infected patients,[49,73] the acceptance and implementation of international standards for other viral nucleic acids has been somewhat slower. Despite the availability of an international standard, quantification values for HBV DNA are still usually reported in copies or picograms per milliliter, not in international units per milliliter. Likewise, the clinical community for HIV-1 continues to express viral load in copies per milliliter, perhaps because this unit of measure is inherently more understandable by clinicians who focus a great deal on virology.

Technical challenges also remain. For example, differences in quantification values of 0.5 to 1.0 $\log_{10}$ still are observed today from different WHO-standardized quantitative HCV assays.[74-77] These differences may reflect inherent performance properties of the technologies employed in the assays, and not in methods for calibration to the international standard. Also, the methods for manufacture and establishment of quantification values for WHO standards have not been tested rigorously. The scientific community anxiously awaits this type of method development.

NON-ANALYTICAL PERFORMANCE CHARACTERISTICS

Development of therapy for disease often closely parallels the elucidation of the disease. Therapies often are directed to subsets of patients who are identified, in part, by signature nucleic acid sequences both in the host and, in the case of infectious disease, in the infectious agent. With increasing constraints on funding for medical care, assays must produce clinically useful and financially grounded results.

Signal amplification technologies, to be useful for clinical decisions, must enable assays to provide inexpensive, automated, and high-throughput results. The assays should use small amounts of readily available specimens, yield multiple answers from a single sample, and utilize different assay platforms (e.g., chips, sequencing columns or other formats, and *in situ* applications). Furthermore, the underlying technologies must be robust enough that assays can be performed in the field, at a bedside, and under conditions found in developing countries. Signal amplification technologies should be readily applied to the detection of emerging targets. None of the signal amplification technologies—or target amplification technologies for that matter—has been developed sufficiently to support all the requirements of all clinical settings.

Manufacturers that incorporate the HC technology (Digene Corporation) and bDNA technology (Bayer HealthCare Diagnostics Division) have developed automated or semi-automated platforms with 96-microwell formats. A lateral-flow strip device has been described for detection of methicillin-resistant *Staphylococcus aureus* using CP technology.[14] It is likely that greater automation for signal amplification technologies will be developed within the next few years. It will be especially interesting to see how these technologies will be adapted to point-of-care applications in diverse clinical environments.

CLINICAL UTILITY

Detection and quantification of disease-associated nucleic acids by signal amplification technologies have touched many faces of medicine. Signal amplification technologies help virologists better understand the natural history and pathogenesis of many clinically important viruses, including HIV-1, HBV, HCV, CMV, and HPV. The technologies provide practitioners with valuable tools for assessing disease progression, determining baselines for initiating treatment, monitoring and assessing the efficacies of therapies and vaccines, predicting treatment outcomes, and detecting emergence of resistant viruses.

While valuable in their own right, signal amplification technologies for viral detection and quantification also enhance the values of other laboratory determinations such as viral genotyping, resistance profiles, serological markers, and CD4 counts. Signal amplification technologies also have been used for detection of bacterial pathogens and bacterial genomic sequences associated with drug resistance, and they have become tools for enhancing the sensitivities of *in situ* hybridization and microarrays.

This section is a review of the clinical utility of signal amplification technologies. We begin by summarizing the diverse applications of these technologies, ranging from routine use in medical practice to applications in veterinary medicine, public health, and basic science. We then focus on the most common applications of signal amplification technologies in medical practice, namely in understanding and treating infectious diseases. For each infectious agent, we begin by describing how signal amplification technologies have aided in elucidating disease pathogenesis — an understanding that is a prerequisite to developing and testing therapeutic compounds. We then discuss how signal amplification has been used in clinical studies and therapeutic trials, from which generalized treatment guidelines are developed. Finally, we describe ways in which these technologies are used continually to update and expand our knowledge of infectious diseases through the study of specialized patient populations.

Diverse Applications of Signal Amplification Technologies

Signal amplification technologies have found applications ranging from quantification of viral and bacterial pathogens in common medical practice to more esoteric applications in histology and basic research (Table 2.4). The bDNA and HC technologies are most widely used to quantify viral nucleic acids and detect bacteria associated with sexually transmitted diseases. bDNA technology is used to measure or detect HIV-1 RNA, CMV DNA, HBV DNA, HCV RNA, and hepatitis G virus (HGV) RNA as well as bacterial nucleic acids, for example, for *Chlamydia trachomatis* or *Neisseria gonorrhoeae*.

bDNA technology also has found places both in histology for detection of HPV DNA and in basic research where it is used in applications such as the identification of novel cytochrome p450 transcripts, quantification of cytokine mRNA, and detection of *Trypanosoma brucei* in human blood. CP technology has been used primarily for detection of bacterial genomic DNA sequences, and has been applied to detection of resistant and non-resistant strains of *Staphylococcus aureus*, *Enterococcus faecalis*, *Enterococcus faecium*, and *Mycobacterium tuberculosis*. HC technology is used in commercial assays to quantify HBV DNA and CMV DNA, and to distinguish and quantify

TABLE 2.4

Example Applications of Signal Amplification Technologies for Detection and/or Quantification of Nucleic Acids

Technology/Application	[References]
Branched DNA (bDNA)	
Quantification of viral nucleic acid (e.g., HIV-1 RNA, CMV DNA, HBV DNA, HCV RNA)	For example [4-6,121]
Determination of prevalence of HGV in patients with cryptogenic cirrhosis or HIV-1 infection	[122-125]
Quantification of HGV RNA in HGV/HIV-1 coinfected patients, patients with end stage liver disease, and liver transplant recipients	[123,125-128]
Detection of HPV-16 and HPV-18 E6/E7 DNA and mRNA in cervical cone biopsies; detection of single copy HPV DNA in whole cells	[8,39]
Detection of *mec A* gene in clinical *Staphylococcus* isolates	[129,130]
Detection of *C. trachomatis* and *N. gonorrhoeae*	[131,132]
Detection and quantification of estrogen and progesterone receptor mRNA in breast carcinomas	[133,134]
Quantification of cytokine mRNA in peripheral blood mononuclear cells	[135]
Quantification of interferon-gamma mRNA levels in unstimulated cells and purified cell populations	[136]
Quantification of tumor necrosis factor mRNA in PBMCs from patients with chronic HCV, correlation of tumor necrosis factor mRNA with plasma viral load in HIV-1 infected subjects	[137,138]
Estimation of insulin RNA splicing rates	[139]
Detection of *T. brucei* in human blood for diagnosis of African trypanosomiasis	[140]
Detection of differential expression of multiple isoforms of rat hepatocellular cytochrome p450	[141]
Cycling Probe (CP)	
Detection of bacterial genomic DNA sequences	[142]
Detection of non-multidrug-resistant and multidrug-resistant *S. aureus*	[70]
Detection of methicillin-resistant *S. aureus*	[14,143]
Detection of *vanA* and *vanB* genes in vancomycin-resistant *E. faecalis* and *E. faecium*	[69]
Detection of *M. tuberculosis* complex	[144]
Hybrid Capture (HC)	
Quantification of HBV DNA and CMV DNA	For example [145-147]
Detection of high risk and/or low risk HPV types	For example [105,148,149]
Detection and typing of herpes simplex virus DNA	[117]
Detection of *C. trachomatis* and *N. gonorrhoeae*	For example [115,116,150]
Detection of urinary tract bacteria	[151]
Detection of *Campylobacter jejuni* in food and poultry viscera	[152]
Invader	
Quantification of intrahepatic cccDNA in chronic HBV patients and those receiving adefovir	[153]
Quantification of HBV cccDNA in liver biopsies and sera	[24]
Detection of mutations associated with resistance to rifampin and isoniazid in *M. tuberculosis*	[154]
Detection and quantification of single nucleotide polymorphisms from unamplified genomic DNA	[155,156]
Genotyping determinations of cytochrome p450 2D6 alleles	[157]
Genotyping of routine clinical patient samples for the Factor V Leiden point mutation	[158]
Multiplexed gene expression analysis	[159]

TABLE 2.4 (continued)
Example Applications of Signal Amplification Technologies for Detection and/or Quantification of Nucleic Acids

Technology/Application	[References]
Tyramide Signal Amplification (TSA), also called catalyzed reporter deposition (CARD)	
Detection of HIV-1 mRNA in formalin-fixed and paraffin-embedded tissues, and in blood smear samples or paraffin-embedded autopsy tissue	[160,161]
Detection of herpes simplex virus DNA in breast milk	[162]
Detection of HPV DNA in formalin-fixed carcinoma cells from uterine cervixes	[163]
Detection of parvovirus B19 DNA in serum	[164]
Detection of mRNA in ES cell-derived cardiomyocytes and in the developing heart	[165]
Analysis of gene amplification in malignant mixed tumor of salivary gland	[166]
Detection of single copy genes in bacterial communities by DNA microarrays	[167]
Direct detection of *16S* rRNA of 4 important bacterial species	[168]
Detection of bacteria in contaminated ground water and in marine and fresh water samples	For example [169-172]
Detection of foot and mouth disease virus RNA in infected cells of cloven-footed animals	[173]
Confirmation of *H. felis* as causative mycoplasmid organism of feline infectious anemia	[174]
Detection of integrated feline leukemia viruses in cat lymphoid tumor cell line	[175]
Enumeration and mapping of integrated murine leukemia viruses	[176]
Detection of catalase mRNA and protein in adult rat brain sections; possible use in study of Alzheimer's and aging	[177]
Localization and semi-quantification of hepatic tamoxifen–DNA adducts in rats exposed to tamoxifen	[178]
Location of T-DNA insertion in transgenic shallots	[179]

DNA from different types of HPV, and to detect *C. trachomatis* and *N. gonorrhoeae* nucleic acids. HC technology also has been used to detect bacterial pathogens that infect the human urinary tract and to detect *Camphylobacter* in food and poultry viscera. Applications for Invader technology range from detection of viral nucleic acids and drug-resistant bacterial pathogens to analysis of human genomic DNA and multiplexed gene expression analysis. TSA or CARD technology also has a broad range of applications, from detection of viral nucleic acids in human tissues and fluids to analysis of viral and bacterial pathogens in water samples and animals.

HIV-1 Infection

Many of the first novel insights into HIV-1 pathogenesis were elucidated through pioneering studies using bDNA technology — the only signal amplification technology applied to quantification of HIV-1 RNA. In the past decade, large studies that utilized bDNA technology clearly demonstrated the link between HIV-1 viral load and disease transmission and progression. An early study[78] examined the relationships of viral load, CD4 count, and disease progression. The research team demonstrated that patients with viral loads $>10^5$ copies/ml within 6 months of seroconversion were 10 times more likely to progress to AIDS within 2 years than those with viral loads $<10^4$ copies/ml.[79] A subsequent study[80] showed that baseline viral load was a stronger predictor of disease progression than CD4 count, but that the two factors together were more predictive of disease progression than either factor alone. The chance of progression to AIDS or death was linearly related to increasing viral load; a patient with a baseline viral load >30,000 copies/ml had a 12.8- to 18.1-fold greater chance of progressing to AIDS or dying within 6 years than patients with viral loads <500 copies/ml. Also, viral load and change in viral load were shown to be predictive factors for the risk of developing opportunistic infections.[81] These findings have contributed to the development of treatment recommendations.

Although the seminal clinical trials of antiviral drugs for treatment of HIV-1 infection were monitored with PCR technology, bDNA technology has been used in a variety of other clinical trials (Table 2.5). Despite the relatively small numbers of patients included in these trials, the findings from these studies have nonetheless contributed to further establishing effective HIV-1 therapy. Moreover, bDNA technology is used in clinical practice to implement guidelines for treatment of HIV-1 infection that rely on viral load quantification in conjunction with CD4 cell counts (Table 2.6). These guidelines, based on analysis of data from multiple clinical trials, reflect consensus opinions and represent the current standard of care.

The goal of therapy for the HIV-1-infected patient is suppression of viral load below the LoDs of the latest generation of quantitative HIV-1 RNA assays: 50 copies/ml for the PCR-based COBAS AMPLICOR HIV-1 MONITOR® test,* v1.5 (Roche Molecular Systems, Pleasanton, California, USA) and 75 copies/ml for the VERSANT HIV-1 RNA 3.0 (bDNA) assay. To date, no studies have shown whether differences in clinical decisions based on the slightly different LoDs of these two assays impact treatment outcomes. Indeed, there has been debate related to the clinical relevance of even wider viral load threshold differences.[82,83] The consensus guidelines may be modified in the future as the assays become even more sensitive and as additional studies shed further light on the clinical relevance of still lower viral threshold values.

CYTOMEGALOVIRUS INFECTION

Two signal amplification technologies are employed in the quantification of CMV DNA: HC technology and a second generation version of bDNA technology. Unlike many other viruses, such as HIV-1, HBV and HCV, CMV is a common virus that usually causes symptoms only when it reaches a critical level. Therefore, measurement of CMV DNA most often is performed in immunocompromised patients who are at risk for CMV disease. Aside from increasing our understanding of CMV infection, reasons to measure CMV DNA include (1) the establishment of critical thresholds for diagnosis of CMV disease, (2) documentation of increasing CMV titers for prediction of CMV disease and decisions regarding pre-emptive therapy, and (3) monitoring of antiviral therapy to determine duration of treatment and evaluate potential resistance.[84]

Assays for measuring CMV DNA are useful both for managing CMV infection and disease in HIV-1 infected patients and identifying post-transplant patients at risk for developing CMV disease and monitoring their therapy (Table 2.5). In HIV-1-infected patients, the two viruses have been recognized as inextricably entwined pathogens because CMV is the most common opportunistic infection in patients with AIDS.[85] In the transplant setting, assays based on signal amplification technologies have been used for early detection of CMV and for establishment of critical viral thresholds predictive of development of CMV disease. Because the virus can replicate quickly post-transplant, both high and increasing levels of CMV DNA are important predictors of development of CMV disease. Hence, assays that are sensitive enough to detect CMV infection early and precise and linear enough to measure increases in CMV DNA levels are critical for the initiation of pre-emptive therapy required to prevent rapid disease progression.

Use of viral load thresholds to guide management of patients at risk for CMV disease has lagged behind the use of such thresholds for management of other viral infections such as HIV-1, HBV, and HCV. This may be due in part to the fact that CMV DNA thresholds predictive of CMV disease appear to vary among different types of transplant recipients, among individual patients, and among assays based on different technologies.[21,55] A much-needed international standard for CMV DNA has yet to be developed. For these reasons, CMV DNA thresholds have not been incorporated into the guidelines for CMV management to date.

---

* COBAS, AMPLICOR, and MONITOR are registered trademarks of a member of the Roche Group, Basel, Switzerland.

TABLE 2.5
Examples of Signal Amplification Technologies in Clinical Trials and Studies of HIV-1 and CMV Infection

Virus	Technology/Patient Population	Study Findings	[References]
HIV	**bDNA Technology**		
	HIV-1-infected adults	Demonstrated interleukin-2 added to nucleoside therapy did not improve immune function	[180]
	HIV-1-infected adults	Established efficacy of recycling stavudine and didanosine in patients requiring salvage therapy	[181]
	HIV-1-infected adults	Showed superiority of triple to dual nucleoside therapy in inducing sustained virologic response	[182]
	HIV-1-infected mothers	Demonstrated nevirapine substantially reduced peripartum mother-to-child transmission in Cameroon	[183]
	HIV-1-infected pediatric patients	Found HIV-1 RNA levels were higher in pediatric patients with AIDS than in moderately symptomatic or asymptomatic patients and viral load gradually decreased with age	[184]
	HIV-1-infected hemophiliacs	Established correlations between lower HIV RNA, higher CD4 cell count, and presence of certain chemokine receptor genotypic variants in hemophiliacs; found patients with CCR2b mutant allele had lower risk of progression to AIDS	[185]
CMV	**bDNA Technology**		
	HIV-1-infected adults	Correlated development of drug resistance and disease progression in patients undergoing therapy for CMV retinitis	[121]
	HIV-1-infected adults	Diagnosed CMV polyradiculopathy; documented declining CMV levels in cerebrospinal fluid in 9/15 patients after mean of 11.5 days of treatment	[186]
	HC Technology		
	HIV-1-infected adults	Established decrease in CMV DNA level as a measure of effectiveness of immune reconstitution in HIV-infected patients receiving HAART but no CMV-specific therapy	[187]
	Renal transplant recipients	Demonstrated 80% reduction in CMV DNA associated with ganciclovir therapy	[188]
	Allogenic stem cell transplant recipients	Showed declining CMV DNA levels in allogenic stem cell recipients treated with myeloablative therapy	[189]
	Lung transplant recipients	Showed CMV DNA was detected on average 36 days earlier than either CMV by culture or by the shell vial assay, and defined definitive viral load cut-off levels that correlate with significant disease	[190]
	Liver transplant recipients	Found independent risk factors for clinically significant CMV infection to include a CMV DNA level of >10 picogram/ml, weekly increases in DNA levels and a critical value (value at which active infection is initially detected) of 13 picogram/ml	[146]
	Solid organ transplant recipients	Showed that CMV DNA levels >50 to 60 picogram/ml were predictive of more severe CMV disease	[20]

TABLE 2.6
Role of Viral Load in Conjunction with CD4 Cell Count in Management of Patients Infected with HIV-1*

Patient Population	Diagnosis	Initiation of Therapy	HIV-1 RNA Monitoring Schedule	Goals of Therapy
Treatment naïve	Use in event of negative or indeterminate anti-HIV test and syndrome consistent with acute HIV-1 infection	Establish baseline viral load using two viral load determinations at least 1 week apart; treat symptomatic HIV-1 disease regardless of viral load and CD4 count; defer therapy if CD4 cell count >350/mm3 and HIV-1 RNA <100,000 copies/ml	2–8 weeks after initiation of therapy; every 3–4 months on therapy	1 $\log_{10}$ drop at 16–24 weeks; eventual suppression of HIV-1 RNA to <50–75 copies/ml ‡
Previously treated†	Not applicable	Consider change of regimen if failure to suppress HIV RNA below 50–75 copies/ml after 1 year or evidence of increasing viral load on two consecutive HIV-1 RNA measurements	Viral load measured at time points determined by clinician; if evidence of virologic failure with limited prior treatment, consider changing regimen; if evidence of virologic failure with extensive prior treatment, may observe on same regimen	Suppression of HIV-1 RNA to <50–75 copies/ml ‡
Pregnant women	Use in event of negative or indeterminate anti-HIV-1 test and syndrome consistent with acute HIV infection	After 10–12 weeks of gestation, treat those not meeting criteria for naïve patients if HIV-1 RNA >1,000 copies/ml; schedule cesarean section at 38 weeks for those with HIV-1 RNA >1,000 copies or untreated patients with viral loads pending	Every trimester	Reduction of HIV-1 RNA to <1,000 copies/ml to prevent vertical transmission
HIV-1-exposed infants	HIV-1 RNA or DNA at 48 hours after birth if negative; repeat at 2 weeks if again negative; repeat at 1–2 months and 3–6 months	Prophylaxis indicated regardless of viral load in mother If HIV-1 infection diagnosed, treat infants under 1 year of age with clinical and immunological symptoms of HIV-1 disease regardless of viral load	Viral load measured at time points determined by clinician	Goal of prophylaxis: to prevent HIV-1 infection; goal of therapy after infection diagnosed: suppression of HIV-1 RNA to <50–75 copies/ml ‡

* Based on guidelines and recommendations available on AIDSinfo Website (http://AIDSinfo.nih.gov), including References 51 and 191.

† Includes both patients who have experienced treatment interruptions and those who are receiving ongoing treatment but are not achieving virologic responses.

‡ Fifty copies/ml by the AMPLICOR HIV-1 MONITOR 1.5 Test and 75 copies/ml by the VERSANT HIV-1 RNA 3.0 Assay (bDNA).

Hepatitis B Virus Infection

Direct detection of HBV DNA by methods such as dot blot or liquid hybridization was employed in early studies of HBV infection. However, in the 1990s, quantitative HBV DNA assays based on signal amplification technologies were developed; they furthered our understanding of HBV infection and its response to therapy (Table 2.7). Studies employing the HC and bDNA technologies also contributed to the development of treatment algorithms for HBV-infected patients.

The original HBV treatment guidelines published by the American Association for the Study of the Liver defined virologic responses to therapy as a decrease in HBV DNA to levels then undetectable by commonly employed signal amplification tests ($<10^5$ copies/ml) and loss of HBeAg in patients who were initially HBeAg-positive.[86] Recently these guidelines were refined further to include more sensitive viral thresholds for response and thresholds for treatment initiation in different HBV-infected populations. Table 2.8 outlines recommendations from the most current guidelines.[87–90]

Although PCR-based assays are recommended[87] for monitoring therapy and determining treatment success of HBV-infected patients, assays based on signal amplification technologies in general have wider dynamic ranges and greater precision and hence may be better suited than most end-point PCR-based assays for determination of baseline viral loads. The upper limits of both the Hybrid Capture II HBV DNA test (1.7×10^9 copies/ml) and the VERSANT HBV DNA 3.0 assay (bDNA) (1×10^8 copies/ml) were higher than that of the PCR-based COBAS AMPLICOR HBV MONITOR test (upper limit of 2×10^5 copies/ml)[6] and hence should quantify baseline viral load in a larger percentage of samples without dilution from untreated HBV-infected populations, especially those who are HBeAg-positive. The real-time PCR-based COBAS HBV TaqMan® analyte-specific reagent test* (Roche) offers the advantage of a wide dynamic range with an upper limit up to 10^9 HBV DNA copies/ml[91] and thus also may be used in the future for determination of baseline viral loads.

Current assays based on signal amplification technologies are often used in conjunction with more sensitive PCR-based assays to monitor patients undergoing therapy.[92] Future generations of assays based on signal amplification technologies will require increased sensitivity for continual monitoring of responses to nucleoside and nucleotide analogues where the virus level quickly drops 2 to 4 $\log_{10}$ units to levels below their current lower limit of quantification.

Hepatitis C Virus Infection

The bDNA technology, the only signal amplification technology used to quantify HCV, has contributed to the general knowledge of the natural history and pathogenesis of untreated HCV infection. The second generation bDNA technology was used in studies that examined stability of viral levels and the relationship of viral levels to liver disease and antibody production. One investigator showed that HCV RNA levels fluctuated on average less than 1 $\log_{10}$ in untreated patients observed over 5 years.[93] Another showed the lack of correlation between HCV RNA level and severity of liver disease.[94] Yet another group showed that intravenous drug users could have low levels of HCV RNA for up to 5 years prior to antibody seroconversion,[95] a finding that underscores the limitations of antibody testing in this population and the necessity of periodic nucleic acid testing. Both second and third generation assays based on bDNA technology have been used in studies of patients considered to have typical chronic HCV infections as well as those considered difficult to treat (Table 2.7).

---

* TaqMan is a registered trademark of a member of the Roche Group, Basel, Switzerland.

TABLE 2.7
Examples of Signal Amplification Technologies in Clinical Trials and Studies of HBV and HCV Infection

Virus	Technology/Patient Population	Study Findings	[References]
HBV	**bDNA Technology**		
	HBV-infected adults	Showed greater efficacy of entecavir as compared to lamivudine in patients chronically infected with HBV	[192]
	HBV-infected adults	Established efficacy of sequential treatment with both lamivudine and IFN in patients who had not previously responded to IFN monotherapy	[193]
	HBV-infected adults with decompensated cirrhosis	Established efficacy of 6 months of lamivudine	[194]
	Adults infected with HBV strains carrying YMDD mutations	Demonstrated HBV DNA levels associated with viral breakthrough	[195]
	HBV-infected pediatric patients	Established efficacy of lamivudine in children at 24 and 48 weeks of therapy	[196]
	HBeAg-positive patients	Showed that HBeAg-positive patients with detectable HBV DNA were more likely to progress to hepatocellular carcinoma than those without detectable HBV DNA, and that the risk for hepatocellular carcinoma increased with increasing HBV DNA levels	[197]
	HBV/HCV coinfected patients	Showed that HCV RNA levels are significantly higher in the HCV monoinfected patients than in HBV/HCV coinfected patients, 3.75×10^6 versus 0.64×10^6 log_{10} equivalents/ml (p <0.05), respectively	[198,199]
	HBV/HCV coinfected patients	Showed that HBV DNA was detected less frequently (33% versus 79%, p<0.05) and at lower titers in HBV/HCV coinfected than monoinfected patients	[200]
	HBV/HCV coinfected patients	Showed that hepatocellular carcinoma was strongly associated with HCV monoinfection and HCV/HBV coinfection	[201]
	HC Technology		
	HBV-infected adults	Demonstrated biphasic decline in HBV DNA in patients receiving 150 mg lamivudine daily	[145]
	HBV-infected adults	Showed ability of 125 mg adefovir to reduce HBV DNA level by more than 1 log_{10} within 28 days	[202]
	HBV-infected adults	Showed ability of 300 mg entricitabine to reduce HBV DNA level by 3.3 log_{10} in 2 months	[203]
	HBV-infected mothers	Documented 50% reduction in vertical transmission of HBV from viremic mothers treated with lamivudine in last trimester of pregnancy compared to untreated cohort	[204]
	Asymptomatic HBV carriers	Demonstrated that hepatitis B surface antigen (HBsAg) and HBV DNA levels in asymptomatic HBV carriers were correlated (r = 0.709, P<0.001), suggesting HBsAg titers could provide an easy means for following HBV replication in carriers if determination of HBV DNA levels was unavailable	[205]

HCV	bDNA Technology		
	HCV-infected adults	Monitored viral load while establishing safety of 300 mg daily dose of amantadine in patients chronically infected with HCV	[206]
	HCV-infected adults	Established superior efficacy of high daily dosage of IFN-α in previous non-responders infected with HCV genotype 1	[207]
	HCV-infected adults	Showed greater $\log_{10}$ decline in HCV RNA within first few days of IFN therapy in HCV genotype non-1 than HCV genotype 1 patients	[208]
	HCV-infected adults	Demonstrated positive predictive value of 83% and negative predictive value of 92% associated with >2 $\log_{10}$ decline in HCV RNA by week 2 of second round of treatment with IFN/ ribavirin in patients who previously relapsed following IFN monotherapy	[209]
	Dialysis patients	Showed low dose IFN in HCV infected dialysis patients decreased viremia and slowed progression of liver disease	[210]
	Transplant recipients	Showed low dose IFN given to post liver transplant patients with recurrent HCV infection resulted in histological improvement	[211]
	Transplant recipients	Demonstrated benefits of peg-IFN/ribavirin following IFN monotherapy	[212]

TABLE 2.8
Role of Viral Load in Managing Patients Infected with HBV*

Patient Population	Treatment Indication by Viral Load	HBV DNA Monitoring Schedule	Goals for Viral Suppression in Conjunction with Other Markers
HBeAg-positive	Treat patients with HBV DNA ≥10^5 copies/ml and elevated alanine aminotransferase (ALT); treatment initiation for patients with HBV DNA ≤10^5 copies/ml and normal ALT requires liver biopsy to determine need for therapy	Every 6–12 months	Suppression of HBV DNA to lowest possible level, loss of HBeAg, conversion to anti-HBe, normalization of ALT, improvement in histology
HBeAg-negative	Treat patients with HBV DNA >10^4 copies/ml	Every 6–12 months	Suppression of HBV DNA to lowest possible level, normalization of ALT
Compensated cirrhosis	Monitor or treat patients with HBV DNA <10^4 copies/ml; treat patients with HBV DNA ≥ 10^4 copies/ml	Every 3 months (on lamivudine every 1–2 months)	Suppression of HBV DNA to undetectable level (< 400 copies/ml), loss of HBsAg
Decompensated cirrhosis	Treat all patients	At least every 3 months (on lamivudine every 1–2 months)	Suppression of HBV DNA to undetectable level (< 400 copies/ml), loss of HBsAg
HBV/HCV coinfected	Determine dominant virus; treat patients with HBV DNA ≥10^5 copies/ml and undetectable HCV RNA and patients with low HBV DNA and detectable HCV RNA	Optimal monitoring schedule not yet established	Suppression of viral load to undetectable levels (<400 copies/ml HBV DNA, <50 IU/ml HCV RNA)
HBV/HIV-1 coinfected	Continue to monitor patients with HBV DNA ≤10^5 copies/ml and normal ALT; consider treatment for all HBeAg-positive and for HBeAg-negative with ALT >1.5 upper limit of normal and HBV DNA >10^5 copies/ml; all HIV-1-infected patients with chronic HBV and elevated transaminase levels should be considered candidates for anti-HBV therapy	Not proposed	Suppression of HBV DNA and HIV-1 RNA to lowest possible levels, improve liver histology, eventual loss of HBeAg and conversion to anti-HBe in e antigen positive patients

* Based on algorithms presented in References 87–90.

Studies in which HCV RNA is detected and/or measured in large populations of chronic hepatitis C patients undergoing therapy led to the development of treatment algorithms. As in the case of algorithms for treatment of HIV-1 infection, algorithms for treatment of HCV infection are largely based on early studies that employed PCR-based assays. These algorithms rely on accurate quantification of HCV RNA along with genotype assignment to determine the initial length of therapy and its possible early termination (Table 2.9).

Because of its linearity and wide dynamic range, the VERSANT HCV RNA 3.0 assay (bDNA) is an excellent choice for implementing treatment algorithms that are dependent on accurate quantification (e.g., in stopping treatment in patients infected with HCV genotype 1 who fail to demonstrate a 2 $\log_{10}$ decline in viral load by week 12 of therapy). Such a metric requires accurate viral load determination in both baseline and week 12 samples. In clinical trials, the (615 to 7.7 × 10^6 IU/ml) dynamic range of the VERSANT HCV RNA 3.0 was shown to be sufficient to quantify neat baseline samples from chronic HCV-infected patients in the 5th to 95th percentiles, whose viral loads ranged from 3.35 ×10^4 to 3.7 ×10^6 IU/ml.[96] By comparison, the dynamic ranges of most end-point PCR-based assays do not extend as high and thus these assays do not quantify without dilution of many of the higher viral load samples.[97]

Dynamic ranges for the PCR-based AMPLICOR HCV MONITOR v 2.0 test (manual and COBAS, Roche), LCx® HCV RNA quantitative assay* (Abbott Diagnostics, Abbott Park, Illinois, USA) and the HCV SuperQuant™† (provided as a service by the National Genetics Institute, Los Angeles, California, USA) are 600 to <5 × 10^5 IU/ml,[64,77,98] 25 to 2.63 ×10^6 IU/ml,[99] and 30 to 1.47 × 10^6 IU/ml,[74,100] respectively. On the other hand, the real-time PCR-based COBAS HCV TaqMan (Roche) has a broad dynamic range of 50 to 2 × 10^8 IU/ml,[101] and could also be used for implementing the 12-week stopping rule.

Treatment algorithms and the viral thresholds they employ are constantly evolving. For example, the viral load threshold of 800,000 IU/ml recommended in 1999 for determining therapy duration and likelihood of response[73] was discarded by 2002 when two new algorithms with different viral load thresholds emerged.[49,102] The VERSANT HCV RNA 3.0 was FDA-cleared in 2003 and thus did not contribute to the development of current algorithms. However, this assay likely will be employed in future studies that will provide additional information on the responses of HCV infections to different therapies and the responses of different patient populations. Such information will undoubtedly lead to further refinement of treatment algorithms.

Human Papillomavirus Infection

Signal amplification technologies for HPV DNA mainly provide qualitative information that distinguishes infections associated with innocuous lesions such as warts from HPV infections associated with cancer. Although most HPV infections are benign or transient, persistent cervical infections by high risk HPV types may lead to cervical cancer. The Hybrid Capture II HPV DNA test distinguishes the low risk HPV-6, -11, -42, -43, and -44 from the high risk HPV-16, -18, -31, -33, -35, -39, -45, -51, -52, -56, -58, -59, and -68. In addition, the Hybrid Capture II detects the 13 high risk HPV types even before the appearance of visible cell changes.[103] A new generation assay based on HC technology that detects HPV and also differentiates and quantifies HPV DNA from specimens containing mixtures of HPV types at different concentrations is being developed.[22]

HPV DNA testing by HC technology has broad applications. It elucidated the epidemiology of HPV infections by demonstrating their prevalence in different populations and has been used for triaging women with abnormal cytologies and as a test of cure (see Table 2.10). HPV DNA testing is a valuable adjunct to screening with cytology or can serve as a stand-alone test. Used alone, it can identify 95 to 100% of cervical intraepithelial neoplasias (CIN 3) and most cervical

---

* LCx is a registered trademark of Abbott Laboratories Corporation, Abbott Park, Illinois, USA.
† SuperQuant is a trademark of the National Genetics Institute, Los Angeles, California, USA.

TABLE 2.9
Role of Viral Load in Managing Patients Infected with HCV

Patient Population*	Duration of Therapy [References]	Prediction of Response [References]	Stopping Rule [References]
HCV genotypes 1 and 4	HCV genotype 1: 48 weeks regardless of baseline viral load [49], by prior guideline 48 weeks for patients with HCV RNA ≥800,000 IU/ml and 24 weeks for patients with HCV RNA ≤800,000 IU/ml [73]; HCV genotype 4: 48 weeks regardless of baseline viral load [213,214], although recent findings suggest 36 weeks of therapy for genotype 4 is sufficient [215]	Baseline viral load ≥800,000 IU/ml and slow rate of response associated with lower response rate [73]; presence of HCV genotype 1 or 4 together with baseline viral load ≥800,000 IU/ml is strongest baseline predictor of non-response [214]	<2 log$_{10}$ drop by week 12 [49]; <2 log$_{10}$ drop by week 12 for HCV genotype 1 only [102]
HCV genotypes 5 and 6	48 weeks regardless of baseline viral load [102]	Little known	Not applied [102]
HCV genotypes 2 and 3	24 weeks independent of viral load [49,73,102,216]	Response appears independent of baseline viral load, although higher relapse rate noted in HCV genotype 3 patients with baseline viral load >600,000 IU/ml [216]	Not applied because of 75–80% response rate
HCV/HIV-1 coinfected	48 weeks independent of viral load or HCV genotype [217]	Baseline viral load >800,000 IU/ml among 70% of coinfected patients and slower rate of viral decline associated with lower response rate	<2 log$_{10}$ drop by week 12 [217]

* For all patient populations described here, the HCV RNA monitoring schedule is quantitative testing at baseline and week 12, and qualitative testing at end of treatment and end of follow-up week 24.

TABLE 2.10

Examples of Signal Amplification Technologies in Clinical Trials and Studies of HPV, *C. trachomatis*, and *N. gonorrhoeae* Infections

Pathogen	Technology/Patient Population	Study Findings	[References]
HPV	**HC Technology**		
	Adult females	Demonstrated detection of HPV DNA by HC was more sensitive than either ThinPrep®* liquid cytology or colposcopy, and therefore could be used for triaging women with cytological determinations of atypical squamous cells of undetermined significance (ASC-US)	[218]
	Adult females	Showed reflex HPV DNA testing of liquid based cytology specimen was less expensive than repeat cytology, and reduced number of colposcopies by 40–60%	[104]
	Adult females	Demonstrated use of HPV DNA as a test of cure by correlating absence of HPV DNA with lower risk of residual or recurrent cervical intraepithelial neoplasia (CIN) 2/3	[104]
	HIV-1-infected adults	Detected HPV DNA in 58% of HIV-1 positive patients versus 6% of control subjects, and high risk HPV genotypes in 40% of HIV-1 positive patients with no difference between homosexual men and other HIV-1 positive patients; results suggest that identification of high risk HPV infection might be warranted in immunosuppressed patients	[149]
	HIV-1-infected adults	Confirmed increased prevalence of latent HPV infection among HIV-1-infected adults but did not show significant effect of imiquimod in its eradication	[219]
	HIV-1-infected females	Detected high risk HPV types in 25% of HIV-1 infected women with normal cytology and 78% of those with abnormal cytology regardless of antiretroviral therapy	[220]
	HIV-1-infected females	Demonstrated significantly higher HPV viral load in women with high grade cervical dysplasia compared to controls and significant decrease in HPV viral load after surgical resection of lesion	[221]
	Females at risk for HIV-1 infection	Established 64% prevalence of HPV infection in HIV positive women versus 28% for HIV negative women	[222]
C. trachomatis, N. gonorrhoeae	**HC Technology**		
	Females at family planning clinics	Established prevalence of 16% for *C. trachomatis* and/or *N. gonorrhoeae* in 1342 women presenting in sexually transmitted disease clinics	[150]
	Males at family planning clinics	Established prevalence of 22% for *C. trachomatis* and/or *N. gonorrhoeae* in 1211 men presenting in sexually transmitted disease clinics	[223]
	Pregnant females with HPV infection	Established higher prevalence of tobacco use (50 vs. 11.5%), bacterial vaginosis (54 vs. 15%) and *C. trachomatis* infection (35 vs. 8%) in pregnant women with HPV infection than in those without	[224]
	Females of reproductive age in northeast Brazil	Established 26% prevalence of HPV and 6% prevalence of *C. trachomatis* infection in 341 women of reproductive age from rural northeast Brazil	[225]

(continued)

TABLE 2.10 (continued)
Examples of Signal Amplification Technologies in Clinical Trials and Studies of HPV, *C. trachomatis*, and *N. gonorrhoeae* Infections

Pathogen	Technology/Patient Population	Study Findings	[References]
C. trachomatis *N. gonorrhaeae*	Females self-referred for routine gynecological care near U.S. Mexican border	Established 8.2% prevalence of *C. trachomatis* infection in 2,270 women living near U.S. Mexican border and its association with young age, new sexual partner, HPV infection, and proximity of clinic to border	[226]
	Females attending clinics for routine gynecologic care	Established higher prevalence of *C. trachomatis* infection among Mexican women overall, but similar prevalence of HPV infection in Mexican versus American women living near U.S. Mexican border	[227]
	Japanese sex workers	Established prevalence of 13% *C. trachomatis* and 4% *N. gonorrhoeae* in 546 Japanese sex workers	[228]

* ThinPrep is a registered trademark of Cytyc Corporation, Marlborough, Massachusetts, USA.

cancers.[103] HPV DNA testing by HC technology shows higher sensitivity and specificity (90 and 70%, respectively) than cytology for managing women with atypical squamous cells of undetermined significance (75 and 60%, respectively); the sensitivity and negative predictive value of both tests combined is near 100%. HPV DNA positivity in the absence of abnormal cytology is not considered a false positive because HPV DNA-positive women are among those at greatest risk for developing abnormal smears and cervical neoplasias.[104] Although the clinical utility of HPV DNA quantification is not clearly established, one large study showed that viral loads of >10 RLU/cut-off by HC technology increased the specificity of the test and its positive predictive value.[105]

HPV DNA screening in combination with cytology for women over 30 years of age has now been incorporated into the clinical management guidelines for obstetricians–gynecologists as a Level B recommendation.[106] Although HPV is an opportunistic infection with a higher prevalence in immunocompromised patients such as those with HIV-1 infection or post-transplant status,[107] routine screening for cervical and anal HPV DNA has not yet been added to the management guidelines for HIV-infected individuals. In the future, HPV DNA screening may be added to certain guidelines as the utility of tests for HPV DNA is further defined.

Chlamydia trachomatis and Neisseria gonorrhoeae

Screening for *C. trachomatis* and *N. gonorrhoeae*, two of the most common sexually transmitted diseases and potential cofactors of HPV infection, has become routine in obstetrics and gynecology clinics, female family planning clinics, and in private physicians' offices for men with urogenital complaints and for both men and women in high risk populations. Prior to the development of nucleic acid tests, screening was dependent on culture methods that required viable organisms and 1 to 2 days to learn results. Although *C. Trachomatis* and *N. gonorrhoeae* can be identified by the unique Hybrid Capture II GC/ID and Hybrid Capture II CT/ID tests, the Hybrid Capture II GC/CT DNA test is used most often to qualitatively detect both organisms. HC technology for detection of *C. trachomatis* and *N. gonorrhoeae* has been used in multicenter trials for direct screening and for establishing the prevalence of infection among people who attend family planning and sexually transmitted disease clinics and among Japanese sex workers. Table 2.10 summarizes these studies.

Screening followed by treatment of individuals with both asymptomatic and symptomatic *C. trachomatis* and *N. gonorrhoeae* infections is the chief mechanism of infection control. These common infections are often silent and are associated with significant pathology, including epididymitis, pelvic inflammatory disease, salpingitis, ectopic pregnancy, adverse outcome of pregnancy, and neonatal infection. For these reasons, the Centers for Disease Control and Prevention recommend *C. trachomatis* screening for all sexually active women under the age of 26.[108] The European guidelines for the management of *C. trachomatis* and *N. gonorrhoeae* infections recommend screening for both organisms.[109,110] Follow-up testing should be performed 3 to 4 weeks after completion of treatment.

CONCLUSIONS

Signal amplification technologies serve as the bases of many assays that are widely used in clinical practice. For the most part, assays based on signal amplification technologies have the sensitivity, specificity, accuracy, standardization, precision, linearity, and dynamic range that enable them to contribute valuable information to a clinical picture. In addition, the inherent designs of some signal amplification technologies (those that include long probes) render them very robust in detecting the nucleic acid target sequences of interest, regardless of common genetic variations.

Today, signal amplification technologies play central roles in patient management. Assays based on these technologies influence the practices of numerous medical specialties including infectious disease, hepatology, immunology, pathology, obstetrics and gynecology, pediatrics, and transplant medicine, and are also useful to epidemiologists and public health laboratories. They are used to detect mutations associated with resistance to therapy in a variety of pathogens, to expand the role of microarrays, and to detect viral nucleic acids in tissues fixed for pathological examination.

The most common applications for signal amplification technologies in medical practice today contribute to understanding and treating infectious diseases. The prevalence and severity of an infectious disease drive the development of therapeutic measures and assays for diagnosis and monitoring. At present, assays based on signal amplification technologies are used primarily for viral load monitoring in HIV-1, HCV, HBV, and CMV infections and for detecting and diagnosing HPV, *C. trachomatis*, and *N. gonorrhoeae* infections.

The global burden of known pathogens for which signal amplification-based assays have been developed is still immense, despite the existence of vaccines and sophisticated therapies for prevention of infection or treatment of the diseases they cause. Therefore, we can expect that resources will continue to be devoted to the development of more effective therapeutics and improved assays. We also can anticipate that new assays will be developed for emerging pathogens that impact public health.

Signal amplification technologies can assess the efficacy of new drugs and refine algorithms for more effective use of existing drugs. The co-development of new therapeutics with assays to monitor their efficacies will lead both to new information about disease etiology and improved predictions of therapeutic outcomes. As more effective therapies and more sensitive assays are developed, current treatment algorithms will be rendered obsolete and new ones will be needed. We should expect ongoing refinement of patient management of currently monitored infections until the global burden of their etiologic pathogens is substantially reduced or eliminated. In the case of pathogens that are never completely eliminated from the body, increasingly sensitive assays will give the physician or scientist a clearer idea of the levels of these pathogens that are clinically relevant.

Emerging applications of signal amplification technologies may extend beyond existing ones as the demands of medicine change. For example, in addition to their roles in diagnosis and monitoring, assays based on signal amplification technologies are beginning to be used for prognostics in the newly emerging field of genomics. In the past, the role of signal amplification technologies in prognostics was limited to assessment of the pathogen. In the future, they will aid in assessing hosts to help physicians identify those who may or may not respond to a given therapy.

Signal amplification technologies must continue to be developed, both to adapt these technologies to automated platforms and to provide more adaptable assays that can be performed in a variety of settings — in developing countries, at the bedside, in a doctor's office, and even at home. The performance and flexibility of signal amplification technologies, indeed of all molecular diagnostic technologies, truly will be challenged to produce ever more refined clinical data required for future patient management.

ACKNOWLEDGMENTS

We thank Joshua Boatright for assistance with literature searches, Patricia Kipnis and Michael Friesenhahn for their editorial input to discussions of analytical performance of assays, Kristina Whitfield (PosterDocs, Oakland, California, USA) for graphics, and Linda Wuestehube (SciScript, Lafayette, California, USA) for literature research and editorial assistance.

REFERENCES

1. Nolte FS. Branched DNA signal amplification for direct quantitation of nucleic acid sequences in clinical specimens. *Adv Clin Chem* 1998; 33: 201–235.
2. Wilber JC. Branched DNA for quantification of viral load. *Immunol Invest* 1997; 10 (1–2): 25–47.

3. Pachl C, Todd JA, Kern DG, Sheridan PJ, Fong SJ, Stempien M, Hoo B, Besemer D, Yeghiazarian T, Irvine B et al. Rapid and precise quantification of HIV-1 RNA in plasma using a branched DNA signal amplification assay. *J Acquir Immune Defic Syndr Hum Retrovirol* 1995; 8 (5): 446–454.

4. Gleaves CA, Welle J, Campbell M, Elbeik T, Ng V, Taylor PE, Kuramoto K, Aceituno S, Lewalski E, Joppa B et al. Multicenter evaluation of the Bayer VERSANT HIV-1 RNA 3.0 assay: analytical and clinical performance. *J Clin Virol* 2002; 25 (2): 205–216.

5. Elbeik T, Surtihadi J, Destree M, Gorlin J, Holodniy M, Jortani SA, Kuramoto K, Ng V, Valdes R, Jr., Valsamakis A et al. Multicenter evaluation of the performance characteristics of the Bayer VERSANT HCV RNA 3.0 assay (bDNA). *J Clin Microbiol* 2004; 42 (2): 563–569.

6. Yao JD, Beld MG, Oon LL, Sherlock CH, Germer J, Menting S, Se Thoe SY, Merrick L, Ziermann R, Surtihadi J et al. Multicenter evaluation of the VERSANT hepatitis B virus DNA 3.0 assay. *J Clin Microbiol* 2004; 42 (2): 800–806.

7. Collins ML, Irvine B, Tyner D, Fine E, Zayati C, Chang C, Horn T, Ahle D, Detmer J, Shen LP et al. A branched DNA signal amplification assay for quantification of nucleic acid targets below 100 molecules/ml. *Nucleic Acids Res* 1997; 25 (15): 2979–2984.

8. Kenny D, Shen LP, Kolberg JA. Detection of viral infection and gene expression in clinical tissue specimens using branched DNA (bDNA) *in situ* hybridization. *J Histochem Cytochem* 2002; 50 (9): 1219–1227.

9. Horn T, Chang C-A, Collins ML. Hybridization properties of the 5-methyl-isocytidine/isoguanosine base pair in synthetic oligonucleotides. *Tetrahedron Lett* 1995; 36: 2033–2036.

10. Geyer CR, Battersby TR, Benner SA. Nucleobase pairing in expanded Watson–Crick-like genetic information systems. *Structure (Camb)* 2003; 11 (12): 1485–1498.

11. Bushnell S, Budde J, Catino T, Cole J, Derti A, Kelso R, Collins ML, Molino G, Sheridan P, Monahan J et al. ProbeDesigner: for the design of probesets for branched DNA (bDNA) signal amplification assays. *Bioinformatics* 1999; 15 (5): 348–355.

12. Duck P, Alvarado-Urbina G, Burdick B, Collier B. Probe amplifier system based on chimeric cycling oligonucleotides. *Biotechniques* 1990; 9 (2): 142–149.

13. Harvey JJ, Lee SP, Chan EK, Kim JH, Hwang ES, Cha CY, Knutson JR, Han MK. Characterization and applications of CataCleave probe in real-time detection assays. *Anal Biochem* 2004; 333 (2): 246–255.

14. Fong WK, Modrusan Z, McNevin JP, Marostenmaki J, Zin B, Bekkaoui F. Rapid solid-phase immunoassay for detection of methicillin-resistant *Staphylococcus aureus* using cycling probe technology. *J Clin Microbiol* 2000; 38 (7): 2525–2529.

15. Hogrefe HH, Hogrefe RI, Walder RY, Walder JA. Kinetic analysis of *Escherichia coli* RNase H using DNA-RNA-DNA/DNA substrates. *J Biol Chem* 1990; 265 (10): 5561–5566.

16. Lorincz AT. Molecular methods for the detection of human papillomavirus infection. *Obstet Gynecol Clin North Am* 1996; 23 (3): 707–730.

17. Iftner T, Villa LL. Human papillomavirus technologies. *J Natl Cancer Inst Monogr* 2003 (31): 80–88.

18. Niesters HG, Krajden M, Cork L, de Medina M, Hill M, Fries E, Osterhaus AD. A multicenter study evaluation of the Digene Hybrid Capture II signal amplification technique for detection of hepatitis B virus DNA in serum samples and testing of EUROHEP standards. *J Clin Microbiol* 2000; 38 (6): 2150–2155.

19. Yuan HJ, Yuen MF, Wong DK, Sum SS, Lai CL. Clinical evaluation of the Digene Hybrid Capture II test and the COBAS AMPLICOR monitor test for determination of hepatitis B virus DNA levels. *J Clin Microbiol* 2004; 42 (8): 3513–3517.

20. Boeckh M, Boivin G. Quantitation of cytomegalovirus: methodologic aspects and clinical applications. *Clin Microbiol Rev* 1998; 11 (3): 533–554.

21. Macartney M, Gane EJ, Portmann B, Williams R. Comparison of a new quantitative cytomegalovirus DNA assay with other detection methods. *Transplantation* 1997; 63 (12): 1803–1807.

22. Lorincz A, Anthony J. Advances in HPV detection by Hybrid Capture. *Pap Report* 2001; 12 (6): 145–154.

23. de Arruda M, Lyamichev VI, Eis PS, Iszczyszyn W, Kwiatkowski RW, Law SM, Olson MC, Rasmussen EB. Invader technology for DNA and RNA analysis: principles and applications. *Expert Rev Mol Diagn* 2002; 2 (5): 487–496.

24. Wong DK, Yuen MF, Yuan H, Sum SS, Hui CK, Hall J, Lai CL. Quantitation of covalently closed circular hepatitis B virus DNA in chronic hepatitis B patients. *Hepatology* 2004; 40 (3): 727–737.

25. Kwiatkowski RW, Lyamichev V, de Arruda M, Neri B. Clinical, genetic, and pharmacogenetic applications of the Invader assay. *Mol Diagn* 1999; 4 (4): 353–364.

26. Eis PS, Olson MC, Takova T, Curtis ML, Olson SM, Vener TI, Ip HS, Vedvik KL, Bartholomay CT, Allawi HT et al. An invasive cleavage assay for direct quantitation of specific RNAs. *Nat Biotechnol* 2001; 19 (7): 673–676.

27. Qian X, Lloyd RV. Recent developments in signal amplification methods for *in situ* hybridization. *Diagn Mol Pathol* 2003; 12 (1): 1–13.

28. Hopman AH, Ramaekers FC, Speel EJ. Rapid synthesis of biotin-, digoxigenin-, trinitrophenyl-, and fluorochrome-labeled tyramides and their application for *in situ* hybridization using CARD amplification. *J Histochem Cytochem* 1998; 46 (6): 771–777.

29. Kohler A, Lauritzen B, Van Noorden CJ. Signal amplification in immunohistochemistry at the light microscopic level using biotinylated tyramide and nanogold–silver staining. *J Histochem Cytochem* 2000; 48 (7): 933–941.

30. Qian X, Bauer RA, Xu HS, Lloyd RV. *In situ* hybridization detection of calcitonin mRNA in routinely fixed, paraffin-embedded tissue sections: a comparison of different types of probes combined with tyramide signal amplification. *Appl Immunohistochem Mol Morphol* 2001; 9 (1): 61–69.

31. Francois P, Bento M, Vaudaux P, Schrenzel J. Comparison of fluorescence and resonance light scattering for highly sensitive microarray detection of bacterial pathogens. *J Microbiol Methods* 2003; 55 (3): 755–762.

32. Bao YP, Huber M, Wei TF, Marla SS, Storhoff JJ, Muller UR. SNP identification in unamplified human genomic DNA with gold nanoparticle probes. *Nucleic Acids Res* 2005; 33 (2): e15.

33. Storhoff JJ, Lucas AD, Garimella V, Bao YP, Muller UR. Homogeneous detection of unamplified genomic DNA sequences based on colorimetric scatter of gold nanoparticle probes. *Nat Biotechnol* 2004; 22 (7): 883–887.

34. Elghanian R, Storhoff JJ, Mucic RC, Letsinger RL, Mirkin CA. Selective colorimetric detection of polynucleotides based on the distance-dependent optical properties of gold nanoparticles. *Science* 1997; 277 (5329): 1078–1081.

35. Kelley SO, Boon EM, Barton JK, Jackson NM, Hill MG. Single-base mismatch detection based on charge transduction through DNA. *Nucleic Acids Res* 1999; 27 (24): 4830–4837.

36. Kelley SO, Jackson NM, Hill MG, Barton JK. Long-range electron transfer through DNA films. *Angew Chem Int Edn Engl* 1999; 38: 941–945.

37. Boon EM, Ceres DM, Drummond TG, Hill MG, Barton JK. Mutation detection by electrocatalysis at DNA-modified electrodes. *Nat Biotechnol* 2000; 18 (10): 1096–1100. Erratum in *Nat Biotechnol* 2000; 18 (12): 1318.

38. National Committee for Clinical Laboratory Standards. *Quantitative Molecular Methods for Infectious Diseases: Approved Guideline.* Document MM6-A. Wayne, PA. 2003.

39. Player AN, Shen LP, Kenny D, Antao VP, Kolberg JA. Single-copy gene detection using branched DNA (bDNA) *in situ* hybridization. *J Histochem Cytochem* 2001; 49 (5): 603–612.

40. Plummer TB, Sperry AC, Xu HS, Lloyd RV. *In situ* hybridization detection of low copy nucleic acid sequences using catalyzed reporter deposition and its usefulness in clinical human papillomavirus typing. *Diagn Mol Pathol* 1998; 7 (2): 76–84.

41. Yang H, Wanner IB, Roper SD, Chaudhari N. An optimized method for *in situ* hybridization with signal amplification that allows the detection of rare mRNAs. *J Histochem Cytochem* 1999; 47 (4): 431–446.

42. van de Corput MP, Dirks RW, van Gijlswijk RP, van Binnendijk E, Hattinger CM, de Paus RA, Landegent JE, Raap AK. Sensitive mRNA detection by fluorescence *in situ* hybridization using horseradish peroxidase-labeled oligodeoxynucleotides and tyramide signal amplification. *J Histochem Cytochem* 1998; 46 (11): 1249–1259.

43. Holmes H, Davis C, Heath A, Hewlett I, Lelie N. An international collaborative study to establish the 1st international standard for HIV-1 RNA for use in nucleic acid-based techniques. *J Virol Methods* 2001; 92 (2): 141–150.

44. Pawlotsky JM, Bouvier-Alias M, Hezode C, Darthuy F, Remire J, Dhumeaux D. Standardization of hepatitis C virus RNA quantification. *Hepatology* 2000; 32 (3): 654–659.

45. Saldanha J, Heath A, Lelie N, Pisani G, Nubling M, Yu M. Calibration of HCV working reagents for NAT assays against the HCV international standard. *Vox Sang* 2000; 78 (4): 217–224.

46. Saldanha J, Gerlich W, Lelie N, Dawson P, Heermann K, Heath A. An international collaborative study to establish a World Health Organization international standard for hepatitis B virus DNA nucleic acid amplification techniques. *Vox Sang* 2001; 80 (1): 63–71.

47. Brambilla D, Leung S, Lew J, Todd J, Herman S, Cronin M, Shapiro DE, Bremer J, Hanson C, Hillyer GV et al. Absolute copy number and relative change in determinations of human immunodeficiency virus type 1 RNA in plasma: effect of an external standard on kit comparisons. *J Clin Microbiol* 1998; 36 (1): 311–314.

48. Klein D. Quantification using real-time PCR technology: applications and limitations. *Trends Mol Med* 2002; 8 (6): 257–260.

49. National Institutes of Health Consensus Development Conference Statement. Management of Hepatitis C. 2002. http://consensus.nih.gov/cons/116/116cdc_intro.htm

50. National Committee for Clinical Laboratory Standards. *Evaluation of the Linearity of Quantitative Measurement Procedures: A Statistical Approach; Approved Guideline.* Document EP6-A. Wayne, PA. 2001.

51. Bartlett JG, Lane HC and Panel on Clinical Practices for Treatment of HIV Infection. *Guidelines for the Use of Antiretroviral Agents in HIV-1-Infected Adults and Adolescents.* April 7, 2005. http://www.aidsinfo.nih.gov/guidelines

52. Pellegrin I, Garrigue I, Binquet C, Chene G, Neau D, Bonot P, Bonnet F, Fleury H, Pellegrin JL. Evaluation of new quantitative assays for diagnosis and monitoring of cytomegalovirus disease in human immunodeficiency virus-positive patients. *J Clin Microbiol* 1999; 37 (10): 3124–3132.

53. Pawlotsky JM, Bastie A, Hezode C, Lonjon I, Darthuy F, Remire J, Dhumeaux D. Routine detection and quantification of hepatitis B virus DNA in clinical laboratories: performance of three commercial assays. *J Virol Methods* 2000; 85 (1–2): 11–21.

54. Caliendo AM, Yen-Lieberman B, Baptista J, Andersen J, Crumpacker C, Schuurman R, Spector SA, Bremer J, Lurain NS. Comparison of molecular tests for detection and quantification of cell-associated cytomegalovirus DNA. *J Clin Microbiol* 2003; 41 (8): 3509–3513.

55. Imbert-Marcille BM, Cantarovich D, Ferre-Aubineau V, Richet B, Soulillou JP, Billaudel S. Usefulness of DNA viral load quantification for cytomegalovirus disease monitoring in renal and pancreas/renal transplant recipients. *Transplantation* 1997; 63 (10): 1476–1481.

56. *Human Retroviruses and AIDS 1997: A Compilation and Analysis of Nucleic Acid and Amino Acid Sequences.* Korber B, Hahn B, Foley B, Mellors JW, Leitner T, Myers G, McCutchan F and Kuiken CL. Eds. Theoretical Biology and Biophysics Group, Los Alamos, NM: Los Alamos National Laboratory. 1997.

57. *Human Retroviruses and AIDS 1996: A Compilation and Analysis of Nucleic Acid and Amino Acid Sequences.* Myers G, Korber BT, Foley BT, Jeang K-T, Mellors JW, Wain-Hobson S, Eds., Theoretical Biology and Biophysics Group, Los Alamos, NM: Los Alamos National Laboratory. 1996.

58. Osmanov S, Pattou C, Walker N, Schwardlander B, Esparza J. Estimated global distribution and regional spread of HIV-1 genetic subtypes in the year 2000. *J Acquir Immune Defic Syndr* 2002; 29 (2): 184–190.

59. Bukh J, Purcell RH, Miller RH. At least 12 genotypes of hepatitis C virus predicted by sequence analysis of the putative E1 gene of isolates collected worldwide. *Proc Natl Acad Sci USA* 1993; 90 (17): 8234–8238.

60. Robertson B, Myers G, Howard C, Brettin T, Bukh J, Gaschen B, Gojobori T, Maertens G, Mizokami M, Nainan O et al. Classification, nomenclature, and database development for hepatitis C virus (HCV) and related viruses: proposals for standardization; International Committee on Virus Taxonomy. *Arch Virol* 1998; 143 (12): 2493–2503.

61. Sablon E, Shapiro F. Hepatitis B and C genotyping: methodologies and implications for patient management. *J Gastroenterol Hepatol* 2004; 19: S329–S337.

62. Michael NL, Herman SA, Kwok S, Dreyer K, Wang J, Christopherson C, Spadoro JP, Young KK, Polonis V, McCutchan FE et al. Development of calibrated viral load standards for group M subtypes of human immunodeficiency virus type 1 and performance of an improved AMPLICOR HIV-1 MONITOR test with isolates of diverse subtypes. *J Clin Microbiol* 1999; 37 (8): 2557–2263.

63. Jagodzinski LL, Wiggins DL, McManis JL, Emery S, Overbaugh J, Robb M, Bodrug S, Michael NI. Use of calibrated viral load standards for group M subtypes of human immunodeficiency virus type 1 to assess the performance of viral RNA quantitation tests. *J Clin Microbiol* 2000; 38 (3): 1247–1249.

64. Lee SC, Antony A, Lee N, Leibow J, Yang JQ, Soviero S, Gutekunst K, Rosenstraus M. Improved version 2.0 qualitative and quantitative AMPLICOR reverse transcription-PCR tests for hepatitis C virus RNA: calibration to international units, enhanced genotype reactivity, and performance characteristics. *J Clin Microbiol* 2000; 38 (11): 4171–4179.

65. Collins ML, Zayati C, Detmer JJ, Daly B, Kolberg JA, Cha TA, Irvine BD, Tucker J, Urdea MS. Preparation and characterization of RNA standards for use in quantitative branched DNA hybridization assays. *Anal Biochem* 1995; 226 (1): 120–129.

66. Yen-Lieberman B, Brambilla D, Jackson B, Bremer J, Coombs R, Cronin M, Herman S, Katzenstein D, Leung S, Lin HJ et al. Evaluation of a quality assurance program for quantitation of human immunodeficiency virus type 1 RNA in plasma by the AIDS Clinical Trials Group virology laboratories. *J Clin Microbiol* 1996; 34 (11): 2695–2701.

67. Lorincz A, Anthony J. A system for nucleic acid detection by signal amplification technology, in Van Dyke K, Van Dyke C, Woodfork K, Eds., *Luminescence Biotechnology: Instruments and Applications.* Boca Raton, FL: CRC Press. 2002.

68. Oliver M, Chuang LM, Chang MS, Chen YT, Pei D, Ranade K, de Witte A, Allen J, Tran N, Curb D et al. High-throughput genotyping of single nucleotide polymorphisms using new biplex Invader technology. *Nucleic Acids Res* 2002; 30 (12): E53.

69. Modrusan Z, Marlowe C, Wheeler D, Pirseyedi M, Bryan RN. CPT-EIA assays for the detection of vancomycin-resistant *vanA* and *vanB* genes in enterococci. *Diagn Microbiol Infect Dis* 2000; 37 (1): 45–50.

70. Merlino J, Rose B, Harbour C. Rapid detection of non-multidrug-resistant and multidrug-resistant methicillin-resistant *Staphylococcus aureus* using cycling probe technology for the *mecA* gene. *Eur J Clin Microbiol Infect Dis* 2003; 22 (5): 322–323.

71. Saldanha J, Lelie N, Heath A. Establishment of the first international standard for nucleic acid amplification technology (NAT) assays for HCV RNA; WHO Collaborative Study Group. *Vox Sang* 1999; 76 (3): 149–158.

72. Saldanha J, Heath A. Collaborative study to calibrate hepatitis C virus genotypes 2–6 against the HCV International Standard 96/790 (genotype 1). *Vox Sang* 2003; 84 (1): 20–27.

73. European Association for the Study of the Liver. Consensus Statement. International Consensus Conference on Hepatitis C. Paris, February 1999. *J Hepatol* 1999; 30 (5): 956–961.

74. Konnick EQ, Erali M, Ashwood ER, Hillyard DR. Performance characteristics of the COBAS Amplicor hepatitis C virus (HCV) monitor, Version 2.0, international unit assay and the National Genetics Institute HCV Superquant assay. *J Clin Microbiol* 2002; 40 (3): 768–773.

75. Germer JJ, Heimgartner PJ, Ilstrup DM, Harmsen WS, Jenkins GD, Patel R. Comparative evaluation of the VERSANT HCV RNA 3.0, QUANTIPLEX HCV RNA 2.0, and COBAS AMPLICOR HCV MONITOR Version 2.0 Assays for quantification of hepatitis C virus RNA in serum. *J Clin Microbiol* 2002; 40 (2): 495–500.

76. Trimoulet P, Halfon P, Pohier E, Khiri H, Chene G, Fleury H. Evaluation of the VERSANT HCV RNA 3.0 assay for quantification of hepatitis C virus RNA in serum. *J Clin Microbiol* 2002; 40 (6): 2031–2036.

77. Beld M, Sentjens R, Rebers S, Weegink C, Weel J, Sol C, Boom R. Performance of the new Bayer VERSANT HCV RNA 3.0 assay for quantitation of hepatitis C virus RNA in plasma and serum: conversion to international units and comparison with the Roche COBAS Amplicor HCV Monitor, Version 2.0, assay. *J Clin Microbiol* 2002; 40 (3): 788–793.

78. Mellors JW, Rinaldo CRJ, Gupta P, White RM, Todd JA, Kingsley LA. Prognosis in HIV-1 infection predicted by the quantity of virus in plasma. *Science* 1996; 272 (5265): 1167–1170. Erratum in *Science* 1997; 275 (5296): 14.

79. Mellors JW, Kingsley LA, Rinaldo CR, Jr., Todd JA, Hoo BS, Kokka RP, Gupta P. Quantitation of HIV-1 RNA in plasma predicts outcome after seroconversion. *Ann Intern Med* 1995; 122 (8): 573–579.

80. Mellors JW, Munoz A, Giorgi JV, Margolick JB, Tassoni CJ, Gupta P, Kingsley LA, Todd JA, Saah AJ, Detels R et al. Plasma viral load and CD4+ lymphocytes as prognostic markers of HIV-1 infection. *Ann Intern Med* 1997; 126: 946–954.

81. Swindells S, Evans S, Zackin R, Goldman M, Haubrich R, Filler SG, Balfour HH, Jr. Predictive value of HIV-1 viral load on risk for opportunistic infection. *J Acquir Immune Defic Syndr* 2002; 30 (2): 154–158.

82. Muir D, White D, King J, Verlander N, Pillay D. Predictive value of the ultrasensitive HIV viral load assay in clinical practice. *J Med Virol* 2000; 61 (4): 411–416.

83. Kempf DJ, Rode RA, Xu Y, Sun E, Heath-Chiozzi ME, Valdes J, Japour AJ, Danner S, Boucher C, Molla A et al. The duration of viral suppression during protease inhibitor therapy for HIV-1 infection is predicted by plasma HIV-1 RNA at the nadir. *AIDS* 1998; 12 (5): F9–F14.

84. Pancholi P, Wu F, Della-Latta P. Rapid detection of cytomegalovirus infection in transplant patients. *Expert Rev Mol Diagn* 2004; 4 (2): 231–242.

85. Whitley RJ. Cytomegalovirus and HIV: inextricably entwined pathogens. *Lancet* 2004; 363 (9427): 2101–2102.

86. Lok AS, McMahon BJ. Chronic hepatitis B. *Hepatology* 2001; 34 (6): 1225–1241.

87. Keeffe EB, Dieterich DT, Han SH, Jacobson IM, Martin P, Schiff ER, Tobias H, Wright TL. A treatment algorithm for the management of chronic hepatitis B virus infection in the United States. *Clin Gastroenterol Hepatol* 2004; 2 (2): 87–106.

88. Lok AS, McMahon BJ. Chronic hepatitis B: update of recommendations. *Hepatology* 2004; 39 (3): 857–861.

89. Brook MG, Gilson R, Wilkins EL. BHIVA Guidelines: coinfection with HIV and chronic hepatitis B virus. *HIV Med* 2003; 4l Suppl 1: 42–51.

90. Soriano V, Miro JM, Garcia-Samaniego J, Torre-Cisneros J, Nunez M, del Romero J, Martin-Carbonero L, Castilla J, Iribarren JA, Quereda C et al. Consensus conference on chronic viral hepatitis and HIV infection: updated Spanish recommendations. *J Viral Hepat* 2004; 11 (1): 2–17.

91. Weiss J, Wu H, Farrenkopf B, Schultz T, Song G, Shah S, Siegel J. Real time TaqMan PCR detection and quantitation of HBV genotypes A–G with the use of an internal quantitation standard. *J Clin Virol* 2004; 30 (1): 86–93.

92. Pichoud C, Berby F, Stuyver L, Petit MA, Trepo C, Zoulim F. Persistence of viral replication after anti-HBe seroconversion during antiviral therapy for chronic hepatitis B. *J Hepatol* 2000; 32 (2): 307–316.

93. Gordon SC, Dailey PJ, Silverman AL, Khan BA, Kodali VP, Wilber JC. Sequential serum hepatitis C viral RNA levels longitudinally assessed by branched DNA signal amplification. *Hepatology* 1998; 28 (6): 1702–1706.

94. Anand BS, Velez M. Assessment of correlation between serum titers of hepatitis C virus and severity of liver disease. *World J Gastroenterol* 2004; 10 (16): 2409–2411.

95. Beld M, Penning M, van Putten M, van den Hoek A, Damen M, Klein MR, Goudsmit J. Low levels of hepatitis C virus RNA in serum, plasma, and peripheral blood mononuclear cells of injecting drug users during long antibody-undetectable periods before seroconversion. *Blood* 1999; 94 (4): 1183–1191.

96. Bayer Corporation. VERSANT HCV RNA 3.0 Assay (bDNA). Package Insert. 2003.

97. Comanor L, Hendricks D. Hepatitis C virus RNA tests: performance attributes and their impact on clinical utility. *Expert Rev Mol Diagn* 2003; 3 (6): 689–702.

98. Jungkind D. Automation of laboratory testing for infectious diseases using the polymerase chain reaction: our past, our present, our future. *J Clin Virol* 2001; 20 (1–2): 1–6.

99. Abbott Laboratories. LCx HCV RNA Quantitative. Package Insert. 2003.

100. National Genetics Institute. Determination of analytical sensitivity of multiplexed and non-multiplexed NGI UltraQual RT PCR assays for the detection of HCV RNA in plasma pools. Product Licensure Application, 1999.

101. Barbeau JM, Goforth J, Caliendo AM, Nolte FS. Performance characteristics of a quantitative TaqMan hepatitis C virus RNA analyte-specific reagent. *J Clin Microbiol* 2004; 42 (8): 3739–3746.

102. Consensus Conference. Treatment of hepatitis C. *Gastroenterol Clin Biol* 2002; 26 (2): B303–B320.

103. Obiso R, Lorincz A. Digene Corporation. *Pharmacogenomics* 2004; 5 (1): 129–132.

104. Lorincz AT. Screening for cervical cancer: new alternatives and research. *Salud Publica Mex* 2003; 45; Suppl 3: S376–S387.

105. Clavel C, Masure M, Bory JP, Putaud I, Mangeonjean C, Lorenzato M, Nazeyrollas P, Gabriel R, Quereux C, Birembaut P. Human papillomavirus testing in primary screening for the detection of high-grade cervical lesions: a study of 7932 women. *Br J Cancer* 2001; 84 (12): 1616–1623.

106. ACOG Practice Bulletin 45. Clinical management guidelines for obstetrician–gynecologists. Cervical cytology screening (replaces Committee Opinion 152, March 1995). *Obstet Gynecol* 2003; 102 (2): 417–427.

107. Chopra KF, Tyring SK. The impact of the human immunodeficiency virus on the human papillomavirus epidemic. *Arch Dermatol* 1997; 133 (5): 629–633.

108. Centers for Disease Control and Prevention. 1998 guidelines for treatment of sexually transmitted diseases. *MMWR* 1998; 47 (RR-1): 1–111.

109. Bignell CJ. European guideline for the management of gonorrhoea. *Int J STD AIDS* 2001; 12; Suppl 3: 27–29.

110. Stary A. European guideline for the management of chlamydial infection. *Int J STD AIDS* 2001; 12; Suppl 3: 30–33.

111. Myerow S, Lee M, Arruda J, Shen L-P, Madej R, Collins M, Miner R, McMullen D, Drew WL, Kolberg J. Development of a branched DNA assay with improved sensitivity for monitoring CMV viral load in immunocompromised patients. Annual Molecular Virology Workshop and Clinical Virology Symposium, Clearwater, FL, 1997.

112. Digene Corporation. Hybrid Capture II HBV DNA Test. Package Insert. 2004.

113. Digene Corporation. Hybrid Capture System CMV DNA Test (Version 2.0). Package Insert. 2002.

114. Wattanamano P, Clayton JL, Kopicko JJ, Kissinger P, Elliot S, Jarrott C, Rangan S, Beilke MA. Comparison of three assays for cytomegalovirus detection in AIDS patients at risk for retinitis. *J Clin Microbiol* 2000; 38 (2): 727–732.

115. Modarress KJ, Cullen AP, Jaffurs WJ, Sr., Troutman GL, Mousavi N, Hubbard RA, Henderson S, Lorincz AT. Detection of *Chlamydia trachomatis* and *Neisseria gonorrhoeae* in swab specimens by the Hybrid Capture II and PACE 2 nucleic acid probe tests. *Sex Transm Dis* 1999; 26 (5): 303–308.

116. van der Pol B, Williams JA, Smith NJ, Batteiger BE, Cullen AP, Erdman H, Edens T, Davis K, Salim-Hammad H, Chou VW et al. Evaluation of the Digene Hybrid Capture II assay with the rapid capture system for detection of *Chlamydia trachomatis* and *Neisseria gonorrhoeae*. *J Clin Microbiol* 2002; 40 (10): 3558–3564.

117. Cullen AP, Long CD, Lorincz AT. Rapid detection and typing of herpes simplex virus DNA in clinical specimens by the Hybrid Capture II signal amplification probe test. *J Clin Microbiol* 1997; 35 (9): 2275–2278.

118. Zeuzem S. Overview of commercial HBV assay systems, in Hamatake RK, Lau JYN, Eds., Hepatitis B and D protocols, Vol. 1. Totowa, NJ: Humana Press; 2004. pp. 3–13.

119. Mazzulli T, Wood S, Chua R, Walmsley S. Evaluation of the Digene Hybrid Capture system for detection and quantitation of human cytomegalovirus viremia in human immunodeficiency virus-infected patients. *J Clin Microbiol* 1996; 34 (12): 2959–2962.

120. Sun CA, Liu JF, Wu DM, Nieh S, Yu CP, Chu TY. Viral load of high-risk human papillomavirus in cervical squamous intraepithelial lesions. *Int J Gynaecol Obstet* 2002; 76 (1): 41–47.

121. Chernoff DN, Miner RC, Hoo BS, Shen L-P, Kelso RJ, Jekic-McMullen D, Lalezari JP, Chou S, Drew WL, Kolberg JA. Quantification of cytomegalovirus DNA in peripheral blood leukocytes by a branched-DNA signal amplification assay. *J Clin Microbiol* 1997; 35 (11): 2740–2744.

122. Martinot M, Marcellin P, Boyer N, Detmer J, Pouteau M, Castelnau C, Degott C, Auperin A, Collins M, Kolberg J et al. Influence of hepatitis G virus infection on the severity of liver disease and response to interferon-alpha in patients with chronic hepatitis C. *Ann Intern Med* 1997; 126 (11): 874–881.

123. Pessoa MG, Terrault NA, Ferrell LD, Kim JP, Kolberg J, Detmer J, Collins ML, Yun AJ, Viele M, Lake JR et al. Hepatitis G virus in patients with cryptogenic liver disease undergoing liver transplantation. *Hepatology* 1997; 25 (5): 1266–1270.

124. Charlton MR, Brandhagen D, Wiesner RH, Gross JB, Jr., Detmer J, Collins M, Kolberg J, Krom RA, Persing DH. Hepatitis G virus infection in patients transplanted for cryptogenic cirrhosis: red flag or red herring? *Transplantation* 1998; 65 (1): 73–76.

125. Lau DT, Miller KD, Detmer J, Kolberg J, Herpin B, Metcalf JA, Davey RT, Hoofnagle JH. Hepatitis G virus and human immunodeficiency virus coinfection: response to interferon- therapy. *J Infect Dis* 1999; 180 (4): 1334–1337.

126. McHutchison JG, Nainan OV, Alter MJ, Sedghi-Vaziri A, Detmer J, Collins M, Kolberg J. Hepatitis C and G co-infection: response to interferon therapy and quantitative changes in serum HGV-RNA. *Hepatology* 1997; 26 (5): 1322–1327.

127. Pessoa MG, Terrault NA, Detmer J, Kolberg J, Collins M, Hassoba HM, Wright TL. Quantitation of hepatitis G and C viruses in the liver: evidence that hepatitis G virus is not hepatotropic. *Hepatology* 1998; 27 (3): 877–880.

128. Brandhagen DJ, Gross JB, Jr., Poterucha JJ, Charlton MR, Detmer J, Kolberg J, Gossard AA, Batts KP, Kim WR, Germer JJ et al. The clinical significance of simultaneous infection with hepatitis G virus in patients with chronic hepatitis C. *Am J Gastroenterol* 1999; 94 (4): 1000–1005.

129. Kolbert CP, Arruda J, Varga-Delmore P, Zheng X, Lewis M, Kolberg J, Persing DH. Branched-DNA assay for detection of the *mecA* gene in oxacillin-resistant and oxacillin-sensitive staphylococci. *J Clin Microbiol* 1998; 36 (9): 2640–2644.

130. Zheng X, Kolbert CP, Varga-Delmore P, Arruda J, Lewis M, Kolberg J, Cockerill FR, Persing DH. Direct *mecA* detection from blood culture bottles by branched-DNA signal amplification. *J Clin Microbiol* 1999; 37 (12): 4192–4193.

131. Kolberg JA, Besemer DJ, Stempien MM, Urdea MS. The specificity of pilin DNA sequences for the detection of pathogenic *Neisseria. Mol Cell Probes* 1989; 3 (1): 59–72.

132. Urdea MS, Kolberg J, Clyne J, Running JA, Besemer D, Warner B, Sanchez-Pescador R. Application of a rapid non-radioisotopic nucleic acid analysis system to the detection of sexually transmitted disease-causing organisms and their associated antimicrobial resistances. *Clin Chem* 1989; 35 (8): 1571–1575.

133. Nargessi RD, Shimizu RM, Xu XM, Connolly J, Zamroud M, Collins ML, Kolberg J. Quantitation of progesterone receptor mRNA in breast carcinoma by branched DNA assay. *Breast Cancer Res Treat* 1998; 50 (1): 57–62.

134. Nargessi RD, Khabbaz NF, Xu XM, Zamroud M, Kolberg J, Collins ML. Quantitation of estrogen receptor mRNA in breast carcinoma by branched DNA assay. *Breast Cancer Res Treat* 1998; 50 (1): 47–55.

135. Shen LP, Sheridan P, Cao WW, Dailey PJ, Salazar-Gonzalez JF, Breen EC, Fahey JL, Urdea MS, Kolberg JA. Quantification of cytokine mRNA in peripheral blood mononuclear cells using branched DNA (bDNA) technology. *J Immunol Methods* 1998; 215 (1–2): 123–134.

136. Breen EC, Salazar-Gonzalez JF, Shen LP, Kolberg JA, Urdea MS, Martinez-Maza O, Fahey JL. Circulating CD8 T cells show increased interferon-gamma mRNA expression in HIV infection. *Cell Immunol* 1997; 178 (1): 91–98.

137. Nelson DR, Lim HL, Marousis CG, Fang JW, Davis GL, Shen L, Urdea MS, Kolberg JA, Lau JY. Activation of tumor necrosis factor-alpha system in chronic hepatitis C virus infection. *Dig Dis Sci* 1997; 42 (12): 2487–2494.

138. Salazar-Gonzalez JF, Martinez-Maza O, Aziz N, Kolberg JA, Yeghiazarian T, Shen LP, Fahey JL. Relationship of plasma HIV-RNA levels and levels of TNF-alpha and immune activation products in HIV infection. *Clin Immunol Immunopathol* 1997; 84 (1): 36–45.

139. Wang J, Shen L, Najafi H, Kolberg J, Matschinsky FM, Urdea M, German M. Regulation of insulin preRNA splicing by glucose. *Proc Natl Acad Sci USA* 1997; 94 (9): 4360–4365.

140. Harris E, Detmer J, Dungan J, Doua F, White T, Kolberg JA, Urdea MS, Agabian N. Detection of *Trypanosoma brucei* spp. in human blood by a nonradioactive branched DNA-based technique. *J Clin Microbiol* 1996; 34 (10): 2401–2407.

141. Hartley DP, Klaassen CD. Detection of chemical-induced differential expression of rat hepatic cytochrome P450 mRNA transcripts using branched DNA signal amplification technology. *Drug Metab Dispos* 2000; 28 (5): 608–616. Erratum in *Drug Metab Dispos* 2000; 28 (8): 1007.

142. Dickinson Laing T, Mah DC, Poirier RT, Bekkaoui F, Lee WE, Bader DE. Genomic DNA detection using cycling probe technology and capillary gel electrophoresis with laser-induced fluorescence. *Mol Cell Probes* 2004; 18 (5): 341–348.

143. Tang T, Badal MY, Ocvirk G, Lee WE, Bader DE, Bekkaoui F, Harrison DJ. Integrated microfluidic electrophoresis system for analysis of genetic materials using signal amplification methods. *Anal Chem* 2002; 74 (4): 725–733.

144. Warnon S, Zammatteo N, Alexandre I, Hans C, Remacle J. Colorimetric detection of the tuberculosis complex using cycling probe technology and hybridization in microplates. *Biotechniques* 2000; 28 (6): 1152–1160.

145. Wolters LM, Hansen BE, Niesters HG, Zeuzem S, Schalm SW, de Man RA. Viral dynamics in chronic hepatitis B patients during lamivudine therapy. *Liver* 2002; 22 (2): 121–126.

146. Norris S, Kosar Y, Donaldson N, Smith HM, Zolfino T, O'Grady JG, Muiesan P, Rela M, Heaton N. Cytomegalovirus infection after liver transplantation: viral load as a guide to treating clinical infection. *Transplantation* 2002; 74 (4): 527–531.

147. Chemaly RF, Yen-Lieberman B, Chapman J, Reilly A, Bekele BN, Gordon SM, Procop GW, Shrestha N, Isada CM, Decamp M et al. Clinical utility of cytomegalovirus viral load in bronchoalveolar lavage in lung transplant recipients. *Am J Transplant* 2005; 5 (3): 544–548.

148. Schiffman M, Solomon D. Findings to date from the ASCUS-LSIL Triage Study (ALTS). *Arch Pathol Lab Med* 2003; 127 (8): 946–949.

149. Drobacheff C, Dupont P, Mougin C, Bourezane Y, Challier B, Fantoli M, Bettinger D, Laurent R. Anal human papillomavirus DNA screening by Hybrid Capture II in human immunodeficiency virus-positive patients with or without anal intercourse. *Eur J Dermatol* 2003; 13 (4): 367–371.

150. Schachter J, Hook EW, 3rd, McCormack WM, Quinn TC, Chernesky M, Chong S, Girdner JI, Dixon PB, DeMeo L, Williams E et al. Ability of the Digene Hybrid Capture II test to identify *Chlamydia trachomatis* and *Neisseria gonorrhoeae* in cervical specimens. *J Clin Microbiol* 1999; 37 (11): 3668–3671.

151. Yehle CO, Patterson WL, Boguslawski SJ, Albarella JP, Yip KF, Carrico RJ. A solution hybridization assay for ribosomal RNA from bacteria using biotinylated DNA probes and enzyme-labeled antibody to DNA:RNA. *Mol Cell Probes* 1987; 1 (2): 177–193.

152. Lamoureux M, MacKay A, Messier S, Fliss I, Blais BW, Holley RA, Simard RE. Detection of *Campylobacter jejuni* in food and poultry viscera using immunomagnetic separation and microtitre hybridization. *J Appl Microbiol* 1997; 83 (5): 641–651.

153. Werle-Lapostolle B, Bowden S, Locarnini S, Wursthorn K, Petersen J, Lau G, Trepo C, Marcellin P, Goodman Z, Delaney WEt et al. Persistence of cccDNA during the natural history of chronic hepatitis B and decline during adefovir dipivoxil therapy. *Gastroenterology* 2004; 126 (7): 1750–1758.

154. Cooksey RC, Holloway BP, Oldenburg MC, Listenbee S, Miller CW. Evaluation of the Invader assay, a linear signal amplification method, for identification of mutations associated with resistance to rifampin and isoniazid in *Mycobacterium tuberculosis*. *Antimicrob Agents Chemother* 2000; 44 (5): 1296–1301.

155. Fors L, Lieder KW, Vavra SH, Kwiatkowski RW. Large-scale SNP scoring from unamplified genomic DNA. *Pharmacogenomics* 2000; 1 (2): 219–229.

156. Olson MC, Takova T, Chehak L, Curtis ML, Olson SM, Kwiatkowski RW. Invader assay for RNA quantitation. *Methods Mol Biol* 2004; 258: 53–69.

157. Neville M, Selzer R, Aizenstein B, Maguire M, Hogan K, Walton R, Welsh K, Neri B, de Arruda M. Characterization of cytochrome p450 2D6 alleles using the Invader™ system. Biotechniques 2002;32(suppl 2):S34-43.

158. Ryan D, Nuccie B, Arvan D. Non-PCR-dependent detection of the factor V Leiden mutation from genomic DNA using a homogeneous Invader microtiter plate assay. *Mol Diagn* 1999; 4 (2): 135–144.

159. Berggren WT, Takova T, Olson MC, Eis PS, Kwiatkowski RW, Smith LM. Multiplexed gene expression analysis using the invader RNA assay with MALDI-TOF mass spectrometry detection. *Anal Chem* 2002; 74 (8): 1745–1750.

160. Nakajima N, Ionescu P, Sato Y, Hashimoto M, Kuroita T, Takahashi H, Yoshikura H, Sata T. *In situ* hybridization AT-tailing with catalyzed signal amplification for sensitive and specific *in situ* detection of human immunodeficiency virus-1 mRNA in formalin-fixed and paraffin-embedded tissues. *Am J Pathol* 2003; 162 (2): 381–389.

161. Murakami T, Hagiwara T, Yamamoto K, Hattori J, Kasami M, Utsumi M, Kaneda T. A novel method for detecting HIV-1 by non-radioactive *in situ* hybridization: application of a peptide nucleic acid probe and catalysed signal amplification. *J Pathol* 2001; 194 (1): 130–135.

162. Kotronias D, Kapranos N. Detection of herpes simplex virus DNA in maternal breast milk by *in situ* hybridization with tyramide signal amplification. *In Vivo* 1999; 13 (6): 463–466.

163. Lizard G, Demares-Poulet MJ, Roignot P, Gambert P. *In situ* hybridization detection of single-copy human papillomavirus on isolated cells, using a catalyzed signal amplification system: *GenPoint*. *Diagn Cytopathol* 2001; 24 (2): 112–116.

164. Zerbini M, Cricca M, Gentilomi G, Venturoli S, Gallinella G, Musiani M. Tyramide signal amplification of biotinylated probe in dot-blot hybridization assay for the detection of parvovirus B19 DNA in serum samples. *Clin Chim Acta* 2000; 302 (1–2): 79–87.

165. Fijnvandraat AC, De Boer PA, Deprez RH, Moorman AF. Non-radioactive *in situ* detection of mRNA in ES cell-derived cardiomyocytes and in the developing heart. *Microsc Res Tech* 2002; 58 (5): 387–394.

166. Tsang YT, Chang YM, Lu X, Rao PH, Lau CC, Wong KK. Amplification of MGC2177, PLAG1, PSMC6P, and LYN in a malignant mixed tumor of salivary gland detected by cDNA microarray with tyramide signal amplification. *Cancer Genet Cytogenet* 2004; 152 (2): 124–128.

167. Denef VJ, Park J, Rodrigues JL, Tsoi TV, Hashsham SA, Tiedje JM. Validation of a more sensitive method for using spotted oligonucleotide DNA microarrays for functional genomics studies on bacterial communities. *Environ Microbiol* 2003; 5 (10): 933–943.

168. Wang D, Zhu L, Jiang D, Ma X, Zhou Y, Cheng J. Direct detection of 16S rRNA using oligonucleotide microarrays assisted by base stacking hybridization and tyramide signal amplification. *J Biochem Biophys Methods* 2004; 59 (2): 109–120.

169. Bakermans C, Madsen EL. Detection in coal tar waste-contaminated groundwater of mRNA transcripts related to naphthalene dioxygenase by fluorescent *in situ* hybridization with tyramide signal amplification. *J Microbiol Methods* 2002; 50 (1): 75–84.

170. Pernthaler A, Pernthaler J, Amann R. Fluorescence *in situ* hybridization and catalyzed reporter deposition for the identification of marine bacteria. *Appl Environ Microbiol* 2002; 68 (6): 3094–3101.

171. Sekar R, Pernthaler A, Pernthaler J, Warnecke F, Posch T, Amann R. An improved protocol for quantification of freshwater *Actinobacteria* by fluorescence *in situ* hybridization. *Appl Environ Microbiol* 2003; 69 (5): 2928–2935.

172. West NJ, Schonhuber WA, Fuller NJ, Amann RI, Rippka R, Post AF, Scanlan DJ. Closely related Prochlorococcus genotypes show remarkably different depth distributions in two oceanic regions as revealed by *in situ* hybridization using 16S rRNA-targeted oligonucleotides. *Microbiology* 2001; 147 (Pt 7): 1731–1744.

173. Zhang Z, Kitching P. A sensitive method for the detection of foot and mouth disease virus by *in situ* hybridisation using biotin-labelled oligodeoxynucleotides and tyramide signal amplification. *J Virol Methods* 2000; 88 (2): 187–192.

174. Berent LM, Messick JB, Cooper SK, Cusick PK. Specific *in situ* hybridization of *Haemobartonella felis* with a DNA probe and tyramide signal amplification. *Vet Pathol* 2000; 37 (1): 47–53.

175. Fujino Y, Satoh H, Hisasue M, Masuda K, Ohno K, Tsujimoto H. Detection of the integrated feline leukemia viruses in a cat lymphoid tumor cell line by fluorescence *in situ* hybridization. *J Hered* 2003; 94 (3): 251–255.

176. Acar H, Copeland NG, Gilbert DJ, Jenkins NA, Largaespada DA. Detection of integrated murine leukemia viruses in a mouse model of acute myeloid leukemia by fluorescence *in situ* hybridization combined with tyramide signal amplification. *Cancer Genet Cytogenet* 2000; 121 (1): 44–51.

177. Schad A, Fahimi HD, Volkl A, Baumgart E. Expression of catalase mRNA and protein in adult rat brain: detection by nonradioactive hybridization with signal amplification by catalyzed reporter deposition (ISH-CARD) and immunohistochemistry (IHC)/immunofluorescence (IF). *J Histochem Cytochem* 2003; 51 (6): 751–760.

178. Divi RL, Dragan YP, Pitot HC, Poirier MC. Immunohistochemical localization and semi-quantitation of hepatic tamoxifen-DNA adducts in rats exposed orally to tamoxifen. *Carcinogenesis* 2001; 22 (10): 1693–1699.

179. Khrustaleva LI, Kik C. Localization of single-copy T-DNA insertion in transgenic shallots (*Allium cepa*) by using ultra-sensitive FISH with tyramide signal amplification. *Plant J* 2001; 25 (6): 699–707.

180. Vogler MA, Teppler H, Gelman R, Valentine F, Lederman MM, Pomerantz RJ, Pollard RB, Cherng DW, Gonzalez CJ, Squires KE et al. Daily low-dose subcutaneous interleukin-2 added to single- or dual-nucleoside therapy in HIV infection does not protect against CD4+ T-cell decline or improve other indices of immune function: results of a randomized controlled clinical trial (ACTG 248). *J Acquir Immune Defic Syndr* 2004; 36 (1): 576–587.

181. Stebbing J, Nelson M, Orkin C, Mandalia S, Bower M, Pozniak A, Gazzard B. A randomized trial to investigate the recycling of stavudine and didanosine with and without hydroxyurea in salvage therapy (RESTART). *J Antimicrob Chemother* 2004; 53 (3): 501–505.

182. Ungsedhapand C, Srasuebkul P, Cardiello P, Ruxrungtham K, Ratanasuwan W, Kroon EDMB, Tongtalung M, Juengprasert N, Ubolyam S, Siangphoe U et al. Three-year durability of dual-nucleoside versus triple-nucleoside therapy in a Thai population with HIV infection. *J Acquir Immune Defic Syndr* 2004; 36 (2): 693–701.

183. Ayouba A, Tene G, Cunin P, Foupouapouognigni Y, Menu E, Kfutwah A, Thonnon J, Scarlatti G, Monny-Lobe M, Eteki N et al. Low rate of mother-to-child transmission of HIV-1 after nevirapine intervention in a pilot public health program in Yaounde, Cameroon. *J Acquir Immune Defic Syndr* 2003; 34 (3): 274–280.

184. Yeghiazarian T, Zhao Y, Read SE, Kabat W, Li X, Hamren S, Sheridan PJ, Wilber JC, Chernoff DN, Yogev R. Quantification of human immunodeficiency virus type 1 (HIV-1) RNA levels in plasma using small volume format branched DNA assays. *J Clin Microbiol* 1998; 36: 2096–2098.

185. Daar ES, Lynn H, Donfield S, Gomperts E, Hilgartner MW, Hoots K, Chernoff D, Winkler C, O'Brien SJ. Effects of plasma HIV RNA, CD4+ T lymphocytes and the chemokine receptors CCR5 and CCR2b on HIV disease progression in hemophiliacs: Hemophilia Growth and Development Study. *J Acquir Immune Defic Syndr* 1999; 21 (4): 317–325.

186. Flood J, Drew WL, Miner R, Jekic-McMullen D, Shen LP, Kolberg J, Garvey J, Follansbee S, Poscher M. Diagnosis of cytomegalovirus (CMV) polyradiculopathy and documentation of *in vivo* anti-CMV activity in cerebrospinal fluid by using branched DNA signal amplification and antigen assays. *J Infect Dis* 1997; 176 (2): 348–352.

187. O'Sullivan CE, Drew WL, McMullen DJ, Miner R, Lee JY, Kaslow RA, Lazar JG, Saag MS. Decrease of cytomegalovirus replication in human immunodeficiency virus infected-patients after treatment with highly active antiretroviral therapy. *J Infect Dis* 1999; 180 (3): 847–849.

188. Barrett-Muir W, Breuer J, Millar C, Thomas J, Jeffries D, Yaqoob M, Aitken C. CMV viral load measurements in whole blood and plasma: which is best following renal transplantation? *Transplantation* 2000; 70 (1): 116–119.

189. Hebart H, Wuchter P, Loeffler J, Gscheidle B, Hamprecht K, Sinzger C, Jahn G, Dietz K, Kanz L, Einsele H. Evaluation of the Murex CMV DNA Hybrid Capture assay (Version 2.0) for early diagnosis of cytomegalovirus infection in recipients of an allogeneic stem cell transplant. *Bone Marrow Transplant* 2001; 28 (2): 213–218.

190. Bhorade SM, Sandesara C, Garrity ER, Vigneswaran WT, Norwick L, Alkan S, Husain AN, McCabe MA, Yeldandi V. Quantification of cytomegalovirus (CMV) viral load by the Hybrid Capture assay allows for early detection of CMV disease in lung transplant recipients. *J Heart Lung Transplant* 2001; 20 (9): 928–934.

191. Perinatal HIV Guidelines Working Group. Recommendations for Use of Antiretroviral Drugs in Pregnant HIV-1-Infected Women for Maternal Health and Interventions to Reduce Perinatal HIV-1 Transmission in the United States, February 24, 2005. http://aidsinfo.nih.gov

192. Lai CL, Rosmawati M, Lao J, Van Vlierberghe H, Anderson FH, Thomas N, Dehertogh D. Entecavir is superior to lamivudine in reducing hepatitis B virus DNA in patients with chronic hepatitis B infection. *Gastroenterology* 2002; 123 (6): 1831–1838.

193. Serfaty L, Thabut D, Zoulim F, Andreani T, Chazouilleres O, Carbonell N, Loria A, Poupon R. Sequential treatment with lamivudine and interferon monotherapies in patients with chronic hepatitis B not responding to interferon alone: results of a pilot study. *Hepatology* 2001; 34 (3): 573–577.

194. Hann HW, Fontana RJ, Wright T, Everson G, Baker A, Schiff ER, Riely C, Anschuetz G, Gardner SD, Brown N et al. A United States compassionate use study of lamivudine treatment in nontransplantation candidates with decompensated hepatitis B virus-related cirrhosis. *Liver Transpl* 2003; 9 (1): 49–56.

195. Yuen MF, Yuan HJ, Sablon E, Wong DK, Chan AO, Wong BC, Lai CL. Long-term follow-up study of Chinese patients with YMDD mutations: significance of hepatitis B virus genotypes and characteristics of biochemical flares. *J Clin Microbiol* 2004; 42 (9): 3932–3936.

196. Jonas MM, Kelley DA, Mizerski J, Badia IB, Areias JA, Schwarz KB, Little NR, Greensmith MJ, Gardner SD, Bell MS et al. Clinical trial of lamivudine in children with chronic hepatitis B. *New Engl J Med* 2002; 346 (22): 1706–1713.

197. Yang HI, Lu SN, Liaw YF, You SL, Sun CA, Wang LY, Hsiao CK, Chen PJ, Chen DS, Chen CJ. Hepatitis B e antigen and the risk of hepatocellular carcinoma. *New Engl J Med* 2002; 347 (3): 168–174.

198. Zampino R, Marrone A, Mangoni ED, Santarpia L, Sica A, Tripodi MF, Utili R, Ruggiero G, Adinolfi LE. Anti-envelope 1 and 2 immune response in chronic hepatitis C patients: effects of hepatitis B virus co-infection and interferon treatment. *J Med Virol* 2004; 73 (1): 33–37.

199. Zarski JP, Bohn B, Bastie A, Pawlotsky JM, Baud M, Bost-Bezeaux F, Tran van Nhieu J, Seigneurin JM, Buffet C, Dhumeaux D. Characteristics of patients with dual infection by hepatitis B and C viruses. *J Hepatol* 1998; 28 (1): 27–33.

200. Dai CY, Yu ML, Chuang WL, Lin ZY, Chen SC, Hsieh MY, Wang LY, Tsai JF, Chang WY. Influence of hepatitis C virus on the profiles of patients with chronic hepatitis B virus infection. *J Gastroenterol Hepatol* 2001; 16 (6): 636–640.

201. Yates SC, Hafez M, Beld M, Lukashov VV, Hassan Z, Carboni G, Khaled H, McMorrow M, Attia M, Goudsmit J. Hepatocellular carcinoma in Egyptians with and without a history of hepatitis B virus infection: association with hepatitis C virus (HCV) infection but not with (HCV) RNA level. *Am J Trop Med Hyg* 1999; 60 (4): 714–720.

202. Gilson RJ, Chopra KB, Newell AM, Murray-Lyon IM, Nelson MR, Rice SJ, Tedder RS, Toole J, Jaffe HS, Weller IV. A placebo-controlled phase I/II study of adefovir dipivoxil in patients with chronic hepatitis B virus infection. *J Viral Hepat* 1999; 6 (5): 387–395.

203. Gish RG, Leung NW, Wright TL, Trinh H, Lang W, Kessler HA, Fang L, Wang LH, Delehanty J, Rigney A et al. Dose range study of pharmacokinetics, safety, and preliminary antiviral activity of emtricitabine in adults with hepatitis B virus infection. *Antimicrob Agents Chemother* 2002; 46 (6): 1734–1740.

204. van Zonneveld M, van Nunen AB, Niesters HG, de Man RA, Schalm SW, Janssen HL. Lamivudine treatment during pregnancy to prevent perinatal transmission of hepatitis B virus infection. *J Viral Hepat* 2003; 10 (4): 294–297.

205. Chen CH, Lee CM, Wang JH, Tung HD, Hung CH, Lu SN. Correlation of quantitative assay of hepatitis B surface antigen and HBV DNA levels in asymptomatic hepatitis B virus carriers. *Eur J Gastroenterol Hepatol* 2004; 16 (11): 1213–1218.

206. Smith JP, Riley TR, 3rd, Bingaman S, Mauger DT. Amantadine therapy for chronic hepatitis C: a dose escalation study. *Am J Gastroenterol* 2004; 99 (6): 1099–1104.

207. Tassopoulos NC, Tsantoulas D, Raptopoulou M, Vassiliadis T, Kanatakis S, Paraskevas E, Vafiadis I, Avgerinos A, Tzathas C, Manolakopoulos S et al. A randomized trial to assess the efficacy of interferon alpha in combination with ribavirin in the treatment of interferon alpha nonresponders with chronic hepatitis C: superior efficacy of high daily dosage of interferon alpha in genotype 1. *J Viral Hepat* 2003; 10 (3): 189–196.

208. Sentjens RE, Weegink CJ, Beld MG, Cooreman MC, Reesink HW. Viral kinetics of hepatitis C virus RNA in patients with chronic hepatitis C treated with 18 MU of interferon alpha daily. *Eur J Gastroenterol Hepatol* 2002; 14 (8): 833–840.

209. Lunel F, Veillon P, Fouchard-Hubert I, Loustaud-Ratti V, Abergel A, Silvain C, Rifflet H, Blanchi A, Causse X, Bacq Y et al. Antiviral effect of ribavirin in early non-responders to interferon monotherapy assessed by kinetics of hepatitis C virus RNA and hepatitis C virus core antigen. *J Hepatol* 2003; 39 (5): 826–833.

210. Huraib S, Iqbal A, Tanimu D, Abdullah A. Sustained virological and histological response with pretransplant interferon therapy in renal transplant patients with chronic viral hepatitis C. *Am J Nephrol* 2001; 21 (6): 435–440.

211. Cotler SJ, Ganger DR, Kaur S, Rosenblate H, Jakate S, Sullivan DG, Ng KW, Gretch DR, Jensen DM. Daily interferon therapy for hepatitis C virus infection in liver transplant recipients. *Transplantation* 2001; 71 (2): 261–266.

212. Beckebaum S, Cicinnati VR, Zhang X, Malago M, Dirsch O, Erim Y, Frilling A, Broelsch CE, Gerken G. Combination therapy with peginterferon alpha-2B and ribavirin in liver transplant recipients with recurrent HCV infection: preliminary results of an open prospective study. *Transplant Proc* 2004; 36 (5): 1489–1491.

213. Castera L, Morice Y, Grando V, Bon C, Deny P, Roulot D. Hepatitis C virus genotype 4: epidemiology and treatment. *Gastroenterol Clin Biol* 2003; 27 (6–7): 596–604.

214. Zeuzem S. Heterogeneous virologic response rates to interferon-based therapy in patients with chronic hepatitis C: who responds less well? *Ann Intern Med* 2004; 140 (5): 370–381.
215. Kamal S, El Tawil A, He Q, Koziel M, Ismail A, Raseneck J, El Sayed K, Hafez A, Madwar M. Peginterferon alpha 2b and ribavirin combination therapy in chronic hepatitis C genotype 4: impact of treatment duration on sustained virologic response (Abstr.). *Hepatology* 2004; 40 (Suppl 1): 322A.
216. Zeuzem S, Hultcrantz R, Bourliere M, Goeser T, Marcellin P, Sanchez-Tapias J, Sarrazin C, Harvey J, Brass C, Albrecht J. Peginterferon alpha-2b plus ribavirin for treatment of chronic hepatitis C in previously untreated patients infected with HCV genotypes 2 or 3. *J Hepatol* 2004; 40 (6): 993–999.
217. Torriani FJ, Rodriguez-Torres M, Rockstroh JK, Lissen E, Gonzalez-Garcia J, Lazzarin A, Carosi G, Sasadeusz J, Katlama C, Montaner J et al. Peginterferon alfa-2a plus ribavirin for chronic hepatitis C virus infection in HIV-infected patients. *New Engl J Med* 2004; 351 (5): 438–450.
218. Atypical Squamous Cells of Undetermined Significance/Low-Grade Squamous Intraepithelial Lesions Triage Study (ALTS) Group. Human papillomavirus testing for triage of women with cytologic evidence of low-grade squamous intraepithelial lesions: baseline data from a randomized trial. *J Natl Cancer Inst* 2000; 92 (5): 397–402.
219. Pelletier F, Drobacheff-Thiebaut C, Aubin F, Venier AG, Mougin C, Laurent R. Effects of imiquimod on latent human papillomavirus anal infection in HIV-infected patients. *Ann Dermatol Venereol* 2004; 131 (11): 947–951.
220. Cubie HA, Seagar AL, Beattie GJ, Monaghan S, Williams AR. A longitudinal study of HPV detection and cervical pathology in HIV infected women. *Sex Transm Infect* 2000; 76 (4): 257–261.
221. Lillo FB, Lodini S, Ferrari D, Stayton C, Taccagni G, Galli L, Lazzarin A, Uberti-Foppa C. Determination of human papillomavirus (HPV) load and type in high-grade cervical lesions surgically resected from HIV-infected women during follow-up of HPV infection. *Clin Infect Dis* 2005; 40 (3): 451–457.
222. Womack SD, Chirenje ZM, Gaffikin L, Blumenthal PD, McGrath JA, Chipato T, Ngwalle S, Munjoma M, Shah KV. HPV-based cervical cancer screening in a population at high risk for HIV infection. *Int J Cancer* 2000; 85 (2): 206–210.
223. Darwin LH, Cullen AP, Crowe SR, Modarress KJ, Willis DE, Payne WJ. Evaluation of the Hybrid Capture 2 CT/GC DNA tests and the GenProbe PACE 2 tests from the same male urethral swab specimens. *Sex Transm Dis* 2002; 29 (10): 576–580.
224. da Silva CS, Adad SJ, Hazarabedian de Souza MA, Macedo-Barcelos AC, Sarreta-Terra AP, Candido-Murta EF. Increased frequency of bacterial vaginosis and *Chlamydia trachomatis* in pregnant women with human papillomavirus infection. *Gynecol Obstet Invest* 2004; 58 (4): 189–193.
225. de Lima Soares V, de Mesquita AM, Cavalcante FG, Silva ZP, Hora V, Diedrich T, de Carvalho Silva P, de Melo PG, Dacal AR, de Carvalho EM et al. Sexually transmitted infections in a female population in rural north-east Brazil: prevalence, morbidity and risk factors. *Trop Med Int Health* 2003; 8 (7): 595–603.
226. Baldwin SB, Djambazov B, Papenfuss M, Abrahamsen M, Denman C, Guernsey de Zapien J, Ortega L, Navarro Henze JL, Hunter J, Rojas M et al. Chlamydial infection in women along the US–Mexico border. *Int J STD AIDS* 2004; 15 (12): 815–821.
227. Giuliano AR, Denman C, Guernsey de Zapien J, Navarro Henze JL, Ortega L, Djambazov B, Mendez Brown de Galaz E, Hatch K. Design and results of the USA–Mexico border human papillomavirus (HPV), cervical dysplasia, and *Chlamydia trachomatis* study. *Rev Panam Salud Publica* 2001; 9 (3): 172–181.
228. Ishi K, Suzuki F, Saito A, Kubota T. Prevalence of human papillomavirus, *Chlamydia trachomatis*, and *Neisseria gonorrhoeae* in commercial sex workers in Japan. *Infect Dis Obstet Gynecol* 2000; 8 (5–6): 235–239.

3 Microarrays: Human Disease Detection and Monitoring

Janet A. Warrington and Thomas B. Broudy

CONTENTS

INTRODUCTION

Starting in the 1970s, a sequencing revolution propelled nucleic acid research into the modern era. Inexpensive, reliable, and automated methods allowed scientists to determine the nucleotide sequences of complete genomes of organisms ranging from bacteria and viruses to plants, animals, and humans. In the wake of this flood of information, we are now faced with the exciting and daunting task of determining how knowledge of billions of nucleotide bases can be put to use in improving human health and treating disease.

High density microarray technology introduced by Stephen P.A. Fodor and colleagues[1–3] allows researchers to take advantage of the sequences of entire genomes to identify important biological information. Microarray technology opens up an entire new world of biology and affords the ability to analyze the nucleotide sequences of genes and to measure gene expression for essentially the complete coding content of the human genome in a single experiment. Whole genome analysis enables the discovery of genetic pathways disrupted by a wide range of diseases including cancer,[4–6] multiple sclerosis,[7] and host responses to infectious agents.[8,9] Across multiple disciplines, whole genome sequencing and gene expression analysis are facilitating the stratification of diseases, prediction of patient outcomes, identification of new drug targets, and better therapeutic choices.

Until recently, geneticists have been limited to using a few hundred markers in disease linkage and association studies. Current microarray technology allows for genotypic analysis of hundreds of thousands of single nucleotide polymorphisms (SNPs) at high density on a genome-wide scale.[10,11] This high resolution coverage enables researchers to conduct linkage analysis and genetic association studies in a manner that was never before practical. These new tools for disease mapping studies have already allowed the pinpointing of genes linked to sudden infant death syndrome,[12] neonatal diabetes,[13,14] bipolar disorder,[15] age-related macular degeneration,[16] and other inherited diseases.[17–19]

As of 2004, over 3,000 scientific publications discussed the use of high density microarrays. While many focus on human studies, microarrays may also be used to study the biology, gene expression, and sequence variation of any organism with a sequenced genome.[20–24]

TECHNOLOGY OVERVIEW

High density microarrays leverage expertise in chemistry, physics, biology, genetics, and semiconductor manufacturing.[1] The integration of semiconductor fabrication techniques, solid phase chemistry, random access combinatorial chemistry, molecular biology, and sophisticated robotics results in a unique photolithographic manufacturing process capable of producing arrays with millions of probes on a single glass chip. This scalability enables consistent high quality and the simultaneous production of hundreds of identical microarrays on a single wafer (Figure 3.1). Commercially available arrays are manufactured at a density of over 100 million probes per wafer. Depending on the demands of the experiment and the number of probes required per array, each wafer can be diced into tens or hundreds of identical individual arrays (Figure 3.2).

Combinatorial theory holds that the number of compounds of length N, composed of Y different subunits, is equal to Y^N. One can synthesize each of Y^N compounds in Y times N steps. Because

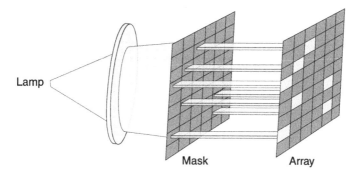

FIGURE 3.1 The photolithographic process begins with coating a 5-inch square wafer with a light-sensitive dimethoxytrityl chemical compound (i.e., protecting group) that prevents coupling between the glass and the first nucleotide of the DNA probe being created. Lithographic masks are used to direct light onto specific locations of the glass surface where DNA probe synthesis then occurs. The surface is flooded with a solution containing nucleoside phosphoramidite monomers of either adenine, thymine, cytosine or guanine. Coupling occurs only in regions on the glass that have been deprotected through illumination. The coupled nucleoside also bears a light-sensitive protecting group, so the cycle can be repeated. In this way, the microarray is built as combinations of probes are simultaneously synthesized through repeated cycles of chemistry.

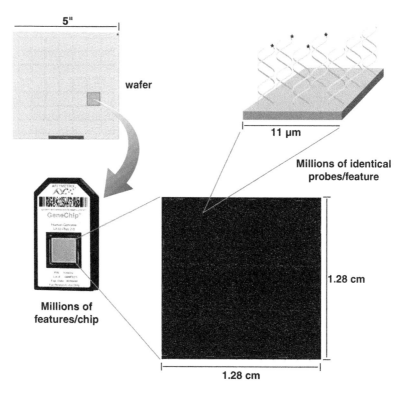

FIGURE 3.2 High density microarray technology leverages techniques from the semiconductor industry to package genetic information onto individual wafers. Each chip contains thousands to millions of features and each feature contains millions of copies of the same probe. This high information capacity enables high performance, akin to higher powered microprocessors yielding faster computer performance.

DNA has four bases, producing all combinations of DNA probes 25 nucleotides in length requires 4 (Y) times 25 (N) or 100 steps. This results in over 10^{15} different probe sequences.

In this way it is possible to construct an array of virtually any size, including arrays that can examine the entire 3.1 billion bases in the human genome, in 100 or fewer steps. Optimization of signal detection through advances in microarray manufacturing, chemistry, labeling methods, and scanning technology will lead to an increase in the amount of genomic information that can be queried on an individual microarray (Figure 3.3).

WHOLE GENOME EXPRESSION ARRAYS

Microarrays provide scientists with tools to study whole genome expression in organisms ranging from humans to flies to bacteria. By comparing genome-wide expression patterns, researchers have a practical method for surveying an entire genome and obtaining a better understanding of the dynamic relationship between gene expression and biological function.

Expression: Probe Set Strategy

Gene expression data reflects the complexity of biology and thus does not lend itself to simple interpretation. To that end, high density gene expression microarrays often use multiple probe pairs — all selected from different regions of the gene or transcript — to provide multiple independent measurements of gene expression for every gene analyzed.[25] Indeed, recent arrays feature over 1.3 million probes used to detect the expression of more than 50,000 transcripts. The use of multiple probes provides redundancy, mitigating common experimental imperfections that might otherwise compromise a complete data set.

One probe from each pair perfectly matches its gene target, while the other contains a single mismatch located directly in the middle of the 25-base probe sequence. Each paired mismatch probe effectively serves as an internal control for its perfect match partner. The mismatch probe hybridizes to nonspecific sequences about as effectively as its counterpart, allowing spurious signals, from cross-hybridization for example, to be efficiently quantified and subtracted from the actual measure of gene expression. The ability to accurately quantify background signals is especially critical for measuring low levels of gene expression.

Following hybridization of labeled target RNA molecules to the microarray, fluorescence values for all of the 11 probe pairs representing an individual transcript are analyzed. The 11 data points are processed through a statistical algorithm that ultimately yields a single measure of gene expression. The multiple probe-pair design assures a minimum sensitivity of 1:100,000, a concentration ratio corresponding to a few copies of transcript per cell.

Expression: Probe Selection and Design

Of all the sequences over 3 billion bases long present in a human genome, selecting the best set of 25-mer probes to specifically measure the expression of a single gene is a challenging task. The ideal probe would specifically detect only its intended sequence and nothing else in the remainder of the genome. To select such probes, candidate sequences are filtered for specificity — making sure they uniquely match their intended genes. The potential for cross-hybridization with similar, but unrelated, sequences is evaluated and minimized. Additionally, the physical nature of an array means that only one hybridization condition can be used for every probe. Thus, selected probe sequences are designed to have similar optimal hybridization properties, such as temperature and salt concentration, in order for the probe array to function efficiently under the single condition.

It is also important to remember that a single gene does not always generate a single transcript. Splice variants and polyadenylation variants can result in transcription of one gene into multiple RNAs, each resulting in a different protein, with potentially vast differences in biological function. When this is the case, meaningful measures of gene expression require probes that can distinguish

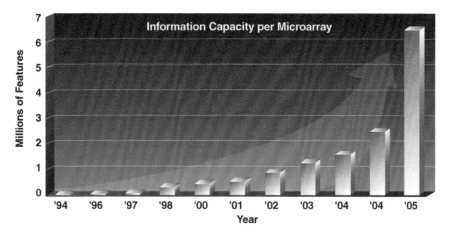

FIGURE 3.3 In 1994, the first commercial high density microarray accommodated had a feature size of 100 μm, accommodating 16,000 features on a 1.28 × 1.28 cm glass surface. By 2005, through technological improvements enhancing signal detection that included manufacturing precision, fluorescent assay development and scanning resolution, feature size was reduced to 5 μm, allowing the same size microarray to feature over 6 million probes. The advances allow the collection of 400 times more information from each microarray. The root of this data explosion lies in the fact that feature shrink increases linearly while the corresponding microarray content grows exponentially. For example, if the diameter of every square on a checkerboard were cut in half, the number of squares on the board would increase by four. Microarray technology has tremendous room for continued growth, submicron lithographic methods are routinely used in semiconductor fabrication, and current R&D efforts are experimenting with 1-μm formats.

between variant transcripts. To obtain a complete picture of the activity of a gene, microarrays use certain probes to distinguish between multiple splice or polyadenylation variants, while others are selected against common regions shared by those variants.

WHOLE GENOME GENOTYPING ARRAYS

Microarrays designed to read SNP-based genotypes operate via the same basic principle as arrays used to measure gene expression levels. For any given SNP of two possible genotypes, A or B, probes are synthesized on the array corresponding to both alleles. Following hybridization of the target to the array, scientists determine whether a SNP is an AA, AB, or BB genotype by determining whether the A allele probes have detected their complementary sequence, whether the B allele probes have detected their complementary sequence, or whether both have detected complementary sequences.[10,11]

Genotyping: Probe Set Strategy

Much like gene expression arrays, every SNP genotype represented on a genotyping array has a perfect match probe as well as a mismatch probe. Once again, the mismatch probe serves as an internal control, accounting for spurious signals and cross-hybridization. Each probe pair is the basic unit used to call a SNP genotype. However, to ensure accurate genotype calls, current GeneChip* microarrays routinely use five different probe pairs for each SNP allele on a single strand of DNA. The first probe pair contains the SNP precisely in the center of the 25-mer sequence. The remaining four probe pairs are positioned or tiled to the right and the left of this central position.

---

* A division of F. Hoffman-LaRoche Ltd., Basel, Switzerland, http://www.roche-diagnostics.com/

Additionally, every SNP is genotyped from both sense and antisense strands of DNA, meaning that 20 probes — 10 probe pairs — are used to genotype a single SNP allele. The same probe set strategy is used for the alternate SNP allele, requiring an additional 20 probes on the array — a total of 40 probes for every SNP. The need for a high density manufacturing technique quickly becomes obvious when genotyping large number of SNPs; nearly 5 million probes are used to genotype every 100,000 SNPs (Figure 3.4).

Genotyping: Probe Selection and Design

When selecting probes to measure gene expression, the 25-base sequence can be chosen from virtually anywhere in the gene; however, because SNPs occur in defined positions, genotyping probes must be selected to complement that precise SNP sequence. In all cases, a 25-base perfect match/mismatch probe pair is designed to contain the complementary SNP sequence in the central position. Each of the additional probe pairs is positioned in one of six other locations relative to the SNP (+1, +3, +4, −1, −2, −4). To identify which probes from which positions function best, each probe is empirically tested on a pilot microarray. Those that function best are selected for further use.

The key to optimizing this type of SNP genotyping is the implementation of a complexity-reducing whole genome sampling assay in which one primer is used to genotype hundreds of thousands of SNPs distributed throughout the genome.[10] Previous SNP mapping efforts have been hampered by the need for locus-specific amplification and for many tens of thousands of PCR primer pairs — an expensive and cumbersome undertaking.

The whole genome sampling assay (WGSA) for microarray-based genotyping requires only one primer to amplify a DNA sequence, making large-scale SNP analysis practical. WGSA uses a simple restriction enzyme to digest genomic DNA, creating various sizes of DNA fragments, each containing their respective SNPs. The same oligonucleotide adapter is then ligated to the genomic DNA fragments, and a primer complementary to the adapter is used for PCR amplification. However, only certain sized fragments are applied to the array, so it is critical to design microarray probes against the SNPs present on the DNA fragments.

Computer modeling of the restriction digest tells researchers which SNPs are present on which DNA fragments, allowing them to then select probes against the SNP sequences that will be applied to the array. For the mapping 100K set, two separate restriction enzymes are used and each enzyme creates a pool of DNA fragments containing over 50,000 SNPs to be genotyped.

HUMAN DISEASE DETECTION AND MONITORING

Perhaps the most compelling application for microarray technology is the potential to advance disease research by exploring the genome in all its dimensions, including whole genome expression and sequence variation studies. The literature is now replete with examples in which these information-rich genetic analysis tools have been used to extend our understanding of the molecular basis of disease and to identify the pathways that modulate it.[26,27] However, the impact on human health is only beginning to be realized. Across multiple disciplines, the technology helps disease stratification, prediction of patient outcomes, and the generation of information to allow better therapeutic choices (Figure 3.5).

Gene Expression

Researchers often use microarrays to compare gene expression levels between disease tissue and control tissue obtained adjacent to disease tissue or tissue from unaffected individuals. The microarray queries the activities of more than 50,000 transcripts in humans and shows which genes in the

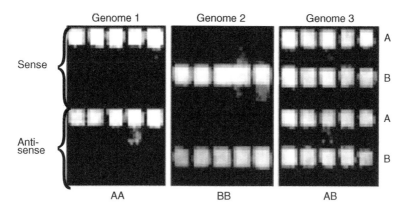

FIGURE 3.4 A scanned microarray readout with fluorescent detection of SNP genotypes. Each row features probes designed to genotype a single SNP that measures either an A allele or B allele from both the sense and antisense sequences. Each panel represents a genome containing different combinations of alleles. For genome 1, each of the 10 probes for allele A (5 sense and 5 antisense probes) have detected the complementary DNA in the sample. Because only A alleles and no B alleles are detected, the SNP under study is referred to as an AA genotype. Similarly, a SNP may be a BB genotype containing two copies of the B allele (genome 2), or an AB genotype containing one copy of each (genome 3).

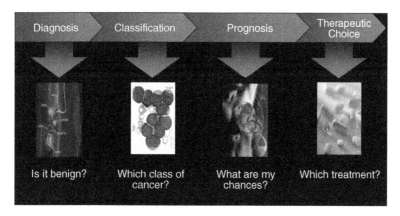

FIGURE 3.5 *(A color version follows page 270)* Rapid detection of differentially expressed genes or genetic mutations such as amplifications, deletions, or loss of heterozygosity, may allow for earlier cancer diagnosis, assessment of cancer predisposing markers, characterization of tumors for tailored chemotherapies, and provision of new insights into the molecular basis of cancer.

disease tissue are expressed at higher or lower levels than those in the unaffected tissue. The genes differentially expressed are called the expression profile or signature genes.

Hundreds of gene expression disease studies offer new hope for diagnosing, classifying, and treating diseases including common and rare cancers. Studies on medulloblastoma,[28] prostate cancer,[29] breast cancer,[30,31] lung cancer,[32] colon cancer,[33] renal cell carcinoma,[34] ovarian cancer,[35] and lymphoma[36,37] are only a few examples of cancers in which gene expression classification systems have been developed. By developing molecular tests, researchers hope to create early detection methods that will allow better treatments and more successful long-term outcomes.

Leukemias are among the cancers most frequently studied with microarray analysis (Figure 3.6). In 1999, Golub et al. were among the first groups to use gene expression profiling to distinguish acute myeloid leukemia (AML) and acute lymphoblastic leukemia (ALL).[38] They showed that different cancers have different sets of genes expressed, and that those gene expression profiles can be used to classify the diseases into their respective clinical subtypes; more importantly, they showed

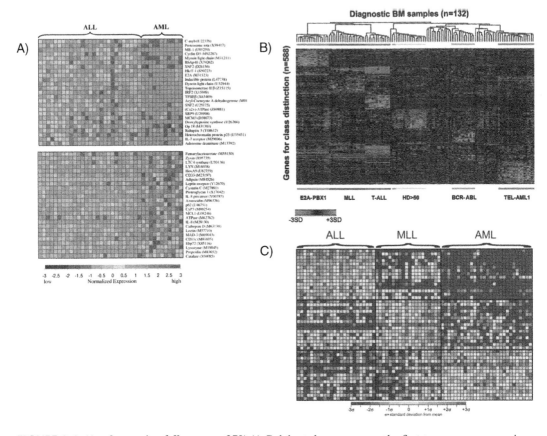

FIGURE 3.6 *(A color version follows page 270)* A) Golub et al. were among the first to use gene expression profiling to classify different kinds of leukemia.[38] B) Ross et al. used expression profiling to subclassify seven distinct subtypes of ALL.[5] C) Armstrong et al. classified patient samples among acute lymphoblastic leukemia, mixed lineage leukemia, and acute myelogenous leukemia with 95% accuracy. Acute lymphoblastic leukemia patients with mixed-lineage leukemia gene (MLL) translocations have particularly poor prognoses.[4]

that expression profiles can also be used to define entirely new subtypes of leukemia. Subsequent studies in which microarrays were used to discriminate seven basic subtypes of ALL have been repeated.[39]

Several additional groups have now reported large-scale classification studies of ALL, describing detailed expression profiles validating and expanding the early results.[4,5] Torsten Haferlach's research team at the Ludwig Maximilians University, Munich, Germany analyzed expression patterns to discriminate eight clinically relevant acute leukemia subgroups.[40] They also confirmed that signatures defined for pediatric ALL could be used to stratify independent adult leukemia patient samples.[6] Haferlach's work is a prime example of the diagnostic value of gene expression signatures across a broad range of leukemias and leukemia related disorders (Figure 3.7), including ALL,[6] AML,[41] and acute promyelocytic leukemia.[42]

A key theme that has emerged is the value of whole genome analysis for the discovery of accurate expression profiles with diagnostic value. Most expression profiles consist of a few dozen or a few hundred markers; finding them requires sifting through as many genes as possible. By using whole genome expression microarrays, researchers have found additional genes correlating with diseases like leukemia, and have been able to use that information to build more discriminating expression profiles.[5]

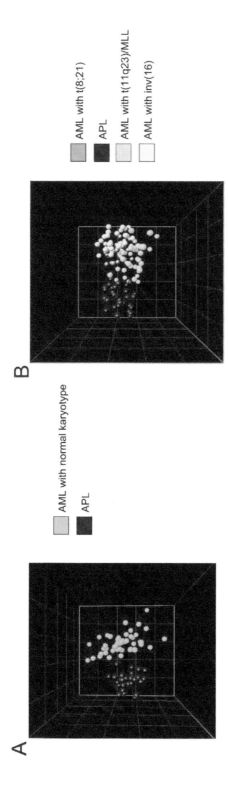

FIGURE 3.7 (*A color version follows page 270*) Haferlach and colleagues used three-dimensional Principal Components Analysis feature space to visualize expression data on genes from the Gene Ontology Biological Process category blood coagulation (accession number GO:0007596).[42] (A) Eighty-three probe sets distinguish patients with acute promyelocytic leukemia (APL) and those with AML with normal karyotypes. (B) Seventy-five probe sets distinguish patients with APL and those with subtypes of AML.

Surrogate Tissue

When studying cancers of the blood or breast, for example, researchers typically have easy, inexpensive, and noninvasive access to the disease tissue, thus meeting the key criteria for simple diagnostics. However, studying other diseases and other types of cancers often presents a far more difficult challenge because tissue access may require invasive procedures. Microarray techniques have allowed researchers to identify "surrogate" gene expression signatures for internal organ malignancies by analyzing molecular changes in tissues such as saliva, airway epithelial cells, blood, and urine.

Using Saliva for Detection of Mouth Cancer

Li et al. developed a microarray-based salivary diagnostic system to identify cancer-associated RNAs from the saliva of oral squamous cell carcinoma (OSCC) patients.[43] They profiled salivary mRNA and discovered 1,679 genes differentially expressed between cancer patient saliva and normal controls. Many of the genes in the expression signature may have little to do directly with cancer etiology, but accurately function as indirect markers for the underlying oral carcinoma.[44] The team focused on seven of the most consistently expressed biomarkers, including transcripts of IL8, IL1B, DUSP1, HA3, OAZ1, S100P, and SAT.

In combination, these markers proved 91% sensitive and 91% specific for detecting cancer in the study of saliva samples from 32 OSCC patients and 32 healthy controls. Results like these offer the promise of a simple early detection screen using an easy-to-obtain surrogate diagnostic fluid, instead of a complicated surgical procedure.

Using Bronchoscopy for Early Detection of Lung Cancer

Cigarette smoke is the major cause of lung cancer. Using high density gene expression arrays, Spira et al. analyzed human airway epithelial cells obtained at bronchoscopy for expression patterns associated with smoking status.[45,46] They found an expression pattern consisting of 97 genes that accurately predicted smokers from never-smokers (Figure 3.8). Remarkably, the research demonstrates that some changes in gene expression are reversible upon smoking cessation, while others are not. For instance, expression levels of smoking-induced genes begin to resemble levels of never-smokers after 2 years of cessation.

Regardless of the number of years elapsed after quitting, the team found 13 genes that never returned to normal expression levels, including a number of potential tumor suppressor genes that remained down-regulated (TU3A and CX3CL1) and a number of possible oncogenes that remained up-regulated (CEACAM6 and HN1). Genetic mutations incurred through smoke exposure may explain these irreversible changes and may also explain why some individuals remain at risk for lung cancer despite having quit cigarette smoking for a lengthy duration.

Using Blood for Detection of Kidney Disease

As part of a Phase II clinical study of renal cell carcinoma (RCC), Twine and colleagues at Wyeth Research (http://www.wyeth.com) profiled gene expression from peripheral blood mononuclear cells (PBMCs) and found a specific set of expressed genes that accurately distinguished RCC cancer patients from controls.[34] The team used high density expression microarrays and multiclass variate analysis to predict patient RCC status with 70% accuracy. Additionally, they found that PBMC expression profiles could distinguish RCC patient samples from other types of solid tumor samples (prostate, head, and neck cancers). These findings have important implications for diagnosis and future clinical pharmacogenomic studies of antitumor therapies. RCC is usually detected by imaging methods, but 30% of apparently nonmetastatic patients undergo relapses after surgery and eventually die of the disease.[47] The need for a simple, noninvasive

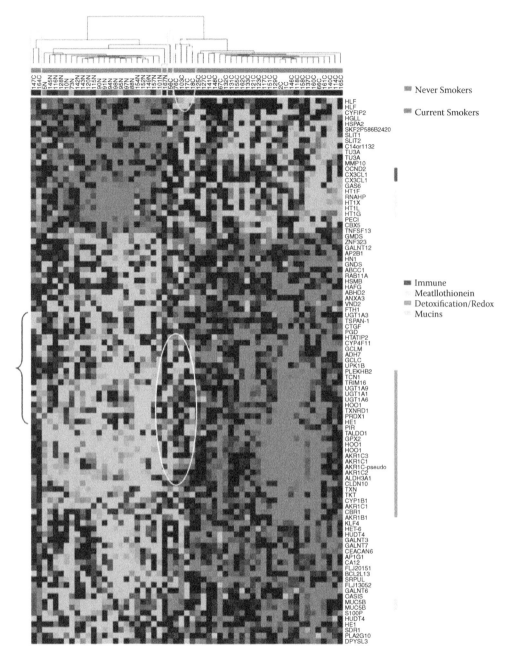

FIGURE 3.8 *(A color version follows page 270)* Hierarchical clustering of 34 current- and 23 never-smoker samples according to expression of 97 genes differentially expressed.[45] While the majority of current smokers and never smokers classify into two distinct groups, there are a few outliers whose gene expression patterns match that of the alternate classification: three current-smokers group with never smokers (yellow rectangle) and one never-smoker (blue box) clusters more closely with current smokers. Some current smokers (yellow oval) did not express redox/xenobiotic gene (white oval) to the same levels as other smokers. The heat map depicts high levels of expression in red and low levels in green.

monitoring method is critical. By using gene expression analysis, researchers can monitor a disease with molecular tools by analyzing a patient's PBMC gene expression and seeing whether it matches that of the RCC profile.

Using Host Response for Detection of Infectious Disease

Similar to the way expression profiles and sets of predictor genes have been used to classify different types of cancer, profiles can also be used to classify host responses to different kinds of infectious diseases (Figure 3.9). For example, when comparing *Chlamydia pneumoniae, Chlamydia trachomatis*, and intracellular *Salmonella typhimurium*, Hess et al. found distinct host response expression profiles in human epithelial cells.[8] Genus- or group-specific transcriptional response patterns likely contribute to the different pathologies of each disease, but could also be used to characterize disease via host response signatures. Nau et al. examined whole genome expression from macrophages exposed to Gram-negative and Gram-positive bacteria.[9] They found distinct host response expression signatures that may provide a basis for diagnosis of clinical Gram-negative infections.

DNA Analysis

For many cancers, improving diagnosis or treatment starts with finding the genetic changes that play a causal role when cells grow out of control, such as gene copy number alterations or loss of heterozygosity (LOH). While both analysis methods have existed for some time, applying high density SNP genotyping microarrays to these studies helps researchers perform such experiments faster and obtain higher resolution.

Genotyping microarrays allow researchers to assess copy number and LOH across an entire genome in only one experiment. By measuring over 100,000 SNPs, arrays detect more detailed copy number changes than conventional comparative genomic hybridization (CGH) and offer many times more markers than microsatellite analysis for LOH studies. This high density combined approach provides researchers with a rapid way to locate difficult-to-detect tumor genes responsible for cancers of the breast,[48–52] bladder,[53,54] prostate,[55,56] bone,[57] mouth,[58] and lung.[49,59,60]

Lung and Breast Cancer

Zhao et al. report using microarrays capable of genotyping 10,000 SNPs to simultaneously detect cancer-specific DNA copy number changes and LOH, two major causes of neoplastic growth. In a study of seven lung cancer cell lines,[49] the team detected twice as many LOH regions as microsatellite analysis. They identified already known deletions as well as 14 previously unknown LOH regions on 9 different chromosomes. In a second study of 18 lung and breast cancer cell lines, the team found copy number amplifications encompassing known proto-oncogenes and deletions encompassing tumor suppressor genes; they were also able to distinguish LOH caused by DNA deletions coupled with gene mutation.[59]

Mouth Cancer

Zhou et al. used a similar strategy to identify four regions of the genome associated with mouth cancer.[58] Their study highlights the need to track both chromosomal copy number changes and loss of heterozygosity. For example, on chromosome 3, the team used copy number analysis to find DNA amplification and deletion mutations shared by premalignant and malignant cells, but finding the mutations associated with malignant cells required looking for LOH. When the team analyzed the genotypic data for LOH, they found two regions of chromosome 3 that mutated in the malignant cells.

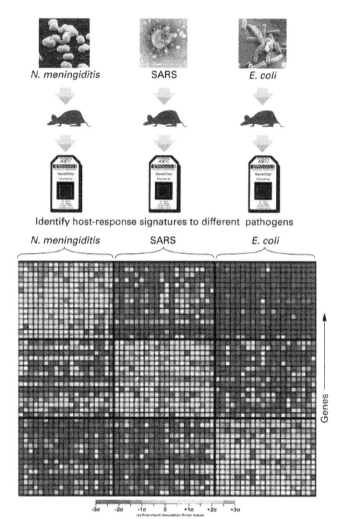

FIGURE 3.9 *(A color version follows page 270)* Host response signatures for pathogen identification. In this animal model, mice were infected with different pathogens and whole genome expression was measured by microarray analysis, resulting in specific host-response signatures.

Bladder Cancer

Hoque et al. extracted DNA from urine sediment and then used genotyping microarrays to identify genetic variations indicative of bladder cancer.[61] Urine is not a true surrogate tissue for bladder cancer because the urine contains actual bladder cells; however, the analysis meets the same easy-to-access, inexpensive, and noninvasive criteria desired for a diagnostic of internal organ disease.

Using the array-based urine assay to examine SNP genotypes among 31 urine DNA samples from patients with bladder tumors and 14 samples from control individuals with no evidence of genitourinary malignancies, the team was able to identify allelic imbalances in 100% of the bladder cancer patients. Importantly, the researchers did not measure any allelic imbalances indicative of cancer in 12 of 13 healthy controls; the one exception suffered from hematuria, a chronic inflammation with possible genetic etiologies. Bladder cancer is typically diagnosed through an invasive cystoscopy procedure; no reliable noninvasive tests that can detect bladder cancers with the sensitivity and specificity demonstrated in the Hoque study are currently available. This new test may be relevant for the 243,000 cases of bladder cancer contracted annually worldwide.

Beyond Cancer

In addition to cancer, gene deletions and amplifications are also major causes of developmental defects such as Down syndrome. Rauch and colleagues used high density genotyping microarrays to study chromosomal amplifications and deletions causing a variety of mental retardation syndromes.[62] The genome-wide analysis of 10,000 SNPs allowed the research team to detect chromosomal aberrations as small as 700 kilobases; they anticipate that this method may allow the identification of further genomic regions that, when deleted, contribute to human pathology.

By combining microarray genotyping with gene expression, researchers can correlate genotypes with changes in gene activity, providing insight into functional consequences of a newly discovered mutation. Likewise, if researchers identify changes in RNA levels with expression arrays, they can check for corresponding genetic changes with genotyping arrays. Advances in microarray information capacity for both gene expression and sequence analysis promise to provide more information about our genome, offering new genetic insights into disease research.

PROGNOSIS

Gene expression profiling of cancers offers important prognostic indications for outcome and recurrence as well as patient response to treatment. Studies have thus far been retrospective, but future maturation and commercialization of the technology promises prospective uses in the clinic and direct impacts on patient care. In diseases like breast cancer, leukemia, and prostate cancer, similar tumor types may have distinct molecular differences.[63] This helps explain why two tumors may look identical histologically, but patient outcomes can be radically different. At a molecular level, the patients have different forms of the disease.

BREAST CANCER

The most important factor in breast cancer prognosis is invasion into axillary lymph nodes. Several studies have identified gene expression classification systems based on lymph node status and estrogen receptor status (another prognostic indicator).[30,31] In 2003, Huang et al. reported using metagene analysis involving multiple aggregates of expression signatures with lymph node status to predict recurrence of disease with 90% accuracy. Scientists at Erasmus University Medical Center, Rotterdam, analyzed expression from 115 lymph node-negative breast cancer patients and discovered a 76-gene signature predictive of developing distant metastases within 5 years.[64] The signature was 93% sensitive and 48% specific in a validation set of 171 patients. The ability to identify cases that need intensive clinical intervention, even if lymph node status is negative, could lead to an improvement in cancer survival.

COLORECTAL CANCER

Frederiksen et al. were among the first groups to classify substages of colorectal cancer using microarray expression profiling.[65] Classifying the disease into its appropriate clinical subtype is important not only for determining prognosis, but also for determining appropriate treatment. The studies compared gene expression from normal mucosa and from each of the Dukes' stages A, B, C, and D. Using a nearest neighbor classifier method, the group could distinguish normal and Dukes' B and C samples with less than 20% error; Dukes' A and D cancers could not be classified correctly.

Within Dukes' B class alone, 25 to 30% of cases develop recurrence and die from the disease; however, no widely adopted method yet exists to predict poor-prognosis cases in order to offer more aggressive therapy. Recent microarray studies by Wang et al. uncovered a 23-gene signature predictive of a 13-fold higher chance of recurrence in Dukes' B patients. The signature was validated in 36 independent patients and predicted poor-prognosis Dukes' B outcomes with 78% accuracy.[66]

LYMPHOMA

Patients with follicular lymphoma survive from 1 to 20 years after diagnosis. In a research program directed by Louis Staudt at the National Cancer Institute (NCI, Bethesda, Maryland, USA, http://cancernet.nci.nih.gov/), Dave et al. measured gene expression from 191 lymphoma biopsy samples and discovered 2 expression signatures, termed immune response-1 and immune response-2, predictive of disease outcome and prognosis.[36] The surprising characteristic is that both signatures predominantly stem from nonmalignant immune cells infiltrating the tumors, not from the cancer cells themselves as may have been expected. A good prognosis signature is affiliated with a mixture of immune cells dominated by T cells, while poor prognosis signature genes are highly expressed by a different group of immune cells that include monocytes (precursors to macrophages) and/or dendritic cells.

Using these signatures, Dave and colleagues were able to subclassify patients into four prognostic categories with average lengths of survival of 13.6, 11.1, 10.8, and 3.9 years. Furthermore, patient stratification offers promise to improve clinical trials. Current trials using survival as an endpoint have been difficult to complete because more than 75% of follicular lymphoma patients survive more than 10 years after diagnosis.

LEUKEMIA

Valk et al. used expression profiling to classify poor-outcome cases of AML that were previously impossible to predict. While the predictive accuracy of each signature varied, some subsets of genes were 100% accurate for predicting particular classes of AML.[67]

PROBABILISTIC DECISION MAKING

RNA profiling and DNA variation information lend themselves to the establishment of reference databases containing disease-related information for affected and unaffected populations. These databases are extremely valuable for combining RNA, DNA, and protein information of an individual with suspected disease. The match to the reference profile combined with other known clinical and demographic data produces probability information that can be used by physician and patient in disease management.

In the case of breast cancer, Nevins et al. report combining multiple gene expression profiles into a 500-gene meta-signature more powerful than any single profile alone for disease subclassification.[68] By integrating clinical data with the genomic information and relying on outcome probabilities, the group made predictive classifications of cancer recurrence and lymph node metastasis around the 85 to 90% level. As the authors noted, the certainty of the prediction must be weighted with equal importance. For example, if a test predicts a 70% recurrence probability, physician and patient may respond differently if the uncertainty is ±40% versus ±1%.

Similarly, Glinsky and colleagues assessed a combination of gene expression profiles, lymph node status, and estrogen receptor markers to create a breast cancer survival algorithm predictive of clinical outcome. The team studied 295 breast cancer cases and found the algorithm provided highly accurate classification into subgroups with distinct 5- and 10-year survival after therapy.[69] By testing for the gene expression survival signature at the time of diagnosis, the researchers suggest that poor-outcome cases may be identified early on and may be more aggressively treated with adjuvant therapy, predicting an estimated increase of ~7000 to 9000 breast cancer survivors after 10-year follow-up in the United States alone.

PHARMACOGENOMICS: EXPRESSION ANALYSIS

Researchers also use microarray expression profiling to identify outcome probabilities predictive of success or failure to drug treatment. Nowhere has this been more apparent than in the field of cancer research in which pharmaceutical companies and academic laboratories have discovered expression signatures that appear to be predictive of treatment outcomes for a number of chemotherapeutics such as Gleevec® (Novartis Pharmaceuticals, East Hanover, New Jersey, USA, http://www.novartis.com/) and Velcade® (Millenium Pharmaceuticals, Cambridge, Massachusetts, USA, http://www.mlnm.com/) used to treat diseases including leukemia,[70,71] myeloma,[72] and breast cancer.[73]

Gleevec

Researchers studying Philadelphia chromosome-positive (Ph+) acute lymphoblastic leukemia used expression profiles to subclassify patients responsive to Gleevec. Hoffman et al. discovered a 95-gene expression signature that distinguished all sensitive samples from resistant samples.[70] Fifty-nine percent of all Ph+ ALL patients are resistant to treatment. Similarly, as part of a recent Phase III clinical trial to predict the success or failure of Gleevec treatment on chronic myelogenous leukemia,[71] patterns were analyzed from patients prior to treatment. A 31-gene no-response signature was found, which predicts a 200-fold higher probability of failed therapy.

Velcade

In the case of myeloma treatment, Mulligan et al. used microarrays to examine expression profiles and collect pharmacogenomic data from patients treated with Velcade.[72] They discovered a pattern consisting of 30 genes that correlate with response or lack of response to therapy. The clinical utility of these biomarkers will be assessed in a Phase III trial.

Pharmacogenomics: DNA Sequence Analysis

High density microarrays representing over 100,000 SNPs enable researchers to readily genotype large populations of responders versus nonresponders to a given drug for phenotypes including efficacy and toxicity. This information may help avoid some of the over 100,000 annual fatalities from adverse drug reactions in the U.S. alone.[74]

In late-stage clinical trials, for example, microarray genotype analysis could be used to stratify patient populations to eliminate poor or toxic responders from key Phase III trials. Such stratification would help ensure maximum effectiveness through clearer statistical differentiation between drug and placebo while also reducing size and cost of trial and improving the odds of drug approval. In addition, once a drug is on the market, patient stratification could be used to accelerate drug expansion into new indications through faster, smaller, more definitive Phase IV trials or establish medical superiority of a late-to-market drug relative to entrenched competitors in an important class of patients. Genome-wide genotype information will also fuel future research.

An example of work now possible is a pharmacogenomic study at the Mayo Clinic (Rochester, Minnesota, USA, http://www.mayoclinic.com), where researchers use genotyping microarrays to investigate the genetic bases for differential responses to antihypertensive drugs in different patients and populations. These scientists hope to identify genes influencing drug response and ultimately tailor antihypertensive therapy for individual patients.

While researchers have made progress discovering complex sets of pharmacogenomic markers, pharmacogenetic studies of individual markers[75] provide insight into the relationship of genetic variation and treatment outcome. For instance, the cytochrome P450 pathway controls the ways a host of drugs are metabolized. Specific known allelic variations may cause an individual to respond poorly or suffer an adverse reaction.[76–78] Genetic variations in critical cancer genes like p53, EGFR, and Neu2 can affect chemotherapy success.

Cytochrome P450 Pharmacogenetic Test

The Roche Diagnostic AmpliChip™ CYP450 Test has been approved for use in the United states and Europe. The test allows identification of certain naturally occurring variations in the drug metabolism genes, CYP2D6 and CYP2C19. These variations affect the rate at which an individual metabolizes many common drugs used to treat cardiovascular disease, high blood pressure, depression, and attention deficit hyperactivity disorder.

The AmpliChip CYP450 microarray has yielded promising results. For example, hoping to find a genotype that would predict a patient's response to risperidone, Peter Wedlund and colleagues at the University of Kentucky Mental Health Research Center used the AmpliChip to investigate the relationship between a patient's CYP450 profile and his response to the schizophrenia drug.[79] This preliminary clinical study resulted in the association of the CYP2D6 poor metabolizer phenotype to risperidone-induced adverse reactions and to treatment discontinuation due to these ADRs.

p53

For the past 25 years, scientists have been elucidating the role of p53 in oncogenesis. Variations in p53 sequences affect the way it functions as a tumor suppressor. Microarray sequence analysis of p53 presents a rapid way to catalog genetic variations that might also impact the way a patient responds to a drug targeted to genetic pathways or genes affected by p53. Vassilev et al. at Hoffmann-La Roche (Nutley, New Jersey, USA, http://www.rocheusa.com/) describe a series of synthetic compounds called "nutlins" that bind to the MDM2 p53 ligand and block the protein–protein interaction, ultimately promoting antitumor activity.[80]

The nutlin compound works by projecting three chemical groups into the MDM2 pockets normally occupied by Phe 19, Trp 23, and Leu 26 of p53 (Figure 3.10). Depending on the genetic variation in the patient's p53 gene, the competitive nutlin inhibition may be variable, potentially requiring a dosage adjustment for optimal treatment. Naturally, using microarrays to examine variations in the MDM2 gene itself, especially in the p53 binding pocket, may help correlate MDM2 genetic variation to nutlin treatment outcome.

Herceptin and Iressa

The real-world use of genetic information in treatment decisions is already a reality for a select few drugs. Using chemotherapeutics like Herceptin® and Iressa®,* scientists have identified single genes, HER2/neu and EGFR, respectively, associated with the differences observed among treatment responders and nonresponders.[81,82] These drugs can now be prescribed in conjunction with genetic tests, allowing a physician to identify patients who stand to benefit most from therapy. Approximately 30% of breast cancer patients benefit from Herceptin therapy, while about 10% of lung cancer cases benefit from Iressa.

The challenge in identifying any of these single genetic markers has been the process of physically pinpointing the gene or mutation among the 3 billion bases in the human genome. When multiple genes or mutations are involved in a variable drug response, far larger clinical studies are needed to adequately power the research. While microarrays provide researchers with high resolution tools to analyze samples collected from such a study, the immense challenges of logistical planning and resource procurement to support large multisite studies remain.

STANDARD CONTROLS AND BEST PRACTICES

As clinical researchers use genomic information and compare their array data within and between laboratories and hospitals, standardized array methodologies and data reporting criteria will be essential. The Microarray Gene Expression Data Society (MGED, http://www.mged.org/) has taken

---

* Herceptin is a registered trademark of Genentech, Inc., South San Francisco, California, USA, http://www.gene.com/ and Iressa is a registered trademark of AstraZeneca International, Wilmslow, Cheshire, United Kingdom, http://www.astrazeneca.com/

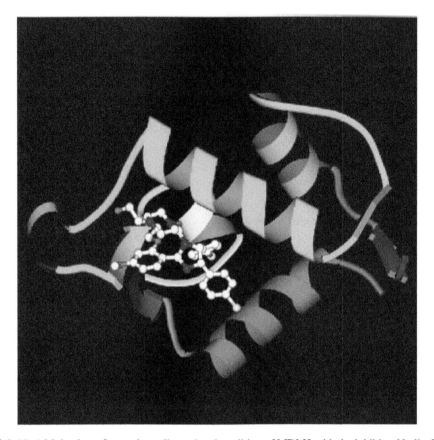

FIGURE 3.10 A Molscript software three-dimensional rendition of MDM2 with the inhibitor Nutlin-2 located in the p53 binding pocket. Genetic variations in the MDM2 binding pocket or the p53 ligand may impact the competitive inhibition of the Nutlin synthetic.[80]

one of the first steps by developing data reporting guidelines,[83,84] enabling scientists to properly compare data from different experiments. However, creating data that can be shared, compared, and developed by the worldwide scientific community requires standard controls and best practices for microarray data generation and interpretation. Toward this goal, a number of collaborative efforts of clinical and laboratory scientists in academia and industry have identified critical areas for standardization, including experimental design, preanalytical methods, technical performance assessment, data analysis and data reporting.

EXPERIMENTAL DESIGN

Successful microarray experimentation begins with appropriate experimental design that involves at least three key focal areas: (1) using sufficient biological replicates, (2) making comparisons of well characterized equivalent tissue types, and (3) standardizing tissue sampling, storage, and assay procedures. The pay-off of adhering to these accepted best practices is high quality data that can readily be compared within and between studies, sites, and dates, creating a more efficient and effective use of resources and precious patient samples.

PREANALYTICAL SAMPLE HANDLING

A preanalytical study by Mutter et al. evaluated the effects of three different tissue storage methods (fresh, frozen and preserved with RNALater*) on the same uterine myometrial tissue specimen by

---

* RNALater is a registered trademark of Ambion Inc., Austin, Texas, USA, http://www.ambion.com/

comparing quantitative microarray expression.[85] The group used analysis of variance (ANOVA) (http://www.physics.csbsju.edu/stats/anova.html) and found very little expression variation among the three cases. When compared by gene functional class, the frequency of outliers was overall no more than what was expected by random chance. This study suggests that if tissue is maintained in a stable environment, for example, via RNALater treatment or snap freezing, a number of methods can be used as a means of routine tissue collection and temporary storage prior to RNA extraction.

The quality and quantity of RNA examined in a microarray experiment is often a critical source of variation as well. Using varying amounts of different quality samples inevitably yields varying results. Adherence to quality control criteria using standard RNA isolation and processing methods yields consistent (between laboratories) and intrinsically comparable data. The same set of criteria can also be used as best practices for data generation in the design and conduct of clinical trials.

Over a dozen biotechnology, pharmaceutical and academic institutions participated in a Best Practices Blood Handling for Array Based Expression Working Group (http://www.affymetrix.com/community/standards/blood_protocol.affx) to develop guidelines for blood handling, stabilization, and RNA extraction. Members of the group conducted a series of longitudinal and variability studies to address handling and stabilization questions, and are evaluating reproducibility and robustness by comparing Ficoll* and other commercially available extraction kits. Initial reports from independent laboratories point to variable incubation times in RNA extraction tubes as a source of variability[86] and as a possible optimization step for high RNA yield.[87]

TECHNICAL PERFORMANCE

In 2004, the National Cancer Institute published a study[88] to assess the comparability of microarray expression data from four distinct laboratories, each following identical protocols, from tissue processing to hybridization and scanning. Intraclass correlation within laboratories was only slightly stronger than between laboratories, and the correlation tended to be weakest for genes expressed at low levels and showing small variation (Figure 3.11).

The multisite collaboration used hierarchical clustering to demonstrate that similar sample types clustered together, independent of the laboratory site where the sample was analyzed. The findings indicate that complete microarray tumor analysis can be performed and compared between multiple independent laboratories for a single study. The data generated from the NCI-funded study are freely available at http://gedp.nci.nih.gov/

Acceptance of microarray data for use in clinical applications requires standardization of technical performance assessment methods across the industry. To this end, over 70 scientists from public, private, and academic institutions have formed the External RNA Controls Consortium (ERCC, http://www.cstl.nist.gov/biotech/Cell&TissueMeasurements/GeneExpression/ERCC.htm) with the goal of developing external RNA spike-in controls and recommended protocols for sample control on gene expression technologies such as microarrays and quantitative real-time reverse transcriptase PCR.[89] The controls will consist of 100 well characterized clones comprised of randomly generated unique sequences as determined by sequence comparison to mouse, rat, human, *Drosophila*, *Escherichia coli*, mosquito sequence databases, and *Bacillus subtillus* and *Arabidopsis thaliana* clones. Control sequences will be tested in the second half of 2006 by multiple labs participating in the ERCC including all major microarray manufacturers, the National Institute of Standards and Technology, the Food and Drug Administration, and the National Cancer Institute. All data will be made public.[90]

---

* Ficoll is a registered trademark of Fluka Chemical Corp., Milwaukee, Wisconsin, USA, https://www.sigmaaldrich.com/Brands/Fluka_Riedel_Home.html

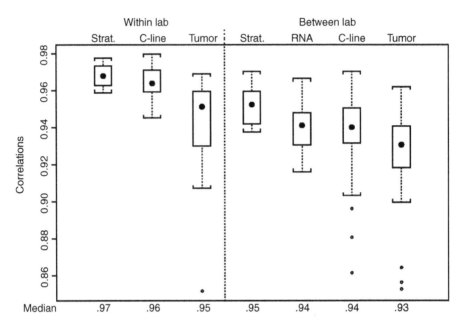

FIGURE 3.11 Left: Within-laboratory variation for expression analysis of purified Stratagene (LaJolla, California, USA, http://www.stratagene.com) reference RNA resulted in a 0.97 median correlation; variation for analysis that included RNA extraction from cell lines correlated at 0.96; and within-laboratory variation from samples requiring sectioning of tissue from a frozen tumor block followed by RNA extraction resulted in a slightly lower correlation of 0.95. Right: Between-laboratory variation revealed a correlation reduction of 0.02 as compared to the within-laboratory variation for the tumors samples, cell lines, and reference RNA. Between-laboratory variability of extracted tumor RNA was analyzed as well, but no within-laboratory replicates on this tumor RNA were performed. All box plots are of Pearson correlations across genes broken down by sample type.

Data Analysis

Microarrays are often designed with multiple probes representing even a single transcript. Depending on the way the multiple probes are evaluated to create the single measure of gene expression, the same microarray data can result in different sets of final data. Differing assumptions, data preprocessing methods, and high level analysis methods can lead to confusion in comparing results.

Microarray researchers often encounter up to 50% variation in comparisons of two different algorithms.[91] For that reason, it is recommended to use multiple methods for comparison and prioritization of expression patterns for identification of robust classifiers. While it is difficult to standardize data analysis methods that are continually improving, the need for consistent analysis is as important as standard experimental design and technical processes.

Developing standardized ways to compare gene expression information from one array to another is essential to this process. However, various statistical and bioinformatics methods are now available for comparing array data generated by probe-level analysis and different programs often provide different results.[92] While avant-garde statistical analysis of microarray data can lead to a number of incomplete or misleading conclusions, microarray users should experiment with various statistical methods to develop the most accurate predictions.

Choe et al. developed a data set to compare expression analysis algorithms such as the affy and gcrma packages from Bioconductor (http://www.bioconductor.org/), scripts adapted from the hdarray library, dChip, MAS 5.0, Perfect Match and SAM.[92] The samples analyzed comprised 1,309 individual cRNAs differing by known relative concentrations (as little as 1.2-fold) between spike-in and control samples in a defined background of 2,551 cRNAs present at identical concentrations.

After evaluating all the software packages, the group described three critical steps for maximizing data accuracy. The first is adjusting the perfect match probe signal with an estimate of nonspecific signal; Choe and colleagues suggest MAS 5.0 (Affymetrix Inc., Santa Clara, California, USA, http://www.affymetrix.com) performs best. The second step is making sure that log-fold changes are distributed around zero. The third step is choosing the most accurate test statistic; the group notes that the regularized t-statistic from CyberT (http://visitor.ics.uci.edu/genex/cybert/index.shtml) is most accurate.

Their work highlights the need to include experienced statisticians in developing experimental designs. Additional critical factors include using sufficient sample numbers to adequately power the study, and, when possible, prospective validation of any findings using new patient samples which serves as an acid test of predictive performance.

DATA REPORTING GUIDELINES

Reporting final microarray data also requires standardization, such that scientists can readily understand assumptions underlying analysis methods and experimental details to determine whether comparing results is logical or meaningful.[91] Implementation of a Minimum Information about a Microarray Experiment (MIAME, http://www.mged.org/Workgroups/MIAME/miame.html) standard[84] guides researchers to interpret data appropriately and provides information for replicating experiments. The key features of the MIAME standard are cited in the MIAME checklist (http://www.mged.org/Workgroups/MIAME/miame_checklist.html) that describes in detail the (1) experiment design, (2) samples used, extract preparation and labeling, (3) hybridization procedures and parameters, (4) measurement data and specifications, and (5) array design.

Standardization has allowed groups to create growing databases such as ArrayExpress at the European Bioinformatics Institute (http://www.ebi.ac.uk/arrayexpress/) and GEO at the National Center for Biotechnology Information (http://www.ncbi.nlm.nih.gov/geo/). Many major journals, including *Nature* and *Cell,* now require microarray data to be submitted in compliance with the MIAME standards. It would benefit the scientific and clinical communities if MIAME guidelines were required for manuscript acceptance for publication.

While compliance with data reporting guidelines has largely been successful, current guidelines do not address essential issues of variability in data generation and interpretation. By developing best practices for the complete microarray workflow, array analysis can be performed according to the guidances and regulations that will be necessary for clinical applications. Microarrays are currently used in dozens of clinical trials, but wide use of the technology in the routine setting of clinical care requires additional standardization.

MICROARRAY DATA INTEGRATION

Microarrays offer the promise of improved healthcare, but to get there, genomic and genetic data must be integrated with clinical information and made readily accessible to practicing physicians. Ideally, the molecular and clinical information will be collected and transmitted through standardized methods and in a standardized format such that the interpretation will be consistent from patient to patient, or from a single patient tested at multiple sites.

A number of initiatives are underway to partner biotechnology, software, database, computer hardware, and diagnostic companies to develop solutions that will integrate genomic research and patient clinical data from disparate databases into a central and organized format. Examples include:

- The cancer Biomedical Informatics Grid (caBIG) initiative sponsored by NCI and launched in 2003 with the purpose of creating a collaborative grid, standardized tools, information, applications, and technologies for cancer researchers world-wide (http://cabig.nci.nih.gov)

- The World Wide Biobank and standardization initiative sponsored by IBM (New York, New York, USA, http://www.ibm.com) to create a database of data for tissue samples, and assemble the informatics to help advance biomedical research and personalized medicine (http://www-03.ibm.com/industries/healthcare/)

The ideal infrastructure will allow clinicians to readily apply DNA and RNA molecular information to diagnosis and treatment of disease, ultimately improving patient care. The integration of clinical information with DNA and RNA data will produce reference databases designed to translate clinical genomics research into life saving, targeted therapies and treatments.

ROAD TO EVIDENCE-BASED MEDICINE

For over a decade, microarrays have revolutionized basic scientific research and redefined our view of the genome and its complexity. These tools not only help to make sense of seemingly endless genome sequences, but allow us to apply this molecular knowledge directly to the patient, helping to develop new drugs and improved therapies. Patients can be diagnosed or treated based on the genetic specifics of their particular diseases, drug tolerances, and immune responses, to yield a more personalized approach to healthcare. While numerous hurdles remain, including challenges surrounding probabilistic decision making, data integration and standardization, evidence-based medicine offers a unique opportunity to combat the current healthcare crisis. With an increasingly older population to care for and ever-increasing costs, microarrays offer the prospect of more efficient, cost-effective, and personalized approaches to human health issues.

REFERENCES

1. Fodor SP, Read JL, Pirrung MC, Stryer L, Lu AT, Solas D. Light-directed, spatially addressable parallel chemical synthesis. *Science* 1991; 251 (4995): 767–773.
2. Fodor SP, Rava RP, Huang XC, Pease AC, Holmes CP, Adams CL. Multiplexed biochemical assays with biological chips. *Nature* 1993; 364 (6437): 555–556.
3. Pease AC, Solas D, Sullivan EJ, Cronin MT, Holmes CP, Fodor SP. Light-generated oligonucleotide arrays for rapid DNA sequence analysis. *Proc Natl Acad Sci USA* 1994; 91 (11): 5022–5026.
4. Armstrong SA, Staunton JE, Silverman LB, Pieters R, den Boer ML, Minden MD, Sallan SE, Lander ES, Golub TR, Korsmeyer SJ. MLL translocations specify a distinct gene expression profile that distinguishes a unique leukemia. *Nat Genet* 2002; 30 (1): 41–47.
5. Ross ME, Zhou X, Song G, Shurtleff SA, Girtman K, Williams WK, Liu HC, Mahfouz R, Raimondi SC, Lenny N et al. Classification of pediatric acute lymphoblastic leukemia by gene expression profiling. *Blood* 2003; 102 (8): 2951–2959.
6. Kohlmann A, Schoch C, Schnittger S, Dugas M, Hiddemann W, Kern W, Haferlach T. Pediatric acute lymphoblastic leukemia (ALL) gene expression signatures classify an independent cohort of adult ALL patients. *Leukemia* 2004; 18 (1): 63–71.
7. Steinman L, Zamvil S. Transcriptional analysis of targets in multiple sclerosis. *Nat Rev Immunol* 2003; 3 (6): 483–492.
8. Hess S, Peters J, Bartling G, Rheinheimer C, Hegde P, Magid-Slav M, Tal-Singer R, Klos A. More than just innate immunity: comparative analysis of *Chlamydophila pneumoniae* and *Chlamydia trachomatis* effects on host-cell gene regulation. *Cell Microbiol* 2003; 5 (11): 785–795.
9. Nau GJ, Schlesinger A, Richmond JF, Young RA. Cumulative Toll-like receptor activation in human macrophages treated with whole bacteria. *J Immunol* 2003; 170 (10): 5203–5209.
10. Kennedy GC, Matsuzaki H, Dong S, Liu WM, Huang J, Liu G, Su X, Cao M, Chen W, Zhang J et al. Large-scale genotyping of complex DNA. *Nat Biotechnol* 2003; 21 (10): 1233–1237.

11. Matsuzaki H, Loi H, Dong S, Tsai YY, Fang J, Law J, Di X, Liu WM, Yang G, Liu G et al. Parallel genotyping of over 10,000 SNPs using a one-primer assay on a high density oligonucleotide array. *Genome Res* 2004; 14 (3): 414–425.

12. Puffenberger EG, Hu-Lince D, Parod JM, Craig DW, Dobrin SE, Conway AR, Donarum EA, Strauss KA, Dunckley T, Cardenas JF et al. Mapping of sudden infant death with dysgenesis of the testes syndrome (SIDDT) by a SNP genome scan and identification of TSPYL loss of function. *Proc Natl Acad Sci USA* 2004; 101 (32): 11689–11694.

13. Sellick GS, Garrett C, Houlston RS. A novel gene for neonatal diabetes maps to chromosome 10p12.1-p13. *Diabetes* 2003; 52 (10): 2636–2638.

14. Sellick GS, Barker KT, Stolte-Dijkstra I, Fleischmann C, R JC, Garrett C, Gloyn AL, Edghill EL, Hattersley AT, Wellauer PK et al. Mutations in PTF1A cause pancreatic and cerebellar agenesis. *Nat Genet* 2004; 36 (12): 1301–1305.

15. Middleton FA, Pato MT, Gentile KL, Morley CP, Zhao X, Eisener AF, Brown A, Petryshen TL, Kirby AN, Medeiros H et al. Genome-wide linkage analysis of bipolar disorder by use of a high density single-nucleotide-polymorphism (SNP) genotyping assay: a comparison with microsatellite marker assays and finding of significant linkage to chromosome 6q22. *Am J Hum Genet* 2004; 74 (5): 886–897.

16. Klein RJ, Zeiss C, Chew EY, Tsai JY, Sackler RS, Haynes C, Henning AK, Sangiovanni JP, Mane SM, Mayne ST et al. Complement factor H polymorphism in age-related macular degeneration. *Science* 2005; 308 (5720) :385–389.

17. Gissen P, Johnson CA, Morgan NV, Stapelbroek JM, Forshew T, Cooper WN, McKiernan PJ, Klomp LW, Morris AA, Wraith JE et al. Mutations in VPS33B, encoding a regulator of SNARE-dependent membrane fusion, cause arthrogryposis-renal dysfunction-cholestasis (ARC) syndrome. *Nat Genet* 2004; 36 (4): 400–404.

18. Janecke AR, Thompson DA, Utermann G, Becker C, Hubner CA, Schmid E, McHenry CL, Nair AR, Ruschendorf F, Heckenlively J et al. Mutations in RDH12 encoding a photoreceptor cell retinol dehydrogenase cause childhood-onset severe retinal dystrophy. *Nat Genet* 2004; 36 (8): 850–854.

19. Uhlenberg B, Schuelke M, Ruschendorf F, Ruf N, Kaindl AM, Henneke M, Thiele H, Stoltenburg-Didinger G, Aksu F, Topaloglu H et al. Mutations in the gene encoding gap junction protein alpha 12 (connexin 46.6) cause Pelizaeus–Merzbacher-like disease. *Am J Hum Genet* 2004; 75 (2): 251–260.

20. Yamada K, Lim J, Dale JM, Chen H, Shinn P, Palm CJ, Southwick AM, Wu HC, Kim C, Nguyen M et al. Empirical analysis of transcriptional activity in the Arabidopsis genome. *Science* 2003; 302 (5646): 842–846.

21. Mitra RM, Gleason CA, Edwards A, Hadfield J, Downie JA, Oldroyd GE, Long SR. A Ca2+/calm-odulin-dependent protein kinase required for symbiotic nodule development: gene identification by transcript-based cloning. *Proc Natl Acad Sci USA* 2004; 101 (13): 4701–4705.

22. Landis GN, Abdueva D, Skvortsov D, Yang J, Rabin BE, Carrick J, Tavare S, Tower J. Similar gene expression patterns characterize aging and oxidative stress in *Drosophila melanogaster*. *Proc Natl Acad Sci USA* 2004; 101 (20): 7663–7668.

23. Wolfgang MC, Jyot J, Goodman AL, Ramphal R, Lory S. *Pseudomonas aeruginosa* regulates flagellin expression as part of a global response to airway fluid from cystic fibrosis patients. *Proc Natl Acad Sci USA* 2004; 101 (17): 6664–6668.

24. Le Roch KG, Zhou Y, Blair PL, Grainger M, Moch JK, Haynes JD, De La Vega P, Holder AA, Batalov S, Carucci DJ et al. Discovery of gene function by expression profiling of the malaria parasite life cycle. *Science* 2003; 301 (5639): 1503–1508.

25. Warrington JA, Dee S, Trulson M. Large-scale genomic analysis using Affymetrix GeneChip® probe arrays, in Schena M, Ed. *Microarray Biochip Technology*. Natick, MA: BioTechniques Books; 2000. pp. 119–148.

26. Ebert BL, Golub TR. Genomic approaches to hematologic malignancies. *Blood* 2004; 104 (4): 923–932.

27. Meyerson M, Franklin WA, Kelley MJ. Molecular classification and molecular genetics of human lung cancers. *Semin Oncol* 2004; 31 (Suppl 1): 4–19.

28. MacDonald TJ, Brown KM, LaFleur B, Peterson K, Lawlor C, Chen Y, Packer RJ, Cogen P, Stephan DA. Expression profiling of medulloblastoma: PDGFRA and the RAS/MAPK pathway as therapeutic targets for metastatic disease. *Nat Genet* 2001; 29 (2): 143–152.

29. Lapointe J, Li C, Higgins JP, van de Rijn M, Bair E, Montgomery K, Ferrari M, Egevad L, Rayford W, Bergerheim U et al. Gene expression profiling identifies clinically relevant subtypes of prostate cancer. *Proc Natl Acad Sci USA* 2004; 101 (3): 811–816.

30. Huang E, Cheng SH, Dressman H, Pittman J, Tsou MH, Horng CF, Bild A, Iversen ES, Liao M, Chen CM et al. Gene expression predictors of breast cancer outcomes. *Lancet* 2003; 361 (9369): 1590–1596.

31. West M, Blanchette C, Dressman H, Huang E, Ishida S, Spang R, Zuzan H, Olson JA, Jr., Marks JR, Nevins JR. Predicting the clinical status of human breast cancer by using gene expression profiles. *Proc Natl Acad Sci USA* 2001; 98 (20): 11462–11467.

32. Beer DG, Kardia SL, Huang CC, Giordano TJ, Levin AM, Misek DE, Lin L, Chen G, Gharib TG, Thomas DG et al. Gene-expression profiles predict survival of patients with lung adenocarcinoma. *Nat Med* 2002; 8 (8): 816–824.

33. Notterman DA, Alon U, Sierk AJ, Levine AJ. Transcriptional gene expression profiles of colorectal adenoma, adenocarcinoma, and normal tissue examined by oligonucleotide arrays. *Cancer Res* 2001; 61 (7): 3124–3130.

34. Twine NC, Stover JA, Marshall B, Dukart G, Hidalgo M, Stadler W, Logan T, Dutcher J, Hudes G, Dorner AJ et al. Disease-associated expression profiles in peripheral blood mononuclear cells from patients with advanced renal cell carcinoma. *Cancer Res* 2003; 63 (18): 6069–6075.

35. Hibbs K, Skubitz KM, Pambuccian SE, Casey RC, Burleson KM, Oegema TR, Jr., Thiele JJ, Grindle SM, Bliss RL, Skubitz AP. Differential gene expression in ovarian carcinoma: identification of potential biomarkers. *Am J Pathol* 2004; 165 (2): 397–414.

36. Dave SS, Wright G, Tan B, Rosenwald A, Gascoyne RD, Chan WC, Fisher RI, Braziel RM, Rimsza LM, Grogan TM et al. Prediction of survival in follicular lymphoma based on molecular features of tumor-infiltrating immune cells. *New Engl J Med* 2004; 351 (21): 2159–2169.

37. Shipp MA, Ross KN, Tamayo P, Weng AP, Kutok JL, Aguiar RC, Gaasenbeek M, Angelo M, Reich M, Pinkus GS et al. Diffuse large B-cell lymphoma outcome prediction by gene-expression profiling and supervised machine learning. *Nat Med* 2002; 8 (1): 68–74.

38. Golub TR, Slonim DK, Tamayo P, Huard C, Gaasenbeek M, Mesirov JP, Coller H, Loh ML, Downing JR, Caligiuri MA et al. Molecular classification of cancer: class discovery and class prediction by gene expression monitoring. *Science* 1999; 286 (5439): 531–537.

39. Yeoh EJ, Ross ME, Shurtleff SA, Williams WK, Patel D, Mahfouz R, Behm FG, Raimondi SC, Relling MV, Patel A et al. Classification, subtype discovery, and prediction of outcome in pediatric acute lymphoblastic leukemia by gene expression profiling. *Cancer Cell* 2002; 1 (2): 133–143.

40. Kohlmann A, Schoch C, Schnittger S, Dugas M, Hiddemann W, Kern W, Haferlach T. Molecular characterization of acute leukemias by use of microarray technology. *Genes Chromosomes Cancer* 2003; 37 (4): 396–405.

41. Kohlmann A, Schoch C, Dugas M, Rauhut S, Weninger F, Schnittger S, Kern W, Haferlach T. Pattern robustness of diagnostic gene expression signatures in leukemia. *Genes Chromosomes Cancer* 2005; 42 (3): 299–307.

42. Haferlach T, Kohlmann A, Schnittger S, Dugas M, Hiddemann W, Kern W, Schoch C. AML M3 and AML M3 variant each have a distinct gene expression signature but also share patterns different from other genetically defined AML subtypes. *Genes Chromosomes Cancer* 2005; 43 (2): 113–127.

43. Li Y, Zhou X, St John MA, Wong DT. RNA profiling of cell-free saliva using microarray technology. *J Dent Res* 2004; 83 (3): 199–203.

44. Li Y, St John MA, Zhou X, Kim Y, Sinha U, Jordan RC, Eisele D, Abemayor E, Elashoff D, Park NH et al. Salivary transcriptome diagnostics for oral cancer detection. *Clin Cancer Res* 2004; 10 (24): 8442–8450.

45. Spira A, Beane J, Shah V, Liu G, Schembri F, Yang X, Palma J, Brody JS. Effects of cigarette smoke on the human airway epithelial cell transcriptome. *Proc Natl Acad Sci USA* 2004; 101 (27): 10143–10148.

46. Shah V, Sridhar S, Beane J, Brody JS, Spira A. SIEGE: Smoking Induced Epithelial Gene Expression Database. *Nucleic Acids Res* 2005; 33 Database Issue: D573–D579.

47. Minasian LM, Motzer RJ, Gluck L, Mazumdar M, Vlamis V, Krown SE. Interferon alfa-2a in advanced renal cell carcinoma: treatment results and survival in 159 patients with long-term follow-up. *J Clin Oncol* 1993; 11 (7): 1368–1375.

48. Huang J, Wei W, Zhang J, Liu G, Bignell GR, Stratton MR, Futreal PA, Wooster R, Jones KW, Shapero MH. Whole genome DNA copy number changes identified by high density oligonucleotide arrays. *Hum Genomics* 2004; 1 (4): 287–299.

49. Zhao X, Li C, Paez JG, Chin K, Janne PA, Chen TH, Girard L, Minna J, Christiani D, Leo C et al. An integrated view of copy number and allelic alterations in the cancer genome using single nucleotide polymorphism arrays. *Cancer Res* 2004; 64 (9): 3060–3071.

50. Schubert EL, Hsu L, Cousens LA, Glogovac J, Self S, Reid BJ, Rabinovitch PS, Porter PL. Single nucleotide polymorphism array analysis of flow-sorted epithelial cells from frozen versus fixed tissues for whole genome analysis of allelic loss in breast cancer. *Am J Pathol* 2002; 160 (1): 73–79.

51. Wang ZC, Lin M, Wei LJ, Li C, Miron A, Lodeiro G, Harris L, Ramaswamy S, Tanenbaum DM, Meyerson M et al. Loss of heterozygosity and its correlation with expression profiles in subclasses of invasive breast cancers. *Cancer Res* 2004; 64 (1): 64–71.

52. Paez JG, Lin M, Beroukhim R, Lee JC, Zhao X, Richter DJ, Gabriel S, Herman P, Sasaki H, Altshuler D et al. Genome coverage and sequence fidelity of phi29 polymerase-based multiple strand displacement whole genome amplification. *Nucleic Acids Res* 2004; 32 (9): e71.

53. Primdahl H, Wikman FP, von der Maase H, Zhou XG, Wolf H, Orntoft TF. Allelic imbalances in human bladder cancer: genome-wide detection with high density single-nucleotide polymorphism arrays. *J Natl Cancer Inst* 2002; 94 (3): 216–223.

54. Hoque MO, Lee CC, Cairns P, Schoenberg M, Sidransky D. Genome-wide genetic characterization of bladder cancer: a comparison of high density single-nucleotide polymorphism arrays and PCR-based microsatellite analysis. *Cancer Res* 2003; 63 (9): 2216–2222.

55. Lieberfarb ME, Lin M, Lechpammer M, Li C, Tanenbaum DM, Febbo PG, Wright RL, Shim J, Kantoff PW, Loda M et al. Genome-wide loss of heterozygosity analysis from laser capture micro-dissected prostate cancer using single nucleotide polymorphic allele (SNP) arrays and a novel bioinformatics platform dChipSNP. *Cancer Res* 2003; 63 (16): 4781–4785.

56. Dumur CI, Dechsukhum C, Ware JL, Cofield SS, Best AM, Wilkinson DS, Garrett CT, Ferreira-Gonzalez A. Genome-wide detection of LOH in prostate cancer using human SNP microarray technology. *Genomics* 2003; 81 (3): 260–269.

57. Wong KK, Tsang YT, Shen J, Cheng RS, Chang YM, Man TK, Lau CC. Allelic imbalance analysis by high-density single-nucleotide polymorphic allele (SNP) array with whole genome amplified DNA. *Nucleic Acids Res* 2004; 32 (9): e69.

58. Zhou X, Mok SC, Chen Z, Li Y, Wong DT. Concurrent analysis of loss of heterozygosity (LOH) and copy number abnormality (CNA) for oral premalignancy progression using the Affymetrix 10K SNP mapping array. *Hum Genet* 2004; 115 (4): 327–330.

59. Janne PA, Li C, Zhao X, Girard L, Chen TH, Minna J, Christiani DC, Johnson BE, Meyerson M. High-resolution single-nucleotide polymorphism array and clustering analysis of loss of heterozygosity in human lung cancer cell lines. *Oncogene* 2004; 23 (15): 2716–2726.

60. Lindblad-Toh K, Tanenbaum DM, Daly MJ, Winchester E, Lui WO, Villapakkam A, Stanton SE, Larsson C, Hudson TJ, Johnson BE et al. Loss-of-heterozygosity analysis of small-cell lung carcinomas using single-nucleotide polymorphism arrays. *Nat Biotechnol* 2000; 18 (9): 1001–1005.

61. Hoque MO, Lee J, Begum S, Yamashita K, Engles JM, Schoenberg M, Westra WH, Sidransky D. High-throughput molecular analysis of urine sediment for the detection of bladder cancer by high-density single-nucleotide polymorphism array. *Cancer Res* 2003; 63 (18): 5723–5726.

62. Rauch A, Ruschendorf F, Huang J, Trautmann U, Becker C, Thiel C, Jones KW, Reis A, Nurnberg P. Molecular karyotyping using an SNP array for genomewide genotyping. *J Med Genet* 2004; 41 (12): 916–922.

63. Ramaswamy S, Golub TR. DNA microarrays in clinical oncology. *J Clin Oncol* 2002; 20 (7): 1932–1941.

64. Wang Y, Klijn JG, Zhang Y, Sieuwerts AM, Look MP, Yang F, Talantov D, Timmermans M, Meijer-van Gelder ME, Yu J et al. Gene expression profiles to predict distant metastasis of lymph-node-negative primary breast cancer. *Lancet* 2005; 365 (9460): 671–679.

65. Frederiksen CM, Knudsen S, Laurberg S, Orntoft TF. Classification of Dukes' B and C colorectal cancers using expression arrays. *J Cancer Res Clin Oncol* 2003; 129 (5): 263–271.

66. Wang Y, Jatkoe T, Zhang Y, Mutch MG, Talantov D, Jiang J, McLeod HL, Atkins D. Gene expression profiles and molecular markers to predict recurrence of Dukes' B colon cancer. *J Clin Oncol* 2004; 22 (9): 1564–1571.

67. Valk PJ, Verhaak RG, Beijen MA, Erpelinck CA, Barjesteh van Waalwijk van Doorn-Khosrovani S, Boer JM, Beverloo HB, Moorhouse MJ, van der Spek PJ, Lowenberg B et al. Prognostically useful gene-expression profiles in acute myeloid leukemia. *New Engl J Med* 2004; 350 (16): 1617–1628.

68. Nevins JR, Huang ES, Dressman H, Pittman J, Huang AT, West M. Towards integrated clinico-genomic models for personalized medicine: combining gene expression signatures and clinical factors in breast cancer outcomes prediction. *Hum Mol Genet* 2003; 12, Spec No 2: R153–R157.

69. Glinsky GV, Higashiyama T, Glinskii AB. Classification of human breast cancer using gene expression profiling as a component of the survival predictor algorithm. *Clin Cancer Res* 2004; 10 (7): 2272–2283.

70. Hofmann WK, de Vos S, Elashoff D, Gschaidmeier H, Hoelzer D, Koeffler HP, Ottmann OG. Relation between resistance of Philadelphia chromosome-positive acute lymphoblastic leukaemia to the tyrosine kinase inhibitor STI571 and gene-expression profiles: a gene-expression study. *Lancet* 2002; 359 (9305): 481–486.

71. McLean LA, Gathmann I, Capdeville R, Polymeropoulos MH, Dressman M. Pharmacogenomic analysis of cytogenetic response in chronic myeloid leukemia patients treated with imatinib. *Clin Cancer Res* 2004; 10 (Pt 1): 155–165.

72. Mulligan G, Kim S, Stec J. Pharmacogenomic Analyses of Myeloma Samples from Bortezomib (VELCADE™) Phase II Clinical Trial. 2002. Presented at American Society of Hematology annual meeting, Philadelphia.

73. Chang JC, Wooten EC, Tsimelzon A, Hilsenbeck SG, Gutierrez MC, Elledge R, Mohsin S, Osborne CK, Chamness GC, Allred DC et al. Gene expression profiling for the prediction of therapeutic response to docetaxel in patients with breast cancer. *Lancet* 2003; 362 (9381): 362–369.

74. Lazarou J, Pomeranz BH, Corey PN. Incidence of adverse drug reactions in hospitalized patients: a meta-analysis of prospective studies. *JAMA* 1998; 279 (15): 1200–1205.

75. Ito RK, Demers LM. Pharmacogenomics and pharmacogenetics: future role of molecular diagnostics in the clinical diagnostic laboratory. *Clin Chem* 2004; 50 (9): 1526–1527.

76. Furman KD, Grimm DR, Mueller T, Holley-Shanks RR, Bertz RJ, Williams LA, Spear BB, Katz DA. Impact of CYP2D6 intermediate metabolizer alleles on single-dose desipramine pharmacokinetics. *Pharmacogenetics* 2004; 14 (5): 279–284.

77. Sachse C, Brockmoller J, Bauer S, Roots I. Cytochrome P450 2D6 variants in a Caucasian population: allele frequencies and phenotypic consequences. *Am J Hum Genet* 1997; 60 (2): 284–295.

78. Bradford LD. CYP2D6 allele frequency in European Caucasians, Asians, Africans and their descendants. *Pharmacogenomics* 2002; 3 (2): 229–243.

79. de Leon J, Susce MT, Pan RM, Fairchild M, Koch WH, Wedlund PJ. The CYP2D6 poor metabolizer phenotype may be associated with risperidone adverse drug reactions and discontinuation. *J Clin Psychiatry* 2005; 66 (1): 15–27.

80. Vassilev LT, Vu BT, Graves B, Carvajal D, Podlaski F, Filipovic Z, Kong N, Kammlott U, Lukacs C, Klein C et al. *In vivo* activation of the p53 pathway by small-molecule antagonists of MDM2. *Science* 2004; 303 (5659): 844–848.

81. Paez JG, Janne PA, Lee JC, Tracy S, Greulich H, Gabriel S, Herman P, Kaye FJ, Lindeman N, Boggon TJ et al. EGFR mutations in lung cancer: correlation with clinical response to gefitinib therapy. *Science* 2004; 304 (5676): 1497–1500.

82. Lynch TJ, Bell DW, Sordella R, Gurubhagavatula S, Okimoto RA, Brannigan BW, Harris PL, Haserlat SM, Supko JG, Haluska FG et al. Activating mutations in the epidermal growth factor receptor underlying responsiveness of non-small-cell lung cancer to gefitinib. *New Engl J Med* 2004; 350 (21): 2129–2139.

83. Spellman PT, Miller M, Stewart J, Troup C, Sarkans U, Chervitz S, Bernhart D, Sherlock G, Ball C, Lepage M et al. Design and implementation of microarray gene expression markup language (MAGE-ML). *Genome Biol* 2002; 3 (9): 46.

84. Brazma A, Hingamp P, Quackenbush J, Sherlock G, Spellman P, Stoeckert C, Aach J, Ansorge W, Ball CA, Causton HC et al. Minimum information about a microarray experiment (MIAME): toward standards for microarray data. *Nat Genet* 2001; 29 (4): 365–371.

85. Mutter GL, Zahrieh D, Liu C, Neuberg D, Finkelstein D, Baker HE, Warrington JA. Comparison of frozen and RNALater solid tissue storage methods for use in RNA expression microarrays. *BMC Genomics* 2004; 5 (1): 88.

86. Thach DC, Lin B, Walter E, Kruzelock R, Rowley RK, Tibbetts C, Stenger DA. Assessment of two methods for handling blood in collection tubes with RNA stabilizing agent for surveillance of gene expression profiles with high density microarrays. *J Immunol Methods* 2003; 283 (1–2): 269–279.

87. Wang J, Robinson JF, Khan HM, Carter DE, McKinney J, Miskie BA, Hegele RA. Optimizing RNA extraction yield from whole blood for microarray gene expression analysis. *Clin Biochem* 2004; 37 (9): 741–744.

88. Dobbin KK, Beer DG, Meyerson M, Yeatman TJ, Gerald WL, Jacobson JW, Conley B, Buetow KH, Heiskanen M, Simon RM et al. Interlaboratory comparability study of cancer gene expression analysis using oligonucleotide microarrays. *Clin Cancer Res* 2005; 11 (Pt 1): 565–572.

89. External RNA Control Consortium Workshop: Specifications for Universal External RNA Spike-In Controls. http://www.cstl.nist.gov/biotech/workshops/ERCC2003/index.html

90. The External RNA Controls Consortium. The External RNA Controls Consortium: a progress report. *Nature Methods* 2005; 2(10): 731–734.

91. Tumor Analysis Best Practices Working Group. Expression profiling: best practices for data generation and interpretation in clinical trials. *Nat Rev Genet* 2004; 5 (3): 229–237.

92. Choe SE, Boutros M, Michelson AM, Church GM, Halfon MS. Preferred analysis methods for Affymetrix GeneChips revealed by a wholly defined control dataset. *Genome Biol* 2005; 6 (2): R16.

4 Bead-Based Flow Systems: From Centralized Laboratories to Genetic Testing in the Field

Darrell P. Chandler

CONTENTS

INTRODUCTION

The convergence of electronics manufacturing techniques with analytical chemistry has produced revolutionary devices and approaches for DNA analysis in clinical samples[1–10] — a trend that will continue to accelerate both life sciences research and applications. From this technological explosion, the concept of a bead-based flow system for genetic testing can easily connote any number of technologies, ranging from disposable bead reagents, pre-packed solid-phase extraction cartridges and multi-pin sample preparation robots to commercial flow cytometers and innumerable microfluidic detection devices. The thesis of this chapter, however, relates to the concept and actual embodiment of a genetic testing system, wherein a system is defined as a single, integrated sample-to-answer device. We further focus our attention on systems that can translate from a high-throughput centralized testing laboratory to a decentralized clinic or bedside, and eventually into nonclinical operational environments. It is ultimately within the longer-term context of technology transfer and deployment that the bead-based and flow principles provide added value beyond the robotic workstations and engineered processes that constitute conventional high-throughput genetic discovery and analysis systems.

BEADS AS A UNIFYING ANALYTICAL PRINCIPLE

Beads (or more generically, microparticles) are at the base of solid phase chromatography for both chemical and biological separations. The concept of bead-based affinity chromatography reaches as far back as 1968,[11] with entire books now devoted to the subject. Indeed, most major life sciences companies (Qiagen Inc., Venlo, Netherlands; Bio-Rad Laboratories, Hercules, California, USA; Applied Biosystems, Foster City, California, USA; Invitrogen Corp., Carlsbad, California, USA; and others) offer numerous bead-based nucleic acid isolation kits, detection reagents, products and/or instruments for molecular biology purposes, many of which are packaged in 96-well plates for high-throughput clinical testing and robotic automation.[12–18]

Some of the additional systems, methods, and techniques for automating genomic testing and analysis in high-throughput contexts were reviewed by Gut.[19] Aside from pre-packed silica cartridges and pipette tips, however, bead-based sample preparation and detection systems face a number of practical problems that can limit their more widespread implementation in high-throughput testing environments. For example, most bead reagents settle, introducing a (high-throughput) mixing and pipetting problem for standard liquid dispensing robots. Many particles (especially silica) aggregate or clump either in solution or on pipette tips with changes in solution pH or buffer conditions. The clumping leads to highly variable pipetting accuracy (especially at low reagent volumes) and biological assays. Some particles have low nucleic acid capture or elution efficiency rates and this makes them problematic for nucleic acid isolation from low biomass or trace samples. Batch-to-batch variations in microparticle manufacture (and surface coating) further exacerbate the mundane yet practical issues of particle transfer, mixing, interaction, and analysis in microtiter plates that may (and perhaps do) limit their utility in high-throughput testing environments.

Of the commercially available, high-throughput sample preparation systems, the MagNA Pure LC product of Roche Applied Science* is one of the most widely distributed and successful platforms now on the market. In this system, paramagnetic particles constitute the base separations technology, which confers a number of automation advantages in a high-throughput testing environment. For example, paramagnetic beads remain in suspension for a relatively long time and are easily dispersed. From an instrumentation perspective, paramagnetic beads are easily concentrated and separated from a bulk solution through an external magnetic field that draws particles to the wall of the reaction vessel or microtiter plate — an engineering solution that avoids vacuum pumps, tubing, additional robotic arms, and centrifugation steps. In this context (high-throughput separations), the most widely used surface chemistries include silica,[20] oligo-dT for mRNA isolation, and various antibodies (e.g., magnetic bead products from Dynal Biotech, Oslo, Norway). At the same time, robotic workstations are the preferred embodiments for massively parallel genetic analyses in centralized testing laboratories, a technology base and implementation strategy that will probably not be supplanted by bead-based flow systems. What, then, are the near-, mid- and long-term values of bead-based flow systems for genetic analysis, especially in light of high-density microarrays and other detection devices that are poised to take advantage of massively parallel (robotic) sample processing capabilities?[21]

If we consider a bead either as a surface that can be custom-coated with reactive moieties or as a packaging strategy with which to embed reagents, we find that beads are conducive to all (possible) operations of a genetic testing system, including sample collection and/or concentration;[22,23] cell lysis;[24,25] nucleic acid purification;[20] nucleic acid amplification;[22,23,26] nucleic acid fragmentation and/or labeling;[27] and detection and reporting.[28–32] This is not to say that beads are required to perform every function within a genetic testing system — only that many of the selective chemistries and functional attributes required of an integrated genetic testing system are compatible with microparticle processing. As such, considering beads as reactive surfaces and reagent delivery

---

*(a division of F. Hoffman-LaRoche Ltd., Basel, Switzerland, http://www.roche-diagnostics.com/)

vehicles leads directly to methods and technologies that can seamlessly integrate both the biochemistry and hardware into fairly generic and flexible genetic testing platforms.

Commercially available particles come in various shapes and sizes, from soft spheres to porous glass or polymer matrices to jagged and irregularly shaped particles in paramagnetic and nonmagnetic varieties. The particle shape, core material, and porosity are typically modified to achieve desired performance criteria, including flow rate, surface area, and/or column packing properties. The original publication (and subsequent patent) by Boom et al.[20] involving silica particles and chaotropic reagents revolutionized nucleic acid isolation protocols and commercial kit development, and forms a basis of many high-throughput and microfluidic sample preparation devices.[1,33–36] The premise of the Boom protocol is nonspecific, electrostatic adsorption of nucleic acids (the charged phosphate backbone) to a silica (glass or diatom) surface under conditions of high ionic strength. Other particle surfaces, including hydroxyapatite, polystyrene, and even streptavidin-coated surfaces can function according to a similar (electrostatic or nonselective) adsorption principle.

Oligo-dT-coated particles, regardless of the underlying core material, are favored reagents for the selective isolation and enrichment of mRNA from fungal, animal, and plant matrices. In this case, sequence-specific oligonucleotides are typically attached to carboxylated beads via a primary amine coupling or to streptavidin-coated beads via a biotin bridge. Unconjugated beads with amino functionalities are also available, providing a range of cross-linking options for both nucleic acid and protein reagents. The principle of sequence-specific nucleic acid isolation has now migrated from sample preparation chemistry to multiplexed detection concepts and instrumentation (see "Coded Beads, in Flow" below). Lectin-coated surfaces are useful for selectively isolating bacteria prior to nucleic acid extraction, while the power and flexibility of antibody-based affinity separations are well known and widespread in all facets of human (and environmental) testing and diagnostic systems.

The introduction of mono-disperse paramagnetic beads (typically <5 μm in diameter) greatly simplified the process of biological affinity separation and purification, especially from small volume samples.[37–39] Of particular utility is the fact that virtually all affinity separation procedures previously developed for sepharose, glass, and polystyrene supports can be transferred to a paramagnetic bead core with fairly standard and straightforward attachment chemistries. The next trend in bead-based biological separations and detection is exploiting the unique properties of nanoparticles (smaller than 1 μm in cross-section) to improve reaction rates and kinetics, and/or provide alternative detection and reporting techniques for genetic analysis systems.[40–45] Within the context of genetic testing, however, beads are still used primarily as immobile support matrices for independent nucleic acid separation and detection in test tube (or microtiter plate) formats, not as part of a mobile (flowing) phase or as a unifying analytical principle at the core of a bead-based flow system. Recent developments in macro- and microfluidics, however, provide several development trajectories for integrated genetic testing systems that can move genetic testing from a high-throughput centralized laboratory into doctors' offices and decentralized operational environments.

BEAD-BASED FLOW ASSAYS

EVOLUTION

The premise of a bead-based genetic testing system is a natural flow from low pressure liquid chromatography and flow injection analysis (FIA), techniques that for the most part were developed and are still used primarily for chemical separations and sensing. While liquid chromatography provided the pumps, valves, fluidics, and computer control systems to move fluids, FIA and its derivatives offered a fundamentally different way of thinking about separation and analysis.[46,47]

FIA considers an injected sample or reagent as a discrete reaction zone that can be transported, manipulated, and/or analyzed in continuous fashion. Sequential injection analysis (SIA) built upon this principle by incorporating bidirectional flow and multipositional valves to sequentially aspirate or stack reagents into discrete diffusion and reaction zones (as opposed to continuous flow in FIA or conventional chromatography). Bead injection analysis (BIA) was born[48,49] via the introduction of a flow cell that allows particles to be automatically trapped, perfused, observed, and released (Figure 4.1), a practice more generically described as renewable surface sensing and separation. The beads serve as renewable surfaces because they are discarded and replaced after each analysis rather than regenerated by chemical stripping procedures.

Several subtle but important operational principles within the concept of a renewable surface lend themselves to biological applications and integrated genetic testing systems in particular. The first relates to the sample-to-detector volume gap or, more formally, the problem of sample concentration. While sufficient numbers of white blood cells (hence, genomic DNA) exist within a single microliter of blood for conducting numerous genetic tests (especially with polymerase chain reaction [PCR] or other amplification techniques), detecting, identifying, and screening for infectious agents may require sample dilution or liquification (tissues, sputum, feces) and/or relatively

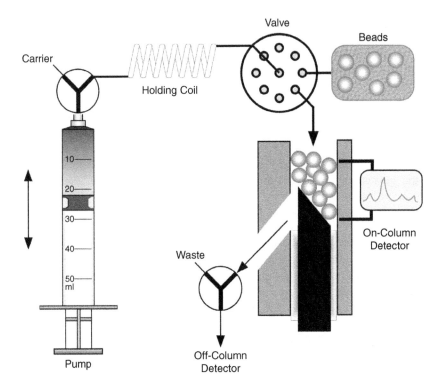

FIGURE 4.1 Basic components of a bead-based flow system. A bidirectional syringe pump and carrier fluid are used to control fluid flow through the system using computer-controlled microvalves and one or more selection valves. The holding coil is an area where reagents can be sequentially aspirated from the selection valve and stacked prior to delivery over the flow cell or detector. A rotating rod renewable surface flow cell for bead trapping and manipulation is illustrated here; other flow cell designs are illustrated in Figure 4.2. Basic flow cell designs include a single inlet and one or more outlets used to control fluid and bead flow. In the device shown, beads will stack against the tip of the rotating rod while fluid escapes through the leaky tolerance between the rod and the flow cell casing. Once the analysis is complete, beads are released by rotating the rod 180 degrees, thereby opening the outlet port and enabling beads to escape (see Figure 4.2C). After release, the flow cell can be automatically cleaned, the rotating rod returned to the "trap" position, and beads loaded for the next analysis. In all cases (e.g., Figure 4.2), target analytes can be detected on-bead within the flow cell itself, eluted to a downstream detector (microarray, PCR device, mass spectrometer), or carried with the beads to a downstream detector (flow cytometer, bead arrays).

large sample volumes (~10 ml for clinical samples, tens or hundreds of liters for pathogen detection in environmental samples) to realize a statistically meaningful assay. Therefore, regardless of whether a perfect detector is available to detect a single DNA molecule, the integrated system principle forces one to simultaneously consider sample processing and detection method efficiencies and the actual concentration of target in the original sample (certainly <100%, perhaps as low as 1 to 10% efficiency). The beauty of a renewable bead surface for sample concentration is that the beads can serve as renewable filters, concentrating whole cells, nucleic acids, proteins or other analytes from a continuous fluid stream.[50–52] Once the targeted analyte is captured and concentrated on or within the beads, the beads can then be utilized as affinity purification matrices to selectively remove undesired sample components that may interfere with subsequent genetic testing and/or detection.[53–58]

As with the sample concentration function, the volumes and types of buffers, reagents, and reactants required to achieve sample purification can vary — restricted only by the dimensions of the flow channels or tubing and the precision of the fluid drive (typical sample purification volumes range from 10 µl to 10 ml per step). Finally, the beads can also be used as sensing surfaces, either immobilized in place or passed on to a downstream detector[29,30,48,59–64] (see "Bead-Based Detectors" below). The simplicity, adaptability, volume range, and availability of flow systems, bead reagents, and detector modalities therefore make BIA and renewable surface concepts very interesting as generic genetic testing platforms for subsequent development into diagnostic devices.

MOVEABLE BARRIERS

Methods for trapping particles within a flow path are varied and depend upon the size and type of particle (Figure 4.2). The original jet ring cell, for example, uses a moveable capillary and solenoid valve to create a small gap between a faceplate (or observation window) and the flow path (Figure 4.2A). Beads stack against the faceplate and perfused fluid moves radially outward from the capillary through the gap.[48] The moveable capillary can be replaced with a moveable rod or other barrier placed perpendicular to or parallel (axial) with the fluid flow (Figure 4.2B). The beads form a packed microcolumn against the end of the barrier and fluid flows around the leaky tolerance machined between the barrier and the flow cell casing. Beads are easily flushed from a moveable barrier system by reversing fluid flow. Both the jet ring cell and moveable barrier lend themselves to relatively large (>40 micron) spherical beads in both flexible (sepharose or agarose) and inflexible (polymer, glass) formulations. Small and nonspherical particles, however, tend to escape from the gap (jet ring cell) or clog the leaky tolerance in the moveable barrier, leading to unacceptable back-pressure and potential cross-contamination between samples.

Moving a bead barrier along an axis is mechanically cumbersome compared to other bead retention methods. One way to simplify bead retention while making use of the leaky tolerance fluidic principle is to machine a chamfered or beveled face on the tip of a rotating barrier, thereby creating a micromechanical stopcock or valve (Figure 4.2C). The rotating rod has also proven effective at trapping nonspherical particles such as the silica resins originally described by Boom et al.[20,65] Even particles smaller than the leaky tolerance (ca. 5 µm in diameter or cross-section) can be effectively manipulated with a rotating rod by first forming a layer of larger particles (ca. 20 µm) to cover the machined gap,[66] a renewable filter principle that should be equally efficacious in the jet ring and moveable piston flow cells. Placing an immovable barrier (with a leaky tolerance) within the flow path, either perpendicular to or parallel with the inlet and outlet ports (Figure 4.2D), eliminates mechanical movement from the flow cell operation. In this design, fluid flow and particle release through and from the flow cell are controlled via an external multiport valve that directs waste to a proper receptacle (the selection valve illustrated in Figure 4.1 or additional three-way valves upstream and downstream from the flow cell).

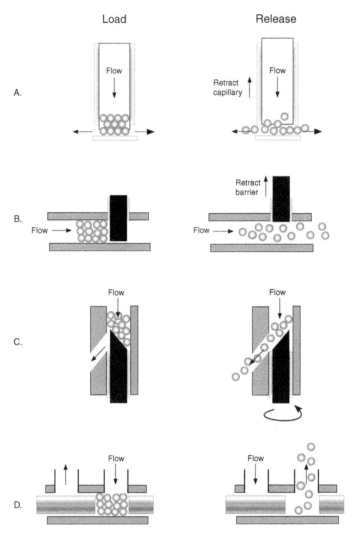

FIGURE 4.2 Renewable microcolumn designs and strategies for particle trapping and manipulation in bead-based flow systems. Only particle trapping and release operations are shown to illustrate the general principles behind each flow cell design. (A) Original jet ring cell; particles are trapped and released via a solenoid-activated capillary column. (B) Solenoid-activated solid barrier (cylindrical rod shown), oriented perpendicular with (as shown) or parallel to the flow path within the flow cell. The fluid flows around the barrier through the leaky tolerance machined between the barrier and flow cell casing. (C) Axial movement of the barrier is replaced with a rotational motion in the rotating rod renewable flow cell while retaining the principle of fluid flow around a leaky tolerance. (D) The leaky tolerance fluidic principle also applies to immoveable barriers oriented perpendicular with or parallel to (as shown) the flow path within the flow cell. (E) Basic frit restriction in a one-inlet, two-outlet design; fluid flow is controlled via external microvalves. (F) Frit restriction within a valve; rotating the valve controls particle trapping and release. (G) Coaxial membrane flow cell. Particles are held within the membrane portion of the flow cell by first positioning an air bubble or air dam at one end and then loading the particles. Closing the inlet and outlet ports and opening the valves that control fluid flow over and around the membrane allow particles to be perfused with reactants (not shown). (H) Solid weirs are common retention devices in microfluidics. (I) Simplified magnetic flow cell. The conductive plate within the flow cell can be replaced by steel wool, mesh, or metal foam structures to hold paramagnetic particles within the entire volume of the flow channel (as in J). The same basic on–off principle applies for other energetic particle trapping methods, including optical tweezers and acoustic standing waves.

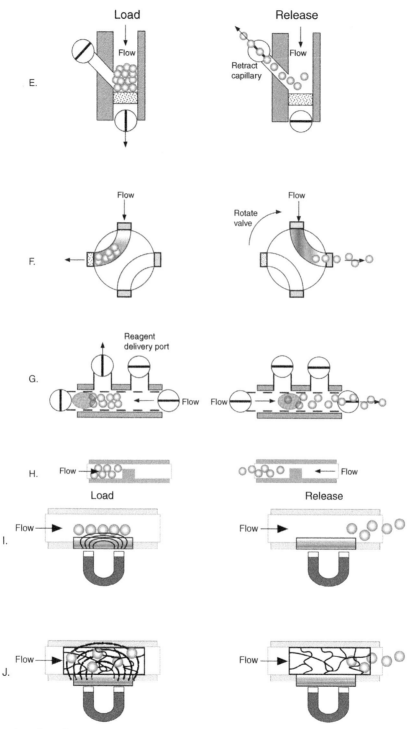

FIGURE 4.2 (continued)

FRITS, MEMBRANES, AND WEIRS

While moveable barriers are efficacious for bead trapping and biological sample manipulation in research settings, they are not yet widespread in commercial instruments or genetic testing devices. Instead, porous frits, membranes, or microfabricated (etched) weirs are preferred bead restriction devices, in part because they are readily available with sub-micron pores or features, they do not contain mechanical components to effect bead manipulation, and they are consistent with fabricating a single-use, disposable product.

Frit restriction can occur either as part of a specialized flow cell in which fluid flow is controlled by external valves (Figure 4.2E) or is embedded as part of a valve mechanism (Figure 4.2F). In a modification of conventional frit design, a coaxial membrane device (Figure 4.2G), external valves, and air bubbles were used to trap and manipulate 5- μm polymer beads in an autonomous pathogen detection system.[67] In this case, the air bubbles serve the function of a frit or barrier by positioning beads within the porous membranes of the flow cell. Finally, microfabricated or etched weirs (Figure 4.2H) are frequently used in microfluidic devices to trap particles[36,68] — a design principle that is also amenable to the manipulation and trapping of nanometer-sized particles and the construction of ordered bead arrays.[69]

The simplicity of frits, membranes and weirs for particle trapping and manipulation is obviously appealing for single-use disposable devices and chip-based fluidics. However, small pores and fluid restrictions are also easily clogged or fouled with soft (pliable) beads, jagged particles, or sample components (tissue homogenates, fats, colloids). Therefore, such bead restriction technologies may be ineffective for some clinical or environmental front-end sample preparation functions, and are presently better suited for the analysis of genetic material from relatively pristine samples (blood, cell cultures) after primary sample treatment (filtration). That said, Weimer et al. scaled these bead trapping principles up to fluidized bed reactors containing 250 g of 3-mm beads for large volume sample concentration at flow velocities of 0.4 l/min.[70] Their work illustrates how simple bead trapping methods can also be applied to field-scale environments and biodetection devices.

MAGNETISM AND ENERGETIC FORCES

Functionalized magnetic particles are used extensively for chemical,[71] nucleic acid,[72] protein,[73] organelle,[74] and whole-cell[75] isolation and analysis. Conventional (high-throughput) biomagnetic separation techniques rely on wall concentration with external magnets to pull paramagnetic particles to the side of a tube so that buffers can be aspirated and exchanged via a manual or robotic pipette.

This principle has also been adopted for electrochemiluminescent (ECL) detection in a renewable magnetic flow cell (Figure 4.2I). However, wall collection of magnetic particles does not allow for efficient mixing or perfusion of target solution with the captured beads, and the external magnetic field strength may not capture beads from dilute, flowing suspensions. This limitation has been addressed by incorporating magnetic grade stainless steel wool, metal-coated fiber or wire, or metal spheres inside a flow cell (Figure 4.2J) to separate colloidal-sized magnetic particles from the flow path.[76] However, steel wool and entangled ferromagnetic fibers bend in response to fluid flow, leading to nonspecific entrapment of the target or sample debris. Likewise, a packed bed of magnetic beads suffers from low porosity and substantial pressure drops over the length of the column,[52,77] similar to some of the challenges associated with trapping beads behind frits, membranes or microfabricated weirs.

One way to avoid the negative effects of flexible magnetic flux conductors within a renewable flow cell is to use highly porous metal foam within the flow path.[50] The foam has an internal structure conceptually similar to a sponge and can be cut to a specific diameter and/or length during the manufacture of the flow cell before inserting the foam within the flow channel. As described by Chandler et al.,[50] the section of tube containing a Ni foam insert was 19.1 mm long × 2.1 mm, with an average pore size of 390 ± 190 μm and a density of 6 to 15%. This strategy provides a very high surface area and magnetic field gradient for bead capture, but the rigid structure does not collapse and encourage physical entrapment of sample or beads. The porosity and large average pore size of metal foam also provide unconstrained fluid flow, low pressure drops, turbulent flow for efficient mixing, efficient passage of nontarget particulates, and new opportunities for repetitive particle trapping and release to maximize analyte–bead interactions.

Such magnetic flux-conducting flow cells can also trap, perfuse, and release nanometer-sized particles (e.g., Miltenyi 50-nm particles [Miltenyi Biotec, Bergisch Gladbach, Germany]) that cannot be captured or concentrated with external block magnets. Unfortunately, magnetic flux conductors of the type described here are not suitable for high-throughput microtiter plates because they would interfere with conventional liquid handling robots and the routine requirements of solution transfer and mixing.

Magnetic separation and flow cells represent only one form of energetic particle trap (as opposed to mechanical particle traps). The notion of turning power on or off to trap and release particles is also embodied in the concepts of optical tweezers,[78,79] ultrasonic standing waves,[80–82] ultrasonic tweezers,[83] dielectrophoresis,[84–86] field-flow fractionation,[87] and combinations thereof.[88] Of particular note is the ability of energetic traps to manipulate nanometer-sized particles[79,82,89] and even control their positions and movements within a three-dimensional lattice or suspension.[90–92] The advent of holographic optical traps now means that large numbers of particles can be individually controlled, patterned, or assembled in three dimensions — technology that is poised to transform fundamental and applied research.[79] From a systems perspective, however, three-dimensional particle manipulation and interactions translate directly into improved kinetics and efficiencies for separation and/or detection.

Acoustic standing waves, in particular, have found use in analytical biotechnology and medicine for cell (or bead) concentration, isolation, and separation; cell lysis; nucleic acid degradation; and even imaging (reviewed by Coakley[80]). Acoustic standing waves are also easily extended to industrial-scale separations and flow rates.[93] In large part, however, nonmagnetic particle trapping techniques are still relatively nascent technologies within the context of genetic testing systems, presently reserved to experimental microdevices and systems at the forefront of microfluidic science.[84,89,94,95]

BEAD-BASED DETECTORS

Although beads are the bases of modern chromatography and biological sample preparation, bead-based genetic detectors are often overlooked or obscured by the tremendous success of and continued interest in PCR, high-density planar microarrays or microtiter plates, and microfabricated electrophoretic chips. Indeed, any or all nucleic acid detectors can (in principle) serve as downstream detection devices for integrated, bead-based sample preparation systems as described above. Building on the previous section describing bead-based flow systems, and assuming the importance of an integrated genetic testing system, this next section will focus on detectors that actually utilize beads as sensor surfaces, thereby completing the conceptual construct of a bead-based flow system.

BEADS IMMOBILIZED IN PLACE

As evident from the section on bead-based flow assays and Figures 4.1 and 4.2, once beads are immobilized in a flow cell any number of reporter or detection modalities is possible in flow or batch fluidic processing. Within the context of biological testing and renewable surface flow cells alone (Figure 4.3A), for example, we find flow cells for absorbance,[96,97] fluorescence,[98,99] electrochemistry,[40,100–103] and other reporter modes.[104,105]

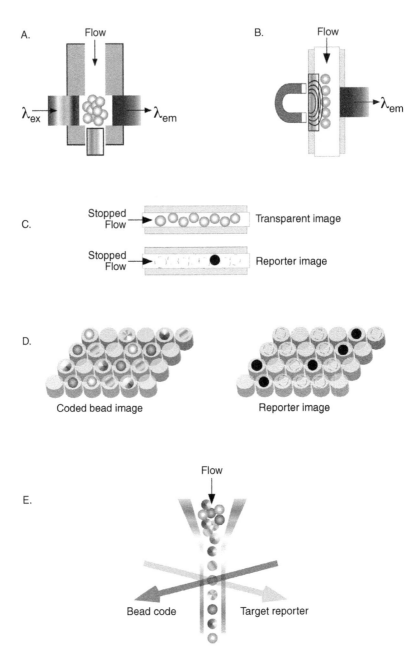

FIGURE 4.3 Bead-based detectors. (A) A generic flow cell (as in Figure 4.2) can be used in conjunction with any number of detection modalities; a typical fluorescence detection scheme is illustrated here. (B) Magnetic electrochemiluminescent flow cell developed by BioVeris holds particles in place magnetically over a conducting plate. An electrochemical reaction ensues on (or very near) the bead surface that emits light for detection when voltage is applied to the plate. (C) An ordered bead array, where homogenous beads are mechanically loaded into capillary tubes and imaged before and after reaction with target solution. Coded beads increase the multiplexing capacity of capillary bead arrays and eliminate the need for mechanical loading in predetermined order. Thus, very high density, self-assembled bead arrays are possible by capturing coded beads in the tips of fiber optic bundles or etched arrays. (D) Inner workings of a typical flow cytometer (E) are likewise amenable to high-throughput analysis of coded beads.

The electrochemiluminescent principle and flow cell (Figure 4.3B), for example, provide the core of the BioVeris (formerly IGEN, Gaithersburg, Maryland, USA) genetic and medical testing systems, with over 36 Food and Drug Administration (FDA)-approved assays and benchtop systems for high-throughput analysis in 384-well plates. The primary difference between ECL and conventional (optical) reporter technologies is that in ECL an applied voltage catalyzes a chemical, light-emitting reaction on (or within approximately 30 angstroms of) the magnetic[61,106] or nonmagnetic[107] bead surface. The specificity of electronic excitation means that background noise is very low, such that ECL has the distinct advantage of "reading through" complex or dirty backgrounds.[60,108] The ability to operate within otherwise filthy backgrounds is a practical benefit that can be used to greatly simplify sample processing functions or requirements associated with integrated systems.

CODED BEADS, IMMOBILIZED IN PLACE

While the simplicity of bulk measurements in a flow cell or test tube lends itself to virtually any analytic problem, one limitation of such devices is their inability to detect and characterize multiple analytes simultaneously in a single sample. Therefore, starting samples must be split into many replicate reaction chambers (e.g., MAG-microarray technology).[109] The split process may produce potential false negatives (splitting a sample to extinction), increased reagent and consumable costs, and/or lower throughput.

One way to circumvent the problem of a split assay is to construct ordered bead arrays in capillaries; target-specific beads are mechanically preloaded in defined order and analyzed in one or more reporter wavelengths or channels (Figure 4.3C).[110,111] Of course, simply fabricating ordered bead arrays in this manner is a challenge[69] and may present some long-term limits on multiplexing capability (e.g., number of ordered beads per capillary) or manufacturing quality control.

A more clever approach to highly multiplexed analysis is to exploit optically or otherwise encoded beads. The "color" (or code) of a single bead is analogous to a spatial address on a planar microarray or physical well within a microtiter plate. While detecting coded particles originated within the realm of flow cytometry (see next section), color-coded particles have been exploited to create randomly ordered or self-assembled high density arrays on the tips of fiber optic bundles[32,112–114] and other chip-based detectors[59,63] (Figure 4.3D). Brenner et al. even stacked coded beads within capillary flow cells (as in Figure 4.3C) for massively parallel signature sequencing and genetic analyses,[115] illustrating how coded beads greatly enhance multiplexing capabilities over ordered bead arrays.[69] In these cases, the bead-based array differs significantly from other microarray platforms, in that each probe in the array is spectrally registered after random distribution into optical detection wells (or fibers), rather than deliberately registered during array fabrication. Walt[32] goes on to describe practical consequences of coded bead arrays relative to competing microarray platforms. From a high-throughput genetic testing perspective, the small spatial scale of coded beads (~ 3 μm) makes it relatively simple to develop and use high density "arrays of arrays" by creating 96 fiberoptic bundles of 50,000 or more fibers each spaced according to the dimensions of a standard 96-well plate.[114]

CODED BEADS, IN FLOW

Cytometry literally means cell (or particle) analysis, and it is in this area that coded beads originated within the realm of detection devices. The basic operational principle in flow cytometry is to deliver particles in single file past a measurement point (Figure 4.3E) — typically a fluorescent or optical detector (although other detection modes are possible).[116]

While flow cytometry is a well established technology (and will not be reviewed here), color-coded beads for multiplexed analysis represent a relatively recent advance in the field.[29] The Luminex detection systems and xMAP technology are the most successful coded bead arrays presently on the market (Luminex Corporation, Austin, Texas, USA), now distributed by most major life sciences suppliers. Early work with Luminex coded bead arrays focused on multiplexing established immunoassays.[29,117–120] The basic technology was quickly adapted to nucleic acid analysis, including an oligonucleotide ligation assay for detecting SNPs; SNP genotyping via solid-phase amplification and minisequencing; and identification of *16S* ribosomal RNA genes.[30,121–123] Luminex xMAP applications now include bead sets for allergens, cancer and cardiac markers, tissue typing, transcription factors, immune and infectious disease profiling, and genetic testing with allele-specific oligonucleotide probes. More recent adaptations of the Luminex detection system include the development of fully automated nucleic acid analysis systems[124] specifically tailored for the analysis of environmental samples.[125]

The power of coded bead arrays relative to other array-based platforms is embodied in several subtle but important respects. First, the sheer scale of bead preparation (10^{10} particles per milliliter for 3- µm particles in a solution containing 20% solids) greatly facilitates uniform and reproducible array fabrication. Second, microparticles react with target analytes with (near) solution-phase kinetics, improving reaction times and (potentially) detection limits. Third, the flow cytometry flow cell is a renewable surface sensor and therefore conceptually and practically consistent with bead-based sample preparation systems as described earlier in this chapter. Fourth (and most importantly), each probe spot on a planar microarray represents a single datum that can be extensively replicated only with considerable time and expense. In contradiction to planar arrays, each particle in a coded bead array likewise represents a datum, but both flow cytometers and self-assembled bead arrays can rapidly and easily replicate 10 to 10^3 such observations for each (corresponding or comparative) planar microarray spot and experimental treatment. Thus, coded bead arrays provide more enhanced replication and sensitivity for statistical modeling and quantitation than other array substrates, a key advantage for developing diagnostic devices for use beyond a centralized testing laboratory (see "From Clinic to Field" later in this chapter).

A principal disadvantage of commercial flow cytometers as detectors for integrated genetic testing systems is their bulky size, despite the fact that the detection flow cell component is very small. The Luminex system represents a modest decrease in overall instrument footprint and has been incorporated into a stand-alone, antibody-based aerosol monitoring system roughly the size of a lecture podium.[126] However, recent developments at Micronics (Redmond, Washington, USA) have converted conventional cytometer fluidics into credit card-sized cartridges (Figure 4.4) for whole blood cytometry and bead-based detection.[28] The extent to which such "personalized" cytometers are or will be reusable (rather than disposable), amenable to massively parallel analysis, or capable of interfacing with renewable surface sample preparation devices in a field setting is still to be determined, even though the conceptual leap to a field-deployable (and potentially autonomous) integrated genetic detection system is obvious.

BEADS AS REPORTERS

Coded beads provide essential information in a multiplexed analysis and are therefore more than simply reactive surfaces. In other words, particles can function as reporter molecules in genetic analyses. Magnetic beads, for example, have been used as reporters in magnetoresistant biosensors, wherein the measured force generated between a reacted particle and a sensor surface serves as a (semi-) quantitative binding signal.[127] Nanoparticles and quantum dots constitute reporters for nucleic acid and protein analyses[45,128,129] in lieu of conventional fluorophores, even in multiplexed formats.[130]

Nanoparticles can also be used in aggregation or agglutination type tests,[43] as reporter molecules for spectroscopic or other detection,[131] and as concentrating probes for matrix-assisted laser desorption/ionization mass spectroscopy (MALDI-MS) analysis.[132,133] As with ECL technology, a key advantage of nanoparticles as reporters is the low intrinsic noise in complex sample backgrounds.

A.

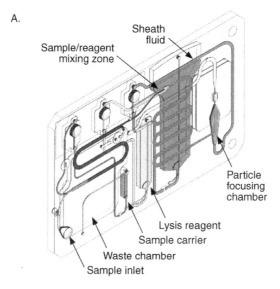

B.

FIGURE 4.4 Microflow cytometer card developed at Micronics (Redmond, Washington). (A) Three-dimensional rendering of a blood cytometry card with on-board fluidics and flow controllers. (B) Disposable cytometer card. (Courtesy of Karen Hedine and C. Fred Battrell, Micronics.)

It is conceivable that integrated bead-based flow systems will someday utilize nanoparticles as a unifying analytical principle within microscale fluidic devices.[2,4,33,134]

FROM CLINIC TO FIELD

As noted above, a number of bead-based genetic testing systems are commercially available. Despite the conceptual linkages among bead- and flow-based sample preparation and detection devices, however, the number of stand-alone, sample-to-answer, bead-based flow systems for use outside a centralized testing laboratory is surprisingly low. Hence, even modest advances in macroengineering and antibody-based testing systems are of tremendous interest.[126,135] The extent to which such integrated, bead-based flow systems are further adapted or developed for genetic testing is indeterminate, but at present such devices are clearly motivated by homeland security, national security, and public health concerns about the dissemination and movement of pathogens in the environment. Regardless of the operational environment, however, a number of practical issues associated with developing an integrated genetic testing system have yet to be addressed.

One way to describe the scientific and engineering issues facing continued development of integrated bead-based flow systems is to consider what it means to perform a descriptive versus diagnostic genetic test. A diagnostic test implies levels of statistical rigor, internal and external standards, quality assurance and quality control, and defined and validated sampling protocols that are not evident or typically considered during fundamental or applied research and engineering. For current bead-based genetic testing devices, these issues are relatively simple to address because genetic tests and associated controls are easily accommodated in replicate or parallel reaction wells (microtiter plates). However, how does one embed internal (and multiplexed) standards, controls, and true biological replicates within a continuous flow path? Certainly, parallel flow channels are relatively simple to etch into microfabricated chips (e.g., the split assay engineering approach), and multisyringe pumps and sample processing schemes are available for conventional liquid chromatography procedures. However, does such parallel processing make biological or engineering sense within a fully integrated, bead-based flow system? Will it be possible to create a diagnostic (as opposed to descriptive) bead-based flow system for use beyond the centralized testing laboratory?

CONCLUSION

From the foregoing discussion, it is relatively simple to see how and why bead-based flow systems are or can be used for genetic testing. The relative obscurity of advanced bead-based genetic testing systems compared to PCR, microarray, and microfluidic devices in the bioengineering research arena is partly an artifact of the "razor blade" consumable mentality that pervades clinical diagnostic markets. Thus, while it is fairly obvious how bead-based flow systems can be engineered and applied for future genetic testing devices, an unresolved question is where and how such systems provide a competitive or compelling market advantage for the technology developer or industry.

The suggestion from the section titled "From Clinic to Field" is that continued development of bead-based flow systems for genetic testing will ultimately find (commercial) value within the larger context of public health and environmental monitoring applications, rather than as instantaneous biosensors or high-throughput testing platforms for centralized laboratories. From this perspective, bead-based flow systems are relatively young and nascent, but the outlook for continued development of integrated, bead-based genetic testing systems is sanguine, in sharp contrast with a recent conclusion regarding Micro-Electro-Mechanical Systems (MEMS)-based subsystems and devices for molecular diagnostics.[136]

ACKNOWLEDGMENTS

The author's work on bead-based flow systems has received historical funding from several programs of the U.S. Department of Energy (DOE). The author gratefully acknowledges the continued support of DOE under the Assessment and Integrated Studies Elements of the Natural and Attenuated Bioremediation Research Program. Argonne National Laboratory is operated for DOE by the University of Chicago under Contract W-31-109-ENG-38.

REFERENCES

1. Breadmore MC, Wolfe KA, Arcibal IG, Leung WK, Dickson D, Giordano BC, Power ME, Ferrance JP, Feldman SH, Norris PM et al. Microchip-based purification of DNA from biological samples. *Anal. Chem.* 2003; 75 (8): 1880–1886.

2. Burns MA, Johnson BN, Brahmasandra SN, Handique K, Webster JR, Krishnan M, Sammarco TS, Man PM, Jones D, Heldsinger D et al. An integrated nanoliter DNA analysis device. *Science* 1998; 282 (5388): 484–487.

3. Giordano BC, Ferrance J, Swedberg S, Hühmer AFR, Landers JP. Polymerase chain reaction in polymeric microchips: DNA amplification in less than 240 seconds. *Anal. Biochem.* 2001; 291 (1): 124–132.

4. Lagally ET, Scherer JR, Blazej RG, Toriello NM, Diep BA, Ramchandani M, Sensabaugh GF, Riley LW, Mathies RA. Integrated portable genetic analysis microsystem for pathogen/infectious disease detection. *Anal. Chem.* 2004; 76 (11): 3162–3170.

5. Landers JP. Molecular diagnostics on electrophoretic microchips. *Anal. Chem.* 2003; 75 (12): 2919–2927.

6. Liu RH, Yang J, Lenigk R, Bonanno J, Grodzinski P. Self-contained, fully integrated biochip for sample preparation, polymerase chain reaction amplification, and DNA microarray detection. *Anal. Chem.* 2004; 76 (7): 1824–1831.

7. Obeid PJ, Christopoulos TK. Continuous-flow DNA and RNA amplification chip combined with laser-induced fluorescence detection. *Anal. Chim. Acta* 2003; 494: 1–9.

8. Schmalzing D, Belenky A, Novotny MA, Koutny L, Salas-Solano O, El-Difrawy S, Adourian A, Matsudaira P, Ehrlich D. Microchip electrophoresis: a method for high-speed SNP detection. *Nucl. Acids Res.* 2000; 28 (9): e43.

9. Tantra R, Manz A. Integrated potentiometric detector for use in chip-based flow cells. *Anal. Chem.* 2000; 72 (13): 2875–2878.

10. Vo-Dinh T, Cullum B. Biosensors and biochips: advances in biological and medical diagnostics. *Fresenius J. Anal. Chem.* 2000; 366 (6–7): 540–551.

11. Cuatrecasas P, Wilchek M, Anfinsen CB. Selective enzyme purification by affinity chromatography. *Proc. Natl. Acad. Sci. USA* 1968; 61 (2): 636–643.

12. Dobrindt U, Agerer F, Michaelis K, Janka A, Buchrieser C, Samuelson M, Svanborg C, Gottschalk G, Karch H, Hacker J. Analysis of genome plasticity in pathogenic and commensal *Escherichia coli* isolates by use of DNA arrays. *J. Bacteriol.* 2003; 185 (6): 1831–1840.

13. Dorris DR, Ramakrishnan R, Trakas D, Dudzik F, Belval R, Zhao C, Nguyen A, Domanus M, Mazumder A. A highly reproducible, linear, and automated sample preparation method for DNA microarrays. *Genome Res.* 2002; 12 (6): 976–984.

14. Hilbert H, Lauber J, Lubenow H, Dusterhoft A. Automated sample-preparation technologies in genome sequencing projects. *DNA Seq.* 2000; 11 (3–4): 193–197.

15. Hong KM, Najjar H, Hawley M, Press RD. Quantitative real-time PCR with automated sample preparation for diagnosis and monitoring of cytomegalovirus infection in bone marrow transplant patients. *Clin. Chem.* 2004; 50 (5): 846–856.

16. Leb V, Stocher M, Valentine-Thon E, Holzl G, Kessler H, Stekel H, Berg J. Fully automated, internally controlled quantification of hepatitis B virus DNA by real-time PCR by use of the MagNA Pure LC and LightCycler instruments. *J. Clin. Microbiol.* 2004; 42 (2): 585–590.

17. Smith K, Diggle MA, Clarke SC. Comparison of commercial DNA extraction kits for extraction of bacterial genomic DNA from whole blood samples. *J. Clin. Microbiol.* 2003; 41 (6): 2440–2443.

18. Wahlberg J, Holmberg A, Bergh S, Hultman T, Uhlen M. Automated magnetic preparation of DNA templates for solid phase sequencing. *Electrophoresis* 1992; 13 (8): 547–551.

19. Gut IG. Automation in genotyping of single nucleotide polymorphisms. *Hum. Mut.* 2001; 17 (6): 475–492.

20. Boom R, Sol CJA, Salimans MMM, Jansen CL, Wertheim-van Dillen PME, van der Noordaa J. Rapid and simple method for purification of nucleic acids. *J. Clin. Microbiol.* 1990; 28 (3): 495–503.

21. Meldrum D. Automation for genomics, part two: sequencers, microarrays, and future trends. *Genome Res.* 2000; 10 (9): 1288–1303.

22. Herbrink P, van den Munckhof HA, Niesters HG, Goessens WH, Stolz E, Quint WG. Solid phase C1q-directed bacterial capture followed by PCR for detection of *Chlamydia trachomatis* in clinical specimens. *J. Clin. Microbiol.* 1995; 33 (2): 283–286.

23. Stark M, Reizenstein E, Uhlen M, Lundeberg J. Immunomagnetic separation and solid-phase detection of *Bordetella pertussis*. *J. Clin. Microbiol.* 1996; 34 (4): 778–784.

24. Hoyt PR, Doktycz MJ. Optimized beadmilling of tissues for high-throughput RNA production and microarray-based analyses. *Anal. Biochem.* 2004; 332 (1): 100–108.
25. Taylor MT, Belgrader P, Furman BJ, Pourahmadi F, Kovacs GTA, Northrup MA. Lysing bacterial spores by sonication through a flexible interface in a microfluidic system. *Anal. Chem.* 2001; 73 (3): 492–496.
26. Augenstein S. Superparamagnetic beads: applications of solid phase RT-PCR. *Am. Biotechnol. Lab.* 1994; 12 (6): 12–14.
27. Kelly JJ, Chernov B, Tovstanovski I, Mirzabekov AD, Bavykin SG. Radical generating coordination complexes as a tool for rapid and effective fluorescent labeling and fragmentation of DNA or RNA for microarray hybridization. *Anal. Biochem.* 2002; 311 (2): 103–118.
28. Battrell F, Hayenga J, Bardell R, Morris C, Graham P, Kesler N, Lancaster C, Schulte T, Weigl B, Saltsman P et al. Cell lysing and cytometry in an integrated microfluidic card system, in Harrison DJ, van den Berg A, Eds. *Micro Total Analysis Systems*. Dordrecht: Kluwer Academic Publishers; 2002.
29. Fulton RJ, McDade RL, Smith PL, Kienker LJ, Kettman Jr. JR. Advanced multiplexed analysis with the FlowMetrix™ system. *Clin. Chem.* 1997; 43 (9): 1749–1756.
30. Iannone MA, Taylor JD, Chen J, Li M-S, Rivers P, Slentz-Kesler KA, Weiner MP. Multiplexed single nucleotide polymorphism genotyping by oligonucleotide ligation and flow cytometry. *Cytometry* 2000; 39 (2): 131–140.
31. Michael KL, Taylor LC, Schultz SL, Walt DR. Randomly ordered addressable high density optical sensor arrays. *Anal. Chem.* 1998; 70 (7): 1242–1248.
32. Walt DR. Bead-based fiberoptic arrays. *Science* 2000; 287 (5452): 451–452.
33. Anderson RC, Su X, Bogdan GJ, Fenton J. A miniature integrated device for automated multistep genetic assays. *Nucl. Acids. Res.* 2000; 28 (12): e60.
34. Dederich DA, Okwuonu G, Garner T, Denn A, Sutton A, Escotto M, Martindale A, Delgado O, Muzny DM, Gibbs RA et al. Glass bead purification of plasmid template DNA for high-throughput sequencing of mammalian genomes. *Nucl. Acids. Res.* 2002; 30 (7): e32.
35. Tian H, Hühmer AFR, Landers JP. Evaluation of silica resins for direct and efficient extraction of DNA from complex biological matrices in a miniaturized format. *Anal. Biochem.* 2000; 283 (2): 175–191.
36. Yuen PK, Kricka LJ, Fortina P, Panaro NJ, Sakazume T, Wilding P. Microchip module for blood sample preparation and nucleic acid amplification reactions. *Genome Res.* 2001; 11 (3): 405–412.
37. Safarík I, Safaríková M. Use of magnetic techniques for the isolation of cells. *J. Chromatogr.* B 1999; 722 (1–2): 33–53.
38. Sinclair B. To bead or not to bead: applications of magnetic bead technology. *Scientist* 1998; 12 (13): 17–24.
39. Uhlen M, Hornes E, Olsvik O. *Advances in Biomagnetic Separation*. Natick, MA: Eaton Publishing; 1994.
40. Cai H, Zhu N, Jiang Y, He P, Fang Y. Cu–Au alloy nanoparticles as oligonucleotide labels for electrochemical stripping detection of DNA hybridization. *Biosens. Bioelectron.* 2003; 18 (11): 1311–1319.
41. Chan WCW, Nie S. Quantum dot bioconjugates for ultrasensitive nonisotopic detection. *Science* 1998; 281 (5385): 2016–2018.
42. Gerion D, Chen F, Kannan B, Fu A, Parak WJ, Chen DJ, Majumdar A, Alivisatos AP. Room-temperature single-nucleotide polymorphism and multiallele DNA detection using fluorescent nano-crystals and microarrays. *Anal. Chem.* 2003; 75 (18): 4766–4772.
43. Ihara T, Tanaka S, Chikaura Y, Jyo A. Preparation of DNA-modified nanoparticles and preliminary study for colorimetric SNP analysis using their selective aggregations. *Nucl. Acids. Res.* 2004; 32 (12): e105.
44. Maruyama A, Ishihara T, Kim J-S, Kim SW, Akaike T. Nanoparticle DNA carrier with poly(L-lysine) grafted polysaccharide copolymer and poly(D,L-lactic acid). *Bioconjugate Chem.* 1997; 8 (5): 735–742.
45. Taylor JR, Fang MM, Nie S. Probing specific sequences on single DNA molecules with bioconjugated fluorescent nanoparticles. *Anal. Chem.* 2000; 72 (9): 1979–1986.
46. Ruzicka J. Discovering flow injection: journey from sample to a live cell and from solution to suspension. *Analyst* 1994; 119 (9): 1925–1934.

47. Ruzicka J, Hansen EH. Flow injection: from beaker to microfluidics. *Anal. Chem.* 2000; 72 (5): 212A–217A.
48. Ruzicka J, Pollema CH, Scudder KM. Jet ring cell: a tool for flow injection spectroscopy and microscopy on a renewable solid support. *Anal. Chem.* 1993; 65 (24): 3566–3570.
49. Ruzicka J, Scampavia L. From flow injection to bead injection. *Anal. Chem.* 1999; 71 (7): 257A–263A.
50. Chandler DP, Brown J, Call DR, Grate JW, Holman DA, Olson L, Stottlemyer MS, Bruckner-Lea CJ. Continuous, automated immunomagnetic separation and microarray detection of *E. coli* O157:H7 from poultry carcass rinse. *Int. J. Food Microbiol.* 2000; 70 (1–2): 143–154.
51. Rossomando EF, Hesla MA, Olson E, Thomas F, Gilbert DC, Herlihy JJ. "Continuous" immunomagnetic capture of *Cryptosporidium* and *Giardia* from surface water, presented at 1995 Water Quality Technology Conference, New Orleans, LA. American Water Works Association. 1996. pp. 848–881.
52. Terranova BE, Burns MA. Continuous cell suspension processing using magnetically stabilized fluidized beds. *Biotechnol. Bioeng.* 1991; 37: 110–120.
53. Jothikumar N, Cliver DO, Mariam TW. Immunomagnetic capture PCR for rapid concentration and detection of hepatitis A virus from environmental samples. *Appl. Environ. Microbiol.* 1998; 64 (2): 504–508.
54. Madonna A, Basile F, Furlong E, Voorhees K. Detection of bacteria from biological mixtures using immunomagnetic separation combined with matrix-assisted laser desorption/ionization time-of-flight mass spectrometry. *Rapid Commun. Mass Spectrom.* 2001; 15(13): 1068–1074.
55. Maher N, Dillon HK, Vermund SH, Unnasch TR. Magnetic bead capture eliminates PCR inhibitors in samples collected from the airborne environment, permitting detection of *Pneumocystis carinii* DNA. *Appl. Environ. Microbiol.* 2001; 67 (1): 449–452.
56. Neto AC, Pasquini C. Flow system for liquid–solid extraction and preconcentration using a renewable extracting solid phase. *Anal. Chim. Acta* 2002; 472: 141–146.
57. Østergaard S, Blankenstein G, Dirac H, Leistiko O. A novel approach to the automation of clinical chemistry by controlled manipulation of magnetic particles. *J. Magn. Magn. Mater.* 1999; 194: 156–162.
58. Watanabe T, Tomita S, Kudo M, Kurokawa M, Orino A, Chiba T. Detection of *Helicobacter pylori* gene by means of immunomagnetic separation-based polymerase chain reaction in feces. *Scand. J. Gastroenterol.* 1998; 33 (11): 1140–1143.
59. Ali MF, Kirby R, Goodey AP, Rodriguez MD, Ellington AD, Neikirk DP, McDevitt JT. DNA hybridization and discrimination of single nucleotide mismatches using chip-based microbead arrays. *Anal. Chem.* 2003; 75 (18): 4732–4739.
60. Bruno JG, Yu H. Immunomagnetic–electrochemiluminescent detection of *Bacillus anthracis* spores in soil matrices. *Appl. Environ. Microbiol.* 1996; 62 (9): 3474–3476.
61. Bruno JG, Kiel JL. Use of magnetic beads in selection and detection of biotoxin aptamers by electrochemiluminescence and enzymatic methods. *BioTechniques* 2002; 32 (1): 178–183.
62. Carella A, Moss M, Provost V, Quinn T. A manual bead assay for the determination of absolute CD4+ and CD8+ lymphocyte counts in human immunodeficiency virus-infected individuals. *Clin. Diagn. Lab. Immunol.* 1995; 2 (5): 623–625.
63. Li AX, Seul M, Cicciarelli J, Yang JC, Iwaki Y. Multiplexed analysis of polymorphisms in the *HLA* gene complex using bead array chips. *Tissue Antigens* 2004; 63 (6): 518–528.
64. Yang X, Li X, Prow TW, Reece LM, Bassett SE, Luxon BA, Herzog NK, Aronson J, Shope RE, Leary JF et al. Immunofluorescence assay and flow cytometry selection of bead-bound aptamers. *Nucl. Acids. Res.* 2003; 31 (10): e54.
65. Bruckner-Lea CJ, Tsukuda T, Dockendorff B, Follensbee JC, Kingsley MT, Ocampo C, Stults JR, Chandler DP. Renewable microcolumns for automated DNA purification and on-line amplification: from sediment samples through PCR. *Anal. Chim. Acta* 2002; 469 (1): 129–140.
66. Grate JW, Bruckner-Lea CJ, Jarrell AE, Chandler DP. Automated sample preparation method for suspension arrays using renewable surface separations with multiplexed flow cytometry fluorescence detection. *Anal. Chim. Acta* 2003; 478: 85–98.
67. Hindson BJ, Brown SB, Marshall GD, McBride MT, Makarewicz AJ, Gutierrez DM, Wolcott DK, Metz TR, Madabhushi RS, Dzenitis JM et al. Development of an automated sample preparation module for environmental monitoring of biowarfare agents. *Anal. Chem.* 2004; 76 (13): 3492–3497.

68. Lettieri GL, Dodge A, Boer G, de Rooij NF, Verpoorte E. A novel microfluidic concept for bioanalysis using freely moving beads trapped in recirculating flows. *Lab Chip* 2003; 3 (1): 34–39.

69. Noda H, Kohara Y, Okano K, Kambara H. Automated bead alignment apparatus using a single bead capturing technique for fabrication of a miniaturized bead-based DNA probe array. *Anal. Chem.* 2003; 75 (13): 3250–3255.

70. Weimer BC, Walsh MK, Beer C, Koka R, Wang X. Solid phase capture of proteins, spores, and bacteria. *Appl. Environ. Microbiol.* 2001; 67 (3): 1300–1307.

71. González-Martínez MA, Puchades R, Maquieira A. On-line immunoanalysis for environmental pollutants: from batch assays to automated sensors. *Trends Anal. Chem.* 1999; 18 (3): 204–218.

72. Day PJR, Flora PS, Fox JE, Walker MR. Immobilization of polynucleotides on magnetic particles. *Biochem. J.* 1991; 278 (Pt 3): 735–740.

73. Pollema CH, Ruzicka J, Christian GD. Sequential injection immunoassay utilizing immunomagnetic beads. *Anal. Chem.* 1992; 64 (13): 1356–1361.

74. Kausch AP, Owen Jr. TP, Narayanswami S, Bruce BD. Organelle isolation by magnetic immunoabsorption. *BioTechniques* 1999; 26 (2): 336–343.

75. Chalmers JJ, Zborowski M, Sun L, Moore L. Flow-through, immunomagnetic cell separation. *Biotechnol. Prog.* 1998; 14 (1): 141–148.

76. Miltenyi S, Muller W, Weichel W, Radbruch A. High gradient magnetic cell separation with MACS. *Cytometry* 1990; 11 (2): 231–238.

77. Miltenyi S. Magnetic separation apparatus. U.S. Patent 5,711,871. January 27, 1998.

78. Block SM. Making light work with optical tweezers. *Nature* 1992; 360 (6403): 493–495.

79. Grier DG. A revoluion in optical manipulation. *Nature* 2003; 424 (6950): 810–816.

80. Coakley WT. Ultrasonic separations in analytical biotechnology. *Trends Biotechnol.* 1997; 15 (12): 506–511.

81. Hawkes JJ, Barrow D, Coakley WT. Microparticle manipulation in millmetre scale ultrasonic standing wave chambers. *Ultrasonics* 1998; 36 (9): 925–931.

82. Sobanski MA, Tucker CR, Thomas NE, Coakley WT. Submicron particle manipulation in an ultrasonic standing wave: applications in detection of clinically important biomolecules. *Bioseparation* 2000; 9 (6): 351–357.

83. Wu JR. Acoustical tweezers. *J. Acoust. Soc. Am.* 1991; 89 (5): 2140–2143.

84. Cummings EB, Singh AK. Dielectrophoresis in microchips containing arrays of insulating posts: theoretical and experimental results. *Anal. Chem.* 2003; 75 (18): 4724–4731.

85. Gascoyne PRC, Vykoukal J. Particle separation by dielectrophoresis. *Electrophoresis* 2002; 23 (13): 1973–1983.

86. Zheng L, Li S, Brody JP, Burke PJ. Manipulating nanoparticles in solution with electrically contacted nanotubes using dielectrophoresis. *Langmuir* 2004; 2 (20): 8612–8619.

87. Yang J, Huang Y, Wang X-B, Becker FF, Gascoyne PRC. Cell separation on microfabricated electrodes using dielectrophoretic/gravitational field flow fractionation. *Anal. Chem.* 1999; 71 (5): 911–918.

88. Masudo T, Okada T. Particle characterization and separation by a coupled acoustic-gravity field. *Anal. Chem.* 2001; 73 (14): 3467–3471.

89. Dürr M, Kentsch J, Müller T, Schnelle T, Stelzle M. Microdevices for manipulation and accumulation of micro- and nanoparticles by dielectrophoresis. *Electrophoresis* 2003; 24 (4): 722–731.

90. Garcés-Chávez V, McGloin D, Melville H, Sibbett W, Dholakia K. Simultaneous micromanipulation in multiple planes using a self-reconstructing light beam. *Nature* 2002; 419 (6903): 145–147.

91. MacDonald MP, Paterson L, Volke-Sepulveda K, Arlt J, Sibbett W, Dholakia K. Creation and manipulation of three-dimensional optically trapped structures. *Science* 2002; 296 (5570): 1101–1103.

92. MacDonald MP, Spalding GC, Dholakia K. Microfluidic sorting in an optical lattice. *Nature* 2003; 426 (6965): 421–424.

93. Gorenflo VM, Smith L, Dedinsky B, Persson B, Piret JM. Scale-up and optimization of an acoustic filter for 200 L/day perfusion of a CHO cell culture. *Biotechnol. Bioeng.* 2002; 80 (4): 438–444.

94. Cheng J, Sheldon EL, Wu L, Heller MJ, O'Connell JP. Isolation of cultured cervical carcinoma cells mixed with peripheral blood cells on a bioelectronic chip. *Anal. Chem.* 1998; 70 (11): 2321–2326.

95. Cheng J, Sheldon EL, Wu L, Uribe A, Gerrue LO, Carrino J, Heller MJ, O'Connell JP. Preparation and hybridization analysis of DNA/RNA from *E. coli* on microfabricated bioelectronic chips. *Nat. Biotechnol.* 1998; 16 (6): 541–546.

96. Ogata Y, Scampavia L, Ruzicka J, Scott CR, Gelb MH, Turecek F. Automated affinity capture–release of biotin-containing conjugates using a lab-on-valve apparatus coupled to UV/visible and electrospray ionization mass spectrometry. *Anal. Chem.* 2002; 74 (18): 4702–4708.

97. Ruzicka J. Bioligand interaction assay by flow injection absorptiometry using a renewable biosensor system enhanced by spectral resolution. *Analyst* 1998; 123: 1617–1623.

98. Hodder PS, Blankenstein G, Ruzicka J. Microfabricated flow chamber for fluorescence-based chemistries and stopped-flow injection cytometry. *Analyst* 1997; 122 (9): 883–887.

99. Pollema CH, Ruzicka J. Flow injection renewable surface immunoassay: a new approach to immunoanalysis with fluorescence detection. *Anal. Chem.* 1994; 66 (11): 1825–1831.

100. Mayer M, Ruzicka J. Flow injection-based renewable electrochemical sensor system. *Anal. Chem.* 1996; 68: 3808–3814.

101. Santandreu M, Solé S, Fabregas E, Alegret S. Development of electrochemical immunosensing systems with renewable surfaces. *Biosens. Bioelectron.* 1998; 13 (1): 7–17.

102. Wang J, Kawde A-N, Erdem Z, Salazar M. Magnetic bead-based label-free electrochemical detection of DNA hybridization. *Analyst* 2001; 126 (11): 2020–2024.

103. Wang J, Xu D, Kawde A-N, Polsky R. Metal nanoparticle-based electrochemical stripping potentiometric detection of DNA hybridization. *Anal. Chem.* 2001; 73 (22): 5576–5581.

104. Hansen EH. Principles and applications of flow injection analysis in biosensors. *J. Mol. Recognit.* 1996; 9 (5–6): 316–325.

105. Wijayawardhana CA, Halsall HB, Heineman WR. Micro volume rotating disk electrode (RDE) amperometric detection for a bead-based immunoassay. *Anal. Chim. Acta* 1999; 399: 3–11.

106. Yu H. Comparative studies of magnetic particle-based solid phase fluorogenic and electrochemiluminescent immunoassay. *J. Immunol. Meth.* 1998; 218 (1–2): 1–8.

107. Miao W, Bard AJ. Electrogenerated chemiluminescence. 77: DNA hybridization detection at high amplification with [Ru(bpy)$_3$]$^{2+}$-containing microspheres. *Anal. Chem.* 2004; 76 (23): 5379–5386.

108. Arai K, Takahashi K, Kusu F. An electrochemiluminescence flow-through cell and its applications to sensitivie immunoassay using *N*-(aminobutyl)-*N*-ethylisoluminol. *Anal. Chem.* 1999; 71 (11): 2237–2240.

109. Matsunaga T, Nakayama H, Okochi M, Takeyama H. Fluorescent detection of cyanobacterial DNA using bacterial magnetic particles on a MAG microarray. *Biotechnol. Bioeng.* 2001; 73 (5): 400–405.

110. Kohara Y, Noda H, Okano K, Kambara H. DNA hybridization using "bead array" probe-attached beads arrayed in a capillary in a predetermined order. *Nucl. Acids Res. Suppl.* 2001; (1): 83–84.

111. Kohara Y, Noda H, Okano K, Kambara H. DNA probes on beads arrayed in a capillary bead array exhibited high hybridization performance. *Nucl. Acids. Res.* 2002; 30 (16): e87.

112. Fan J-B, Yeakley JM, Bibikova M, Chudin E, Wickham E, Chen J, Doucet D, Rigault P, Zhang B, Shen R et al. A versatile assay for high-throughput gene expression profiling on universal array matrices. *Genome Res.* 2004; 14 (5): 878–885.

113. Ferguson JA, Boles TC, Adams CP, Walt DR. A fiberoptic DNA biosensor microarray for the analysis of gene expression. *Nat. Biotechnol.* 1996; 14 (13): 1681–1684.

114. Gunderson KL, Kruglyak S, Graige MS, Garcia F, Kermani BG, Zhao C, Che D, Dickinson T, Wickham E, Bierle J et al. Decoding randomly ordered DNA arrays. *Genome Res.* 2004; 14 (5): 870–877.

115. Brenner S, Johnson M, Bridgham J, Golda G, Lloyd DH, Johnson D, Luo S, McCurdy S, Foy M, Ewan M et al. Gene expression analysis by massively parallel signature sequencing (MPSS) on microbead arrays. *Nat. Biotechnol.* 2000; 18 (6): 630–634.

116. Burger D, Gershman R. Acousto-optic laser scanning cytometer. *Cytometry* 1988; 9 (2): 101–110.

117. Carson RT, Vignali DAA. Simultaneous quantitation of 15 cytokines using a multiplexed flow cytometric assay. *J. Immunol. Methods* 1999; 227 (1–2): 41–52.

118. Park MK, Briles DE, Nahm MH. A latex bead-based flow cytometric immunoassay capable of simultaneous typing of multiple pneumococcal serotypes multibead assay. *Clin. Diagn. Lab. Immunol.* 2000; 7 (3): 486–489.

119. Tarnok A, Hambsch J, Chen R, Varro R. Cytometric bead array to measure six cytokines in twenty-five microliters of serum. *Clin. Chem.* 2003; 49(6, Pt 2): 1000–1002.

120. Vignali DA. Multiplexed particle-based flow cytometric assays. *J. Immunol. Meth.* 2000; 243 (1–2): 243–255.

121. Lowe M, Spiro A, Zhang YZ, Getts R. Multiplexed, particle-based detection of DNA using flow cytometry with three DNA dendrimers for signal amplification. *Cytometry* 2004; 60A (2): 135–144.
122. Nolan JP, Sklar LA. Suspension array technology: evolution of the flat array paradigm. *Trends Biotechnol.* 2002; 20 (1): 9–12.
123. Shapero MH, Leuther KK, Nguyen A, Scott M, Jones KW. SNP genotyping by multiplexed solid phase amplification and fluorescent minisequencing. *Genome Res.* 2001; 11 (11): 1926–1934.
124. Chandler DP, Jarrell AE. Enhanced nucleic acid capture and flow cytometry detection with peptide nucleic acid probes and tunable surface microparticles. *Anal. Biochem.* 2003; 312 (2): 182–190.
125. Chandler DP, Jarrell AE. Automated purification and suspension array detection of 16S rRNA from soil and sediment extracts using tunable surface microparticles. *Appl. Environ. Microbiol.* 2004; 70 (5): 2621–2631.
126. McBride MT, Gammon S, Pitesky M, O'Brien TW, Smith T, Aldrich J, Langlois RG, Colston B, Venkateswaran KS. Multiplexed liquid arrays for simultaneous detection of simulants of biological warfare agents. *Anal. Chem.* 2003; 75 (8): 1924–1930.
127. Baselt DR, Lee GU, Natesan M, Metzger SW, Sheehan PE, Colton RJ. A biosensor based on magnetoresistance technology. *Biosens. Bioelectron.* 1998; 13 (7–8): 731–739.
128. Nam J-M, Thaxton CS, Mirkin CA. Nanoparticle-based bio bar codes for the ultrasensitive detection of proteins. *Science* 2003; 301 (5641): 1884–1886.
129. Taton TA, Mirkin CA, Letsinger RL. Scanometric DNA array detection with nanoparticle probes. *Science* 2000; 289 (5485): 1757–1760.
130. Han M, Gao X, Su JZ, Nie S. Quantum dot-tagged microbeads for multiplexed optical coding of biomolecules. *Nat. Biotechnol.* 2001; 19 (7): 631–635.
131. Cao YC, Jin R, Mirkin CA. Nanoparticles with Raman spectroscopic fingerprints for DNA and RNA detection. *Science* 2002; 297 (5586): 1536–1540.
132. Teng C-H, Ho K-C, Lin Y-S, Chen Y-C. Gold nanoparticles as selective and concentrating probes for samples in MALDI MS analysis. *Anal. Chem.* 2004; 76 (15): 4337–4342.
133. Schürenberg M, Dreisewerd K, Hillenkamp F. Laser desorption/ionization mass spectrometry of peptides and proteins with particle suspension matrices. *Anal. Chem.* 1999; 71 (1): 221–229.
134. Khandurina J, McKnight TE, Jacobson SC, Waters LC, Foote RS, Ramsey JM. Integrated system for rapid PCR-based DNA analysis in microfluidic devices. *Anal. Chem* 2000; 72 (13): 2995–3000.
135. McBride MT, Masquelier D, Hindson BJ, Makarewicz AJ, Brown S, Burris K, Metz T, Langlois RG, Tsang KW, Bryan R et al. Autonomous detection of aerosolized *Bacillus anthracis* and *Yersinia pestis*. *Anal. Chem.* 2003; 75 (20): 5293–5299.
136. Huang Y, Mather EL, Bell JL, Madou M. MEMS-based sample preparation for molecular diagnostics. *Anal. Bioanal. Chem.* 2002; 372 (1): 49–65.

5 *In Situ* Hybridization

Xiang Qian and Ricardo V. Lloyd

CONTENTS

INTRODUCTION

In situ hybridization (ISH) is a powerful technique that allows visualization of specific gene products in cells and tissues through microscopy procedures. The identification of gene expression patterns in tissues can provide critical information for understanding gene function.[1–7] ISH also allows for morphological identification of viruses, bacteria and fungi.[8–15] The method involves a hybridization reaction between a labeled nucleotide probe and complementary target RNA or DNA sequences. The hybrids can be detected by autoradiographic emulsion for radioactively (isotope) labeled probes[16–19] or by histochemical chromogen development for nonisotopically labeled probes.[19–26] ISH localizes target sequences while preserving cell integrity within heterogeneous tissues, permitting meaningful anatomic interpretations.[1–7]

Although isotopic methods (such as labels with ^{3}H, ^{32}P, ^{35}S, ^{125}I) create problems with long turn-around times, exposure risk, and waste disposal, they remain very useful because of their greater sensitivity.[16–19] Nonisotopic methods are considered better for routine use of ISH in diagnostic pathology.[20–27] However, nonisotopic ISH can be limited by lower sensitivity.[27–30] ISH detection limits on tissue sections are in the range of 40 KB of target DNA and 10 to 20 copies of mRNA or viral DNA per cell.[6–7] Increased absolute amounts of hybridized probes have been used successfully to improve ISH detection sensitivity, e.g., cocktails of oligonucleotides or multiple cRNA probes.[18–20]

Several strategies have been developed to improve ISH sensitivity by the application of PCR-based target amplification[27–29] and signal amplification (CARD).[30–33] Application of nucleic acid target and signal amplification techniques to ISH allows detection of as few as one or two copies of specific DNA molecules in cell preparations and provides precise localization, yielding positive signals retained within the subcellular compartments. This chapter reviews the basic principles and approaches of ISH, discusses recent technical advances, the use of ISH for the diagnosis of specific diseases, and its uses in research.

IN SITU HYBRIDIZATION TECHNIQUES

BASIC PRINCIPLES

The basic requirements for a probe are specificity for the sequence of interest and labeling with a reporter to allow appropriate detection. The "melting" temperature (Tm) of a hybrid is the point at which 50% of the double-stranded nucleic acid chains are separated. The optimal temperature for hybridization is 15 to 25°C below the Tm.

Various formulas can be used to calculate Tm, depending on probe lengths and types of hybrids. For DNA–DNA hybrids with probes longer than 22 bases, the following formula can be used:

$$Tm = 81.5 + 16.5 \log (Na) + 0.41 \ (\%GC) - 0.62 \ (\% \ formamide)$$
$$- \ 500/length \ of \ base \ pairs \ of \ probe$$

RNA–RNA hybrids are generally 10 to 15°C more stable than either DNA–DNA or DNA–RNA, and therefore require more stringent conditions for hybridization and post-hybridization washing.[5,34] A wide variety of probes can be used for ISH, and the appropriate type of probe is determined to a large extent by the application.[18–27]

TISSUE PREPARATIONS

ISH has been applied to cell specimens (smears, cytospins, or cell pellets) and tissue sections (frozen, paraffin, semithin, and ultrathin plastic sections). Tissue processing, including storage and fixation, should be optimized to detect intracellular nucleic acids. Both intact cells and fresh frozen tissues (stored at –70°C) are ideal for ISH because they contain better preserved nucleotide sequences. Formalin-fixed and paraffin-embedded (FFPE) archival tissues can be used for ISH after storage for several years.[35,36] They provide almost limitless sources of materials for study using a variety of molecular techniques.

Ideal fixation for ISH should preserve both RNA and DNA and tissue morphology. Cross-linking fixatives such as paraformaldehyde, formalin, and glutaraldehyde are most commonly used. Bouin's solution contains picric acid and poorly preserves nucleic acids.[37] The nature of the fixation is a very important consideration when proteolytic digestion conditions are optimized. Fixed frozen sections or cytospin cells require fairly mild proteolytic digestion. FFPE archival tissues with better morphology require more rigorous proteolysis. The slides should be pretreated with a suitable coating solution such as 3-aminopropyltrimethoxysilane or poly-L-lysine to ensure adherence of tissue sections to the glass slides.

PRETREATMENT OF CELLS AND TISSUES

A series of pretreatment steps before hybridization increases the efficiency of hybridization and reduces nonspecific background staining. Some steps are routinely used in most ISH protocols.[38–44] Protease treatment (e.g., proteinase K) is considered one of the most important steps to increase the accessibility of the target nucleic acid, especially for nonisotopic probes with paraffin sections.[38–40] The concentrations of proteinase K (1 to 50 μg/ml) and the length of treatment (5 to 30 min) depend on tissue type, fixative, and length of fixation.

Prolonged incubation will lead to over-digested tissues, resulting in loss of signal and morphologic integrity. Under-digestion leads to suboptimal probe penetration. Both situations lead to relative failure of the hybridization reaction because the probe and target nucleic acids are not brought together under optimal conditions. Sometimes acceptable results cannot be obtained despite optimization of proteolytic digestion. Coupling of digestion to other unmasking techniques such as treatment with sodium bisulfite, sodium thiocyanate, or hydrochloric acid may increase the hybridization signal.[40–42] Acetylation of sections using 0.25% acetic anhydride/0.1 M triethanolamine can reduce charged probes binding nonspecifically to tissues. Acetic anhydride also reduces the nonspecific binding of unrelated digoxigenin (DIG)-labeled probes to neuroendocrine cells.[22]

The presence of endogenous biotin or alkaline phosphatase (AP) should be anticipated in some tissues when using nonisotopic probes. Endogenous AP can be inhibited by treatment of sections with 0.2 N HCl and levamisole and endogenous biotin by biotin-blocking agents. Microwave pretreatment can be used for FFPE archival tissues to improve staining sensitivity.[43,44] The mechanisms by which these steps improve nucleic acid unmasking are not clearly understood, although extraction of histone proteins from the cell nucleus is thought to be one important factor. Following unmasking, post-fixation such as with paraformaldehyde may be useful to prevent loss of material from the slide and for further preservation of morphology.

SELECTION OF OPTIMAL PROBES

The choice of optimal probes for ISH must consider specificity and sensitivity, ease of tissue penetration, stability of hybrids, and reproducibility of the technique. Optimal probe sizes for tissue penetration are probably 200 to 500 base pairs (bp).[7,26]

Double-Stranded DNA Probes

Specific sequences of DNA or cDNA derived by reverse transcription of mRNA are cloned into vectors such as bacteriophages, plasmids, and cosmids.[26,45,46] The amplified sequences are extracted and labeled using nick translation or random primer methods. These probes have high levels of specificity and sensitivity because of their lengths and the numbers of incorporated reporter molecules. They require denaturation to produce single-stranded DNA before hybridization and may require cleavage into smaller sequences to allow optimal access to fixed tissues.

Single-Stranded Antisense RNA Probes

Single-stranded antisense RNA probes (riboprobes) are prepared by *in vitro* transcription using the cDNA sequences as templates.[23,26,47] The cDNA insert is subcloned into a transcription vector and flanked by two different RNA polymerase (e.g., T7, Sp6) initiation sites, thus enabling either the sense or antisense strand to be synthesized. The sense RNA probe is usually performed as a negative control.

Treatment with ribonuclease A (RNaseA) after hybridization with riboprobes will reduce nonspecific background signal because the enzyme digests single-stranded but not double-stranded RNA hybrids. Riboprobes are very sensitive and useful for detection of low copy numbers of expressed genes.[18,20] They may also give rise to higher background than DNA probes because of increased nonspecific binding.[22]

Single-Stranded DNA Probes

Single-stranded DNA probes can be generated by PCR using a DNA template and only antisense primer producing single-stranded DNA, which is directly labeled during synthesis by the incorporation of nucleotide conjugated to a reporter molecule.[45] These probes have similar advantages to the oligonucleotide probes except they are much larger, probably in the 200 to 500 bp size range.

Synthetic Oligodeoxynucleotide Probes

With the increasing numbers of cloned and sequenced genes, oligoprobes can be generated from cDNA maps in the literature or GeneBank and synthesized rapidly and inexpensively.[48] Single-stranded DNA molecules ranging in size from 20 to 50 bases probably penetrate cells more readily and can produce excellent hybridization signals. Oligoprobes are commonly 3′ or 5′ ends and "tailings" labeled by using the enzyme terminal transferase with relatively few incorporated label molecules,[4,24] and are relatively less sensitive than longer cDNA or cRNA probes.[20,21,25]

Oligoprobes are generally considered most suitable for detection of relatively abundant expressed genes such as hormone mRNAs.[13,25,29] This disadvantage can be overcome by using a "cocktail" of multiple oligoprobes that are complementary to different regions of the target molecules. Careful selection of oligoprobes with low or no homologies to other nucleotide sequences is most important to ascertain the specificity of ISH.[48] Because of their short lengths, they readily gain access to appropriate sites in fixed tissues, but if very short (fewer than 12 bp), they may bind nonspecifically. The hybrids may also be more easily disrupted in post-hybridization washes, leading to false negative results.

Oligoriboprobes

Oligoriboprobes may be generated in the same way as standard riboprobes using a short DNA template or by combining single-stranded oligonucleotides with a bacteriophage promoter. Although they have the advantages of both access and stability of hybrids, they are probably more susceptible to degradation by RNases because of their short lengths.

CHOICE OF LABEL MOLECULES

The choice of label molecules is governed to some extent by personal preference and, in some situations, commercial availability. Isotopic (radioactive) labeling is now largely confined to RNA detection, particularly where nonisotopic alternatives are unsuccessful.[16-20 35] The most popular label is ^{35}S because it combines reasonable specific activity with relatively high morphological resolution.

Many laboratories now use nonisotopic labels exclusively.[20-26] The most common are biotin and digoxigenin but many others — for example, fluorescein — are available.[7] One advantage of using a fluorochrome as a hapten is that both direct fluorescent detection and indirect immunological chromogen detection can be used with the same probe. The flexibility of these nonisotopic labels coupled with their high morphological resolution has led to a dramatic increase in their use for ISH.

Isotopic Labels

Isotopic labeling may be used to detect low copy sequences and multiple nucleotide sequences, or further to combine with immunohistochemistry (IHC). Hybridization signals are detected by autoradiography, using either liquid emulsion or x-ray film and quantified by using silver grain counting or semiquantified by densitometry.[5,49] These applications are rare in diagnostic practice.

Disadvantages of isotopic probes include radioactive hazards and short half-life, as well as the fact that they are time-consuming. The choice of isotope usually reflects a compromise between the quality of resolution and the time of exposure. ^{32}P gives a rapid result and poor resolution; ^{3}H provides the best resolution, but with long exposure. ^{35}S is widely used since it produces a reasonable resolution in acceptable exposure time.[16] ^{33}P has recently been reported to produce results superior to ^{35}S.[50] Other isotopes are rarely used.

Nonisotopic Labels

Nonisotopic ISH methods are now commonly used.[20–28] The first nonisotopic label was biotin, because of the high sensitivity of streptavidin detection systems. However, the widespread presence of endogenous biotin and the limited success of blocking methods stimulated the development of a range of other labels. Digoxigenin (DIG) is a derivative of the digoxin cardiac glycoside and can be used for probe labeling. DIG-labeled probes were more recently introduced and have become widely used, with higher sensitivity and less background staining than biotinylated probes.[22–24]

The absence of DIG in mammalian cells is a particular advantage when studying tissues such as liver or kidney that may contain endogenous biotin. The signals can be visualized by using an anti-DIG antibody fragment conjugated to AP or horseradish peroxidase (HRP), with respective substrates that yield insoluble-colored products.[7,20] Nonisotopic probes are generally considered less sensitive than corresponding isotopic probes, and the hybridization results are difficult to quantify.

Other less common methods of nonisotopic probe labeling include direct conjugation to AP, labeling with bromodeoxyuridine (BrdU) or phenytoin, and chemical modifications including mercuration, sulfonation, and the addition of 2-acetylaminofluorene or dinitrophenol.[33,51–53] Probes labeled with different haptens may allow the simultaneous detection of multiple nucleotide targets in a single experiment.[7]

DENATURATION AND HYBRIDIZATION

Once the appropriate probe has been labeled and purified and the target nucleic acid exposed within the cell or tissue of interest, the probe and target must be brought together in such a way that specific hybridization can occur.[34] For DNA detection with DNA probes, both probe and target molecules must be denatured at 95°C. This can be achieved either separately or by co-denaturation.

The choice between these two methods is largely based on personal preference, although separate denaturation is most commonly used in molecular cytogenetics for fluorescent *in situ* hybridization (FISH); co-denaturation is most commonly used to analyze tissue sections.[8–15] The main advantage of co-denaturation is the reduction of the number of practical steps. However, some argue that morphological preservation is less optimal than with separate probe and target denaturation. Riboprobes and oligonucleotide probes are single-stranded, as is cellular RNA. Therefore, denaturation is not essential for RNA detection but improves the sensitivity of riboprobe detection of RNA, possibly by removing the secondary structure of RNA probe and target.

If double-stranded DNA is heated above its melting temperature (determined by its length and sequence), the two strands separate. The temperature at which this occurs can be altered by the inclusion of organic solvents in the denaturation–hybridization solution. The most common solvent is formamide, which destabilizes the double-stranded structure of DNA at a given temperature, thus reducing the effective melting temperature of the hybrids. This reduces the need for high temperature incubation and consequently leads to better preservation of morphology.

After the probe and target molecules have been made single-stranded, all that is required for annealing to take place is for the probe and target molecules to be brought together (for RNA detection and separate denaturation of DNA) or for the incubation temperature to be reduced below the melting temperature of the required hybrids (for DNA detection by co-denaturation). At this point, the specificity (or stringency) of the hybridization reaction is determined. Thus, if hybridization is carried out at too high a temperature, no probe annealing occurs and the probe and targets remain single-stranded. If hybridization is carried out at too low a temperature, probe and target molecules that are not perfectly matched will be allowed to anneal, thus reducing the specificity of the reaction.

The appropriate hybridization temperature is determined by experimentation. Alternatively, the same effect can be achieved by altering the chemical constitution of the hybridization solution while keeping the temperature constant. Thus, increasing the formamide concentration (which destabilizes mismatched hybrids) has the same effect as hybridization at a higher temperature. Other parameters that affect the specificity of the reaction are the concentration of monovalent cation (usually Na$^+$), the lengths of the probe molecules, and the probe concentration. Thus, a reduction in salt concentration increases the specificity of the reaction, as does lengthening the probe.

Increasing the probe concentration drives the reaction in favor of the formation of probe–target hybrids, thus speeding the reaction up, but may also lead to nonspecific background staining. Generally, a probe concentration of 1 to 2 ng/μl is optimum for nonisotopic labeling, but should be determined by experimentation. Hybridization time is also generally set by experimentation. However, as a rule, the time required is dependent on how repetitive the target sequence is. Thus, highly repetitive targets or targets present in high copy numbers generally require short hybridization times (2 hr), whereas low copy number targets require overnight hybridization.[1,2]

DETECTION OF HYBRIDS

Following hybridization, the first step is to remove unbound hybridization probe. This is generally carried out in a saline solution in which probe–target hybrids are stable (standard saline citrate, SSC). At this point, the specificity of the reaction can again be manipulated, although only an increase in specificity is possible at this stage. This can be done by (a) washing in solutions containing lower salt concentrations; (b) higher formamide concentrations than the hybridization solution; (c) using a higher temperature than that at which hybridization was carried out, which increases specificity by dissociating imperfectly matched hybrids.

An example of these conditions is: hybridization in 60% formamide, 2× SSC at 37°C followed by washing in 60% formamide, 2× SSC at 42°C; and hybridization in 50% formamide, 2× SSC at 37°C followed by washing in 0.5× SSC at 37°C.[5,20] Formamide is added to decrease the melting temperatures of hybrids. High temperatures, high formamide concentrations, and low ionic strength provide highly stringent conditions. In general, high degrees of specificity can be obtained by increasing the stringency.

Once the appropriate level of specificity has been achieved, the presence of probe–target hybrids can be demonstrated by detection of the probe label molecules. For isotopic labels, this is achieved by using dip-slide emulsion techniques. For nonisotopic labels, the two basic choices are fluorescent ISH (FISH) or nonfluorescent chromogenic ISH (CISH) detection. Probes labeled directly with fluorochromes or enzymes can be detected directly by immediate fluorescence microscopy or by incubation in substrate solution, respectively.

CHROMOGENIC *IN SITU* HYBRIDIZATION

Indirect detection methods using nonisotopic labeled probes followed by chromogenic (enzymatic) detection have been applied mainly for identification of virus, bacterial, and fungal infections and for detection of mRNA and DNA in tissue sections.[6–15] The probes can be labeled with DIG or biotin and then detected by sequential incubations with mouse anti-DIG and goat anti-mouse-HRP/diaminobenzidine (DAB) reaction or streptavidin–HRP/DAB, respectively. Immunohistochemical detection may use any of the standard systems, but is most often based on AP-labeled antibodies with nitroblue tetrazolium or 5-bromo-4-chloro-3-indolyl phosphate (NBT/BCIP) as chromogens, resulting in a blue–black precipitate at binding sites.[7–10]

AP-based systems are generally more sensitive than those using peroxidase but peroxidase substrates tend to give greater resolution. This approach is advantageous in surgical pathology with accurate morphological correlation, particularly in the assessment of paraffin sections.[1–4] Application of CISH using conventional ISH probes to detect gene amplification and/or deletion and chromosome translocation has been limited by the low ratio of signal-to-background staining. Interpretation of CISH is performed using a standard light microscope and permits simultaneous evaluation of gene copies and tissue morphology on the same slide. Large regions of the tissue section can be scanned rapidly in CISH using a conventional counterstain such as hematoxylin or nuclear fast red.[54,55] Amplification systems based on peroxidase-catalyzed deposition of biotinylated tyramine allow enhancement of the sensitivity of peroxidase-based systems and can also be applied to AP-based detection.[27–33]

The major advantages of CISH with precipitating enzyme reactions include the stability of the resulting precipitate and thus the possibility of permanently storing cell and tissue preparations. The combination of precipitates with routine stains, enabling the use of a standard brightfield microscope for the analysis, is an additional advantage, in particular in a setting where histopathological diagnostic analyses must be performed. For optimal nucleic acid detection *in situ*, enzymatic precipitation reactions must possess both high sensitivity and precise localization properties. Moreover, rapid staining reactions resulting in stable reaction products with contrasting colors are preferred.

Despite numerous efforts, the most efficient results have been achieved with HRP (molecular weight 40 kDa) and AP (molecular weight 100 kDa) enzymatic systems. In case the biological material examined contains endogenous enzyme activity or pseudo-peroxidase activity (such as hemoglobin in erythrocytes), the activity must be blocked to prevent the formation of unacceptable background.[56] Inhibition or elimination of these enzyme activities can be realized by a number of different treatments including acetic acid, HCl, or levamisole for endogenous AP and treatment with, for example, periodate and borohydride, sodium nitroferricyanide, phenylhydrazine, azide, and/or hydrogen peroxide, or HCl in case of endogenous HRP (e.g., in erythrocytes, neutrophils, and macrophages).

In some cases, it may be better to change the enzyme system used rather than attempt to remove excessive endogenous enzyme activity.[57] After performing the enzyme reactions and prior to embedding, the cell preparations can be lightly counterstained with, for example, hematoxylin and/or eosin, methyl green, neutral red, or nuclear fast red for bright-field microscopic analysis, and with fluorescent counterstains in the case of fluorescence microscopic analysis. Counterstaining can be omitted if reflection/contrast microscopy is utilized. Enzymatic systems that produce chemiluminescent signals can also be used for ISH probe detection.

Double or Triple Chromogenic *In Situ* Hybridization

In performing double-ISH methods for the simultaneous detection of multiple target nucleic acids such as two mRNAs, DNA and mRNA,[58–61] the two probes may be hybridized simultaneously or sequentially and the signals detected simultaneously or sequentially. For combined ISH/IHC, ISH is generally performed before IHC because it reduces the chances of RNase contamination. Combinations of nonisotopic and isotopic ISH methods are mainly used for detection of two mRNAs in the same tissue sections.[18,59]

After histochemical detection for nonisotopic signals, the slides are subject to an autoradiographic approach to determine radioactive signals. A combination of two nonisotopic labeled probes, mainly biotin and DIG conjugated to different enzymes (AP or HRP) or fluorophores followed by respective detection systems, allow a simultaneous localization of multiple mRNA and genomic DNA within tissue sections or even single cells.[7,61] Colloidal gold with different, nonoverlapping particle sizes (6 to 30 nm) greatly facilitates multiple labeling for ISH analysis at the electron microscopy (EM) level.[62] Colloidal gold particles 5 to 15 nm in size are excellent for post-embedding cytochemistry. ISH signal by EM can be evaluated quantitatively by counting colloidal gold particles. A double EM *in situ* hybridization was performed using a DIG-labeled probe and a biotin-labeled probe. The hybrids were revealed, respectively, with specific antibodies coupled to colloidal gold particles of different sizes (10 and 15 nm).[63]

Chromosome-specific DNA probes labeled with biotin, DIG, or fluorescein were hybridized simultaneously and then detected by enzyme cytochemistry using one AP reaction and two subsequent temporally discrete HRP incubations. For triple-color detection on single cell preparations, the combination of the enzyme precipitates HRP/diaminobenzidine (DAB, brown color), AP/fast red (FR, red color) and HRP/tetramethylbenzidine (TMB, green color) resulted in accurate detection of DNA targets.[61]

Embedding of the preparations in a thin cross-linked protein layer further stabilized the enzyme reaction products. For ISH on tissue sections, however, this detection procedure showed some limitations with respect to both the stability of the AP/FR and HRP/TMB precipitates and the sequences of immunochemical layers in multiple-target procedures. For this reason, the AP/FR reaction was replaced by the AP/new fuchsin (NF, red color) reaction and the washing steps after the HRP/TMB reaction were restricted to the use of pH 6.0 phosphate buffer. Furthermore, to improve the efficiency of the ISH reaction, AP/NF was applied in an avidin–biotin complex detection system and, to avoid target shielding in the triple-target ISH, the third primary antibody was applied prior to the second enzyme cytochemical reaction.[7,61]

FLUORESCENT *IN SITU* HYBRIDIZATION

Fluorescent *in situ* hybridization (FISH) is commonly used in molecular cytogenetics. It is particularly useful for interphase cytogenetics where multicolor hybridization requires high resolution. Applications of FISH include karyotype analysis, gene mapping, DNA replication, and DNA recombination.[64–67] To perform the procedure, a permeabilization step is required to facilitate penetration of the fluorescence-labeled probe and a denaturation step is necessary to convert double-stranded DNA in tissue sections to single-stranded DNA to allow re-annealing of the probe and target genome.

FISH facilitates the precise detection and localization of specific DNA sequences in intact cells and chromosomes, and so is extensively applied to the histochemical mapping of specific nucleic acid sequences on chromosomes in metaphase and to interphase nuclei. FISH allows quantitative analysis of various sequences in interphase nuclei with specific numerical and structural chromosomal abnormalities.[65] It can also be employed for simultaneous detection of multiple mRNA species in a single tissue section or cell.

The identification of specific nucleic acid sequences with FISH reveals sites of mRNA processing, transport, and cytoplasmic localization.[66,67] The sensitivity of FISH can be increased to detect low-abundance mRNA by amplifying the signal through a combination of tyramide signal amplification (TSA) and enzyme-labeled fluorescent AP substrate.[68] FISH and digital imaging microscopy were modified to allow quantitative detection of single RNA molecules.[67] Oligodeoxynucleotide probes can be synthesized with five fluorochromes per molecule. The rates of transcription initiation and termination and mRNA processing may be determined by positioning probes along the transcription unit.[66]

The use of fluorescence detection protocols for ISH provides a number of advantages, including easy and rapid detection of fluorochrome-labeled probes, high sensitivity with low endogenous background, high resolution, multiple-target analyses with different fluorochromes, and the possibility of quantifying signal intensity. A large number of fluorochromes with high quantum yields and good spectral separation properties are now available for use in *direct* and the more sensitive *indirect* [conjugated to (strept)avidin and antibody molecules] ISH detection procedures.

The number of nucleic acid targets that can be detected in DAPI (4′,6-diamidino-2-phenylindole) stained chromosomes has been extended by FISH. The Vectashield® mounting medium (Vector Laboratories, Burlingame, California, USA) has been recommended specifically for FISH. Development of a variety of optical filter sets coupled with the availability of diverse fluorochromes has expanded the limited spectral band width (approximately 350 to 800 nm) that exists for fluorescence imaging. Most FISH applications, however, are still based on the use of two or three fluorochromes (blue, green, and red).

If further increase of the FISH detection sensitivity or efficiency is needed, the fluorescence detection procedures can be extended with catalyzed reporter deposition (CARD) signal amplification.[69] Painting all chromosomes in a single experiment simplifies chromosome karyotyping and provides for rapid diagnosis of aberrations such as the identification of rearrangements, imbalanced material, and marker chromosomes in, for example, tumor specimens.[65]

Despite advantages, FISH also displays disadvantages, including fading of the fluorescence signals upon exposure to light, significant autofluorescence, for example, in FFPE tissue sections, difficulties of combining FISH with routine histopathological stains for morphological examination, and, to a lesser extent, economical and practical considerations related to the cost of the microscopic and imaging equipment needed. To obtain permanent preparations with nonfading FISH signals, enzyme cytochemical and immunogold-silver detection systems have been utilized.[70]

Body cavity effusion specimens are more suitable for FISH than paraffin-embedded tissue sections, because nuclei remain intact, thereby avoiding the nuclear sectioning problem inherent in FISH performed on standard surgical specimens.[71] One advantage of FISH is its easy-to-engage multicolor detection; CISH is restricted to single or dual colors due to the limitations of current chromogenic detection technology. Many factors affect the detection of fluorescent signals including the type and duration of fixation, artifacts caused by sectioning of tissue blocks, thickness of sections, storage of embedded tissues or sections, and probe penetration. Because visual evaluation of large numbers of cells and enumeration of hybridization signals are tedious and time consuming, FISH analysis with dot counting can be expedited via an automated system.

FISH is also used to tag specific RNAs with fluorescent moieties without interfering with cell vitality, providing insight into the spatial organization of specific RNA transcripts in cell nuclei. Various methods have been developed to deliver probes into a living cell, such as *in vivo* hybridization of fluorochrome-labeled 2′-*O*-methyl oligoribonucleotide (2′-OMe RNA) probes to endogenous RNAs and microscopic imaging of the tagged RNAs in living cells. The 2′-OMe RNA probes are not degraded by nucleases, form stable hybrids with structured RNAs, and do not interfere with cell vitality.[65,66]

CONTROLS FOR *IN SITU* HYBRIDIZATION

Controls are critical for all ISH assays to assess whether the signal does indeed represent specific hybridization to the target sequence. A variety of controls can be used: (1) pretreatment of tissues with RNase or DNase, depending on the target tested; (2) omission of the specific probes in the hybridization reaction; (3) using an unrelated or sense probe; (4) competition studies with unlabeled probes before adding labeled probes for hybridization; and (5) combining ISH with immunostaining to localize the translated protein product in the same cells. Loss of target RNA or DNA may result in false-negative results, particularly for paraffin-embedded tissues. β-actin[72] poly(dT),[73] or ribosomal RNA[74] and Alu DNA probes[75] are used for the integrity of target RNA and DNA.

The probe should be characterized in Northern or Southern blot studies and on known positive and negative tissues. Competitive hybridization with an excess of unlabeled probe or prehybridization with the complementary sequence should reduce signal intensity. Substitution of sense probe should result in no signal. Technical problems may arise during the assay for various reasons, such as nonspecific reaction of probe binding to targets encoded by as yet unsequenced genes within the genome. The application of irrelevant oligonucleotide probes may detect nonspecific binding of nucleotide sequences to specific cell types, such as gut neuroendocrine cells.[76]

RECENT TECHNICAL ADVANCES IN *IN SITU* HYBRIDIZATION

Recent advances of ISH can be divided into two categories: (1) combination use of various ISH approaches for simultaneous detection of multiple target nucleic acids[58-61] and (2) increasing ISH sensitivity by amplification of either target nucleic acid sequences prior to ISH or signal detection after hybridization is completed.[27-33] Current approaches to increase ISH sensitivity include target amplification (*in situ* PCR, primed labeling, self-sustained sequence replication), signal amplification (TSA, branched DNA amplification), and probe amplification (rolling circle amplification). Application of some of these techniques has extended the utility of ISH in the diagnostic pathology field and in research because of its ability to detect targets with low copy numbers of DNA and RNA.

TARGET AMPLIFICATION

PCR with ISH can amplify specific DNA (*in situ* PCR) or RNA (*in situ* RT-PCR) sequences inside single cells or tissue sections and increase the copy numbers to levels readily detectable by conventional ISH methods.[77-79] In theory, *in situ* PCR techniques are straightforward and should be reproducible — like conventional PCR. In practice, however, *in situ* PCR is associated with many problems, for example, low amplification efficiency and poor reproducibility.

The results of direct *in situ* PCR can be influenced by incorporation of labeled nucleotides into nonspecific PCR products and the generation of nonspecific PCR products resulting from mispriming, fragmented DNA undergoing repair by DNA polymerase, repair artifacts, priming of nonspecific DNA or cDNA fragments, or endogenous priming artifacts. Rigorous use of controls is required to allow adequate interpretation of *in situ* PCR results and to use indirect *in situ* PCR to increase the specificity of the amplified nucleic acids. *In situ* PCR has been mainly applied to detect DNA sequences not easily detected by conventional ISH, including human single-copy genes, rearranged cellular genes, and chromosomal translocations — and also to map low-copy number genomic sequences in metaphase chromosomes.

The employment of *in situ* PCR to detect low copy number viral genes, especially HIV and hepatitis C, has led to significant discoveries about viral infectious diseases.[80,81] Successful amplification of mRNA by *in situ* RT-PCR includes hormone, receptor, and oncogene transcripts and is still limited primarily to cell preparations and frozen sections.[29,82-84] Very few applications in paraffin sections have been reported.[78]

Novel target amplification ISH techniques, such as self-sustained sequence replication (IS-3SR)[85] and primed *in situ* labeling (PRINS)[86,87] are still in the research stage and have been applied successfully only to experimental cytocentrifuge preparations and not to archival tissue sections. Extension of its use for mRNA detection and application on tissue sections from paraffin-embedded specimens is required to render it more reliable and acceptable. Similar drawbacks of these novel methods have been encountered as for *in situ* PCR, including false positive results and background staining due to mispriming and artifacts related to the incorporation of labeled nucleotides into nonspecific DNA.[88]

SIGNAL AMPLIFICATION

Catalyzed Reporter Deposition Technique

CARD is based on the deposition of activated biotinylated tyramine onto electron-rich moieties such as tyrosine, phenylalanine, or tryptophan at and near the sites of HRP activity. The biotin sites on the bound tyramide act as further binding sites for streptavidin–biotin complexes or enzyme- and fluorochrome-labeled streptavidin, for example.[30–33] CARD allows a 10- to 100-fold increase in sensitivity of ISH signals when compared to conventional avidin–biotin complex procedures, without production of increased background.

Visualization of deposited tyramides can be performed either directly after the CARD reaction with fluorescence microscopy if fluorochrome-labeled tyramides are used or indirectly via fluorescence or bright field microscopy if biotin, digoxigenin, dinitrophenyl, or trinitrophenyl is used as a hapten. The tyramides can act as further binding sites for anti-hapten antibodies or streptavidin conjugates.[33] The main advantage of using CARD signal amplification for ISH is that it is performed after probe hybridization and stringent washings, so that the specificity of the probe hybridization is not compromised. The practical limitations of CARD relate to a number of factors. Because of its high sensitivity, CARD has the potential to amplify nonspecific background signal, which may result in an unfavorable signal-to-noise ratio. Endogenous peroxidases in human tissue can catalyze the CARD reaction.

To avoid this unwanted reaction, endogenous peroxidase must be blocked or quenched and appropriate positive and negative controls should be used. CARD was applied successfully to detect both repetitive and single-copy DNA target sequences in cytospin preparations with high efficiency. By combining streptavidin–nanogold/silver ISH with CARD, ISH sensitivity can allow detection of single-copy sequences in FFPE sections.[89]

Branched DNA *In Situ* Hybridization

Branched DNA (bDNA) technology has been recently adapted to ISH for improving nucleic acid detection. This technique increases the signal but not the target.[90–92] To reduce potential nonspecific hybridization, non-natural nucleotides 5-methyl-2′-deoxyisocytidine (isoC) and 2′-deoxyisoguanosine (isoG) are included in the target, preamplifier, amplifier, and AP-conjugated label probes. AP-labeled probes then catalyze the reaction of the NBT-BCIP substrate.

The advantage of this method is its specificity. Although both bDNA and CARD improve the sensitivity of ISH by increasing the numbers of binding sites for reporter molecules, they use different mechanisms to achieve signal amplification. Because bDNA ISH is based on the sequential hybridization of synthetic DNA probes, it does not require any DNA or RNA polymerase activity or repeated cycling through elevated temperatures.[91] Avoiding high temperature incubation is important for applications in which preservation of intricate cell morphology is required. Another feature is that bDNA ISH does not use an avidin–biotin signal amplification system and hence is not affected by the binding of avidin-conjugated reporter molecules to endogenous biotin.

By providing greater flexibility in probe design and avoiding potential interference by endogenous biotin, the nonisotopic and rapid bDNA ISH method can be adapted to generate chromogenic and/or fluorescent signals, and should be amenable to automation and quantification. The sensitivity of the bDNA ISH method is similar to that of CARD and is sufficient to detect relatively low-abundance targets — as few as one or two copies of human papillomavirus (HPV) type 16 DNA in SiHa cells.[92] The disadvantages of bDNA ISH include less sensitivity compared to target and probe amplification methods such as *in situ* PCR.

Probe Amplification

A padlock probe is composed of a single-stranded linear oligonucleotide of about 90 base pairs (bp) that hybridizes to targets of 30 bases. The 30-base target-binding region of the probe is split into two 15-base segments placed in opposite orientation at each end of the linear probe. The remaining 60 bases form an unhybridized single-stranded loop. Posthybridization DNA ligation connects the two ends of the probe in the middle of the 30-base binding region. The unbound 60-base loop facilitates probe circularization and permits 20 bases to serve as a primer recognition site for DNA polymerase to replicate the circle.[93]

Rolling circle amplification (RCA) generates a localized signal via isothermal amplification of an oligonucleotide circle, which can be performed using padlock probes as templates. The polymerase progresses continuously around the loop until the 100 bases have been replicated hundreds or thousands of times. Incorporating a labeled nucleotide during the RCA reaction produces sufficient signal for easy visualization of the target. One advantage of RCA is a significant increase in the signal achieved through a single round of enzymatic amplification of a nucleic acid substrate, leading to improvements in relative discrimination as well as absolute amounts of measurable signal.[94]

RCA *in situ* is useful for discriminating alleles, determining gene copy numbers, and quantifying gene expression in single cells.[94,95] Because of its exquisite sensitivity, *in situ* RCA may add an entirely new dimension to the fields of genomics, pathology, and cytogenetics. Because the procedure is technically complicated, application of RCA *in situ* in FFPE tissues has not been uniformly successful to date. Future work is required to confirm its real potential in FFPE tissue.

APPLICATIONS OF *IN SITU* HYBRIDIZATION

ISH methods have found many applications in basic clinical research and in diagnostic pathology.[1–6] Diagnostic applications are emerging in the fields of (1) identification of gene expression by the detection of mRNA (usually where levels of gene product are below the limits of immunohistochemical detection); (2) diagnosis of infections, particularly viral; (c) cell cycles and apoptosis; and (4) interphase cytogenetics as an alternative to classical cytogenetics in the investigation of fetal abnormalities and tumors.

Infectious Diseases

ISH has been used for identification of foreign genes, including bacteria, fungi, and viruses in tissue sections. Detection of infectious agents with sensitive nonisotopic ISH methods provides valuable information about the etiologies of infectious diseases because many infectious agents can be visualized readily by ISH methods.

Virus Infections

Viral infections have been widely investigated. The abilities to detect human immunodeficiency virus (HIV), cytomegalovirus (CMV), human papillomavirus (HPV), herpes simplex virus (HSV), hepatitis B virus (HBV), Epstein–Barr virus (EBV), parvovirus B19, and polyoma viruses BK/JC are only some of the diagnostic applications of ISH methods[96–104] (see Figure 5.1, panels A through E). More than 50 types of HPV exist and probes specific for the various types are available to assess infection by particular types associated with neoplastic development.[103] HPV types 16 and 18 are more likely to be associated with malignant progression.[104] The test has provided some insight into the epidemiology in that similar patterns of infections were identified in two cohorts separated by 25 years.[103–105] However, the role played by HPV in other lesions including Bowen's disease, squamous cell papilloma of bronchus, and larynx, is unclear.[106–109]

EBV has been implicated in the pathogenesis of a wide range of human lymphoid and epithelial tumors.[110–112] Detecting viral DNA in latent infection with ISH is difficult because of low copy

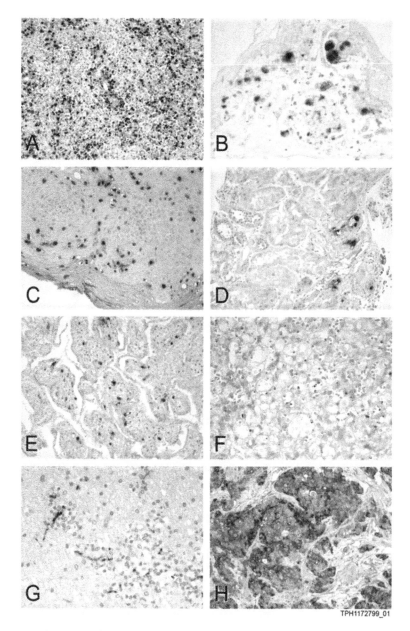

TPH1172799_01

FIGURE 5.1 *(A color version follows page 270)* Examples of ISH using chromogenic methods. A. Detection of Epstein-Barr virus (EBV) in a B cell lymphoma with diffuse large cell type using EBERs oligonucleotide probes visualized by alkaline phosphatase (AP) with nitroblue tetrazolium and 5-bromo-4-chloro-3-indolyl phosphate (NBT/BCIP) as chromogen substrate. B. ISH for adenovirus (ADV) using ADV cDNA probe with AP and NBT/BCIP. C. ISH for human papillomavirus (HPV) type 6/11 showing viral DNA in the nuclei of condyloma acuminatum using DNA probe with AP and NBT/BCIP. D. ISH positive signals for BK polyoma virus are seen in the nuclei of tubular epithelium from a renal transplant case using DNA probe with AP and NBT/BCIP. E. ISH positive cells of parvovirus B19 (PVB19) present in nucleated red cells of placenta using cDNA probes with AP and NBT/BCIP. F. Detection of *Cryptococcus* by ISH in a lung biopsy. The organisms are detected using oligonucleotide probes targeted at the ribosomal RNA visualized by AP and NBT/BCIP. G. ISH for *Aspergillus* showing hyphal forms. Oligonucleotide probes with AP and NBT/BCIP. H. ISH detecting albumin mRNA in a hepatocellular carcinoma metastatic to the scapula using riboprobes with AP and NBT/BCIP.

numbers per cell. However, small nuclear RNAs (EBERs) encoded by the virus are highly expressed and detectable by nonisotopic ISH. EBV has been detected in some cases of Hodgkin's disease and in a variety of non-Hodgkin's lymphomas including Burkitt's lymphoma, a minority of B-cell non-Hodgkin's lymphomas, certain T-cell lymphomas, and oral hairy leukoplakias.[110,111] The presence of EBV in undifferentiated nasopharyngeal carcinomas is well proven.[112] ISH is useful for confirming infectious mononucleosis in atypical cases and renal disease associated with EBV in patients with negative EBV serologies.

Hybridization to viral DNA and mRNAs in liver has aided the understanding of the complexity of hepatitis B virus infection. Detection of HBV DNA by ISH is superior to IHC detection for hepatitis B surface antigen (HBsAg) with a higher degree of sensitivity.[100] Recently, ISH has been applied to the detection of hepatitis C virus (HCV) RNA.[15,80] Most studies applied these techniques to FFPE tissue due to its availability and the good preservation of morphology, which allows for better localization of HCV. Few studies have applied the technique to frozen tissue.

Most studies showed that HCV-positive cells were usually isolated with occasional clustering, and fewer than 20% of cells were positive.[113] HCV RNA was shown in the cytoplasms of hepatocytes with occasional signals detected in mononuclear cells, bile duct epithelium, and sinusoidal cells. The relationship between the detection of HCV and cell damage is important. The level of HCV positivity appeared to correlate with serum aminotransferase, suggesting that HCV may cause damage to the liver cells directly in the absence of overt morphological changes.[114]

ISH has been also applied to the identification of cytomegalovirus (CMV) in viral encephalitis and chronic encephalitis associated with acquired immune deficiency syndrome (AIDS)[97,115] and in chronic encephalitis with epilepsy.[116] Pulmonary CMV involvement has been detected on cytological specimens[117] and the extent of systemic disease documented where the presentation indicated isolated oophoritis.[118] CMV hepatitis has been demonstrated in liver allografts,[119] and gastrointestinal biopsies have been used to detect infections in cardiac transplant recipients.[120] However, for CMV, ISH may not exhibit significantly greater sensitivity than histology and IHC techniques. Other human herpesviruses (HHVs) may also be detected, including herpes simplex (HSV) in lymphadenitis and endometritis,[121] HHV-6 in erythroderma associated with an infectious mononucleosis-like syndrome, in a variety of lymphoproliferative states, and in AIDS-associated retinitis.[122,123] The presence of HHV-8 in Kaposi's sarcoma has been demonstrated using PCR-ISH.[124] Evidence suggests that ISH may not be as sensitive as IHC in the detection of HSV, but proper validation should be performed with the newer amplified techniques. ISH may still have an advantage over IHC in early HSV infection.[125]

Identification of Bacteria and Fungi

A rapid method for the identification of bacterial cells using 16S rRNA-directed, fluorescently tagged oligonucleotide probes has been developed.[10,126] The results of tests using a variety of Gram-positive and Gram-negative microorganisms are presented. *Staphylococcus epidermidis* and *Staphylococcus aureus* are the most common causes of medical device-associated infections. The microbiological diagnosis of these infections may occasionally be unclear, so detection and identification of *S. aureus* and *S. epidermidis* by ISH with fluorochrome- or fluorophore-labeled oligonucleotide probes specific for staphylococcal 16S rRNA has proven useful.[127]

Helicobacter pylori, the causative agent of chronic gastritis and peptic ulcers, is also associated with gastric cancer. Eradication of *H. pylori* infection may be difficult to confirm. Specificity of ISH for *H. pylori* was proven by the lack of hybridization on sections with Gram-negative and Gram-positive bacteria other than *H. pylori* and various controls.[128] FISH is a highly sensitive and reliable method for detecting macrolide-resistant *H. pylori* in FFPE biopsy specimens, which represents the routine method of processing tissue obtained upon gastroscopy.

The specific identification of yeast and yeast-like organisms in tissue sections can sometimes be difficult because several common species have overlapping histologic features. Aspergillosis produces significant mortality in immunosuppressed patients. Rapid diagnosis is often required to initiate appropriate therapy. The histology of *Aspergillus* species may overlap with a variety of fungi, so diagnosis often relies on fungal cultures that can take weeks to complete.

ISH targeting *Aspergillus* 5S ribosomal RNA (rRNA) identified 41 localized aspergillomas in the lung, brain, sinonasal tract, and ear, and two cases of invasive aspergillosis involving pleura and soft tissue of the scapular region.[8,9] The diagnosis of *Pneumocystis carinii* by ISH was performed on FFPE human lung tissues and detected with the avidin–biotin peroxidase complex method. The reactions were positive in all 12 cases of *P. carinii* pneumonia, but in none of the infections with other pathogenic agents, including viruses (6 cases), mycobacteria (4 cases), protozoa (4 cases), and fungi (8 cases).[129]

The reactivity and specificity of this method were comparable with IHC results with a monoclonal anti-human *P. carinii* antibody. *Blastomyces dermatitidis*, *Coccidioides immitis*, *Cryptococcus neoformans*, *Histoplasma capsulatum*, and *Sporothrix schenckii* can also be detected with ISH.[8] Probes for both the 18S and 28S ribosomal RNAs were tested against 98 archived FFPE tissue specimens, each of which had culture-proven involvement by one of these organisms. ISH was uniformly positive with all species-specific probes, yielding 100% specificity. It also achieved a higher positive predictive value (100% in all cases) compared with Grocott methenamine silver (GMS) staining (83.3 to 100%). Four cases with rare organisms present (4% of cases tested) were detected by ISH but not by GMS staining. These results show that ISH provides a rapid and accurate technique for the identification of fungal organisms in histologic tissue sections (Figure 5.1, panels F and G).

CYTOGENETICS

ISH now serves as a practical technique in cytogenetics and molecular genetics. FISH is most commonly used.[130–133] Probes are usually directly fluorescein labeled or labeled with biotin or digoxigenin, followed by a second signal detection coupled with fluorochrome.[26,134] The primary applications of FISH in cytogenetic analysis include chromosomal gene mapping, characterization of genetic aberrations, identification of genetic abnormalities associated with genetic disease or neoplastic disorders, and detection of viral genomes in interphase nuclei or metaphase chromosomes.[130–133] Synchronization of cell cultures by BrdU or thymidine is used to increase the frequency of chromosomes at the different stages of condensation (metaphase, prometaphase, or prophase) for ISH analysis.[5] The discrete hybridization signals (fluorescent spots) can be visualized by fluorescent microscopy. FISH using probes with nonisotopic labeling has successfully localized single copy gene sequences.[134–136]

Interphase FISH can be applied to imprints from fresh tissue or to paraffin sections after appropriate pretreatment. Centromeric, telomeric, and locus DNA-sequence-specific probes can be used to identify aneuploidy or gene mutations.[130–134] Several protocols combine molecular cytogenetics with classic karyotyping.[133]

Multicolor FISH assays are indispensable for obtaining precise descriptions of complex chromosomal rearrangements. Routine application of such techniques to human chromosomes started in 1996 with the simultaneous use of all 24 human whole-chromosome painting probes in multiplex-FISH and spectral karyotyping (SKY). Since then, different approaches for chromosomal differentiation based on multicolor-FISH assays have been described,[136] predominantly to characterize marker chromosomes identified in conventional banding analysis. Their characterization is of high clinical impact and is the requisite condition for further molecular investigations aimed at the identification of disease-related genes.

FISH tests are now available for several well known genetic disorders caused by deletions or duplications that are otherwise undetectable by classical cytogenetic analysis. These disorders include Williams syndrome, Prader–Willi syndrome, and the 22q11.2 deletion syndrome, also known as DiGeorge or velocardiofacial syndrome.[137] Ewing's sarcoma and alveolar rhabdomyosarcoma, are now essentially defined by their respective translocations. This may also be true for synovial sarcoma because the t(X;18) translocation was not detected in a panel of 23 nonsynovial sarcomas.[132,134]

New chromosomal translocations have also been described for extraskeletal myxoid chondrosarcoma [t(9;17)(q22;q11)] and alveolar soft part sarcoma ([t(X;17)(p11.2;q25)]).[138] FISH studies have demonstrated that 60 to 75% of myeloma patients have illegitimate rearrangements involving the immunoglobulin heavy-chain gene at 14q32.[131,132] Loss of chromosome 13 has been associated with a poor prognosis in myeloma. Cytogenetic studies showed a significantly lower incidence of monosomy 13 in monoclonal gammopathy of undetermined significance (MGUS) compared with myeloma and a significantly higher incidence of this deletion in post-MGUS myeloma compared with *de novo* myeloma.[139,140] These observations suggest that this chromosomal alteration may confer a growth or survival advantage to the malignant plasma cell and may be involved in the transformation of MGUS to myeloma. As these and other putative myeloma-specific translocations are characterized, it will continue to be important to determine their sensitivities and specificities for particular diseases and the relative importance of molecular versus histologic classification of each myeloma.

Ewing's sarcoma primitive neuroectodermal tumor (ES/PNET) is characterized by rearrangements of the EWS gene located at 22q12 and over-expression of MIC2 (CD99) antigen. In 85% of ES cases, the t(11;22)(q24;q12) generates chimeric fusion transcripts between the EWS and the FLI1 gene, whereas in the remaining cases, the EWS gene is rearranged with different partners of the ES oncogene family.[141,142]

In addition to classical cytogenetic analysis, FISH and reverse transcriptase PCR (RT-PCR) can be used to demonstrate 22q12 rearrangements that are pathognomonic for ES/PNET.[134,141] To visualize 22q12 rearrangements in individual cells, DNA probes flanking the EWS-R1 breakpoint region on chromosome 22 can be hybridized in double-target FISH experiments on tumor cell preparations. Intact chromosomes 22 are indicated by spatial juxtaposition of signals from distinct DNA probes, whereas rearrangements of the EWS gene separate the hybridization signals (Figure 5.2). In addition to 22q12 rearrangements, numerical aberrations of chromosomes 8 and 12 can be observed in about 50% of ES and deletions at the short arm of chromosome 1 and der (16)t(1;16)(q12;q11.2) chromosomes in about 20% of the cases. Numerical aberrations, deletions at 1p36.3, and at t(1;16) were detected by using double-target FISH on touch, cytospin, and chromosome preparations, on frozen and paraffin sections and on isolated deparaffinized nuclei. Numerical aberrations of chromosomes 8 and 12 did not show prognostic impacts. However, deletions at 1p36.3 and imbalances between the long and short arms of chromosome 1 were associated with adverse clinical outcomes in a group of Ewing's sarcoma patients with localized disease.[142]

GENE EXPRESSION

ISH methods are powerful tools for the analysis of gene expression in normal and pathologic tissues.[1-6] A major advantage of ISH is its ability to localize mRNA at the cellular level in heterogeneous tissues, thus expanding the results of other molecular techniques such as Northern blot hybridization for specific gene analysis. ISH methods have been of particular value for the study of mRNA-encoding oncogenes, growth factors and their receptors, hormones and hormone receptors, cytokines, structural proteins, and enzymes (collagenase and others).[18-25]

ISH methods have many practical applications in tumor pathology. The correlation of oncogene expression with prognosis is being investigated in neuroblastomas and epithelial neoplasms such as colon, lung, prostate, and breast carcinomas. Detection of genes encoding cell structural proteins including tumor-associated markers represents potential for the application of ISH methods in pathologic diagnosis.[143] For example, nonisotopic ISH methods for localization of immunoglobulin light chain mRNAs in hyperplastic and neoplastic lymphoproliferative disorders[144] albumin mRNA

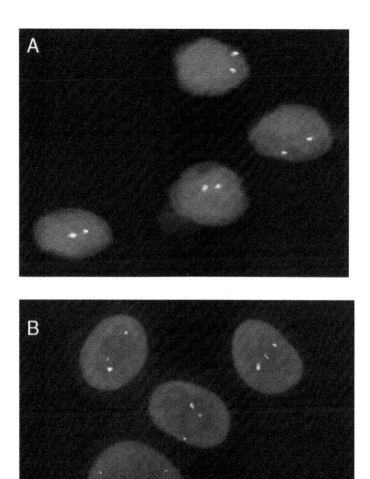

FIGURE 5.2 *(A color version follows page 270)* Interphase FISH analysis of normal tissues and Ewing's sarcomas from FFPE sections. A. Normal interphase cell nucleus without EWS translocation showing green and red FISH signals in juxtaposition (fusion). B. Tumor cell nucleus with EWS translocation showing separation of green and red FISH signals indicating EWS gene breakage (translocation).

to distinguish between hepatocellular and metastatic carcinomas in the liver (Figure 5.1, panel H), and chromogranin–secretogranin mRNAs in the classification of neuroendocrine tumors[145,146] are used in some diagnostic pathology laboratories.[143]

ISH methods used to identify cells or tumors on the basis of their specific mRNA contents are different from IHC, which is dependent on the protein contents of cells. Thus, ISH identifies the gene products from *de novo* synthesis, rather than nonspecific uptakes of proteins by cells that may result in false-positive immunostaining results.[147] ISH analysis has also been extensively used in studies of endocrine tumors. For example, a significant number of small cell lung carcinomas with few secretory granules are commonly negative for chromogranin proteins, but the mRNA may be detected by ISH methods.[25]

Studies of gene expression by ISH in endocrine tumors, including chromogranin A, thyroglobulin, estrogen receptor proteins, parathyroid hormones, and calcitonin gene-related peptides, have contributed to our understanding of the biologies and pathophysiologies of various endocrine disorders.[145,146]

Proliferative capacity is an important determinant of tumor biological behavior. The expression of histone genes is a fundamental step constituting the process of cell proliferation. Histone H3 mRNA accumulates in the cytoplasm during S phase and then decreases as cells approach G2 phase. Histone H3 mRNA levels are dependent on transcriptional and post-transcriptional mechanisms, with transcription rates increasing 10-fold at the onset of S phase. they are down-regulated at the cessation of cell proliferation and during quiescence.[148–150] Histone H3 mRNA level is a specific marker of S phase cells. The principal advantage of histone H3 mRNA determination by ISH is that the results are tightly coupled with *de novo* DNA synthesis. The ISH technique permits proliferative capacity to be assessed rapidly, retrospectively, accurately, and easily in FFPE tissues.

CONCLUSIONS AND FUTURE PROSPECTS

Factors that influence the sensitivity of ISH include the type, quality, and method of detection of the hybridization probe, the effects of tissue fixation on target mRNA preservation and accessibility to probe, the efficiency of hybrid formation, the stability of hybrids formed *in situ* during posthybridization treatments, and background noise masking or confusing the hybridization signals. Strategies designed to optimize these variables can improve the sensitivity and specificity of ISH. Reliable detection and localization of mRNAs using ISH in histological sections require the use of optimally fixed tissues to minimize nucleic acid degradation. RNA degradation in fresh tissues is well known.

Storage of paraffin sections of formalin-fixed tissues for weeks or longer prior to ISH diminishes the intensity of the radioactive signal. The decrease in signal intensity is not dependent on the type of the tissue. Diminished signal intensity is more pronounced in the case of mRNA present in abundant quantity. Such decrease is less consistent for mRNA present in low to moderate amounts. For these reasons, it is recommended to use freshly prepared sections of formalin-fixed tissues. If storage is unavoidable, slides should be stored at −70°C. Antigenicity is also diminished in paraffin sections stored for prolonged periods at room temperature.

The relative insensitivity of nonisotopic ISH has been cited as an important limiting factor in many applications, especially in diagnostic pathology.[23–33] Improving the sensitivity of ISH has been a goal of many investigators. The two general approaches to achieve this are amplification of target sequences and signal enhancement. Currently, nucleic acid amplification has become both a necessity and a routine procedure in many aspects of molecular biology. Although extremely versatile and sensitive, *in situ* PCR has limitations such as poor preservation of morphologies of some biological structures and lack of reliability as a quantitative method. Clinical applications of *in situ* PCR must await the resolution of some of the current limitations.

Isothermal methods such as IS-3SR and RCA are extremely sensitive qualitative methods. Amplifying the signal rather than the target or probe is suitable for quantitative interpretations, as in bDNA ISH, which addresses both the detection and quantification of the target, providing valuable information in the molecular diagnosis of pathogens. CARD with a combination of enzyme and substrate can generate and deposit large numbers of molecules at a detection site. Therefore, CARD ISH results in considerable improvement in sensitivity. In addition, brighter fluorescent polystyrene microspheres and enzyme-labeled fluorescence signal amplification technologies now in development may improve detection in the future.[151]

One new approach has been the use of molecular beacons (MBs) for visualizing real-time mRNA hybridization in single living cells.[152] MBs represent a new class of nucleic acid probes that become fluorescent when they bind to a complementary sequence. Because of the sequence specificity of nucleic acid hybridization and marked sensitivity, MBs can be used to detect low copy numbers of DNA or RNA molecules.[153]

More sensitive techniques for bright-field detection of hybridization products are being developed. A bright-field assay has been recently reported for assessment of Her-2/neu gene amplification using Gold Enhance gold-based autometallography, CARD, and a biotinylated labeled probe that approached the sensitivity of the FISH technique.[154] These new developments indicate that highly sensitive and specific probes and amplification systems for ISH will continue to improve analyses.

REFERENCES

1. Jin L, Lloyd RV. *In situ* hybridization: methods and applications. *J Clin Lab Anal* 1997; 11 (1): 2–9.
2. Jin L, Qian X, Lloyd RV. *In situ* hybridization: detection of DNA and RNA, in Lloyd RV, Ed. *Morphology Methods: Cell and Molecular Biology Techniques.* Totowa: Humana Press; 2001. pp. 27–47.
3. Hougaard DM, Hansen H, Larsson LI. Nonradioactive *in situ* hybridization for mRNA with emphasis on the use of oligodeoxynucleotide probes. *Histochem Cell Biol* 1997, Oct–Nov; 108 (4–5): 335–44.
4. Raap AK. Advances in fluorescence *in situ* hybridization. *Mutat Res* 1998, May; 400 (1–2): 287–98.
5. Wilcox JN. Fundamental principles of *in situ* hybridization. *J Histochem Cytochem* 1993; 41 (12): 1725–1733.
6. Denijn M, Schuurman HJ, Jacobse KC, De Weger RA. *In situ* hybridization: a valuable tool in diagnostic pathology. *APMIS* 1992; 100 (8): 669–681.
7. Speel EJ. Detection and amplification systems for sensitive, multiple-target DNA and RNA *in situ* hybridization: looking inside cells with a spectrum of colors. *Histochem Cell Biol* 1999; 112 (2): 89–113.
8. Hayden RT, Qian X, Roberts GD, Lloyd RV. *In situ* hybridization for the identification of yeast-like organisms in tissue section. *Diagn Mol Pathol* 2001;10 (1): 15–23.
9. Montone KT, Litzky LA. Rapid method for detection of Aspergillus 5S ribosomal RNA using a genus-specific oligonucleotide probe. *Am J Clin Pathol* 1995; 103 (1): 48–51.
10. Hayden RT, Uhl JR, Qian X, Hopkins MK, Aubry MC, Limper AH, Lloyd RV, Cockerill FR. Direct detection of *Legionella* species from bronchoalveolar lavage and open lung biopsy specimens: comparison of LightCycler PCR, *in situ* hybridization, direct fluorescence antigen detection, and culture. *J Clin Microbiol* 2001; 39 (7): 2618–2626.
11. Pollanen R, Vuopala S, Lehto VP. Detection of human papillomavirus infection by nonisotopic *in situ* hybridisation in condylomatous and CIN lesions. *J Clin Pathol* 1993; 46 (10): 936–939.
12. Herrington CS, Anderson SM, Graham AK, McGee JO. The discrimination of high-risk HPV types by *in situ* hybridization and the polymerase chain reaction. *Histochem J* 1993; 25 (3): 191–198.
13. Ambinder RF, Mann RB. Epstein–Barr-encoded RNA *in situ* hybridization: diagnostic applications. *Hum Pathol* 1994; 25 (6): 602–605.
14. Anagnostopoulos I, Hummel M. Epstein–Barr virus in tumours. *Histopathology* 1996; 29 (4): 297–315.
15. Qian X, Guerrero RB, Plummer TB, Alves VF, Lloyd RV. Detection of hepatitis C virus RNA in formalin-fixed paraffin-embedded sections with digoxigenin-labeled cRNA probes. *Diagn Mol Pathol* 2004; 13 (1): 9–14.
16. Komminoth P, Merk FB, Leav I, Wolfe HJ, Roth J. Comparison of ^{35}S- and digoxigenin-labeled RNA and oligonucleotide probes for *in situ* hybridization. Expression of mRNA of the seminal vesicle secretion protein II and androgen receptor genes in the rat prostate. *Histochemistry* 1992; 98 (4): 217–228.
17. Dagerlind A, Friberg K, Bean AJ, Hokfelt T. Sensitive mRNA detection using unfixed tissue: combined radioactive and nonradioactive *in situ* hybridization histochemistry. *Histochemistry* 1992; 98 (1): 39–49.
18. Miller MA, Kolb PE, Raskind MA. A method for simultaneous detection of multiple mRNAs using digoxigenin and radioisotopic cRNA probes. *J Histochem Cytochem* 1993; 41 (12): 1741–1750.
19. Steel JH, Jeffery RE, Longcroft JM, Rogers LA, Poulsom R. Comparison of isotopic and nonisotopic labelling for *in situ* hybridisation of various mRNA targets with cRNA probes. *Eur J Histochem* 1998; 42 (2): 143–150.

20. Qian X, Bauer RA, Xu HS, Lloyd RV. *In situ* hybridization detection of calcitonin mRNA in routinely fixed, paraffin-embedded tissue sections: a comparison of different types of probes combined with tyramide signal amplification. *Appl Immunohistochem Mol Morphol* 2001; 9 (1): 61–69.

21. Szakacs JG, Livingston SK. mRNA *in situ* hybridization using biotinylated oligonucleotide probes: implications for the diagnostic laboratory. *Ann Clin Lab Sci* 1994; 24 (4): 324–338.

22. Panoskaltsis–Mortari A, Bucy RP. *In situ* hybridization with digoxigenin-labeled RNA probes: facts and artifacts. *Biotechniques* 1995; 18 (2): 300–307.

23. Tata AM. An *in situ* hybridization protocol to detect rare mRNA expressed in neural tissue using biotin-labeled oligonucleotide probes. *Brain Res Protoc* 2001; 6 (3): 178–184.

24. Durrant I, Fowler S. Nonradioactive oligonucleotide probe labeling. *Methods Mol Biol* 1994; 31: 163–175.

25. Lloyd RV, Jin L. *In situ* hybridization analysis of chromogranin A and B mRNAs in neuroendocrine tumors with digoxigenin-labeled oligonucleotide probe cocktails. *Diagn Mol Pathol* 1995; 4 (2): 143–151.

26. Roche Applied Science. Procedures for labeling DNA, RNA, and oligonucleotides with digoxigenin, biotin or fluorochromes, in *Nonradioactive* in Situ *Hybridization Application Manual,* 2nd ed. Indianapolis: Roche Applied Science; 1996. pp. 34–56.

27. Qian X, Lloyd RV. Recent developments in signal amplification methods for *in situ* hybridization. *Diagn Mol Pathol* 2003; 12 (1): 1–13.

28. Komminoth P, Werner M. Target and signal amplification: approaches to increase the sensitivity of *in situ* hybridization. *Histochem Cell Biol* 1997; 108 (4–5): 325–333.

29. Jin, L., Qian, X., Lloyd, R.V. Comparison of mRNA expression detected by *in situ* polymerase chain reaction and *in situ* hybridization in endocrine cells. *Cell Vision* 1995; 2: 314–321.

30. Plummer TB, Sperry AC, Xu HS, Lloyd RV. *In situ* hybridization detection of low copy nucleic acid sequences using catalyzed reporter deposition and its usefulness in clinical human papillomavirus typing. *Diagn Mol Pathol* 1998; 7 (2): 76–84.

31. Speel EJ, Hopman AH, Komminoth P. Amplification methods to increase the sensitivity of *in situ* hybridization: play card(s). *J Histochem Cytochem* 1999; 47 (3): 281–288.

32. Speel EJ, Saremaslani P, Roth J, Hopman AH, Komminoth P. Improved mRNA *in situ* hybridization on formaldehyde-fixed and paraffin-embedded tissue using signal amplification with different haptenized tyramides. *Histochem Cell Biol* 1998; 110 (6): 571–577.

33. Hopman AH, Ramaekers FC, Speel EJ. Rapid synthesis of biotin-, digoxigenin-, trinitrophenyl-, and fluorochrome-labeled tyramides and their application for *in situ* hybridization using CARD amplification. *J Histochem Cytochem* 1998; 46 (6): 771–777.

34. Gall JG, Pardue ML. Formation and detection of RNA–DNA hybrid molecules in cytological preparations. *Proc Natl Acad Sci USA* 1969; 63 (2): 378–383.

35. Biffo S. *In situ* hybridization: optimization of the techniques for collecting and fixing the specimens. *Liver* 1992; 12 (Pt 2): 227–229.

36. Lisowski AR, English ML, Opsahl AC, Bunch RT, Blomme EA. Effect of the storage period of paraffin sections on the detection of mRNAs by *in situ* hybridization. *J Histochem Cytochem* 2001; 49 (7): 927–928.

37. Weiss LM, Chen YY. Effects of different fixatives on detection of nucleic acids from paraffin-embedded tissues by in situ hybridization using oligonucleotide probes. *J Histochem Cytochem* 1991; 39 (9): 1237–1242.

38. Oliver KR, Heavens RP, Sirinathsinghji DJ. Quantitative comparison of pretreatment regimens used to sensitize *in situ* hybridization using oligonucleotide probes on paraffin-embedded brain tissue. *J Histochem Cytochem* 1997; 45 (12): 1707–1713.

39. Nitta H, Kishimoto J, Grogan TM. Application of automated mRNA *in situ* hybridization for formalin-fixed, paraffin-embedded mouse skin sections: effects of heat and enzyme pretreatment on mRNA signal detection. *Appl Immunohistochem Mol Morphol* 2003; 11 (2): 183–187.

40. Chin SF, Daigo Y, Huang HE, Iyer NG, Callagy G, Kranjac T, Gonzalez M, Sangan T, Earl H, Caldas C. A simple and reliable pretreatment protocol facilitates fluorescent *in situ* hybridisation on tissue microarrays of paraffin wax embedded tumour samples. *Mol Pathol* 2003; 56 (5): 275–279.

41. Carr NJ, Talbot IC. *In situ* end labelling: effect of proteolytic enzyme pretreatment and hydrochloric acid. *Mol Pathol* 1997; 50 (3): 160–163.

42. Ensinger C, Obrist P, Rogatsch H, Ramoni A, Schafer G, Mikuz G. Improved technique for investigations on archival formalin-fixed, paraffin-embedded tumors by interphase *in situ* hybridisation. *Anticancer Res* 1997; 17 (6D): 4633–4637.

43. Oliver KR, Wainwright A, Heavens RP, Sirinathsinghji DJ. Retrieval of cellular mRNA in paraffin-embedded human brain using hydrated autoclaving. *J Neurosci Methods* 1997; 77 (2): 169–174.

44. Sperry A, Jin L, Lloyd RV. Microwave treatment enhances detection of RNA and DNA by *in situ* hybridization. *Diagn Mol Pathol* 1996; 5 (4): 291–296.

45. Emanuel JR. Simple and efficient system for synthesis of non-radioactive nucleic acid hybridization probes using PCR. *Nucleic Acids Res* 1991; 19 (10): 2790.

46. Temsamani J, Agrawal S. Enzymatic labeling of nucleic acids. *Mol Biotechnol* 1996; 5 (3): 223–232.

47. Witkiewicz H, Bolander ME, Edwards DR. Improved design of riboprobes from pBluescript® and related vectors for *in situ* hybridization. *Biotechniques* 1993; 14 (3): 458–463.

48. Mitsuhashi M. Basic requirements for designing optimal oligonucleotide probe sequences, technical report, part 1. *J Clin Lab Anal* 1996; 10 (5): 277–284.

49. Baskin DG, Stahl WL. Fundamentals of quantitative autoradiography by computer densitometry for *in situ* hybridization, with emphasis on ^{33}P. *J Histochem Cytochem* 1993; 41 (12): 1767–1776.

50. Durrant I, Dacre B, Cunningham M. Evaluation of novel formulations of ^{35}S- and ^{33}P-labeled nucleotides for *in situ* hybridization. *Histochem J* 1995; 27 (1): 89–93.

51. Eizuru Y, Minamishima Y, Matsumoto T, Hamakado T, Mizukoshi M, Nabeshima K, Koono M, Yoshida A, Yoshida H, Kikuchi M. Application of *in situ* hybridization with a novel phenytoin-labeled probe to conventional formalin-fixed, paraffin-embedded tissue sections. *J Virol Methods* 1995; 52 (3): 309–316.

52. Long S, Rebagliati M. Sensitive two-color whole-mount *in situ* hybridizations using digoxygenin- and dinitrophenol-labeled RNA probes. *Biotechniques* 2002; 32 (3): 494–496.

53. Harper SJ, Bailey E, McKeen CM, Stewart AS, Pringle JH, Feehally J, Brown T. A comparative study of digoxigenin, 2,4-dinitrophenyl, and alkaline phosphatase as deoxyoligonucleotide labels in non-radioisotopic *in situ* hybridisation. *J Clin Pathol* 1997; 50 (8): 686–690.

54. Marquez A, Wu R, Zhao J, Tao J, Shi Z. Evaluation of epidermal growth factor receptor (EGFR) by chromogenic *in situ* hybridization (CISH™) and immunohistochemistry (IHC) in archival gliomas using bright field microscopy. *Diagn Mol Pathol* 2004; 13 (1): 1–8.

55. Wixom CR, Albers EA, Weidner N. Her2 amplification: correlation of chromogenic *in situ* hybridization with immunohistochemistry and fluorescence *in situ* hybridization. *Appl Immunohistochem Mol Morphol* 2004; 12 (3): 248–251.

56. Trabandt A, Gay RE, Sikhatme VP, Gay S. Enzymatic detection systems for non-isotopic *in situ* hybridization using biotinylated cDNA probes. *Histochem J* 1995; 27 (4): 280–290.

57. Koller U, Stockinger H, Majdic O, Bettelheim P, Knapp W. A rapid and simple immunoperoxidase staining procedure for blood and bone marrow samples. *J Immunol Methods* 1986; 86 (1): 75–81.

58. Kriegsmann J, Keyszer G, Geiler T, Gay RE, Gay S. A new double labeling technique for combined *in situ* hybridization and immunohistochemical analysis. *Lab Invest* 1994; 71 (6): 911–917.

59. Trembleau A, Roche D, Calas A. Combination of non-radioactive and radioactive *in situ* hybridization with immunohistochemistry: a new method allowing the simultaneous detection of two mRNAs and one antigen in the same brain tissue section. *J Histochem Cytochem* 1993; 41 (4): 489–498.

60. Young WS 3rd, Hsu AC. Observations on the simultaneous use of digoxigenin- and radio-labeled oligodeoxyribonucleotide probes for hybridization histochemistry. *Neuropeptides* 1991; 18 (2): 75.

61. Hopman AH, Claessen S, Speel EJ. Multicolour bright field *in situ* hybridisation on tissue sections. *Histochem Cell Biol* 1997; 108 (4–5): 291–298.

62. Sibon OC, Cremers FF, Humbel BM, Boonstra J, Verkleij AJ. Localization of nuclear RNA by pre- and post-embedding *in situ* hybridization using different gold probes. *Histochem J* 1995; 27 (1): 35–45.

63. Siffroi JP, Alfonsi MF, Dadoune JP. Co–localization of HP1 and TP1 transcripts in human spermatids by double electron microscopy *in situ* hybridization. *Int J Androl* 1999; 22 (2): 83–90.

64. Isola J, Tanner M, Forsyth A, Cooke TG, Watters AD, Bartlett JM. Interlaboratory comparison of HER–2 oncogene amplification as detected by chromogenic and fluorescence *in situ* hybridization. *Clin Cancer Res* 2004; 10 (14): 4793–4798.

65. Fan YS, Rizkalla K. Comprehensive cytogenetic analysis including multicolor spectral karyotyping and interphase fluorescence *in situ* hybridization in lymphoma diagnosis: summary of 154 cases. *Cancer Genet Cytogenet* 2003; 143 (1): 73–79.

66. Dirks RW, Molenaar C, Tanke HJ. Visualizing RNA molecules inside the nucleus of living cells. *Methods* 2003; 29 (1): 51–57.

67. Femino AM, Fay FS, Fogarty K, Singer RH. Visualization of single RNA transcripts *in situ*. *Science* 1998; 280 (5363): 585–590.

68. Breininger JF, Baskin DG. Fluorescence *in situ* hybridization of scarce leptin receptor mRNA using the enzyme-labeled fluorescent substrate method and tyramide signal amplification. *J Histochem Cytochem* 2000; 48 (12): 1593–1599.

69. Schriml LM, Padilla–Nash HM, Coleman A, Moen P, Nash WG, Menninger J, Jones G, Ried T, Dean M. Tyramide signal amplification (TSA)–FISH applied to mapping PCR-labeled probes less than 1 kb in size. *Biotechniques* 1999; 27 (3): 608–613.

70. Zehbe I, Hacker GW, Su H, Hauser–Kronberger C, Hainfeld JF, Tubbs R. Sensitive *in situ* hybridization with catalyzed reporter deposition, streptavidin–nanogold, and silver acetate autometallography: detection of single-copy human papillomavirus. *Am J Pathol* 1997; 150 (5): 1553–1561.

71. Florentine BD, Sanchez B, Raza A, Frankel K, Martin SE, Kovacs B, Felix JC. Detection of hyperdiploid malignant cells in body cavity effusions by fluoresence *in situ* hybridization on ThinPrep slides. *Cancer* 1997; 81 (5): 299–308.

72. Choi YJ. *In situ* hybridization using a biotinylated DNA probe on formalin-fixed liver biopsies with hepatitis B virus infections: *in situ* hybridization superior to immunochemistry. *Mod Pathol* 1990; 3 (3): 343–347.

73. Pringle JH, Primrose L, Kind CN, Talbot IC, Lauder I. *In situ* hybridization demonstration of polyadenylated RNA sequences in formalin-fixed paraffin sections using a biotinylated oligonucleotide poly d(T) probe. *J Pathol* 1989; 158 (4): 279–286.

74. Yoshii A, Koji T, Ohsawa N, Nakane PK. *In situ* localization of ribosomal RNAs is a reliable reference for hybridizable RNA in tissue sections. *J Histochem Cytochem* 1995; 43 (3): 321–327.

75. Yulug IG, Yulug A, Fisher EM. The frequency and position of Alu repeats in cDNAs, as determined by database searching. *Genomics* 1995; 27 (3): 544–548.

76. Pagani A, Cerrato M, Bussolati G. Nonspecific *in situ* hybridization reaction in neuroendocrine cells and tumors of the gastrointestinal tract using oligonucleotide probes. *Diagn Mol Pathol* 1993; 2 (2): 125–130.

77. Nuovo GJ. Co–labeling using *in situ* PCR: a review. *J Histochem Cytochem* 2001; 49 (11): 1329–1339.

78. Martinez A, Miller MJ, Quinn K, Unsworth EJ, Ebina M, Cuttitta F. Non-radioactive localization of nucleic acids by direct *in situ* PCR and *in situ* RT-PCR in paraffin-embedded sections. *J Histochem Cytochem* 1995; 43 (8): 739–747.

79. Bagasra O, Seshamma T, Hansen J, Bobroski L, Saikumari P, Pestaner JP, Pomerantz RJ. Application of *in situ* PCR methods in molecular biology. I. Details of methodology for general use. *Cell Vision* 1994; 1: 324–335.

80. Alzahrani AJ, Vallely PJ, McMahon RF. Development of a novel nested *in situ* PCR-ISH method for detection of hepatitis C virus RNA in liver tissue. *J Virol Methods* 2002; 99 (1–2): 53–61.

81. Strappe PM, Wang TH, McKenzie CA, Lowrie S, Simmonds P, Bell JE. *In situ* polymerase chain reaction amplification of HIV-1 DNA in brain tissue. *J Virol Methods* 1998; 70 (2): 119–127.

82. Long AA, Komminoth P. *In situ* polymerase chain reaction and its applications to the study of endocrine diseases. *Endocr Pathol* 1995; 6: 167–171.

83. Embleton MJ, Gorochov G, Jones PT, Winter G. In-cell PCR from mRNA: amplifying and linking the rearranged immunoglobulin heavy and light chain V genes within single cells. *Nucleic Acids Res* 1992; 20 (15): 3831–3837.

84. Steele A, Uckan D, Steele P, Chamizo W, Washington K, Koutsonikolis A, Good RA. RT *in situ* PCR for the detection of mRNA transcripts of Fas-L in the immune-privileged placental environment. *Cell Vision* 1998; 5 (1): 13–19.

85. Hofler H, Putz B, Mueller JD, Neubert W, Sutter G, Gais P. *In situ* amplification of measles virus RNA by the self-sustained sequence replication reaction. *Lab Invest* 1995; 73 (4): 577–585.

86. Wilkens L, Tchinda J, Komminoth P, Werner M. Single- and double-color oligonucleotide primed *in situ* labeling (PRINS): applications in pathology. *Histochem Cell Biol* 1997; 108 (4–5): 439–446.

87. Coullin P, Roy L, Pellestor F, Candelier JJ, Bed–Hom B, Guillier–Gencik Z, Bernheim A. PRINS, the other *in situ* DNA labeling method useful in cellular biology. *Am J Med Genet* 2002; 107 (2): 127–135.

88. Komminoth P, Long AA. *In situ* polymerase chain reaction: overview of methods, applications and limitations of a new molecular technique. *Virchows Arch B Cell Pathol Incl Mol Pathol* 1993; 64 (2): 67–73.

89. Hacker GW. High performance nanogold–silver *in situ* hybridisation. *Eur J Histochem* 1998; 42 (2): 111–120.

90. Bushnell S, Budde J, Catino T, Cole J, Derti A, Kelso R, Collins ML, Molino G, Sheridan P, Monahan J, Urdea M. ProbeDesigner: for the design of probesets for branched DNA (bDNA) signal amplification assays. *Bioinformatics* 1999; 15 (5): 348–355.

91. Urdea MS. Branched DNA signal amplification. *Biotechnology* 1994; 12 (9): 926–928.

92. Player AN, Shen LP, Kenny D, Antao VP, Kolberg JA. Single-copy gene detection using branched DNA (bDNA) *in situ* hybridization. *J Histochem Cytochem* 2001; 49 (5): 603–612.

93. Baner J, Nilsson M, Mendel–Hartvig M, Landegren U. Signal amplification of padlock probes by rolling circle replication. *Nucleic Acids Res* 1998; 26 (22): 5073–5078.

94. Christian AT, Pattee MS, Attix CM, Reed BE, Sorensen KJ, Tucker JD. Detection of DNA point mutations and mRNA expression levels by rolling circle amplification in individual cells. *Proc Natl Acad Sci USA* 2001; 98 (25): 14238–14243.

95. Zhou Y, Calciano M, Hamann S, Leamon JH, Strugnell T, Christian MW, Lizardi PM. *In situ* detection of messenger RNA using digoxigenin-labeled oligonucleotides and rolling circle amplification. *Exp Mol Pathol* 2001; 70 (3): 281–288.

96. Murakami T, Hagiwara T, Yamamoto K, Hattori J, Kasami M, Utsumi M, Kaneda T. A novel method for detecting HIV–1 by non-radioactive *in situ* hybridization: application of a peptide nucleic acid probe and catalysed signal amplification. *J Pathol* 2001; 194 (1): 130–135.

97. Musiani M, Zerbini M, Venturoli S, Gentilomi G, Borghi V, Pietrosemoli P, Pecorari M, La Placa M. Rapid diagnosis of cytomegalovirus encephalitis in patients with AIDS using *in situ* hybridisation. *J Clin Pathol* 1994; 47 (10): 886–891.

98. Aksamit AJ Jr. Nonradioactive *in situ* hybridization in progressive multifocal leukoencephalopathy. *Mayo Clin Proc* 1993; 68 (9): 899–910.

99. Gaffey MJ, Ben–Ezra JM, Weiss LM. Herpes simplex lymphadenitis. *Am J Clin Pathol* 1991; 95 (5): 709–714.

100. Loriot MA, Marcellin P, Walker F, Boyer N, Degott C, Randrianatoavina I, Benhamou JP, Erlinger S. Persistence of hepatitis B virus DNA in serum and liver from patients with chronic hepatitis B after loss of HBsAg. *J Hepatol* 1997; 27 (2): 251–258.

101. Walters C, Powe DG, Padfield CJ, Fagan DG. Detection of parvovirus B19 in macerated fetal tissue using *in situ* hybridisation. *J Clin Pathol* 1997; 50 (9): 749–754.

102. Randhawa PS, Finkelstein S, Scantlebury V, Shapiro R, Vivas C, Jordan M, Picken MM, Demetris AJ. Human polyoma virus-associated interstitial nephritis in the allograft kidney. *Transplantation* 1999; 67 (1): 103–109.

103. zur Hausen H. Papillomavirus infections — a major cause of human cancers. *Biochim Biophys Acta* 1996; 1288 (2): F55–F78.

104. Stoler MH, Rhodes CR, Whitbeck A, Wolinsky SM, Chow LT, Broker TR. Human papillomavirus type 16 and 18 gene expression in cervical neoplasias. *Hum Pathol* 1992; 23 (2): 117–128.

105. Anderson SM, Brooke PK, Van Eyck SL, Noell H, Frable WJ. Distribution of human papillomavirus types in genital lesions from two temporally distinct populations determined by *in situ* hybridization. *Hum Pathol* 1993; 24 (5): 547–553.

106. Collina G, Rossi E, Bettelli S, Cook MG, Cesinaro AM, Trentini GP. Detection of human papillomavirus in extragenital Bowen's disease using *in situ* hybridization and polymerase chain reaction. *Am J Dermatopathol* 1995; 17 (3): 236–241.

107. Welt A, Hummel M, Niedobitek G, Stein H. Human papillomavirus infection is not associated with bronchial carcinoma: evaluation by *in situ* hybridization and the polymerase chain reaction. *J Pathol* 1997; 181 (3): 276–280.

108. Popper HH, el–Shabrawi Y, Wockel W, Hofler G, Kenner L, Juttner–Smolle FM, Pongratz MG. Prognostic importance of human papilloma virus typing in squamous cell papilloma of the bronchus: comparison of *in situ* hybridization and the polymerase chain reaction. *Hum Pathol* 1994; 25 (11): 1191–1197.

109. Gorgoulis VG, Zacharatos P, Kotsinas A, Kyroudi A, Rassidakis AN, Ikonomopoulos JA, Barbatis C, Herrington CS, Kittas C. Human papilloma virus (HPV) is possibly involved in laryngeal but not in lung carcinogenesis. *Hum Pathol* 1999; 30 (3): 274–283.

110. Hummel M, Anagnostopoulos I, Dallenbach F, Korbjuhn P, Dimmler C, Stein H. EBV infection patterns in Hodgkin's disease and normal lymphoid tissue: expression and cellular localization of EBV gene products. *Br J Haematol* 1992; 82 (4): 689–694.

111. Niedobitek G. Patterns of Epstein–Barr virus infection in non-Hodgkin's lymphomas. *J Pathol* 1995; 175 (3): 259–261.

112. Wu TC, Mann RB, Epstein JI, MacMahon E, Lee WA, Charache P, Hayward SD, Kurman RJ, Hayward GS, Ambinder RF. Abundant expression of EBER1 small nuclear RNA in nasopharyngeal carcinoma: a morphologically distinctive target for detection of Epstein–Barr virus in formalin-fixed paraffin-embedded carcinoma specimens. *Am J Pathol* 1991; 138 (6): 1461–1469.

113. Lau JY, Krawczynski K, Negro F, Gonzalez–Peralta RP. *In situ* detection of hepatitis C virus — a critical appraisal. *J Hepatol* 1996; 24 (2 Suppl): 43–51.

114. Chang M, Marquardt AP, Wood BL, Williams O, Cotler SJ, Taylor SL, Carithers RL Jr, Gretch DR. *In situ* distribution of hepatitis C virus replicative intermediate RNA in hepatic tissue and its correlation with liver disease. *J Virol* 2000; 74 (2): 944–955.

115. Balluz IM, Farrell MA, Kay E, Staunton MJ, Keating JN, Sheils O, Cosby SL, Mabruk MJ, Sheahan BJ, Atkins GJ. Colocalisation of human immunodeficiency virus and human cytomegalovirus infection in brain autopsy tissue from AIDS patients. *Ir J Med Sci* 1996; 165 (2): 133–138.

116. Jay V, Becker LE, Otsubo H, Cortez M, Hwang P, Hoffman HJ, Zielenska M. Chronic encephalitis and epilepsy (Rasmussen's encephalitis): detection of cytomegalovirus and herpes simplex virus 1 by the polymerase chain reaction and *in situ* hybridization. *Neurology* 1995; 45 (1): 108–117.

117. Iwa N, Sasaki M, Yutani C, Wakasa K. Detection of cytomegalovirus DNA in pulmonary specimens: confirmation by *in situ* hybridization in two cases. *Diagn Cytopathol* 1992; 8 (4): 357–360.

118. Sharma TM, Nadasdy T, Leech RW, Kingma DW, Johnson LD, Hanson–Painton O. *In situ* DNA hybridization study of 'primary' cytomegalovirus (CMV) oophoritis. *Acta Obstet Gynecol Scand* 1994; 73 (5): 429–431.

119. Colina F, Juca NT, Moreno E, Ballestin C, Farina J, Nevado M, Lumbreras C, Gomez–Sanz R. Histological diagnosis of cytomegalovirus hepatitis in liver allografts. *J Clin Pathol* 1995; 48 (4): 351–357.

120. Muir SW, Murray J, Farquharson MA, Wheatley DJ, McPhaden AR. Detection of cytomegalovirus in upper gastrointestinal biopsies from heart transplant recipients: comparison of light microscopy, immunocytochemistry, *in situ* hybridisation, and nested PCR. *J Clin Pathol* 1998; 51 (11): 807–811.

121. Remadi S, Finci V, Ismail A, Zacharie S, Vassilakos P. Herpetic endometritis after pregnancy. *Pathol Res Pract* 1995; 191 (1): 31–34.

122. Borisch B, Ellinger K, Neipel F, Fleckenstein B, Kirchner T, Ott MM, Muller–Hermelink HK. Lymphadenitis and lymphoproliferative lesions associated with the human herpes virus-6 (HHV-6). *Virchows Arch B Cell Pathol Incl Mol Pathol* 1991; 61 (3): 179–187.

123. Fillet AM, Reux I, Joberty C, Fournier JG, Hauw JJ, Le Hoang P, Bricaire F, Huraux JM, Agut H. Detection of human herpes virus 6 in AIDS-associated retinitis by means of *in situ* hybridization, polymerase chain reaction and immunohistochemistry. *J Med Virol* 1996; 49 (4): 289–295.

124. Boshoff C, Schulz TF, Kennedy MM, Graham AK, Fisher C, Thomas A, McGee JO, Weiss RA, O'Leary JJ. Kaposi's sarcoma-associated herpes virus infects endothelial and spindle cells. *Nat Med* 1995; 1 (12): 1274–1278.

125. Kobayashi TK. Comparison of immunocytochemistry and *in situ* hybridization in the cytodiagnosis of genital herpetic infection. *Diagn Cytopathol* 1992; 8 (1): 53–60.

126. Moter A, Gobel UB. Fluorescence *in situ* hybridization (FISH) for direct visualization of microorganisms. *J Microbiol Methods* 2000; 41 (2): 85–112.

127. Oliveira K, Procop GW, Wilson D, Coull J, Stender H. Rapid identification of *Staphylococcus aureus* directly from blood cultures by fluorescence *in situ* hybridization with peptide nucleic acid probes. *J Clin Microbiol* 2002; 40 (1): 247–251.

128. Bashir MS, Lewis FA, Quirke P, Lee A, Dixon MF. *In situ* hybridisation for the identification of *Helicobacter pylori* in paraffin wax-embedded tissue. *J Clin Pathol* 1994; 47 (9): 862–864.

129. Hayashi Y, Watanabe J, Nakata K, Fukayama M, Ikeda H. A novel diagnostic method of *Pneumocystis carinii*. *In situ* hybridization of ribosomal ribonucleic acid with biotinylated oligonucleotide probes. *Lab Invest* 1990; 63 (4): 576–580.

130. Patel AS, Hawkins AL, Griffin CA. Cytogenetics and cancer. *Curr Opin Oncol* 2000; 12 (1): 62–67.

131. Gozzetti A, Le Beau MM. Fluorescence *in situ* hybridization: uses and limitations. *Semin Hematol* 2000; 37 (4): 320–333.

132. Huang NF, Gupta M, Varghese S, Rao S, Luke S. Detection of numerical chromosomal abnormalities in epithelial ovarian neoplasms by fluorescence *in situ* hybridization (FISH) and a review of the current literature. *Appl Immunohistochem Mol Morphol* 2002; 10 (2): 187–193.

133. Liehr T, Starke H, Weise A, Lehrer H, Claussen U. Multicolor FISH probe sets and their applications. *Histol Histopathol* 2004; 19 (1): 229–237.

134. Qian X, Jin L, Shearer BM, Ketterling RP, Jalal SM, Lloyd RV. Molecular diagnosis of Ewing's sarcoma/primitive neuroectodermal tumor in formalin-fixed paraffin-embedded tissues by RT–PCR and fluorescence *in situ* hybridization. *Diagn Mol Pathol* 2005; 14 (1): 23–28.

135. Arnould L, Denoux Y, MacGrogan G, Penault–Llorca F, Fiche M, Treilleux I, Mathieu MC, Vincent–Salomon A, Vilain MO, Couturier J. Agreement between chromogenic *in situ* hybridisation (CISH) and FISH in the determination of HER2 status in breast cancer. *Br J Cancer* 2003; 88 (10): 1587–1591.

136. Lee C, Gisselsson D, Jin C, Nordgren A, Ferguson DO, Blennow E, Fletcher JA, Morton CC. Limitations of chromosome classification by multicolor karyotyping. *Am J Hum Genet* 2001; 68 (4): 1043–1047.

137. Bearden CE, Wang PP, Simon TJ. Williams syndrome cognitive profile also characterizes velocardio-facial/DiGeorge syndrome. *Am J Med Genet* 2002; 114 (6): 689–692.

138. Sandberg AA. Genetics of chondrosarcoma and related tumors. *Curr Opin Oncol* 2004; 16 (4): 342–354.

139. Swansbury J. Lymphoid disorders other than common acute lymphoblastic leukemia: background. *Methods Mol Biol* 2003; 220: 93–110.

140. Avet–Loiseau H, Li JY, Morineau N, Facon T, Brigaudeau C, Harousseau JL, Grosbois B, Bataille R. Monosomy 13 is associated with the transition of monoclonal gammopathy of undetermined significance to multiple myeloma. *Blood* 1999; 94 (8): 2583–2589.

141. Meier VS, Kuhne T, Jundt G, Gudat F. Molecular diagnosis of Ewing tumors: improved detection of EWS-FLI-1 and EWS-ERG chimeric transcripts and rapid determination of exon combinations. *Diagn Mol Pathol* 1998; 7 (1): 29–35.

142. Arvand A, Denny CT. Biology of EWS/ETS fusions in Ewing's family tumors. *Oncogene* 2001; 20 (40): 5747–5754.

143. DeLellis RA. *In situ* hybridization techniques for the analysis of gene expression: applications in tumor pathology. *Hum Pathol* 1994; 25 (6): 580–585.

144. Weiss LM, Movahed LA, Chen YY, Shin SS, Stroup RM, Bui N, Estess P, Bindl JM. Detection of immunoglobulin light-chain mRNA in lymphoid tissues using a practical *in situ* hybridization method. *Am J Pathol* 1990; 137 (4): 979–988.

145. Jin L, Chandler WF, Smart JB, England BG, Lloyd RV. Differentiation of human pituitary adenomas determines the pattern of chromogranin/secretogranin messenger ribonucleic acid expression. *J Clin Endocrinol Metab* 1993; 76 (3): 728–735.

146. Lloyd RV. Molecular probes and endocrine diseases. *Am J Surg Pathol* 1990; 14, Suppl 1: 34–44.

147. Kriegsmann J, Keyszer GM, Geiler T, Brauer R, Gay RE, Gay S. Expression of vascular cell adhesion molecule-1 mRNA and protein in rheumatoid synovium demonstrated by *in situ* hybridization and immunohistochemistry. *Lab Invest* 1995; 72 (2): 209–214.

148. Muskhelishvili L, Latendresse JR, Kodell RL, Henderson EB. Evaluation of cell proliferation in rat tissues with BrdU, PCNA, Ki-67 (MIB-5) immunohistochemistry and *in situ* hybridization for histone mRNA. *J Histochem Cytochem* 2003; 51 (12): 1681–1688.

149. Arakura N, Hayama M, Honda T, Matsuzawa K, Akamatsu T, Ota H. Histone H3 mRNA *in situ* hybridization for identifying proliferating cells in human pancreas, with special reference to the ductal system. *Histochem J* 2001; 33 (3): 183–191.

150. Kotelnikov V, Cass L, Coon JS, Spaulding D, Preisler HD. Accuracy of histone H3 messenger RNA *in situ* hybridization for the assessment of cell proliferation in human tissues. *Clin Cancer Res* 1997; 3 (5): 669–673.

151. Hakala H, Heinonen P, Iitia A, Lonnberg H. Detection of oligonucleotide hybridization on a single microparticle by time-resolved fluorometry: hybridization assays on polymer particles obtained by direct solid phase assembly of the oligonucleotide probes. *Bioconjug Chem* 1997; 8 (3): 378–384.

152. Antony T, Subramaniam V. Molecular beacons: nucleic acid hybridization and emerging applications. *J Biomol Struct Dyn* 2001; 19 (3): 497–504.

153. Yao G, Fang X, Yokota H, Yanagida T, Tan W. Monitoring molecular beacon DNA probe hybridization at the single molecule level. *Chemistry* 2003; 9 (22): 5686–5692.

154. Tubbs R, Pettay J, Skacel M, Powell R, Stoler M, Roche P, Hainfeld J. Gold–facilitated *in situ* hybridization: a bright field autometallographic alternative to fluorescence *in situ* hybridization for detection of HER–2/neu gene amplification. *Am J Pathol* 2002; 160 (5): 1589–1595.

6 Detection Techniques for Single Nucleotide Polymorphisms

W. Mathias Howell, Johan Stenberg, Chatarina Larsson, Mats Nilsson, and Ulf Landegren

CONTENTS

INTRODUCTION

Today's physicians have an expanding repertoire of molecular analytical options to assist in diagnosis and treatment of human disease. Medical research strives to provide tools to optimize patient care according to a variety of molecular measures including variations found in patients' genomes. As more is understood about how subtle genetic variations affect disease states and responses to drug treatment, the more important it becomes to extract genetic information in an easy and inexpensive way. An impressive number of alternative technologies for collecting genotypic data from patient samples currently exists; however, most of the techniques can be clearly understood based on a few basic principles.

This chapter will discuss in some detail methods for detection of single nucleotide polymorphisms (SNPs). For completeness we start with a basic description including SNP abundance and distribution, followed by a brief introduction of concepts such as haplotypes and linkage disequilibrium. The underlying anatomies of common SNP genotyping techniques are presented with emphasis on modes of allele discrimination, methods of signal detection, and the physical arrangements of the assays. We then present a summary of how different commercial platforms combine these three components along with the advantages and disadvantages of each strategy. Finally, we explore the current and potential applications of these technologies.

SINGLE NUCLEOTIDE POLYMORPHISM

A working definition of single nucleotide polymorphism or SNP (Figure 6.1) builds on several closely related terms encountered in the study of genetic variation. *Polymorphism* refers to any position (*locus*) in a genome where an alternative version of the sequence (*allele*) can be observed, on average, when comparing any 100 randomly chosen chromosomes from a given population. Genetic variants found less frequently, that is, loci where the frequency of the less common allele falls below this 1% criterion, are usually referred to as rare variants.[1] This frequency definition is a common source of confusion in discussions of genetics and human disease.

In a classical sense, mutations are aberrant DNA sequences that are necessary and sufficient to cause disease. The Online Mendelian Inheritance in Man (OMIM) database (http://www.ncbi.nlm.nih.gov/omim/) describes almost 1400 diseases with causes that fit this definition. In the context of genetic variation, however, a mutation is considered a DNA variant found infrequently in a population. Millions of recorded point mutations have no known links to diseases. Conversely, many polymorphisms are known to predispose individuals to diseases (reviewed by Zondervan and Cardon[2]). Thus, not all mutations are bad and not all polymorphisms are innocuous.

What distinguishes SNPs from other forms of genetic polymorphism is that the alleles at an SNP locus differ in DNA sequence by a single base. Although it is theoretically possible for an SNP locus to be tetra-allelic, the vast majority of SNPs have only two alleles. Curiously, roughly two thirds of human SNPs involve a single nucleotide substitution of a T for a C (Figure 6.2),

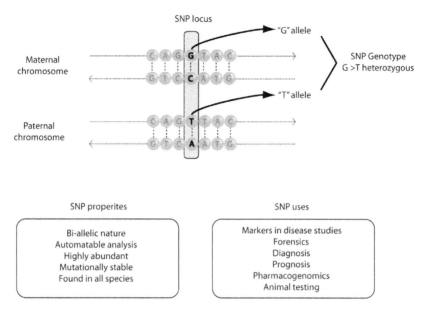

FIGURE 6.1 Fundamentals of single nucleotide polymorphisms (SNPs). A locus is a particular position in the genome. An allele is the particular nucleotide present at the locus. A genotype is the combination of alleles present in an individual.

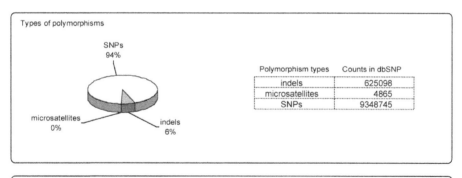

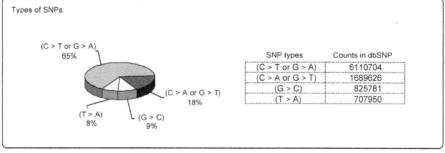

FIGURE 6.2 Summary of polymorphism data as present in dbSNP (build 124) for SNPs, insertions and deletions (indels), and microsatellites. For each section, distributions are represented in a pie chart, and the actual numbers presented in the adjacent table. Under types of SNPs, only four types of substitutions are presented. The reason for combining (C > T) and (G > A) polymorphisms is that in genomic DNA, a substitution of C for a T in one strand results in the G to A substitution in the complementary strand. Thus, the counts for (C > T) and (G > A) are combined to represent type of polymorphism present rather than polarity of DNA strand upon which the SNP was discovered. Similar reasoning is applied to (C > A) and (G > T) substitutions.

probably arising through a de-amination mechanism known to occur frequently at CpG dinucle-otides.[3] The allele with a higher frequency is termed the *major allele*; the other is designated the *minor allele*. Because the human genome is diploid in nature, an individual can be either *homozy-gous* (having two copies of the same allele at an SNP locus) or *heterozygous* (having one copy of each). The particular combination of alleles that an individual has at a particular locus is called a *genotype*.

DISCOVERY, ABUNDANCE, AND DISTRIBUTION

With a complete sequence of the human genome as a reference,[4–6] major efforts have been asserted toward producing detailed maps of genome variation. Many SNPs have been defined by re-analysis of data generated from the human genome sequencing projects. Comparative analysis of DNA sequences from overlapping genomic clones,[5,7,8] shotgun sequencing assemblies,[7,9] and libraries of cDNA sequences[10–14] have been good sources for finding SNPs. A large contributor to the public discovery effort is the SNP Consortium or TSC (http://snp.cshl.org). Using some of the aforemen-tioned strategies, this coalition of industry and academic partners, in a single publication in 2001,[7] released 1.4 million putative SNPs into the public domain. To date, the number has increased to a total of 1.8 million.

These SNPs and additional polymorphisms from many other private and public groups can be found in an online database dbSNP (http://www.ncbi.nlm.nih.gov/projects/SNP/). This collection of sequence variants is freely accessible and contains information about SNPs and other types of genetic polymorphisms relevant to the study of human disease, for example, microsatellites and insertions/deletions (*indels*).[15] Single nucleotide polymorphisms are by far the most abundant (Figure 6.2). As of the writing of this chapter, there are almost 9.5 million SNP entries in dbSNP (build 124). Roughly half have been validated from at least two independent sources or using two genotyping methods. About a million of these SNP entries include allele frequency information from at least one population. HGVBase (http://hgvbase.cgb.ki.se)[16,17] is another online repository of polymorphisms that is highly curated and focuses on the relationship of phenotype and DNA variation. Additionally, a number of SNP and mutation databases centered on specific genes or diseases can also be found on the Web (Table 6.1).

An estimated 11 million SNPs are thought to reside in the human genome.[18] When averaged across the entire genome, the result is one SNP for every 300 bases. Discovery efforts indicate, however, that SNPs appear to vary widely in distribution among different genomic regions.[5] When SNPs are placed in the context of a chromosomal map, they occur roughly four times more frequently in non-coding sequences than in coding sequences.[19,20]

Within genes, there are fewer SNPs at splice junctions than in exons. Within exons, SNPs are much more likely to reflect "silent" or *synonymous* changes rather than *non-synonymous* changes that when translated would change the amino acid sequence of the encoded protein.[21–24] These trends make sense from a biological perspective, as sequence changes in introns and other non-coding regions are much less likely to adversely affect normal cell function than genetic variants that affect mRNA splicing or protein sequences. Other genes appear to be under diversifying selection, meaning that variation in gene sequence is biologically beneficial, resulting in greater numbers of polymorphisms within such genes. For example, the diversity of polymorphism found in the genes encoding the human leukocyte antigens (HLAs)[25] contributes to the repertoire of proteins and peptides that can be effectively recognized and presented during an immune response.

LINKAGE DISEQUILIBRIUM, HAPLOTYPES, AND HAPMAP PROJECT

The phenomenon of linkage disequilibrium (LD) has been widely discussed in the SNP literature.[26] Basically, LD is a statistical measure of relationships between alleles at neighboring polymorphic loci along a chromosome. Although alleles located close to one another tend to be co-inherited as

TABLE 6.1
Online Databases Containing Entries on Human Genetic Variation

Type of Database	Name	Description	Location
Public	dbSNP — a database of single nucleotide polymorphisms	Major online repository for DNA polymorphisms including SNPs, microsatellites, and small insertion/deletions	http://www.ncbi.nlm.nih.gov/SNP/
	HGVbase — human genome variation database	Large database of polymorphisms with emphasis on genotype/phenotype relationships	http://hgvbase.cgb.ki.se/
	GBrowse — SNP resource	Database of SNPs discovered by SNP Consortium, now hosted on International HapMap project website	http://www.hapmap.org/cgi-perl/gbrowse/gbrowse
	ALFRED — allele frequency database	Resource of gene frequency data on human populations supported by U.S. National Science Foundation	http://alfred.med.yale.edu/alfred/index.asp
	CEPH — genotype database	Database of genotypes for genetic markers typed on CEPH reference family resource for linkage mapping	http://www.cephb.fr/cephdb
	GeneSNPs	Database sponsored by Environmental Genome Project that integrates gene, sequence, and polymorphism data into individually annotated gene models	http://www.genome.utah.edu/genesnps/
Commercial	Perlegen genotype browser	Database of SNPs, linkage disequilibrium bins, and haplotype blocks across all three populations examined in the study by Hinds et al. [34]	http://genome.perlegen.com/browser/index.html
	Human SNP database	Online resource of SNPs created in collaboration with Affymetrix, Inc., sponsored and supported in part by pharmaceutical companies and the National Institute of Standards and Technology	http://www.broad.mit.edu/snp/human/
	SNPBrowser	Applied Biosystems software and database over SNPs integrated with assay information	www.allsnps.com/snpbrowser.
Mutation	HGMD — human gene mutation database	Database of various types of mutation within coding regions of human nuclear genes causing inherited disease	http://archive.uwcm.ac.uk/uwcm/mg/hgmd0.html
	Mutation database website	Database sponsored by HUGO Mutation Database Initiative (MDI) that hosts links to numerous mutation and DNA variation databases	http://www2.ebi.ac.uk/mutations/cotton/
	Genome Web — human mutation database	Extensive collection of links to human mutation and SNP databases	http://www.hgmp.mrc.ac.uk/GenomeWeb/human-gen-db-mutation.html
Clinical resource	EDDNAL — European Directory of DNA Diagnostic Laboratories	Directory of non-profit European laboratories offering diagnostic services for numerous diseases	http://www.eddnal.com/
	PharmGKB — pharmacogenetics and pharmacogenomics knowledge base	Integrated resource about how variation in human genes leads to variation in drug response	http://www.pharmgkb.org/

a unit, alleles at loci that are far apart on a chromosome tend to be shuffled as an effect of recombination during meiosis. LD can be thought of as a measurement of the probability that two or more alleles occur together because they are linked as compared to co-occurrence because of chance. There are several alternative methods for calculating LD based on different statistical models of inference.[27-33]

The existence of LD is the underlying reason that it is possible to identify discrete collections of alleles from nearby loci that tend to occur together. These clusters or combinations of alleles are termed *haplotypes*.[34,35] Since recombination is spotty and occurs with varying frequency along the chromosomes,[36-38] it creates many islands of chromosomal segments where LD is strong and clusters of common haplotypes can be identified.[39-43] These conserved regions are called haplotype blocks.

In addition to recombination, the sizes and compositions of haplotype blocks are also influenced by other forces such as natural selection or genetic drift.[2] A current international effort called the HapMap Project is underway to identify common haplotype blocks across genomes in four human populations.[44] Once completed, haplotype frequency and distribution information can be used to guide optimal selection of SNPs for studies designed to identify genetic factors influencing human disease.

SNPs IN INHERITED DISEASE

The demands on SNP genotyping techniques vary greatly, depending on intended applications, for example, studies designed to seek the genetic determinants of disease where hereditary components are implicated. A classic approach is to perform an association study (reviewed in Reference 45). In this design, a group of individuals affected with a disease of interest (cases), and a group of people from the general population (controls) are selected. The cases and controls are chosen to be as closely matched as possible to avoid stratification due to genetic and environmental differences that are irrelevant for the investigated trait.

One or more SNPs located in genes suspected to be biologically relevant for the disease are chosen and genotyping is performed in the two groups. Statistically significant differences in allele or genotype frequencies of the two groups are taken as evidence of disease association. A positive association can be attributed to either the tested SNP allele directly influencing the risk of disease or indirect action via linkage disequilibrium with a nearby disease-influencing genetic variant.[46] Typically, a limited number of SNPs are genotyped in a relatively small number of individuals.

Many common diseases have been studied with this traditional approach[47] — with precious little success.[48-50] One possible explanation is that many common diseases are caused by combinations of alleles at a number of loci that act in concert to influence disease. Under such conditions, analysis of individual loci would be insufficient to identify genetic risk factors of that disease. The common disease/common variant (CD/CV) theory[51] proposes that some or all common diseases are due to relatively few high frequency risk alleles present at a few major disease loci. Since common alleles are traceable through haplotypes, the CD/CV theory has inspired much of the work of the HapMap Project and fueled interest in the possibility of performing whole genome association studies.[52,53] Rather than relying on biological guesswork, this "hypothesis-free" approach involves performing a large number of association studies on SNPs covering the entire human genome. Information from the HapMap Project will be useful for such an approach; SNPs to include in the study could be chosen to scan most of the common haplotypes using the most informative SNPs.[54]

The sensibility of performing whole genome association studies is a source of great controversy,[55-58] but the demand for higher throughput–lower cost genotyping has made a dramatic impact on SNP profiling techniques. At least four companies have responded and are developing assays specifically for whole genome analysis: Affymetrix (Santa Clara, California, USA, http://www.affymetrix.com/), Centurion DNA Array; Illumina (San Diego, California, USA, http://www.illumina.com/), Linkage IV Panel; Perlegen (Santa Clara, California, USA, http://www.perlegen.com/), Whole Genome Scanning Collection; and ParAllele BioScience (South San Francisco, California, USA,

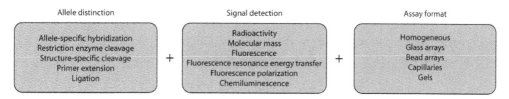

FIGURE 6.3 The anatomy of SNP genotyping protocols. Genotyping platforms often involve combinations of one or more of these allele distinction techniques, along with a signal detection strategy using one of these assay formats.

http://www.parallelebio.com), MegAllele™ Mapping Set. These systems allow the prospect of genotyping many thousands to possibly millions of SNPs simultaneously and in multiplex at a fraction of the previous cost.

SNP DETECTION TECHNIQUES

As noted earlier, genetic studies vary greatly in design and consequently impose very different requirements on SNP typing methods. Fortunately, researchers have a generous variety of innovative SNP genotyping solutions to choose from, each with its own advantages and disadvantages. All these techniques can be thought of as particular assemblies of three fundamental components (Figure 6.3):

1. Allele distinction
2. Signal detection
3. Assay format

In brief, allele distinction refers to the reaction principle by which alleles are distinguished. Signal detection is the way in which allele-specific products are visualized, and the assay format is simply the milieu in which the reactions take place. In the following sections, these key building blocks are described individually, setting the stage for an understanding of the architectures of commercial SNP genotyping systems.

In addition, the reaction format of most SNP detection techniques relies on amplification of the target molecule prior to the allele discrimination step, while other techniques amplify signals solely during the detection step. Target amplification serves to both reduce complexity of the genetic sample to be interrogated and to enhance subsequent detection signals via the increased quantity of targets. If allele distinction is performed directly on the genomic material, the intent of a signal amplification step is to increase the specific genotyping signal relative to background. In both cases, the use of the polymerase chain reaction (PCR) is by far the most common means of enhancing target concentration and/or signal intensity.

ALLELE DISTINCTION TECHNIQUES

A central aspect of any genotyping method is the technique used to distinguish different allelic variants. In the case of SNP genotyping, this requires that single base differences in a certain position can be identified and scored. The five fundamentally different techniques currently in use for this purpose are (1) allele-specific hybridization, (2) restriction enzyme cleavage, (3) enzymatic ligation, (4) primer extension by polymerases, and (5) structure-specific cleavage (Figure 6.4).

Allele-Specific Hybridization

Allele-specific hybridization techniques take advantage of the fact that the stability of the duplex formed between two nucleic acid molecules is sensitive to mismatches in the duplex where the two

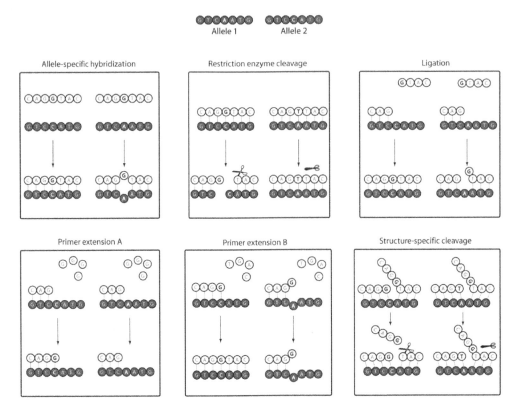

FIGURE 6.4 Principles of allele distinction. Target molecules are represented by dark circles, probe sequences by grey circles. Two different mechanisms for allele distinction by primer extension are presented. Primer extension A depicts the single base pair extension approach where polymerization results in incorporation of an allele-specific nucleotide. Primer extension B refers to allele-specific primer extension in which polymerization occurs only with primers that are complementary at the polymorphic position.

opposing bases do not form a Watson–Crick base-pair. This is utilized by having oligonucleotide probes form duplexes with the target molecules. Perfectly matched duplexes exhibit greater stability, and thus require higher temperatures to separate, than duplexes with mismatched positions. The existence of duplexes can be monitored by different means to reveal the alleles present in a sample.

Hybridization modifiers such as minor-groove binders (MGBs)[59] and locked nucleic acids (LNAs)[60] have been incorporated into a variety of genotyping platforms. These modifications increase the stability of DNA duplexes, thereby enabling the use of shorter oligonucleotides in analyses which in turn leads to greater differences in allele-specific hybridization reactions.

Restriction Digestion

Restriction enzymes cleave DNA molecules at specific sequences. The substitution of a single nucleotide for another frequently abolishes or creates restriction enzyme recognition sequences, thereby altering the ability of the restriction enzyme to cleave. Thus, single nucleotide differences may be detected as a change in the set of fragments resulting from a restriction enzyme cleavage reaction.[61]

Enzymatic Ligation

Ligation-based distinction techniques utilize the ability of DNA ligase enzymes to discriminate between matched and mismatched ligation substrates. In these methods, oligonucleotide probes are hybridized to the target molecule so that a 5' end and a 3' end are positioned immediately next to each other. If the ends are perfectly matched to the target, a ligase is able to join the two ends. If either end is mismatched (in particular the 3' end), ligation is greatly inhibited.[62]

Enzymatic Polymerization

Primer extension is used for allele distinction in several SNP genotyping strategies. A DNA polymerase is used to interrogate the base at the variable position. Two different approaches are possible here. In the first approach (Figure 6.4, Primer Extension A), a primer is designed to hybridize to the target with its 3' end complementary to a position immediately 3' of the nucleotide to be interrogated, so that the nucleotide incorporated in the primer by the polymerase will reflect the identity of the nucleotide present at the variable position.[63]

The nucleotide incorporated can then be determined by adding only one type of nucleotide at a time,[64,65] by using differentially labeled nucleotides[66,67] or by using a mix of nucleotides (some terminating or absent) so that the length of the extension product depends on what nucleotide can be incorporated.[68,69] In the second approach, allele-specific primers are designed to be extended only if they are correctly matched to the target at the variable position[70,71] or, in an alternative approach, only if mismatched.[72]

Structure-Specific Cleavage

The structure-specific cleavage or "invader" technique uses the capacities of certain classes of enzymes to cleave nucleic acid molecules at the site of a specific structure formed when a DNA strand hybridizes to a target nucleic acid molecule in such a way that its 3' end partially displaces another nucleic acid. The displaced strand is then cleaved at the position corresponding to the 3' end of the displacing or invading strand. The target strand and the displaced strand must be matched at the invasion site for this cleavage to occur, thus allowing allele discrimination.[73]

SIGNAL DETECTION

Rather than detecting hybridizing or enzymatically modified DNA directly, most SNP screening techniques measure a change in a surrogate molecule that is modified during or as a result of the allele distinction process. This is typically accomplished through addition of labels or tags to DNA probes, which can be followed through changes in one of many physical properties. Physical properties used in SNP detection strategies include radioactivity, molecular mass, and a variety of luminescence phenomena such as fluorescence, fluorescence resonance energy transfer (FRET), fluorescence polarization (FP), and chemiluminescence.

Radioactivity

One of the earliest approaches to detecting DNA molecules was through attachment of radioactive tags. Different isotopes such as ^3H, ^{32}P, or ^{35}S can be readily incorporated into DNA via polymerase or kinase reactions and the labeled products can be detected by autoradiography or other radiation measuring devices such as scintillation counters or phosphoimagers. Although radiolabeling is very sensitive, commercial platforms have adopted more benign alternatives for DNA detection.

Mass Detection

Mass spectrometry (MS) can be used to measure small differences in molecular weights of allele-specific reaction products. MS is also the only signal detection technique that does not require labeling of DNA prior to detection. The MS technique termed matrix-assisted laser desorption/ionization time-of-flight mass spectrometry (MALDI-TOF) has been successfully implemented in a number of SNP detection schemes (reviewed in Reference 74). In these approaches, small amounts of allele-specific reaction product and matrix are co-deposited onto a target plate. A short laser pulse serves to desorb the DNA sample/matrix into the gas phase and the ionized products are accelerated toward the detector.

The mass of the product can be inferred from the relative time it takes for the product to travel from the point of ionization to the detector. The resolution of this type of mass measure allows for even single base differences to be distinguished with ease in short DNA fragments. In addition, the dynamic range of MALDI-TOF has been extended by incorporation of mass tags into the allele-specific products.[75] Mass tags are small molecules with defined masses that can be designed to differ by only a few Daltons. Libraries of mass tags can be used to sort allele-specific products from several SNP loci in parallel.

Fluorescence Detection

Fluorescence detection involves the measurement of emitted light. In principle, a fluorescent dye or *fluorophore* is excited with an appropriate wavelength of light and induces an electron to enter an excited state. As the electron returns to the more stable ground state, excess energy is dissipated as fluorescence or longer-lived phosphorescence, depending on the type of fluorophore used. The emitted light is always of a lower energy/longer wavelength than that of the excitation source. Fluorophores having specific excitation and emission spectra that span a wide range of wavelengths are commonly available. By selecting fluorophores with dissimilar emission spectra, differently labeled DNA molecules can be resolved on the bases of their fluorescent signatures.

Many fluorescent detection reagents have been developed to stain or tag DNA molecules for SNP genotyping. For example, SYBR® Green I from Molecular Probes (Eugene, Oregon, USA, http://probes.invitrogen.com/) is an intercalating dye that undergoes a 10,000-fold increase in fluorescence when bound to double-stranded DNA. This dye has proven useful in assays that detect the presence or absence of DNA duplexes as part of a genotyping procedure.[76] Dyes with other desirable characteristics such as a large *Stokes shift* (difference between peak excitation and emission wavelengths), a high *extinction coefficient* (ability to capture excitation photons), and a high *quantum yield* (efficiency of converting between incoming and outgoing photons) are under continual development, exemplified by the Bodipy® and Alexa Flour® dye series also from Molecular Probes. Quantum dots or qdots represent a relatively new class of fluorescent labels composed of Zn or Cd nanoparticles encapsulated in a biocompatible, hydrophilic shell. Among the advantages of qdots over other fluorophores is that they are highly resistant to *photobleaching* (fluorophore destruction due to over-excitation), have relatively narrow emission spectra, and can all be excited at a fixed wavelength of light.

Fluorescence Resonance Energy Transfer

Fluorescence resonance energy transfer (FRET) is a phenomenon that occurs between appropriately matched pairs of fluorophores. The process involves the transfer of excitation energy from a *donor* fluorophore to an *acceptor* fluorophore located in close proximity. The donor does not emit fluorescence; rather it gives its energy to the acceptor fluorophore via dipole–dipole interaction. Upon energy transfer, the acceptor enters an excited state and can then dissipate the energy according to its own spectral properties. For FRET to occur, the emission spectrum of the donor must overlap the excitation spectrum of the acceptor.

In addition, the efficiency of energy transfer is greatly influenced by the spatial relationship between the donor and acceptor. In general, the closer the donor is to the acceptor, the more efficient is the energy transfer. Many SNP detection strategies have been developed in which a FRET reaction changes as a response to the allele discrimination event (reviewed in Reference 77; see Figure 6.5).

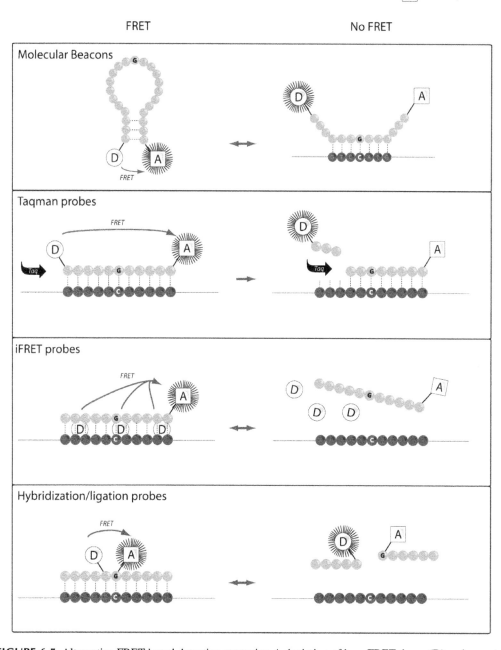

FIGURE 6.5 Alternative FRET-based detection strategies. A depiction of how FRET donor (D) and acceptor (A) molecules (sometimes referred to as fluorophore and quencher moieties) are separated as part of the allele discrimination event for a number of different platforms. Radiating lines depict fluorescent emission.

Most methods rely on covalent attachment or linkage of donor and acceptor fluorophores to a probe DNA molecule. One variation known as induced fluorescent energy transfer (iFRET™)[78] circumvents the necessity for physical attachment of the donor by employing the intercalating SYBR Green I dye as a donor.

Fluorescence Polarization

Another aspect of luminescence that has been used successfully for SNP detection is measuring the polarization of light emitted by fluorophores excited with polarized light. A stationary fluorophore, when excited with plane-polarized light, will emit fluorescence with the same plane of polarization. If, however, the fluorophore is free to tumble in solution before light is emitted, then emission polarization is reduced or lost. Reduction or increase of size of a fluorophore-labeled DNA molecule as part of an allele-specific reaction will influence the rate of tumbling, and thereby result in loss or gain of fluorescence polarization (FP) that is detectable with appropriate instrumentation (reviewed in Reference 79).

Chemiluminescence

Light produced via a chemical or biochemical reaction is referred to as chemiluminescence. The process often involves the enzymatic conversion of a substrate from one form to another, with part of the catalysis involving emission of light. A prominent example in SNP typing is the activity of the luciferase enzyme on the luciferin substrate that produces a greenish yellow fluorescent emission.

Assay Formats

Genetic material can be organized in a number of ways prior to analysis. Solution-based analyses can take place in tubes, microtiter plates, or capillaries. For assays involving immobilized genetic material, a wide variety of solid supports including various types of beads, glass slides, membranes, and gel matrices are available. Active movement of reactants between different formats often involves pipetting, electrophoresis, magnetism, or fluidic manipulation.

Since target amplification-based and signal amplification-based techniques, microarrays, and bead flow systems are discussed in depth in other chapters of this book, the following section will discuss only a few examples of the integration of these principles and components into the sorting procedures of various SNP genotyping techniques.

Sorting via Charge, Length, and Mass

DNA has an inherent negative charge due to its chemical composition. Thus, when placed in an electrical field, DNA tends to migrate toward the positive pole. This property is used in electrophoresis in which DNA molecules are electrically drawn through a gel matrix. The matrix, usually composed of agarose or polyacrylamide, serves to separate DNA fragments based on length and strand conformation. Slab-gel and capillary electrophoresis are common formats for many molecular biological techniques including SNP genotyping.

Similar to electrophoresis, mass spectrometry uses the charges on ionized DNA molecules to accelerate the molecules within an electric field. DNA fragments are further sorted and identified according to differences in molecular mass. In addition to relying on the physical properties inherent to DNA, tags influencing mass, length, or electrophoretic mobility can be attached to probes to further enhance separation of allele-specific signals.

Sorting via Arrays

Arranging clusters of DNA molecules in a two-dimensional space is a highly effective means of separating reactions and signals. In the case of microarrays, DNA can be synthesized separately and spotted to particular locations on a glass slide or synthesis can take place directly on a surface. The spots or *features* can be made to contain a series of probes that correspond to a number of different SNP loci and subsequent allele distinction occurs in these features. Alternatively, generic arrays can be created.

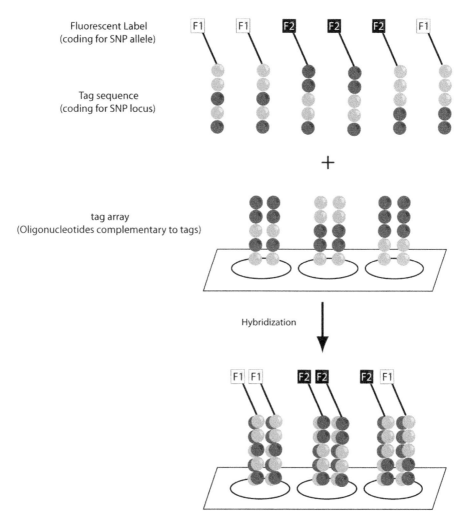

FIGURE 6.6 Principles of a tag array. Multiplex SNP detection techniques such as the SNPStream, Golden-Gate, and MIP assays result in production of DNA molecules bearing a tag sequence coding for the SNP locus, and a fluorescent label coding for the allele detected. These products are then hybridized to a complementary tag array and genotypes are scored by comparison of fluorescence at each feature of the tag array.

Instead of representations of target or probe sequences, a generic or tag array involves the immobilization of a series of unique sequences to the surface. Complementary sequences to the immobilized tags can be built into SNP detection probes. After an allele discrimination reaction, probes can be sorted on the tag array via selective hybridization of tail sequences to corresponding array tags (Figure 6.6). Thus, the same array can be used to assay many different SNPs as the specificity lies in design of the probe rather than the array.

Other types of supports for arrays are microparticles or beads. They are often small silica or plastic particles that have been surface functionalized for the attachment of DNA. Once DNA is stuck to a bead, it is possible to interrogate the beads in fluidic suspensions for determination of the identity of the genetic material on the bead. It is also possible to create a two-dimensional structure of beads by capture of the beads at appropriate locations on a surface. Both specific and tag array formats have been successfully implemented with beads.

Beads have a number of useful properties. They have large surface-to-mass ratios, facilitating attachment of large numbers of DNA molecules. Furthermore, reactions involving beads with immobilized DNA behave more like solution-phase assays than the linear arrays because the beads float around and come into contact with more solution. The beads can be labeled and aid in decoding the signals from the genotyping procedure.

Sorting via Optical Spectra

Sorting of mixtures of DNA can also be accomplished through spectral analysis. By labeling DNA with different fluorescent dyes, examination of a mixed sample over different portions of the emission spectra will allow inference as the sum totality of the contents of the tube. Otherwise referred to as *spectral multiplexing*, this type of signal separation permits simultaneous detection of several different allele-specific products. The use of optical spectral separation can be combined with microarrays or bead arrays to increase the information content of assays performed in those formats.

SNP GENOTYPING PLATFORMS

Drawing on the common elements of allele distinction, signal detection, and reaction format, a diversified assortment of SNP typing platforms have been commercialized and put into use in clinical labs around the world. A number of commercially available genotyping platforms are listed in Table 6.2. In the following section, a number of SNP typing strategies are described along with details regarding nuances of each technique. Variations of these techniques are also described to illustrate modifications commonly encountered in applying the various methods.

TaqMan® (Hybridization–FRET–Solution)

This assay from Applied Biosystems (Foster City, California, USA, http://www.appliedbiosystems.com/) involves a pair of PCR primers and two detection probes.[80,81] The PCR primers are designed to amplify a short DNA fragment containing the polymorphic locus. Two probes, one for each allele and labeled with different colored FRET pairs of fluorophores, are designed to hybridize across the polymorphic position. PCR conditions are chosen so that allele-specific hybridization occurs between probe and PCR target sequences. During the extension step of the PCR cycle, *Taq* polymerase adds nucleotides as it reads from the target strand. Upon encountering a correctly annealed probe, the nuclease activity of *Taq* polymerase cleaves the probe.

Digestion of the probe causes the separation of the FRET donor and acceptor, resulting in an increase of fluorescence from the donor. During PCR, this process is cycled, resulting in accumulation of donor fluorescence throughout the PCR process. The fractional point (cycle number) at which the fluorescence signal from the digested probe exceeds the assay background is termed the *Ct value*. Collection of data in real time allows for quantitative analysis over a broad range of target DNA concentrations, but for simple scoring of SNP genotypes, endpoint analysis is sufficient since the presence or absence of each allele can be inferred via comparison of total fluorescence from each probe.[82]

The major advantage of the TaqMan assay is its homogeneous format. After patient DNA and reaction mix are added to a reaction vessel, the tube is sealed and never has to be opened again, reducing the chances of cross-contamination. Also, Applied Biosystems has provided a number of conveniences for users. Hundreds of thousands of analytically pre-validated assays along with searchable mapping software over the chromosomal locations of the SNPs are available online.

A related strategy for homogeneous SNP detection using fluorescence involves molecular beacon (MB) probes.[83,84] An MB probe consists of an oligonucleotide bearing a member of a FRET pair at each end. The middle of the MB probe encodes the allele recognition sequence. The ends of the oligonucleotide are designed to be weakly self-complementary. In the absence of target, the

TABLE 6.2
Classification of Common SNP Genotyping Platforms

Allele Distinction	Detection Mechanism	Reaction Format	Platform and Provider	Website
Hybridization	FRET	Solution	TaqMan, Applied Biosystems	http://www.appliedbiosystems.com/
Enzymatic polymerization	Chemiluminescence	Solution	Pyrosequencing, Biotage	http://www.pyrosequencing.com/
Hybridization	iFRET	Array	DASH, DynaMetrix	http://www.dynametrix-ltd.com/
Structure-specific cleavage	Fluorescence	Solution	Invader, Third Wave Technologies	http://www.twt.com/
Enzymatic polymerization	Mass spectrosopy	Solution	Homogenous MassExtend, Sequenom	www.sequenom.com/
Enzymatic polymerization	Mass spectroscopy	Solution	MASScode, Qiagen	http://www1.qiagen.com/
Enzymatic polymerization	Fluorescence	Microarray	SNPstream, Beckman Coulter	http://www.beckman.com/
Enzymatic polymerization	Fluorescence	Microarray	APEX, Asper Biotech	http://www.asperbio.com/
Ligation	Fluorescence	Electrophoresis	SNPlex, Applied Biosystems	http://www.appliedbiosystems.com/
Enzymatic polymerization	Fluorescence	Bead array	GoldenGate, Illumina	http://www.illumina.com/
Ligation	Fluorescence	Microarray	MIP, ParAllele	http://www.parallelebio.com/
Ligation	Fluorescence	Electrophoresis	SNPWave, Keygene	http://www.keygene.com/
Hybridization	Fluorescence	Microarray	GeneChip, Affymetrix	http://www.affymetrix.com/
Enzymatic polymerization	Fluorescence	Beadarray	WGG, Illumina	http://www.illumina.com/

MB probe forms a stem–loop structure bringing the FRET pair close together and thus enabling energy transfer. Upon hybridization of the MB probe to a target, the stem structure is opened, separating the FRET pair. An increase in donor fluorescence is an indication of hybridization of the MB probe to its target.

PYROSEQUENCING (POLYMERIZATION–CHEMILUMINESCENCE–SOLUTION)

When a polymerase enzyme extends a primer by the incorporation of a nucleotide added as a nucleotide triphosphate, a pyrophosphate (PPi) molecule is released. By adding a single kind of deoxyribonucleotide triphosphate at a time to a primer extension reaction and monitoring the release of PPi once incorporation occurs, it is possible to extract sequence information from a sample. This is utilized in the pyrosequencing assay[85] currently marketed by Biotage (Uppsala, Sweden, http://www.biotage.com/) in which several cycles of nucleotide addition are performed and the PPi release is monitored in real time. The detection is based on the conversion of PPi to ATP and ATP to emitted light in an enzymatic chain reaction.

Different chemistries and ways of removing unincorporated deoxyribonucleotide triphosphates are in use.[86] Like many other assays, pyrosequencing requires amplification of the genomic target regions prior to analysis, and the assay has a very limited capacity for multiplexing. However, its ability to extract stretches of DNA sequence information from a sample makes it useful also for purposes other than SNP genotyping.

DASH™ (HYBRIDIZATION–FLUORESCENCE–PLATE ARRAY)

Dynamic allele-specific hybridization (DASH), supported by DynaMetrix Ltd. (Leicester, United Kingdom, www.dynametrix-ltd.com/) is based on analysis of DNA melting curves.[76,87] The assay involves PCR amplification of the SNP locus and subsequent immobilization to a well of a streptavidin-coated microtiter plate via a biotin attached to one of the PCR primers.

Following removal of excess PCR reagents and the non-biotinylated strand, a detection probe is hybridized over the polymorphic positions on the solid phase targets. The SYBR Green I dye is added in solution and allowed to bind to the DNA duplexes. The samples are heated through a temperature range while fluorescence is continually monitored. As the melting temperature (Tm) of each allele-specific duplex is surpassed, the DNA strands separate, releasing the dye and causing a dramatic drop in fluorescence. The DNA duplexes formed with targets containing the 100% complementary or "matched" allele are more stable than duplexes formed with the alternate or "mismatched" allele.

A plot of fluorescence versus temperature thus reveals the Tms of the duplexes in the sample, and genotypes are inferred by comparison of their relative Tms. The most recent rendition of the assay termed DASH-2[88] involves fluorescence detection via iFRET™ detection[78] as facilitated by attachment of a FRET acceptor to a detection probe and the use of a macroarray format created by robotic spotting of samples on a streptavidin-coated membrane.

INVADER® (CLEAVAGE–FLUORESCENCE–SOLUTION)

The Invader assay from Third Wave Molecular Diagnostics (Madison, Wisconsin, USA, http://www.twt.com/) utilizes structure-specific endonucleolytic cleavage to achieve SNP geno-typing.[89,90] A locus-specific invader oligonucleotide is used to partially displace or invade one of two allele-specific probes hybridized to a PCR-amplified target sequence. These allele-specific probes differ in the base positions opposite the SNP to be analyzed. The 3′ end of the locus-specific oligonucleotide displaces the allele-specific probe at the SNP position, and the specificity of the endonuclease results in the cleavage of only those probes that are correctly matched to the target.

The cleavage that occurs thus reflects the alleles of the SNP and is commonly detected by fluorescence. In the original set-up, the allele-specific probes are differentially labeled with fluorophores at their 5' ends and they have quenchers 3' of the cleavage site. Thus, cleavage results in a pattern of fluorescence that reveals the alleles of the SNP. Reaction conditions can be chosen so that probes will repeatedly hybridize to and dissociate from the target molecule, resulting in multiple cleavage events per target in an isothermal reaction.

In another set-up known as the serial invasive signal amplification reaction (SISAR), the allele-specific probes are unlabeled, and secondary detection probes are used instead.[91] The fragments cleaved off the primary detection probes are used to invade corresponding secondary probes that are labeled as previously. This procedure results in increased signal amplification and allows the detection of sequences directly from limited amounts of genomic DNA. Also, the expensive dually labeled probes are generic and a single pair of such probes may be used to genotype many SNPs, using many different pairs of primary probes. Use of one pair of detection probes, however, does limit the number of SNPs that can be scored in each genotyping reaction.

MassEXTEND® (Polymerization–Mass Spectrometry–Solution)

The homogeneous MassEXTEND (hME) assay[68] is an SNP genotyping strategy implemented on the MassARRAY™ system[92] of Sequenom (San Diego, California, USA, http://www.sequenom.com/). The hME assay involves initial PCR amplification of target DNA containing the polymorphic site. Following an enzymatic step to deactivate excess PCR reagents, a locus-specific oligonucleotide is added and annealed adjacent to the SNP position. DNA polymerase and a cocktail of three dNTPs (nucleotides) and one selected ddNTP (terminating nucleotide) are added to the sample. The annealed oligonucleotide acts as a primer and is extended by the DNA polymerase until a terminating nucleotide is incorporated. Allelic differences in the targets give rise to extension products of different lengths. After the extension reactions, the samples are purified with a resin and arrayed onto a silicon chip (384 SpectroCHIP™ bioarray) preloaded with matrix. The SpectroCHIP is then read by the MALDI-TOF instrument and genotype calls are recorded by measuring the masses of the extension products observed in the sample compared to a theoretical mass based on the allele-specific differences in length expected from the extension reaction.

The hME assay is used to illustrate SNP genotyping by multiple-base extension and MS detection as Sequenom offers a turn-key solution for performing SNP genotyping including chemistries, sample preparation robotics, MALDI-TOF instrumentation, and analysis software. The company also has validated assays for several hundred thousand SNPs and the current system offers up to 12-fold multiplexing.

The predecessor of the MassEXTEND is called the primer oligo base extension (PROBE) assay.[93] The systems differ in several ways. The PROBE procedure involves immobilization of PCR products to streptavidin-coated beads via a biotin label present in one PCR primer. The extension reaction occurs on the solid phase target and sample purification is via ethanol precipitation.

Another MS-based method called the GOOD assay[94,95] avoids the need for column purification or ethanol precipitation. A chemical modification and enzymatic digestion strategy enables preparation of samples for analysis in as few as three addition reactions.[96] The PROBE and GOOD assays and several other SNP genotyping platforms based on MALDI-TOF are extensively described in a recent review.[97]

Masscode™ (Polymerization–Mass Spectrometry–Solution)

The Masscode SNP allele discrimination assay[98] marketed by Qiagen (Venlo, Netherlands, http://www1.qiagen.com/) is based on allele-specific PCR and a series of Masscode tags for allele detection using MS. In principle, two rounds of PCR are performed; first a locus-specific PCR is used to amplify the SNP-containing fragment, then an allele-specific PCR incorporates a unique Masscode tag to each allele-specific product. The mass tags are photocleavable. After purification of the amplicons, the tags are released by treating with light and injected into a mass spectrometer. Alleles are scored via analysis of mass signatures. As of 2001, 30 unique Masscode tags had been characterized, enabling multiplex analysis of up to 15 SNPs.

SNPSTREAM® (POLYMERIZATION–FLUORESCENCE–MICROARRAY)

The GenomeLab™ SNPstream genotyping system of Beckman Coulter (Fullerton, California, USA, www.beckman.com/) is based on the SNP-IT™ assay[99,100] originally marketed by Orchid Cellmark (Princeton, New Jersey, http://www.orchid.com/). Essentially, the method entails single-base extension (SBE) or "minisequencing" followed by fluorescence detection on tag arrays. The assay procedure starts by multiplex PCR amplification of up to 12 different SNP loci. After purification, locus-specific primers are added and annealed adjacent to the SNP position. Each primer contains a unique tag sequence. A cyclic, allele-specific extension reaction ensues in which fluorescence-labeled ddNTPs are added to the extension primers.

The extension products are sorted via their tag sequences on a complementary tag array, followed by placement into an array scanner. Alleles are called by measuring the color and intensity of fluorescence at each position on the tag array. Currently the 4 × 4 tag arrays are printed in a 384-well format, increasing the potential throughput of SNP genotypes called per scan.

Another assay involving minisequencing on an array is the arrayed primer extension (APEX) assay[101] of Asper Biotech (Tartu, Estonia, http://www.asperbio.com/). Rather than using a tag array as described above, extension reactions are performed directly on probes immobilized to the microarray surface.

SNPLEX™ (LIGATION–FLUORESCENCE–ELECTROPHORESIS)

The SNPlex assay[102] recently released by Applied Biosystems is a multiplexed ligation-based assay performed on fragmented genomic DNA. The allele discrimination procedure basically involves differential ligation of one locus-specific oligonucleotide with a pair of allele-specific oligonucleotides. Each allele-specific oligonucleotide encodes a unique tag termed a ZipCode™. Following a clean-up step, successfully ligated products are amplified by PCR and immobilized to a streptavidin-coated plate via a biotin label on one of the PCR primers. A second clean-up step serves to remove excess PCR reagents and the non-biotinylated strand of the PCR product. A set of Zip-Chute™ detection probes is added to the plate and allowed to hybridize.

Each ZipChute probe contains an oligonucleotide that is complementary to one of the ZipCode sequences, a fluorescent label, and a unique mobility shift tag. Non-hybridized probes are washed away. The hybridized ZipChute probes are then eluted and run on a capillary electrophoresis instrument. Since each probe migrates at a different rate, alleles are scored by measuring the sizes of the fluorescent bands during electrophoresis. The assay reagents have been optimized for 48-plex reactions.

GOLDENGATE™ (POLYMERIZATION/LIGATION–FLUORESCENCE–BEAD ARRAY)

In the GoldenGate™ assay developed by Illumina, allele-specific extension and ligation are combined to achieve a high performance multiplexed genotyping method.[103] Sample genomic DNA is bound to paramagnetic particles and locus- and allele-specific oligonucleotide probes are hybridized to the genomic DNA. For each SNP analyzed, there are two allele-specific probes and one locus-specific probe.

The allele-specific probes are designed to hybridize to the target so that their 3′ terminal nucleotides are complementary to either variant of the SNP. The locus-specific probe is designed to hybridize some distance downstream of the allele-specific probe. By adding a polymerase and a ligase, only allele-specific probes that are matched at their 3' ends are extended and can be ligated to a locus-specific probe hybridized downstream of the primer. Undesired reaction byproducts and remaining non-hybridized probes are removed by washing.

All ligation products can be amplified with a standard set of amplification primers, while each locus-specific probe is also equipped with a unique address sequence. Allele information is thus encoded in the reaction product as one of two primer sequences, while locus information is encoded as an address sequence. Following PCR amplification of the reacted probes with differentially labeled primers, allele information is represented by one of two fluorescent molecules. The amplification products are then rendered single-stranded by a solid phase extraction and hybridized to a generic oligonucleotide array, the BeadArray™ platform, containing oligonucleotides complementary to each of the address sequences used. This way, locus information is converted into position information, and reading the fluorescence of each position of the array reveals the genotype at the corresponding loci.

PADLOCK PROBE AND MIP ASSAYS
(POLYMERIZATION/LIGATION–FLUORESCENCE–ARRAY/ELECTROPHORESIS)

The oligonucleotide ligation assay mentioned previously has been further developed in the padlock probe assay.[104] Allele-specific and locus-specific probes are designed so that they are connected at their distal ends to form a single probe that becomes circularized upon ligation. Its shape resembles a padlock locked around its target molecule. Two such allele-specific padlock probes are used for each SNP to be genotyped.

The circular nature of reacted probes allows them to be distinguished from unreacted or dimerized probes, and the padlock probe method requires no washes. Ligation is performed directly on genomic DNA and the reacted probes are amplified to generate sufficient signal for detection. Locus- and allele-specific information can be encoded in the parts of the padlock probes that are not target-complementary via tag sequences for array hybridization and for detection of allele information using differentially labeled fluorescent probes.[105] The circular probes may also be amplified using the circle-to-circle amplification technique.[106] In the SNPWave™ assay (Keygene, Wageningen, Netherlands, http://www.keygene.com/), allele and locus information is instead encoded in the length of the padlock probe and decoding is achieved through physical separation by capillary gel electrophoresis.[107]

A variation of the padlock probe assay is the molecular inversion probe (MIP) assay in which each SNP is interrogated by a single probe and allele distinction is achieved by polymerization instead of ligation.[65] The probes are designed so that the ends hybridize to the target, leaving a single nucleotide gap at the position of the SNP. In separate reactions for each nucleotide, this gap is filled with an allele-specific nucleotide by a polymerase, allowing the probe to be circularized by ligation. Next, cleavage of the probes between the PCR primer sequences results in "inversion" of the ligation product, releasing the probes from the target sequence for PCR amplification. Following amplification, the PCR products are hybridized to tag oligonucleotide arrays as described above. Molecular inversion probes have been used by ParAllele BioScience to achieve 10,000-plex SNP genotyping.[108]

GENECHIP® (HYBRIDIZATION–FLUORESCENCE–MICROARRAY)

A primary example of the use of microarrays for SNP detection is the GeneChip® probe assay from Affymetrix. Pre-selected sets of probe sequences for 10,000[109] or 100,000[110] SNPs are synthesized directly on glass slides using a photolithographic technique.[111] SNPs for inclusion on the arrays were chosen from information from the human genome sequencing project. Thus the precise locations of the SNPs and the surrounding DNA sequences are known.

The method involves a complexity reduction step to prepare patient DNA samples for hybridization to the array. Whole genomic DNA from the patient is digested with a chosen set of restriction enzymes to produce 400- to 800-bp fragments of DNA. The restriction fragments containing the SNP target loci are of optimal length for PCR. Adapter cassettes encoding a universal PCR primer sequence are ligated to both ends of each fragment. A single PCR primer sequence is then used for multiplex amplification of the set of SNP targets. The PCR products are digested with an endonuclease to produce shorter fragments and are end-labeled with biotin. Finally, the digestion mix is hybridized to the array and labeled using a fluorescence staining procedure. The arrays are placed into a scanner and sequence variants are determined from examination of the fluorescence patterns across the surface of the array.

This approach provides several advantages. From a relatively small amount of DNA (250 ng), it is possible to retrieve genotypes from up to 100,000 SNPs.[110] Also, a single PCR primer is used for amplifying all SNPs, reducing the number of assay-specific reagents needed per SNP studied. The investment per study thus resides mainly in the production and analysis of the array. This assay is most suitable for large scale genotyping. Also, custom SNPs to be added to the chip must conform to the entire strategy which may or may not be a trivial consideration.

WHOLE GENOME GENOTYPING (POLYMERIZATION–FLUORESCENCE–BEADARRAY™)

Illumina recently published a radically new genotyping strategy called the whole genome genotyping (WWG) assay.[112] In this approach an initial whole genome amplification reaction similar to the ϕ29 method described in the "Quantity of DNA" section below yields a >1000-fold amplification of 100- to 200-bp DNA fragments covering the entire genome. Following purification, the amplified products are hybridized to an array. The oligonucleotides on the array contain a complementary sequence to one SNP loci with a 3′ allele-specific nucleotide. DNA polymerase and a mixture of labeled nucleotides are added to perform allele-specific primer extension reactions.

After a clean-up step and staining procedure, genotypes are scored by measuring fluorescence in an array scanner. A major advantage of the WGG is that the assay performs well in the full complexity of genomic DNA. This allows great flexibility in selection of SNPs to be analyzed, while the array format maintains the ability to perform highly multiplexed reactions.

CAVEAT

Admittedly, this list of techniques is far from exhaustive. It is meant to give readers a foundation for understanding how common genotyping platforms work. Further descriptions of the presented techniques can be found in the primary literature. Expanded coverage over the breadth of SNP genotyping methodologies can be found in one of many high-quality reviews.[113–122]

CHOOSING A METHOD: A FEW CLINICAL CONSIDERATIONS

A number of factors are important in selecting a method for scoring SNP genotypes. Non-trivial matters, especially in clinical settings, include approval of methods and tests by regulatory agencies and accessibility to equipment. Although technology is changing, the basics of PCR machines, electrophoresis equipment, and DNA sequencers form the core activities of many clinical laboratories. It is no surprise that methods such as restriction fragment length polymorphism (RFLP) and sequencing account for a large proportion of techniques used to genotype SNPs in patients today.

The cost of importing a new genotyping platform into an existing lab involves several different sources. The initial investment for instrumentation, potential robotic systems, software, and maintenance is one source. A second is the running cost of assays including reagents, plastics, enzymes, oligonucleotides, and other consumables involved in performing individual genotyping reactions. The third and often underestimated cost is for labor. Beyond the hours involved in performing the assay, the expense involved in training staff to design, perform, and interpret assay data as well as repeat inconclusive assays is often omitted from expense calculations. Meaningful comparisons of cost are thus extremely difficult; few publications report comprehensive analyses of the expenses involved.

Aside from these practical matters, the choice of SNP genotyping technology is highly application-dependent. For example, characteristics such as sensitivity and specificity may be prioritized over cost and throughput in diagnostic tests. For other applications such as whole genome association studies, somewhat lower accuracy rates may be tolerated but throughput must be high to finish a project within a reasonable time frame and cost per genotype must be kept to a minimum. Several other criteria worth considering are noted below.

QUANTITATIVE DATA

For some applications determining the presence or absence of alleles is insufficient. The number of copies of an SNP locus is the clinically relevant fact. For example, gene copy number changes are important elements of malignancy in diseases such as breast cancer[123] and they can influence susceptibility to infectious diseases such as HIV/AIDS.[124] In such cases, both genotypic variation and gene copy number can impact the risk or severity of disease. Another clinically relevant example of the importance of quantification is in monitoring the success of bone marrow transplantation. This can be done by analyzing polymorphisms in different fractions of patient blood.

Genetic markers in which the donor and recipient differ in genotype are chosen and blood samples are taken repeatedly after transplantation. The stable presence of both donor and recipient genotypes (*chimerism*) is taken as a sign of successful transplantation. Reversion of genotype toward that of the recipient is an indication of graft rejection. Traditionally, various repeat polymorphisms have been used for this type of analysis (reviewed in Reference 125), but recent publications discuss the use of sets of SNPs.[126] Examples of techniques that are better suited for quantitative measures are pyrosequencing and the Sequenom MassARRAY™ system.

QUANTITY OF DNA

All SNP genotyping methods are highly affected by the quality and quantity of DNA used in the analysis. Although genetic material can be extracted from a variety of sources including biopsies and tissue cultures, one very common source is patient blood. From 1 ml of blood, roughly 25 µg of DNA can be retrieved. For many SNP genotyping platforms, a minimum of 1 to 10 ng of genomic DNA (corresponding to 150 to 1500 copies of any single copy gene) is required. Considering possible loss of DNA during pipetting, storage, dilution, and assay optimization, DNA extracted from 1 ml of blood would thus be sufficient for roughly 2000 genotyping assays on typical genotyping platforms. This is more than enough for genotyping a small number of loci; however, it may prove insufficient for large scale projects like whole genome association studies.

For some purposes, generation of immortalized cell lines from patients can be a suitable means of ensuring access to biological material. A number of *in vitro* approaches can extend the amount of genomic DNA starting from a minimal amount of genetic material. The first is the whole genome amplification (WGA) method. Although several different strategies have been proposed (reviewed in Reference 127), one method that has been successfully implemented into SNP genotyping is multiple displacement amplification (MDA).[128,129] In this isothermal technique, a series of random hexamers (six-base oligonucleotides) are mixed with genomic DNA along with ϕ29 DNA polymerase. When the hexamers anneal to the genome, they serve as primers for replication by the polymerase.

Primer extension progresses along the target strand, displacing any DNA molecules encountered along the way. Displaced strands then serve as templates for hexamer annealing and extension, and the process of DNA duplication continues. Starting from as few as 1 to 10 copies of genomic DNA, up to 20 to 30 μg of amplified product may be produced.[129] The fidelity of genotyping results from MDA-amplified samples has been demonstrated in a number of publications.[130–132]

Instead of amplifying the entire genome, directed amplification of selected portions of the genome may be more appropriate for some applications. PCR is commonly used for amplification of a single target sequence or a few sequences in parallel. However, for high levels of parallelism, a large number of separate amplification reactions must be performed.

As briefly noted previously, the so-called adaptor ligation method uses restriction endonucleases to digest a genomic sample into fragments of a known distribution. Common adaptor oligonucleotides are then ligated to the ends of the generated fragments. A subsequent PCR amplification with a single pair of primers can be adjusted to preferentially amplify fragments within a certain range of sizes. This way, a subset of the genome is amplified and a reduction of sequence complexity is achieved.[109,133] A variant of the method uses adaptors, one common primer, and a single specific primer per target sequence to achieve amplification of specific genomic sequences.[134,135]

If restriction digestion is carried out with type II S restriction enzymes, different end sequences are produced for different fragments. This way, adaptor sequences can be selected to circularize a certain population of fragments. These fragments can then be amplified and the procedure repeated, allowing the selective amplification of a single fragment from a genomic DNA sample.[136] This method may also be multiplexed by using several adaptor sequences, but incompatibilities of end sequences limit the extent to which this can be performed.

None of the methods above achieves the ideal goal of amplifying a large set of arbitrary sequences without the production of undesired amplification products. In the recently described selector amplification method, restriction fragments are made single-stranded by heat denaturation, circularized by hybridization to specific selector probes, and ligated to a common oligonucleotide, thus incorporating a common sequence motif in the single-stranded circles formed from the selected fragments. The selector probe end sequences are 10 to 20 nt in length to achieve specificity. Furthermore, structure-specific cleavage, using the same mechanism as in the Invader assay, can be utilized to remove a 5' portion of the fragment prior to ligation, allowing the creation of circles of similar lengths. All circles can then be amplified by PCR using a single primer pair for multiplex amplification for hundreds to thousands of selected fragments of interest.[137]

Known versus Unknown Variations

The vast majority of techniques described here are applied to SNPs where sequence and mapping information are known. Other methods are more adept at looking for sequence variants where exact information is unavailable. For example, if a particular gene has been implicated in a given disease, scanning the exons and promoters for variation could be desirable from a diagnostic point of view.

A number of methods can speed the search for such unknown sequence variants. Single-strand conformation polymorphism (SSCP),[138] denaturing gradient gel electrophoresis (DGGE),[139] and denaturing high performance liquid chromatography (dHPLC)[140] are examples of such methods. Although these three techniques can identify DNA fragments bearing sequence differences, the position or nature of genetic change is difficult to characterize based on results of these methods alone. Therefore, these methods are combined with or in fact replaced by sequencing techniques to derive exact sequence information.

Speed to Results

Another aspect to consider when selecting a genotyping platform is amount of time required to obtain the results. For point-of-care applications, the homogeneous PCR-based methods are quickest for typing a limited number of genetic variants in one or a few patients. In addition to the homogenous TaqMan and molecular beacon assays mentioned previously, methods based on allele-specific PCR have been developed for SNP genotyping.

The principles of the amplification refractory mutation system[70] have been adapted for fluorescence detection using intercalating dyes as in the Tm shifting assay[141] or through strategies involving FRET reactions.[142–146] These methods require both thermocycling and fluorescent detection, and many commercially available devices combine these two functions. The LightCycler® marketed by Roche Diagnostics (division of F. Hoffmann–La Roche Ltd, Basel, Switzerland, http://www.roche-diagnostics.com/) and the SmartCycler® by Cepheid (Sunnyvale, California, USA, http://www.cepheid.com/) perform PCR with real time fluorescence detection in as few as 20 minutes.

POSSIBILITIES AND PERSPECTIVES

Poised with a broad range of tools and technologies, biomedical research is entering a phase of exploring the information embodied within the human genome. Starting from extensive knowledge about the basic architecture and sequence of our genetic blueprint, major efforts are underway to determine the impact of genetic variation on risk for developing disease or responding to medical therapy.[147] For the field of pharmacogenomics, analysis of SNPs will probably not become the sole basis for personalized medicine, but rather an important component to the sum of molecular measurements that together reveal an individual's medical condition. A physician's evaluation will involve a totality of molecular information together with other information to determine the cause and identity of the disease and the best course of treatment.

The continued advancement of SNP genotyping technologies opens new and exciting possibilities in molecular medicine. One such opportunity is investigation of polymorphic differences directly within cells or tissue samples. SNP genotyping *in situ* can be important if the material to be analyzed is characterized by significant cell-to-cell variation as in tumor-derived samples. When such heterogeneity is encountered, conventional analysis in solution may fail to reveal heterogeneity. Using solution-phase methods for detection of the distribution of cells with specific sequence variants cannot be evaluated unless samples are micro-dissected.

Average measurements over many cells also fail to show the presence of distinct populations of cells that differ in target content. To access this kind of information, single cells must be defined and analyzed. The most widespread technique for analyzing the distribution of DNA and RNA sequences is *in situ* hybridization (ISH), which can be combined with fluorescent detection and is then known as fluorescence *in situ* hybridization (FISH). Until now, no reliable method has been available for *in situ* analysis of single copy molecules for SNP discrimination. So-called padlock probes have been applied recently for *in situ* analysis of mitochondrial DNA,[148,149] offering new possibilities for diagnostics.

As described earlier, padlock probes are linear oligonucleotides whose ends, upon hybridization next to each other on a target strand, can be joined by ligation. This creates a circularized DNA strand, wound around a target.[104] Circularized probes can then be used for template synthesis of extensive rolling-circle replication products anchored at targets and easily detected using fluorescent probes hybridized to the repeated sequence.

The robust target-primed rolling circle amplification (RCA) approach has been used on mitochondrial DNA targets in both cultured cells (Figure 6.7) and fresh frozen tissue sections.[148] Target-primed RCA should also be applicable for studies of gene expression by enabling *in situ* studies of allele-specific transcripts and the expression of highly similar members of gene families and splice variants. Another possible application is localized detection of infectious microbial agents. The procedure is suitable for use in pathological examinations by providing short analysis time without a need for expensive laboratory equipment other than a fluorescence microscope.

Adaptation of SNP genotyping platforms is also used to analyze other molecular genomic variables such as gene copy number[150] and allele-specific expression of transcripts[92] due, for example, to promoter polymorphisms. Epigenetic factors modulating gene expression such as

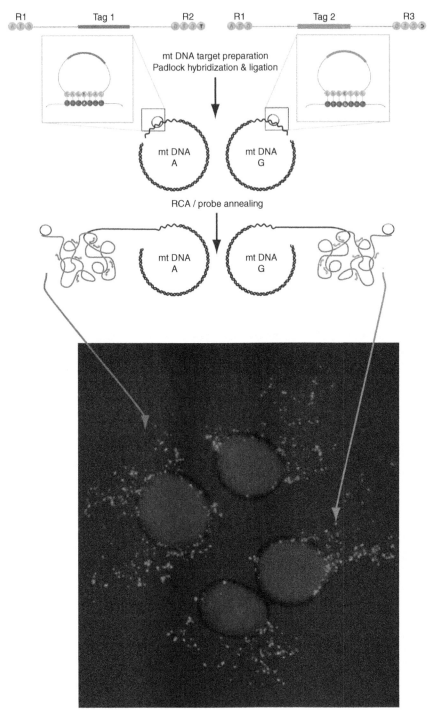

FIGURE 6.7 *(A color version follows page 270) In situ* genotyping using padlock probes. R1 are the locus-specific recognition sequences of the padlock probe; R2 and R3 are the allele-specific recognition sequences. The drawing (top) depicts the basic steps of the procedure, and the picture (bottom) presents an example of genotyping of a heteroplasmic cell line called G55 1.1. In these cells, roughly 65% of mitochondrial genomes contain the A3243G (MELAS) mutation.

methylation and histone hyperacetylation[151] have been studied by technologies similar to those used for SNP genotyping.

Infectious disease diagnostics are also of potentially great value; however, this type of application introduces new problems. Many pathogenic organisms mutate rapidly as part of their biological strategies to maintain virulence. Methods for SNP genotyping based on recognition of specific genetic sequences encounter difficulties when applied to viruses and bacteria that evolve rapidly. Assay specificity and selectivity must be balanced with the ability to detect slight variations in sequence in order to successfully detect or distinguish variants of the organism if present.

One further technological development that could drastically affect the way we approach analysis of the genome is the emergence of powerful and economical sequencing techniques. Rather than looking at single base positions, the strategy would involve decoding the entire genome or extensive selected regions thereof. Complete information of the genetic sequence of an individual would thus become available, obviating the need for genotyping specific single base positions. Research activities aimed at achieving "$1000" genome sequencing[152] are under way, making cataloguing entire genome sequence information in populations a foreseeable possibility. In the meantime, many of the technologies described here will certainly assist in harvesting the fruits of the human genome initiative and play key roles in biomedical research for years to come.

REFERENCES

1. Brookes AJ. The essence of SNPs. *Gene* 1999; 234 (2): 177–186.
2. Zondervan KT, Cardon LR. The complex interplay among factors that influence allelic association. *Nat Rev Genet* 2004; 5 (2): 89–100.
3. Holliday R, Grigg GW. DNA methylation and mutation. *Mutat Res* 1993; 285 (1): 61–67.
4. International Human Genome Sequencing Consortium. Initial sequencing and analysis of the human genome. *Nature* 2001; 409 (6822): 860–921.
5. Venter JC, Adams MD, Myers EW, Li PW, Mural RJ, Sutton GG, Smith HO, Yandell M, Evans CA, Holt RA et al. Sequence of the human genome. *Science* 2001; 291 (5507): 1304–1351.
6. International Human Genome Sequencing Consortium. Finishing the euchromatic sequence of the human genome. *Nature* 2004; 431 (7011): 931–945.
7. International SNP Map Working Group. A map of human genome sequence variation containing 1.42 million single nucleotide polymorphisms. *Nature* 2001; 409 (6822): 928–933.
8. Taillon-Miller P, Gu Z, Li Q, Hillier L, Kwok PY. Overlapping genomic sequences: a treasure trove of single-nucleotide polymorphisms. *Genome Res* 1998; 8 (7): 748–754.
9. Altshuler D, Pollara VJ, Cowles CR, Van Etten WJ, Baldwin J, Linton L, Lander ES. An SNP map of the human genome generated by reduced representation shotgun sequencing. *Nature* 2000; 407 (6803): 513–516.
10. Picoult-Newberg L, Ideker TE, Pohl MG, Taylor SL, Donaldson MA, Nickerson DA, Boyce-Jacino M. Mining SNPs from EST databases. *Genome Res* 1999; 9 (2): 167–174.
11. Irizarry K, Kustanovich V, Li C, Brown N, Nelson S, Wong W, Lee CJ. Genome-wide analysis of single-nucleotide polymorphisms in human expressed sequences. *Nat Genet* 2000; 26 (2): 233–236.
12. Garg K, Green P, Nickerson DA. Identification of candidate coding region single nucleotide polymorphisms in 165 human genes using assembled expressed sequence tags. *Genome Res* 1999; 9 (11): 1087–1092.
13. Marth GT, Korf I, Yandell MD, Yeh RT, Gu Z, Zakeri H, Stitziel NO, Hillier L, Kwok PY, Gish WR. A general approach to single-nucleotide polymorphism discovery. *Nat Genet* 1999; 23 (4): 452–456.
14. Buetow KH, Edmonson MN, Cassidy AB. Reliable identification of large numbers of candidate SNPs from public EST data. *Nat Genet* 1999; 21 (3): 323–325.
15. Schlotterer C. The evolution of molecular markers: just a matter of fashion? *Nat Rev Genet* 2004; 5 (1): 63–69.
16. Fredman D, Siegfried M, Yuan YP, Bork P, Lehvaslaiho H, Brookes AJ. HGVbase: a human sequence variation database emphasizing data quality and a broad spectrum of data sources. *Nucleic Acids Res* 2002; 30 (1): 387–391.

17. Fredman D, Munns G, Rios D, Sjoholm F, Siegfried M, Lenhard B, Lehvaslaiho H, Brookes AJ. HGVbase: a curated resource describing human DNA variation and phenotype relationships. *Nucleic Acids Res* 2004; 32: D516–D519.

18. Kruglyak L, Nickerson DA. Variation is the spice of life. *Nat Genet* 2001; 27 (3): 234–236.

19. Nickerson DA, Taylor SL, Weiss KM, Clark AG, Hutchinson RG, Stengard J, Salomaa V, Vartiainen E, Boerwinkle E, Sing CF. DNA sequence diversity in a 9.7-kb region of the human lipoprotein lipase gene. *Nat Genet* 1998; 19 (3): 233–240.

20. Li WH, Sadler LA. Low nucleotide diversity in man. *Genetics* 1991; 129 (2): 513–523.

21. Halushka MK, Fan JB, Bentley K, Hsie L, Shen N, Weder A, Cooper R, Lipshutz R, Chakravarti A. Patterns of single-nucleotide polymorphisms in candidate genes for blood pressure homeostasis. *Nat Genet* 1999; 22 (3): 239–247.

22. Cargill M, Altshuler D, Ireland J, Sklar P, Ardlie K, Patil N, Shaw N, Lane CR, Lim EP, Kalyanaraman N et al. Characterization of single-nucleotide polymorphisms in coding regions of human genes. *Nat Genet* 1999; 22 (3): 231–238.

23. Livingston RJ, von Niederhausern A, Jegga AG, Crawford DC, Carlson CS, Rieder MJ, Gowrisankar S, Aronow BJ, Weiss RB, Nickerson DA. Pattern of sequence variation across 213 environmental response genes. *Genome Res* 2004; 14 (10A): 1821–1831.

24. Costas J, Salas A, Phillips C, Carracedo A. Human genome-wide screen of haplotype-like blocks of reduced diversity. *Gene* 2005; 349: 219–225.

25. Horton R, Niblett D, Milne S, Palmer S, Tubby B, Trowsdale J, Beck S. Large-scale sequence comparisons reveal unusually high levels of variation in the HLA-DQB1 locus in the class II region of the human MHC. *J Mol Biol* 1998; 282 (1): 71–97.

26. Reich DE, Cargill M, Bolk S, Ireland J, Sabeti PC, Richter DJ, Lavery T, Kouyoumjian R, Farhadian SF, Ward R et al. Linkage disequilibrium in the human genome. *Nature* 2001; 411 (6834): 199–204.

27. Clark AG. Inference of haplotypes from PCR-amplified samples of diploid populations. *Mol Biol Evol* 1990; 7 (2): 111–122.

28. Excoffier L, Slatkin M. Maximum likelihood estimation of molecular haplotype frequencies in a diploid population. *Mol Biol Evol* 1995; 12 (5): 921–927.

29. Hawley ME, Kidd KK. HAPLO: a program using the EM algorithm to estimate the frequencies of multi-site haplotypes. *J Hered* 1995; 86 (5): 409–411.

30. Long JC, Williams RC, Urbanek M. An EM algorithm and testing strategy for multiple-locus haplotypes. *Am J Hum Genet* 1995; 56 (3): 799–810.

31. Stephens M, Smith NJ, Donnelly P. A new statistical method for haplotype reconstruction from population data. *Am J Hum Genet* 2001; 68 (4): 978–989.

32. Stephens M, Donnelly P. A comparison of Bayesian methods for haplotype reconstruction from population genotype data. *Am J Hum Genet* 2003; 73 (5): 1162–1169.

33. Weale ME. A survey of current software for haplotype phase inference. *Hum Genomics* 2004; 1 (2): 141–144.

34. Hinds DA, Stuve LL, Nilsen GB, Halperin E, Eskin E, Ballinger DG, Frazer KA, Cox DR. Whole-genome patterns of common DNA variation in three human populations. *Science* 2005; 307 (5712): 1072–1079.

35. Crawford DC, Nickerson DA. Definition and clinical importance of haplotypes. *Annu Rev Med* 2005; 56: 303–320.

36. Winckler W, Myers SR, Richter DJ, Onofrio RC, McDonald GJ, Bontrop RE, McVean GA, Gabriel SB, Reich D, Donnelly P et al. Comparison of fine-scale recombination rates in humans and chimpanzees. *Science* 2005; 308 (5718): 107–111.

37. Zhang K, Akey JM, Wang N, Xiong M, Chakraborty R, Jin L. Randomly distributed crossovers may generate block-like patterns of linkage disequilibrium: an act of genetic drift. *Hum Genet* 2003; 113 (1): 51–59.

38. Crawford DC, Bhangale T, Li N, Hellenthal G, Rieder MJ, Nickerson DA, Stephens M. Evidence for substantial fine-scale variation in recombination rates across the human genome. *Nat Genet* 2004; 36 (7): 700–706.

39. Daly MJ, Rioux JD, Schaffner SF, Hudson TJ, Lander ES. High-resolution haplotype structure in the human genome. *Nat Genet* 2001; 29 (2): 229–232.

40. Gabriel SB, Schaffner SF, Nguyen H, Moore JM, Roy J, Blumenstiel B, Higgins J, DeFelice M, Lochner A, Faggart M et al. The structure of haplotype blocks in the human genome. *Science* 2002; 296 (5576): 2225–2229.

41. Phillips MS, Lawrence R, Sachidanandam R, Morris AP, Balding DJ, Donaldson MA, Studebaker JF, Ankener WM, Alfisi SV, Kuo FS et al. Chromosome-wide distribution of haplotype blocks and the role of recombination hot spots. *Nat Genet* 2003; 33 (3): 382–387.

42. Wall JD, Pritchard JK. Assessing the performance of the haplotype block model of linkage disequilibrium. *Am J Hum Genet* 2003; 73 (3): 502–515.

43. Patil N, Berno AJ, Hinds DA, Barrett WA, Doshi JM, Hacker CR, Kautzer CR, Lee DH, Marjoribanks C, McDonough DP et al. Blocks of limited haplotype diversity revealed by high-resolution scanning of human chromosome 21. *Science* 2001; 294 (5547): 1719–1723.

44. International HapMap Consortium. The International HapMap Project. *Nature* 2003; 426 (6968): 789–796.

45. Newton–Cheh C, Hirschhorn JN. Genetic association studies of complex traits: design and analysis issues. *Mutat Res* 2005; 573 (1–2): 54–69.

46. Cardon LR, Bell JI. Association study designs for complex diseases. *Nat Rev Genet* 2001; 2 (2): 91–99.

47. Botstein D, Risch N. Discovering genotypes underlying human phenotypes: past successes for Mendelian disease, future approaches for complex disease. *Nat Genet* 2003; 33 Suppl: 228–237.

48. Lohmueller KE, Pearce CL, Pike M, Lander ES, Hirschhorn JN. Meta-analysis of genetic association studies supports a contribution of common variants to susceptibility to common disease. *Nat Genet* 2003; 33 (2): 177–182.

49. Tabor HK, Risch NJ, Myers RM. Opinion: candidate-gene approaches for studying complex genetic traits: practical considerations. *Nat Rev Genet* 2002; 3 (5): 391–397.

50. Hirschhorn JN, Lohmueller K, Byrne E, Hirschhorn K. A comprehensive review of genetic association studies. *Genet Med* 2002; 4 (2): 45–61.

51. Reich DE, Lander ES. On the allelic spectrum of human disease. *Trends Genet* 2001; 17 (9): 502–510.

52. Risch N, Merikangas K. The future of genetic studies of complex human diseases. *Science* 1996; 273 (5281): 1516–1517.

53. Carlson CS, Eberle MA, Kruglyak L, Nickerson DA. Mapping complex disease loci in whole-genome association studies. *Nature* 2004; 429 (6990): 446–452.

54. Johnson GC, Esposito L, Barratt BJ, Smith AN, Heward J, Di Genova G, Ueda H, Cordell HJ, Eaves IA, Dudbridge F et al. Haplotype tagging for the identification of common disease genes. *Nat Genet* 2001; 29 (2): 233–237.

55. Kwok PY. Genomics: genetic association by whole-genome analysis? *Science* 2001; 294 (5547): 1669–1670.

56. Altshuler D, Daly M, Kruglyak L. Guilt by association. *Nat Genet* 2000; 26 (2): 135–137.

57. Hirschhorn JN, Daly MJ. Genome-wide association studies for common diseases and complex traits. *Nat Rev Genet* 2005; 6 (2): 95–108.

58. Wang WY, Barratt BJ, Clayton DG, Todd JA. Genome-wide association studies: theoretical and practical concerns. *Nat Rev Genet* 2005; 6 (2): 109–118.

59. Kutyavin IV, Afonina IA, Mills A, Gorn VV, Lukhtanov EA, Belousov ES, Singer MJ, Walburger DK, Lokhov SG, Gall AA et al. 3' minor groove binder DNA probes increase sequence specificity at PCR extension temperatures. *Nucleic Acids Res* 2000; 28 (2): 655–661.

60. Vester B, Wengel J. LNA (locked nucleic acid): high-affinity targeting of complementary RNA and DNA. *Biochemistry* 2004; 43 (42): 13233–13241.

61. Kan YW, Dozy AM. Polymorphism of DNA sequence adjacent to human -globin structural gene: relationship to sickle mutation. *Proc Natl Acad Sci USA* 1978; 75 (11): 5631–5635.

62. Landegren U, Kaiser R, Sanders J, Hood L. A ligase-mediated gene detection technique. *Science* 1988; 241 (4869): 1077–1080.

63. Syvanen AC, Aalto-Setala K, Harju L, Kontula K, Soderlund H. A primer-guided nucleotide incorporation assay in the genotyping of apolipoprotein E. *Genomics* 1990; 8 (4): 684–692.

64. Nyren P., Pettersson B., Uhlen M. Solid Phase DNA Minisequencing by an enzymatic luminometric inorganic pyrophosphate detection assay. *Anal Biochem* 1993; 208 (1): 171–175.

65. Hardenbol P, Baner J, Jain M, Nilsson M, Namsaraev EA, Karlin-Neumann GA, Fakhrai-Rad H, Ronaghi M, Willis TD, Landegren U et al. Multiplexed genotyping with sequence-tagged molecular inversion probes. *Nat Biotechnol* 2003; 21 (6): 673–678.

66. Chen X, Zehnbauer B, Gnirke A, Kwok PY. Fluorescence energy transfer detection as a homogeneous DNA diagnostic method. *Proc Natl Acad Sci USA* 1997; 94 (20): 10756–10761.

67. Fan JB, Chen X, Halushka MK, Berno A, Huang X, Ryder T, Lipshutz RJ, Lockhart DJ, Chakravarti A. Parallel genotyping of human SNPs using generic high-density oligonucleotide tag arrays. *Genome Res* 2000; 10 (6): 853–860.

68. Rodi CP, Darnhofer-Patel B, Stanssens P, Zabeau M, van den Boom D. A strategy for the rapid discovery of disease markers using the MassARRAY™ system. *Biotechniques* 2002; 32 (Suppl 2): S62–S69.

69. Blondal T, Waage BG, Smarason SV, Jonsson F, Fjalldal SB, Stefansson K, Gulcher J, Smith AV. A novel MALDI-TOF-based methodology for genotyping single nucleotide polymorphisms. *Nucleic Acids Res* 2003; 31 (24): e155.

70. Newton CR, Heptinstall LE, Summers C, Super M, Schwarz M, Anwar R, Graham A, Smith JC, Markham AF. Amplification refractory mutation system for prenatal diagnosis and carrier assessment in cystic fibrosis. *Lancet* 1989; 2 (8678–8679): 1481–1483.

71. Pastinen T, Raitio M, Lindroos K, Tainola P, Peltonen L, Syvanen AC. A system for specific, high-throughput genotyping by allele-specific primer extension on microarrays. *Genome Res* 2000; 10 (7): 1031–1042.

72. Cahill P, Bakis M, Hurley J, Kamath V, Nielsen W, Weymouth D, Dupuis J, Doucette–Stamm L, Smith DR. Exo-proofreading, a versatile SNP scoring technology. *Genome Res* 2003; 13 (5): 925–931.

73. Lyamichev V, Brow MA, Dahlberg JE. Structure-specific endonucleolytic cleavage of nucleic acids by eubacterial DNA polymerases. *Science* 1993; 260 (5109): 778–783.

74. Tost J, Gut IG. Genotyping single nucleotide polymorphisms by MALDI mass spectrometry in clinical applications. *Clin Biochem* 2005; 38 (4): 335–350.

75. Fei Z, Ono T, Smith LM. MALDI-TOF mass spectrometric typing of single nucleotide polymorphisms with mass-tagged ddNTPs. *Nucleic Acids Res* 1998; 26 (11): 2827–2828.

76. Howell WM, Jobs M, Gyllensten U, Brookes AJ. Dynamic allele-specific hybridization: a new method for scoring single nucleotide polymorphisms. *Nat Biotechnol* 1999; 17 (1): 87–88.

77. De Angelis DA. Why FRET over genomics? *Physiol Genomics* 1999; 1 (2): 93–99.

78. Howell WM, Jobs M, Brookes AJ. iFRET: an improved fluorescence system for DNA-melting analysis. *Genome Res* 2002; 12 (9): 1401–1407.

79. Kwok PY. SNP genotyping with fluorescence polarization detection. *Hum Mutat* 2002; 19 (4): 315–323.

80. Livak KJ, Flood SJ, Marmaro J, Giusti W, Deetz K. Oligonucleotides with fluorescent dyes at opposite ends provide a quenched probe system useful for detecting PCR product and nucleic acid hybridization. *PCR Methods Appl* 1995; 4 (6): 357–362.

81. Livak KJ. Allelic discrimination using fluorogenic probes and the 5′ nuclease assay. *Genet Anal* 1999; 14 (5–6): 143–149.

82. Lee LG, Livak KJ, Mullah B, Graham RJ, Vinayak RS, Woudenberg TM. Seven-color, homogeneous detection of six PCR products. *Biotechniques* 1999; 27 (2): 342–349.

83. Tyagi S, Kramer FR. Molecular beacons: probes that fluoresce upon hybridization. *Nat Biotechnol* 1996; 14 (3): 303–308.

84. Tyagi S, Bratu DP, Kramer FR. Multicolor molecular beacons for allele discrimination. *Nat Biotechnol* 1998; 16 (1): 49–53.

85. Ronaghi M, Karamohamed S, Pettersson B, Uhlen M, Nyren P. Real-time DNA sequencing using detection of pyrophosphate release. *Anal Biochem* 1996; 242 (1): 84–89.

86. Langaee T, Ronaghi M. Genetic variation analyses by pyrosequencing. *Mutat Res* 2005; 573 (1–2): 96–102.

87. Prince JA, Feuk L, Howell WM, Jobs M, Emahazion T, Blennow K, Brookes AJ. Robust and accurate single nucleotide polymorphism genotyping by dynamic allele-specific hybridization (DASH): design criteria and assay validation. *Genome Res* 2001; 11 (1): 152–162.

88. Jobs M, Howell WM, Stromqvist L, Mayr T, Brookes AJ. DASH-2: flexible, low-cost, and high-throughput SNP genotyping by dynamic allele-specific hybridization on membrane arrays. *Genome Res* 2003; 13 (5): 916–924.

89. Lyamichev V, Mast AL, Hall JG, Prudent JR, Kaiser MW, Takova T, Kwiatkowski RW, Sander TJ, de Arruda M, Arco DA et al. Polymorphism identification and quantitative detection of genomic DNA by invasive cleavage of oligonucleotide probes. *Nat Biotechnol* 1999; 17 (3): 292–296.

90. Olivier M. The invader assay for SNP genotyping. *Mutat Res* 2005; 573 (1–2): 103–110.

91. Olivier M, Chuang LM, Chang MS, Chen YT, Pei D, Ranade K, de Witte A, Allen J, Tran N, Curb D et al. High-throughput genotyping of single nucleotide polymorphisms using new biplex invader technology. *Nucleic Acids Res* 2002; 30 (12): e53.

92. Jurinke C, Denissenko MF, Oeth P, Ehrich M, van den Boom D, Cantor CR. A single nucleotide polymorphism based approach for the identification and characterization of gene expression modulation using MassARRAY. *Mutat Res* 2005; 573 (1–2): 83–95.

93. Braun A, Little DP, Koster H. Detecting *CFTR* gene mutations by using primer oligo base extension and mass spectrometry. *Clin Chem* 1997; 43 (7): 1151–1158.

94. Sauer S, Lechner D, Berlin K, Lehrach H, Escary JL, Fox N, Gut IG. A novel procedure for efficient genotyping of single nucleotide polymorphisms. *Nucleic Acids Res* 2000; 28 (5): e13.

95. Sauer S, Lechner D, Berlin K, Plancon C, Heuermann A, Lehrach H, Gut IG. Full flexibility genotyping of single nucleotide polymorphisms by the GOOD assay. *Nucleic Acids Res* 2000; 28 (23): e100.

96. Sauer S, Gelfand DH, Boussicault F, Bauer K, Reichert F, Gut IG. Facile method for automated genotyping of single nucleotide polymorphisms by mass spectrometry. *Nucleic Acids Res* 2002; 30 (5): e22.

97. Tost J, Gut IG. Genotyping single nucleotide polymorphisms by mass spectrometry. *Mass Spectrom Rev* 2002; 21 (6): 388–418.

98. Kokoris M, Dix K, Moynihan K, Mathis J, Erwin B, Grass P, Hines B, Duesterhoeft A. High-throughput SNP genotyping with the MASScode system. *Mol Diagn* 2000; 5 (4): 329–340.

99. Nikiforov TT, Rendle RB, Goelet P, Rogers YH, Kotewicz ML, Anderson S, Trainor GL, Knapp MR. Genetic bit analysis: a solid phase method for typing single nucleotide polymorphisms. *Nucleic Acids Res* 1994; 22 (20): 4167–4175.

100. Bell PA, Chaturvedi S, Gelfand CA, Huang CY, Kochersperger M, Kopla R, Modica F, Pohl M, Varde S, Zhao R et al. SNPstream® UHT: ultra-high throughput SNP genotyping for pharmacogenomics and drug discovery. *Biotechniques* 2002; 32 (Suppl 2): S70–S77.

101. Shumaker JM, Metspalu A, Caskey CT. Mutation detection by solid phase primer extension. *Hum Mutat* 1996; 7 (4): 346–354.

102. de la Vega FM, Lazaruk KD, Rhodes MD, Wenz MH. Assessment of two flexible and compatible SNP genotyping platforms: TaqMan® SNP genotyping assays and the SNPlex genotyping system. *Mutat Res* 2005; 573 (1–2): 111–135.

103. Oliphant A, Barker DL, Stuelpnagel JR, Chee MS. BeadArray™ technology: enabling an accurate, cost-effective approach to high-throughput genotyping. *Biotechniques* 2002; 32 (Suppl 2): S56–S61.

104. Nilsson M, Malmgren H, Samiotaki M, Kwiatkowski M, Chowdhary BP, Landegren U. Padlock probes: circularizing oligonucleotides for localized DNA detection. *Science* 1994; 265 (5181): 2085–2088.

105. Banér J, Isaksson A, Waldenstrom E, Jarvius J, Landegren U, Nilsson M. Parallel gene analysis with allele-specific padlock probes and tag microarrays. *Nucleic Acids Res* 2003; 31 (17): e103.

106. Dahl F, Baner J, Gullberg M, Mendel-Hartvig M, Landegren U, Nilsson M. Circle-to-circle amplification for precise and sensitive DNA analysis. *Proc Natl Acad Sci USA* 2004; 101 (13): 4548–4553.

107. van Eijk MJ, Broekhof JL, van der Poel HJ, Hogers RC, Schneiders H, Kamerbeek J, Verstege E, van Aart JW, Geerlings H, Buntjer JB et al. SNPWave™: a flexible multiplexed SNP genotyping technology. *Nucleic Acids Res* 2004; 32 (4): e47.

108. Hardenbol P, Yu F, Belmont J, Mackenzie J, Bruckner C, Brundage T, Boudreau A, Chow S, Eberle J, Erbilgin A et al. Highly multiplexed molecular inversion probe genotyping: over 10,000 targeted SNPs genotyped in a single tube assay. *Genome Res* 2005; 15 (2): 269–275.

109. Kennedy GC, Matsuzaki H, Dong S, Liu WM, Huang J, Liu G, Su X, Cao M, Chen W, Zhang J et al. Large-scale genotyping of complex DNA. *Nat Biotechnol* 2003; 21 (10): 1233–1237.

110. Matsuzaki H, Dong S, Loi H, Di X, Liu G, Hubbell E, Law J, Berntsen T, Chadha M, Hui H et al. Genotyping over 100,000 SNPs on a pair of oligonucleotide arrays. *Nat Methods* 2004; 1 (2): 109–111.

111. Fodor SP, Read JL, Pirrung MC, Stryer L, Lu AT, Solas D. Light-directed, spatially addressable parallel chemical synthesis. *Science* 1991; 251 (4995): 767–773.

112. Gunderson KL, Steemers FJ, Lee G, Mendoza LG, Chee MS. A genome-wide scalable SNP genotyping assay using microarray technology. *Nat Genet* 2005; 37 (5): 549–554.

113. Nakatani K. Chemistry challenges in SNP typing. *ChemBioChem* 2004; 5 (12): 1623–1633.

114. Chen X, Sullivan PF. Single nucleotide polymorphism genotyping: biochemistry, protocol, cost and throughput. *Pharmacogenomics J* 2003; 3 (2): 77–96.

115. Kwok PY. High-throughput genotyping assay approaches. *Pharmacogenomics* 2000; 1 (1): 95–100.

116. Landegren U, Nilsson M, Kwok PY. Reading bits of genetic information: methods for single-nucleotide polymorphism analysis. *Genome Res* 1998; 8 (8): 769–776.

117. Tsuchihashi Z, Dracopoli NC. Progress in high throughput SNP genotyping methods. *Pharmacogenomics J* 2002; 2 (2): 103–110.

118. Brennan MD. High throughput genotyping technologies for pharmacogenomics. *Am J Pharmacogenomics* 2001; 1 (4): 295–302.

119. Gut IG. Automation in genotyping of single nucleotide polymorphisms. *Hum Mutat* 2001; 17 (6): 475–492.

120. Syvanen AC. Accessing genetic variation: genotyping single nucleotide polymorphisms. *Nat Rev Genet* 2001; 2 (12): 930–942.

121. Shi MM. Technologies for individual genotyping: detection of genetic polymorphisms in drug targets and disease genes. *Am J Pharmacogenomics* 2002; 2 (3): 197–205.

122. Drabek J. A commented dictionary of techniques for genotyping. *Electrophoresis* 2001; 22 (6): 1024–1045.

123. Slamon DJ, Clark GM, Wong SG, Levin WJ, Ullrich A, McGuire WL. Human breast cancer: correlation of relapse and survival with amplification of the HER-2/neu oncogene. *Science* 1987; 235 (4785): 177–182.

124. Gonzalez E, Kulkarni H, Bolivar H, Mangano A, Sanchez R, Catano G, Nibbs RJ, Freedman BI, Quinones MP, Bamshad MJ et al. The influence of *CCL3L1* gene-containing segmental duplications on HIV-1/AIDS susceptibility. *Science* 2005; 307 (5714): 1434–1440.

125. Leclair B, Fregeau CJ, Aye MT, Fourney RM. DNA typing for bone marrow engraftment follow-up after allogeneic transplant: a comparative study of current technologies. *Bone Marrow Transplant* 1995; 16 (1): 43–55.

126. Oliver DH, Thompson RE, Griffin CA, Eshleman JR. Use of single nucleotide polymorphisms (SNP) and real-time polymerase chain reaction for bone marrow engraftment analysis. *J Mol Diagn* 2000; 2 (4): 202–208.

127. Lasken RS, Egholm M. Whole genome amplification: abundant supplies of DNA from precious samples or clinical specimens. *Trends Biotechnol* 2003; 21 (12): 531–535.

128. Lizardi PM, Huang X, Zhu Z, Bray-Ward P, Thomas DC, Ward DC. Mutation detection and single-molecule counting using isothermal rolling-circle amplification. *Nat Genet* 1998; 19 (3): 225–232.

129. Dean FB, Hosono S, Fang L, Wu X, Faruqi AF, Bray-Ward P, Sun Z, Zong Q, Du Y, Du J et al. Comprehensive human genome amplification using multiple displacement amplification. *Proc Natl Acad Sci USA* 2002; 99 (8): 5261–5266.

130. Lovmar L, Fredriksson M, Liljedahl U, Sigurdsson S, Syvanen AC. Quantitative evaluation by minisequencing and microarrays reveals accurate multiplexed SNP genotyping of whole genome amplified DNA. *Nucleic Acids Res* 2003; 31 (21): e129.

131. Paez JG, Lin M, Beroukhim R, Lee JC, Zhao X, Richter DJ, Gabriel S, Herman P, Sasaki H, Altshuler D et al. Genome coverage and sequence fidelity of 29 polymerase-based multiple strand displacement whole genome amplification. *Nucleic Acids Res* 2004; 32 (9): e71.

132. Pask R, Rance HE, Barratt BJ, Nutland S, Smyth DJ, Sebastian M, Twells RC, Smith A, Lam AC, Smink LJ et al. Investigating the utility of combining 29 whole genome amplification and highly multiplexed single nucleotide polymorphism BeadArray™ genotyping. *BMC Biotechnol* 2004; 4 (1): e15.

133. Matsuzaki H, Loi H, Dong S, Tsai YY, Fang J, Law J, Di X, Liu WM, Yang G, Liu G et al. Parallel genotyping of over 10,000 SNPs using a one-primer assay on a high-density oligonucleotide array. *Genome Res* 2004; 14 (3): 414–425.

134. Broude NE, Zhang L, Woodward K, Englert D, Cantor CR. Multiplex allele-specific target amplification based on PCR suppression. *Proc Natl Acad Sci USA* 2001; 98 (1): 206–211.

135. Shapero MH, Zhang J, Loraine A, Liu W, Di X, Liu G, Jones KW. MARA: a novel approach for highly multiplexed locus-specific SNP genotyping using high-density DNA oligonucleotide arrays. *Nucleic Acids Res* 2004; 32 (22): e181.

136. Callow MJ, Drmanac S, Drmanac R. Selective DNA amplification from complex genomes using universal double-sided adapters. *Nucleic Acids Res* 2004; 32 (2): e21.

137. Dahl F, Gullberg M, Stenberg J, Landegren U, Nilsson M. Multiplex amplification enabled by selective circularization of large sets of genomic DNA fragments. *Nucleic Acids Res* 2005; 33 (8): e71.

138. Orita M, Suzuki Y, Sekiya T, Hayashi K. Rapid and sensitive detection of point mutations and DNA polymorphisms using the polymerase chain reaction. *Genomics* 1989; 5 (4): 874–879.

139. Myers RM, Lumelsky N, Lerman LS, Maniatis T. Detection of single base substitutions in total genomic DNA. *Nature* 1985; 313 (6002): 495–498.

140. Underhill PA, Jin L, Lin AA, Mehdi SQ, Jenkins T, Vollrath D, Davis RW, Cavalli-Sforza LL, Oefner PJ. Detection of numerous Y chromosome biallelic polymorphisms by denaturing high-performance liquid chromatography. *Genome Res* 1997; 7 (10): 996–1005.

141. Germer S, Higuchi R. Single-tube genotyping without oligonucleotide probes. *Genome Res* 1999; 9 (1): 72–78.

142. Bernard PS, Wittwer CT. Homogeneous amplification and variant detection by fluorescent hybridization probes. *Clin Chem* 2000; 46 (2): 147–148.

143. Solinas A, Brown LJ, McKeen C, Mellor JM, Nicol J, Thelwell N, Brown T. Duplex scorpion primers in SNP analysis and FRET applications. *Nucleic Acids Res* 2001; 29 (20): e96.

144. Nazarenko IA, Bhatnagar SK, Hohman RJ. A closed tube format for amplification and detection of DNA based on energy transfer. *Nucleic Acids Res* 1997; 25 (12): 2516–2521.

145. Myakishev MV, Khripin Y, Hu S, Hamer DH. High-throughput SNP genotyping by allele-specific PCR with universal energy-transfer-labeled primers. *Genome Res* 2001; 11 (1): 163–169.

146. Whitcombe D, Theaker J, Guy SP, Brown T, Little S. Detection of PCR products using self-probing amplicons and fluorescence. *Nat Biotechnol* 1999; 17 (8): 804–807.

147. Collins FS, Guttmacher AE. Genetics moves into the medical mainstream. *JAMA* 2001; 286 (18): 2322–2324.

148. Larsson C, Koch J, Nygren A, Janssen G, Raap AK, Landegren U, Nilsson M. *In situ* genotyping individual DNA molecules by target-primed rolling-circle amplification of padlock probes. *Nat Methods* 2004; 1 (3): 227–232.

149. Christian AT, Pattee MS, Attix CM, Reed BE, Sorensen KJ, Tucker JD. Detection of DNA point mutations and mRNA expression levels by rolling circle amplification in individual cells. *Proc Natl Acad Sci USA* 2001; 98 (25): 14238–14243.

150. Bignell GR, Huang J, Greshock J, Watt S, Butler A, West S, Grigorova M, Jones KW, Wei W, Stratton MR et al. High-resolution analysis of DNA copy number using oligonucleotide microarrays. *Genome Res* 2004; 14 (2): 287–295.

151. Li L, Shi H, Yiannoutsos C, Huang TH, Nephew KP. Epigenetic hypothesis tests for methylation and acetylation in a triple microarray system. *J Comput Biol* 2005; 12 (3): 370–390.

152. Collins FS, Green ED, Guttmacher AE, Guyer MS. A vision for the future of genomics research. *Nature* 2003; 422 (6934): 835–847.

7 RNA Expression Profiling

Payman Hanifi-Moghaddam, Curt W. Burger,
Theo J.M. Helmerhorst, and Leen J. Blok

CONTENTS

INTRODUCTION

Generating gene expression profiles has become relatively easy in recent years, but publishing these "simple" profiles has become exceedingly difficult. It is no longer sufficient to generate long lists of genes. From the millions of data points generated, one must be able to extract meaningful information about the system under study. This task has not turned out to be easy, because no single method will analyze and interpret all variations of genome-wide expression data.

In this chapter, we will provide the reader with some basic background information about types of microarrays available, a practical overview to help navigate among the ever-increasing numbers of available tools, and the techniques of filtering relevant information from the thousands of data points generated from microarray experiments.

PRINCIPLE

Although many protocols and types of systems are available, the basis of a successful microarray experiment lies in the extraction of fully intact and highly pure RNA from biological samples. The mRNA within the total RNA preparation is then copied, while incorporating either fluorescent nucleotides (Cy3 and Cy5) or a tag that is later stained with a fluorophore. The labeled RNA is then hybridized to the microarray, after which excess labeled mRNA is washed away. Subsequently, the information encoded in the microarray is scanned by laser light.

OLIGONUCLEOTIDE AND cDNA MICROARRAYS

Two different microarray methods are used generally to measure gene expression profiles.

Oligonucleotide Microarrays

This type of array is produced by Affymetrix (Santa Clara, California, USA) and is named the GeneChip®. The array contains hundreds of thousands of ordered, single-strand, synthetic oligonucleotides that are typically 25 bases in length. These oligonucleotides have been synthesized *in situ* on a glass surface and are designed to have uniform hybridization temperature profiles and RNA-binding affinities. A gene is generally detected by using 16 of these oligonucleotide probes, and each microarray measures the RNA abundance of thousands of different genes in one single RNA sample (no reference sample is used). The resulting data represent absolute levels of RNA, although this absolute measurement does not correlate exactly with the mRNA concentration in micrograms per microliter.

Spotted cDNA Microarrays

This type of array contains ordered DNA probes that were either synthesized or created by polymerase chain reaction (PCR). They usually correspond to cDNA sequences (but genomic sequences are also used) linked to the surfaces of glass slides. Usually there is one or sometimes two probes per gene, and each probe has its own preferred hybridization temperature. The resulting data represents the relative concentration of a certain transcript under a condition compared to a reference condition that may serve as a control.

ADVANTAGES AND DISADVANTAGES OF OLIGONUCLEOTIDE AND cDNA MICROARRAYS

Oligonucleotide microarrays

The probes are designed from an understanding of sequence information and are directly synthesized on the glass surface of the hybridization chamber. Detection of hybridization is based on a physical interaction between the cRNA and a number of probes (usually 16). Moreover, the probes are relatively short (25 mers); this makes it easier to choose a unique sequence representing one specific gene in a whole family of related genes.

Spotted cDNA Oligonucleotide Microarrays

A synthetic oligonucleotide (60 to 70 mers) is bound to a glass surface. Most of the process can be automated (oligonucleotide concentrations are therefore more comparable), resulting in fewer sample mix-ups and smaller losses of samples. The main advantage is that this technique is not difficult to applys and is thus becoming increasingly available. The main disadvantages are the high cost and the fact that one can run out of a given batch of synthesized oligonucleotides.

Spotted cDNA PCR Probe Microarrays

The main advantage of these arrays is that they are available to individual research laboratories. The cDNA probes can easily be generated by PCR amplification of plasmid DNA, and the resulting product can be purified over inexpensive purification columns and spotted on fairly inexpensive glass slides. Furthermore, one can be very flexible in what is spotted on each microarray. This technique can, for example, be applied to make a small array of genes that are relevant to a certain disease type. The disadvantage is the 10 to 20% drop-out rate of failed PCR reactions, failed spots, and misidentified clones.

COMPARABILITY OF EXPRESSION MEASUREMENTS MADE BY OLIGONUCLEOTIDE AND cDNA MICROARRAYS

Oosterhoff et al.[1,2] performed comparisons of different microarrays and found this a very unrewarding task. The most important problem in comparisons is that the arrays are not the same and have only limited overlaps in genes that are studied. Comparing microarrays works best if as many parameters as possible are the same. If possible, the same cell lines, same time frame of experimentation, same method of isolating RNA and, most importantly, the same microarrays should be used. Our laboratory has now embraced a single microarray system (Affymetrix) and this has enhanced inter-experiment comparability. Furthermore, when we tried to confirm our microarray data using Northern blotting or RT-PCR, we found that the oligonucleotide arrays worked best for us for high and low expression level genes.

FUNDAMENTALS OF MICROARRAY EXPERIMENTS

A good microarray experimental design is the first step toward obtaining interpretable data and should contain the following elements:

1. A well thought-through biological question should be investigated. It may seem that control experiments unnecessarily make an investigation much more expensive; this is

not true. Omitting controls will harm the interpretability of results and journal reviewers will ask for comparisons with controls.

2. The experimental treatment of samples should be as little affected by systematic and experimental errors as possible. For example, treatment of cultured cells with hormones dissolved in different solvents introduces an extra parameter that should be controlled.

3. Biological samples should be of the best quality and purity. Human tissue samples obtained at surgery should, for example, be snap-frozen in liquid nitrogen and stored below −80°C. Alternatively the tissues can be stored for a limited time in RNA*later*® (Ambion, Inc., Austin, Texas, USA).[3]

4. Isolated RNA should be of superb quality. It is not a question whether the RNA is as intact as possible; RNA should always be 100% intact. Variations in RNA quality will appear in the final analysis and are likely to lead to wrong answers.

5. An experiment should be in agreement with the Minimum Information about a Microarray Experiment (MIAME) guidelines[4] of the international Microarray Gene Expression Data (MGED) Society.[5] The guidelines are specific about experimental design, including the number of replicates, and this allows researchers to interpret one another's data more easily. The MGED Society has effectively developed data reporting guidelines, but has not addressed issues of data generation and interpretation. The latter are more intimately coupled to specific experimental platforms. In addition, microarray manufacturers such as Affymetrix have implemented MIAME-compliant data output in their new software releases.

6. One should take great care in choosing appropriate statistical methods for low level analysis (image analysis, data quality check, and data normalization) and high level analysis (estimation of magnitude and significance of differential gene expression) to reach biological conclusions.

On the basis of our experiences and those of others, we cannot stress strongly enough the importance of great experimental care, well characterized and rigorous analysis, and the need for appropriate follow-up and verification in the performance of microarray expression experiments.

SOURCES OF VARIATIONS IN MICROARRAY EXPERIMENTS

There are three sources of variation in a microarray experiment. The first is biological variation among different organisms from the same species. Variations can be substantial, especially in conducting array experiments on human tissues. This relates to the fact that such experiments are usually poorly controlled. Differences in age, body mass index (BMI), feeding status, and time of surgery are usually difficult to control (Figure 7.1A). For inbred laboratory animals, many of these variations disappear and variation among cell lines is virtually gone (Figure 7.1B). However, if one compares cell line experiments conducted at different points in time, variation can increase again due to different passage numbers or, for example, the use of different preparations of fetal calf serum.

The second source of technical variation is introduced during dissection of tissue and the extraction, labeling, and hybridization of RNA samples to the array. We have performed microarrays on endometrial tissues from women treated with different hormones. Upon surgery, the patients' uteri were dissected and immediately frozen in liquid nitrogen. In order to make sure that only endometrial tissue was evaluated, cryosections were produced and only those sections containing 100% of the desired tissue were used to isolate RNA (Figure 7.2).

Finally, measurement variation may be associated with reading the fluorescence signals that may be affected by factors such as dust on the array or in the scanner.

FIGURE 7.1 Oligonucleotide microarray results from endometrial samples from human patients (A) and human endometrial cancer cell lines (B). RNA was isolated and quality controlled; cRNA was generated and subsequently tagged for fluorescence staining. The labeled RNA was then hybridized to the microarray (U133plus 2 array; Affymetrix), after which excess of labeled mRNA was washed away. Panel A represents the expression profiles of 650 selected genes in the endometria of seven post-menopausal women. Panel B represents the expression profiles of 3500 selected genes in six independent control samples of the ECC1-PRAB-72 human endometrial carcinoma cell line.

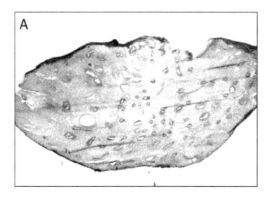

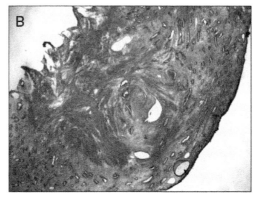

FIGURE 7.2 Pure endometrium used for gene expression profiling (A), and endometrium contaminated with myometrium (B) that was not used for RNA extraction for gene expression profiling. Macroscopically dissected endometrium was flash-frozen in liquid nitrogen. Subsequently sections were prepared using a cryostat. Every tenth section was stained and evaluated for pure endometrium. Only those sections that were in between two stained sections that contained 100% endometrium were used for RNA isolation for profiling. This way we were sure that only 100% endometrium was profiled.

IMPROVEMENT OF MICROARRAY EXPERIMENTS

The following general guidelines are intended to improve the quality of microarray experiments:

1. When the microarray data are very precise, it may be worthwhile to consider adding more biological samples rather than performing technical replications.
2. Technical replications are generally necessary when measurements show high variability.
3. In cases of a high degree of variability among individual samples, the samples should be pooled.

We have found that all handling and tissue dissection must be conducted with great care, precision, and accuracy to obtain the highest quality, most meaningful, and reproducible results in expression measurements of human tissues.

EXPERIMENTAL DESIGNS FOR cDNA MICROARRAYS

Most of the considerations noted above apply to both oligonucleotide and cDNA microarrays. However, because cDNA microarrays involve comparing two RNA samples in one experiment, special attention should be paid to the designs of these experiments:

Direct comparison — The comparison of most interest should be made directly on one array, for example, control versus treatment.

Dye swap — On array 1, the control sample is assigned to the red dye and the treatment sample is assigned to the green dye. On array 2, the dye assignments are reversed.

Reference sample design — In most experimental designs, a sample of interest is compared to a standard reference sample. Ideally the reference sample should remain the same for different experiments conducted over many years. Therefore, the reference material should be plentiful and remain stable over time. The use of a reference sample has certain advantages and disadvantages.

One disadvantage is that half the measurements of an experiment are made on a reference sample that is presumably of little or no interest. As a consequence, technical variation is increased compared to the level that can be achieved with direct comparisons. The main advantage is that the path connecting any two samples is never longer (or shorter) than two steps; thus all comparisons are made with equal efficiency. Reference designs can be extended (as long as the reference sample is available) to assay large numbers of samples collected over a long period. This reduces the possibility of laboratory error and increases the efficiency of sample handling in large projects.

DATA HANDLING

NORMALIZATION OF DATA

Although a number of normalization methods are available, it is difficult to decide which method performs best. Furthermore, normalization techniques for one microarray technology do not always apply to others. As a first step for both oligonucleotide and cDNA microarrays, one needs to decide which set of genes to use for normalization. Yang et al.[6] suggested three approaches:

Use all genes on the array.
Use only the constantly expressed (housekeeping) genes.
Use spiked (exogenous control genes added during hybridization) or control genes.

Normalization of Oligonucleotide Microarrays

Before a comparison of two arrays can be made, normalization must correct for variations between the two experiments caused by technical and biological factors. Normalization can be performed using either data from user-selected genes (for example, a group of housekeeping genes) or data from all genes on the array. Using data from all genes for normalization will introduce less bias.

The simplest approach to normalizing Affymetrix data is to re-scale each chip in an experiment by its total intensity. Variants of this approach scaling by trimmed mean intensity or by median intensity, are widely available in commercial software.[7] This approach does not deal particularly well with cases involving non-linear relationships between different arrays. For its new 133 series chips, Affymetrix introduced a new approach using a set of 100 so-called housekeeping genes. The chips are re-scaled so the average values of these housekeeping genes are equal across all chips.[8] It is believed that, in practice, these approaches are adequate for about 80% of chips.

Bolstad et al.[9] proposed three other methods for normalizing signal intensity levels of oligonucleotide array data. The first two methods, cyclic loss and contrast, are extensions of accepted normalization methods often used with cDNA microarray data. The third method based on quantiles is quicker and simpler.[9] It relies on the assumption that the distribution of gene abundances is nearly the same in all samples. For convenience, we take the pooled distribution of probe intensities on all chips. Then, for normalization of each chip, we compute for each value the quantile of that value in the distribution of probe intensities. For more information, access http://discover.nci.nih.gov/microarrayAnalysis/Affymetrix.Preprocessing.jsp.[8] We found that quantile normalization is able to reduce variance without increasing bias (Figure 7.3).

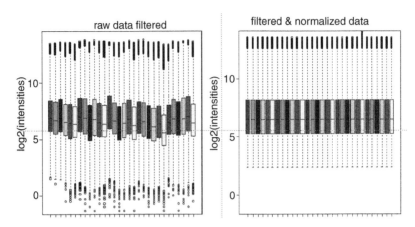

FIGURE 7.3 Quantile normalization. Thirty human samples (X-axis) were used for expression profiling. The log of measured expression intensities is depicted on the Y-axis. The mean ±SD (standard deviation) of one quantile is depicted in the boxes. On the left are the raw data; on the right the normalized data.

Normalization of cDNA Microarrays

The process of normalization of cDNA microarray data aims to remove systematic errors by balancing the fluorescence intensities of the two labeling dyes. The dye bias can come from various sources, including differences in dye labeling efficiencies or heat and light sensitivity, as well as from the scanner settings for scanning two channels at the same time.

Before deciding on the normalization method, the linearity of data (log ratio) should be checked. Linearity means that in the scatter plot of channel 1 (red color) *versus* channel 2 (green color), the relationship between the channels is linear. It is often more informative to produce a scatter plot of the log-transformed intensities, because the lowest intensities are better represented in such a plot. If the data are linear, such procedures as median centering can be applied. If the data is non-linear, "lowess smoothing"[10] or another local method should be applied.

In general, the normalization methods include (1) within-slide normalization, (2) paired-slide normalization for dye swap experiments, and (3) multiple slide normalization. For comprehensive descriptions of different normalization methods for cDNA arrays, see Yang et al.[10]

After normalization, the data for each gene are typically reported as an expression ratio or as the logarithm of the expression ratio. The expression ratio is simply the normalized value of the expression level for a particular gene in the query sample divided by its normalized value for the control.[11] Genes that are up-regulated by a factor of 2 have an expression ratio of 2, whereas those down-regulated by the same factor have an expression ratio of one half (0.5). As a consequence, down-regulated genes are situated between 1 and 0. The advantage of using the logarithm to base 2 of the expression ratio is simple to understand. A gene up-regulated by a factor of 2 has a $\log_2$ (ratio) of 1, whereas a gene down-regulated by a factor of 2 has a $\log_2$ (ratio) of −1. A gene expressed at a constant level (with a ratio of 1) has a $\log_2$ (ratio) of 0.[11]

It should be noted that using only expression ratios for data analysis has certain disadvantages. Although ratios can help to reveal some patterns in the data, they remove all information about the absolute gene expression levels.[11] Various parameters depend on the measured intensity, including the confidence limits placed on any microarray measurement. Although most of the techniques developed for analysis of microarray data use ratios, many of them can be adapted for use with measured intensities.[11]

IDENTIFYING SIGNIFICANTLY DIFFERENTIALLY EXPRESSED GENES

Regardless of the experiment performed, one outcome is the identification of genes whose expression levels are changed significantly between one or more pairs of samples in the data set. Identification of differentially expressed genes involves many considerations. There may be two or more experimental conditions; the conditions may be independent or related (time series). Many replicates may be present and they may be biological (samples from different animals) or technical (repeated hybridizations of the same samples). Reflecting this variety, many different methods are used for identifying significant changes.

Most of the early transcriptional profiling experiments used a fixed fold-change cut-off (generally two-fold) to identify the genes exhibiting the most significant variations.[12–14] Replication is essential in an experimental design because it allows an estimate of variability. The ability to assess such variability allows identification of biologically reproducible changes in gene expression levels. It has been suggested that three replications of a microarray experiment are needed to obtain a good estimate of the variation in gene expression for statistical calculations and to increase data reliability.[15]

As noted earlier, the two types of replication are biological and technical. Due to tissue heterogeneity or variability of expression among the same tissue types in a disease sample group, biological replication is particularly important for expression profiling. Technical replication eliminates many technical variations introduced during an experiment and provides a precise measurement of gene expression for a particular sample. Unfortunately, technical replication will not resolve the problem of biological variation. Therefore it is usually preferable to have biological rather than technical replications.[16,17]

Identification of differentially expressed genes consists of two parts: choosing a statistical test and determining the significance of the observation. Various tests can detect significant changes among repeated measurements of a variable in two groups. A program known as statistical analysis of microarray (SAM) has been developed to identify differentially expressed genes and to estimate the probability that a given gene identified as differentially expressed is in fact a false positive (false discovery rate). SAM can be used to determine fold-change cut-offs for significantly expressed genes.[18] Other methods for determining differentially expressed genes include the Student's T test and its variants,[19] analysis of variance (ANOVA),[20] Bayesian method,[21,22] and the Mann–Whitney test.[23] When replicates are insufficient to allow accurate estimates of experimental variances, modeling methods that use pooled variance estimates may be helpful.[24,25]

FROM GENES TO BIOLOGY

EXPERIMENTAL DESIGN

Two experimental designs lend themselves to different types of statistical and visual analysis that will help interpret data: the cross-sectional study and the time series study.[26]

Cross-Sectional Study

The goal of a cross-sectional study is to identify genes or patterns of expression for diagnostic purposes or to identify relevant biochemical pathways. A very small number of genes are identified and correlated and therefore are predictive of the biological variables of interest. The gene or pattern selection must be validated by predictive computer modeling (internal cross-validation) or, preferably, validated prospectively on new data.[27,28]

Time Series Study

In time series expression experiments, a temporal process is measured; a snapshot of the expression of genes in different samples is measured in cross-sectional expression experiments. Time series data exhibit a strong autocorrelation between successive points. The action of the gene as a function of time can quickly be assessed, while static data from the sample population (e.g., housekeeping genes) are assumed to be identically distributed. Time series profiles act as biomarkers that can be used to predict the underlying mechanisms of control of gene expression.[29]

TYPES OF DATA ANALYSIS

The current methodology for analyzing microarray data sets can be divided into unsupervised and supervised approaches.

Unsupervised Approach

The unsupervised approach attempts to characterize internal structures or relationships in a data set without any previous knowledge or classification scheme. Common categories of unsupervised techniques include (1) feature determination or determination of genes with interesting properties without looking for a particular pattern, for example, principal components analysis (PCA),[30,31] and (2) cluster determination or determination of groups of genes or samples with similar patterns of gene expression; different clustering methods can be applied to microarray data analysis.[11,32–37]

Unsupervised methods are used in cancer studies to determine whether discrete subsets of a disease entity can be defined based on unique expression profiles. Examples of such "class discovery" studies include those by Alizadeh et al.[38] and Bittner et al.[39] Alizadeh looked at diffuse B cell lymphomas and identified a subtype with a distinct expression pattern correlating with particular clinical implications such as expected survival time. Bittner et al. investigated human genes in melanoma cells and found a group associated with lower invasive ability, reduced motility, and (possibly) lower death rates. Class discovery by gene expression data has been attempted through a variety of clustering techniques.[35,40]

Supervised Approach

Another intriguing type of data analysis is training a classifier algorithm using the expression profiles of predefined sample groups, so that the classifier can assign any new sample to a predefined group.[41] Also known as supervised data analysis, this approach holds great promise in clinical diagnostics and has been used successfully in recent studies.[42–45]

Supervised methods include a training phase run on samples whose classes are already known and a testing phase in which the algorithm generalizes from the training data to predict the classifications of previously unseen samples.[24] Choosing a prediction method requires selecting from a vast range of techniques.

One widely used method is the support vector machine (SVM)[46,47] that uses a training set in which genes known to be related, for example, by function, are provided as positive examples and genes known not to be members of that class are negative examples.[11] Both types are combined into a set of training examples used by the SVM to learn to distinguish members of the class from non-members on the basis of expression data. Having learned the expression features of the class, the SVM can then be used to recognize and classify unseen genes on the basis of their expression. In this way, SVM uses biological information to determine expression features that are characteristic of a group in order to assign genes to that group.[11]

ONE STEP FURTHER: GENERATING A NETWORK

The complex functions of a living cell are carried out through the concerted activities of many genes and gene products. As a result, analyzing the microarray data in a pathway perspective could lead to a higher level of understanding of the system.[41] To fully understand a cellular system, we must combine data on proteins, metabolites, and gene expression. However, methods for profiling proteins and metabolites are not yet as widespread as methods for profiling gene expression.

Genetic networks are models that represent relationships of gene activities. The detailed molecular mechanisms of how the product of one gene affects the expression of another gene are often unknown but the eventual effect can easily be observed in a gene expression experiment. Various methods have been proposed for constructing a network from these kinds of microarray data. The simplest methods for modeling the interactions of genes are Boolean networks in which a 1 or 0 is used to express simply whether a gene is or is not induced. The induction of each gene is a deterministic function of the state of a set of other genes.[39] These representations are easy to compute and require a minimum number of parameters to be estimated but may be too simplified.[48,49] Readers are referred to three excellent reviews on these and other methods for reverse engineering of networks.[50–52]

Different approaches are to study the expression data in a pathway perspective or genetic network models using text mining and expert knowledge. In the first approach, one can map expression data onto metabolic or regulatory pathways and evaluate which pathways are most affected by transcriptional changes in whole genome expression experiments. Statistical methods like the Fisher exact test are used to score biochemical pathways according to the probability that a certain number of genes in a pathway significantly altered in a given experiment more than by chance alone. Results from multiple experiments can be compared, reducing the analysis of a full set of differentially expressed genes to a limited number genes acting in pathways of interest. Examples of such programs are GO-Mapper,[53] Ease,[54] GoMiner,[55] and the FatiGo module of GEPAS.[56]

The basic idea for the construction of a genetic network based on literature mining and expert knowledge is the assumption that if two genes are co-mentioned in a MEDLINE record, there is an underlying biological relationship. The extracted information can be stored in a database that can be integrated with other public databases such as the complete Kyoto Encyclopedia of Genes and Genomes,[57] the Database of Interacting Proteins,[58] the Bimolecular Interaction Network Database,[59] and Gene Ontology.[60]

The integrated database would contain biological knowledge represented in a formalized form with the focus on how proteins, cellular processes, and small molecules interact with, modify, and regulate each other.[61] Examples of programs that help interpret experimental results in the context of pathways, gene regulation networks, and protein interaction maps are PathwayAssists,[62] OmniViz,[63] and Ingenuity.[64] For an overview of programs for pathway construction and analysis, see http://ihome.cuhk.edu.hk/~b400559/arraysoft_pathway.html.[65] Figure 7.4 shows an example of a genetic network constructed with steroid hormone-regulated genes in an endometrium cancer cell line.

RNA AND SNP MICROARRAYS: OUTCOME PREDICTIONS IN OTHER CANCER TYPES

RNA MICROARRAYS

Several studies claim that clinical outcomes for cancer patients can be predicted using gene expression profiles of the primary tumors at diagnosis.[42,43,45,66–74] In essence, this approach involves the application of the (un)supervised classifications of gene expression profiles of tumors in the search for specific reporter genes for a preset parameter of interest — which is survival. These searches[42,43,45,66–73] lead to identification of gene expression patterns and a number of key genes in

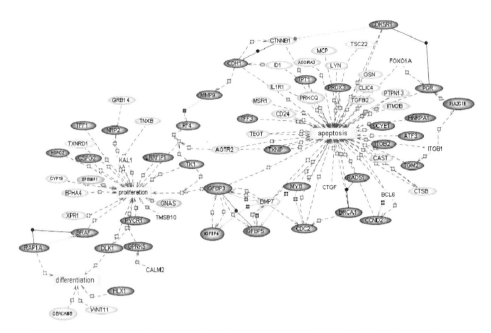

FIGURE 7.4 Genetic network of genes regulated by estrogen in the ECC1 cell line. Three cellular processes depicted are apoptosis, proliferation, and differentiation. The genes are indicated by dark gray circles (up-regulated), light gray circles (down-regulated), and no circles (unchanged expression). Types of connections: effect on expression (dashed gray line, gray box); effect on regulation (solid light gray line, light gray box); effect on binding (solid dark gray line, dark gray circle).

different subgroups of patients, for example, one subgroup with a high likelihood of dying from the disease and the other indicating increased long-term survival. To estimate the accuracy of a classification method, these studies use a training validation approach in which a training set is used to identify the molecular signature and a validation set is used to estimate the proportion of misclassifications.

Gene expression profiling of primary tumors is a much better predictor of patient survival than clinicopathological parameters.[42,66] Importantly, gene expression profiles are highly discriminative in biopsies that show inconclusive morphological features and prevent a pathologist from obtaining good information. For follicular lymphoma (FL), Glas et al.[66] showed that expression profiling accurately classified 93% of FL samples; in up to 30% of all cases, histologic grading and clinical criteria proved insufficient. Another example comes from Croner et al.[74] who calculated prediction rates for lymphatic metastasis using conventional clinicopathologic parameters, gene expression data, and a combination of both. They observed that the conventional estimates were improved by 9 to 12% when array data were added.

The major shortcoming of the outcome predictor system based on gene expression profiles was recently revealed by Michiels and colleagues.[75] Upon reanalyzing seven of the largest microarray studies in an attempt to predict cancer outcomes, they found that five of the seven studies did not classify patients better than chance. They concluded that published estimates often seem excessively optimistic, probably due to opportune selection bias in the analysis mode or in the validation processes.

SNP Microarrays

Single nucleotide polymorphism (SNP) microarrays provide three types of genetic information: loss of heterozygosity (LOH), chromosomal copy number changes, and genotype information. The genetic changes — copy number alterations or losses of heterozygosity — cause a cell to grow wildly out of control and activation of oncogenes or inactivation of tumor-suppressing genes often occurs. Most studies have used LOH and copy number methods to find genes associated with cancer or obtain refinements of tumor classifications.[76–90]

The application of genotype information provided by an SNP array in outcome prediction is limited to marker gene analysis and pharmacogenomics. An example of an SNP as a prognostic marker is the association of the SNPs in *CYP3A4* with severity of disease and long-term outcomes in prostate cancer (*CYP3A4* is involved in the oxidation of testosterone to hydroxytestosterone).[91,92] SNPs are widely applied in pharmacogenomics — the study of inter-individual differences in drug responses.[93] The two current approaches are (1) the search for SNPs associated with severe adverse effects that can be used to screen for subjects who should not receive the drug in question (e.g., *TPMT* genes and hematopoietic malignancies)[94,95] and (2) the identification of markers that predict drug efficacy (e.g., Cytochrome P-450).[96] Most marker gene studies that examine functional SNPs are performed on genes with known functions. Therefore, only very limited numbers of genes have been adequately studied.

COMBINING DATA FROM DIFFERENT SOURCES

Accurate tumor classification and optimal outcome prediction in our opinion require gene expression data and the inclusion of tumor genotype information. Distinct gene expression profiles partly arise from their associated genetic alterations (e.g. mutations and gene and chromosome deletions or amplifications). One example of the limitation of the sole use of gene expression profiles is that some prognostic indicators may escape detection by expression profiling because they are qualitative rather than quantitative, that is, genetic (allelic) differences may result in relevant changes in some genes that escape expression screening. They could encode proteins with different activity or stability that affects tumor progression or response to treatment.

Sole use of genotypic data also involves limitations. For example, it is not possible to determine the hierarchical effects of different genetic mutations on the gene expression signatures of tumors. In this respect, monitoring of SNPs (individual differences at a single base pair that mark a particular genetic variation in the population) would constitute an appealing complementary approach to gene expression profiling.[97] Valk et al.[98] showed that integration of gene expression profiling with cytogenetic data allowed a comprehensive classification of acute myeloid leukemia that includes previously identified genetically defined subgroups and a novel cluster with an adverse prognosis.

In conclusion, several challenges lie ahead. We must combine array results with other relevant independent results in order to build sensitive and specific classifiers that will help us answer specific questions. Furthermore, these laboratory-invented methods to improve patient outcome and well-being will have to be incorporated into daily medical practice and that will not be easy. What we need in the near future is a new specialization within the medical profession that will function in close collaboration with pathology: medical bioinformatics.

REFERENCES

1. Oosterhoff JK, Grootegoed JA, Blok LJ. Expression profiling of androgen-dependent and -independent LNCaP cells: EGF versus androgen signalling. *Endocr Relat Cancer* 2005; 12 (1): 135–148.
2. Oosterhoff JK, Kuhne LC, Grootegoed JA, Blok LJ. EGF signalling in prostate cancer cell lines is inhibited by a high expression level of the endocytosis protein REPS2. *Int J Cancer* 2005; 113 (4): 561–567.
3. Mutter GL, Zahrieh D, Liu C, Neuberg D, Finkelstein D, Baker HE, Warrington JA. Comparison of frozen and RNALater solid tissue storage methods for use in RNA expression microarrays. *BMC Genomics* 2004; 5 (1): 88.
4. www.mged.org/Workgroups/MIAME/miame.html

5. Brazma A, Hingamp P, Quackenbush J, Sherlock G, Spellman P, Stoeckert C, Aach J, Ansorge W, Ball CA, Causton HC, Gaasterland T, Glenisson P, Holstege FC, Kim IF, Markowitz V, Matese JC, Parkinson H, Robinson A, Sarkans U, Schulze-Kremer S, Stewart J, Taylor R, Vilo J, Vingron M. Minimum information about a microarray experiment (MIAME): toward standards for microarray data. *Nat Genet* 2001; 29 (4): 365–371.
6. Yang YH, Buckley MJ, Speed TP. Analysis of cDNA microarray images. *Brief Bioinform* 2001; 2 (4): 341–349.
7. www.bea.ki.se/staff/reimers/Web.Pages/Affymetrix.Normalization.htm
8. www.discover.nci.nih.gov/microarrayAnalysis/Affymetrix.Preprocessing.jsp
9. Bolstad BM, Irizarry RA, Astrand M, Speed TP. A comparison of normalization methods for high density oligonucleotide array data based on variance and bias. *Bioinformatics* 2003; 19 (2): 185–93.
10. Yang YH, Dudoit S, Luu P, Lin DM, Peng V, Ngai J, Speed TP. Normalization for cDNA microarray data: a robust composite method addressing single and multiple slide systematic variation. *Nucleic Acids Res* 2002; 30 (4): e15.
11. Quackenbush J. Computational analysis of microarray data. *Nat Rev Gen* 2001; 2 (6): 418–427.
12. DeRisi JL, Iyer VR, Brown PO. Exploring the metabolic and genetic control of gene expression on a genomic scale. *Science* 1997; 278 (5338): 680–686.
13. Schena M, Shalon D, Heller R, Chai A, Brown PO, Davis RW. Parallel human genome analysis: microarray-based expression monitoring of 1000 genes. *Proc Natl Acad Sci USA* 1996; 93 (20): 10614–10619.
14. Wodicka L, Dong H, Mittmann M, Ho MH, Lockhart DJ. Genome-wide expression monitoring in *Saccharomyces cerevisiae. Nat Biotechnol* 1997; 15 (13): 1359–1367.
15. Lee ML, Kuo FC, Whitmore GA, Sklar J. Importance of replication in microarray gene expression studies: statistical methods and evidence from repetitive cDNA hybridizations. *Proc Natl Acad Sci USA* 2000; 97 (18): 9834–9839.
16. Yang YH, Speed T. Design issues for cDNA microarray experiments. *Nat Rev Genet* 2002; 3 (8): 579–588.
17. Simon RM, Dobbin K. Experimental design of DNA microarray experiments. *Biotechniques* 2003; Suppl: 16–21.
18. Storey JD, Tibshirani R. SAM thresholding and false discovery rates for detecting differential gene expression in DNA microarraysm in *Analysis of Gene Expression Data: Methods and Software*, Parmigiani G, Garrett ES, Irizarry RA, et al., Eds. Springer: New York, 2003.
19. Lonnstedt I, Speed, TP. Replicated microarray data. *Stat Sin* 2002; 12: 31–46.
20. Kerr MK, Martin M, Churchill GA. Analysis of variance for gene expression microarray data. *J Comput Biol* 2000; 7 (6): 819–837.
21. Long AD, Mangalam HJ, Chan BY, Tolleri L, Hatfield GW, Baldi P. Improved statistical inference from DNA microarray data using analysis of variance and a Bayesian statistical framework: analysis of global gene expression in *Escherichia coli* K12. *J Biol Chem* 2001; 276 (23): 19937–19944.
22. Baldi P, Long AD. A Bayesian framework for the analysis of microarray expression data: regularized *t* -test and statistical inferences of gene changes. *Bioinformatics* 2001; 17 (6): 509–519.
23. Wu TD. Analysing gene expression data from DNA microarrays to identify candidate genes. *J Pathol* 2001; 195 (1): 53–65.
24. Slonim DK. From patterns to pathways: gene expression data analysis comes of age. *Nat Genet* 2002; 32, Suppl: 502–508.
25. Li C, Hung Wong W. Model-based analysis of oligonucleotide arrays: model validation, design issues and standard error application. *Genome Biol* 2001; 2 (8): 32.
26. Tumor Analysis Best Practices Working Group. Expression profiling: best practices for data generation and interpretation in clinical trials. *Nat Rev Genet* 2004; 5 (3): 229–237.
27. Ambroise C, McLachlan GJ. Selection bias in gene extraction on the basis of microarray gene-expression data. *Proc Natl Acad Sci USA* 2002; 99 (10): 6562–6566.
28. West M, Blanchette C, Dressman H, Huang E, Ishida S, Spang R, Zuzan H, Olson JA Jr, Marks JR, Nevins JR. Predicting the clinical status of human breast cancer by using gene expression profiles. *Proc Natl Acad Sci USA* 2001; 98 (20): 11462–11467.

29. Almon RR, Chen J, Snyder G, DuBois DC, Jusko WJ, Hoffman EP. *In vivo* multi-tissue corticosteroid microarray time series available online at Public Expression Profile Resource (PEPR). *Pharmacogenomics* 2003; 4 (6): 791–799.

30. Raychaudhuri S, Stuart JM, Altman RB. Principal components analysis to summarize microarray experiments: application to sporulation time series. *Pac Symp Biocomput* 2000; 5: 455–466.

31. Landgrebe J, Wurst W, Welzl G. Permutation-validated principal components analysis of microarray data. *Genome Biol* 2002; 3 (4): 19.

32. Sokal RR, Sneath PH. *Principles of Numerical Taxonomy.* Freeman: San Francisco, 1963, p. 8.

33. Wen X, Fuhrman S, Michaels GS, Carr DB, Smith S, Barker JL, Somogyi R. Large-scale temporal gene expression mapping of central nervous system development. *Proc Natl Acad Sci USA* 1998; 95 (1): 334–339.

34. Michaels GS, Carr DB, Askenazi M, Fuhrman S, Wen X, Somogyi R. Cluster analysis and data visualization of large-scale gene expression data. *Pac Symp Biocomput* 1998; 32-53.

35. Eisen MB, Spellman PT, Brown PO, Botstein D. Cluster analysis and display of genome-wide expression patterns. *Proc Natl Acad Sci USA* 1998; 95 (25): 14863–14868.

36. Heyer LJ, Kruglyak S, Yooseph S. Exploring expression data: identification and analysis of coexpressed genes. *Genome Res* 1999; 9 (11): 1106–1115.

37. Quackenbush J. Extracting meaning from functional genomics experiments. *Toxicol Appl Pharmacol* 2005; 207 (2 suppl): 195–199.

38. Alizadeh AA, Eisen MB, Davis RE, Ma C, Lossos IS, Rosenwald A, Boldrick JC, Sabet H, Tran T, Yu X, Powell JI, Yang L, Marti GE, Moore T, Hudson J Jr, Lu L, Lewis DB, Tibshirani R, Sherlock G, Chan WC, Greiner TC, Weisenburger DD, Armitage JO, Warnke R, Levy R, Wilson W, Grever MR, Byrd JC, Botstein D, Brown PO, Staudt LM. Distinct types of diffuse large B-cell lymphoma identified by gene expression profiling. *Nature* 2000; 403 (6769): 503–511.

39. Bittner M, Meltzer P, Chen Y, Jiang Y, Seftor E, Hendrix M, Radmacher M, Simon R, Yakhini Z, Ben-Dor A, Sampas N, Dougherty E, Wang E, Marincola F, Gooden C, Lueders J, Glatfelter A, Pollock P, Carpten J, Gillanders E, Leja D, Dietrich K, Beaudry C, Berens M, Alberts D, Sondak V. Molecular classification of cutaneous malignant melanoma by gene expression profiling. *Nature* 2000; 406 (6795): 536–540.

40. Alon U, Barkai N, Notterman DA, Gish K, Ybarra S, Mack D, Levine AJ. Broad patterns of gene expression revealed by clustering analysis of tumor and normal colon tissues probed by oligonucleotide arrays. *Proc Natl Acad Sci USA* 1999; 96 (12): 6745–6750.

41. Leung YF, Cavalieri D. Fundamentals of cDNA microarray data analysis. *Trends Genet* 2003; 19 (11): 649–659.

42. Pomeroy SL, Tamayo P, Gaasenbeek M, Sturla LM, Angelo M, McLaughlin ME, Kim JY, Goumnerova LC, Black PM, Lau C, Allen JC, Zagzag D, Olson JM, Curran T, Wetmore C, Biegel JA, Poggio T, Mukherjee S, Rifkin R, Califano A, Stolovitzky G, Louis DN, Mesirov JP, Lander ES, Golub TR. Prediction of central nervous system embryonal tumour outcome based on gene expression. *Nature* 2002; 415 (6870): 436–442.

43. Shipp MA, Ross KN, Tamayo P, Weng AP, Kutok JL, Aguiar RC, Gaasenbeek M, Angelo M, Reich M, Pinkus GS, Ray TS, Koval MA, Last KW, Norton A, Lister TA, Mesirov J, Neuberg DS, Lander ES, Aster JC, Golub TR. Diffuse large B-cell lymphoma outcome prediction by gene-expression profiling and supervised machine learning. *Nat Med* 2002; 8 (1): 68–74.

44. Khan J, Wei JS, Ringner M, Saal LH, Ladanyi M, Westermann F, Berthold F, Schwab M, Antonescu CR, Peterson C, Meltzer PS. Classification and diagnostic prediction of cancers using gene expression profiling and artificial neural networks. *Nat Med* 2001; 7 (6): 673–679.

45. vant Veer LJ, Dai H, van de Vijver MJ, He YD, Hart AA, Mao M, Peterse HL, van der Kooy K, Marton MJ, Witteveen AT, Schreiber GJ, Kerkhoven RM, Roberts C, Linsley PS, Bernards R, Friend SH. Gene expression profiling predicts clinical outcome of breast cancer. *Nature* 2002; 415 (6871): 530–536.

46. Brown MP, Grundy WN, Lin D, Cristianini N, Sugnet CW, Furey TS, Ares M Jr, Haussler D. Knowledge-based analysis of microarray gene expression data by using support vector machines. *Proc Natl Acad Sci USA* 2000; 97 (1): 262–267.

47. Furey TS, Cristianini N, Duffy N, Bednarski DW, Schummer M, Haussler D. Support vector machine classification and validation of cancer tissue samples using microarray expression data. *Bioinformatics* 2000; 16 (10): 906–914.

48. Liang S, Fuhrman S, Somogyi R. Reveal, a general reverse engineering algorithm for inference of genetic network architectures. *Pac Symp Biocomput* 1998; 3: 18–29.

49. Ideker TE, Thorsson V, Karp RM. Discovery of regulatory interactions through perturbation: inference and experimental design. *Pac Symp Biocomput* 2000; 5: 305–316.

50. de Jong H. Modeling and simulation of genetic regulatory systems: a literature review. *J Comput Biol* 2002; 9 (1): 67–103.

51. D'haeseleer P, Liang S, Somogyi R. Genetic network inference: from co-expression clustering to reverse engineering. *Bioinformatics* 2000; 16 (8): 707–726.

52. van Someren EP, Wessels LF, Backer E, Reinders MJ. Genetic network modeling. *Pharmacogenomics* 2002; 3 (4): 507–525.

53. Smid M, Dorssers LC. GO-Mapper: functional analysis of gene expression data using the expression level as a score to evaluate gene ontology terms. *Bioinformatics* 2004; 20 (16): 2618–2625.

54. Hosack DA, Dennis G Jr, Sherman BT, Lane HC, Lempicki RA. Identifying biological themes within lists of genes with EASE. *Genome Biol* 2003; 4 (10): R70.

55. Zeeberg BR, Feng W, Wang G, Wang MD, Fojo AT, Sunshine M, Narasimhan S, Kane DW, Reinhold WC, Lababidi S, Bussey KJ, Riss J, Barrett JC, Weinstein JN. GoMiner: a resource for biological interpretation of genomic and proteomic data. *Genome Biol* 2003; 4 (4): R28.

56. Herrero J, Al-Shahrour F, Diaz-Uriarte R, Mateos A, Vaquerizas JM, Santoyo J, Dopazo J. GEPAS: A web-based resource for microarray gene expression data analysis. *Nucleic Acids Res* 2003; 31 (13): 3461–3467.

57. www.genome.ad.jp/kegg

58. http://dip.doe-mbi.ucla.edu

59. http://bind.mshri.on.ca

60. www.geneontology.org

61. Hanifi-Moghaddam P, Gielen SC, Kloosterboer HJ, De Gooyer ME, Sijbers AM, van Gool AJ, Smid M, Moorhouse M, van Wijk FH, Burger CW, Blok LJ. Molecular portrait of the progestagenic and estrogenic actions of tibolone: behavior of cellular networks in response to tibolone. *J Clin Endocrinol Metab* 2005; 90 (2): 973–983.

62. www.ariadnegenomics.com

63. www.omniviz.com

64. www.ingenuity.com

65. http://ihome.cuhk.edu.hk/~b400559/arraysoft_pathway.html

66. Glas AM, Kersten MJ, Delahaye LJ, Witteveen AT, Kibbelaar RE, Velds A, Wessels LF, Joosten P, Kerkhoven RM, Bernards R, van Krieken JH, Kluin PM, van't Veer LJ, de Jong D. Gene expression profiling in follicular lymphoma to assess clinical aggressiveness and to guide the choice of treatment. *Blood* 2005; 105 (1): 301–307.

67. Kitahara O, Katagiri T, Tsunoda T, Harima Y, Nakamura Y. Classification of sensitivity or resistance of cervical cancers to ionizing radiation according to expression profiles of 62 genes selected by cDNA microarray analysis. *Neoplasia* 2002; 4 (4): 295–303.

68. Rosenwald A, Wright G, Chan WC, Connors JM, Campo E, Fisher RI, Gascoyne RD, Muller-Hermelink HK, Smeland EB, Giltnane JM, Hurt EM, Zhao H, Averett L, Yang L, Wilson WH, Jaffe ES, Simon R, Klausner RD, Powell J, Duffey PL, Longo DL, Greiner TC, Weisenburger DD, Sanger WG, Dave BJ, Lynch JC, Vose J, Armitage JO, Montserrat E, Lopez-Guillermo A, Grogan TM, Miller TP, LeBlanc M, Ott G, Kvaloy S, Delabie J, Holte H, Krajci P, Stokke T, Staudt LM; Lymphoma/Leukemia Molecular Profiling Project: use of molecular profiling to predict survival after chemotherapy for diffuse large-B-cell lymphoma. *New Engl J Med* 2002; 346 (25): 1937–1947.

69. Yeoh EJ, Ross ME, Shurtleff SA, Williams WK, Patel D, Mahfouz R, Behm FG, Raimondi SC, Relling MV, Patel A, Cheng C, Campana D, Wilkins D, Zhou X, Li J, Liu H, Pui CH, Evans WE, Naeve C, Wong L, Downing JR. Classification, subtype discovery, and prediction of outcome in pediatric acute lymphoblastic leukemia by gene expression profiling. *Cancer Cell* 2002; 1 (2): 133–143.

70. Beer DG, Kardia SL, Huang CC, Giordano TJ, Levin AM, Misek DE, Lin L, Chen G, Gharib TG, Thomas DG, Lizyness ML, Kuick R, Hayasaka S, Taylor JM, Iannettoni MD, Orringer MB, Hanash S. Gene-expression profiles predict survival of patients with lung adenocarcinoma. *Nat Med* 2002; 8 (8): 816–824.

71. Bhattacharjee A, Richards WG, Staunton J, Li C, Monti S, Vasa P, Ladd C, Beheshti J, Bueno R, Gillette M, Loda M, Weber G, Mark EJ, Lander ES, Wong W, Johnson BE, Golub TR, Sugarbaker DJ, Meyerson M. Classification of human lung carcinomas by mRNA expression profiling reveals distinct adenocarcinoma subclasses. *Proc Natl Acad Sci USA* 2001; 98 (24): 13790–13795.

72. Ramaswamy S, Ross KN, Lander ES, Golub TR. A molecular signature of metastasis in primary solid tumors. *Nat Genet* 2003; 33 (1): 49–54.

73. Iizuka N, Oka M, Yamada-Okabe H, Nishida M, Maeda Y, Mori N, Takao T, Tamesa T, Tangoku A, Tabuchi H, Hamada K, Nakayama H, Ishitsuka H, Miyamoto T, Hirabayashi A, Uchimura S, Hamamoto Y. Oligonucleotide microarray for prediction of early intrahepatic recurrence of hepatocellular carcinoma after curative resection. *Lancet* 2003; 361 (9361): 923–929.

74. Croner RS, Peters A, Brueckl WM, Matzel KE, Klein-Hitpass L, Brabletz T, Papadopoulos T, Hohenberger W, Reingruber B, Lausen B. Microarray versus conventional prediction of lymph node metastasis in colorectal carcinoma. *Cancer* 2005; 104 (2): 395–404.

75. Michiels S, Koscielny S, Hill C. Prediction of cancer outcome with microarrays: a multiple random validation strategy. *Lancet* 2005; 365 (9458): 488–492.

76. Lindblad-Toh K, Tanenbaum DM, Daly MJ, Winchester E, Lui WO, Villapakkam A, Stanton SE, Larsson C, Hudson TJ, Johnson BE, Lander ES, Meyerson M. Loss-of-heterozygosity analysis of small-cell lung carcinomas using single-nucleotide polymorphism arrays. *Nat Biotechnol* 2000; 18 (9): 1001–1005.

77. Lindblad-Toh K, Winchester E, Daly MJ, Wang DG, Hirschhorn JN, Laviolette JP, Ardlie K, Reich DE, Robinson E, Sklar P, Shah N, Thomas D, Fan JB, Gingeras T, Warrington J, Patil N, Hudson TJ, Lander ES. Large-scale discovery and genotyping of single-nucleotide polymorphisms in the mouse. *Nat Genet* 2000; 24 (4): 381–386.

78. Huang J, Wei W, Zhang J, Liu G, Bignell GR, Stratton MR, Futreal PA, Wooster R, Jones KW, Shapero MH. Whole genome DNA copy number changes identified by high density oligonucleotide arrays. *Hum Genomics* 2004; 1 (4): 287–299.

79. Zhao X, Li C, Paez JG, Chin K, Janne PA, Chen TH, Girard L, Minna J, Christiani D, Leo C, Gray JW, Sellers WR, Meyerson M. An integrated view of copy number and allelic alterations in the cancer genome using single nucleotide polymorphism arrays. *Cancer Res* 2004; 64 (9): 3060–3071.

80. Schubert EL, Hsu L, Cousens LA, Glogovac J, Self S, Reid BJ, Rabinovitch PS, Porter PL. Single nucleotide polymorphism array analysis of flow-sorted epithelial cells from frozen versus fixed tissues for whole genome analysis of allelic loss in breast cancer. *Am J Pathol* 2002; 160 (1): 73–79.

81. Wang ZC, Lin M, Wei LJ, Li C, Miron A, Lodeiro G, Harris L, Ramaswamy S, Tanenbaum DM, Meyerson M, Iglehart JD, Richardson A. Loss of heterozygosity and its correlation with expression profiles in subclasses of invasive breast cancers. *Cancer Res* 2004; 64 (1): 64–71.

82. Paez JG, Lin M, Beroukhim R, Lee JC, Zhao X, Richter DJ, Gabriel S, Herman P, Sasaki H, Altshuler D, Li C, Meyerson M, Sellers WR. Genome coverage and sequence fidelity of φ29 polymerase-based multiple strand displacement whole genome amplification. *Nucleic Acids Res* 2004; 32 (9): e71.

83. Primdahl H, Wikman FP, von der Maase H, Zhou XG, Wolf H, Orntoft TF. Allelic imbalances in human bladder cancer: genome-wide detection with high-density single-nucleotide polymorphism arrays. *J Natl Cancer Inst* 2002; 94 (3): 216–223.

84. Hoque MO, Lee CC, Cairns P, Schoenberg M, Sidransky D. Genome-wide genetic characterization of bladder cancer: a comparison of high-density single-nucleotide polymorphism arrays and PCR-based microsatellite analysis. *Cancer Res* 2003; 63 (9): 2216–2222.

85. Lieberfarb ME, Lin M, Lechpammer M, Li C, Tanenbaum DM, Febbo PG, Wright RL, Shim J, Kantoff PW, Loda M, Meyerson M, Sellers WR. Genome-wide loss of heterozygosity analysis from laser capture microdissected prostate cancer using single nucleotide polymorphic allele (SNP) arrays and a novel bioinformatics platform dChipSNP. *Cancer Res* 2003; 63 (16): 4781–4785.

86. Dumur CI, Dechsukhum C, Ware JL, Cofield SS, Best AM, Wilkinson DS, Garrett CT, Ferreira-Gonzalez A. Genome-wide detection of LOH in prostate cancer using human SNP microarray technology. *Genomics* 2003; 81 (3): 260–269.

87. Wong KK, Tsang YT, Shen J, Cheng RS, Chang YM, Man TK, Lau CC. Allelic imbalance analysis by high-density single-nucleotide polymorphic allele (SNP) array with whole genome amplified DNA. *Nucleic Acids Res* 2004; 32 (9): e69.

88. Zhou X, Mok SC, Chen Z, Li Y, Wong DT. Concurrent analysis of loss of heterozygosity (LOH) and copy number abnormality (CNA) for oral premalignancy progression using the Affymetrix 10K SNP mapping array. *Hum Genet* 2004; 115 (4): 327–330.

89. Janne PA, Li C, Zhao X, Girard L, Chen TH, Minna J, Christiani DC, Johnson BE, Meyerson M. High-resolution single-nucleotide polymorphism array and clustering analysis of loss of heterozygosity in human lung cancer cell lines. *Oncogene* 2004; 23 (15): 2716–2726.

90. Rauch A, Ruschendorf F, Huang J, Trautmann U, Becker C, Thiel C, Jones KW, Reis A, Nurnberg P. Molecular karyotyping using an SNP array for genomewide genotyping. *J Med Genet* 2004; 41 (12): 916–922.

91. Rebbeck TR, Jaffe JM, Walker AH, Wein AJ, Malkowicz SB. Modification of clinical presentation of prostate tumors by a novel genetic variant in CYP3A4. *J Natl Cancer Inst* 1998; 90 (16): 1225–1229. Erratum in *J Natl Cancer Inst* 1999; 91 (12): 1082.

92. Paris PL, Kupelian PA, Hall JM, Williams TL, Levin H, Klein EA, Casey G, Witte JS. Association between a *CYP3A4* genetic variant and clinical presentation in African-American prostate cancer patients. *Cancer Epidemiol Biomarkers Prev* 1999; 8 (10): 901–905.

93. Kalow W, Tang BK, Endrenyi L. Hypothesis: comparisons of inter- and intra-individual variations can substitute for twin studies in drug research. *Pharmacogenetics* 1998; 8 (4): 283–289.

94. Krynetski EY, Tai HL, Yates CR, Fessing MY, Loennechen T, Schuetz JD, Relling MV, Evans WE. Genetic polymorphism of thiopurine S-methyltransferase: clinical importance and molecular mechanisms. *Pharmacogenetics* 1996; 6 (4): 279–290.

95. Black AJ, McLeod HL, Capell HA, Powrie RH, Matowe LK, Pritchard SC, Collie-Duguid ES, Reid DM. Thiopurine methyltransferase genotype predicts therapy-limiting severe toxicity from azathioprine. *Ann Intern Med* 1998; 129 (9): 716–718.

96. Kroemer HK, Eichelbaum M. "It's the genes, stupid": Molecular bases and clinical consequences of genetic cytochrome P450 2D6 polymorphism. *Life Sci* 1995; 56 (26): 2285–2298.

97. Cargill M, Altshuler D, Ireland J, Sklar P, Ardlie K, Patil N, Shaw N, Lane CR, Lim EP, Kalyanaraman N, Nemesh J, Ziaugra L, Friedland L, Rolfe A, Warrington J, Lipshutz R, Daley GQ, Lander ES. Characterization of single-nucleotide polymorphisms in coding regions of human genes. *Nat Genet* 1999; 22 (3): 231–238. Erratum in *Nat Genet* 1999; 23 (3): 373.

98. Valk PJ, Verhaak RG, Beijen MA, Erpelinck CA, Barjesteh van Waalwijk van Doorn-Khosrovani S, Boer JM, Beverloo HB, Moorhouse MJ, van der Spek PJ, Lowenberg B, Delwel R. Prognostically useful gene-expression profiles in acute myeloid leukemia. *New Engl J Med* 2004; 350 (16): 1617–1628.

8 Evolution of the Molecular Microbiology Laboratory

Gary W. Procop and Belinda Yen-Lieberman

CONTENTS

INTRODUCTION

Molecular methods, once only research tools, have now permeated to some degree every area of the clinical microbiology laboratory. Assays approved by the U.S. Food and Drug Administration (FDA) are available to detect fastidious organisms like *Neisseria gonorrhoeae* and *Chlamydia trachomatis*, as well as commonly encountered, easily cultivated bacteria such as Group B *Streptococcus* and methicillin-resistant *Staphylococcus aureus*.

The degree to which these methods are used varies, based often on the experience and comfort levels of laboratory medical technologists. However, other factors such as the availability of high quality, commercial diagnostic kits, the cost of such kits, and the reimbursement for tests performed for clinical care also influence the use of this technology. All indications suggest that the migration of microbiologic assays from traditional to molecular-based methods will continue.

The purpose of this chapter is to examine in part the current state of molecular diagnostics in the clinical microbiology laboratory, but moreover to predict what may happen in the not-too-distant future and to suggest what may facilitate the introduction or further implementation of molecular methods into the clinical microbiology laboratory.

A molecular microbiology laboratory uses a variety of techniques. Every method, regardless of its level of complexity, requires some type of specimen preparation. The preparation used prior to nucleic acid amplification varies from simple lysis procedures to complete nucleic acid extractions. The most common molecular methods currently used in the clinical microbiology laboratory can be broadly separated into signal amplification assays, nucleic acid (target) amplification assays — many of which are migrating to real-time or homogeneous formats — and often post-amplification analyses such as DNA sequencing. In addition, molecular typing is often used to determine whether strains of the same species are indistinguishable or different. These categories will be discussed in detail in this chapter.

MARKET MATURATION FOR MOLECULAR MICROBIOLOGY

Neither of the authors of this chapter is an economist. However, studies of the maturation of new markets are well documented. As laboratorians, we actively gain knowledge concerning such maturation in our field as we practice our professions throughout our lives. When a new economic niche becomes available or is discovered, a number of competitors rapidly emerge with products for consumers. For example, the availability of products for testing for the presence of West Nile virus soon followed the introduction of this zoonotic pathogen into North America. In the field of laboratory medicine, these new products are often diagnostic kits or other devices that industrial marketing organizations would have us believe everyone uses and attempt to convince us that we cannot possibly live without. The products produced vary in quality, costs, ease of use and other parameters.

As the markets for these products mature with time, the best products tend to emerge from a competition that is hopefully competently evaluated in a non-biased manner in the peer-reviewed literature. The market then usually resolves into stable competition between two or three superior vendors.

We are in the early stages of this type of competition in the molecular microbiology market. Recent advances in nucleic acid amplification technologies, namely homogeneous formats and advances in DNA sequencing, have made these tools more cost-effective, easier to use, and more readily available than ever. Because of these changes, molecular techniques are used both for detection and characterization of fastidious and/or unusual pathogens and also for the more rapid detection and characterization of commonly occurring microorganisms.

Manufacturer competition is currently stratified from vary large, well known vendors to upstart biotechnology companies that have developed novel approaches to current problems. It is anticipated that the novel technologies of many of these small biotechnology companies may be purchased by larger manufacturers that will incorporate them into their own portfolios of products. Conversely, some small companies will not survive in this marketplace due to the high level of competition, and their methods will not be used for diagnostic products. The intense competition at this stage of market maturation often drives down the costs of the assays as individual vendors seek ways to expand their customer bases.

Many of these vendors are competing for the same subset of consumers. For example, consumers who seek Group B *Streptococcus* detection have more than one kit to choose from. This environment produces competitive pressure on a manufacturer to produce technically sound assays that are easy to use and available at reasonable prices. These kits may be approved by the FDA or produced as an analyte-specific reagent (ASR) kit. If an ASR kit is used, the responsibility for validation of the performance of the assay is shifted from the manufacturer to the laboratory, as with currently used, laboratory-validated ("home-brew") assays. The pursuit of FDA approval versus the manufacturing of ASR kits is the choice of the manufacturer and is part of strategic planning for the use of finite resources, since FDA submission is costly. The future of ASR kits will largely be dictated by decisions made by the FDA. If the current situation persists, we predict these kits will become more widely used, and, as cost-competitive kits are produced for particular organisms, they will begin to replace many of the currently used, laboratory-validated assays. This should produce better standardization of test results because many of the variables (e.g., different vendors used for primer and probe synthesis) that differ among laboratories performing the same types of assays will be removed. In addition, we anticipate continued progress in clinical applications of DNA sequencing technologies, as instruments and software become more user friendly and microarrays for the assessment of infectious diseases become more routine.

FUTURE OF SPECIMEN PREPARATION AND NUCLEIC ACID EXTRACTION

Specimen preparation is one of the most important pre-analytic phases of testing under the direct control of a laboratory director. Another pre-analytic component critical to the optimal performance of an assay is the quality of specimens collected. Although quality is primarily the responsibility of the healthcare provider submitting the specimen, the laboratorian should provide guidelines and direction to healthcare providers and feedback when inadequate specimens are submitted for testing. The quality of a specimen affects the likelihood or probability that an assay will produce the correct result. For example, a deep respiratory specimen (sputum or bronchoalveolar lavage [BAL]) from a patient with *Mycobacterium tuberculosis* is more likely to produce a positive result by PCR, culture, or acid-fast staining, than an unsatisfactory upper respiratory specimen (saliva) from the same patient. The quality of some specimens, such as blood, may be easier to control than other specimens, such as deep respiratory specimens.

Several studies have revealed specimen parameters that may be used to predict the usefulness of subsequent molecular testing, but many more such studies are needed. For example, Tang and Persing have shown that none of 209 cerebrospinal fluid (CSF) specimens that had normal leukocyte and protein levels were positive for the presence of herpes simplex virus (HSV) by polymerase chain reaction (PCR).[1] From studies such as these, one may argue that unnecessary, expensive molecular testing may be curtailed based on specimen screening parameters. Such approaches have been used successfully for many years in traditional microbiology laboratories. It is likely that these approaches will also be taken in the future to achieve optimal predictive value from molecular assays, as data emerges from thoughtfully designed trials.

Assay requirements vary according to the different types of assays performed and often are recommended by manufacturers. Requirements for obtaining the best test results, however, include (1) high quality clinical specimens, (2) a work environment that will minimize specimen contamination, (3) a skilled workforce, and (4) timely reporting to the clinician of easily interpreted test results. A discussion of the specimen requirements for individual assays is beyond the scope of this text, but the predictive value of any assay will be improved if the clinician collects the specimen from a site likely to contain nucleic acid from the microorganism of interest and has the specimen transported to the laboratory in an appropriate manner.

The methods of nucleic acid extraction have evolved in conjunction with the evolution of nucleic acid amplification. Most recently, a number of rapid, efficient extraction methods have been developed to complement the rapid-cycle amplification and detection afforded by real-time PCR and other rapid nucleic acid amplification techniques. These methods have been modified to address small numbers of specimens, with the intention that individual, important clinical specimens will not have to wait for batch extraction and testing. The methods of nucleic acid preparation and extraction vary from the simple lysis of the target microorganism to a complete nucleic acid extraction resulting in a very pure nucleic acid product free of inhibitors of nucleic acid amplification.

The rapid lysis methods used are effective if a relatively high quantity of microorganisms is present. Such methods are inexpensive and very fast — taking possibly less than a minute per lysis. Lysis methods result in a specimen that does not consist of pure nucleic acids derived from the clinical specimen, but rather a lysed potpourri of everything present in the clinical specimen, such as mucus, blood contents, bacterial and fungal cell wall components, human and other cell membranes, proteins, and other constituents. It is obvious, once one considers these factors, that molecular test results from specimens prepared in such a manner will be less sensitive compared with those performed on the far purer nucleic acids obtained following a nucleic acid extraction. Conversely, lysis methods are often excellent and inexpensive ways to prepare bacteria or fungi for molecular testing following culture. These methods have been used successfully to detect mucosal colonization by bacteria, e.g., vaginal colonization by Group B *Streptococcus* and nasal colonization by *S. aureus*, wherein the results of PCR are often comparable with culture. Lysis specimen preparation methods would not be considered optimal processing prior to a nucleic acid amplification assay if highly sensitive results are needed.

A large number of high quality extraction kits and systems are commercially available. These vary in cost, hands-on-time, and the quality of the extracted nucleic acid product. Although the quality of the nucleic acid products varies within a range, there is a strong tendency for the development of reliable, reproducible nucleic acid extraction procedures. Several comparative studies have examined different extraction methods for the recovery of nucleic acids from organisms of interest from the specimen matrices from which they are normally recovered. For example, we compared the recovery of *Legionella pneumophila* DNA from BAL and sputum specimens using two automated and three manual methods.[2] We used a known quantity of *Legionella* in specimen matrices to examine the different methods of extraction. How these five assays performed when a complex specimen matrix (sputum) was examined is demonstrated in Table 8.1.

The results from such studies provide useful data that should help the laboratorian decide which extraction methods may be optimal for particular clinical specimens. Other issues such as cost, ease-of-use, and other parameters should be considered in addition to technical performance. Similarly, comparative studies have been performed with clinical specimens from infected patients when specimen quantity was sufficient to evaluate several methods. For example, redundant or excess urine samples from patients with BK or JC polyomaviruria were used to evaluate the extraction of the DNA from these viruses from urine specimens.[3] We recommend that prior to the implementation of a molecular assay, the laboratorian review the literature and concentrate efforts on methods that are most versatile (may be used for other assays), cost-effective, and above all will provide an extraction product that complements the quality of the molecular assay performed.

TABLE 8.1
Concentrations of *Legionella* DNA Extracted by Various Methods[a]

Specimen Type	Nucleic Acid Extraction Method and DNA Concentration				
	ViralXpress™ (cfu/ml)	QIAamp® DNA Mini Kit (cfu/ml)	High Pure PCR Template Preparation Kit (cfu/ml)	MagNA Pure (cfu/ml)	NucliSens® Extractor (cfu/ml)
Sputum (median concentration of *Legionella* DNA)	13,635	46,380	133,900	526,200	171,800
Bronchoalveolar lavage (mean concentration of *Legionella* DNA)	23,599	102,345	41,090	596,805	474,575

[a] A certain amount of *Legionella pneumophila* was added to all sputum specimens and a different certain amount was added to all BAL specimens. The starting DNA concentration and the specimen matrix were held constant and the method of extraction varied. The median concentrations recovered from the sputum and the mean concentrations from the BAL are shown. We refer readers interested in additional details to the published article.[2]

The future will find a variety of nucleic acid extraction procedures, the vast majority of which will produce high quality nucleic acid products free of inhibitors. As these products are evaluated over time and demonstrate consistently high quality nucleic acid products with highly effective removal of clinically relevant inhibitors, it is possible that regulations requiring the presence of internal amplification controls may become unnecessary.

Automation is another substantial advance in nucleic acid extraction methods that has occurred over the past decade and continues. Although manual methods of nucleic acid extraction may provide high quality products, they are labor intensive and involve several areas (e.g., pipetting) wherein human error may be introduced. Both of these issues are addressed to a large extent through automated platforms. Several automated and semiautomated methods are available. These vary from high throughput robots to semiautomated and automated systems designed to process small numbers of specimens. The high throughput extractors contribute significantly to labor savings, reproducibility, and the reduction of repetitive use-associated injuries such as carpal tunnel syndrome. These instruments, however, are costly and would likely only be effectively used by large laboratories that process large numbers of specimens.

At the other end of the spectrum, several automated and semiautomated nucleic acid extractors are capable of producing high quality end products and are designed to process small numbers of specimens. Examples of the types of automated and semiautomated platforms currently available are listed on Table 8.2. A typical run on one of these types of instruments may contain one to eight specimens. Such extractors answer the need for high quality extractions that do not depend on batch size, and would be most effectively used when linked with a rapid nucleic acid amplification test that produces important clinical and cost-saving information. For example, rapid extraction may be linked to reverse transcriptase polymerase chain reaction (RT-PCR) for the detection of enteroviruses in children with meningitis. The rapid and accurate diagnosis of children with enteroviral meningitis has been shown to result in decreased length of hospital stays, fewer ancillary test procedures, and the discontinuation of unnecessary antimicrobial agents.[4]

Another example (Figure 8.1 and Figure 8.2) is the rapid detection or exclusion of *M. tuberculosis* complex from a patient whose respiratory specimen contains acid-fast bacilli.[5] The molecular detection of *M. tuberculosis*, which may be merely confirmatory to an astute clinician, nevertheless removes any ambiguity that may be associated with the diagnosis and directs the continuation of therapy. Conversely, the detection of non-tuberculous mycobacteria changes the diagnosis entirely, removes the necessity for respiratory isolation, and may alter therapy. (Members of the MAI complex are treated very differently from members of the *M. tuberculosis* complex.)

The use of such diagnostic and cost-effective nucleic acid amplification assays is only possible in routine practice when linked to user-friendly, technically sound nucleic acid extraction methods such as those described. The future in this area will see the development of a variety of automated and semiautomated specimen preparation robots. The degree of technical excellence will be comparable, and consumers will make choices based on cost, throughput, and laboratory utilization. Many of the robotic systems and their general characteristics are listed in Table 8.2.

SIGNAL AMPLIFICATION ASSAYS

The signal amplification methods described combine some type of nucleic acid probe with the generation of a signal. The signal is often amplified through an enzymatic reaction. A signal from fluorescence *in situ* hybridization (FISH), however, may be directly observed microscopically following the hybridization of a fluorescently labeled probe to its complementary nucleic acid target. A variety of chemical and enzymatic modifications may be applied to FISH or chromogenic *in situ* hybridization if further amplification of the signal is necessary, but these matters are beyond the scope of this text. Alternatively, the hybridization may take place in a tube or microtiter plate and the signal detected with an instrument such as a fluorometer.

TABLE 8.2
Currently Available Automated and Semiautomated Extraction Systems

System and Manufacturer	Hands-on Time	Disposables	Method	Throughput	Comments
NucliSens Extractor (bioMerieux)	+	Many	Boom	Approx. 10 samples/hr	Excellent NA product; expensive, largely secondary to disposables
NucliSens MiniMag (bioMerieux)	+++	Few	Boom, MB	Approx. 12 samples/35 to 40 min	Labor intense; almost a manual system
NucliSens EasyMag (bioMerieux)	+	Few	MB	1 to 24 samples/hr	Combination of the best of bioMerieux NucliSens platforms; small size
MagNA Pure (Roche)	+	Few-to-Moderate	MB	Approx. 32 samples/90 min	Different NA kits available (e.g., mRNA); post-elution functions allow for dispensation into COBAS A-ring, LightCycler capillary tubes, or standard tubes; large size
AmpliPrep (Roche)	+	Moderate	MB	72 samples/ approx. 1.45 hr	FDA approved for HIV-1 and HCV extraction; various methods available; for use with COBAS AMPLICOR analyzer
MagNA Pure Compact	+	Few	MB	1 to 8 samples/20 to 40 min	Various NA kits available; small size
Qiagen BioRobot M48	+	Few	MB	6 to 48 samples; 6 in approx. 20 min; 48 in 3 to 3.5 hr	Various NA kits available; small size
Qiagen BioRobot EZ1	+	Few	Silica gel membrane	1 to 6 samples in 15 to 20 min	Various NA kits available; pre-programed protocol cards lead to additional expense
Corbett CAS-1820 X-Tractor Gene	+	Few	Various	96 samples in approx. 90 min	Open platform; various protocols available
Abbott m1000 Generic Extractor	+	Few	MB	Approx. 48 samples in 2 hr	Various NA kits available; large size

+ = minimal. ++ = moderate. +++ = extensive. NA = nucleic acid. MB = magnetic beads (compositions vary by vendor).

Table constructed from lecture information presented by R.L. Hodinka (Man versus Robot: Approaches to Nucleic Acid Extraction), Clinical Virology Symposium, Clearwater Beach, Florida, USA, May 8–11, 2005.

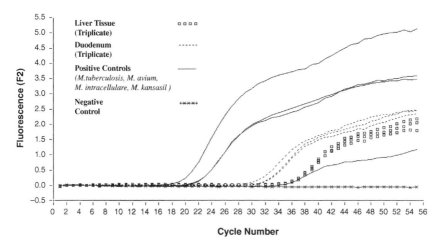

FIGURE 8.1 *(A color version follows page 270)* The positive amplification in this pan-mycobacteria PCR demonstrates the presence of mycobacterial DNA in the liver and duodenum of this patient.

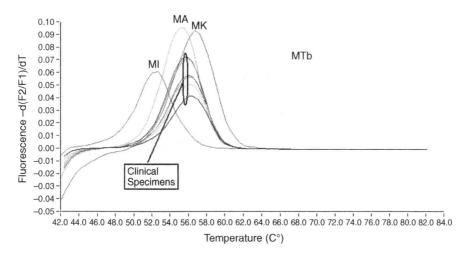

FIGURE 8.2 *(A color version follows page 270)* The post-amplification melt curve analysis of this assay differentiates *M. tuberculosis* from non-tuberculous mycobacteria. The patient is infected with a mycobacterium that is definitely not *M. tuberculosis*; it has a melt curve suggestive of *M. avium*. DNA sequence-based identification revealed *M. avium*.

Signal amplification is usually considered to be less analytically sensitive than nucleic acid amplification. One exception may be FISH, wherein a single copy of a target molecule may be identified in a large population of other cells. This application of FISH is more useful for anatomic pathology specimens or in research when a minor subpopulation of cells may be of clinical importance. Signal amplification technologies have several advantages over nucleic acid target amplification methods. These procedures are far less likely to produce false-positive results secondary to contamination compared with traditional nucleic acid amplification assays. However, signal contamination or the bleed-over of a strong signal from one reaction chamber into an adjacent well has been reported.[6] We will briefly discuss four types of signal amplification methods in this section: nucleic acid probes, Hybrid Capture, branched chain DNA, and *in situ* hybridization.

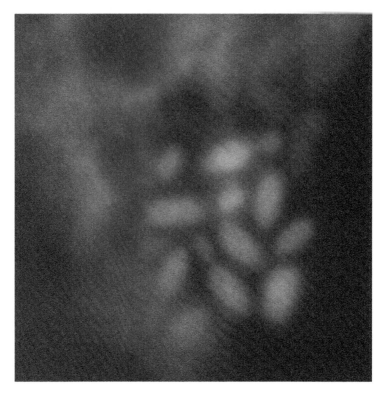

FIGURE 8.3 *(A color version follows page 270)* Yeast cells in a positive blood culture hybridize with a *Candida albicans* PNA FISH probe (AdvanDx, Inc., Woburn, Massachusetts, USA, http://www.advandx.com/) and demonstrate fluorescence that thereby identifies them as *C. albicans*. 1000× magnification.

NUCLEIC ACID PROBES

Nucleic acid probes were the first molecular assays to become commonplace in many laboratories. Commercially available AccuProbes® (Gen-Probe, Inc., San Diego, California, USA, http://www.gen-probe.com/) have been used for many years and continue to be used for the detection of numerous microorganisms. In brief, DNA probes hybridize with the target rRNA of the microorganism of interest. This is followed by an application of Gen-Probe's proprietary chemistry that allows differentiation of the hybridized from non-hybridized probe. Light is produced when a stable DNA probe:rRNA hybrid is formed and is measured with a luminometer. Such assays are available for the direct detection of bacteria in clinical specimens (e.g., Group A or B *Streptococcus*) or as a rapid and specific method of identifying microorganisms in culture.[7–9]

The latter application is an important means of rapidly and accurately identifying commonly encountered mycobacteria and systemic dimorphic fungi in culture. These assays are expected to continue to be used for such applications in the future. However, real-time PCR assays will compete with genetic probes for the direct analysis of clinical specimens. For example, a real-time PCR for Group A *Streptococcus* has already been described and is substantially more sensitive than culture.[10] In our experience, the sensitivity and specificity of the Group A *Streptococcus* AccuProbe on direct clinical specimens are comparable to those of cultures. The use of individual genetic probes for the identification of mycobacteria and systemic dimorphic fungi will have to compete with microarrays, nucleic acid sequencing, and/or reverse hybridization assays in the future, as these latter methods become more cost-competitive and user-friendly.

Hybrid Capture

The Hybrid Capture® system (Digene Corporation, Gaithersburg, Maryland, USA, http://www.digene.com/) is a signal amplification technology that consists of the capture (retention) of a DNA:RNA molecular complex (hybrid) in a tube or microtiter plate. Following specimen processing to prepare the DNA, RNA probes are used that are complementary to the DNA molecule for the microorganism of interest. If a DNA:RNA hybrid is formed, it is captured by an antibody that coats the walls of the test tube or microtiter plate and recognizes this unique nucleic acid complex. After this, another antibody is added which also recognizes the DNA:RNA complex, but this antibody is labeled with a reporter molecule (alkaline phosphatase) that can generate a chemiluminescent signal in the form of light.

An amount of relative light units (RLU) above a certain threshold is considered a positive reaction. Because the amount of light correlates with the amount of DNA:RNA complex captured, quantitative data may be generated with the use of external standards for calibration. Currently, the Hybrid Capture 2 assay (hc2) is the most commonly used, commercially available assay for the detection of high risk human papillomavirus (HPV) subtypes.

The high risk HPV assay has become an important supplemental assay for women with cytologic diagnoses of atypical squamous cells of undetermined significance (ASCUS) and as a routine assay to accompany cervical cytology for women over 30 years of age.[11,12] We believe that this assay will continue to be used in the future.

In addition, Hybrid Capture assays are available for the qualitative detection of *Neisseria gonorrhoeae* and *Chlamydia trachomatis* and the quantitative detection of hepatitis B virus (HBV) and cytomegalovirus (CMV).[13–17] Hybrid Capture technology will find competition with real-time PCR for the quantitative and qualitative detection of these pathogens and other pathogens, now and in the future.

Branched DNA

Branched DNA technology is used for the detection and quantification of pathogens such as human immune-deficiency virus (HIV), HBV, and hepatitis C virus (HCV). This technology consists of a variety of branched DNA (bDNA) probes associated with signal amplification reporter molecules. In brief, the organism-specific bDNA oligonucleotide probes hybridize to the nucleic acid sequences of the target organisms of interest and the complex is thereby captured onto a solid substrate. bDNA oligonucleotide reporter molecules conjugated to reporter enzymes are added and a chemiluminescent signal is generated after the addition of the appropriate substrate.

As with Hybrid Capture, the amount of signal is proportional to the amount of target nucleic acid. Therefore, quantitative results can be generated when this assay is performed with quantitative standards.[18,19] Applications of this technology have demonstrated high reproducibility and a wide linear range.[20–24] The future will likely hold some applications for bDNA, but the extent to which these assays will penetrate the marketplace and compete with alternate technologies such as quantitative real-time PCR remains to be seen.

In Situ Hybridization

In situ hybridization has been used successfully in molecular pathology for many years to detect chromosomal translocations and gene amplifications and to identify infectious agents.[25] Only recently have these techniques been shown to be potentially useful in the routine clinical microbiology laboratory, and *in situ* hybridization assays for infectious agents become commercially available.

The rapid identification of microorganisms from blood culture bottles that produce positive signals for a general screen has been the most popular use of this technology to date[26–31] (Figure 8.3). It is possible that *in situ* hybridization may retain this new niche position in the clinical microbiology laboratory of the future, but to do so, it will have to compete with several other technologies, such as real-time PCR.

NUCLEIC ACID AMPLIFICATION

The market for clinical assays for microbiology, as noted above, is early in the process of maturation, and much is predicted to occur before complete maturation is achieved. PCR-based assays currently dominate this market, but other methods of effective nucleic acid amplification. Strand displacement amplification (SDA), nucleic acid sequence-based amplification (NASBA), and transcription-mediated amplification (TMA) will continue to occupy part, if not a greater percentage, of the competitive marketplace. Regardless of the type of nucleic acid amplification used, the strong migration of these assays to real-time formats is expected to continue.

The use of nucleic acid amplification testing (NAT) in the clinical laboratory has changed, continues to change, and will continue to change in the future. Early in the use of such testing, the technical complexity and poor turn-around time limited the use of this technology to the detection of fastidious pathogens that could not be sufficiently detected by traditional means. Many years ago, semiautomated platforms brought the molecular detection of *N. gonorrhoeae* and *C. trachomatis* into many laboratories. The necessity of determining viral loads of patients infected with HIV to monitor their responses to antiretroviral therapy made it necessary to provide quantitative RT-PCR services on a routine basis. The most recent change that has affected the implementation of new molecular assays in the clinical laboratory has been the introduction of the real-time or homogeneous assays. These assays provide more timely results than traditional NAT assays, and because they are performed in closed systems, they present significantly diminished chances of amplicon contamination.[32] The small volumes employed make this technology utilizing relatively expensive fluorescently labeled oligonucleotide probes cost-competitive with cultures and other strategies of microorganism detection. However, the sensitivity of these assays may be limited in some instances by the low volume of specimen extract used. Typically, only 2 to 5 μl of specimen extract are present in a 20-μl LightCycler (Roche Applied Science, http://www.roche-applied-science.com) reaction. Therefore, if very low copies of the target organism exist in the extract of a clinical specimen, it may be a matter of statistical sampling to determine whether the DNA molecule of interest is pipetted into a test tube or remains in the extract eluant, present but untested. For this reason and in such situations, there has been a movement toward testing larger volumes of specimen extract by real-time PCR.

Reischl and colleagues effectively demonstrated this by comparing the results of five 20-μl PCR reactions with a single 100-μl PCR reaction for the detection of *M. tuberculosis* in clinical specimens from patients suspected of having tuberculosis.[33] Not surprisingly, the assays that used 100 μl total volume were uniformly positive, whereas fewer than five positive reactions usually resulted when the five 20-μl volume reactions were tested, demonstrating the necessity of testing larger volumes of extract. This is significant: by performing a 100-μl reaction, this study would have always characterized the test specimen (and therefore the patient) correctly, whereas a 20-μl reaction would have produced a significant number of false negative reactions. The use of small volume reactions may be appropriate for certain clinical specimens if the number of microorganisms is expected to be high enough to provide sufficient positive and negative predictive values for an assay. However, in the future, larger volume reactions may be favored for the detection of certain pathogens.

Several real-time assays have been shown to have excellent sensitivity and expanded dynamic ranges. Because of this expansion of dynamic range, the idea of separate less sensitive quantitative assays preceded by more a sensitive qualitative assay is becoming a thing of the past. For example, the recently released quantitative HCV TaqMan® (Roche Diagnostics, http://www.roche-diagnostics.com) assay has been demonstrated to have a 6-log dynamic range and would be expected to function as both a qualitative and quantitative assay.[34] A single, highly sensitive, quantitative real-time assay will likely replace the current quantitative and qualitative assays in the future.

Commercially available kits are available for the detection of several pathogenic microorganisms. These are either FDA approved or are ASR kits. The ASR kits are the basic component parts for the assay (primers, probes, and other master mix components). The laboratory that uses an ASR kit is responsible for the appropriate performance or validation of the assay when it is used for testing clinical specimens, just as it is responsible for the validation of any laboratory-designed (home-brew) assays. The long-term role of ASR and RUO (research use only) kits in the diagnostic arena is unclear and subject to the discretion of the FDA. However, in the foreseeable future the number and use of these kits are expected to increase at an exponential rate. Suppliers are currently vying for market share and are in direct competition with each other and also with the home-brew assays already in place in many laboratories throughout the country.

Because real-time NAT technology has become cost-effective, a number of kits have emerged and a large variety of studies examined the use of this technology for more timely detection of microorganisms that are not difficult to grow and were traditionally identified by culture. The two FDA-approved real-time PCR kits are used for the rapid detection of nasal colonization by methicillin-resistant *S. aureus* (MRSA) and the vaginal or rectal colonization of pregnant women by Group B *Streptococcus* — neither organism is difficult to cultivate. The use of real-time PCR for screening for these bacteria, however, affords a more rapid time to detection and avoids the need for subculture and traditional identification of suspect isolates present in the complex mixture of normal flora. How these tests will be used in the future remains to be seen, but we predict that in many institutions these assays will replace traditional approaches, particularly when only a qualitative answer is desired (e.g., presence or absence of MRSA) and additional susceptibility testing is not needed.

We, however, favor a merging of traditional and molecular methods in the clinical microbiology laboratory of the future and have had success with this approach in our own laboratory. For example, rather than simply testing a vaginal or rectal swab for the presence of Group B *Streptococcus* by real-time PCR (recommended by the manufacturer), we pre-incubate the collection swab in LIM broth to achieve growth (i.e., biologic amplification of the bacteria prior to PCR) which results in better sensitivity when the LIM broth is tested because more target organism is present. Furthermore, the LIM broth may be subcultured and antimicrobial susceptibility testing may be performed on the isolate for women who are allergic to penicillin. The merger of traditional microbiologic methods and newer molecular methods may be the best "new" technology of the future.

POST-AMPLIFICATION ANALYSIS

Southern blotting is the traditional method of post-amplification analysis. This is done to assure that the amplified product, usually first assessed by gel electrophoresis, is the intended product and not an aberrant product of similar size. Although useful, this technique is rarely employed in the modern clinical laboratory. It has largely been replaced by enzyme immunoassay detection and, more recently, by fluorescently labeled oligonucleotide probes that function in homogeneous amplification assays. The specific detection of products of nucleic acid amplification using enzyme immunoassays is not covered in detail here, but will likely continue to be used in the near future. This type of detection technology, however, may be replaced by homogeneous detection chemistries as nucleic acid amplification assays migrate to real-time platforms.

Melting Curves

Following real-time PCR, the products of amplification may be examined by post-amplification melt curve analysis. Melt curves may be performed when non-specific DNA binding dyes (e.g.,

SYBR Green) are used for amplicon detection and when specific hybridization probes are used. The melt curves generated by these dyes are useful in differentiating amplicons of different sizes and/or significantly different nucleotide compositions. For example, SYBR Green melting curves have been used following a broad-range PCR reaction to differentiate the four species of *Plasmodium* responsible for human malaria.[35] Similarly, the use of melt curves has been reported following a multiplex real-time PCR reaction for the detection and differentiation of influenza A, influenza B, and respiratory syncytial virus.[36]

The melt curves generated by specific hybridization probes are significantly different from those generated by non-specific DNA binding dyes. In addition, there are important differences in the abilities of different types of hybridization probes to generate melt curves as well as differences in the quality of the melt curves generated, all of which depend on the type of hybridization probe chemistry used. TaqMan probes are hydrolyzed during the real-time PCR reaction and therefore are not available after PCR for the generation of post-amplification melt curves.

Molecular beacons generate melt curves, but it has been our experience that these provide limited discrimination compared with the next probes described. Florescence resonance energy transfer (FRET) probes and eclipse probes are hybridization probes that generate the most useful melt curves. The curves generated by these probes are very sensitive and may be used to detect even single nucleotide polymorphisms in a probe hybridization site. For example, commercially available kits are used to detect the single nucleotide polymorphism responsible for Factor V Leiden coagulopathy. In the microbiology laboratory, these applications have been used to differentiate a variety of closely related microorganisms.

We have used FRET real-time PCR and post-amplification melt curve analysis to detect and differentiate the most commonly occurring *Bartonella* species, to distinguish *M. tuberculosis* complex from non-tuberculous mycobacteria, and to differentiate the BK and JC polyoma viruses (Figure 8.4). Others have used these features to differentiate HSV-1 from HSV-2.[37,38] We predict in the future such assays will become more commonly used to differentiate microorganisms that are closely related.

One cautionary note, however: Stevenson et al. have shown that there may be a compromised sensitivity in such assays for the organism for which the hybridization probe is <100% complementary, which is the basis for separating closely related microorganisms by melt curve analysis.[37]

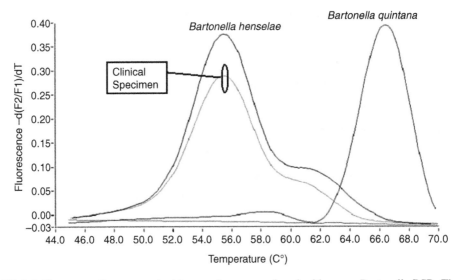

FIGURE 8.4 The extract from an excised heart valve was analyzed with a pan-*Bartonella* PCR. The post-amplification melt curve analysis was used to identify the infecting species as *B. henselae*.

For example, if the FRET probe is 100% homologous with the target DNA from HSV-1, but contains one or more nucleotide polymorphisms at the hybridization site for HSV-2, then it would be expected that the limit of detection would be better for HSV-1 compared with HSV-2. This may become a factor in the decision to use this technology or an alternate if low copy numbers are expected and if the presence of a microorganism may be missed because of a compromised sensitivity. This would become particularly important if the sensitivity of the assay varied to such a degree that the specimen was mischaracterized with regard to the presence or absence of the microorganism of interest (if a truly positive specimen was characterized as negative because of a lack of sensitivity).

DNA SEQUENCING

DNA sequencing, predominantly Sanger's sequencing or sequencing by termination, was once a tool used almost exclusively for research and discovery. This technology is now commonly available in many large microbiology and reference laboratories. It is used for a variety of clinical indications, but several other applications are possible.

One of the most popular uses is for the characterization of bacteria and fungi that are difficult to identify by phenotypic means. The 16S ribosomal DNA gene is a commonly used target that helps characterize a difficult-to-identify bacterium by placing it in a taxonomic category based on DNA sequence homology. In many instances, the 16S rDNA sequence may identify the bacteria of interest to the species level. Other genetic targets such as *rpoB* and *hsp* genes are alternative genetic targets and may be useful for the specific characterization of organisms in a particular group. For example, it has been demonstrated that the *rpoB* gene is likely superior to the 16S rDNA gene for the identification of *Corynebacterium* species.[39]

One limitation of sequencing a specific gene such as *rpoB* is that primers specific for the group of organisms under study must be used. Therefore, the optimal approach may begin with a generic 16S rDNA PCR followed by DNA sequencing to, if not definitively identify the bacterium, then at least to categorize it to the genus level. If the identification is not definitive, then one could resort to sequencing the product of a more specific gene such as *rpoB*.

In addition to taxonomic uses, sequencing has become the standard of care for determining mutations associated with HIV resistance to antiretroviral drugs. Similarly, DNA sequencing may be used to characterize cytomegalovirus isolates that are resistant to commonly used antiviral agents.[40,41] DNA sequencing may also be used following RT-PCR to determine the genotype of the hepatitis C virus, an important indicator in the prediction of a patient's response to therapy.[42,43]

DNA sequencing for the analysis of an amplified product is now a common method of post-amplification analysis. Over the past decade, DNA sequencing has become more user-friendly and now achieves a higher level of accuracy compared with former techniques. Nevertheless, this technology is complicated and requires users to have experience with sequence alignment and/or editing software and genetic databases. We predict the continued and expanded use of DNA sequencing in the clinical laboratory. Likely targets, in addition to those already discussed, include human genetic polymorphisms that may reveal increased susceptibility of patients for contracting infectious diseases.

Sequencing by Synthesis

A relatively new method of DNA sequencing known as sequencing by synthesis, in contrast to traditional Sanger sequencing or sequencing by termination, is now available.[44–48] This method, also known as Pyrosequencing® (Biotage Uppsala, Sweden), utilizes a novel and proprietary chemistry that consists of a mixture of enzymes that when used with the appropriate instrumentation detects the incorporation of the next single nucleotide added to the strand of DNA being synthesized.

The technology has a number of advantages in that it does not require the use of a gel, it is performed in a common microtiter plate, and is easy to perform. The disadvantage is that the length of sequence generated is small in comparison to sequences generated by Sanger sequencing. Approximately 30 bases constituted the longest sequence that we could generate with confidence, but there is some evidence that the length may be extended to 70 bases or more with modifications of the chemistry. This technology is particularly useful when variable regions that contain particular genetic information are identified. It is very useful for SNP analysis and may be used to identify single nucleotide polymorphisms associated with genetic diseases or resistance to antimicrobial agents.

In addition, Pyrosequencing has been used to interrogate variable regions that confer taxonomic information. In this regard, we have used it successfully for the identification of mycobacteria and nocardia and the differentiation of BK and JC polyoma viruses. (Figure 8.5) Similar to traditional 16S sequencing, but with less discriminating power, Pyrosequencing of a portion or portions of the 16S ribosomal DNA gene may be used to help categorize bacteria of unknown etiology.

Preliminary studies have been used to help categorize the fungi that are most commonly encountered in clinical specimens. DNA sequencing will continue to be used as a research tool to study microorganisms of interest and human genetic polymorphisms associated with increased susceptibility to infectious diseases. These assays, whether performed by traditional (Sanger) DNA sequencing or by the interrogation of relatively compact genetic elements by Pyrosequencing, will be present in molecular microbiology laboratories for many years to come.

REVERSE HYBRIDIZATION

The Southern blot was used to determine whether a product amplified by PCR that had the appropriate size as determined by gel electrophoresis was indeed the specific target of interest. This cumbersome and complex method required the transfer of the amplicon from the gel to a nitrocellulose membrane, and then probe hybridization was attempted using a specific, labeled oligonucleotide probe. Such assays took at least 2 days to perform and significantly hampered the turnaround times of clinical assays.

Reverse hybridization, either line probe or line blot assay, utilizes this technology in a more user-friendly, rapid format. It is termed reverse hybridization because the probes of interest are immobilized on the nitrocellulose membrane rather than the amplicon, which was the case with the Southern blot. With reverse hybridization, numerous probes may be immobilized along a nitrocellulose strip and then the amplicon generated by PCR or RT-PCR applied to the nitrocellulose membrane containing various probes. A reaction, which is usually chromogenic in nature, is then performed to determine where the hybridization occurs along the nitrocellulose strip. This may then be compared with the reference standard indicating where along the nitrocellulose strip the

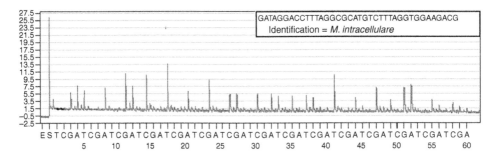

FIGURE 8.5 This pyrogram was generated following a broad-range mycobacteria PCR. The extract was from a cytology specimen that contained acid-fast bacilli, but was formalin fixed and therefore could not be cultured. The pyrosequencing of a portion of the hypervariable region A was used to make the definitive species identification.

probes of interest have been placed. The specific identification of the amplicon may be determined in this manner. Not surprisingly, weak hybridizations may occasionally occur between the amplicon and probes that are highly but not completely complementary, which may result in the presence of "ghost bands" that may interfere with interpretation.

This type of technology has been used to determine the most commonly occurring mutations associated with HIV resistance, to determine HCV genotypes, and to differentiate high and low risk HPV subtypes.[49–55] In addition to molecular virology, reverse hybridization has been used following broad range PCR to identify commonly occurring mycobacteria and fungi.[56–58] Advantages of this technology are that it is easy to use and often demonstrates a clearly discernible endpoint. Disadvantages are the inability to detect novel mutations that would be detected by DNA sequencing and the occasional "ghost bands" mentioned earlier.

Reverse hybridization assays are also relatively expensive, which has hampered their routine use for the identification of mycobacteria and fungi. We predict that reverse hybridization assays will continue to be used in microbiology laboratories in the future. Their use, however, depends to a large degree on cost and competitive technology. If costs decrease, such assays may become more routinely used for differentiating commonly occurring microorganisms and would represent competitive assays to DNA sequencing for the molecular identification of commonly occurring mycobacteria and fungi.

MICROARRAYS

Microarrays have served the scientific community well with regard to gene discovery and studies of genetic expression. These devices detect a large number of signals simultaneously, and may be used to detect DNA or RNA. The detection of mRNA signals is useful for determining which genes are expressed and which genes are silenced in response to various stimuli.

These assays, largely due to cost and the complexity of data management, have not made their way into routine use in clinical laboratories. However a DNA microarray that examines the genetic polymorphisms associated with drug metabolism within the cytochrome p450 system has been approved for use by the FDA. In addition to the commonly used fluorometric system, many other types of microarrays, several of which are being developed by small biotechnology companies, hold great promise in their discriminating abilities.

These microarrays may generate either fluorescent or electrical signals (bioelectric arrays) and in the future are likely to be employed for both the identification and characterization of microorganisms and genetic polymorphisms associated with increased susceptibilities of humans to infectious diseases. For example, microarrays have been used to differentiate with complete certainty the commonly encountered mycobacteria associated with human disease and simultaneously used to detect genetic polymorphisms associated with rifampin and kanamycin drug resistance.[59]

The primary limitations of using microarrays are cost, and for large microarrays, data management and interpretation. Issues of non-specific hybridization and the inclusion of appropriate quality controls also must be addressed for clinical applications. In the future, as costs decline and competition in this area increases, clinically useful microarray assays may become available for routine use in clinical microbiology laboratories. These technologies will compete directly with DNA sequencing for the rapid identification of microorganisms and for the detection of genetic polymorphisms associated with antimicrobial drug resistance and human susceptibility to infection.

OTHER METHODS OF POST-AMPLIFICATION ANALYSIS

Several other post-amplification analysis methods are likely to be used in the future. Older technologies such as enzyme immunoassays for the detection of PCR products continue to prove useful for the testing of clinical specimens. The Invader® assays (Third Wave Molecular Diagnostics, Madison, Wisconsin, USA, http://www.twt.com/) use a proprietary technology capable of differentiating single nucleotide polymorphisms.[60,61] This technology has been used to detect mutation-associated rifampin and isoniazid resistance in *M. tuberculosis*.[61] It could also be employed for a variety of other uses, from the differentiation of HCV and HPV subtypes to the differentiation of microorganisms in taxonomically related groups.

Another post-amplification technology that holds great promise is Luminex (Luminex Corporation, Austin, Texas, USA, http://www.luminexcorp.com/). It has the ability to resolve complex multiplex reactions using oligonucleotide probes that are covalently linked to tiny color-coded beads called microspheres. The nature of the amplicon is determined by the oligonucleotide with which it hybridizes, as determined by the color of the associated bead. Antibodies in place of the oligonucleotide probe may be used for the detection of antigen. Inflammatory markers and bacterial genomic DNA, and bacterial and viral proteins among many other applications have been analyzed using this platform.[63–66] More recently, this technology has been used to differentiate species within the genus *Trichosporon*[67] and holds the same promise as the microarray for the clinical laboratory.

MICROBIAL TYPING

Molecular epidemiologic tools (e.g., strain typing) are important and useful tools for hospital epidemiologists. The premier epidemiologists use such tools in conjunction with traditional epidemiologic methods to characterize and curtail epidemics that may occur on a large scale, such as food-borne epidemics that affect an entire nation, or are localized, such as epidemics occurring in a single intensive care unit. These methods of genetic typing, of which there are a considerable number, seek to characterize phenotypically similar organisms based on similarities and differences in their genetic materials.

The variety and specifics are beyond the scope of this chapter, and each laboratorian has his or her own favorite method. Issues with the use and interpretation of these assays often concern standardization and methods used to determine strain differentiation (indistinguishable versus different). Other important issues are availability, cost, and the need for a high level of technical expertise to perform these assays. These issues are particularly important for microbiology laboratories in community hospitals where financial support is limited and molecular technical expertise may be lacking. However, local epidemiology still plays a critical role in the control of outbreaks of infections. In such settings, these assays may be used only once or twice a year and the return on investment for installing such technology may be appropriately questioned by administrators. These laboratories would likely do better to outsource microbial typing to a reliable reference laboratory and limit typing to infections suspected to be related, as determined by an infection control committee.

In addition to traditional typing methods such as pulsed field gel electrophoresis, several alternative techniques may be easier to perform than traditional methods. Some methods such as rep-PCR, are commercially available and user friendly.[68,69] Unfortunately, some methods are expensive. The determination of typing methods that will be most common in the laboratories of the future remains unclear. While numerous questions regarding microbial typing, such as reimbursement for tests and budgets that will support such testing remain open, these tools are important to health and will to some degree be utilized by all laboratories in the future.

SUMMARY

The future of molecular microbiology is bright. Traditional research tools are being modified into diagnostic clinical assays. The skills of molecular biologists and the applications of the techniques are being brought to bear in the routine diagnosis and treatment of patients with infectious diseases in a same-day, real-time manner. These applications are not the ultimate answers in laboratory medicine, and are only tools that may be used appropriately or inappropriately. The tests are best performed by microbiologists who understand the technical aspects of the assays and also the clinical implications of the results. Judicious use of molecular diagnostic tests as defined by the outcomes of carefully designed trials and inputs from clinical colleagues will improve patient care through more rapid diagnostics and evidence-directed (rather than empiric) use of therapeutics.

REFERENCES

1. Tang YW, Hibbs JR, Tau KR, Qian Q, Skarhus HA, Smith TF, Persing DH. Effective use of polymerase chain reaction for diagnosis of central nervous system infections. *Clin Infect Dis* 1999; 29 (4): 803–806.
2. Wilson D, Yen-Lieberman B, Reischl U, Warshawsky I, Procop GW. Comparison of five methods for extraction of *Legionella pneumophila* from respiratory specimens. *J Clin Microbiol* 2004; 42 (12): 5913–5916.
3. Tang YW, Sefers SE, Li H, Kohn DJ, Procop GW. Comparative evaluation of three commercial systems for nucleic acid extraction from urine specimens. *J Clin Microbiol* 2005; 43 (9): 4830–4835.
4. Ramers C, Billman G, Hartin M, Ho S, Sawyer MH. Impact of a diagnostic cerebrospinal fluid enterovirus polymerase chain reaction test on patient management. *JAMA* 02000; 283 (20): 2680–2685.
5. Kaul KL. Molecular detection of *Mycobacterium tuberculosis*: impact on patient care. *Clin Chem* 2001; 47 (8): 1553–1558.
6. de Cremoux P, Coste J, Sastre-Garau X, Thioux M, Bouillac C, Labbe S, Cartier I, Ziol M, Dosda A, Le Gales C et al. Efficiency of the Hybrid Capture 2 HPV DNA test in cervical cancer screening: study by the French Society of Clinical Cytology. *Am J Clin Pathol* 2003; 120 (4): 492–499.
7. Overman SB, Eley DD, Jacobs BE, Ribes JA. Evaluation of methods to increase the sensitivity and timeliness of detection of *Streptococcus agalactiae* in pregnant women. *J Clin Microbiol* 2002; 40 (11): 4329–4331.
8. Bourbeau PP, Heiter BJ, Figdore M. Use of Gen-Probe AccuProbe Group B streptococcus test to detect group B streptococci in broth cultures of vaginal–anorectal specimens from pregnant women: comparison with traditional culture method. *J Clin Microbiol* 1997; 35 (1): 144–147.
9. Bull TJ, Shanson DC. Evaluation of a commercial chemiluminescent gene probe system "AccuProbe" for the rapid differentiation of mycobacteria, including MAIC X isolated from blood and other sites from patients with AIDS. *J Hosp Infect* 1992; 21 (2): 143–149.
10. Uhl JR, Adamson SC, Vetter EA, Schleck CD, Harmsen WS, Iverson LK, Santrach PJ, Henry NK, Cockerill FR. Comparison of LightCycler PCR, rapid antigen immunoassay, and culture for detection of group A streptococci from throat swabs. *J Clin Microbiol* 2003; 41 (1): 242–249.
11. ACOG Practice Bulletin. Clinical management guidelines for obstetrician–gynecologists: human papillomavirus. No. 61, April 2005. *Obstet Gynecol* 2005; 105 (4): 905–918.
12. Holmes J, Hemmett L, Garfield S. The cost-effectiveness of human papillomavirus screening for cervical cancer: a review of recent modelling studies. *Eur J Health Econ* 2005; 6 (1): 30–37.
13. Darwin LH, Cullen AP, Crowe SR, Modarress KJ, Willis DE, Payne WJ. Evaluation of the Hybrid Capture 2 CT/GC DNA tests and the GenProbe PACE 2 tests from the same male urethral swab specimens. *Sex Transm Dis* 2002; 29 (10): 576–580.
14. Darwin LH, Cullen AP, Arthur PM, Long CD, Smith KR, Girdner JL, Hook EW, 3rd, Quinn TC, Lorincz AT. Comparison of Digene Hybrid Capture 2 and conventional culture for detection of *Chlamydia trachomatis* and *Neisseria gonorrhoeae* in cervical specimens. *J Clin Microbiol* 2002; 40 (2): 641–644.
15. Schachter J, Hook EW, 3rd, McCormack WM, Quinn TC, Chernesky M, Chong S, Girdner JI, Dixon PB, DeMeo L, Williams E and others. Ability of the Digene Hybrid Capture II test to identify *Chlamydia trachomatis* and *Neisseria gonorrhoeae* in cervical specimens. *J Clin Microbiol* 1999; 37 (11): 3668–3671.

16. Lau GK, Leung YH, Fong DY, Au WY, Kwong YL, Lie A, Hou JL, Wen YM, Nanj A, Liang R. High hepatitis B virus (HBV) DNA viral load as the most important risk factor for HBV reactivation in patients positive for HBV surface antigen undergoing autologous hematopoietic cell transplantation. *Blood* 2002; 99 (7): 2324–2330.

17. Bhorade SM, Sandesara C, Garrity ER, Vigneswaran WT, Norwick L, Alkan S, Husain AN, McCabe MA, Yeldandi V. Quantification of cytomegalovirus (CMV) viral load by the Hybrid Capture assay allows for early detection of CMV disease in lung transplant recipients. *J Heart Lung Transplant* 2001; 20 (9): 928–934.

18. Sanchez-Pescador R SM, Urdea MS. Rapid chemiluminescent nucleic acid assays for detection of TEM-1 beta-lactamase-mediated penicillin resistance in *Neisseria gonorrhoeae* and other bacteria. *J Clin Microbiol* 1988; 26: 1934–1938.

19. Urdea MS HT, Fultz TJ, Anderson M, Running JA, Hamren S et al. Branched DNA amplification multimers for the sensitive, direct detection of human hepatitis viruses. *Nucleic Acids Symp Ser* 1991; 24: 197–200.

20. Gleaves CA, Welle J, Campbell M, Elbeik T, Ng V, Taylor PE, Kuramoto K, Aceituno S, Lewalski E, Joppa B et al. Multicenter evaluation of the Bayer VERSANT HIV-1 RNA 3.0 assay: analytical and clinical performance. *J Clin Virol* 2002; 25 (2): 205–216.

21. Elbeik T, Alvord WG, Trichavaroj R, de Souza M, Dewar R, Brown A, Chernoff D, Michael NL, Nassos P, Hadley K et al. Comparative analysis of HIV-1 viral load assays on subtype quantification: Bayer Versant HIV-1 RNA 3.0 versus Roche Amplicor HIV-1 Monitor version 1.5. *J Acquir Immune Defic Syndr* 2002; 29 (4): 330–339.

22. Peter JB, Sevall JS. Molecular-based methods for quantifying HIV viral load. *AIDS Patient Care Stds* 2004; 18 (2): 75–79.

23. Elbeik T, Surtihadi J, Destree M, Gorlin J, Holodniy M, Jortani SA, Kuramoto K, Ng V, Valdes R, Jr., Valsamakis A et al. Multicenter evaluation of the performance characteristics of the Bayer Versant HCV RNA 3.0 assay (bDNA). *J Clin Microbiol* 2004; 42 (2): 563–569.

24. Yao JD, Beld MG, Oon LL, Sherlock CH, Germer J, Menting S, Se Thoe SY, Merrick L, Ziermann R, Surtihadi J et al. Multicenter evaluation of the Versant hepatitis B virus DNA 3.0 assay. *J Clin Microbiol* 2004; 42 (2): 800–806.

25. McNicol AM, Farquharson MA. *In situ* hybridization and its diagnostic applications in pathology. *J Pathol* 1997; 182 (3): 250–261.

26. Jansen GJ, Mooibroek M, Idema J, Harmsen HJM, Welling GW, Degener JE. Rapid identification of bacteria in blood cultures by using fluorescently labeled oligonucleotide probes. *J Clin Microbiol* 2000; 38 (2): 814–817.

27. Kempf VA, K. Trebesius, I.B. Autenrieth. Fluorescent *in situ* hybridization allows rapid identification of microorganisms in blood cultures. *J Clin Microbiol* 2000; 38 (2): 830–838.

28. Hogardt M, Trebesius K, Geiger AM, Hornef M, Rosenecker J, Heesemann J. Specific and rapid detection by fluorescent in situ hybridization of bacteria in clinical samples obtained from cystic fibrosis patients. *J Clin Microbiol* 2000; 38 (2): 818–825.

29. Oliveira K, Brecher SM, Durbin A, Shapiro DS, Schwartz DR, De Girolami PC, Dakos J, Procop GW, Wilson D, Hanna CS et al. Direct identification of *Staphylococcus aureus* from positive blood culture bottles. *J Clin Microbiol* 2003; 41 (2): 889–891.

30. Oliveira K, Procop GW, Wilson D, Coull J, Stender H. Rapid identification of *Staphylococcus aureus* directly from blood cultures by fluorescence *in situ* hybridization with peptide nucleic acid probes. *J Clin Microbiol* 2002; 40 (1): 247–251.

31. Rigby S, Procop GW, Haase G, Wilson D, Hall G, Kurtzman C, Oliveira K, Von Oy S, Hyldig-Nielsen JJ, Coull J et al. Fluorescence *in situ* hybridization with peptide nucleic acid probes for rapid identification of *Candida albicans* directly from blood culture bottles. *J Clin Microbiol* 2002; 40 (6): 2182–2186.

32. Cockerill FR, 3rd. Application of rapid-cycle real-time polymerase chain reaction for diagnostic testing in the clinical microbiology laboratory. *Arch Pathol Lab Med* 2003; 127 (9): 1112–1120.

33. Burggraf S, Reischl U, Malik N, Bollwein M, Naumann L, Olgemoller B. Comparison of an internally controlled, large-volume LightCycler assay for detection of *Mycobacterium tuberculosis* in clinical samples with the COBAS AMPLICOR assay. *J Clin Microbiol* 2005; 43 (4): 1564–1569.

34. Forman MS, Valsamakis A. Verification of an assay for quantification of hepatitis C virus RNA by use of an analyte-specific reagent and two different extraction methods. *J Clin Microbiol* 2004; 42 (8): 3581–3588.

35. Mangold KA, Manson RU, Koay ES, Stephens L, Regner M, Thomson RB, Jr., Peterson LR, Kaul KL. Real-time PCR for detection and identification of Plasmodium spp. *J Clin Microbiol* 2005; 43 (5): 2435–2440.

36. Boivin G, Cote S, Dery P, De Serres G, Bergeron MG. Multiplex real-time PCR assay for detection of influenza and human respiratory syncytial viruses. *J Clin Microbiol* 2004; 42 (1): 45–51.

37. Stevenson J, Hymas W, Hillyard D. Effect of sequence polymorphisms on performance of two real-time PCR assays for detection of herpes simplex virus. *J Clin Microbiol* 2005; 43 (5): 2391–2398.

38. Burrows J, Nitsche A, Bayly B, Walker E, Higgins G, Kok T. Detection and subtyping of herpes simplex virus in clinical samples by LightCycler PCR, enzyme immunoassay and cell culture. *BMC Microbiol* 2002; 2 (1): 12.

39. Khamis A, Raoult D, La Scola B. Comparison between *rpoB* and 16S rRNA gene sequencing for molecular identification of 168 clinical isolates of Corynebacterium. *J Clin Microbiol* 2005; 43 (4): 1934–1936.

40. Scott GM, Isaacs MA, Zeng F, Kesson AM, Rawlinson WD. Cytomegalovirus antiviral resistance associated with treatment induced UL97 (protein kinase) and UL54 (DNA polymerase) mutations. *J Med Virol* 2004; 74 (1): 85–93.

41. Weinberg A, Jabs DA, Chou S, Martin BK, Lurain NS, Forman MS, Crumpacker C. Mutations conferring foscarnet resistance in a cohort of patients with acquired immunodeficiency syndrome and cytomegalovirus retinitis. *J Infect Dis* 2003; 187 (5): 777–784.

42. Zekri AR, El-Din HM, Bahnassy AA, El-Shehabi AM, El-Leethy H, Omar A, Khaled HM. TRUGENE sequencing versus INNO-LiPA for sub-genotyping of HCV genotype-4. *J Med Virol* 2005; 75 (3): 412–420.

43. Lole KS, Jha JA, Shrotri SP, Tandon BN, Prasad VG, Arankalle VA. Comparison of hepatitis C virus genotyping by 5′ noncoding region- and core-based reverse transcriptase PCR assay with sequencing and use of the assay for determining subtype distribution in *India*. *J Clin Microbiol* 2003; 41 (11): 5240–5244.

44. Diggle MA, Clarke SC. Pyrosequencing: sequence typing at the speed of light. *Mol Biotechnol* 2004; 28 (2): 129–137.

45. Ronaghi M, Elahi E. Pyrosequencing for microbial typing. *J Chromatogr B Analyt Technol Biomed Life Sci* 2002; 782 (1–2): 67–72.

46. Franca LT, Carrilho E, Kist TB. A review of DNA sequencing techniques. *Q Rev Biophys* 2002; 35 (2): 169–200.

47. Fakhrai-Rad H, Pourmand N, Ronaghi M. Pyrosequencing: an accurate detection platform for single nucleotide polymorphisms. *Hum Mutat* 2002; 19 (5): 479–485.

48. Ronaghi M. Pyrosequencing sheds light on DNA sequencing. *Genome Res* 2001; 11 (1): 3–11.

49. Stuyver L, Wyseur A, Rombout A, Louwagie J, Scarcez T, Verhofstede C, Rimland D, Schinazi RF, Rossau R. Line probe assay for rapid detection of drug-selected mutations in the human immunodeficiency virus type 1 reverse transcriptase gene. *Antimicrob Agents Chemother* 1997; 41 (2): 284–291.

50. Puchhammer-Stöckl E, Schmied B, Mandl CW, Vetter N, Heinz FX. Comparison of line probe assay (LiPA) and sequence analysis for detection of HIV-1 drug resistance. *J Med Virol* 1999; 57 (3): 283–289.

51. Castelain S, Zawadzki P, Khorsi H, Darchis JP, Capron D, Sueur JM, Eb F, Duverlie G. Comparison of hepatitis C virus serotyping and genotyping in French patients. *Clin Diagn Virol* 1997; 7 (3): 159–165.

52. Servais J, Lambert C, Fontaine E, Plesseria JM, Robert I, Arendt V, Staub T, Schneider F, Hemmer R, Burtonboy G et al. Comparison of DNA sequencing and a line probe assay for detection of human immunodeficiency virus type 1 drug resistance mutations in patients failing highly active antiretroviral therapy. *J Clin Microbiol* 2001; 39 (2): 454–459.

53. van Doorn LJ, Quint W, Kleter B, Molijn A, Colau B, Martin MT, Kravang I, Torrez-Martinez N, Peyton CL, Wheeler CM. Genotyping of human papillomavirus in liquid cytology cervical specimens by the PGMY line blot assay and the SPF(10) line probe assay. *J Clin Microbiol* 2002; 40 (3): 979–983.

54. Levi JE, Kleter B, Quint WG, Fink MC, Canto CL, Matsubara R, Linhares I, Segurado A, Vanderborght B, Neto JE et al. High prevalence of human papillomavirus (HPV) infections and high frequency of multiple HPV genotypes in human immunodeficiency virus-infected women in Brazil. *J Clin Microbiol* 2002; 40 (9): 3341–3345.

55. Zheng X, Pang M, Chan A, Roberto A, Warner D, Yen-Lieberman B. Direct comparison of hepatitis C virus genotypes tested by INNO-LiPA HCV II and TRUGENE HCV genotyping methods. *J Clin Virol* 2003; 28 (2): 214–216.

56. Tortoli E, Mariottini A, Mazzarelli G. Evaluation of INNO-LiPA MYCOBACTERIA v2: improved reverse hybridization multiple DNA probe assay for mycobacterial identification. *J Clin Microbiol* 2003; 41 (9): 4418–4420.

57. Miller N, Infante S, Cleary T. Evaluation of the LiPA MYCOBACTERIA assay for identification of mycobacterial species from BACTEC 12B bottles. *J Clin Microbiol* 2000; 38 (5): 1915–1919.

58. Martin C, Roberts D, van Der Weide M, Rossau R, Jannes G, Smith T, Maher M. Development of a PCR-based line probe assay for identification of fungal pathogens. *J Clin Microbiol* 2000; 38 (10): 3735–3742.

59. Troesch A, Nguyen H, Miyada CG, Desvarenne S, Gingeras TR, Kaplan PM, Cros P, Mabilat C. Mycobacterium species identification and rifampin resistance testing with high-density DNA probe arrays. *J Clin Microbiol* 1999; 37 (1): 49–55.

60. Mein CA, Barratt BJ, Dunn MG, Siegmund T, Smith AN, Esposito L, Nutland S, Stevens HE, Wilson AJ, Phillips MS et al. Evaluation of single nucleotide polymorphism typing with Invader on PCR amplicons and its automation. *Genome Res* 2000; 10 (3): 330–343.

61. Olivier M. The Invader assay for SNP genotyping. *Mutat Res* 2005; 573 (1–2): 103–110.

62. Cooksey RC, Holloway BP, Oldenburg MC, Listenbee S, Miller CW. Evaluation of the Invader assay, a linear signal amplification method, for identification of mutations associated with resistance to rifampin and isoniazid in *Mycobacterium tuberculosis*. *Antimicrob Agents Chemother* 2000; 44 (5): 1296–1301.

63. Dunbar SA, Vander Zee CA, Oliver KG, Karem KL, Jacobson JW. Quantitative, multiplexed detection of bacterial pathogens: DNA and protein applications of the Luminex LabMAP™ system. *J Microbiol Methods* 2003; 53 (2): 245–252.

64. Skogstrand K, Thorsen P, Norgaard-Pedersen B, Schendel DE, Sorensen LC, Hougaard DM. Simultaneous measurement of 25 inflammatory markers and neurotrophins in neonatal dried blood spots by immunoassay with xMAP technology. *Clin Chem* 2005; 51 (10): 1854–1866.

65. Giavedoni LD. Simultaneous detection of multiple cytokines and chemokines from nonhuman primates using luminex technology. *J Immunol Methods* 2005; 301 (1–2): 89–101.

66. Rao RS, Visuri SR, McBride MT, Albala JS, Matthews DL, Coleman MA. Comparison of multiplexed techniques for detection of bacterial and viral proteins. *J Proteome Res* 2004; 3 (4): 736–742.

67. Diaz MR, Fell JW. High-throughput detection of pathogenic yeasts of the genus trichosporon. *J Clin Microbiol* 2004; 42 (8): 3696–3706.

68. Rasschaert G, Houf K, Imberechts H, Grijspeerdt K, De Zutter L, Heyndrickx M. Comparison of five repetitive-sequence-based PCR typing methods for molecular discrimination of *Salmonella enterica* isolates. *J Clin Microbiol* 2005; 43 (8): 3615–3623.

69. Pounder JI, Williams S, Hansen D, Healy M, Reece K, Woods GL. Repetitive-sequence-PCR-based DNA fingerprinting using the Diversilab system for identification of commonly encountered dermatophytes. *J Clin Microbiol* 2005; 43 (5): 2141–2147.

Part II

Disease Applications of Nucleic Acid Tests

9 Bacterial Sexually Transmitted Diseases

Julius Schachter and Stephen A. Morse

CONTENTS

INTRODUCTION

Nucleic acid amplification (NAA) technologies have exerted a major impact on the field of sexually transmitted diseases (STDs). Better etiologic diagnoses, changed perceptions regarding the accuracy of clinical diagnoses, greater understanding of the epidemiology of STDs, and innovative public health and STD control programs have resulted from the use of nucleic acid amplification tests (NAATs).

It is likely that these tests produced their greatest impacts in the area of chlamydial infections. *Chlamydia trachomatis* is the most common sexually transmitted bacterial agent.[1] A consideration of the characteristics of these infections would predict that NAATs would exert major impacts. The infection is highly prevalent, difficult to diagnose, easy to treat, and most often asymptomatic in both males and females. It has serious and costly long term consequences as well. Programs of screening females for *C. trachomatis* and treating those found to be infected prevents pelvic inflammatory disease (PID) and presumably its late sequelae of tubal factor infertility and ectopic pregnancy as well as transmission from infected pregnant women to their newborns.[2,3] To a slightly lesser degree, this description also fits infections caused by *Neisseria gonorrhoeae*.

The need (as well as the market) for diagnostic tests for these infections was so great that major manufacturers made efforts to develop nonculture diagnostic and screening tests. The first to be introduced were antigen detection methods such as direct fluorescent antibody (DFA) and enzyme immunoassays (EIAs). These were followed by direct nucleic acid hybridization assays[4] that were rendered obsolete when the more sensitive NAATs became available.[5] The NAATs have evolved and have become more sensitive. The ways in which NAATs are used continue to evolve; for example, we have seen an increased use of vaginal swabs as specimens. The U.S. Food and Drug Administration (FDA) recently cleared the use of vaginal swabs for screening with the APTIMA® Combo 2 test (Gen-Probe Inc., San Diego, California, USA).

In this chapter, when considering detection of *N. gonorrhoeae* or *C. trachomatis*, we will restrict our discussions solely to commercially available NAATs. The literature dealing with these tests is extensive and allows for comparative evaluations. While many researchers have developed "home brew" tests, it is difficult to evaluate much of that work on a comparative basis. Some have applied NAA technology for purposes other than diagnosis, e.g., typing, but we will limit our discussion to diagnostic tests.

For some STDs, companies have perceived that the potential sizes of their markets are too small for them to commit resources to developing and marketing tests, but NAATs have been used to great advantage even in these circumstances. One example is the diagnosis of genital ulcer disease (GUD). Use of NAATs demonstrated unequivocally that clinical diagnosis was often inaccurate, and that herpes simplex virus was often a major but previously unappreciated contributor to the etiology of GUD in many parts of the world. As is customary, we will discuss the use of NAATs for the diagnosis of GUD separately from those used to detect etiologic agents that are more often found causing discharges from the lower genital tract.

LOWER GENITAL TRACT DISEASE-CAUSING AGENTS

CHLAMYDIA TRACHOMATIS AND NEISSERIA GONORRHOEAE

Chlamydia trachomatis and *Neisseria gonorrhoeae* are the two most common causes of bacterial STDs.[1] Both cause urethritis and epididymitis in males and mucopurulent endocervicitis and acute salpingitis in women (with potential sequelae of tubal factor infertility and ectopic pregnancy).[6] Both organisms can be transmitted to newborns. Infections are often asymptomatic (chlamydiae more so than gonorrhea) and are found at higher rates in sexually active adolescents. Their epidemiology and clinical consequences are generally similar enough to have led manufacturers to develop combined tests to detect both infections from a single specimen. Therefore, we will discuss these two agents together.

NAATs are the most sensitive diagnostic tests available for detecting both *C. trachomatis* and *N. gonorrhoeae* infections. They represent major improvements over the isolation of *C. trachomatis* in cell culture — the most sensitive method in the pre-NAAT era.[4,5] Under ideal circumstances, in high prevalence settings (as in STD clinics) with expert laboratories, *N. gonorrhoeae* culturing is sensitive and NAATs offer only a marginal improvement. However, in most screening settings, culture performance is far from optimal, and the NAAT sensitivity advantage is greater.

The ability to use NAATs with noninvasively collected specimens (first-catch urine [FCU] from both sexes and vaginal swabs) enables these tests to be employed in screening programs to detect asymptomatic individuals who represent the majority of prevalent infections. Thus, NAATs are useful for diagnosis, screening, epidemiologic studies. and public health purposes. Specimens collected from other anatomic sites beyond the typical cervical, urethral, and urine specimens, for example, rectal and oropharyngeal specimens, can be subjected to further evaluation.

The way we use NAATs has been complicated by the Centers for Disease Control and Prevention (CDC) guidelines that recommend confirmation of positive results when a positive predictive value is below 90%.[7] This recommendation assumes a firm understanding of the specificity of these tests;

however, we really do not know how to accurately measure NAAT specificity. Further, the methods suggested for the confirmation of positive results are either not easily implemented or their accuracy is uncertain. The issues related to confirmatory testing will be discussed later in this chapter.

EVALUATION OF NUCLEIC ACID AMPLIFICATION TESTS

NAATs provided a challenge to the microbiologists involved in the first evaluations of these tests for *C. trachomatis* because many more positive results were obtained with NAATs than with previous existing diagnostic methods. Indeed, if DFA, EIA, direct probe and tissue culture (TC) were all used on the same patient specimens, they would collectively identify fewer infected patients than the NAATs would. This problem is a generic issue in diagnostic test evaluation and not unique to chlamydial infections. It is nonetheless most relevant in the context of *C. trachomatis* because it was the first organism for which NAATs became widely available. The question was how to determine whether the positive results unique to the new test represented a true increment in sensitivity (i.e., detection of more truly infected individuals) or whether a problem with specificity may have, for example, produced false positive (FP) results.

LACK OF "GOLD STANDARD"

A "gold standard" is a test (or group of tests) that correctly identifies specimens as truly positive or truly negative. When a new test is evaluated, it is compared to the gold standard to define its performance characteristics. The problem for *C. trachomatis* test evaluation was that no gold standard existed. When NAATs were first introduced they were compared to culture tests that were then the best available.

To further evaluate NAAT results and ensure that the increment in positive results was not due to FPs, investigators used a procedure called discrepant analysis (DA).[8,9] To be considered a true positive (TP), a specimen had to be either TC positive or positive in two different nonculture tests. Thus, the specimens that were initially positive only in the NAAT were retested first by a relatively insensitive DFA test (DFA could yield positive results if a lot of organisms were present; however, most (if not all) were dead so that the TC would be negative). Approximately 50% of the excess NAAT positives could be explained on this basis, but certain specimens were positive only in NAAT and negative by TC and DFA and most were positive when a second NAAT for a different target was performed. These specimens that yielded two positive results in different NAATs were also accepted as TPs. These were the specimens containing the least amount of target, thus requiring NAATs to detect them. In DA, the specimens that appeared negative in the first run were never retested with another procedure because that could not result in the two positive results required to accept a TP result.

The DA procedure has been criticized as based on selective testing, in which (1) the experiment was used to identify positive specimens and (2) because the resultant sensitivity and specificity were upwardly biased.[10] This is because the only movement in classification of results by the retest would be from an FP to a TP result. Of course, the goal of the researchers was to determine whether the increase in positive results represented newly detected TPs or not, but the issue involves more practical considerations. When the first NAATs were introduced, no other tests were capable of detecting the very small number of organisms (about 10^1 to 10^2) that could be detected by these new tests; EIAs and DFAs require $\geq 10^4$ organisms to yield positive signals.

There is no question that the results generated by DA are upwardly biased. The ultimate questions are (1) how important is that difference and (2) what options exist for the analysis of test performance?[11] The bias in specificity was very small.[12] Only the specimens that initially tested positive by the NAAT alone were retested. The results of the second tests moved the number of confirmed positive results out of the FP cell into the TP cell. Specimens that were initially both TC positive and NAAT positive were never retested. The theoretical possibility that some of those initial results could in fact be FPs exists, of course, but this number is too small to be meaningful. Similarly, mathematical calculations can show that some of the confirmed positives are really FPs, even though they tested positive with two different NAATs with different targets, but this number must be very small indeed. Thus, without DA, we face greater misclassification bias because confirmable and presumably TP NAAT results would be considered FPs.

The only large multicenter trial in which specimens from the same patient were analyzed by multiple tests and different methods of analysis used with the same data found that specificity changed only slightly with DA.[13] An effort to evaluate the two then-commercially available NAATs in parallel was sponsored by the CDC. When these tests were compared using a third, newly introduced NAAT to resolve differences, the specificities of the two NAATs (LCR [ligase chain reaction] and PCR [polymerase chain reaction]) were found to be equivalent to that of TC (generally assumed to be 100% specific) at 99.6%.

This probably means that, with existing technologies, we have reached the limit of our ability to assess specificity, with results approaching 100%. It is a given that some FP results arise from any test, but with the use of NAATs in expert laboratories, we are now approaching the level where FPs are more likely due to labeling errors and mistakes in specimen collection or processing than problems with the technology.

The DA bias in sensitivity is somewhat larger, but here too it is not a major issue because the sensitivity values of NAATs generated in any of these evaluations are relative and not absolute. They are only useful in comparing different tests. Certainly, the results seen after DA are preferable to simply accepting the increase in number of positive results as FPs. The sensitivity depends on the number of specimens that test positive for the organism of interest with the new test compared to the total number of positives generated by the gold standard. If a new test represents a real increment in sensitivity, more infections are detected than could have been detected with the older tests that comprise the gold standard. Sensitivity represents a moving target because new and better tests identify positive specimens that would have tested negative with the earlier generation of less sensitive tests. Test evaluators have become well aware of this from experience.

Head-to-head evaluations of two nonamplified tests (e.g., DFA versus TC) will generate higher values for sensitivity than would be seen if these tests were compared to NAATs. In fact, in the first evaluations of nonculture tests, the sensitivity of culture was taken as 100%. The literature or package inserts from diagnostic tests of 10 to 15 years ago include some sensitivity claims that are higher than those of more recently introduced NAATs, which in reality are far more sensitive. Because sensitivity appears to be a moving target, we may well want to consider another term to express how many more truly infected individuals are identified by the new tests or simply what percentage increment the new tests offer when compared to the old tests (e.g., 20% more infections detected than by EIA).

Specificity remains a very important issue because it determines how useful a test is in screening low prevalence populations. Too many FPs and the positive predictive values of tests in low prevalence populations could be relatively small. One general rule is that routine screening is not cost-effective and will not be performed in populations where the prevalence is <3%. With the 99.6% specificity found in the CDC study, a test with 90% sensitivity would have a positive predictive value of 88% in that population. That would generally be considered acceptable.

NAATs actually produce very few FP results, and specificities after resolution typically run on the order of 99.7%. Thus, while one accepts that FP results can exist (the clinician should know this and understand that all positive results should not simply be accepted; some may be questioned) confirmatory tests should not be needed. The determination of which confirmatory test to use would actually represent a major technical problem. The only test that would be equally sensitive to a NAAT would be another NAAT, but not all NAATs have the same sensitivities. In addition, the transport media for the different tests are not the same so this issue introduces specimen collection and processing problems.

Confirmatory testing assumes that repeating a positive result means that the initial positive result was correct, a TP; and that failure to repeat the initial positive result means that the initial result was incorrect, in other words a FP. It is not clear why the first assumption is always correct. Why shouldn't some FP results repeat? But the second assumption is clearly false. A specimen with low target concentration may provide erratic results because the specimen is not homogeneous or because the efficiency of amplification varies. A low target concentration is understandable from a clinical perspective because pathogen load varies between the incubation period and the acute or chronic stages of infection.

We have no completely acceptable way to evaluate new and superior tests. Indeed, even today, no combination of two or even three tests (even three NAATs) will identify all infected individuals. In large studies, a few specimens tested positive in TC and negative in multiple NAATs. However, many more TC-negative specimens tested positive with NAATs. A dramatic increase in sensitivity with a new test requires a focus on the apparent FP results.

The criticisms of DA fail to deal with some of the real issues faced in evaluation of a new diagnostic test.[11] All the statistical manipulations and approximations will not correct for a single test that detects 10 to 20% more infected individuals than all the comparison tests. We cannot wait for more improved tests to be developed before we accept the first one in a new generation of superior tests. If the comparative tests have a sensitivity threshold of 10^4 organisms and a new test has a threshold of 10^1 organisms, what result would be expected if 10 to 20% of specimens actually contained 10^1 to 10^4 organisms? If DA had not been applied to the increased number of positives detected in the first evaluations of NAATs, it is likely those tests never would have come to market and we would have lost valuable tools. Had those excess positives been dismissed as FPs, performance profiles of the tests would have been too poor to permit their routine use.

If we were to choose an appropriate way to evaluate NAATs today, we would insist that any NAAT be compared to two other NAATs, as a simple head-to-head comparison is still inadequate because it would generate too many apparent FP results. The different technologies are not equally effective in detecting low levels of organisms, but the differences are reduced when a third NAAT is used.[13]

PERFORMANCE OF NUCLEIC ACID AMPLIFICATION TESTS

The first NAAT to come to market was PCR.[8] When it was first evaluated in comparison with culture using urethral swabs and urine specimens from men attending STD clinics, it was found to be far more sensitive than TC. Evaluations of PCR usually noted that the sensitivity was in the 90% range compared to about 65% for culture. Results using cervical swab specimens were more variable. Some investigators found satisfactory performance with cervical swabs, similar to performance with urethral swab specimens from men, while others found markedly lower sensitivities.[14]

At the University of California, San Francisco (UCSF) laboratory, we too found a sensitivity of only 60% for PCR from cervical swabs as compared to TC. In retrospect, the differences among the various studies are fairly easy to explain. It has become clear that the initial PCR test was sensitive to inhibitors present in cervical swabs. If laboratories processed the specimens quickly, the inhibitor was present and PCR testing yielded many false negatives (FNs), thus reducing sensitivity. However, if specimens were stored frozen or kept for several days in a refrigerator before processing, the inhibitors, being labile, diminished, resulting in higher sensitivity for the PCR assay. We now know that specimen dilution, a freeze-thaw cycle, or overnight storage at refrigerator temperatures will reduce inhibitor problems.[15]

The next NAAT to become commercially available was LCR. In fairly large, multicentered trials, sensitivities in the order of 90 to 95% were found for urethral swabs, male first-catch urine (FCU) specimens, and cervical swabs.[9,16] In each case, a 30 to 50% increase was noted in the number of positive specimens detected by LCR compared to TC, which had a sensitivity ranging from 50 to 70%, depending on the specimen tested. Transcription-mediated amplification (TMA)

TABLE 9.1
Additional *C. trachomatis* Infections Detectable by Changing to NAATs from Other Diagnostic Tests

Change from:	Increase in infections detected (%)
Culture	30
Pace-2, better EIAs	30
DFA	35
First generation EIAs	50
Rapid tests	90

Based on evaluations performed in authors' laboratory. Reprinted with permission from Schachter J. *Expert Rev Mol Diagn* 2001, 1(2), 137–144.

and strand displacement amplification (SDA) are two other NAATs that arrived later and evaluations revealed they were at least as good as PCR and LCR.[17,18] In general, the NAATs performed in a similar manner.

It has become obvious as more and more tests were evaluated that sensitivity figures generated for these tests have been moving targets since the introduction of DFA tests for *C. trachomatis* in the 1980s.[4] The total number of positive results obtained in the early evaluations of EIA and DFA versus culture were far fewer than the number of positives that would have been identified in the same populations by use of NAATs.

In some ways, it may be more realistic to discuss the increments in positivity seen with the different tests. Table 9.1 presents the differences noted in percentages of positive specimens obtained with the varying technologies based on comparative evaluations performed at the UCSF laboratory. By extrapolation, a laboratory using rapid tests would essentially double the number of positives by using an NAAT while a laboratory changing from better EIA, DFA, TC, or nucleic acid hybridization (e.g., PACE 2® of Gen-Probe Inc.) to NAAT would see a 30 to 50% increase in the number of positive specimens.

The current generation of NAATs provides the best technologies for clinical laboratories and clinicians interested in chlamydial diagnosis. They are more sensitive, easier to use, and provide answers in a reasonable period. Their major drawback is that they are relatively expensive, although probably less expensive than TC in most settings. Still, if NAATs were used widely, laboratory costs would increase. This perception of cost has limited the acceptance of NAATs. It must be stressed that looking only at laboratory costs is short-sighted. Certainly cost-benefit analyses will show that in many settings the preventable cost of chlamydia-induced disease and its complications is greater than the cost of screening programs using NAATs.[19]

Even if one considers only the economics of testing, the ability to use FCU specimens and vaginal swabs eliminates the costs associated with pelvic exams and makes the total cost of specimen collection and testing favorable for NAATs.[20] Having tests with similar performance characteristics is beneficial because laboratories can choose on the basis of price (hopefully competition will drive prices down), ease of use, and other issues, and not be concerned about test performance.

Approaches to reducing the costs of NAATs have been developed. Pooling strategies (a number of specimens are pooled and tested with retesting on an individual basis if the pool tests positive) have been evaluated[21] and found to save costs. In low prevalence populations, pooling is an appropriate and effective strategy. Few positive specimens are lost and FPs are relatively uncommon. We evaluated pooling of a number of swab specimens from genital tracts, ocular specimens from trachoma patients, and urine specimens. When pooled results were compared with those obtained by testing individual specimens, we confirmed the results published by others. However, on occasion, we failed to obtain positive results with a pool when some of the individual specimens tested positive. Whether this is due to inhibitors in some of the specimens comprising the pool or simply

TABLE 9.2

Targets Used in *N. gonorrhoeae* and *C. trachomatis* NAATs

Test	Target	Size	Detection
	N. gonorrhoeae		
LCx	Opa 1	?	MEIA
Amplicor	M-Ngo PII	201 bp	Hybridization EIA
ProbeTec	Chromosomal pilin gene-inverting protein homologue	<50 bp	Fluorescent probe
Aptima Combo 2	16s rRNA	120 bp	Hybridization chemiluminescence
Amp-CT	16S rRNA, different region	153 bp	Hybridization chemiluminescence
Hybrid Capture 2	Cryptic plasmid and chromosomal gene	4200 bp	Signal amplification EIA
	C. trachomatis		
LCx	Cryptic plasmid	145 bp	MEIA
Amplicor	Cryptic plasmid	207 bp	Hybridization EIA
ProbeTec	Cryptic plasmid	<50 bp	Fluorescent probe
Aptima Combo 2	23s rRNA	120 bp	Hybridization chemiluminescence
Amp-CT	16S rRNA	153 bp	Hybridization chemiluminescence
Hybrid Capture 2	Cryptic plasmid and chromosomal gene	7500 bp	Signal amplification EIA

dilution of target is uncertain. Clearly, this can happen and must be taken into consideration, but the sensitivity loss is less than 5%.

Three NAATs, each using a different amplification technology, are commercially available. In order of their introduction, they are a PCR assay (Amplicor, Roche Diagnostics, Indianapolis, Indiana, USA), a TMA assay (originally AMP-CT, now replaced by the next generation known as the APTIMA Combo 2, Gen-Probe Incorporated, San Diego, California, USA), and an SDA assay (ProbeTec®, Becton Dickinson, Sparks, Maryland, USA). Abbott Laboratories' LCR assay known as LCx has been withdrawn by the manufacturer. Some NAATs target nucleic acid sequences present in multiple copies (i.e., a plasmid that has 7 to 10 copies per cell or RNA sequence present in thousands of copies per cell) that are further amplified many times by enzymes to make detection easier. Targets for these assays are shown in Table 9.2.

The Hybrid Capture 2® (hc2, Digene Corporation, Gaithersburg, Maryland, USA) is a combination test designed to detect *C. trachomatis* and *N. gonorrhoeae* in a single specimen. It is a nucleic acid hybridization test that uses signal amplification to increase sensitivity. This test is more sensitive than *C. trachomatis* culture and is at least as sensitive as culture for gonococci.[22] Its use of signal amplification made it more sensitive than the PACE 2[23] which detects multiple copies of ribosomal RNA.

We evaluated all these assays, and in our hands the best performer in terms of sensitivity and specificity is the new APTIMA Combo 2. It is a multiplex assay based on TMA that simultaneously detects both pathogens and uses target capture to evade inhibitors. Unfortunately, it is not very user-friendly. Sensitivity and specificity for *C. trachomatis* were 94.2 and 97.6%, respectively, with cervical swab specimens and 94.7 and 98.9%, respectively, with FCU specimens. Sensitivity and specificity for *N. gonorrhoeae* were 99.2 and 98.7%, respectively, with swabs and 91.3 and 99.3%, respectively, with FCU.[24]

In this chapter, we have chosen not to list sensitivity and specificity results generated in the many studies evaluating available diagnostic tests for these two organisms because the numbers would be misleading. While most studies are internally consistent, the methods are not comparable across studies. For example, studies employing the ProbeTec assay (Becton Dickinson) generate sensitivities around 90%, only slightly below the APTIMA Combo 2 results cited above. Yet, when we analyzed results from a study that used both tests on multiple specimens from the same patients, the APTIMA Combo 2 detected approximately 15% more positive specimens than did the ProbeTec.[25] In summary, the NAATs correctly identified more infected individuals than we ever could have identified in the past.

A question concerns whether NAATs can be used when test results have medical–legal consequences such as in cases of incest, child abuse, or rape. It is likely that answers will vary from country to country, but in the United States, culture must be used as a diagnostic test if the results are to be used as evidence. This obviously is a compromised standard, as a TC for chlamydia is not readily available and it is certainly not as sensitive as NAATs. The issue will require educating those responsible for establishing legal requirements about the enhanced sensitivity of NAATs and the possibility of directly typing a strain from a specimen by obtaining the DNA sequences for the major outer membrane protein (MOMP) gene or other genes.

Reliability of NAATs has become an issue. These tests represent major technological advances. They are complicated and contain many reagents. What was not foreseen was the considerable difficulty in maintaining quality control in the manufacture of the reagents. Product recalls and warnings about test performance have been issued. The most publicity surrounded the use of LCx and the fact that some positive results, particularly those for *N. gonorrhoeae*, could not be repeated. The initial focus was on low level positive results (where the signal was just above the cut-off for determining a positive result); the assumption was that highly positive signals were not suspect. It was uncertain whether the low-level positive results that could not be repeated reflected problems with the test, stability of the specimens, or the distribution of target.

Because we tend to do a lot of retesting and our internal quality control measures exceed those suggested by manufacturers, the experiences of the UCSF laboratory may be of interest. For a long time we retested all positive specimens and confirmed them by using another NAAT.[25] We did not expect complete concordance of positive results as we have learned that each NAAT will detect some positive specimens that would not test positive with other NAATs. In other words, no single test could reach 100% sensitivity.

As a matter of routine, we retested all positive LCx specimens. In one long-term study, the prevalence of *C. trachomatis* was about 5.7%; 251 of 4392 specimens tested were positive and 9.6% (24 of 251) failed to repeat. Twenty specimens revealed low signal to cut-off ratios between 1 and 2 and one specimen with a high signal ratio >3 also failed to repeat. This was somewhat disconcerting, but obviously reflected a common problem that was recognized by the manufacturer. At the same time, we used the Amplicor PCR assay to confirm some of the LCx positive specimens and found 89.3% (50 of 56) specimens positive; the six failures were with LCx low-level positive reactions.

An obvious interpretation of these failures to repeat could be that they were FPs, but it was also plausible that some low-level positives may not have confirmed because of varying distribution of target in the specimens. When we tested the specimens that were initially positive and failed to repeat, we found an irregular pattern of positive and negative results. After three repeat tests, >90% of initial positive results could be confirmed. When we seeded specimens near the threshold level of the tests, we found the same pattern. We also looked at the reproducibility of negative results by repeat testing of 32 negative specimens (288 total tests); only 0.35% (1 of 288) gave a positive signal.

A similar reproducibility problem has been described with the BD ProbeTec system. Only 80.8% (21 of 26) of low-positive samples in the *C. trachomatis* test and 33.3% (4 of 12) of low-positive samples in the *N. gonorrhoeae* test retested positive.[26] We observed that multiple testing actually confirmed >90% of the positive results with all the NAATs for *C. trachomatis*. The same was not true for *N. gonorrhoeae*. Some of the specificity problems with *N. gonorrhoeae* reflected an unfortunate choice of targets that were also present in other Neisseria species. This led to the development of retesting algorithms and other confirmatory testing strategies.[27,28] In our hands, the APTIMA Combo 2 test for *N. gonorrhoeae* revealed no specificity problems.

Noninvasive Specimens

What is the specimen of choice for the diagnosis of genital chlamydial infection? Obviously swab specimens are standard, but noninvasive specimens (e.g., FCU specimens from males and females and vaginal swabs) allow for self-collection. This, in turn, facilitates the collection of specimens from asymptomatic individuals and thus enables the screening of at-risk populations and determination of population-based estimates of prevalence and incidence. Much research is being done in this area but it is still difficult to provide satisfactory answers to all of the questions. However, the trend is clear. It is likely that FCU will be the specimen of choice for males and vaginal swabs will be the choice for women and this may apply to diagnostic as well as screening tests.

EIAs have been used with FCUs from symptomatic men with some success, but were relatively insensitive for asymptomatic men and both insensitive and nonspecific with FCU from women.[4] When NAATs were performed on FCU from both symptomatic and asymptomatic men, the tests were found to be highly sensitive and specific (mid-90% sensitivity).[8,16] It was immediately obvious that NAATs on FCU would allow screening of asymptomatic males. This is a particularly important finding because it is impossible to perform population-based or other studies on large numbers of asymptomatic men if testing requires intraurethral swabbing to collect specimens. The results suggest that FCU from a man, symptomatic or not, is the specimen of choice for diagnosis of urethral chlamydial infection.

It was somewhat surprising to find that FCU from females could also be used for the diagnosis of chlamydial infection. The first study in 1994 found that FCU specimens from women were almost as sensitive as cervical swabs for *C. trachomatis*.[29] Similarly, LCx testing of urine specimens for *N. gonorrhoeae* yielded positive results for 94.4% (51 of 54) of women with positive cervical or urethral cultures.[30]

These findings opened the possibility of broad-based testing of asymptomatic women, and most importantly, routine testing of women without the need for performing pelvic exams. This dramatically reduces the cost of testing[20] and increases opportunities for the collection of specimens away from traditional medical and clinical settings.

All the NAATs work reasonably well with male urethral swabs, FCU specimens, and female endocervical swabs. However, some variation has been noted in results of studies using female FCU, but some of that may reflect factors such as technical difficulties in processing the specimen, lack of standardized protocols for urine collection, dilution of target when different volumes are collected or processed, stability, different storage and transport conditions, and the degradation of target or inhibition of amplification.

The first inkling of problems involving urine as a specimen came from a study reporting poor results with LCx in the diagnosis of chlamydial infection in pregnant women.[31] The assumption was that hormones present in the urine inhibited the LCR. In fact, what evolved is a base of knowledge that says that the accuracy of urine processing varies from laboratory to laboratory. This was confirmed with LCx quite readily when it was shown that phosphate present in urine was an inhibitor of the LCR.[32] Thus, leaving excess urine on the sediment after centrifugation will produce FN results. Incorrect supernatant aspiration after centrifugation can result in loss of target, also causing FN findings.[33]

In multicenter trials, we found that inappropriate handling of urine specimens resulted in decreased sensitivity of all NAATs evaluated.[34] In most of these trials, one or more laboratories noted NAAT sensitivity with urine at least 10 to 20% lower than values seen at the other sites. Close examination of laboratory procedures usually identified flaws in specimen processing. Our first experience with this problem was in a study on acquisition of STDs. We trained a highly experienced collaborating laboratory how to correctly perform the LCx test. When we compared our results with split specimens (half tested on-site at the collaborating laboratory and half tested in our laboratory), the collaborating laboratory noted 0.8% (4 of 500) of specimens positive whereas our laboratory found 3.6% (18 of 500). A review of the protocols, correction of processing procedures, and repeat testing using frozen aliquots resulted in total agreement.

In other multicenter trials, we found that when technicians missed on-site training for urine processing for a new test, their sensitivities were lower. In other studies, the overly aggressive aspiration of supernatant after centrifugation of the urine disturbed the sediment and decreased sensitivity. In all these trials, we had opportunities to retest frozen specimens that initially tested negative and found that they were actually positive. The lesson learned is somewhat disconcerting. If this is what happens in world-class laboratories, one wonders what is going on in routine clinical laboratories that simply test and report results. They have no access to information that indicates their procedures may be inadequate and that they may be generating falsely negative results.

Fortunately, however, we may well have a superior alternative to FCU specimens in testing women. Self- or physician-collected vaginal swabs have been found essentially equivalent to cervical swabs.[35–37] A few positive specimens were detected using vaginal swabs that were negative with cervical specimens, representing contamination from urethral infection (in some studies, approximately 15 to 20% of women have infections of the urethra but no detectable cervical infections).[5] The vaginal swab misses some specimens that are positive only with cervical specimens, but given the trade-off with urethral infection, the sensitivity is similar to that found with endocervical specimens. No pelvic examination is needed and the specimens are less expensive to handle than FCU (no centrifugation step). The stability issues and variability introduced by the centrifugation of urine are not problems with vaginal swabs. Thus, vaginal swabs and perhaps introital swabs are likely to be specimens of choice in routine screening of asymptomatic women. Patients do as good a job of collecting specimens as do clinicians. This is important as it opens up many possibilities for testing strategies.

A number of studies have found vaginal swabs to be the best specimens. In one such study, the total positive rates for *C. trachomatis* and *N. gonorrhoeae* were 11.6 and 2.4%, respectively. The proportions of positives identified by specimen type were for *C. trachomatis* and *N. gonorrhoeae*, respectively, endocervical, 65 and 40%; urine, 72 and 24%; and vaginal, 81 and 72%. The proportions of positives when specimen results were combined were, for *C. trachomatis* and *N. gonorrhoeae*, respectively, cervical plus urine, 86 and 49%; cervical plus vaginal, 91 and 93%; and vaginal plus urine, 94 and 79%.

Self-collected vaginal swabs identified the highest number of infected individuals based on use of a single specimen, with the combination of cervical and vaginal specimens identifying the highest number of infected individuals.[38] Similarly, LCx testing of self-collected vaginal swab specimens was significantly more sensitive for the diagnosis of gonorrhea in women than LCx testing or culture of cervical specimens (100 versus 84.6%).[39]

The use of different sampling sites introduced an interesting problem for test evaluators: whether to use as the denominator in the calculations of sensitivity the infected patient (proven to have a chlamydial infection at any site) or a specimen-specific evaluation. In a very real sense, the comparison of results obtained with a single specimen gives a superior feel for performance of any specific test. Using the infected female patient (for example, testing both urethra and cervix; a woman testing positive at either site is considered a positive patient) as the denominator decreases the sensitivity for any test, but gives a better reflection of the real world use of the test (Table 9.3). The reason for lower sensitivity is that invariably some women are infected at only one of the sampled sites, so that specimens from the other site will test negative (true negative [TN] on a specimen basis but FN on a patient basis).

INHIBITORS OF NUCLEIC ACID AMPLIFICATION TESTS

It is clear that NAATs have many potential inhibitors.[10] The concern is real, but the effect is probably exaggerated. Even in the presence of some defined inhibition, many positive specimens are still identified. Inhibitors occur at relatively low levels in most specimens. For example, in a large

TABLE 9.3
Calculation of Sensitivity Based on Specimen of Infected Patient

	TC Result	+	+	−	Sensitivity of Cervical Swab by	
				+		
NAAT Result		−	+	(DFA+)	NAAT	TC
Cervical swab		5	60	35	95% (95 of 100)	65% (65 of 100)
Additional positives detected by urethral swab		2	10	8	79% (95 of 120)	54% (65 of 120)

Reprinted with permission from Schachter J. *Expert Rev Mol Diagn* 2001, 1(2), 137–144.

multicenter evaluation of the Amplicor PCR assay, 2.4% of specimens were inhibitory, with the highest rate (5.4%) found with FCU from symptomatic males.[40] NAAT inhibition appears most marked with some FCU specimens. We do not know exactly what compounds are inhibitory. The important fact evolving from studies of inhibitors is that they are relatively labile and can be effectively minimized. Dilution (1:10 removes inhibitors at the likely cost of an occasional lost positive result), overnight storage at refrigerator temperature, and subjecting specimens to a freeze–thaw cycle have been shown to reduce inhibition.[21]

The use of controls for inhibition (i.e., amplification control) should be subjected to a cost–benefit analysis. If there is an internal amplification control, and there is no reduction in throughput (a separate amplification control-tube or well reduces the number of specimens processed by 50%) then such controls are likely to be reasonable. Because the enzymes involved in amplification are potentially sensitive to inhibitors, a number of issues have been raised about inhibition. However, no unanimity has been reached regarding the approaches to take or even whether amplification control is needed.

If we accept the need for an amplification control, what should that control be? Certainly the presence of an internal control provides evidence of success or failure of amplification for that specific target, but we need more information as to how these controls should be designed and used. Some manufacturers use Chlamydia- or Neisseria-specific targets; others use irrelevant targets meant to provide checks for any amplification. Another question is the level of inhibition that should be detected. Assays for inhibition depend on the concentration of the amplification control. At the moment, we do not have a clear rationale for selecting the concentration level.

WHAT PACKAGE INSERTS MEAN

Just as researchers are faced with problems in evaluation of NAATs, regulatory authorities are faced with problems in determining what data to accept for package inserts. Initially, the use of second NAATs to confirm NAAT positive results with specimens that were TC negative was not accepted. Thus, only TC positive and DFA positive results were considered to be TPs; if the NAAT results agreed, they were accepted as TPs but excess NAAT positives (i.e., TC and DFA negatives) were considered FPs. This meant that even if a specimen tested positive for the chlamydial plasmid and for another target gene, such as MOMP, it was not considered a TP specimen.

Researchers were not subject to these constraints when they carried out the evaluations. They used the alternate target NAATs in their evaluations of possible FPs and determined that a great majority of them were TPs. Thus, the specificities cited in the literature are much higher than those included in package inserts. An insert may contain a footnote indicating that many FPs were shown to contain multiple chlamydial genes, but they are still considered FPs. Obviously, microbiologists disagreed and accepted these as TPs (Table 9.4).

The problem, of course, is that anyone simply looking at a package insert and seeing a specificity of 93% (as may be found for a single specimen) would say the test is not very useful. It certainly could not be used to screen low-prevalence populations if 7% of the uninfected are going to yield

TABLE 9.4
Specificities of NAATs Cited in Package Inserts versus Specificity if Alternate Target NAATs Had Been Allowed

Test	Overall Specificity per Package Insert	Pos by Alternate NAAT	Revised Specificity
Amplicor	97.1% (8136 of 8378)	153 of 242	98.9% (8288 of 8378)
LCx	97.1% (2848 of 2932)	80 of 84	99.9% (2928 of 2932)
Aptima	98.0% (3593 of 3665)	62 of 72	99.7% (3655 of 3665)
ProbeTec	96.6% (3480 of 3603)	71 of 123	98.6% (3551 of 3603)

Reprinted with permission from Schachter J. *Expert Rev Mol Diagn* 2001, 1(2), 137–144.

FPs. The fact that about 0.4% are actually FPs obtained with male urines more closely reflects reality.[13] Thus, anyone wanting to understand the performance profiles of a test must look very carefully at the package insert to see what comparison tests were used and how the specificity was calculated. If FPs were shown to contain multiple chlamydial genes, they are not FPs even if the regulators insist they be treated as such.

It is also important to determine whether sensitivity is expressed on the basis of a single specimen evaluation or by using an infected patient as the standard. If the infected patient serves as the basis, then all sensitivity results will be somewhat lower than they would be if the evaluation was performed on a per-site basis (Table 9.3).

One unfortunate aspect of package inserts is that they do not change over time. Thus, it is possible to see sensitivity claims close to 90% for the antigen detection methods of the 1980s and see similar or even somewhat lower numbers for the NAATs of the late 1990s, even though the NAATs are actually 30 to 40% more sensitive than the earlier tests. This is simply a reflection of a secular trend in the sensitivity of the tests that are available.

FUTURE OF NUCLEIC ACID AMPLIFICATION TESTS IN DIAGNOSIS OF CHLAMYDIAL AND GONOCOCCAL INFECTIONS

We will see more innovative approaches to exploit NAA technology. In one creative experiment, home collection kits were given to high school students along with instructions for self-collection of specimens and mailing. More than a 10-fold increase in the proportion tested was found among students who mailed in self-collected specimens compared to students referred to clinicians' offices for tests.[41] High school screening and intervention programs have exploited the use of NAATs and noninvasive specimens.[42] It has also been possible to measure incidence of infection.[43] One dramatic result of this technology was a population-based study that concluded that in 1997 and 1998, the estimated number of undiagnosed gonococcal and chlamydial infections prevalent in Baltimore adults aged 18 to 35 years (N = 9241; SE = 2441) exceeded the number of infections (N = 4566) diagnosed and reported to the Baltimore City Health Department in 1998.[44]

MYCOPLASMA GENITALIUM

M. genitalium is an organism that provides an excellent example of the value of NAA technology to research and perhaps to important public health considerations. *M. genitalium* was first isolated from the urethras of 2 of 13 men with urethritis.[45] However, further studies on its role in genital tract disease were thwarted by the lack of a useful diagnostic method and an effective culture system. Several groups developed PCR methods that were used in studies that implicated *M. genitalium* in the etiology of non-gonococcal urethritis (NGU).[46-50]

Mena et al.[51] used PCR to detect *M. genitalium* in 97 men with urethritis and in 184 asymptomatic control subjects at an STD clinic in New Orleans. *M. genitalium* was found in 25% of symptomatic and 7% of asymptomatic men who were negative for *C. trachomatis* and *N. gonorrhoeae*. *M. genitalium* co-infection rates among men with chlamydial and gonococcal urethritis were 35 and 14%, respectively.

Mycoplasma-associated NGU is treatable.[52] Thus, the use of PCR has allowed researchers to generate evidence as to the cause of nonchlamydial NGU, a common condition in STD clinics. Certainly *M. genitalium* is sexually transmitted, but its role in disease of the female genital tract remains to be elucidated. A commercially available NAAT would conceivably help address these questions and gain wide use in STD clinic settings. In addition, further studies are needed to determine the course of untreated *M. genitalium* infection and its sequelae.

TRICHOMONAS VAGINALIS

Trichomonas vaginalis is considered the world's most common curable STD.[1] It is an important cause of adverse pregnancy outcomes and has been associated with increased HIV transmission. Although not a bacterium, it is often considered together with the bacterial STDs because trichomoniasis is one of the common causes of vaginal discharge. Like the bacterial STDs, the condition is often asymptomatic in males and females. The commonly used diagnostic tests, wet mount and culture, are relatively insensitive. Thus, *T. vaginalis* is a likely candidate for study using NAATs.

Several groups have used PCR assays for *T. vaginalis* during studies on STD prevalence in selected male and female populations.[53–55] Predictably, the NAAT was found to be more sensitive than the other diagnostic tests and higher prevalences of *T. vaginalis* were found. Vaginal swabs constitute useful specimens and real-time PCR has been successfully used with urine specimens.[56] This could have broad public health implications. However, the ultimate significance of the high prevalence found in some male populations is uncertain.[57]

GENITAL ULCER DISEASE

Genital ulcerative diseases are common in all countries. Because specific diagnosis is difficult, many people do not seek medical attention and reporting is incomplete. Thus, the true incidence of genital ulcers is unknown. The proportion of patients treated for STDs who have genital ulcerations depends on several factors including (1) the prevalence of other STDs; (2) circumcision rates; (3) diagnostic and treatment facilities; and (4) specific sexual practices.[58]

Patients with genital ulcers comprise 18 to 70% of STD patients seen in clinics of East and Southeast Asia, Africa, and India.[58–61] An intermediate proportion of patients seen in areas of Latin America and the Caribbean have genital ulcer disease. In Sweden, the United Kingdom, and the United States, as few as 3 to 4% of patients in some STD clinics have ulcerative lesions of the genitalia.[58]

Studies of genital ulcer disease show differences in etiologies that may partially reflect differences in study methods, but also genuine geographical and temporal differences in the incidence of specific genital pathogens. The primary agents of genital ulcer disease and the diseases they cause are *Treponema pallidum* (syphilis); *Haemophilus ducreyi* (chancroid); *C. trachomatis* (lymphogranuloma venereum); *Calymmatobacterium granulomatis* (donovanosis); and herpes simplex virus types 1 and 2 (genital herpes).

Clinical diagnosis often has been used to establish the etiology of a genital ulcer in areas where genital ulcer disease is prevalent or laboratory facilities are absent or minimal. Unfortunately, the etiologic diagnosis based on the presence or absence of specific clinical signs has a relatively low sensitivity in both developing[62] and developed countries.[63] The sensitivity of a clinical diagnosis can also be affected by factors such as differences in disease expression between men and women and atypical presentation due to concomitant HIV infection.[64]

The etiologic diagnosis of genital ulcers using standard laboratory methods such as microscopy, serology, culture, and animal inoculation is difficult, imprecise, and often costly.[62,65-67] A number of factors are responsible: the inability to cultivate *T. pallidum in vitro*; the requirement for cell culture to grow *C. trachomatis*, *C. granulomatis*, and herpes simplex viruses; multiple etiologic agents in the same ulcer; and difficulties in differentiating current ulcer etiology from past infections in highly endemic areas using serological methods.[62] For all of these reasons, a syndromic approach has been recommended for the management of genital ulcers.[60,68]

Under ideal circumstances, a current knowledge of the local etiology of genital ulcers will be necessary to avoid over-treatment and the selection of antibiotic-resistant organisms.[60,68]

NAATs could solve many problems associated with the etiologic diagnosis of genital ulcers. PCR assays for detecting the presence of nucleic acid from agents such as *T. pallidum*,[69-71] *H. ducreyi*,[72] and herpes simplex virus[73,74] were first published in the early 1990s and used primarily in research settings. Roche Molecular Systems pioneered the development of a multiplex PCR assay for the three major causes of genital ulcers, i.e., *T. pallidum*, *H. ducreyi*, and herpes simplex virus.[75] Unfortunately, market size probably dictated the decision against commercialization of this assay. However, a detailed method has been published to make it widely available to the research community.[76] The use of NAATs for the diagnosis of genital ulcer disease will be discussed from an etiologic perspective; herpes simplex virus has been included because it is one of the agents detected by the multiplex PCR assay for genital ulcer pathogens.[75]

GENERAL CONSIDERATIONS

Specimen Collection and Processing

After thorough cleaning of each lesion with a gauze pad, material is collected from the base of the lesion with a cotton-[62] or Dacron-tipped[59] swab. When multiple ulcers are present, the specimen is collected from the largest. Because many genital ulcer specimens contain PCR inhibitors such as blood, it has been necessary to extract the DNA prior to amplification.

Numerous studies have shown that a significant proportion of specimens are negative when assayed by the multiplex PCR assay. Among 31 separate collections of genital ulcer specimens assayed by multiplex PCR (Table 9.6), the proportion of negative specimens ranged from 6 to 68% (average 23%). These negative specimens are most likely due to inadequate specimen collection, particularly when additional swab specimens are being collected. However, low target concentration (can also result from the collection of multiple specimens) or agents other than *T. pallidum*, *H. ducreyi*, and herpes simplex virus are also important factors.

Amplicon Detection

The method used for amplicon detection has changed as PCR assays have become more widely used in studies on the etiology of genital ulcer disease. The earliest assays employed agarose gel electrophoresis with or without Southern blotting to identify the amplicon.[72,77-82] The multiplex PCR assay uses the same methodology employed in the AMPLICOR *C. trachomatis* assay for the detection of the amplicons by enzyme-linked immunosorbent assay (ELISA).[75,76] This method is somewhat labor-intensive and time-consuming for use with a multiplex assay, involving replicates of each of the three target agents plus an internal control. The use of different fluorescent-labeled primers for the amplification of each target and analysis on an ABI Prism 310 genetic analyzer or GeneScan-500 (Applied Biosystems, Foster City, California, USA) has allowed easier differentiation of peaks indicative of each amplification product.[83]

TABLE 9.5
PCR Assays for *Haemophilus ducreyi*[a]

Assay	Primer	Target	Amplicon Size	Detection	[References]
1	**Forward:** 5'-CCCCGACACTTTTACACGCGCT **Reverse:** 5'-GCCAGCCCAGTGACGCCGATGCC	*recD*	1.1-kb	agarose gel	[77–79]
2[b]	**Forward:** 5'-biotinyl-CAAGTCGAACGGTAGCACGAAG **Reverse:** 5'-biotinyl-TTCTGTGACTAACGTCAATCAATTTTG **Probe:** 5'-CCGAAAGGTCCCACCCTTTAATCCGA	16s rRNA gene	439 bp	capture probe + ELISA	[75,76]
3	**Forward:** 5'-AGGCAGTTTGTTAATAGCA **Reverse:** (1) 5'-ATAGAAGAAACTCAGAGATGA (2) 5'-AGAACTCAGAGATGAGTTTG	16s rRNA gene	585 bp 590 bp	agarose gel	[80]
4[c]	**Forward:** 5'-GACAAGTCGGAATACATCT **Reverse:** (1) 5'-TATGCGCGAGGCATATTG (2) 5'-GATTACTCAGACTTTCTACTTTAG	*rrs-rrl* ISR	277 bp 197 bp	agarose gel	[81]
5	**Forward:** 5'-AGGTGCTGCATGGCTGTC **Reverse:** 5'-CTAGCGATTCCGACTTCA **Probe:** 5'-ATGTAGTGATGGGAAC	16s rRNA gene	303 bp	agarose gel + Southern blot	[72]
6[d]	**Outer forward:** 5'-TGTGACAGGTGCTGCATGGCTGTC **Outer reverse:** 5'-ACTTCAGCCTTAGCGATCATTAGCG **Inner forward:** 5'-biotinyl-CTCAGACGTTGAGCTG **Inner reverse:** 5'-digoxigenin-ATGTAA/GTGATGGAAC	16s rRNA gene	? ?	ELISA	[84]
7[a]	**Forward:** 5'-TET-CAAGTCGAACGGTAGCACGAAG **Reverse:** 5'-FAM-TTCTGTGACTAACGTCAATCAATTTTG	16s rRNA gene	439 bp	ABI 310	[83]
8	**Forward:** 5'-ATGGTACAGGTTTAGATGATGCCTTAGATG **Reverse:** 5'-AACTACGCGTGCTTTAATTTGTGCTTCATC **Probe:** 5'-AGCCGGTACGGTTGAATTAGACAACCCATA	*groEL*	505 bp	Southern blot	[82]

a. Probes are shown for assays employing solid phase detection methodologies.

b. Multiplex assay for *Treponema pallidum*, *Haemophilus ducreyi*, and herpes simplex virus.

c. One-tube PCR assay.

d. One-tube nested PCR assay.

CHANCROID

A number of PCR assays have been developed for the specific detection of *H. ducreyi* DNA (Table 9.5). Most of the assays amplify sequences within the 16S rRNA gene.[72,75,76,80,83,84] Others amplify sequences within *rec*D,[75–79] *rrs-rrl* ISR,[81] and *gro*EL.[72] All these assays appear to be specific for *H. ducreyi* and do not amplify sequences from other genital pathogens or humans.

Culture of *H. ducreyi* on various complex media is considered the gold standard against which other diagnostic tests for chancroid are compared.[64,85] Before the development of PCR assays, the sensitivity of culture had been determined among patients with clinical diagnoses of chancroid.[85] Culture sensitivities ranging from 53 to 85% were reported.[85] It is likely that this variation reflects, in part, the diagnostic acumen of clinicians and the relative sensitivities of different culture media[62,63,85] and suggests that potentially more sensitive assays such as PCR would identify more positive patients than culture.

Compared to culture of *H. ducreyi*, multiplex PCR had a sensitivity of 95 to 98.4%.[62,75] The multiplex PCR-positive, culture-negative specimens were tested by a second confirmatory PCR[75] and the resolved specificities were calculated to be 99.6 to 100%.[62,75] Analogous to the situation with NAATs for *C. trachomatis*, when compared to PCR, the sensitivity of *H. ducreyi* culture was only 74.2 to 75%.

PCR has been used to determine the prevalence of *H. ducreyi* DNA in genital ulcer specimens obtained from different geographical locations (Table 9.6). Overall, the results show that the prevalence of chancroid varies geographically. The highest rates are seen in Africa and the Caribbean while the lowest rates occur in the United States, Thailand, and China. Rates of chancroid have decreased dramatically in the United States over time. Chancroid also appears to be decreasing in some areas of Africa as well and may be associated with the implementation of STD control programs for reducing HIV transmission.

One advantage of the multiplex PCR assay is that it can easily identify genital ulcers containing multiple etiological agents. This was most evident in *H. ducreyi*-positive specimens from Africa in which up to 17% also contained herpes simplex DNA, *T. pallidum* DNA, or both.

SYPHILIS

If left untreated, syphilis is a chronic disease that spreads throughout the body hematogenously and can produce manifestations in virtually any organ system.[96] The infectious clinical manifestations, i.e., primary and secondary syphilis, are transient events. The primary lesion, the chancre, appears at the site where treponemal invasion of the dermis first occurred, usually on or near the genitals. However, it may occur on any skin or mucous membrane. The disease can be transmitted vertically *in utero* resulting in congenital syphilis. *T. pallidum* subsp. pallidum, the causative agent, is an ideal candidate for the development of a PCR assay because it has never been cultured successfully on artificial media, and does not take up the Gram stain.[97]

A number of PCR assays have been developed for the detection of *T. pallidum* subsp. pallidum DNA (Table 9.7). However, the targets are more variable than those used to develop assays for *H. ducreyi*. Three assays use different primers based on sequences within the gene encoding the 47-kDA membrane protein[70,75,76,98]; other assays have targeted sequences within the TmpA gene,[69] 4D gene,[69] *bmp* gene,[71] 16S rRNA gene,[88] and the *pol*A gene.[99] Some of the assays[75,76] have been tested against DNA from a large number of microorganisms including other etiologies of genital ulcers and those responsible for other STDs and found to be specific for *T. pallidum*. However, due to the extensive genetic homology, none of the PCR assays for syphilis will distinguish between the pathogenic treponemes of humans: *T. pallidum* subsp. pallidum, *T. pallidum* subsp. pertenue, *T. pallidum* subsp. endemicum, and *T. carateum*. While infections from all these organisms will cause serologic tests for syphilis to be reactive, only *T. pallidum* subsp. pallidum is sexually transmitted.

TABLE 9.6
Prevalence of *H. ducreyi*, *T. pallidum*, and HSV Determined by PCR in Genital Ulcer Specimens from Different Geographic Locations

Location	Year	Number Analyzed	Number Positive (Percent Positive)							PCR Technique [References]	Study [References]
			H. ducreyi Alone	*H. ducreyi* Co-Infection	*T. pallidum*	*T. pallidum* Co-Infection	HSV Alone	HSV Co-Infection	Unknown Etiology		
United States											
New Orleans, LA	1992	101[a]	35 (35)	2 (2)	15 (15)	0	28 (28)	2 (2)	21 (31)	[75,76]	[75]
New Orleans, LA	1993	97[a]	17 (18)	2 (2)	34 (35)	2 (2)	29 (30)	4 (4)	13 (13)	[75,76]	[75]
New Orleans, LA	1994	100[a]	9 (9)	0	23 (23)	1 (1)	38 (38)	1 (1)	29 (29)	[75,76]	[75]
Atlanta, GA	1995–1996	95[a]	3 (3)	4 (4)	12 (13)	4 (4)	39 (27)	5 (3)	7 (7)	[75,76]	Flemming et al. (unpubl.)
Jackson, MS	1994–1995	143[a]	47 (33)	9 (6)	16 (11)	11 (8)	65 (68)	8 (8)	29 (20)	[75,76,79]	[86]
Birmingham, AL	1996	50[a]	0	0	13 (26)	0	25 (50)	0	12 (24)	[75,76]	[87]
Chicago, IL	1996	49[a]	6 (12)	0	4 (8)	1 (2)	24 (49)	0	14 (29)	[75,76]	[87]
Cincinnati, OH	1996	52[a]	0	0	1 (2)	0	41 (79)	0	10 (19)	[75,76]	[87]
Dallas, TX	1996	52[a]	0	0	6 (12)	2 (4)	35 (67)	0	9 (17)	[75,76]	[87]
Houston, TX	1996	51[a]	0	0	1 (2)	1 (2)	38 (75)	0	11 (22)	[75,76]	[87]
Los Angeles, CA	1996	54[a]	0	0	0	0	41 (76)	0	13 (24)	[75,76]	[87]
Memphis, TN	1996	50[a]	9 (18)	1 (2)	15 (30)	6 (12)	14 (28)	1 (2)	5 (10)	[75,76]	[87]
New York, NY	1996	55[a]	0	0	1 (2)	1 (2)	36 (65)	0	17 (31)	[75,76]	[87]
Philadelphia, PA	1996	50[a]	0	0	3 (6)	1 (2)	38 (76)	0	8 (16)	[75,76]	[87]
St. Louis, MO	1996	53[a]	0	0	7 (13)	1 (2)	28 (53)	0	17 (32)	[75,76]	[87]
Africa											
Dakar, Senegal	1992	39[a]	19 (49)	3 (8)	3 (8)	3 (8)	ND[b]	ND	10 (26)	[79,82]	[88]
Nairobi, Kenya	1993	100	62 (62)	ND	ND	ND	ND	ND	ND	[72]	[72]
Nairobi, Kenya	1995	152	79 (52)	ND	ND	ND	ND	ND	ND	[82]	[82]
Malawi	1995[c]	156	ND	ND	26 (17)	ND	ND	ND	130 (83)	[70]	[67][e]
Maseru, Lesotho	1993–1994	100[a]	41 (41)	17 (17)	16 (16)	7 (7)	17 (17)	9 (9)	6 (6)	[75,76]	[62]
Johannesburg, RSA[b]	1995[d]	83[d]	69 (83)	ND	ND	ND	ND	ND	ND	[84]	[84]
Central African Republic	1994	66[a]	10 (15)	7 (11)	16 (24)	5 (8)	15 (23)	10 (16)	15 (23)	[75,76]	Lewis et al. (unpubl.)
Morocco	1995–1996	46[a]	23 (49)	1 (2)	ND	ND	ND	ND	15 (34)	[79]	Ryan et al. (unpubl.)
Harare, Zimbabwe	1998	74	35 (47)	ND	ND	ND	ND	ND	ND	[80]	[80]
Antananarivo, Madagascar	1997	196[a]	61 (31)	3 (2)	51 (26)	5 (3)	15 (8)	4 (2)	62 (32)	[75,76]	[89]

TABLE 9.6 (continued)
Prevalence of *H. ducreyi*, *T. pallidum*, and HSV Determined by PCR in Genital Ulcer Specimens from Different Geographic Locations

| Location | Year | Number Analyzed | Number Positive (Percent Positive) | | | | | | | [References] | |
			H. ducreyi Alone	*H. ducreyi* Co-Infection	*T. pallidum*	*T. pallidum* Co-Infection	HSV Alone	HSV Co-Infection	Unknown Etiology	PCR Technique	Study
Cape Town, RSA	1993–1994	180[a]	27 (15)	4 (2)	53 (29)	5 (3)	40 (22)	3 (2)	54 (30)	[75,76]	[90]
Johannesburg, RSA	1993–1994	159[a]	31 (20)	3 (2)	16 (10)	3 (3)	72 (45)	6 (4)	34 (21)	[75,76]	[90]
Durban, RSA	1993–1994	199[a]	74 (37)	32 (16)	18 (9)	11 (5)	45 (23)	27 (14)	29 (15)	[75,76]	[90]
Carletonville, RSA	1993–1994	232[a]	146 (63)	15 (6)	16 (7)	8 (3)	28 (12)	12 (5)	25 (11)	[83]	[83]
Carletonville, RSA	1998	186[a]	67 (36)	27 (15)	10 (5)	13 (7)	39 (21)	28 (15)	39 (21)	[83]	[83]
Welkom, RSA	1997	273[a]	95 (35)	3 (1)	110 (4)	2 (1)	68 (25)	1 (<1)	97 (36)	[83]	Chen et al. (unpubl.)
Botswana	2002	147[a]	1 (1)	0	2 (1)	0	84 (57)	0	60 (41)	[83]	Chen et al. (unpubl.)
Mali	2002	37[a]	1 (3)	0	0	0	11 (30)	0	25 (68)	[83]	Chen et al. (unpubl.)
Ethiopia	2003	77[a]	0	3 (4)	6 (8)	1 (1)	52 (68)	3 (4)	15 (20)	[83]	Chen et al. (unpubl.)
Mozambique	2003	137[a]	5 (4)	4 (3)	7 (5)	3 (2)	83 (61)	5 (4)	36 (26)	[83]	Chen et al. (unpubl.)
Asia											
Chiang Mai, Thailand	1995–1996	38[a]	0	0	0	1 (1)	31 (82)	1 (3)	6 (16)	[75,76]	[91]
Pune, India	1994	302[a]	69 (23)	15 (5)	29 (10)	12 (4)	79 (26)	16 (5)	104 (34)	[75,76]	[59]
Shanghai, China	2000–2001	80[a]	0	0	44 (55)	16 (20)	5 (6)	16 (20)	15 (19)	[76]	[92]
Chengdu, China	2000–2001	147[a]	0	0	34 (23)	12 (8)	38 (26)	12 (8)	63 (43)	[76]	[92]
Caribbean/Latin America											
Nassau, Bahamas	1992	47[a]	7 (15)	0	ND	ND	ND	ND	10 (21)	[79]	[93]
Kingston, Jamaica	1996	304[a]	54 (18)	18 (6)	22 (7)	9 (3)	138 (45)	20 (7)	67 (22)	[75,76]	[94]
Santo Domingo, Dominican Republic	1995–1996	81[a]	21 (26)	0	4 (5)	0	35 (43)	0	21 (26)	[75,76]	[95]
Lima, Peru	1994–1995	63[a]	0	3 (5)	6 (10)	0	24 (38)	3 (5)	30 (48)	[75,76]	[95]

a. Samples from consecutive patients with genital ulcers.

b. ND = not determined. RSA = Republic of South Africa.

c. Date of publication; date of specimen collection was not published.

d. Samples from patients with clinical diagnoses of chancroid.

e. Study only concerned syphilis.

TABLE 9.7
PCR Assays for *Treponema pallidum*

Assay	Primer	Target	Amplicon Size	Detection	[References]
1	Forward: 5'-biotinyl-GAAGTTGTCCCAGTTGCGGTT Reverse: 5'-biotinyl-CAGAGCCATCAGCCCTTTTCA Probe: 5'-CGGGCTCTCCATGCTGCTTACCTTA	47-KdA membrane protein gene	260 bp	Capture probe + ELISA	[75,76][b]
2	Forward: 5'-GCAGTGTCTGCAATCCACCGT Reverse: 5'-CGAGAAATTACGTGTCTTCAT	TmpA gene	599 bp	Southern blot	[69]
	Forward: 5'-TGTACCGGACGCTCGTGCCATT Reverse: 5'-TGGCCTTTCCCAACGTCCTCAG	4D gene	428 bp	Southern blot	[69]
3	Forward: 5'-GACAATGCTCACTGAGGATAGT Reverse: 5'-ACGCACAGAAACGAATTCCTTG	47-KdA membrane protein gene	658 bp	Dot blot	[70]
4	Forward: 5'-CTCAGCACTGCTGAGCGTAG Reverse: 5'-AACGCCTCCATCGTCAGACC Inner forward: 5'-CAGGTAACGGATGCTGAAGT Inner reverse: 5'-CGTGGCAGTAACCGCAGTCT	*bmp* gene	617 bp 500 bp	Agarose gel	[71][c]
5	Forward: 5'-CTCTTTTGGACGTAGGTCTTTGAG Reverse: 5'-TTACGTGTTACCGCGGCTGG	16S rRNA gene	366 bp	Agarose gel	[98][d]
	Forward: 5'-CGTGTGGTATCAACTATGG Reverse: 5'-TCAACCGTGTACTCAGTGC	47-kDA membrane protein gene	310 bp	Agarose gel	[98]
6	Forward: 5'-FAM-TGCGCGTGTGCGAATGGTGTGGTC Reverse: 5'-TET-CACAGTGCTCAAAAACGCCTGCACG	*polA*	377 bp	ABI 310	[99]
	Forward: 5'-FAM-CGTCTGGTCGATGTGCAAATGAGTG Reverse: 5'-TET-TGCACATGTACACTGAGTTGACTCGG		393 bp	ABI 310	[99]

a. Probes shown for assays employing solid phase detection methodologies.

b. Multiplex assay for *Treponema pallidum*, *Haemophilus ducreyi*, and herpes simplex virus.

c. Nested PCR assay.

d. Reverse transcriptase PCR assay.

The rabbit infectivity test (RIT) is the gold standard for the direct detection of *T. pallidum* in clinical specimens.[97] This method has an analytic sensitivity of about a single organism when repeated passages in rabbits are used. However, it is expensive, time-consuming (takes about 1 to 2 months to complete), and requires access to an animal facility. Thus, microscopy (dark-field examination, DFA) and serology (nontreponemal and treponemal test results) have been used for evaluating the clinical performance of PCR assays for *T. pallidum*. Orle et al.[75] compared the detection of *T. pallidum* DNA in 295 genital ulcer specimens by multiplex PCR with dark-field microscopy. After resolution of the discrepant results using another NAAT, a total of 80 specimens were positive. The sensitivity of multiplex PCR was 91.2% (73 of 80) while the sensitivity of dark-field microscopy was 81.2% (65 of 80). The specificity of multiplex PCR was 99.1% (213 of 215) and that of dark-field 100% (215 of 215).

In another study, Mertz et al.[87] reported that only 49% (31 of 64) of persons tested positive for *T. pallidum* by multiplex PCR in 10 U.S. cities were positive by dark-field microscopy. Dark-field microscopy requires considerable skill and its performance has been shown to vary among microscopists. Thus, it is not surprising that multiplex PCR is more sensitive than dark-field microscopy for the detection of *T. pallidum*.

Detection of *T. pallidum* can also be accomplished by DFA staining performed with a fluorescein-conjugated monoclonal anti-*T. pallidum* antibody.[100] Jethwa et al.[67] compared the performances of PCR and DFA for the detection of *T. pallidum* in touch preparations from 156 patients with genital ulcer disease. After microscopic examination of the DFA slides, the cover slips were removed and the DNA removed with 0.5% sodium dodecylsulfate and a swab. After the PCR amplification step, amplicons were detected by dot blot hybridization. Overall, 14.1% (22 of 156) specimens were positive by both DFA and PCR while 127 were negative by both methods. PCR was positive in four DFA-negative specimens and negative in three DFA-positive specimens. Thus, there was good concordance between the two methods.

A comparison of the performance of PCR with the results of serological tests for syphilis is problematic. For example, only 80 to 87% of those diagnosed with primary syphilis have positive nontreponemal test results.[97] A high prevalence of untreated (latent) or recently treated syphilis is present in highly endemic regions. Thus, patients presenting with genital ulcers may not have primary syphilis in spite of positive nontreponemal tests.[62,101] Orle et al.[75] compared multiplex PCR results for *T. pallidum* with nontreponemal serologic test results (VDRL [Venereal Disease Research Laboratories] or RPR [rapid plasma reagin] test) in 296 consecutive men with genital ulcers in New Orleans. The concordance of the assays was 88%. When the RPR and VDRL test results were analyzed separately, multiplex PCR detected 11 specimens containing *T. pallidum* that were negative by the RPR test and 15 specimens that were negative by the VDRL test. Thirteen specimens tested positive by serology were negative via both multiplex PCR and dark-field microscopy.

Morse et al.[62] tested genital ulcer specimens by multiplex PCR from 105 consecutive patients seen at a genitourinary medicine clinic in Maseru, Lesotho. Nontreponemal (RPR) and treponemal (fluorescent treponemal antigen-antibody absorption test [FTA-ABS]) serologic test results were available from 98 patients. When serological test results were compared to those obtained by multiplex PCR, 6% (2 of 35) with negative serologic test results, 0% (0 of 1) with only a reactive RPR test result, 57% (13 of 23) with reactive RPR and FTA-ABS tests and 21% (8 of 39) with only reactive FTA-ABS tests had *T. pallidum* DNA detected by multiplex PCR. A positive multiplex PCR result was strongly associated (P = 0.003) with both reactive RPR and FTA-ABS tests in those with indeterminate diagnoses. Overall, FTA-ABS tests were reactive in 91% (21 of 23) of patients with positive multiplex PCR results for *T. pallidum*; in contrast, RPR tests were only reactive in 56.5% (13 of 23) of these 23 patients (P = 0.007). *T. pallidum* multiplex PCR results were negative in 67% (42 of 63) of patients with reactive RPR or FTA-ABS test results.

Ballard et al.[101] obtained similar results. They reported that *T. pallidum*-specific DNA sequences were detected by multiplex PCR in 18.8% (163 of 868) of patients with genital ulcer disease. Overall, the sensitivity of the RPR test was 69% and that of the FTA-ABS test was 87%. Thus, in areas of high syphilis endemicity, a single nontreponemal test may miss 6 to 21% of patients with primary syphilis.

The prevalence of *T. pallidum* DNA in genital ulcer specimens obtained from different geographic locations was determined by multiplex PCR (Table 9.6). Syphilis has been a significant cause of genital ulcers in most of the locations sampled. A smaller proportion of patients positive for *T. pallidum* DNA also had second organisms identified in their ulcers than was observed in patients with chancroid. Recently, however, rates of syphilis have declined in specimens collected in Sub-Saharan Africa. Some of the decline may be due to implementation of STD control programs or to increases in HIV infection.

As noted previously, *T. pallidum* subsp. pallidum can disseminate hematogenously and infect virtually any organ. Several studies with PCR were intended to diagnose other forms of syphilis. All these studies involved small numbers of patients and should be repeated with larger numbers. PCR has also been used to detect the presence of *T. pallidum* DNA in blood from a small number of patients with incubating, primary, and secondary syphilis.[102] The results have been encouraging although problems with inhibitors have not been overcome.[102] A larger study is needed to evaluate whether blood can be used for the diagnosis of syphilis by PCR.

Equally encouraging was a report by Liu et al.[103] who used a real-time semiquantitative PCR to show that the number of spirochetes in the blood of patients with syphilis varied with the stage of syphilis and ranged from 200 to 100,000 organisms per milliliter of blood. However, the number of spirochetes in ulcer specimens from primary syphilis patients was greater and ranged from 22,000 to 5.7 million treponemes per milliliter of specimen collection fluid.

PCR has also been used with serum or cerebrospinal fluid (CSF) specimens for the diagnosis of early latent[70] neurosyphilis[69,71] and meningitis.[70] It has also been used with formalin-fixed tissue in the diagnosis of cardiovascular syphilis.[104] PCR may prove to be a useful method to aid in the diagnosis of congenital syphilis. It has been used to test serum and CSF from symptomatic infants and amniotic fluid from mothers.[65,66] While the data are encouraging, these studies must be repeated with larger numbers of patients.

Genital Herpes

Herpes simplex is a major cause of genital ulcer disease throughout the world. Herpes simplex virus type 2 (HSV-2) is the predominant cause of genital herpes. Herpes simplex type 1 (HSV-1) has been associated with genital herpes, but to a lesser degree.[105,106] Cell culture is the gold standard for the diagnosis of genital herpes,[107] but the sensitivity of culture is affected by the stage of the lesion[108] and by specimen transport conditions. The culture can also be time-consuming — the median time to cytopathic effect (CPE) is 2 to 3 days, and approximately 5% of specimens require 7 days to develop cytopathic effect (CPE).[107]

PCR is a viable technology for the diagnosis of genital herpes. Numerous publications describe the use of PCR for the diagnosis of herpes virus infections. However, in this chapter the use of PCR for HSV diagnosis will be discussed in the context of genital ulcer disease. A number of these assays have been used for the diagnosis of genital herpes (Table 9.8) as well as multiplex assays that include *T. pallidum* and *H. ducreyi*. All these HSV PCR assays are based on the amplification of sequences within the glycoprotein B gene. Different strategies have been used for the differentiation of HSV-1 and -2.

Orle et al.[75,76] used a generic HSV capture probe and ELISA to detect HSV-1 and -2. Lai et al.[83] identified the HSV amplicon by its characteristic size and then performed a separate real-time PCR assay that could differentiate HSV-1 from -2. More recently, Chen and Ballard[109] developed a quadraplex real-time PCR assay that could simultaneously identify HSV-1, HSV-2, *T. pallidum*, and *H. ducreyi*.

The performance of multiplex PCR in the detection of HSV-1 or -2 DNA has been compared with that of cell culture. Orle at al.[75] isolated HSV in 28 of 99 genital ulcer (nonvesicular) specimens

TABLE 9.8
PCR Assays for Herpes Simplex Virus Types -1 and -2

Assay	Primer	Target	Amplicon Size	Detection	[References]
1	**HSV-1 + HSV-2**				
	Forward: 5'-biotinyl-TTCAAGGCCACCATGTACTACAAAGACGT	Glycoprotein B gene	431 bp	Capture probe + ELISA	[75,76][b]
	Reverse: 5'-biotinyl-GCCGTAAAACGGGGACATGTACACAAAGT				
	Probe: 5'-GGTCTCGTGGTCGTCCCGGTGAAA[a]				
2	**HSV-1 + -2**				
	Forward: 5'-TET-TTCAAGGCCACCATGTACTACAAAGACGT	Glycoprotein B gene	431 bp	ABI 310	[83][b]
	Reverse: 5'-GCCGTAAAACGGGGACATGTACACAAAGT				
3	**HSV-1**				
	Forward: 5'-CTGTTCTCGTTCCTCACTGCCT	Glycoprotein B gene	52 bp	Smart Cycler	[83][d]
	Reverse: 5'-CGTTTTTGTGTATTGGTGCG				
	Probe: 5'-FAM-CCCTGGACACCCTCTTCGTCGTCAG-BH[c]				
	HSV-2				
	Forward: 5'-CGGCTCCCCTGCTCTAGATA	Glycoprotein B gene	112 bp	Smart Cycler	
	Reverse: 5'-CTTGGCAGCACAACTTTGG				
	Probe: 5'-TET-GCGGGCGTTCGTTTGTCTGGT-BH[a]				

a. Probes shown for assays employing solid phase detection methodologies.

b. Multiplex assay for *Treponema pallidum*, *Haemophilus ducreyi*, and herpes simplex virus.

c. Black hole quencher.

d. Real time PCR assay.

tested, of which 27 were HSV-2 and 1 was HSV-1. Multiplex PCR detected HSV DNA in all 28 specimens that were positive by culture and detected 11 additional specimens that were negative by culture; using a second NAAT, all 39 positive specimens resolved as positive for the presence of HSV.

The sensitivity of the multiplex PCR relative to culture was 100%, compared with a sensitivity of culture of 71.8%. The specificity of both multiplex PCR and culture was 100%. In another study, Morse et al.[62] examined 105 nonvesicular genital ulcer patients in Lesotho; multiplex PCR identified 15 of 17 culture-positive specimens and showed positive in an additional 10 culture-negative specimens. All 10 of the culture-negative specimens were positive in a confirmatory PCR assay. The resolved sensitivity and specificity of the multiplex PCR were 93% and 100%, respectively. The sensitivity of culture relative to multiplex PCR was 60%.

The performance of multiplex PCR for HSV has also been compared to performance of a rapid enzyme immunoassay (HerpChek, Dupont, Wilmington, Delaware, USA). When compared with multiplex PCR, the HerpChek test was found to be 68.5% sensitive and 99.5% specific.[59] In another study,[95] HerpChek had a sensitivity of 70.4% and a specificity of 85.3% in comparison with multiplex PCR. Thus, with nonvesicular lesions, multiplex PCR is more sensitive and rapid than culture and more sensitive than HerpChek for the diagnosis of genital herpes.

Multiplex PCR has been used to determine the prevalence of herpes simplex virus DNA in nonvesicular genital ulcer specimens from different geographic locations (Table 9.6). Genital herpes is a major cause of genital ulcers in most of the locations sampled. A significant number of individuals from southern Africa and other areas with high prevalence of HIV had other agents detected in their ulcers. In the context of HIV infection, lesions due to HSV are often atypical, making clinical diagnosis difficult and overuse of antibiotics likely.

Donovanosis

C. granulomatis is the causative agent of donovanosis (also known as granuloma inguinale), an uncommon but important cause of genital ulcer disease. Donovanosis is often considered a tropical sexually transmitted infection, but in the preantibiotic era it was found throughout the world. It is now reported to be endemic in southern India, Papua New Guinea, southern Africa, parts of South America, and northern and central Australia.[110]

C. granulomatis is a Gram-negative, intracellular, pleomorphic, encapsulated bacillus. In specimens from infected individuals, it is seen in the cytoplasm of large mononuclear cells (Donovan bodies), and less commonly in polymorphonuclear leukocytes.[110] It is difficult to cultivate *in vitro*, but two new methods involving co-culture in fresh human monocytes and in a human epithelial cell line have recently been reported.

Confirmation of the diagnosis is currently based on the morphological examination of smears or biopsy specimens for the presence of the causative organisms (Donovan bodies) within large mononuclear cells. However, the sensitivity of these cytological and histological techniques varies between 60 and 80%.[111] A 334-bp sequence within the *pho*E gene of *C. granulomatis* was shown to be very similar to those from *Klebsiella pneumoniae*, *K. ozaenae*, and *K. rhinoscleromatis*.[112] A comparison of the sequences showed two base changes unique to *C. granulomatis*.[112]

These differences were used to develop a PCR assay that can be used on simple swab or biopsy specimens (Table 9.9).[113] The assay has been modified so that it can be used routinely in clinical laboratories.[114] After digesting the PCR products with *Hae*III (undigested *C. granulomatis* product; cleaved *Klebsiella* spp. products releasing biotinylated fragment), magnetic beads with covalently linked oligonucleotides are used to capture the cleaved and uncleaved PCR products. The *C. granulomatis* amplicon is detected colorimetrically after the addition of an avidin-labeled enzyme. This assay must be extensively validated using specimens from donovanosis patients from several geographic areas.

TABLE 9.9
PCR Assays for *Calymmatobacterium granulomatis*

Assay	Primer	Target	Amplicon Size	Detection	[References]
1	**Forward:** 5'-GATCTGACCCTGCAGTACC **Reverse:** 5'-CAGACCGAAGTCGAACTGATACTG	*pho*E	188 bp	HaeIII digestion + agarose gel	[113]
2	**Forward:** 5'-biotinyl-GATCTGACCCTGCAGTACC **Reverse:** 5'-CAGACCGAAGTCGAACTGATACTG **Probe:** 5'-NH₂CCAGCAGGTTCTGATCGTT ᵃ	*pho*E	188 bp	HaeIII digestion + magnetic bead capture + avidin–HRP + substrate	[113]

a. Capture probes shown for assays employing solid phase detection methodologies.
b. HRP = horseradish peroxidase.

LYMPHOGRANULOMA VENEREUM

Lymphogranuloma venereum (LGV) is caused by *C. trachomatis*; three serovars (L1, L2, L3) are responsible for the majority of cases. LGV is a chronic disease with acute and late manifestations. A primary stage causing a small genital lesion appears in a minority of patients. Most common is the secondary stage characterized by acute inguinal (and femoral) lymphadenitis with bubo formation. The laboratory diagnosis of LGV has usually been based on positive chlamydial serology or isolation of *C. trachomatis* from an infected site.[115]

PCR has been used to detect the presence of chlamydial DNA in genital ulcer specimens,[59,62,89,93] but the presence of chlamydial DNA in an ulcer specimen is not sufficient evidence for a diagnosis of LGV. Bauwens et al.[93] verified that genital ulcer disease cases were caused by one of the serovars of *C. trachomatis* that cause LGV, by amplifying and sequencing an approximately 1-kb portion of the MOMP gene from Chlamydia-positive PCR samples. MOMP sequences vary among the different *C. trachomatis* serovars, and these differences can be used to genotype the chlamydial strain. Use of similar probes identified an outbreak of LGV proctitis in men who have sex with men (MSM) in Rotterdam.[116]

SUMMARY

NAATs have been developed for the major etiologic agents associated with genital discharges and ulcers. It is important to note that several of these agents are difficult to culture or have not been cultured *in vitro*. Furthermore, current laboratory tests are relatively insensitive. NAATs only for *C. trachomatis* and *N. gonorrhoeae* are commercially available as of this writing. This may reflect the perception that the commercial market for other STD agents is limited. Nevertheless, assays for *T. pallidum*, *H. ducreyi*, and HSV have been employed in multiplex formats by reference and research laboratories to determine the etiology of genital ulcer disease in many countries.

The etiology exhibits both geographical and temporal variations. This information has been important in evaluating the performance of syndromic algorithms for the management of genital ulcer disease. Improvements are still needed in sample collection and preparation in order to reduce the proportion of PCR-negative specimens. Perhaps additional studies are warranted to identify additional causes of genital ulcers.

We should also see better front end procedures (e.g., automated specimen processing) and improved throughput for the commercial NAATs, making them more efficient and moving toward totally automated testing. Improvements should reduce the costs of the tests. Hopefully, we would have better rapid tests that would allow more effective management of infected patients by reducing delays in treatment and ensuring that infected patients would not be lost to follow-up.[117]

Technological advances should provide faster, simpler, isothermic NAATs that will be even more suitable for diagnosis and screening. More multiplexing will be seen to allow detection of multiple pathogens with a single test.

REFERENCES

1. Gerbase AC, Rowley JT, Heymann DH, Berkley SF, Piot P. Global prevalence and incidence estimates of selected curable STDs. *Sex Transm Infect* 1998, 74 (Suppl 1), S12–S16.
2. Scholes D, Stergachis A, Heidrich FE, Andrilla H, Holmes KK, Stamm WE. Prevention of pelvic inflammatory disease by screening for cervical chlamydial infection. *New Engl J Med* 1996, 334 (21), 1362–1366.
3. Schachter J, Sweet RL, Grossman M, Landers D, Robbie M, Bishop E. Experience with the routine use of erythromycin for chlamydial infections in pregnancy. *New Engl J Med* 1986, 314 (5), 276--279.

4. Schachter J, Stamm WE. Chlamydia, in Murray PR, Baron EJ, Pfaller MA, Tenover FC, Yolken RH, Eds., *Manual of Clinical Microbiology*, 7th ed. Washington: American Society for Microbiology; 1999. pp. 795–806.

5. Schachter J. NAATs to diagnose *Chlamydia trachomatis* genital infection: a promise still unfulfilled. *Expert Rev Mol Diagn* 2001, 1 (2), 137–144.

6. Stamm WE. *Chlamydia trachomatis* infections of the adult, in Holmes KK, Sparling PF, March P-A, Lemon SM, Stamm WE, Piot P, Wasserheit JN, Eds., *Sexually Transmitted Diseases*, 3rd ed. New York: McGraw Hill; 1999. pp. 407–422.

7. Centers for Disease Control and Prevention. Sexually transmitted diseases: treatment guidelines 2002. *MMWR* 2002, 51 (RR-6), 1–82.

8. Jaschek G, Gaydos CA, Welsh LE, Quinn TC. Direct detection of *Chlamydia trachomatis* in urine specimens from symptomatic and asymptomatic men by using a rapid polymerase chain reaction assay. *J Clin Microbiol* 1993, 31 (5), 1209–1212.

9. Schachter J, Stamm WE, Quinn TC, Andrews WW, Burczak JD, Lee HH. Ligase chain reaction to detect *Chlamydia trachomatis* infection of the cervix. *J Clin Microbiol* 1994, 32 (10), 2540–2543.

10. Hadgu A. The discrepancy in discrepant analysis. *Lancet* 1996, 348 (9027), 592–593.

11. Schachter J. Two different worlds we live in. *Clin Infect Dis* 1998, 27 (5), 1181–1185.

12. Green TA, Black CM, Johnson RE. Evaluation of bias in diagnostic test sensitivity and specificity estimates computed by discrepant analysis. *J Clin Microbiol* 1998, 36 (2), 375–381.

13. Johnson RE, Green TA, Schachter J, Jones RB, Hook EW, Black CM, Martin DH, St. Louis ME, Stamm WE. Evaluation of nucleic acid amplification tests as reference tests for *Chlamydia trachomatis* infections in asymptomatic men. *J Clin Microbiol* 2000, 38 (12), 4382–4386.

14. Bauwens JE, Clark AM, Stamm WE. Diagnosis of *Chlamydia trachomatis* endocervical infections by a commercial polymerase chain reaction assay. *J Clin Microbiol* 1993, 31 (11), 3023–3027.

15. Mahony J, Chong S, Jang D, Luinstra K, Faught M, Dalby D, Sellors J, Chernesky M. Urine specimens from pregnant and nonpregnant women inhibitory to amplification of *Chlamydia trachomatis* nucleic acid by PCR, ligase chain reaction, and transcription-mediated amplification: identification of urinary substances associated with inhibition and removal of inhibitory activity. *J Clin Microbiol* 1998, 36 (11), 3122–3126.

16. Chernesky MA, Lee H, Schachter J, Burczak JD, Stamm WE, McCormack WM, Quinn TC. Diagnosis of *Chlamydia trachomatis* urethral infection in symptomatic and asymptomatic men by testing first-void urine in a ligase chain reaction assay. *J Infect Dis* 1994, 170 (5), 1308–1311.

17. Crotchfelt KA, Pare B, Gaydos C, Quinn TC. Detection of *Chlamydia trachomatis* by the Gen-Probe amplified *Chlamydia trachomatis* assay (AMP CT) in urine specimens from men and women and endocervical specimens from women. *J Clin Microbiol* 1998, 36 (2), 391–394.

18. Van Der Pol B, Ferrero DV, Buck-Barrington L, Hook E, Lenderman C, Quinn T, Gaydos CA, Lovchik J, Schachter J, Moncada J, Hall G, Tuohy MJ, Jones RB. Multicenter evaluation of the BDProbeTec ET system for detection of *Chlamydia trachomatis* and *Neisseria gonorrhoeae* in urine specimens, female endocervical swabs and male urethral swabs. *J Clin Microbiol* 2001, 39 (3), 1008–1016.

19. Howell MR, Quinn TC, Brathwaite W, Gaydos CA. Screening women for *Chlamydia trachomatis* in family planning clinics: the cost-effectiveness of DNA amplification assays. *Sex Transm Dis* 1998, 25 (2), 108–117.

20. Shafer MA, Pantell RH, Schachter J. Is the routine pelvic examination needed with the advent of urine-based screening for sexually transmitted diseases? *Arch Pediatr Adolesc Med* 1999, 153 (2), 119–125.

21. Kacena KA, Quinn SB, Howell MR, Madico GE, Quinn TC, Gaydos CA. Pooling urine samples for ligase chain reaction screening for genital *Chlamydia trachomatis* infection in asymptomatic women. *J Clin Microbiol* 1998, 36 (2), 481–485.

22. Schachter J, Hook EW 3rd, McCormack WM, Quinn TC, Chernesky M, Chong S, Girdner JI, Dixon PB, DeMeo L, Williams E, Cullen A, Lorincz A. Ability of the Digene Hybrid Capture II test to identify *Chlamydia trachomatis* and *Neisseria gonorrhoeae* in cervical specimens. *J Clin Microbiol* 1999, 37 (11), 3668–3371.

23. Darwin LH, Cullen AP, Crowe SR, Modarress KJ, Willis DE, Payne WJ. Evaluation of the Hybrid Capture 2 CT/GC DNA tests and the GenProbe PACE 2 tests from the same male urethral swab specimens. *Sex Transm Dis* 2002, 29 (10), 576–580.

24. Gaydos CA, Quinn TC, Willis D, Weissfeld A, Hook EW, Martin DH, Ferrero D V, Schachter J. Performance of the APTIMA Combo 2 assay for detection of *Chlamydia trachomatis* and *Neisseria gonorrhoeae* in female urine and endocervical swab specimens. *J Clin Microbiol* 2003, 41 (1), 304–309.

25. Schachter J, Hook EW, Martin DH, Willis D, Fine P, Fuller D, Jordan J, Janda WM, Chernesky M. Confirming positive results of nucleic acid amplification tests (NAATs) for *Chlamydia trachomatis*: all NAATs are not created equal. *J Clin Microbiol* 2005, 43 (3), 1372–1373.

26. Culler EE, Caliendo AM, Nolte FS. Reproducibility of positive test results in the BDProbeTec ET system for detection of *Chlamydia trachomatis* and *Neisseria gonorrhoeae*. *J Clin Microbiol* 2003, 41 (8), 3911–3914.

27. Van der Pol B. Identifying an expanded grey zone that enhances the specificity of the COBAS AMPLICOR CT/NG Test for *Neisseria gonorrhoeae*. Presented at First Joint Meeting of the American Sexually Transmitted Diseases Association and the Medical Society for the Study of Venereal Diseases, Baltimore, May 3–7, 2000.

28. Tabrizi SN, Chen S, Cohenford MA, Lentrichia BB, Coffman E, Shultz T, Tapsall JW, Garland SM. Evaluation of real time polymerase chain reaction assays for confirmation of *Neisseria gonorrhoeae* in clinical samples tested positive in the Roche COBAS AMPLICOR assay. *Sex Transm Infect* 2004, 80 (1), 68–71.

29. Lee HH, Chernesky MA, Schachter J, Burczak JD, Andrews WW, Muldoon S, Leckie G, Stamm WE. Diagnosis of *Chlamydia trachomatis* genitourinary infection in women by ligase chain reaction assay of urine specimens. *Lancet* 1995, 345 (8944), 213–216.

30. Smith KR, Ching S, Lee H, Ohhashi Y, Hu HY, Fisher HC 3rd, Hook EW 3rd. Evaluation of ligase chain reaction for use with urine for identification of *Neisseria gonorrhoeae* in females attending a sexually transmitted disease clinic. *J Clin Microbiol* 1995, 33 (2), 455–457.

31. Jensen IP, Thorsen P, Moller BR. Sensitivity of ligase chain reaction assay of urine from pregnant women for *Chlamydia trachomatis* [letter]. *Lancet* 1997, 349 (9048), 329–330.

32. Notomi T, Ikeda Y, Okadome A, Nagayama A. The inhibitory effect of phosphate on the ligase chain reaction used for detecting *Chlamydia trachomatis*. *J Clin Pathol* 1998, 51 (4), 306–308.

33. Martin DH, Cammarata C. Potential causes of decreased sensitivity of the Abbott ligase chain reaction (LCR) assay for *Chlamydia trachomatis* in urine specimens (Abstr 385), in Proceedings of 13th Meeting of International Society for Sexually Transmitted Diseases Research, Denver, July 11–14, 1999.

34. Schachter J, McCormack WM, Chernesky MA, Martin DH, Van Der Pol B, Rice PA, Hook EW 3rd, Stamm WE, Quinn TC, Chow JM. Vaginal swabs are appropriate specimens for diagnosis of genital tract infection with *Chlamydia trachomatis*. *J Clin Microbiol* 2003, 41 (8), 3784–3789.

35. Wiesenfeld HC, Heine RP, Rideout A, Macio I, DiBiasi F, Sweet RL. The vaginal introitus: a novel site for *Chlamydia trachomatis* testing in women. *Am J Obstet Gynecol* 1996, 174 (5), 1542–1546.

36. Hook EW 3rd, Smith K, Mullen C, Stephens J, Rinehardt L, Pate MS, Lee HH. Diagnosis of genitourinary *Chlamydia trachomatis* infections by using the ligase chain reaction on patient-obtained vaginal swabs. *J Clin Microbiol* 1997, 35 (8), 2133–2135.

37. Stary A, Najim B, Lee HH. Vulval swabs as alternative specimens for ligase chain reaction detection of genital chlamydial infection in women. *J Clin Microbiol* 1997, 35 (4), 836–838.

38. Shafer M-A, Moncada J, Boyer CB, Betsinger K, Flinn SD, Schachter J. Comparing first-void urine specimens, self-collected vaginal swabs, and endocervical specimens to detect *Chlamydia trachomatis* and *Neisseria gonorrhoeae* by a nucleic acid amplification test. *J Clin Microbiol* 2003, 41 (9), 4395–4399.

39. Hook EW 3rd, Ching SF, Stephens J, Hardy KF, Smith KR, Lee HH. Diagnosis of *Neisseria gonorrhoeae* infections in women by using the ligase chain reaction on patient-obtained vaginal swabs. *J Clin Microbiol* 1997, 35 (8), 2129–2132.

40. Van Der Pol B, Quinn TC, Gaydos CA, Crotchfelt K, Schachter J, Moncada J, Jungkind D, Martin DH, Turner B, Peyton C, Jones RB. Multicenter evaluation of the AMPLICOR and automated COBAS AMPLICOR CT/NG tests for detection of *Chlamydia trachomatis*. *J Clin Microbiol* 2000, 38 (3), 1105–1112.

41. Ostergaard L, Andersen B, Moller JK, Olesen F. Home sampling versus conventional swab sampling for screening of *Chlamydia trachomatis* in women: a cluster-randomized one-year follow-up study. *Clin Infect Dis* 2000, 31 (4), 951–957.

42. Cohen DA, Nsuami M, Etame RB, Tropez-Sims S, Abdalian S, Farley TA, Martin DH. A school-based chlamydia control program using DNA amplification technology. *Pediatrics* 1998, 101 (1), e1.

43. Burstein GR, Gaydos CA, Diener-West M, Howell MR, Zenilman JM, Quinn TC. Incident *Chlamdyia trachomatis* infections among inner-city adolescent females. *JAMA* 1998, 280 (6), 521–526.

44. Turner CF, Rogers SM, Miller HG, Miller WC, Gribble JN, Chromy JR, Leone PA, Cooley PC, Quinn TC, Zenilman JM. Untreated gonococcal and chlamydial infection in a probability sample of adults. *JAMA* 2002, 287 (6), 726–733.

45. Tully JG, Taylor-Robinson D, Cole RM, Rose DL. A newly discovered mycoplasma in the human urogenital tract. *Lancet* 1981, 1 (8233), 1288–1291.

46. Palmer HM, Gilroy CB, Furr PM, Taylor-Robinson D. Development and evaluation of the polymerase chain reaction to detect *Mycoplasma genitalium*. *FEMS Microbiol Lett* 1991, 61 (2–3), 199–203.

47. Jensen JS, Uldum SA, Sondergard-Andersen J, Vuust J, Lind K. Polymerase chain reaction for detection of *Mycoplasma genitalium* in clinical samples. *J Clin Microbiol* 1991, 29 (1), 46–50.

48. Totten PA, Schwartz MA, Sjostrom KE, Kenny GE, Handsfield HH, Weiss JB, Whittington WL. Association of *Mycoplasma genitalium* with nongonococcal urethritis in heterosexual men. *J Infect Dis* 2001, 183 (2), 269–276.

49. Horner P, Thomas B, Gilroy CB, Egger M, Taylor-Robinson D. Role of *Mycoplasma genitalium* and *Ureaplasma urealyticum* in acute and chronic nongonococcal urethritis. *Clin Infect Dis* 2001, 32 (7), 995–1003.

50. Dutro SM, Hebb JK, Garin CA, Hughes JP, Kenny GE, Totten PA. Development and performance of a microwell plate-based polymerase chain reaction assay for *Mycoplasma genitalium*. *Sex Transm Dis* 2003, 30 (10), 756–763.

51. Mena L, Wang X, Mroczkowski TF, Martin DH. *Mycoplasma genitalium* infections in asymptomatic men and men with urethritis attending a sexually transmitted diseases clinic in New Orleans. *Clin Infect Dis* 2002, 35 (10), 1167–1173.

52. Gambini D, Decleva I, Lupica L, Ghislanzoni M, Cusini M, Alessi E. *Mycoplasma genitalium* in males with nongonococcal urethritis: prevalence and clinical efficacy of eradication. *Sex Transm Dis* 2000, 27 (4), 226–229.

53. Madico G, Quinn TC, Rompalo A, McKee KT Jr, Gaydos CA. Diagnosis of *Trichomonas vaginalis* infection by PCR using vaginal swab samples. *J Clin Microbiol* 1998, 36 (11), 3205–3210.

54. Lawing LF, Hedges SR, Schwebke JR. Detection of trichomoniasis in vaginal and urine specimens from women by culture and PCR. *J Clin Microbiol* 2000, 38 (10), 3585–3588.

55. Wiesenfeld HC, Lowry DL, Heine RP, Krohn MA, Bittner H, Kellinger K, Shultz M, Sweet RL. Self-collection of vaginal swabs for the detection of chlamydia, gonorrhea, and trichomoniasis: opportunity to encourage sexually transmitted disease testing among adolescents. *Sex Transm Dis* 2001, 28 (6), 321–325.

56. Hardick J, Yang S, Lin S, Duncan D, Gaydos C. Use of the Roche LightCycler instrument in a real-time PCR for *Trichomonas vaginalis* in urine samples from females and males. *J Clin Microbiol* 2003, 41 (12), 5619–5622.

57. Schwebke JR, Hook EW 3rd. High rates of *Trichomonas vaginalis* among men attending a sexually transmitted diseases clinic: implications for screening and urethritis management. *J Infect Dis* 2003, 188 (3), 465–468.

58. Weisner P, Brown S, Kraus S, Perine P. Genital ulcers, in Holmes KK, Mardh P-A, Eds. *International Perspectives on Neglected Sexually Transmitted Diseases*. New York: McGraw-Hill, 1983. pp. 219–234.

59. Risbud A, Chan-Tack K, Gadkari D, Gangakhedkar RR, Shepherd ME, Bollinger R, Mehendale S, Gaydos C, Divekar A, Rompalo A, Quinn TC. The etiology of genital ulcer disease by multiplex polymerase chain reaction and relationship to HIV infection among patients attending sexually transmitted disease clinics in Pune, India. *Sex Transm Dis* 1999, 26 (1), 55–62.

60. Htun Y, Morse SA, Dangor Y, Fehler G, Radebe F, Trees DL, Beck-Sague CM, Ballard RC. Comparison of clinically directed, disease specific, and syndromic protocols for the management of genital ulcer disease in Lesotho. *Sex Transm Infect* 1998, 74 (Suppl 1), S23–S28.

61. Kibukamusoke JW. Venereal disease in East Africa. *Trans R Soc Trop Med Hyg* 1965, 59, 642–648.

62. Morse SA, Trees DL, Htun Y, Radebe F, Orle KA, Dangor Y, Beck-Sague CM, Schmid S, Fehler G, Weiss JB, Ballard RC. Comparison of clinical diagnosis and standard laboratory and molecular methods for the diagnosis of genital ulcer disease in Lesotho: association with human immunodeficiency virus infection. *J Infect Dis* 1997, 175 (3), 583–589.

63. DiCarlo RP, Martin DH. The clinical diagnosis of genital ulcer disease in men. *Clin Infect Dis* 1997, 25 (2), 292–298.

64. Ballard R, Morse SA. Chancroid, in Morse SA, Ballard R, Holmes KK, Moreland AA, Eds., *Atlas of Sexually Transmitted Diseases*, 3rd ed. London: Mosby; 2003. pp. 53–71.

65. Sanchez PJ, Wendel GD Jr, Grimprel E, Goldberg M, Hall M, Arencibia-Mireles O, Radolf JD, Norgard MV. Evaluation of molecular methodologies and rabbit infectivity testing for the diagnosis of congenital syphilis and neonatal central nervous system invasion by *Treponema pallidum*. *J Infect Dis* 1993, 167 (1), 148–157.

66. Grimprel E, Sanchez PJ, Wendel GD, Burstain JM, McCracken GH Jr, Radolf JD, Norgard MV. Use of polymerase chain reaction and rabbit infectivity testing to detect *Treponema pallidum* in amniotic fluid, fetal and neonatal sera, and cerebrospinal fluid. *J Clin Microbiol* 1991, 29 (8), 1711–1718.

67. Jethwa HS, Schmitz JL, Dallabetta G, Behets F, Hoffman I, Hamilton H, Lule G, Cohen M, Folds JD. Comparison of molecular and microscopic techniques for detection of *Treponema pallidum* in genital ulcers. *J Clin Microbiol* 1995, 33 (1), 180–183.

68. Ndowa F, Ballard R. Syndromic management, in Morse SA, Ballard R, Holmes KK, Moreland AA, Eds., *Atlas of Sexually Transmitted Diseases*, 3rd ed. London: Mosby; 2003. pp. 365–373.

69. Hay PE, Clarke JR, Strugnell RA, Taylor-Robinson D, Goldmeier D. Use of the polymerase chain reaction to detect DNA sequences specific to pathogenic treponemes in cerebrospinal fluid. *FEMS Microbiol Lett* 1990, 56 (3), 233–238.

70. Burstain JM, Grimprel E, Lukehart SA, Norgard M V, Radolf JD. Sensitive detection of *Treponema pallidum* by using the polymerase chain reaction. *J Clin Microbiol* 1991, 29 (1), 62–69.

71. Noordhoek GT, Wolters EC, de Jonge ME, van Embden JD. Detection by polymerase chain reaction of *Treponema pallidum* DNA in cerebrospinal fluid from neurosyphilis patients before and after antibiotic treatment. *J Clin Microbiol* 1991, 29 (9), 1976–1984.

72. Chui L, Albritton W, Paster B, Maclean I, Marusyk R. Development of the polymerase chain reaction for diagnosis of chancroid. *J Clin Microbiol* 1993, 31 (3), 659–664.

73. Kimura H, Shibata M, Kuzushima K, Nishikawa K, Nishiyama Y, Morishima T. Detection and direct typing of herpes simplex virus by polymerase chain reaction. *Med Microbiol Immunol (Berl)* 1990, 179 (4), 177–184.

74. Kimura H, Futamura M, Kito H, Ando T, Goto M, Kuzushima K, Shibata M, Morishima T. Detection of viral DNA in neonatal herpes simplex virus infections: frequent and prolonged presence in serum and cerebrospinal fluid. *J Infect Dis* 1991, 164 (2), 289–293.

75. Orle KA, Gates CA, Martin DH, Body BA, Weiss JB. Simultaneous PCR detection of *Haemophilus ducreyi*, *Treponema pallidum*, and herpes simplex virus types 1 and 2 from genital ulcers. *J Clin Microbiol* 1996, 34 (1), 49–54.

76. Orle KA, Weiss JB. Detection of *Treponema pallidum*, *Haemophilus ducreyi* and herpes simplex virus by multiplex PCR, in Peeling RW, Sparling F, Eds., *Sexually Transmitted Diseases: Methods and Protocols*. Totowa, NJ: Humana Press, 1999. pp. 67–69.

77. Johnson SR, Martin DH, Cammarata C, Morse SA. Development of a polymerase chain reaction assay for the detection of *Haemophilus ducreyi*. *Sex Transm Dis* 1994, 21 (1), 13–23.

78. Johnson SR, Martin DH, Cammarata C, Morse SA. Alterations in sample preparation increase sensitivity of PCR assay for diagnosis of chancroid. *J Clin Microbiol* 1995, 33 (4), 1036–1038.

79. Totten PA, Kuypers J, Morse SA. *Haemophilus ducreyi* detection by PCR, in Peeling RW, Sparling F, Eds., *Sexually Transmitted Diseases: Methods and Protocols*. Totowa, NJ: Humana Press, 1999. pp. 47–65.

80. Roesel DJ, Gwanzura L, Mason PR, Joffe M, Katzenstein DA. Polymerase chain reaction detection of *Haemophilus ducreyi* DNA. *Sex Transm Infect* 1998, 74 (1), 63–65.

81. Gu XX, Rossau R, Jannes G, Ballard R, Laga M, Van Dyck E. The *rrs* (16S)-*rrl* (23S) ribosomal intergenic spacer region as a target for the detection of *Haemophilus ducreyi* by a heminested PCR assay. Microbiology 1998, 144 (Pt 4), 1013–1019.

82. Parsons LM, Waring AL, Otido J, Shayegani M. Laboratory diagnosis of chancroid using species-specific primers from *Haemophilus ducreyi* groEL and the polymerase chain reaction. *Diagn Microbiol Infect Dis* 1995, 23 (3), 89–98.

83. Lai W, Chen CY, Morse SA, Htun Y, Fehler HG, Liu H, Ballard RC. Increasing relative prevalence of HSV-2 infection among men with genital ulcers from a mining community in South Africa. *Sex Transm Infect* 2003, 79 (3), 202–207.

84. West B, Wilson SM, Changalucha J, Patel S, Mayaud P, Ballard RC, Mabey D. Simplified PCR for detection of *Haemophilus ducreyi* and diagnosis of chancroid. *J Clin Microbiol* 1995, 33 (4), 787–790.

85. Morse SA. Chancroid and *Haemophilus ducreyi*. *Clin Microbiol Rev* 1989, 2 (2), 137–157.

86. Mertz KJ, Weiss JB, Webb RM, Levine WC, Lewis JS, Orle KA, Totten PA, Overbaugh J, Morse SA, Currier MM, Fishbein M, St Louis ME. An investigation of genital ulcers in Jackson, Mississippi, with use of a multiplex polymerase chain reaction assay: high prevalence of chancroid and human immunodeficiency virus infection. *J Infect Dis* 1998, 178 (4), 1060–1066.

87. Mertz KJ, Trees D, Levine WC, Lewis JS, Litchfield B, Pettus KS, Morse SA, St Louis ME, Weiss JB, Schwebke J, Dickes J, Kee R, Reynolds J, Hutcheson D, Green D, Dyer I, Richwald GA, Novotny J, Weisfuse I, Goldberg M, O'Donnell JA, Knaup R. Etiology of genital ulcers and prevalence of human immunodeficiency virus coinfection in 10 U.S. cities. *J Infect Dis* 1998, 178 (6), 1795–1798.

88. Totten PA, Kuypers JM, Chen CY, Alfa MJ, Parsons LM, Dutro SM, Morse SA, Kiviat NB. Etiology of genital ulcer disease in Dakar, Senegal and comparison of PCR and serologic assays for detection of *Haemophilus ducreyi*. *J Clin Microbiol* 2000, 38 (1), 268–273.

89. Behets FM, Andriamiadana J, Randrianasolo D, Randriamanga R, Rasamilalao D, Chen CY, Weiss JB, Morse SA, Dallabetta G, Cohen MS. Chancroid, primary syphilis, genital herpes, and lymphogranuloma venereum in Antananarivo, Madagascar. *J Infect Dis* 1999, 180 (4), 1382–1385.

90. Chen CY, Ballard RC, Beck-Sague CM, Dangor Y, Radebe F, Schmid S, Weiss JB, Tshabalala V, Fehler G, Htun Y, Morse SA. Human immunodeficiency virus infection and genital ulcer disease in South Africa: the herpetic connection. *Sex Transm Dis* 2000, 27 (1), 21–29.

91. Beyrer C, Jitwatcharanan K, Natpratan C, Kaewvichit R, Nelson KE, Chen CY, Weiss JB, Morse SA. Molecular methods for the diagnosis of genital ulcer disease in a sexually transmitted disease clinic population in northern Thailand: predominance of herpes simplex virus infection. *J Infect Dis* 1998, 178 (1), 243–246.

92. Wang Q-Q, Mabey D, Peeling RW, Tan M-L, Jian D-M, Yang P, Zhong M-Y, Wang G-J. Validation of syndromic algorithm for the management of genital ulcer diseases in China. *Int J STD AIDS* 2002, 13 (7), 469–474.

93. Bauwens JE, Orlander H, Gomez MP, Lampe M, Morse S, Stamm WE, Cone R, Ashley R, Swenson P, Holmes KK. Epidemic Lymphogranuloma venereum during epidemics of crack cocaine use and HIV infection in the Bahamas. *Sex Transm Dis* 2002, 29 (5), 253–258.

94. Behets FM, Brathwaite AR, Hylton-Kong T, Chen CY, Hoffman I, Weiss JB, Morse SA, Dallabetta G, Cohen MS, Figueroa JP. Genital ulcers: etiology, clinical diagnosis, and associated human immunodeficiency virus infection in Kingston, Jamaica. *Clin Infect Dis* 1999, 28 (5), 1086–1090.

95. Sanchez J, Volquez C, Totten PA, Campos PE, Ryan C, Catlin M, Hasbun J, Rosado De Quinones M, Sanchez C, De Lister MB, Weiss JB, Ashley R, Holmes KK. The etiology and management of genital ulcers in the Dominican Republic and Peru. *Sex Transm Dis* 2002, 29 (10), 559–567.

96. Clark EG, Dunbolt N. The Oslo study of the natural course of untreated syphilis: an epidemiological investigation based on a re-study of the Boeck–Bruusgaard material. *Med Clin North Am* 1964, 48, 613–623.

97. Cox D, Liu H, Moreland AA, Levine W. Syphilis, in Morse SA, Ballard R, Holmes KK, Moreland AA, Eds., *Atlas of Sexually Transmitted Diseases*, 3rd ed. London: Mosby, 2003. pp. 23–51.

98. Centurion-Lara A, Castro C, Shaffer JM, Van Voorhis WC, Marra CM, Lukehart SA. Detection of *Treponema pallidum* by a sensitive reverse transcriptase PCR. *J Clin Microbiol* 1997, 35 (6), 1348–1352.

99. Liu H, Rodes B, Chen CY, Steiner B. New tests for syphilis: rational design of a PCR method for detection of *Treponema pallidum* in clinical specimens using unique regions of the DNA polymerase I gene. *J Clin Microbiol* 2001, 39 (5), 1941–1946.

100. Ito F, Hunter EF, George RW, Pope V, Larsen SA. Specific immunofluorescent staining of pathogenic treponemes with a monoclonal antibody. *J Clin Microbiol* 1992, 30 (4), 831–838.

101. Ballard RC, Htun Y, Fehler G, Chen C-Y, Morse S. Interpretation of serological tests for primary syphilis in the era of HIV infection and multiplex PCR for genital ulcer disease. *Int J STD AIDS* 2001, 12 (Suppl. 2), 43–44.

102. Sutton MY, Liu H, Steiner B, Pillay A, Mickey T, Finelli L, Morse S, Markowitz LE, St Louis ME. Molecular subtyping of *Treponema pallidum* in an Arizona county with increasing syphilis morbidity: use of specimens from ulcers and blood. *J Infect Dis* 2001, 183 (11), 1601–1606.

103. Liu H, McCaustland K, Holloway B. Evaluating the concentration of *Treponema pallidum* in blood and body fluid using a semi-quantitative PCR method. *Int J STD AIDS* 2001, 12 (Suppl. 2), 142–143.

104. O'Regan AW, Castro C, Lukehart SA, Kasznica JM, Rice PA, Joyce-Brady MF. Barking up the wrong tree? Use of polymerase chain reaction to diagnose syphilitic aortitis. *Thorax* 2002, 57 (10), 917–918.

105. Lafferty WE, Downey L, Celum C, Wald A. Herpes simplex virus type 1 as a cause of genital herpes: impact on surveillance and prevention. *J Infect Dis* 2000, 181 (4), 1454–1457.

106. Tran T, Druce JD, Catton MC, Kelly H, Birch CJ. Changing epidemiology of genital herpes simplex virus infection in Melbourne, Australia, between 1980 and 2003. *Sex Transm Infect* 2004, 80 (4), 277–279.

107. Ashley RL. Laboratory techniques in the diagnosis of herpes simplex infection. *Genitourin Med* 1993, 69 (3), 174–183.

108. Cone RW, Hobson AC, Palmer J, Remington M, Corey L. Extended duration of herpes simplex virus DNA in genital lesions detected by the polymerase chain reaction. *J Infect Dis* 1991, 164 (4), 757–760.

109. Chen C-Y, Ballard RC. Simultaneous detection of four etiologic agents of genital ulcer disease by multiplex PCR and real-time multiplex PCR assays, in Wilson JD, Ed., Proceedings of BASSH/ASTDA Spring Meeting. London: Royal Society of Medicine Press, 2004. p. 32.

110. Bowden F. Donovanosis, in Morse SA, Ballard R, Holmes KK, Moreland AA, Eds., *Atlas of Sexually Transmitted Diseases,* 3rd ed. London: Mosby, 2003. pp. 97–108.

111. Richens J. Diagnosis and treatment of donovanosis (granuloma inguinale). *Genitourin Med* 1991, 67 (6), 441–452.

112. Bastian I, Bowden FJ. Amplification of Klebsiella-like sequences from biopsy samples from patients with donovanosis. *Clin Infect Dis* 1996, 23 (6), 1328–1330.

113. Carter J, Bowden FJ, Sriprakash KS, Bastian I, Kemp DJ. Diagnostic polymerase chain reaction for donovanosis. *Clin Infect Dis* 1999, 28 (5), 1168–1169.

114. Carter JS, Kemp DJ. A colorimetric detection system for *Calymmatobacterium granulomatis. Sex Transm Infect* 2000, 76 (2), 134–136.

115. Van Dyck E, Piot P. Laboratory techniques in the investigation of chancroid, lymphogranuloma venereum and donovanosis. *Genitourin Med* 1992, 68 (2), 130–133.

116. Nieuwenhuis RF, Ossewaarde JM, Gotz HM, Dees J, Thio HB, Thomeer MGJ, den Hollander JC, Neumann MHA, van der Meijdeil WI. Resurgence of lymphogranuloma venereum in Western Europe: an outbreak of *Chlamydia trachomatis* serovar L2 proctitis in The Netherlands among men who have sex with men. *Clin Infect Dis* 2004, 39 (7), 996–1003.

117. Gift TL, Pate MS, Hook EW 3rd, Kassler WJ. The rapid test paradox: when fewer cases detected leads to more cases treated: decision analysis of tests for *Chlamydia trachomatis. Sex Transm Dis* 1999, 26 (4), 232–240.

10 Sexually Transmissible Viral Pathogens: Human Papillomaviruses and Herpes Simplex Viruses

Attila T. Lorincz and Jennifer S. Smith

CONTENTS

INTRODUCTION

Many viruses find access to human tissues by multiple routes; thus, classifying them strictly according to mode of transmission is not an easy or clear-cut exercise. However, several viruses are transmitted mainly by sexual activity and are regarded primarily as sexually transmissible infections (STIs) with the capacity to produce sexually transmitted disease (STD). Of these, the most important agents of human disease are the human immunodeficiency virus (HIV), human papillomavirus (HPV), and human herpesvirus (HHV) herpes simplex virus (HSV type 2). HIV is covered in Chapter 11 by Niesters and colleagues; this chapter will focus specifically on HPV and HSV infections.

HUMAN PAPILLOMAVIRUSES

Spectrum of Disease

The HPV genome is an 8-Kb double-stranded DNA circle consisting of a control region and a relatively small number of genes coding for remarkably multifunctional proteins. Almost 200 genetically diverse HPV types infect epithelial cells at a variety of different body sites.[2] Mucosa-tropic HPVs predominantly infect genital, anal, oral, and nasopharyngeal skin and cause the majority of human HPV diseases including warts, papules, diverse flat lesions, and cervical, anal, vulvar, vaginal, penile, oral, tonsillar, and other carcinomas.

Cancer is typically the result of persistent multidecade infection by a specific group of HPV types referred to alternatively as the high-risk (HR), oncogenic or carcinogenic types.[3] Careful analyses indicate that more than 99% of cervical cancers have detectable HPV DNA,[4] although in "real world" settings the better HPV DNA tests achieve a clinical sensitivity of only 95 to 98%,[5] due presumably to sampling issues or slight variations in laboratory technique.[4,6–8]

Many studies, including the landmark investigation by the late Jan Walboomers,[4] have demonstrated that invasive cervical cancer in the absence of infection by HR HPV is very rare. In fact, the clinical negative predictive value (NPV) of a highly accurate and well validated HPV DNA test can reach 99.95% or better in a screening population of women over 30 years of age.[9,10] The relevant meaning of NPV here is the percentage of women found negative by the HPV DNA test who do not have cervical cancer or precancer; alternatively expressed, there is an approximately 0.05% chance that a woman negative for HR HPV has precancer or cancer.

Unfortunately, HPV diagnostics have confused some clinicians and laboratorians because HPV acquisition typically results in highly prevalent STI in the short term, causing genital warts and various flat lesions. According to old teachings, a dual STI–carcinogenic action was erroneously regarded as improbable. Luckily, the STD and carcinogenic effects of the virus are temporally very distinct and are also related to different HPV types (low-risk types such as HPV 6 and 11 causing genital warts; high-risk types causing cervical disease including carcinoma).

The STI phase, which is most common in young adults, is of minor clinical consequence because 95% of infections either clear rapidly or are suppressed by the immune system below a subpathogenic threshold. However, a HR HPV infection that persists over a span of 5 to 20 years is a serious condition because it predisposes the host to a dramatically increased risk of carcinoma of 100-fold or greater. By comparison, the relative risks of smoking and lung cancer are on the order of 20 to 40-fold. Interestingly, smoking modestly increases the risk of cervical cancer in women infected by HR HPV DNA.[3,11]

Cervical Cancer Screening by Cytology

The global incidence of cervical cancer is more than 493,000 cases per year; it caused approximately 274,000 deaths in 2002[12] (see also Globocan 2002, http://www.iarc.fr/ENG/Databases/index.php). Most cases of this disease arise in less developed countries. The Papanicolaou (Pap) test, named after the famous pathologist, has been the mainstay of cervical cancer prevention for the past 50 years and has produced a dramatic reduction in cervical cancer incidence and death in many developed countries. The incidence and mortality of cervical cancer in the United States declined more than 70% after the introduction of Pap screening (Figure 10.1).

In recent years, the declines have been more modest, dropping from about 20,000 new invasive cervical cancer cases in 1977 to approximately 10,000 new cases and about 4,000 deaths in 2005.[13] It is reasonable to state that cervical cancer prevention is a success story in more developed countries where quality Pap screening programs have been implemented. Unfortunately, this success is tarnished by the general lack of quality screening services in resource-poor regions, where the incidence of cervical cancer is increasing and may grow to one million new cases annually by the year 2050 if something creative is not done.[14] Clearly, considerable effort is needed to reduce the rate of this totally preventable cancer to acceptable levels in less developed as well as developed countries.

Published data show that the clinical sensitivity of conventional cytology is low — in the range of 50 to 60%.[15,16] Most Pap testing in the United States is now performed via liquid cytology (e.g., ThinPrep), which appears to show moderately higher clinical sensitivity than conventional cytology — in the range of 60 to 85%[17,18] — but potentially lower specificity. However, none of these cytology tests appears to approach the 95 to 100% clinical sensitivity required for a truly effective screening program.

Cytological diagnoses are subjective and widely variable in real world settings, depending strongly on the experiences, biases, and quality control capacities of the human interpreters. While the Pap test has been shown to be reasonably effective in detecting cytological abnormalities, it has not been found effective for detecting evidence of HPV infection. This point may seem intuitively obvious to microbiologists experienced with ultrasensitive nucleic acid tests (NAT) like the polymerase chain reaction (PCR) but it is surprising how often a Pap procedure has been mistaken as an HPV test by clinicians because the koilocytes seen on Pap tests were commonly reported in the past as definitive indicators of HPV infection. Since koilocytotic changes are usually in the context of benign low-grade lesions, many clinicians came to interpret Pap reports with morphology consistent with an HPV effect to mean that high-grade disease was absent — a serious misinterpretation. The limitation in sensitivity of cytology to detect cervical disease has been addressed in part by repeat testing for most women (annually in the United States or every few years in Europe) and the repetition has proven very costly.

RATIONALES FOR DIAGNOSTICS

In contrast to Pap cytology, extensive data show that HPV DNA is a highly sensitive marker for women at risk of neoplasia over a span of up to a decade.[3-6,19-21] The main drawback of the HR

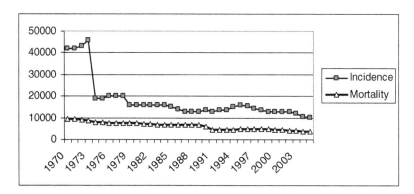

FIGURE 10.1 Cervical cancer incidence and mortality in the United States. (Compiled from annual reports published by the American Cancer Society in *CA Cancer J Clin.*)

HPV DNA test is that it is less specific than a Pap cytology test,[5] although the higher cost of the HPV test is also often raised as an important issue. At present, HPV DNA testing is widely accepted as an adjunct to cervical cytology in cancer prevention programs for triage of atypical squamous cells of undetermined significance (ASC-US), but in the future HR HPV detection may likely become the primary screening test, with adjunctive or reflex cytology.

The specific current clinical uses for HPV DNA testing include triage of patients with ASC-US diagnoses on their Pap tests; triage of low-grade squamous intraepithelial lesion (LSIL) Pap results in adolescent and post-menopausal women; resolution of discordant cytology–colposcopy–histology findings; follow-up after treatment for test of cure; follow-up after a negative or normal colposcopy; or population screening as an adjunct to the Pap test — effectively summarized in several seminal publications by Wright and others.[22–24]

The clinical value of HPV DNA triage for ASC-US in particular is now clear and convincing, as described in the recent ASCUS/LSIL study (ALTS).[17,22] HPV DNA testing has recently become the standard of care for ASC-US triage in the United States; most of the >2.0 million women diagnosed with ASC-US each year are tested for HPV DNA.

Cost effectiveness studies have demonstrated that HR HPV DNA testing as a reflex to ASC-US saves costs for the overall healthcare system.[25] Similarly, HR HPV DNA testing to screen women over the age of 30 years can save overall healthcare costs if it is used to extend cytology screening intervals for HPV DNA negative women.[26,27]

It is important to understand certain natural history attributes of HPV infection in relation to cervical carcinogenesis. HR HPV infection is the earliest known causal step on the pathway to cervical cancer. A positive HR HPV result can define at least two secular groups of women at a higher risk of serious cervical disease: women who have high probabilities of current cervical intraepithelial neoplasia grade 2 or 3 (CIN 2/3, also called high-grade CIN or high-grade disease) and women who are at significantly elevated risk for future high-grade disease and cancer. For the former group, a single positive HR HPV DNA test carries about a 5 to 15% probability of detecting CIN 2/3 or cancer at colposcopy. Women with persistent HPV infection may face a 15 to 30% or greater probability for detection of CIN 2/3 or cancer within the next several years.[5,28–34]

RISK GROUPS

In 1992, genital HPV was initially broadly categorized into low-risk (LR), intermediate-risk (IR), and high-risk (HR) types.[19] However, shortly after the types were distinguished, the IR and HR HPVs were combined into only one broad HR group for diagnostic purposes, to reduce HPV test costs and avoid confusion in clinical circles. Thus, the broader HR group contains a diversity of HPVs with widely different relative risks for cervical cancer. HPVs 16 and 18 in particular, along with HPV 45, were originally classified as HR HPVs (see Figure 10.2 and Table 10.1). HPVs 31, 33, 35, 39, 51, and 52 were initially regarded as IR types[19] HPV 56 was placed in the original smaller HR group but more recent data indicate that it is a low prevalence IR type.[35] The classification error appears to have been caused by the very small number of cervical cancers positive for this type in the original study.[19]

In 2005, 13 HPV types were classified as high-risk based on an expert group meeting at the International Agency for Research on Cancer: HPV 16, 18, 31, 33, 35, 39, 45, 51, 52, 56, 58, 59, and 66.[36] HPV 16 and 18 (associated with ~70% of invasive cervical cancers) have been singled out recently by the Schiffman team (National Cancer Institute [NCI], Bethesda, Maryland, http://www.cancer.gov/) as causing particularly risky infections that can strongly predispose women to cervical squamous carcinomas and adenocarcinomas, respectively (Table 10.2).[37,38] New HPVs have been added to the HR group, such as types 26, 73, 82, and others,[35,39–41] although the exact risk order of these types remains to be refined. In contrast to LR HPVs, HR HPVs establish dramatically elevated risks of cervical cancer and high-grade CIN, on the order of 100- to 200-fold higher than the risks observed for HR HPV-negative controls.

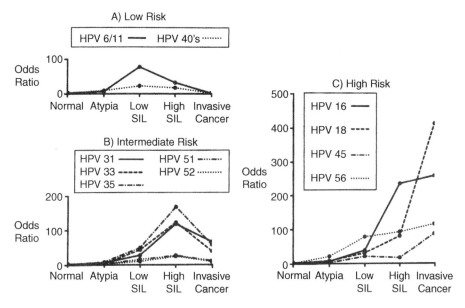

FIGURE 10.2 Graphic profiles of odds ratios for (A) low-risk, (B) intermediate-risk, and (C) high-risk human papillomavirus (HPV). Each graph is drawn to the same scale; hence, the pronounced variations in strengths of association are accentuated by the differences in the heights of the Y axes. SIL = squamous intraepithelial lesions. (Source: Lorincz AT et al. *Obstet Gynecol* 1992; 79 (3): 328–337. Figure 1. With permission.)

TABLE 10.1
Distribution of HPV 16, 18, and 45 Genotypes as Percentages of HR HPV Positives among HPV-Positive Women with Normal, CIN 2/3, and Cancer Diagnoses

Diagnosis (Number Positive for Any HPV)	HPV 16% (Number)	HPV 18/45% (Number)	HPV 16/18/45% (Number)
Normal (101)	16 (16)	7 (7)	23 (23)
CIN 2/3 (228)	54 (123)	6 (14)	60 (137)
Cancer (137)	53 (72)	28 (39)	81 (111)

CIN = cervical intraepithelial neoplasia. HPV = human papillomavirus. These HR HPVs form a special group because they are much more prevalent in high-grade disease than other HR HPV types. (Adapted from Lorincz et al., 1992[19], Table 3.)

It is not believed that LR HPVs lead to cervical cancer. However, LR HPV DNA is present sporadically in some cases of high-grade CIN and very rarely in cervical carcinomas, most likely as passenger viral DNA along with detectable or undetectable HR HPV DNA. LR HPVs are predominantly associated with genital warts (condylomata acuminata) and low-grade cervical disease such as CIN 1 (Figure 10.2).

DETECTION METHODS

The currently preferred methods for detecting HPV are all DNA tests, the most common of which are *in situ* (ISH) hybridization, PCR, and the Hybrid Capture 2® test (hc2, Digene Corporation, Gaithersburg, Maryland, USA, http://www.digene.com/). Each test has its particular strengths and weakness. ISH localizes the HPV DNA within the nuclei of infected cells, preserving limited aspects of the original morphology and distinguishing some integrated from episomal HPV infections.[42–44] The test provides satisfactory visual images of HPV-infected tissues and cells.

TABLE 10.2
Positive Predictive Values (PPVs) for CIN 3 among Women with Various
Pap or HPV Diagnoses in Portland Study by Khan et al., 2005[37]

Portland Study Screening Test Results	PPV for CIN 3
NIL	0.8
ASC-US	4.2
LSIL	10.9
HPV 16[a]	20.1
HPV 18[a]	15.4
HR HPV other than 16/18	1.8
HR HPV negative	0.5

ASC-US = atypical squamous cells of undetermined significance. CIN = cervical intraepithelial neoplasia. HPV = human papillomavirus. HR = high risk. LSIL = low-grade squamous intraepithelial lesion. NIL = no intraepithelial lesion. PPV = positive predictive value.

[a] HPV 16 and 18 positivity has a PPV that appears greater than an LSIL Pap diagnosis.

While ISH has good specificity, it suffers from some cross-reactivity among related HPV types. However, its main disadvantage is poor analytical and clinical sensitivity. For cells to react as positive, they must have more than 10 genomic copies of HPV DNA. Many older studies and three recent ones show that ISH has a clinical sensitivity of only 50 to 75% for CIN 2/3.[45–47] For example, in a study by Hesselink, 23% (7 of 30) high-grade CIN lesions were negative by the Inform® HPV DNA ISH test (Ventana Medical Systems, Tucson, Arizona, USA, http://www.ventanamed.com/), whereas all the lesions tested positive with general primer GP 5+/6+ PCR and hc2.

Another contrast between ISH and hc2 or PCR is that ISH identifies only current disease; it does not reliably identify increased risk for future CIN 3 among cytologically normal women because many such women have lower levels of HPV DNA and no abnormal cells to stain on a Pap slide for ISH.

PCR has very high analytical sensitivity, with some expert PCR labs able to detect fewer than 10 copies of HPV genomic DNA, typically in a few microliters of input specimen.[48–50] Some PCR procedures also appear to provide accurate HPV genotyping. Despite more than 15 years of PCR research on HPV detection, most available validation data are analytical and based primarily on plasmid reconstructions in artificial matrices, sometimes in combination with simple studies that compare two or more PCR tests that themselves have not been clinically validated.

The clinical sensitivity of PCR for detecting high-grade cervical disease may not be as great as initially expected. Despite an extensive HPV PCR publication list, few papers show an ability of the PCR test to detect more than about 95% of high-grade CIN and cancer in a large, well designed study. More typically, the clinical sensitivity of HPV PCR is reported as 75 to 95%,[18,33,51–64] with a median clinical sensitivity of 82% reported in 16 recent papers (Figure 10.3a).

It appears that HPV PCR may suffer from several negative effects specific to clinical specimens, for example, random partial inhibition not revealed by a beta-globin or other internal control. This inhibition persists even in purified specimens and may attenuate the very high analytical sensitivity of the test. The presence of multiple HPV types in 20 to 30% of specimens can cause competition effects and may suppress the detection of the lower-level HPV types, resulting in irreproducible genotyping.[65] L1 region PCR primers may produce negative results in up to 5% of cervical cancers that have integrated and deleted HPV genomes lacking this region.[66] Of particular interest is the comparative PCR versus hc2 data from the large ALTS study on 278 cases of CIN 3 and cancer.[51] A prototype PCR test employing PGMY09/11 primers attained clinical sensitivity and specificity levels of 87.4 and 55.6%, while the corresponding values for hc2 were 92.5 and 51.1%, respectively.

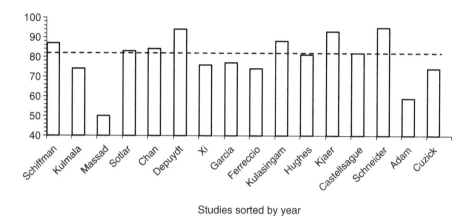

FIGURE 10.3a Clinical sensitivity (%) of PCR for CIN 2/3 and cancer as shown by major studies from 1999–2005.[18,33,51–64] The median sensitivity (82%) is shown by dashed horizontal line.

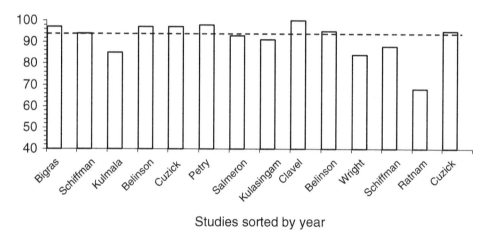

FIGURE 10.3b Clinical sensitivity (%) of hc2 for CIN 2/3 and cancer as shown by major studies from 1999–2005.[9,10,18,29,51,52,64,69–74,84] The median sensitivity (94%) is shown by dashed horizontal line.

The only HPV DNA test currently approved for routine clinical use by the U.S. Food and Drug Administration is hc2. The test is a relatively simple, high-throughput semi-automated procedure with approved claims for use in adjunctive screening with the Pap test in women over the age of 30 years and for ASC-US triage in women of all ages.[67] The hc2 can detect one or more of 13 HR HPV types (16, 18, 31, 33, 35, 39, 45, 51, 52, 56, 58, 59, and 68) at a level of 1 pg/ml each, corresponding to 5000 HPV genomes per test well.[68] All hc2 technology operates on the principle of signal amplification and thus requires minimal specimen preparation. Two methods are preferred for collecting specimens. One employs the co-collection approach using a small conical-shaped nylon bristle brush that is rotated three times in the cervical os and then placed into a specimen transport medium (STM) tube (Digene Corporation, Gaithersburg, Maryland, USA, http://www.digene.com/). The other type of specimen is obtained from a liquid cytology tube; hc2 is FDA-approved for reflex testing from PreservCyt® fluid (PC, Cytyc Corporation, Marlborough, Massachusetts, USA, http://www.cytyc.com/).

The clinical sensitivity of hc2 is high, the test consistently appears more sensitive than other HPV DNA detection methods, and it exhibits a median clinical sensitivity for CIN 2/3 of 94% (Figure 10.3b).[5,7,9,10,18,29,64,69–74] The test exhibits robust performance, with repeat test kappa values typically in the 0.85 to 0.95 range. Reproducibility is typically better from STM than from PC specimens and for the former specimen type, no retesting is required.[75,76]

For PreservCyt specimens, the FDA has approved a retest zone for initial relative light units (RLU) cut-off values of 1.0 to 2.5. In this case, one or two additional tests must be performed to confirm or contradict the original result. A drawback of hc2 is that the test exhibits some cross-reactivity to other HPV types including both LR and non-targeted HR HPVs not present in the probe cocktail.[77,78] This cross-reactivity is of concern although it produces a relatively minor effect on the clinical specificity of the hc2 test. Only about 10 to 15% of test positives are attributable to non-targeted HPV types and many of these test positives occur in women with CIN. The hc2 has a moderate analytical sensitivity of about 5,000 copies per test or approximately 100,000 HPV genomes per milliliter of specimen (typical volume for an STM specimen collection). This moderate analytical sensitivity nevertheless translates to high clinical sensitivity as shown in extensive validation studies.[5,29,69-74]

TEST VALIDATION ISSUES

Sensitivity is a term that often lacks clarity and, when used without further qualifiers, may be misleading or even meaningless. In much of the HPV literature to date, the difference between clinical and analytical sensitivity is not considered. The omission becomes an important error when it is perpetuated by researchers who emphasize analytical performance and overlook the subtleties of HPV natural history.

Analytical sensitivity involves a lower limit of detection, usually reported as a genome or copy number per milliliter or per aliquot. Clearly, a test that can detect 100 copies is more analytically sensitive than a test that can detect 5,000 copies. Clinical sensitivity typically refers to the detection of a specific disease endpoint, for example, high-grade disease (CIN2/3 or cancer) caused by HPV; this parameter is reported as a proportion or percentage.

If there is a critical viral load threshold at which high-grade disease occurs and the threshold is fairly high relative to the analytical sensitivity of the various tests (perhaps 100,000 viral genomes per milliliter of clinical specimen), then the analytically more sensitive test may not be the clinically more sensitive test and could ultimately be less sensitive due to other issues such as inhibition effects, genome deletions, etc. This appears to be the situation with HPV disease[79] and it must be considered by developers of diagnostic tests. On the other hand, a test that has too much analytical sensitivity may produce poor clinical specificity if it detects low levels of viral DNA not related to clinical disease or risk of cancer.

It is obvious from the literature that suppositions about diagnostic test performance predicted from theoretical considerations have a limited bearing on real-world performance. Furthermore, analytical validation studies alone or in combination with small or otherwise inadequate clinical validation studies do not improve this situation in a meaningful way. The complexity of carcinogenesis and HPV natural history make it imperative that tests for HPV intended for routine clinical use must undergo extensive clinical validation studies in realistic use settings.

CLINICAL STUDIES SUPPORTING UTILITY OF DNA SCREENING

HPV DNA testing for ASC-US triage and related uses and for test of cure is well established in the literature. Many comprehensive papers and reviews[5,17,22,80,81] have been published and will not be discussed here. In contrast, HPV DNA population screening is still a controversial issue. Many large clinical studies of HPV screening, mostly via hc2 have been performed and most enrolled 5,000 to 15,000 women. These include studies in the United Kingdom,[29] Germany,[10] France,[9,82] The Netherlands,[83] Switzerland,[84] Finland,[85] Canada,[74] Costa Rica,[73] Mexico,[69] South Africa,[86] and China.[71]

The studies without exception concluded that HPV DNA testing was much more sensitive in detecting high-grade CIN and cancer than Pap cytology, although the latter was almost always the

TABLE 10.3
Average HPV Prevalence by hc2 for 58,000 Women Aged 30 Years or Older

Pap Diagnosis	hc2 HPV-Negative (%)	hc2 HPV-Positive (%)
Pap <ASC-US	89	6
Pap ASC-US+	2	3
Any Pap	91	9

ASC-US = atypical squamous cells of undetermined significance. hc2 = Hybrid Capture 2 HPV DNA test. HPV = human papillomavirus. Pap = Papanicolaou smear. Adapted from underlying raw data presented in six studies.[10,29,71–73,197] Percentages based on population total.

more specific test. A multi-center randomized controlled trial from Italy with 16,700 women in each arm[86a] reached conclusions very similar to those of the studies listed above. Cumulatively, the studies examined more than 100,000 women and included more than 1,500 cases of CIN 2/3 and cancer. Overall, HR HPV DNA prevalence was 9% in a subset of 58,000 women aged 30 or older (Table 10.3). Stratification of the women according to cytological diagnosis revealed that 6% were positive with a Pap smear diagnosis "less than ASC-US." In other words, women who had Pap smears within normal limits or reactive atypias not suspect for AS-CUS (considered normal) had an average rate of hc2 positivity of about 6%.

Recent data on 125,000 women compiled by the Kinney team of the Kaiser clinics in California revealed a 3.7% hc2 HR HPV DNA prevalence in cytologically normal women over age 30.[87] In a similar but smaller screening study from the Netherlands (8,132 women), HPV positivity by hc2 was 6.8% at the usual test cut-off of 1.0 RLU/CO but prevalence dropped to only 3.4% at a test RLU/CO of 3.0. This improved specificity was accompanied by a decrease in sensitivity for CIN 2/3 from 97 to 92% (Hesselink et al., unpublished data).

Figure 10.4a is a scatterplot of clinical sensitivity data from the above studies. The median sensitivity across the screening studies was 68% for Pap cytology alone, whereas the median clinical sensitivity for Pap plus hc2 was 100%. The clinical specificity of these tests is shown in Figure 10.4b. It is clear that the median specificity of HPV testing plus Pap cytology (90%) is lower than the median specificity of Pap alone (96%). Cost effectiveness calculations show that such a drop in specificity can be accommodated in current screening programs by extending the screening intervals.[26,27,88]

The results of these large HPV DNA screening studies provide compelling evidence for the clinical utility of HPV DNA testing as an adjunct to cervical cytology for routine screening in women older than age 30 and perhaps in younger women in certain settings. The wide diversity of study locations indicates that the data may be generalizable to many screening settings worldwide, but the studies exhibit several weaknesses. One criticism is that none of the currently published studies is a randomized controlled trial (RCT) with cervical cancer as an endpoint. The ethics of cancer as a study endpoint merit much debate, but meanwhile several RCT studies with CIN 3 as an endpoint are under way with encouraging initial results. It has been argued by Sir Bradford Hill, an epidemiologist of considerable renown,[89] that it is not necessary to wait for RCT data before remediating toxic hazards or implementing obviously beneficial new technologies; in fact it may be unethical to wait. New tests are almost invariably implemented well in advance of RCT data because doing otherwise would unacceptably delay the availability of beneficial technologies to patients in need.

Another criticism of the screening studies relates to the lack of true assessment of all potential CIN 2/3 because individual tests and possibly even combinations of tests may still miss a large proportion of true high-grade neoplasia on the cervix. This effect is known as verification bias and it tends to make tests appear more sensitive and more specific than they actually are. However, many researchers (Belinson,[70,71] Petry,[10] and Bigras[84]) conducted careful assessments of all (Belin-

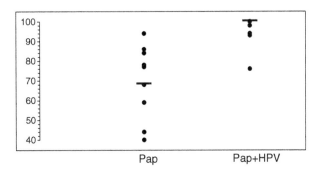

FIGURE 10.4a Clinical sensitivity (%) of Pap cytology and hc2 HPV DNA screening in major studies from 1999–2005. Median sensitivity (shown by short horizontal bar) across these studies was 68% for Pap alone[9,10,18,29,64,69,71–74,84] and 100% for Pap plus hc2.[9,10,29,69,71–74,84]

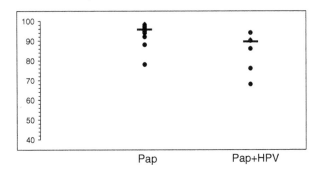

FIGURE 10.4b Clinical specificity (%) of Pap cytology and hc2 HPV DNA screening. Median specificity (shown by short horizontal bar) across these studies was 96% for Pap alone[9,10,18,29,64,69,71–74,84] and 90% for Pap plus hc2.[10,29,69,71–74]

son) or large percentages of test-negative women by colposcopy and biopsy and detected very little high-grade CIN in HR HPV DNA-negative women; the combined Pap and HPV tests detected essentially all prevalent CIN 2/3 and cancers.

PERSISTENCE AND RISK OF FUTURE CIN 2/3 AND CANCER

HPV testing can detect prevalent high-grade CIN and can also identify women at risk of future disease. Studies by many authors demonstrated that women persistently infected by HR HPV were at very high risk for the development of high-grade CIN.[30–34] A study of particular note was conducted at the Kaiser Permanente clinics in Portland, Oregon and involved 23,000 women over a period of 10 years. HR HPV DNA was tested in cervical specimens collected at baseline in 1989 (the baseline was the only hc2 HPV test considered in this analysis) and then the women were followed by routine annually scheduled Pap cytology.

After 10 years of follow-up, all accumulated data were examined[31] and 171 cases of CIN 3 were detected over that period. Of those cases, only about 33% were predicted by minimally abnormal cytology performed at or near the baseline. The rest of the cervical disease apparently developed over the ensuing 10 years and was detected by repeat Pap testing and colposcopy. In contrast, one baseline HR HPV test set identified a population of positive women from whom approximately 70% of the CIN 3 developed. Thus, a woman who is HPV-positive has significant future risk for CIN 3 even if her cytology is normal.

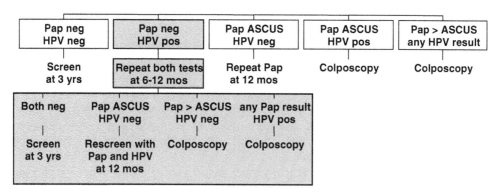

FIGURE 10.5 Screening algorithm for adjunctive HPV DNA testing in women aged 30 or more as presented by Wright et al.[24] The recommendation in the three boxes on the top right side of the figure are the same as the ASCUS guidelines for abnormal Pap smears (ASCUS and above),[93] but the recommendation on the left for Pap- and HPV-negative women is to rescreen every 3 years. The recommendation (second box from left, top) for Pap-negative/HPV-positive women is to repeat both tests in 6 to 12 months. Depending on the results, a woman would be rescreened in 3 years if both Pap and HPV results were negative; rescreened in 12 months if the results were Pap-ASCUS- and HPV-negative; and scheduled for prompt colposcopy if the HR HPV test was positive or the Pap was diagnosed as indicating LSIL, HSIL, or other serious lesion.

In another cohort study conducted in Denmark,[33] 10,700 women 20 to 29 years of age were followed for 2 years by cytology and HPV testing by GP5+/6+ PCR. Women who were HPV negative, both at a given moment and at follow-up 2 years later were by convention assigned a relative risk (RR) of 1.0. Women who were positive for HR HPV at the given moment and at follow-up 2 years later had an RR of 692 for high-grade CIN. Those who were positive for the same HPV type at a given moment and again at follow-up 2 years later had an RR of 813. These are extraordinarily elevated RR values, showing how important it is to follow up HR HPV-positive women because of their high risk of developing disease in the future. The potential utility of genotyping for persistence must be evaluated by additional studies and cost effectiveness analyses before it can be implemented into routine clinical use.

MEDICAL SOCIETY GUIDELINES

The American Cancer Society (ACS, http://www.cancer.org/docroot/home/index.asp) promulgated guidelines for adjunctive HPV testing[90] that recognize the value of HPV information to clinicians. ACS provides recommendations to screen with combined Pap cytology and HPV DNA testing every 3 years, promote counseling and education on HPV issues, and develop more specific consensus screening guidelines. Relevant guidelines were developed by Wright et al.[24] (Figure 10.5).

The American College of Obstetrics and Gynecology (ACOG, http://www.acog.org/) has a very similar set of guidelines — basically the use of a combination of cervical cytology and HPV DNA screening for women aged 30 years and older. Women who test negative with both tests should be rescreened no more frequently than every 3 years.[91–93]

HERPES SIMPLEX VIRUSES

The focus of this section will be on comparisons of two methods for the direct detection of genital herpetic infections: viral culture (the current gold standard) and recently developed PCR-based and hybridization methods. Data will be presented on the relative sensitivities and specificities of culture versus PCR-based methods for the detection of genital HSV infection. Issues related to the clinical relevance of the use of PCR-based methods for HSV viral DNA detection will be also discussed.

VIRUS CHARACTERISTICS

Herpes simplex type-1 (HSV-1) and type-2 (HSV-2) DNA viruses preferentially infect skin and mucosal tissues. Herpetic viruses generally establish latency in sensory ganglia following initial acquisition, causing an infection that persists for life. Individuals with HSV-2 and HSV-1 herpetic viruses have recurrent outbreaks and episodic shedding of virus throughout their lives.[94]

HSV-2 infection is the primary cause of genital herpes and the most common cause of genital ulceration throughout the world, and, like HPV infection, it is one of the most prevalent.[95,96] Individuals with HSV-2 infections do not necessarily develop clinical disease, but most intermittently shed HSV from their genital tracts.[97]

HSV-1 infection is generally transmitted via non-sexual contact during childhood and adolescence, and if symptomatic, is usually characterized by oral or facial lesions.[94,96] However, recent data from developed countries (United States, United Kingdom, Israel, and Sweden) indicate that HSV-1 may account for up to 50% of genital herpes.[98] Possible causes of the notable increase in the proportion of first episode genital herpes due to HSV-1 infection include changes in sexual behavior (higher prevalence of oral–genital contact), increased condom use, reducing genital HSV-2 transmission, and increased susceptibility to genital HSV-1 infections due to currently lower HSV-1 oral–labial acquisition in adolescence.[99]

DNA virus sequence analyses of herpes virus infections indicate that HSV-2 and HSV-1 DNA sequences share significant nucleic acid homologies, characterized by closely related genomic sequences.[100] Genital infections due to HSV-2 and HSV-1 are largely asymptomatic. Thus individuals unknowingly transmit HSV infections to their sexual partners and pregnant women can transmit infections to their infants during delivery.[101,102] Although the overall clinical presentation of genital ulcers caused by HSV-2 and HSV-1 are clinically indistinguishable, several noteworthy differences exist in the clinical management and natural history of HSV-2 and HSV-1 infections due to differences in viral recurrences and shedding patterns. In terms of clinical symptoms, individuals with genital HSV-2 infections have been found to experience more herpetic recurrences and are more likely to shed virus in the absence of symptoms than those with HSV-1 infections.[103–105] Based on these differences, recent diagnostic efforts have focused on means to differentiate HSV-1 and HSV-2 infections in order to improve clinical management and provide tailored counseling to patients with genital herpes.

WORLDWIDE PREVALENCE AND CLINICAL DISEASES RELATED TO GENITAL HERPES INFECTION

Although endemic in many areas of the world, genital HSV infections are highly under-recognized and under-diagnosed in many countries. Reliable data using type-specific serological assays to ascertain HSV-2 seroprevalence provide epidemiological measures of the population burden of genital herpes.[96] However, genital herpes prevalence may be under-estimated in areas where the prevalence of HSV-1 genital herpes is noteworthy. HSV-2 seropositivity is most common in certain areas of Africa (30 to 82% seroprevalence) and parts of the Americas (33 to 50%).

HSV-2 seroprevalence in Western and Southern Europe is usually lower than in Northern Europe and in the United States and tends to be lower in Asia than in all other geographic areas.[96] HSV-2 is associated with substantial morbidity; it causes painful and recurrent genital lesions[106] and is commonly associated with non-classical symptoms such as painful urination, limb pain, and genital discharge.[94,107] HSV-2 can infect cervical squamous epithelial tissue in the squamo-columnar junction. It can also cause serious but relatively rare systemic complications such as encephalitis and meningitis.[107]

The acquisition of genital herpes among women during pregnancy has been associated with a higher risk of prenatal morbidity (spontaneous abortion, prematurity).[108–111] Maternal-to-infant HSV-2 transmission may also result in severe pregnancy complications of neonatal herpes.[111,112]

Neonatal herpes is associated with high levels of infant mortality and morbidity, including neurological complications. Although available data from the United States indicate that the prevalence of neonatal herpes is as high as 1 in 1000 live births,[113] reliable data are limited on the prevalence of neonatal herpes in Africa, Asia, and South America.

The diagnosis of genital herpes infection has become more important as increasing evidence indicates a higher risk of HIV acquisition and shedding among HSV-2 seropositive individuals.[114,115] Individuals dually seropositive for HSV-2 and HIV have also been found to have higher risks of both cervical[116,117] and rectal HIV shedding.[118] Similarly, concomitant HIV infection has been found to change the natural history of HSV-2, causing more frequent HSV-2 viral shedding and reactivation, also for patients receiving highly active antiretroviral therapy (HAART).[118,119] In addition, HIV viral replication may be higher during HSV reactivation.[120] As the epidemiological evidence becomes more convincing, myriad biological explanations may reveal why HSV may increase HIV risk.[121]

LABORATORY DIAGNOSIS

Methods for Culture, Direct Detection, and Diagnosis of Genital Herpes Infection

A number of HSV diagnostic methods are commonly used for the direct detection of herpes simplex viruses, each with varying sensitivities.[122,123] Viral culture detects cytopathic effects and remains the gold standard, despite a generally low sensitivity when compared to newly developed, most sensitive PCR-based methods[124,125] (Table 10.4).

Antigen detection tests are also generally less sensitive than HSV ascertainment via conventional culture.[126] Tzanck smear preparation is a morphologic assay used in the past that cannot differentiate HSV and varicella zoster virus (VZV) infections and is less sensitive than viral culture. The direct fluorescent antibody test provides quick results but its optimum use generally requires the presence of clinical lesions.[127] Cytological diagnosis, electron microscopy, and early hybridization assays have also been used to directly detect HSV genital infection, although these tests have also been hampered by relatively low sensitivities.[122,128] An hc2 signal amplification probe test for the rapid detection and typing of herpes simplex virus DNA has also been described, with results indicating high sensitivity (93%) and specificity (100%).[129]

Although not the focus of this chapter, recent HSV-2 type-specific serological assays, based on glycoprotein Gg-2 or purified antigen permit reliable type-specific HSV-2 and HSV-1 serum antibody differentiation.[130] HSV-2 serological diagnosis, however, does not directly detect genital herpetic infection. The presence of IgG antibodies reflects lifetime exposure to HSV-2. Because the acquisition of genital herpes in the late stages of pregnancy appears to be the biggest risk factor for neonatal herpes, the use of type-specific serological screening in pregnant women with no history of prior HSV infection is being discussed as a potential prevention strategy for neonatal herpes.[131]

In order to provide differential diagnosis of HSV-1 versus HSV-2 genital infections, two main diagnostic measures may be used: (1) PCR-based methods and (2) type-specific serological assays for differentiation of HSV-1 and HSV-2 type infections. After culture-based testing, it is also possible to obtain HSV-1 and HSV-2 typing results in a separate confirmatory step using type-specific antigens with fluorescence or enzyme immunoassays.[122]

PCR Techniques for Type-Specific Detection

PCR-based technology has notably increased our understanding of the role of herpetic infections in the pathogenesis of diseases of the central nervous system (CNS). In fact, PCR-based methods are the current gold standard for the diagnosis of CNS diseases caused by herpes viruses (i.e., encephalitis) because these diseases cannot be easily diagnosed by standard culture based methods due to sampling difficulties.[132] HSV genomes have been detected in spinal fluid of the central nervous system[133,134] and to a lesser extent in the brain.[135] PCR has also been used to detect HSV viral DNA in human blood,[136] cornea,[137] skin,[138] semen,[139] and saliva[140] samples.

COMPARISONS OF PCR-BASED METHODS AND CURRENT GOLD STANDARD CULTURE FOR DIRECT DETECTION

Relative Analytical Sensitivity and Programmatic Implications

Several studies have directly compared the prevalence of HSV detection using traditional culture-based methods and PCR (Table 10.4), concluding that PCR assays are more sensitive than conventional culture detection for the ascertainment of genital HSV infections. Due to the use of hybridization, the specificity of PCR-based assays is generally high.[122] Higher prevalence of HSV detected in clinical specimens by PCR than those revealed by culture have been consistent across studies, in individuals with clinical lesions,[141,142] and in those with asymptomatic infections.[143,144]

In clinical practice, the relative higher sensitivity of PCR-based methods may provide a noteworthy advantage over culture-based detection. For example, PCR has been shown to improve the ability to detect herpetic ulcers in relatively late stages (those characterized by ulcerative or crusted lesions).[107,145] When compared with culture techniques, more sensitive PCR techniques can also improve the ability to detect recurrent HSV infections, generally characterized by lower viral titers than first episode genital herpes.[146]

False negative results from culture have also been attributed to problems with the quality of sample storage and shipment,[142] and thus PCR-based techniques, due to their greater sensitivity, may be optimal for the ascertainment of HSV DNA in samples collected in difficult field or storage conditions or subject to longer durations of storage.[142,145] Quantitative stability of HSV DNA levels by PCR has also been shown in specimens stored for long periods.[147]

PCR-based methods have indeed increased our ability to ascertain asymptomatic HSV infections that are generally characterized by lower prevalences of virions. The clinical relevance of the greater diagnostic sensitivity must be established in terms of identifying individuals at higher risk of HSV transmission. However, the greater sensitivity of PCR and other hybridization methods could lead to the earlier identification and treatment of HSV-infected individuals, and thus theoretically contribute to a decrease in HSV transmission on a population level.

Resolving the issue of a PCR-positive/culture-negative sample is not without its complexities; the specificity of PCR results requires confirmation. Indeed, difficulties in the identification of a true gold standard for evaluating diagnostic methods are not limited to the discussion of direct detection diagnostic methods for HSV.

Real-time PCR-based methods have been recently developed for HSV DNA amplification and may produce lower false positive rates due to contamination than conventional PCR-based assays.[148] The quantitative assays also have shorter turn-around times than conventional PCR-based assays,[149] and a recent real-time PCR has been developed to differentiate HSV-1 and HSV-2 infections.[150]

Use of PCR to Achieve Improved Management of Patients with Genital Ulcers

For patients presenting with genital ulcerative diseases, PCR-based methods have recently been recommended for use by the Centers for Disease Control (CDC) for the differentiation of genital ulcerative conditions (Meeting on Directions in Type-Specific HSV Serologic Testing, March 31, 2005). A review of data on the prevalence of herpes, *H. ducreyi*, and *T. pallidum* in studies of genital ulcers indicates that a significant proportion of genital ulcerative lesions are caused by herpes viruses (see Chapter 9 by Schachter and Morse).

Individuals with genital HSV-2 infections experience more herpetic recurrences and are more likely to shed virus in the absence of symptoms than those with HSV-1 infections.[104,105,151] Thus, diagnostic methods that permit the differentiation of genital lesions caused by HSV-2 versus HSV-1 may be very useful for counseling patients with genital herpes.

TABLE 10.4
Comparison of HSV Type-2 Detection by Culture and PCR Assays

[References]	Study Population	HSV Prevalence	
		Culture	PCR
Hardy [144]	11 pregnant women with no signs of genital herpes	90.9% (10 of 11)	100% (11 of 11)
Kimura [180]	15 oral lesion samples from children with herpetic gingivostomatitis	86.7% (13 of 15)	100% (15 of 15)
Cone [145]	3 women with recurrent genital herpes	100% (75% of 4 episodes)	100% (100% of 4 episodes)
Rogers [182]	68 women undergoing HSV routine testing	42% (33 of 79) = HSV+; 10.1% (8 of 79) = HSV-1+; 31.6% (25 of 79) = HSV-2+	51.9% = HSV+
Cone [182]	275 women with herpetic lesions	53%	100% of culture+/HerpChek- results were PCR+; 93% of culture-/HerpChek+ results were PCR+
Cone [143]	100 asymptomatic pregnant women	0%	9% overall; HSV-1: 3%; HSV-2: 5%; HSV-1 and 2: 1%
Diaz-Mitoma [165]	40 patients treated with acyclovir or placebo	Placebo: HSV-2, 32%; overall: HSV-2, 13%	Placebo: HSV-2: 67%; overall: HSV-2: 30%
Guibinga [166]	216 clinical specimens	N = 107 (49.5%)	105 (51%); PCR sensitivity = 98.1%
Orle [183]	99 specimens	N = 28 (28.3%); sensitivity = 78.1%; specificity = 100%	39.3% (N = 39); sensitivity = 100%; specificity = 100%
Bogges [184]	9 HSV-2 seropositive pregnant women	2.3%	13%
Cullen [129]	104 ulcerative genital lesions	35.6% (37 of 104) culture+; sensitivity = 84.1%; specificity = 100%	39.42 % (41 of 104)) hc2+; sensitivity = 93.2%; specificity = 100%
Morse [185]	27 genital ulcer specimens from patients with genital ulcer disease	62.9% (17 of 27); resolved sensitivity = 60%; specificity = 100%	92.5% (25 of 27); resolved sensitivity = 93%; resolved specificity = 100%
Safrin [186]	246 patients with mucocutaneous lesions	HSV: 47.2% (116 of 246); HSV-1: 7.3% (18 of 246); HSV-2: 39.8% (96 of 246)	HSV: 59.8% (147 of 246); HSV-1: 9.8% (24 of 246); HSV-2: 50% (123 of 246)
Slomka [187]	194 genitourinary clinic patients (126 female, 86 male)	HSV: 48% (93 of 194) sensitivity versus PCR = 80.9%; specificity versus PCR = 100%	HSV: 59% (115 of 194)
Do Nascimento [188]	41 AIDS patients	HSV: 58.5% (24 of 41)	HSV: 87.8% (36 of 41)
Beards [189]	132 patients with rashes	HSV: 7.3%	HSV=12.9%

TABLE 10.4 (continued)
Comparison of HSV Type-2 Detection by Culture and PCR Assays

[References]	Study Population	HSV Prevalence Culture	HSV Prevalence PCR
Coyle [190]	126 patients with suspected clinical HSV-1 and HSV-2 infections	HSV: 16.4%; HSV-1: 48.6% (18 of 37); HSV-2: 10.8% (4 of 37); sensitivity of culture versus PCR = 100%; specificity of culture versus PCR = 100%	HSV: 27.6% (37 of 134); HSV-1: 73% (27 of 37); HSV-2: 27% (10 of 37)
Madhaven [191]	54 ophthalmic specimens	HSV: 9.3% (5 of 54)	HSV: 51.9% (28 of 54)
Ryncarz [192]	335 genital samples from patients with suspected genital lesions	HSV-2: 48.3% (162 of 335)	HSV-2: 77.9% (261 of 335), Taq man PCR; 82.3% (276 of 335), qualitative competive PCR
Waldhuber [141]	131 patients with genital ulcers	HSV-2: 22.1% (29 of 131); HSV-1: 8.4% (11 of 131); sensitivity of culture = 67%; specificity of culture = 97% versus PCR	HSV-2: 39.7% (52 of 131); HSV-1: 6.1% (8 of 131)
Espy [193]	500 specimens (288 genital; 192 dermal; 20 ocular)	30% (150 of 500)	57.4% (287 of 500)
Espy [194]	200 specimens (160 genital, 38 dermal, 2 ocular)	24.5% (69 of 200)	44% (88 of 200)
Bruisten [195]	Specimens from 372 patients with genital ulcer disease	HSV-1: 5.6% HSV-2: 35%	HSV-1: 7.8% HSV-2: 48%
Marshall [196]	100 patient sample swabs	32%	36%
Espy [197]	198 specimens; (152 genital, 46 dermal) from patients suspected of having HSV infections	35.4% (N = 70)	46.2% (N = 92)
Markoulatos [174]	55 samples from oral, genital and skin sites from patients with clinical evidence of HSV	HSV-2: 100% (20 of 20); HSV-1: 78.6% (22 of 28)	HSV-2: 100% (20 of 20); HSV-1: 100% (28 of 28)
Burrows [198]	262 specimens including 106 cutaneous, 20 ocular, 106 anogenital, 1 respiratory, and 29 from unspecified sites	57% (36 of 63) of HSV+ specimens	99% (66 of 67)

Druce [199]	656 specimens	4.7% (31 of 656) virus isolation and IF-positive samples combined	25.8% (169 of 656)
Scoular et al. [200]	236 patients with clinical features suggestive of genital herpes	HSV: 37.2% (88 of 236)	HSV: 46.2% (109/236)
Van Doornam [201]	668 patients with genital samples, throat swabs from transplant patients, dermal or oral specimens from patients suspected having VZV or HSV infections; 161 STD patients with genital ulcers	HSV-1: 19% (127 of 168); HSV-2: 10.8% (72 of 668); sensitivity versus PCR: HSV-1 = 86%; HSV-2 = 73%	HSV-1: 21.4% (143 of 668); HSV-2: 14.5% (97 of 668); sensitivity versus culture: HSV-1: 96.9%; HSV-2: 98.6%
Wald [142]	296 HSV-infected persons (137 women; 159 men) with >36,000 samples of mucosal secretions	3%	12.1%
Mengelle [202]	313 samples (70 dermal, 81 cerebrospinal fluid, 47 ocular, 42 anogenital, 34 throat swabs, 33 oral samples, 3 whole blood, 2 biopsies, and 1 bronchoalveolar lavage); 226 included in PCR-cell culture comparisons	9.3% HSV (21 of 226)	15% (34 of 226)
Ramaswamy [203]	Genital swabs from 233 consecutive genitourinary medicine attendees with suspected genital herpes	34% (79 of 233)	57% (132 of 233)

[a] HerpChek (NEA-106, DuPont Medical Products, Wilmington, Delaware, USA, http://www2.dupont.com/DuPont_Home/en_US/index.html

Potential of Using Direct Testing for Neonatal Herpes Prevention

In an effort to eliminate the risk of neonatal herpes, efforts to prevent maternal HSV viral shedding at the time of delivery are needed. Improving the sensitivities of diagnostic tests to ascertain genital HSV shedding in pregnant women at the time of delivery could potentially help identify infants at higher risk of neonatal herpes.[152] For example, one study of 612 pregnant women in Brazil found a low incidence of asymptomatic HSV infection at the time of delivery using viral culture-based methods.[153] More sensitive PCR-based assays have consistently found higher prevalences of HSV shedding at the time of delivery[143] when compared with culture-based methods.

An infant's risk of neonatal herpes is associated with a number of factors, including the quantity and duration of HSV viral excretion to the fetus[154] and the type of genital infection (HSV-1 or HSV-2). It is noteworthy that higher levels of HSV have been isolated in the genital tracts of women with primary rather than recurrent HSV genital infections[155] and among women with recurrent HSV-2 versus HSV-1 infections.[146] Thus, the risk of neonatal herpes has been characterized as highest among women who report no symptoms of genital HSV and no previous history of genital herpes.[154]

Further studies should compare the prevalence of HSV excretion in women antepartum with that at the time of delivery in pregnant women using culture and PCR-based technologies. Despite the higher sensitivity of HSV PCR than that of culture, significant differences in the prevalence of HSV using PCR-based methods have been found in different laboratories. Laboratory variations must be considered when interpreting PCR results. Despite this caveat, there has been a recent call for bedside HSV DNA testing of women at time of delivery, as was done earlier to reduce Group B streptococcal maternal–infant transmission.[152]

TRANSMISSION

Like neonatal HSV transmission described above, genital transmission has been shown to depend upon a number of factors including sexual behavior and the duration of sexual relationships, condom use, and knowledge of partners' HSV status.[113] Further research is needed to determine the relative importance of increasing the sensitivity of an HSV detection assay (by the use of PCR-based methods versus less sensitive and culture-based methods) for clinical use to predict individuals at higher risk of HSV transmission.

In contrast to culture-based methods, PCR does not require infectious viral particles for HSV DNA viral detection. PCR-based assays for the detection of type-specific HSV-1 and HSV-2 DNA (similar to HPV PCR-based assays described earlier in this chapter) do not differentiate infectious and non-infectious herpetic viruses. As expected, culture-positive specimens have been shown to have higher HSV DNA PCR-detected viral levels, and conversely culture-negative/HSV DNA-positive samples are generally characterized by lower HSV titers.[145] Because larger quantities of HSV in the genital tract are associated with higher likelihoods of HSV transmission (both between sexual partners and from mother to child), the clinical significance of the higher sensitivity of PCR and how it relates to potential clinically relevant increased risk of sexual and maternal-to-infant HSV transmission remains unclear. For example, the clinical utility of PCR-based techniques for predicting higher likelihoods of neonatal herpes in high-risk infants who are HSV culture-negative/HSV PCR-positive is unclear. Another unclear issue is to what extent higher HSV viral shedding correlates with higher probabilities of HSV transmission.[156] PCR-based techniques may indeed help to better identify individuals at higher risk of HSV transmission or infants with increased risk of neonatal herpes, but this remains to be proven. The advent of real-time PCR (including quantification of HSV load in genetic tissues at time of delivery) will increase our understanding of the clinical significance of detectable excreted HSV viral levels to the risk of both sexual and maternal–infant transmission. Thus, the clinical relevance of using PCR to identify higher HSV transmission probabilities between sexual partners and from mothers to infants during delivery[157] merits further study.

TIMING OF REPORTING OF LABORATORY RESULTS

Time issues for reporting laboratory results to patients should be considered for clinical management. Traditional methods of culture detection generally require a relatively long time to report results (2 to 4 days); confirmation of negative results generally takes approximately 2 weeks.[149] In contrast, detection time for PCR-based techniques for HSV detection is shorter (1 to 3 days). However, a need exists, as stated earlier, for even more rapid and simpler diagnostic tests.[129] For example, to ascertain which pregnant women may be shedding HSV at time of delivery. Modified or shell vial cultures have been developed that may provide answers in 2 to 6 hours, but they are characterized by low sensitivities similar to standard culture methods.[122]

Because genital HSV cervico–vaginal specimens may be collected using self-sampling techniques, further research is also needed on the potential for home-based sampling for future HSV diagnostic efforts[158] — as more extensively evaluated for HPV self-collection techniques.[70]

TECHNICAL DIFFERENCES OF PCR- AND CULTURE-BASED ASCERTAINMENTS

PCR is relatively labor intensive and more prone to false positive results than culture-based methods. Furthermore, due to requirements for high levels of technical expertise, laboratory equipment, and quality control, PCR technology is not widely available and has been generally limited to specialized laboratories. Significant efforts should be made to reduce the risk of contamination with PCR-based technology (i.e., negative controls, separating pre- and post-PCR equipment and procedures).[159,160] Some laboratories use negative controls with each marked clinical specimen to avoid the potential for false positive reactions. PCR-based methods additionally require the extraction of DNA from clinical specimens, which may require significant work. Previous studies have quantified differences in laboratory methods to extract HSV DNA successfully.[161]

New automated extraction methods currently under development[162] are intended to optimize DNA extraction from collected samples and thus decrease overall cost. However, because the cost of HSV DNA testing can vary widely among laboratories and is currently limited to in-house techniques rather than standardized commercial diagnostic kits, cost should be considered in deciding on the use of PCR-based technology in different clinical settings. PCR-based ascertainment methods are also characterized by significant costs for equipment and technician time. Reimbursement systems for the diagnosis of HSV infections also vary and will influence decisions concerning which diagnostic methods are ultimately chosen.

HSV PCR allows for the detection of both exogenous and endogenous herpetic DNA viral sequences. In contrast to HSV, infections HPV does not lyse cells and its genomic DNA is seldom detected extracellularly. Hence the beta-globin human gene can be used to ascertain the quality of the HPV DNA in a collected genital specimen. However, because exogenous HSV DNA viral sequences are commonly detected, a significant proportion of specimens are HSV DNA-positive/beta-globin gene-negative, making the use of the beta-globin gene unreliable for HSV DNA detection. Currently, there is no standard use of a marker to determine the quality of collected exfoliated cell samples as input into HSV PCR systems.

Most PCR-based assays require the detection of HSV-1 and HSV-2 in separate reactions and separate test tubes.[163] However, PCR-based assay detection systems that simultaneously detect both HSV-1 and HSV-2[149] have recently been successful. The specificity of assay detection for this real-time PCR-based assay for both types was confirmed by the non-amplification of other herpes infections, including CMV, VZV, HHV-6, and HHV-8. One issue of potential concern, however, is the extent of possible competition between HSV-2 and HSV-1 amplification, potentially causing low levels of one type not to be detected in the case of high levels of competition by the other type, with subsequent false negative results.[149] However, some multiplex primers sets have been shown to have equivalent sensitivity for the detection of individual virus types as compared to single-primer PCR-based detection.[164]

NATURAL HISTORY OF HSV DNA AND EVALUATION OF HSV-2
SUPPRESSIVE THERAPY

PCR-based methods allow a more accurate picture of the natural history of herpes viral shedding, reactivation, and herpetic lesions.[165] This represents a clear advantage for research purposes. For example, PCR allows the detection of specific HSV genes (those with potential transforming abilities)[166] and the potential quantification of HPV DNA viral titers.

Acyclovir (GlaxoSmithKline Biologicals, http://www.gsk.com/index.htm) has been successfully used since the early 1980s to treat primary genital herpes and has been shown to decrease notably the duration of herpetic lesions and clinical severity.[98] Epidemiological studies have shown that episodic acyclovir treatment significantly reduces HSV shedding and recurrent infections.[167] However, acyclovir treatment does not completely eradicate HSV shedding, and systematic HSV therapy has also not been shown to prevent latency of herpetic infections or the frequency of recurrences.[157] The higher level of PCR sensitivity (than culture) may allow the ascertainment of the treatment effectiveness of acyclovir over follow-up.[168] PCR-based technology may thus have a unique use for studying the effects of antiviral treatment on HSV viral recurrences in prospective studies of treated patients.

The rates of acyclovir resistance for HSV-2 can be as high as 7%, and culture does have a distinct advantage over PCR-based methods for the testing of antiviral susceptibility. If the genetic mutations responsible for acyclovir resistance can be determined in the future, PCR-based tests may be particularly useful for the identification of these mutations and thus provide an important role for susceptibility testing.[169]

The use of real-time PCR in research studies has allowed the quantification of HSV viral shedding. This information, for example, was useful in showing that suppressive treatment of discordant monogamous couples with valacyclovir (GlaxoSmithKline Biologicals)[170] both lowered HSV infection transmission by ~48% and significantly decreased the quantities of HSV genomes detected when compared to controls.[113,171]

FUTURE DIAGNOSTIC APPROACHES

As developments of future HSV diagnostics move forward, microarray techniques will likely increase in use. These arrays represent the potential to incorporate thousands of gene probes for the simultaneous detection of multiple virus types.[164] For example, microarray chip techniques are currently being developed to ascertain specific DNA sequences for the detection of HSV type-2 PCR-amplified viruses[164] and for viral transcription.[172]

The use of multiplex PCR-based assays for the detection of multiple infections within the same sample will also certainly become more common.[173] Future multiplex PCR assays for detecting panels of different infectious agents include (1) specific herpes viruses such as HSV-1, HSV-2, and varicella zoster viruses[174] for the detection of acute central nervous system diseases, (2) genital ulcer diseases including herpes, *H. ducreyi*, and *T. pallidum*, and (3) simultaneous assessment of different sexually transmitted infections, such as HSV, HPV, and chlamydia.[175]

COMMENTS

In recent years, the public has shown heightened awareness of both human papillomavirus and herpes simplex virus infections, largely due to increased media attention paid to early clinical and ongoing Phase III clinical trials of prophylactic HPV and HSV vaccines. These two vaccine candidates aim to prevent the acquisition of HPV[176,177] and HSV[178] infections in young, unexposed adolescents, with ongoing clinical studies targeted at females. Given the likely future implementation of these vaccines in the public sector, assuring quality clinical diagnostic testing for both HPV and HSV viral infections will become even more important as a facet of future clinical care.

REFERENCES

1. Lorincz A. Human papillomaviruses, in Lennette EH, Lennette DA, Lennette ET, Eds., *Diagnostic Procedures for Viral, Rickettsial, and Chlamydial Infections*, 7th ed. Washington: American Public Health Association, 1995. pp. 465–480.
2. de Villiers E-M, Fauquet C, Broker TR, Bernard H-U, zur Hausen H. Classification of papillomaviruses. *Virology* 2004; 324 (1): 17–27.
3. Bosch FX, Lorincz A, Munoz N, Meijer CJLM, Shah KV. The causal relation between human papillomavirus and cervical cancer. *J Clin Pathol* 2002; 55 (4): 244–265.
4. Walboomers JMM, Jacobs MV, Manos MM, Bosch FX, Kummer JA, Shah KV, Snijders PJF, Peto J, Meijer CJLM, Munoz N. Human papillomavirus is a necessary cause of invasive cervical cancer worldwide. *J Pathol* 1999; 189 (1): 12–19.
5. Lorincz AT, Richart RM. Human papillomavirus DNA testing as an adjunct to cytology in cervical screening programs. *Arch Pathol Lab Med* 2003; 127 (8): 959–968.
6. Bosch FX, Manos MM, Munoz N, Sherman M, Jansen AM, Peto J, Schiffman MH, Moreno V, Kurman R, Shah KV. Prevalence of human papillomavirus in cervical cancer: a worldwide perspective. *J Natl Cancer Inst* 1995; 87 (11): 796–802.
7. Herrero R, Hildesheim A, Bratti C, Sherman ME, Hutchinson M, Morales J, Balmaceda I, Greenberg MD, Alfaro M, Burk RD, Wacholder S, Plummer M, Schiffman M. Population-based study of human papillomavirus infection and cervical neoplasia in rural Costa Rica. *J Natl Cancer Inst* 2000; 92 (6): 464–474.
8. Thomas DB, Qin Q, Kuypers J, Kiviat N, Ashley RL, Koetsawang A, Ray RM, Koetsawang S. Human papillomaviruses and cervical cancer in Bangkok. II. Risk factors for *in situ* and invasive squamous cell cervical carcinomas. *Am J Epidemiol* 2001; 153 (8): 732–739.
9. Clavel C, Masure M, Bory J-P, Putaud I, Mangeonjean C, Lorenzato F, Nazeyrollas P, Gabriel R, Quereux C, Birembaut P. Human papillomavirus testing in primary screening for the detection of high-grade cervical lesions: a study of 7932 women. *Br J Cancer* 2001; 84 (12): 1616–1623.
10. Petry K-U, Menton S, Menton M, van Loenen-Frosch F, de Carvalho Gomes H, Holz B, Schopp B, Garbrecht-Buettner S, Davies P, Boehmer G, van den Akker E, Iftner T. Inclusion of HPV testing in routine cervical cancer screening for women above 29 years in Germany: results for 8468 patients. *Br J Cancer* 2003; 88 (10): 1570–1577.
11. Hildesheim A, Herrero R, Castle PE, Wacholder S, Bratti MC, Sherman ME, Lorincz AT, Burk RD, Morales J, Rodriguez AC, Helgesen K, Alfaro M, Hutchinson M, Balmaceda I, Greenberg M, Schiffman M. HPV co-factors related to the development of cervical cancer: results from a population-based study in Costa Rica. *Br J Cancer* 2001; 84 (9): 1219–1226.
12. Parkin DM, Bray F, Ferlay J, Pisani P. Global cancer statistics, 2002. *CA Cancer J Clin* 2005; 55 (2): 74–108.
13. Jemal A, Murray T, Ward E, Samuels A, Tiwari RC, Ghafoor A, Feuer EJ, Thun MJ. Cancer statistics 2005. *CA Cancer J Clin* 2005; 55 (1): 10–30.
14. Monsonego J. HPV infections and cervical cancer prevention. Priorities and new directions: highlights of EUROGIN International Expert Meeting, Nice, France, October 21–23, 2004. *Gynecol Oncol* 2005; 96 (3): 830–839.
15. Fahey MT, Irwig L, Macaskill P. Meta-analysis of Pap test accuracy. *Am J Epidemiol* 1995; 141 (7): 680–689.
16. Nanda K, McCrory DC, Myers ER, Bastian LA, Hasselblad V, Hickey JD, Matchar DB. Accuracy of the Papanicolaou test in screening for and follow-up of cervical cytologic abnormalities: a systematic review. *Ann Intern Med* 2000;132 (10): 810–819.
17. Solomon D, Schiffman M, Tarone B. Comparison of three management strategies for patients with atypical squamous cells of undetermined significance (ASCUS): baseline results from a randomized trial. *J Natl Cancer Inst* 2001; 93 (4): 293–299.
18. Kulasingam SL, Hughes JP, Kiviat NB, Mao C, Weiss NS, Kuypers JM, Koutsky LA. Evaluation of human papillomavirus testing in primary screening for cervical abnormalities: comparison of sensitivity, specificity, and frequency of referral. *JAMA* 2002; 288 (14): 1749–1757.

19. Lorincz AT, Reid R, Jenson AB, Greenberg MD, Lancaster W, Kurman RJ. Human papillomavirus infection of the cervix: relative risk associations of 15 common anogenital types. *Obstet Gynecol* 1992; 79 (3): 328–337.

20. Schiffman MH, Bauer HM, Hoover RN, Glass AG, Cadell DM, Rush BB, Scott DR, Sherman ME, Kurman RJ, Wacholder S, Stanton CK, Manos MM. Epidemiologic evidence showing that human papillomavirus infection causes most cervical intraepithelial neoplasia. *J Natl Cancer Inst* 1993; 85 (12): 958–964.

21. Franco EL, Villa LL, Sobrinho JP, Prado JM, Rousseau M-C, Desy M, Rohan TE. Epidemiology of acquisition and clearance of cervical human papillomavirus infection in women from a high-risk area for cervical cancer. *J Infect Dis* 1999; 180 (5): 1415–1423.

22. Wright TC, Jr., Cox JT, Massad LS, Twiggs LB. 2001 consensus guidelines for the management of women with cervical cytological abnormalities. *JAMA* 2002; 287 (16): 2120–2129.

23. O'Meara AT. Present standards for cervical cancer screening. *Curr Opin Oncol* 2002; 14 (5): 505–511.

24. Wright TC, Jr., Schiffman M, Solomon D, Cox JT, Garcia F, Goldie S, Hatch K, Noller KL, Roach N, Runowicz C, Saslow D. Interim guidance for the use of human papillomavirus DNA testing as an adjunct to cervical cytology for screening. *Obstet Gynecol* 2004; 103 (2): 304–309.

25. Kim JJ, Wright TC, Goldie SJ. Cost-effectiveness of alternative triage strategies for atypical squamous cells of undetermined significance. *JAMA* 2002; 287 (18): 2382–2390.

26. Goldie SJ, Kim JJ, Wright TC. Cost-effectiveness of human papillomavirus DNA testing for cervical cancer screening in women aged 30 years or more. *Obstet Gynecol* 2004; 103 (4): 619–631.

27. Mandelblatt J, Lawrence W, Yi B, King J. The balance of harms, benefits, and costs of screening for cervical cancer in older women: the case for continued screening. *Arch Intern Med* 2004; 164 (3): 245–247; rebuttal: 248.

28. Elfgren K, Rylander E, Radberg T, Strander B, Strand A, Paajanen K, Sjoberg I, Ryd W, Silins I, Dillner J. Colposcopic and histopathologic evaluation of women participating in population-based screening for human papillomavirus deoxyribonucleic acid persistence. *Am J Obstet Gynecol* 2005; 193 (3, Pt 1): 650–657.

29. Cuzick J, Szarewski A, Cubie H, Hulman G, Kitchener H, Luesley D, McGoogan E, Menon U, Terry G, Edwards R, Brooks C, Desai M, Gie C, Ho L, Jacobs I, Pickles C, Sasieni P. Management of women who test positive for high-risk types of human papillomavirus: the HART study. *Lancet* 2003; 362 (9399): 1871–1876.

30. Bory J-P, Cucherousset J, Lorenzato M, Gabriel R, Quereux C, Birembaut P, Clavel C. Recurrent human papillomavirus infection detected with the hybrid capture II assay selects women with normal cervical smears at risk for developing high grade cervical lesions: a longitudinal study of 3091 women. *Int J Cancer* 2002; 102 (5): 519–525.

31. Sherman ME, Lorincz AT, Scott DR, Wacholder S, Castle PE, Glass AG, Mielzynska-Lohnas I, Rush BB, Schiffman M. Baseline cytology, human papillomavirus testing, and risk for cervical neoplasia: a 10-year cohort analysis. *J Natl Cancer Inst* 2003; 95 (1): 46–52.

32. Dalstein V, Riethmuller D, Pretet J-L, Le Bail Carval K, Sautiere J-L, Carbillet J-P, Kantelip B, Schaal J-P, Mougin C. Persistence and load of high-risk HPV are predictors for development of high-grade cervical lesions: a longitudinal French cohort study. *Int J Cancer* 2003; 106 (3): 396–403.

33. Kjaer SK, van den Brule AJC, Paull G, Svare EI, Sherman ME, Thomsen BL, Suntum M, Bock JE, Poll PA, Meijer CJLM. Type specific persistence of high risk human papillomavirus (HPV) as indicator of high grade cervical squamous intraepithelial lesions in young women: population based prospective follow-up study. *Br Med J* 2002; 325 (7364): 572–578.

34. Rozendaal L, Westerga J, van der Linden JC, Walboomers JMM, Voorhorst FJ, Risse EKJ, Boon ME, Meijer CJLM. PCR-based high risk HPV testing is superior to neural network based screening for predicting incident CIN III in women with normal cytology and borderline changes. *J Clin Pathol* 2000; 53 (8): 606–611.

35. Franceschi S, Clifford GM. Re: A study of the impact of adding HPV types to cervical cancer screening and triage tests. *J Natl Cancer Inst* 2005; 97 (12): 938–941.

36. Cogliano V, Baan R, Straif K, Grosse Y, Secretan B, El Ghissassi F. Carcinogenicity of human papillomaviruses. *Lancet Oncol* 2005; 6 (4): 204.

37. Khan MJ, Castle PE, Lorincz AT, Wacholder S, Sherman M, Scott DR, Rush BB, Glass AG, Schiffman M. The elevated 10-year risk of cervical precancer and cancer in women with human papillomavirus (HPV) type 16 or 18 and the possible utility of type-specific HPV testing in clinical practice. *J Natl Cancer Inst* 2005; 97 (14): 1072–1079.

38. Castle PE, Solomon D, Schiffman M, Wheeler CM, for the ALTS Group. Human papillomavirus type 16 infections and 2-year absolute risk of cervical precancer in women with equivocal or mild cytologic abnormalities. *J Natl Cancer Inst* 2005; 97 (14): 1066–1071.

39. Munoz N, Bosch FX, de Sanjose S, Herrero R, Castellsague X, Shah KV, Snijders PJF, Meijer CJLM. Epidemiologic classification of human papillomavirus types associated with cervical cancer. *New Engl J Med* 2003; 348 (6): 518–527.

40. Schiffman M, Khan MJ, Solomon D, Herrero R, Wacholder S, Hildesheim A, Rodriguez AC, Bratti MC, Wheeler CM, Burk RD. Study of the impact of adding HPV types to cervical cancer screening and triage tests. *J Natl Cancer Inst* 2005; 97 (2): 147–150.

41. Clifford GM, Gallus S, Herrero R, Munoz N, Snijders PJF, Vacarella S, Anh PTH, Ferreccio C, Hieu NT, Matos E, Molano M, Rajkumar R, Ronco G, de Sanjose S, Shin HR, Sukvirach S, Thomas JO, Tunsakul S, Meijer CJLM, Franceschi S. Worldwide distribution of human papillomavirus types in cytologically normal women in the International Agency for Research on Cancer HPV prevalence surveys: a pooled analysis. *Lancet* 2005; 366 (9490): 991–998.

42. Cooper K, Herrington CS, Graham AK, Evans MF, McGee JO. *In situ* evidence for HPV 16, 18, 33 integration in cervical squamous cell cancer in Britain and South Africa. *J Clin Pathol* 1991; 44 (5): 406–409.

43. Wolber RA, Clement PB. *In situ* DNA hybridization of cervical small cell carcinoma and adenocarcinoma using biotin-labeled human papillomavirus probes. *Mod Pathol* 1991; 4 (1): 96–100.

44. Samama B, Plas-Roser S, Schaeffer C, Chateau D, Fabre M, Boehm N. HPV DNA detection by *in situ* hybridization with catalyzed signal amplification on thin-layer cervical smears. *J Histochem Cytochem* 2002; 50 (10): 1417–1420.

45. Hesselink AT, van den Brule AJC, Brink AATP, Berkhof J, van Kemenade FJ, Verheijen RHM, Snijders PJF. Comparison of Hybrid Capture 2 with *in situ* hybridization for the detection of high-risk human papillomavirus in liquid-based cervical samples. *Cancer Cytopathol* 2004; 102 (1): 11–18.

46. Bewtra C, Xie Q, Soundararajan S, Gatalica Z, Hatcher L. Genital human papillomavirus testing by *in situ* hybridization in liquid atypical cytologic materials and follow-up biopsies. *Acta Cytol* 2005; 49 (2): 127–131.

47. Davis-Devine S, Day SJ, Freund GG. Test performance comparison of Inform HPV and Hybrid Capture 2 high-risk HPV DNA tests using the SurePath liquid-based Pap test as the collection method. *Am J Clin Pathol* 2005; 124 (1): 24–30.

48. Quint WGV, Scholte G, van Doom LJ, Kleter B, Smits PHM, Lindeman J. Comparative analysis of human papillomavirus infections in cervical scrapes and biopsy specimens by general SPF10 PCR and HPV genotyping. *J Pathol* 2001; 194 (1): 51–58.

49. Coutlee F, Gravitt P, Kornegay J, Hankins C, Richardson H, Lapointe N, Voyer H, Canadian Women's HIV Study Group, Franco E. Use of PGMY primers in L1 consensus PCR improves detection of human papillomavirus DNA in genital samples. *J Clin Microbiol* 2002; 40 (3): 902–907.

50. Kornegay JR, Shepard AP, Hankins C, Franco E, Lapointe N, Richardson H, Canadian Women's HIV Study Group, Coutlee F. Nonisotopic detection of human papillomavirus DNA in clinical specimens using a consensus PCR and a generic probe mix in an enzyme-linked immunosorbent assay format. *J Clin Microbiol* 2001; 39 (10): 3530–3536.

51. Schiffman M, Castle PE, Solomon D, Stoler M, Wheeler C. A study of the impact of adding HPV types to cervical cancer screening and triage tests. *J Natl Cancer Inst* 2005; 97 (12): 939–941. Authors' response to letter by Franceschi et al.: 938–939.

52. Kulmala S-M, Syrjanen S, Shabalova I, Petrovichev N, Kozachenko V, Podistov J, Ivanchenko O, Zakharenko S, Nerovina R, Kljukina L, Branovskaja M, Grunberga V, Juschenko A, Tosi P, Santopietro R, Syrjanen K. Human papillomavirus testing with the Hybrid Capture 2 assay and PCR as screening tools. *J Clin Microbiol* 2004; 42 (6): 2470–2475.

53. Massad LS, Schneider MF, Watts DH, Strickler HD, Melnick S, Palefsky J, Anastos K, Levine AM, Minkoff H. HPV testing for triage of HIV-infected women with Papanicolaou smears read as atypical squamous cells of uncertain significance. *J Women's Health* 2004; 13 (2): 147–153.

54. Sotlar K, Diemer D, Dethleffs A, Hack Y, Stubner A, Vollmer N, Menton S, Menton M, Dietz K, Wallwiener D, Kandolf R, Bultmann B. Detection and typing of human papillomavirus by E6 nested multiplex PCR. *J Clin Microbiol* 2004; 42 (7): 3176–3184.

55. Chan JK, Monk BJ, Brewer C, Keefe KA, Osann K, McMeekin S, Rose GS, Youssef M, Wilczynski SP, Meyskens FL, Berman ML. HPV infection and number of lifetime sexual partners are strong predictors for 'natural' regression of CIN 2 and 3. *Br J Cancer* 2003; 89 (6): 1062–1066.

56. Depuydt CE, Vereecken AJ, Salembier GM, Vanbrabant AS, Boels LA, van Herck E, Arbyn M, Segers K, Bogers JJ. Thin-layer liquid-based cervical cytology and PCR for detecting and typing human papillomavirus DNA in Flemish women. *Br J Cancer* 2003; 88 (4): 560–566.

57. Xi LF, Toure P, Critchlow CW, Hawes SE, Dembele B, Sow PS, Kiviat NB. Prevalence of specific types of human papillomavirus and cervical squamous intraepithelial lesions in consecutive, previously unscreened, West African women over 35 years of age. *Int J Cancer* 2003; 103 (6): 803–809.

58. Garcia F, Barker B, Santos C, Mendez Brown E, Nuno T, Giuliano A, Davis J. Cross-sectional study of patient- and physician-collected cervical cytology and human papillomavirus. *Obstet Gynecol* 2003; 102 (2): 266–272.

59. Ferreccio C, Bratti MC, Sherman ME, Herrero R, Wacholder S, Hildesheim A, Burk RD, Hutchinson M, Alfaro M, Greenberg MD, Morales J, Rodriguez AC, Schussler J, Schiffman M. A comparison of single and combined visual, cytologic, and virologic tests as screening strategies in a region at high risk of cervical cancer. *Cancer Epidemiol Biomarkers Prev* 2003; 12 (9): 815–823.

60. Hughes SA, Sun D, Gibson C, Bellerose B, Rushing L, Chen H, Harlow BL, Genest DR, Sheets EE, Crum CP. Managing atypical squamous cells of undetermined significance (ASCUS): human papillomavirus testing, ASCUS subtyping, or follow-up cytology? *Am J Obstet Gynecol* 2002; 186 (3): 396–403.

61. Castellsague X, Menendez C, Loscertales M-P, Kornegay JR, dos Santos F, Gomez-Olive FX, Lloveras B, Abarca N, Vaz N, Barreto A, Bosch FX, Alonso P. Human papillomavirus genotypes in rural Mozambique. *Lancet* 2001; 359 (9291): 1429–1430.

62. Schneider A, Hoyer H, Lotz B, Leistritz S, Kuhne-Heid R, Nindl I, Muller B, Haerting J, Durst M. Screening for high-grade cervical intra-epithelial neoplasia and cancer by testing for high-risk HPV, routine cytology or colposcopy. *Int J Cancer* 2000; 89 (6): 529–534.

63. Adam E, Berkova Z, Daxnerova Z, Icenogle J, Reeves WC, Kaufman RH. Papillomavirus detection: demographic and behavioral characteristics influencing the identification of cervical disease. *Am J Obstet Gynecol* 2000; 182 (2): 257–264.

64. Cuzick J, Beverley E, Ho L, Terry G, Sapper H, Mielzynska I, Lorincz A, Chan W-K, Krausz T, Soutter P. HPV testing in primary screening of older women. *Br J Cancer* 1999; 81 (3): 554–558.

65. Qu W, Jiang G, Cruz Y, Chang CJ, Ho GYF, Klein RS, Burk RD. PCR detection of human papillomavirus: comparison between MY09/MY11 and GP5+/GP6+ primer systems. *J Clin Microbiol* 1997; 35 (6): 1304–1310.

66. Karlsen F, Kalantari M, Jenkins A, Pettersen E, Kristensen G, Holm R, Johansson B, Hagmar B. Use of multiple PCR primer sets for optimal detection of human papillomavirus. *J Clin Microbiol* 1996; 34 (9): 2095–2100.

67. Lorincz AT. Hybrid Capture™ method for detection of human papillomavirus DNA in clinical specimens: a tool for clinical management of equivocal Pap smears and for population screening. *J Obstet Gynaecol Res* 1996; 22 (6): 629–636.

68. Lorincz A, Anthony J. Advances in HPV detection by Hybrid Capture®. *Pap Rep* 2001; 12 (6): 145–154.

69. Salmeron J, Lazcano-Ponce E, Lorincz A, Hernandez M, Hernandez P, Leyva A, Uribe M, Manzanares H, Antunez A, Carmona E, Ronnett BM, Sherman ME, Bishai D, Ferris D, Flores Y, Yunes E, Shah KV. Comparison of HPV-based assays with Papanicolaou smears for cervical cancer screening in Morelos State, Mexico. *Cancer Causes Control* 2003; 14 (6): 505–512.

70. Belinson JL, Qiao YL, Pretorius RG, Zhang WH, Rong SD, Huang MN, Zhao FH, Wu LY, Ren SD, Huang RD, Washington MF, Pan QJ, Li L, Fife D. Shanxi Province cervical cancer screening study II: self-sampling for high-risk human papillomavirus compared to direct sampling for human papillomavirus and liquid based cervical cytology. *Int J Gynecol Cancer* 2003; 13 (6): 819–826.

71. Belinson J, Qiao YL, Pretorius R, Zhang WH, Elson P, Li L, Pan QJ, Fischer C, Lorincz A, Zahniser D. Shanxi Province cervical cancer screening study: a cross-sectional comparative trial of multiple techniques to detect cervical neoplasia. *Gynecol Oncol* 2001; 83 (2): 439–444.

72. Wright TC, Jr., Denny L, Kuhn L, Pollack A, Lorincz A. HPV DNA testing of self-collected vaginal samples compared with cytologic screening to detect cervical cancer. *JAMA* 2000; 283 (1): 81–86.

73. Schiffman M, Herrero R, Hildesheim A, Sherman ME, Bratti M, Wacholder S, Alfaro M, Hutchinson M, Morales J, Greenberg MD, Lorincz AT. HPV DNA testing in cervical cancer screening: results from women in a high-risk province of Costa Rica. *JAMA* 2000; 283 (1): 87–93.

74. Ratnam S, Franco EL, Ferenczy A. Human papillomavirus testing for primary screening of cervical cancer precursors. *Cancer Epidemiol Biomarkers Prev* 2000; 9 (9): 945–951.

75. Carozzi FM, Del Mistro A, Confortini M, Sani C, Puliti D, Trevisan R, De Marco L, Tos AG, Girlando S, Dalla Palma P, Pellegrini A, Schiboni ML, Crucitti P, Pierotti P, Vignato A, Ronco G. Reproducibility of HPV DNA testing by Hybrid Capture 2 in a screening setting: intralaboratory and interlaboratory quality control in seven laboratories participating in the same clinical trial. *Am J Clin Pathol* 2005; 124 (5): 1–6.

76. Cubie HA, Moore C, Waller M, Moss S. Development of a quality assurance programme for HPV testing with the UK NHS cervical screening LBC/HPV studies. *J Clin Virol* 2005; 33 (4): 287–292.

77. Peyton CL, Schiffman M, Lorincz AT, Hunt WC, Mielzynska I, Bratti C, Eaton S, Hildesheim A, Morera LA, Rodriguez AC, Sherman ME, Wheeler CM. Comparison of PCR- and Hybrid Capture-based HPV detection systems using multiple cervical specimen collection strategies. *J Clin Microbiol* 1998; 36 (11): 3248–3254. Erratum: 1999; 37 (2): 478.

78. Terry G, Ho L, Londesborough P, Cuzick J, Mielzynska-Lohnas I, Lorincz A. Detection of high-risk HPV types by the Hybrid Capture 2 test. *J Med Virol* 2001; 65 (1): 155–162.

79. Snijders PJF, van den Brule AJC, Meijer CJLM. The clinical relevance of human papillomavirus testing: relationship between analytical and clinical sensitivity. *J Pathol* 2003; 201 (1): 1–6.

80. Arbyn M, Buntinx F, Van Ranst M, Paraskevaidis E, Martin-Hirsch P, Dillner J. Virologic versus cytologic triage of women with equivocal Pap smears: a meta-analysis of the accuracy to detect high-grade intraepithelial neoplasia. *J Natl Cancer Inst* 2004; 96 (4): 280–293.

81. Paraskevaidis E, Arbyn M, Sotiriadis A, Diakomanolis E, Martin-Hirsch P, Koliopoulos G, Markydimas G, Tofoski J, Roukos DH. The role of HPV DNA testing in the follow-up period after treatment for CIN: a systematic review of the literature. *Cancer Treat Rev* 2004; 30 (2): 205–211.

82. Clavel C, Masure M, Bory J-P, Putaud I, Mangeonjean C, Lorenzato M, Gabriel R, Quereux C, Birembaut P. Hybrid Capture II-based human papillomavirus detection, a sensitive test to detect in routine high-grade cervical lesions: a preliminary study on 1518 women. *Br J Cancer* 1999; 80 (9): 1306–1311.

83. Bulkmans NWJ, Rozendaal L, Voorhorst FJ, Snijders PJF, Meijer CJLM. Long-term protective effect of high-risk human papillomavirus testing in population-based cervical screening. *Br J Cancer* 2005; 92 (9): 1800–1802.

84. Bigras G, de Marval F. The probability for a Pap test to be abnormal is directly proportional to HPV viral load: results from a Swiss study comparing HPV testing and liquid-based cytology to detect cervical cancer precursors in 12,842 women. *Br J Cancer* 2005; 93 (5): 575–581.

85. Nieminen P, Vuorma S, Viikki M, Hakama M, Anttila A. Comparison of HPV test versus conventional and automation-assisted Pap screening as potential screening tools for preventing cervical cancer. *BJOG* 2004; 111 (8): 842–848.

86. Kuhn L, Denny L, Pollack A, Lorincz A, Richart RM, Wright TC. Human papillomavirus DNA testing for cervical cancer screening in low-resource settings. *J Natl Cancer Inst* 2000; 92 (10): 818–825.

87. Fetterman B, Shaber R, Pawlick G, Kinney W. 2005. Lessons from practice: the first hundred thousand Pap and HPV cotests for general population screening.
Abstr C-02. Poster presented at 22nd International Papillomavirus Conference, Vancouver, April 30–May 6, 2005.

88. Mandelblatt JS, Lawrence WF, Womack SM, Jacobson D, Yi B, Hwang Y, Gold K, Barter J, Shah K. Benefits and costs of using HPV testing to screen for cervical cancer. *JAMA* 2002; 287 (18): 2372–2381.

89. Hill AB. The environment and disease: association or causation? President's address. *Proc Royal Soc Med* 1965; 58: 295–300.

90. Saslow D, Runowicz CD, Solomon D, Moscicki A-B, Smith RA, Eyre IIJ, Cohen C. American Cancer Society guidelines for the early detection of cervical neoplasia and cancer. *CA Cancer J Clin* 2002; 52 (6): 342–362. Erratum in 2003; 53 (2): 126–127.

91. ACOG. Clinical management guidelines for obstetrician–gynecologists: cervical cytology screening. *Obstet Gynecol* 2003; 102 (2): 417–427. Practice Bulletin Number 45, August 2003; replaces Committee Opinion 152, March 1995.

92. ACOG. Clinical management guidelines for obstetrician–gynecologists: human papillomavirus. *Obstet Gynecol* 2005; 105 (4): 905–918. ACOG Practice Bulletin 61, April 2005.

93. ACOG. Clinical management guidelines for obstetrician–gynecologists: management of abnormal cervical cytology and histology. *Obstet Gynecol* 2005; 106 (3): 645–664. ACOG Practice Bulletin 66, September 2005.

94. Corey L, Wald A. Genital herpes, in Holmes KK, Mardh P-A, Sparling PF, Lemon SM, Stamm WE, Piot P, Wasserheit JN, Eds., *Sexually Transmitted Diseases*. 3rd ed., New York: McGraw-Hill, 1999. pp. 285–312.

95. Nahmias AJ, Lee FK, Beckman-Nahmias S. Sero-epidemiological and -sociological patterns of herpes simplex virus infection in the world. *Scand J Infect Dis Suppl* 1990; 69: 19–36.

96. Smith JS, Robinson NJ. Age-specific prevalence of infection with herpes simplex virus types 2 and 1: a global review. *J Infect Dis* 2002; 186 (Suppl 1): S3–S28.

97. Wald A, Corey L, Cone R, Hobson A, Davis G, Zeh J. Frequent genital herpes simplex virus 2 shedding in immunocompetent women: effect of acyclovir treatment. *J Clin Invest* 1997; 99 (5): 1092–1097.

98. Brugha R, Keersmaekers K, Renton A, Meheus A. Genital herpes infection: a review. *Int J Epidemiol* 1997; 26 (4): 698–709.

99. Tran T, Druce JD, Catton MC, Kelly H, Birch CJ. Changing epidemiology of genital herpes simplex virus infection in Melbourne, Australia, between 1980 and 2003. *Sex Transm Infect* 2004; 80 (4): 277–279.

100. Dolan A, Jamieson FE, Cunningham C, Barnett BC, McGeoch DJ. The genome sequence of herpes simplex virus type 2. *J Virol* 1998; 72 (3): 2010–2021.

101. Mertz GJ, Coombs RW, Ashley R, Jourden J, Remington M, Winter C, Fahnlander A, Guinan M, Ducey H, Corey L. Transmission of genital herpes in couples with one symptomatic and one asymptomatic partner: a prospective study. *J Infect Dis* 1998; 157 (6): 1169–1177.

102. Koelle DM, Wald A. Herpes simplex virus: the importance of asymptomatic shedding. *J Antimicrob Chemother* 2000; 45: 1–8.

103. Benedetti JK, Zeh J, Corey L. Clinical reactivation of genital herpes simplex virus infection decreases in frequency over time. *Ann Intern Med* 1999; 131 (1): 14–20.

104. Wald A, Zeh J, Selke S, Ashley RL, Corey L. Virologic characteristics of subclinical and symptomatic genital herpes infections. *New Engl J Med* 1995; 333 (12): 770–775.

105. Wald A, Zeh J, Selke S, Warren T, Ashley R, Corey L. Genital shedding of herpes simplex virus among men. *J Infect Dis* 2002; 186: S34–S39.

106. Corey L, Handsfield HH. Genital herpes and public health: addressing a global problem. *JAMA* 2000; 283 (6): 791–794.

107. Cusini M, Ghislanzoni M. The importance of diagnosing genital herpes. *J Antimicrob Chemother* 2001; 47: 9–16.

108. Whitley R, Arvin A, Prober C, Corey L, Burchett S, Plotkin S, Starr S, Jacobs R, Powell D, Nahmias A, Sumaya C, Edwards K, Alford C, Caddell G, Soong S-J. Predictors of morbidity and mortality in neonates with herpes simplex virus infections. *New Engl J Med* 1991; 324 (7): 450–457.

109. Brown ZA, Benedetti J, Selke S, Ashley R, Watts DH, Corey L. Asymptomatic maternal shedding of herpes simplex virus at the onset of labor: relationship to preterm labor. *Obstet Gynecol* 1996; 87 (4): 483–488.

110. Brown ZA, Selke S, Zeh J, Kopelman J, Maslow A, Ashley RL, Watts H, Berry S, Herd M, Corey L. The acquisition of herpes simplex virus during pregnancy. *New Engl J Med* 1997; 337 (8): 509–515.

111. Gutierrez KM, Falkovitz Halpern MS, Maldonado Y, Arvin AM. The epidemiology of neonatal herpes simplex virus infections in California from 1985 to 1995. *J Infect Dis* 1999; 180 (1): 199–202.

112. Whitley R. Neonatal herpes simplex virus infection. *Curr Opin Infect Dis* 2004; 17 (3): 243–246.

113. Sacks SL, Griffiths PD, Corey L, Cohen C, Cunningham A, Dusheiko GM, Self S, Spruance S, Stanberry LR, Wald A, Whitley RJ. HSV-2 transmission. *Antiviral Res* 2004; 63: S27–S35.

114. Keet IPM, Lee FK, van Griensven GJP, Lange JMA, Nahmias A, Coutinho RA. Herpes simplex virus type 2 and other genital ulcerative infections as a risk factor for HIV-1 acquisition. *Genitourin Med* 1990; 66 (5): 330–333.

115. Wald A, Link K. Risk of human immunodeficiency virus infection in herpes simplex virus type 1-seropositive persons: a meta-analysis. *J Infect Dis* 2002; 185 (1): 45–52.

116. Augenbraun M, Feldman J, Chirgwin K, Zenilman J, Clarke L, DeHovitz J, Landesman S, Minkoff H. Increased genital shedding of herpes simplex virus type 2 in HIV-seropositive women. *Ann Intern Med* 1995; 123 (11): 845–847.

117. Mostad SB, Kreiss JK, Ryncarz A, Chohan B, Mandaliya K, Ndinya-Achola J, Bwayo JJ, Corey L. Cervical shedding of herpes simplex virus and cytomegalovirus throughout the menstrual cycle in women infected with human immunodeficiency virus type 1. *Am J Obstet Gynecol* 2000; 183 (4): 948–955.

118. Lowhagen GB, Bergbrant IM, Bergstrom T, Ryd W, Voog E. PCR detection of Epstein–Barr virus, herpes simplex virus and human papillomavirus from the anal mucosa in HIV-seropositive and HIV-seronegative homosexual men. *Int J STD AIDS* 1999; 10 (9): 615–618.

119. Posavad CM, Wald A, Kuntz S, Huang ML, Selke S, Krantz E, Corey L. Frequent reactivation of herpes simplex virus among HIV-1-infected patients treated with highly active antiretroviral therapy. *J Infect Dis* 2004; 190 (4): 693–696.

120. Schacker T. The role of HSV in the transmission and progression of HIV. *Herpes* 2001; 8 (2): 46–49.

121. Wald A, Corey L. How does herpes simplex virus type 2 influence human immunodeficiency virus infection and pathogenesis? *J Infect Dis* 2003; 187 (10): 1509–1512.

122. Ashley RL. Laboratory techniques in the diagnosis of herpes simplex infection. *Genitourin Med* 1993; 69 (3): 174–183.

123. Aurelian L. Herpes simplex viruses, in Specter S, Lancz G, Eds., *Clinical Virology Manual*, 2nd ed., Amsterdam: Elsevier; 1992. pp. 473–499.

124. Cohen PR. Tests for detecting herpes simplex virus and varicella-zoster virus infections. *Dermatol Clin* 1994; 12 (1): 51–68.

125. Koutsky LA, Stevens CE, Holmes KK, Ashley RL, Kiviat NB, Critchlow CW, Corey L. Underdiagnosis of genital herpes by current clinical and viral isolation procedures. *New Engl J Med* 1992; 326 (23): 1533–1539.

126. Warford AL, Levy RA, Rekrut KA. Evaluation of a commercial enzyme-linked immunosorbent assay for detection of herpes simplex virus antigen. *J Clin Microbiol* 1984; 20 (3): 490–493.

127. Lafferty WE, Krofft S, Remington M, Giddings R, Winter C, Cent A, Corey L. Diagnosis of herpes simplex virus by direct immunofluorescence and viral isolation from samples of external genital lesions in a high-prevalence population. *J Clin Microbiol* 1987; 25 (2): 323–326.

128. Redfield DC, Richman DD, Albanil S, Oxman MN, Wahl GM. Detection of herpes simplex virus in clinical specimens by DNA hybridization. *Diagn Microbiol Infect Dis* 1983; 1 (2): 117–128.

129. Cullen AP, Long CD, Lorincz AT. Rapid detection and typing of herpes simplex virus DNA in clinical specimens by the Hybrid Capture II signal amplification probe test. *J Clin Microbiol* 1997; 35 (9): 2275–2278.

130. Ashley RL. Performance and use of HSV type-specific serology test kits. *Herpes* 2002; 9 (2): 38–45.

131. Cleary KL, Pare E, Stamilio D, Macones GA. Type-specific screening for asymptomatic herpes infection in pregnancy: a decision analysis. *BJOG* 2005; 112 (6): 731–736.

132. Sauerbrei A, Wutzler P. Laboratory diagnosis of central nervous system infections caused by herpes viruses. *J Clin Virol* 2002; 25: S45–S51.

133. Kimura H, Futamura M, Kito H, Ando T, Goto M, Kuzushima K, Shibata M, Morishima T. Detection of viral DNA in neonatal herpes simplex virus infections: frequent and prolonged presence in serum and cerebrospinal fluid. *J Infect Dis* 1991; 164 (2): 289–293.

134. Rowley AH, Whitley RJ, Lakeman FD, Wolinsky SM. Rapid detection of herpes simplex virus DNA in cerebrospinal fluid of patients with herpes simplex encephalitis. *Lancet* 1990; 335 (8687): 440–441.

135. Baringer JR, Pisani P. Herpes simplex virus genomes in human nervous system tissue analyzed by polymerase chain reaction. *Ann Neurol* 1994; 36 (6): 823–829.

136. Malm G, Forsgren M. Neonatal herpes simplex virus infections: HSV DNA in cerebrospinal fluid and serum. *Arch Dis Child Fetal Neonatal Ed* 1999; 81 (1): F24–F29.

137. Openshaw H, McNeill JI, Lin XH, Niland J, Cantin EM. Herpes simplex virus DNA in normal corneas: persistence without viral shedding from ganglia. *J Med Virol* 1995; 46 (1): 75–80.

138. Miura S, Smith CC, Burnett JW, Aurelian L. Detection of viral DNA within skin of healed recurrent herpes simplex infection and erythema multiforme lesions. *J Invest Dermatol* 1992; 98 (1): 68–72.

139. Wald A, Matson P, Ryncarz A, Corey L. Detection of herpes simplex virus DNA in semen of men with genital HSV-2 infection. *Sex Transm Dis* 1999; 26 (1): 1–3.

140. Robinson PA, High AS, Hume WJ. Rapid detection of human herpes simplex virus type 1 in saliva. *Arch Oral Biol* 1992; 37 (10): 797–806.

141. Waldhuber MG, Denham I, Wadey C, Leong-Shaw W, Cross GF. Detection of herpes simplex virus in genital specimens by type-specific polymerase chain reaction. *Int J STD AIDS* 1999; 10 (2): 89–92.

142. Wald A, Huang M-L, Carrell D, Selke S, Corey L. Polymerase chain reaction for detection of herpes simplex virus (HSV) DNA on mucosal surfaces: comparison with HSV isolation in cell culture. *J Infect Dis* 2003; 188 (9): 1345–1351.

143. Cone RW, Hobson AC, Brown Z, Ashley R, Berry S, Winter C, Corey L. Frequent detection of genital herpes simplex virus DNA by polymerase chain reaction among pregnant women. *JAMA* 1994; 272 (10): 792–796.

144. Hardy DA, Arvin AM, Yasukawa LL, Bronzan RN, Lewinshohn DM, Hensleigh PA, Prober CG. Use of polymerase chain reaction for successful identification of asymptomatic genital infection with herpes simplex virus in pregnant women at delivery. *J Infect Dis* 1990; 162 (5): 1031–1035.

145. Cone RW, Hobson AC, Palmer J, Remington M, Corey L. Extended duration of herpes simplex virus DNA in genital lesions detected by the polymerase chain reaction. *J Infect Dis* 1991; 164 (4): 757–760.

146. Kinghorn GR. Genital herpes: natural history and treatment of acute episodes. *J Med Virol* 1993; Suppl. 1: 33–38.

147. Jerome KR, Huang M-L, Wald A, Selke S, Corey L. Quantitative stability of DNA after extended storage of clinical specimens as determined by real-time PCR. *J Clin Microbiol* 2002; 40 (7): 2609–2611.

148. Kessler HH, Muhlbauer G, Rinner B, Stelzl E, Berger A, Dorr H-W, Santner B, Marth E, Rabenau H. Detection of herpes simplex virus DNA by real-time PCR. *J Clin Microbiol* 2000; 38 (7): 2638–2642.

149. Adelson ME, Feola M, Trama J, Tilton RC, Mordechai E. Simultaneous detection of herpes simplex virus types 1 and 2 by real-time PCR and pyrosequencing. *J Clin Virol* 2005; 33 (1): 25–34.

150. Corey L, Huang M-L, Selke S, Wald A. Differentiation of herpes simplex virus types 1 and 2 in clinical samples by a real-time Taqman PCR assay. *J Med Virol* 2005; 76 (3): 350–355.

151. Lafferty WE, Coombs RW, Benedetti J, Critchlow C, Corey L. Recurrences after oral and genital herpes simplex virus infection: influence of site of infection and viral type. *New Engl J Med* 1987; 316 (23): 1444–1449.

152. Kimberlin DW, Whitley RJ. Neonatal herpes: what have we learned. *Semin Pediatr Infect Dis* 2005; 16 (1): 7–16.

153. Weinberg A, Canto CLM, Pannuti CS, Kwang WN, Garcia SAL, Zugaib M. Herpes simplex virus type 2 infection in pregnancy: asymptomatic viral excretion at delivery and seroepidemiologic survey of two socioeconomically distinct populations in Sao Paulo, Brazil. *Rev Inst Med Trop Sao Paulo* 1993; 35 (3): 285–290.

154. Stagno S, Whitley RJ. Herpesvirus infections in neonates and children: cytomegalovirus and herpes simplex virus, in Holmes KK, Mardh P-A, Sparling PF, Lemon SM, Stamm WE, Piot P, Wasserheit JN, Eds., *Sexually Transmitted Diseases*, 3rd ed., New York: McGraw-Hill; 1999. pp. 1191–1212.

155. Koelle DM, Benedetti J, Langenberg A. Asymptomatic reactivation of herpes simplex virus in women after the first episode of genital herpes. *Ann Intern Med* 1992; 116 (6): 433–437.

156. Sacks SL, Griffiths PD, Corey L, Cohen C, Cunningham A, Dusheiko GM, Self S, Spruance S, Stanberry LR, Wald A, Whitley RJ. Introduction: is viral shedding a surrogate marker for transmission of genital herpes? *Antiviral Res* 2004; 63: S3–S10.

157. Barton SE, Munday PE, Patel RJ. Asymptomatic shedding of herpes simplex virus from the genital tract: uncertainty and its consequences for patient management. *Int J STD AIDS* 1996; 7 (4): 229–232.

158. Stanberry LR. Asymptomatic herpes simplex virus shedding and Russian roulette. *Clin Infect Dis* 2000; 30 (2): 268–269.

159. Atkins JT. HSV PCR for CNS infections: pearls and pitfalls. *Pediatr Infect Dis J* 1999; 18 (9): 823–824.

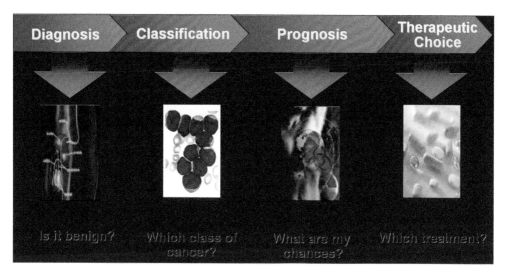

FIGURE 3.5 Rapid detection of differentially expressed genes or genetic mutations such as amplifications, deletions, or loss of heterozygosity (LOH), may allow for earlier cancer diagnosis, assessment of cancer predisposing markers, characterization of tumors for tailored chemotherapies, and provision of new insights into the molecular basis of cancer.

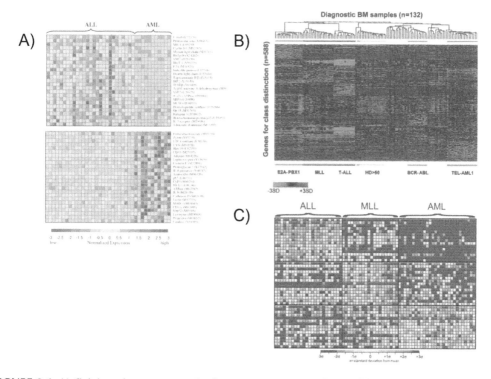

FIGURE 3.6 A) Golub et al. were among the first to use gene expression profiling to classify different kinds of leukemia.[38] B) Ross et al. used expression profiling to subclassify seven distinct subtypes of ALL.[5] C) Armstrong et al. classified patient samples among acute lymphoblastic leukemia, mixed lineage leukemia, and acute myelogenous leukemia with 95% accuracy. Acute lymphoblastic leukemia patients with mixed-lineage leukemia gene (MLL) translocations have particularly poor prognoses.[4]

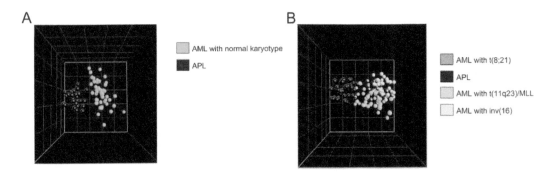

FIGURE 3.7 Haferlach and colleagues used three-dimensional Principal Components Analysis feature space to visualize expression data on genes from the Gene Ontology Biological Process category blood coagulation (accession number GO:0007596).[42] (A) Eighty-three probe sets distinguish patients with acute promyelocytic leukemia (APL) and those with AML with normal karyotypes. (B) Seventy-five probe sets distinguish patients with APL and those with subtypes of AML.

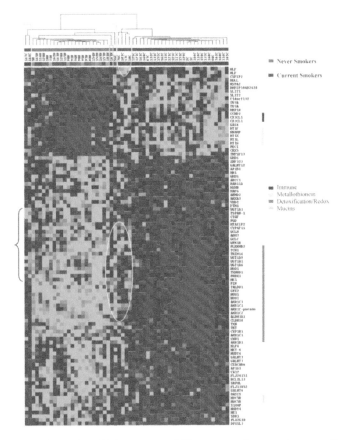

FIGURE 3.8 Hierarchical clustering of 34 current- and 23 never-smoker samples according to expression of 97 genes differentially expressed.[45] While the majority of current smokers and never smokers classify into two distinct groups, there are a few outliers whose gene expression patterns match that of the alternate classification: three current-smokers group with never smokers (yellow rectangle) and one never-smoker (blue box) clusters more closely with current smokers. Some current smokers (yellow oval) did not express redox/xenobiotic gene (white oval) to the same levels as other smokers. The heat map depicts high levels of expression in red and low levels in green.

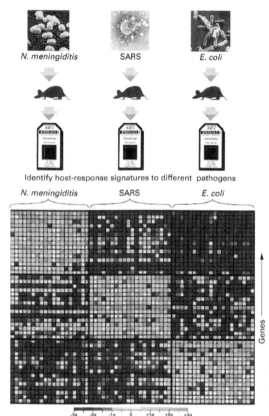

FIGURE 3.9 Host response signatures for pathogen identification. In this animal model, mice were infected with different pathogens and whole genome expression was measured by microarray analysis, resulting in specific host-response signatures.

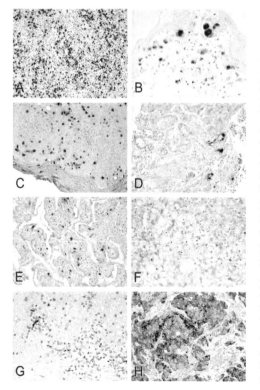

FIGURE 5.1 Examples of ISH using chromogenic methods. A. Detection of Epstein-Barr virus (EBV) in a B cell lymphoma with diffuse large cell type using EBERs oligonucleotide probes visualized by alkaline phosphatase (AP) with nitroblue tetrazolium and 5-bromo-4-chloro-3-indolyl phosphate (NBT/BCIP) as chromogen substrate. B. ISH for adenovirus (ADV) using ADV cDNA probe with AP and NBT/BCIP. C. ISH for human papillomavirus (HPV) type 6/11 showing viral DNA in the nuclei of condyloma acuminatum using DNA probe with AP and NBT/BCIP. D. ISH positive signals for BK polyoma virus are seen in the nuclei of tubular epithelium from a renal transplant case using DNA probe with AP and NBT/BCIP. E. ISH positive cells of parvovirus B19 (PVB19) present in nucleated red cells of placenta using cDNA probes with AP and NBT/BCIP. F. Detection of *Cryptococcus* by ISH in a lung biopsy. The organisms are detected using oligonucleotide probes targeted at the ribosomal RNA visualized by AP and NBT/BCIP. G. ISH for *Aspergillus* showing hyphal forms. Oligonucleotide probes with AP and NBT/BCIP. H. ISH detecting albumin mRNA in a hepatocellular carcinoma metastatic to the scapula using riboprobes with AP and NBT/BCIP.

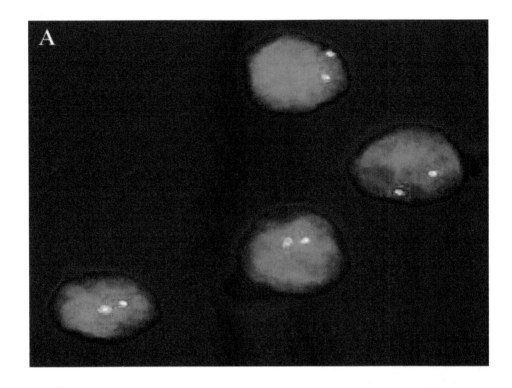

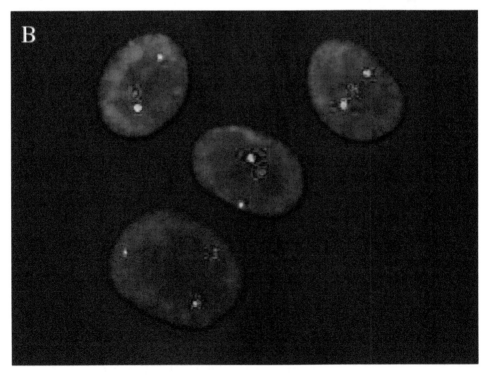

FIGURE 5.2 Interphase FISH analysis of normal tissues and Ewing's sarcomas from FFPE sections. A. Normal interphase cell nucleus without EWS translocation showing green and red FISH signals in juxtaposition (fusion). B. Tumor cell nucleus with EWS translocation showing separation of green and red FISH signals indicating EWS gene breakage (translocation).

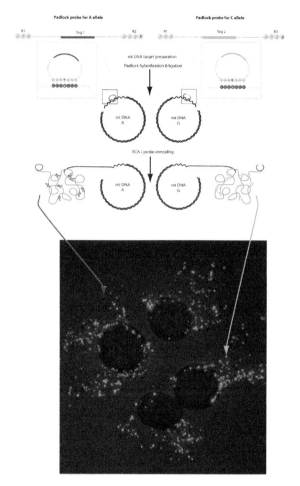

FIGURE 6.7 *In situ* genotyping using padlock probes. R1 are the locus-specific recognition sequences of the padlock probe; R2 and R3 are the allele-specific recognition sequences. The drawing (top) depicts the basic steps of the procedure, and the picture (bottom) presents an example of genotyping of a heteroplasmic cell line called G55 1.1. In these cells, roughly 65% of mitochondrial genomes contain the A3243G (MELAS) mutation.

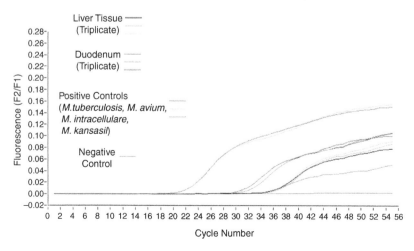

FIGURE 8.1 The positive amplification in this pan-mycobacteria PCR demonstrates the presence of myco-bacterial DNA in the liver and duodenum of this patient.

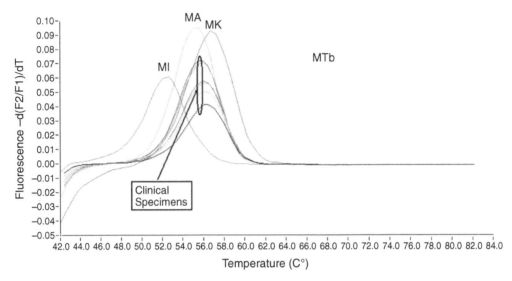

FIGURE 8.2 The post-amplification melt curve analysis of this assay differentiates *M. tuberculosis* from non-tuberculous mycobacteria. The patient is infected with a mycobacterium that is definitely not *M. tuberculosis*; it has a melt curve suggestive of *M. avium*. DNA sequence-based identification revealed *M. avium*. (See legend on Figure 8.1.)

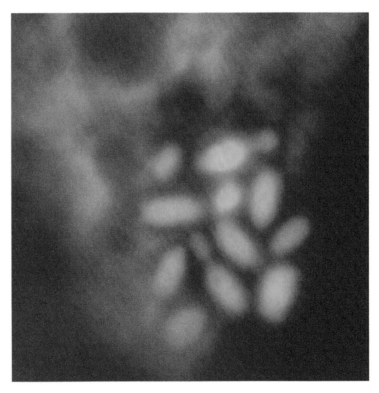

FIGURE 8.3 Yeast cells in a positive blood culture hybridize with a *Candida albicans* PNA FISH probe (AdvanDx, Inc., Woburn, Massachusetts, USA< http://www.advandx.com/) and demonstrate fluorescence that thereby identifies them as *C. albicans*. 1000× magnification.

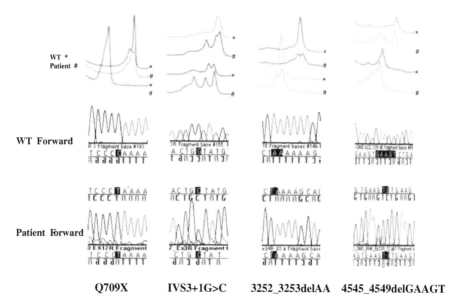

Q709X   IVS3+1G>C   3252_3253delAA   4545_4549delGAAGT

FIGURE 15.8 DMD *dystrophin* gene point mutation analysis. Wild-type DHPLC pattern (top) and patient DHPLC pattern (bottom) at two temperatures for each exon for different mutations. Sequence analysis showing confirmation of each mutation is shown below. DNA from patients having definite DMD or BMD but no deletion or duplication mutation detected by multiplex PCR or Southern blot was subjected to DHPLC to scan for other types of mutations. The DHPLC profile is shown for the wild-type compared to the patient. Following DHPLC, samples demonstrating a heteroduplex (sequence variant) were subjected to sequencing in both forward and reverse direction compared with the wild-type sequence. Four types of mutations are shown: a nonsense mutation Q709X resulting in the change of a glutamine to a stop codon at codon 709; a splicing mutation IVS3 + 1G > C that represents a change of a guanine to a cytosine at the first nucleotide position in intron 3; a deletion of two adenines at cDNA nucleotide position 3252; and a 5-bp deletion at cDNA position 4545.

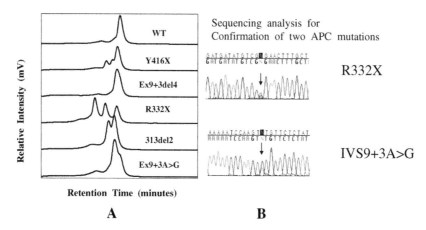

FIGURE 15.11 A: Scanning *APC* gene for mutations using DHPLC. DHPLC patterns of patients with various mutations in exon 9 of the *APC* gene identified by DHPLC. The top profile shows the wild-type pattern. All variants shown below have a pattern shifted to the left and include nonsense, deletions, and splicing mutations. The left and right panels represent the same samples analyzed under different denaturing conditions, showing differences in the heteroduplex patterns and pointing to the necessity for well defined conditions for analysis. B: Sequence analysis confirmation of two *APC* gene mutations in exon 9 shown in Figure A: R332X and IVS9+3A>G. Arrows point to sequence changes.

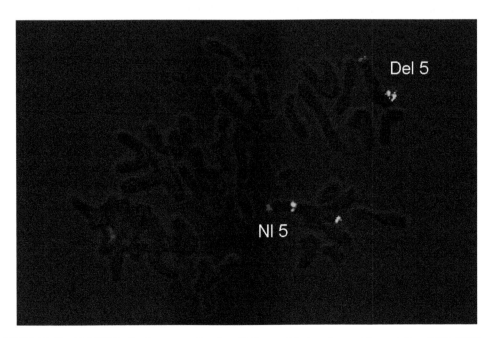

FIGURE 15.13 *APC* FISH. Confirmation of entire *APC* gene deletion detected by real-time PCR. FISH was performed using a biotin-labeled BAC RP11-107C15 as a probe. Arrow shows no hybridization of BAC RP11-107C15 probe, indicating entire *APC* gene deletion. The 5p telomere (green signals) and the 5q telomere (red signals) were used as control probes. Nl 5 indicates normal chromosome 5.

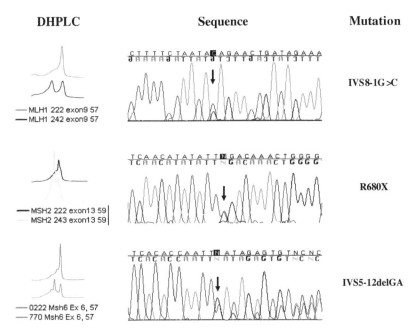

FIGURE 15.19 Identification of HNPCC by DHPLC and sequencing. Similar to the strategy for FAP testing, HNPCC analysis uses DHPLC to identify a variant and sequencing to define it. DHPLC patterns are shown to the left. The wild-type profile is shown directly above the patient profile. Also shown are a splicing mutation in *MLH1* (upper panel), a nonsense mutation in *MSH2* (middle panel), and a 2-base deletion creating a frameshift in *MSH6* (lower panel).

160. Burggraf S, Olgemoller B. Simple technique for internal control of real-time amplification assays. *Clin Chem* 2004; 50 (5): 819–825.
161. Kok T, Wati S, Bayly B, Devonshire-Gill D, Higgins G. Comparison of six nucleic acid extraction methods for detection of viral DNA or RNA sequences in four different non-serum specimen types. *J Clin Virol* 2000; 16 (2): 59–63.
162. Issa NC, Espy MJ, Uhl JR, Harmsen WS, Mandrekar JN, Gullerud RE, Davis MD, Smith TF. Comparison of specimen processing and nucleic acid extraction by the swab extraction tube system versus the MagNA Pure LC system for laboratory diagnosis of herpes simplex virus infections by LightCycler PCR. *J Clin Microbiol* 2005; 43 (3): 1059–1063.
163. Vesanen M, Piiparinen H, Kallio A, Vaheri A. Detection of herpes simplex virus DNA in cerebrospinal fluid samples using the polymerase chain reaction and microplate hybridization. *J Virol Methods* 1996; 59 (1–2): 1–11.
164. Boriskin YS, Rice PS, Stabler RA, Hinds J, Al Ghusein H, Vass K, Butcher PD. DNA microarrays for virus detection in cases of central nervous system infection. *J Clin Microbiol* 2004; 42 (12): 5811–5818.
165. Diaz-Mitoma F, Ruben M, Sacks S, MacPherson P, Caissie G. Detection of viral DNA to evaluate outcome of antiviral treatment of patients with recurrent genital herpes. *J Clin Microbiol* 1996; 34 (3): 657–663.
166. Guibinga GH, Coutlee F, Kessous A, Hankins C, Lapointe N, Richer G, Tousignant J. Detection of a transforming fragment of herpes simplex virus type 2 in clinical specimens by PCR. *J Clin Microbiol* 1996; 34 (7): 1654–1659.
167. Sacks SL, Griffiths PD, Corey L, Cohen C, Cunningham A, Dusheiko GM, Self S, Spruance S, Stanberry LR, Wald A, Whitley RJ. HSV shedding. *Antiviral Res* 2004; 63: S19–S26.
168. Kinghorn GR. Limiting the spread of genital herpes. *Scand J Infect Dis* 1996; 100: 20–25.
169. Bacon TH, Levin MJ, Leary JJ, Sarisky RT, Sutton D. Herpes simplex virus resistance to acyclovir and penciclovir after two decades of antiviral therapy. *Clin Microbiol Rev* 2003; 16 (1): 114–128.
170. Corey L, Ashley R. Prevention of herpes simplex virus type 2 transmission with antiviral therapy. *Herpes* 2004; 11:170A–174A.
171. Corey L, Wald A, Patel R, Sacks SL, Tyring SK, Warren T, Douglas JM, Jr., Paavonen J, Morrow RA, Beutner KR, Stratchounsky LS, Mertz G, Keene ON, Watson HA, Tait D, Vargas-Cortes M. Once-daily valacyclovir to reduce the risk of transmission of genital herpes. *New Engl J Med* 2004; 350 (1): 11–20.
172. Stingley SW, Garcia Ramirez JJ, Aguilar SA, Simmen K, Sandri-Goldin RM, Ghazal P, Wagner EK. Global analysis of herpes simplex virus type 1 transcription using an oligonucleotide-based DNA microarray. *J Virol* 2000; 74 (21): 9916–9927.
173. Tenorio A, Echevarria JE, Casas I, Echevarria JM, Tabares E. Detection and typing of human herpesviruses by multiplex polymerase chain reaction. *J Virol Methods* 1993; 44 (2–3): 261–269.
174. Markoulatos P, Georgopoulou A, Kotsovassilis C, Karabogia-Karaphillides P, Spyrou N. Detection and typing of HSV-1, HSV-2, and VZV by a multiplex polymerase chain reaction. *J Clin Lab Anal* 2000; 14 (5): 214–219.
175. Mitrani-Rosenbaum S, Tsvieli R, Lavie O, Boldes R, Anteby E, Shimonovitch S, Lazarovitch T, Friedmann A. Simultaneous detection of three common sexually transmitted agents by polymerase chain reaction. *Am J Obstet Gynecol* 1994;1994 (3): 784–790.
176. Koutsky LA, Ault KA, Wheeler CM, Brown DR, Barr E, Alvarez FB, Chiacchierini LM, Jansen KU. A controlled trial of a human papillomavirus type 16 vaccine. *New Engl J Med* 2002; 347 (21): 1645–1651.
177. Harper DM, Franco EL, Wheeler C, Ferris DG, Jenkins D, Schuind A, Zahaf T, Innis B, Naud P, De Carvalho NS, Roteli-Martins CM, Teixeira J, Blatter MM, Korn AP, Quint W, Dubin G. Efficacy of a bivalent L1 virus-like particle vaccine in prevention of infection with human papillomavirus types 16 and 18 in young women: a randomised controlled trial. *Lancet* 2004; 364 (9447): 1757–1765.
178. Stanberry LR, Spruance SL, Cunningham AL, Bernstein DI, Mindel A, Sacks S, Tyring S, Aoki FY, Slaoui M, Denis M, Vandepapeliere P, Dubin G. Glycoprotein-D adjuvant vaccine to prevent genital herpes. *New Engl J Med* 2002; 347 (21): 1652–1661.

179. Flores Y, Bishai D, Lazcano E, Shah K, Lorincz A, Hernandez M, Salmeron J. Improving cervical cancer screening in Mexico: results from the Morelos HPV study. *Salud Publica Mex* 2003; 45: S388–S398.

180. Kimura H, Shibata M, Kuzushima K, Nishikawa K, Nishiyama Y, Morishima T. Detection and direct typing of herpes simplex virus by polymerase chain reaction. *Med Microbiol Immunol (Berl)* 1990; 179 (4): 177–184.

181. Rogers BB, Josephson SL, Mak SK, Sweeney PJ. Polymerase chain reaction amplification of herpes simplex virus DNA from clinical samples. *Obstet Gynecol* 1992; 79 (3): 464–469.

182. Cone RW, Swenson PD, Hobson AC, Remington M, Corey L. Herpes simplex virus detection from genital lesions: a comparative study using antigen detection (HerpChek) and culture. *J Clin Microbiol* 1993; 31 (7): 1774–1776.

183. Orle KA, Gates CA, Martin DH, Body BA, Weiss JB. Simultaneous PCR detection of *Haemophilus ducreyi*, *Treponema pallidum*, and herpes simplex virus types 1 and 2 from genital ulcers. *J Clin Microbiol* 1996; 34 (1): 49–54.

184. Boggess KA, Watts DH, Hobson AC, Ashley RL, Brown ZA, Corey L. Herpes simplex virus type 2 detection by culture and polymerase chain reaction and relationship to genital symptoms and cervical antibody status during the third trimester of pregnancy. *Am J Obstet Gynecol* 1997; 176 (2): 443–451.

185. Morse SA, Trees DL, Htun Y, Radebe F, Orle KA, Dangor Y, Beck-Sague CM, Schmid S, Fehler G, Weiss JB, Ballard RC. Comparison of clinical diagnosis and standard laboratory and molecular methods for the diagnosis of genital ulcer disease in Lesotho: association with human immunodeficiency virus infection. *J Infect Dis* 1997; 175 (3): 583–589.

186. Safrin S, Shaw H, Bolan G, Cuan J, Chiang CS. Comparison of virus culture and the polymerase chain reaction for diagnosis of mucocutaneous herpes simplex virus infection. *Sex Transm Dis* 1997; 24 (3): 176–180.

187. Slomka MJ, Emery L, Munday PE, Moulsdale M, Brown DWG. A comparison of PCR with virus isolation and direct antigen detection for diagnosis and typing of genital herpes. *J Med Virol* 1998; 55 (2): 177–183.

188. do Nascimento MC, Sumita LM, de Souza VA, Pannuti CS. Detection and direct typing of herpes simplex virus in perianal ulcers of patients with AIDS by PCR. *J Clin Microbiol* 1998; 36 (3): 848–849.

189. Beards G, Graham C, Pillay D. Investigation of vesicular rashes for HSV and VZV by PCR. *J Med Virol* 1998; 54 (3): 155–157.

190. Coyle PV, Desai A, Wyatt D, McCaughey C, O'Neill HJ. A comparison of virus isolation, indirect immunofluorescence and nested multiplex polymerase chain reaction for the diagnosis of primary and recurrent herpes simplex type 1 and type 2 infections. *J Virol Methods* 1999; 83 (1–2): 75–82.

191. Madhavan HN, Priya K, Anand AR, Therese KL. Detection of herpes simplex virus (HSV) genome using polymerase chain reaction (PCR) in clinical samples: comparison of PCR with standard laboratory methods for the detection of HSV. *J Clin Virol* 1999; 14 (2): 145–151.

192. Ryncarz AJ, Goddard J, Wald A, Huang M-L, Roizman B, Corey L. Development of a high-throughput quantitative assay for detecting herpes simplex virus DNA in clinical samples. *J Clin Microbiol* 1999; 37 (6): 1941–1947.

193. Espy MJ, Ross TK, Teo R, Svien KA, Wold AD, Uhl JR, Smith TF. Evaluation of LightCycler PCR for implementation of laboratory diagnosis of herpes simplex virus infections. *J Clin Microbiol* 2000; 38 (8): 3116–3118.

194. Espy MJ, Uhl JR, Mitchell PS, Thorvilson JN, Svien KA, Wold AD, Smith TF. Diagnosis of herpes simplex virus infections in the clinical laboratory by LightCycler PCR. *J Clin Microbiol* 2000; 38 (2): 795–799.

195. Bruisten SM, Cairo I, Fennema H, Pijl A, Buimer M, Peerbooms PGH, Van Dyck E, Meijer A, Ossewaarde JM, van Doornum GJJ. Diagnosing genital ulcer disease in a clinic for sexually transmitted diseases in Amsterdam, The Netherlands. *J Clin Microbiol* 2001; 39 (2): 601–605.

196. Marshall DS, Linfert DR, Draghi A, McCarter YS, Tsongalis GJ. Identification of herpes simplex virus genital infection: comparison of a multiplex PCR assay and traditional viral isolation techniques. *Mod Pathol* 2001; 14 (3): 152–156.

197. Espy MJ, Rys PN, Wold AD, Uhl JR, Sloan LM, Jenkins GD, Ilstrup DM, Cockerill FR, III, Patel R, Rosenblatt JE, Smith TF. Detection of herpes simplex virus DNA in genital and dermal specimens by LightCycler PCR after extraction using the IsoQuick, MagNA Pure, and BioRobot 9604 methods. *J Clin Microbiol* 2001; 39 (6): 2233–2236.

198. Burrows J, Nitsche A, Bayly B, Walker E, Higgins G, Kok T. Detection and subtyping of Herpes simplex virus in clinical samples by LightCycler PCR, enzyme immunoassay and cell culture. *BMC Microbiology* 2002; 2 (1): 12.

199. Druce J, Catton M, Chibo D, Minerds K, Tyssen D, Kostecki R, Maskill B, Leong-Shaw W, Gerrard M, Birch C. Utility of a multiplex PCR assay for detecting herpes virus DNA in clinical samples. *J Clin Microbiol* 2002; 40 (5): 1728–1732.

200. Scoular A, Gillespie G, Carman WF. Polymerase chain reaction for diagnosis of genital herpes in a genitourinary medicine clinic. *Sex Transm Infect* 2002; 78 (1): 21–25.

201. van Doornum GJJ, Guldemeester J, Osterhaus ADME, Niesters HGM. Diagnosing herpes virus infections by real-time amplification and rapid culture. *J Clin Microbiol* 2003; 41 (2): 576–580.

202. Mengelle C, Sandres-Saune K, Miedouge M, Mansuy J-M, Bouquies C, Isopet J. Use of two real-time polymerase chain reactions (PCRs) to detect herpes simplex type 1 and 2-DNA after automated extraction of nucleic acid. *J Med Virol* 2004; 74 (3): 459–462.

203. Ramaswamy M, McDonald C, Smith M, Thomas D, Maxwell S, Tenant-Flowers M, Geretti AM. Diagnosis of genital herpes by real time PCR in routine clinical practice. *Sex Transm Infect* 2004; 80 (5): 406–410.

11 Blood-Borne Viruses in Clinical and Diagnostic Virology

Hubert G.M. Niesters, Martin Schutten, and Elisabeth Puchhammer-Stöckl

CONTENTS

INTRODUCTION

The ability to detect nucleic acids in blood has made a large impact on screening of blood and blood products, in particular a reduction in the transmission of human immunodeficiency virus type 1 (HIV-1) and hepatitis C virus (HCV). Commercial nucleic acid tests (NATs) for these blood-borne viruses have become readily available. Antiviral treatment regimens for both HIV and HCV were introduced and algorithms to select and monitor patients became part of disease and patient management. Additional amplification assays were developed in clinical virology laboratories and technological changes from manual methods to automation were introduced. Finally, the introduction of real-time amplification technologies enabled laboratories to focus their attention toward more targets and other clinically important viruses and to routinely introduce semi-quantitative assays with short turn-around times.

Although it seems logical to focus attention initially on viruses that cannot be detected easily by classical virological methods (like HCV, HIV-1, hepatitis B virus [HBV], and human papillomavirus [HPV]), the detection of viral nucleic acids is now available for most if not all viruses, although mostly through "home brew" assays. Transition from a research and specialized clinical laboratory setting to a routine clinical diagnostic setting is ongoing for assays developed by laboratories and for commercial assays that generally have relatively short life spans. Furthermore, quality control procedures for these nucleic acid detection technologies are more strongly implemented than they were a decade ago. Because of their ability to quantify both RNA and DNA, efforts to standardize the tests are actively being pursued.

The ability to detect and quantify viral nucleic acids easily from whole blood or plasma has made it possible to use these new technologies to detect viruses important to specific patient groups and, in particular, for solid organ and bone marrow transplant patients. The tests for herpes viruses such as Epstein–Barr virus (EBV) and cytomegalovirus (CMV) and for the adenoviruses (AdVs) have proven to be of enormous value in reducing morbidity and mortality in these patient groups.

This chapter will focus on clinical aspects of the application of tests for blood-borne viruses although consensus conferences (in particular for HIV-1, HBV, and HCV) may require changes to current guidelines after more clinical data become available. However, the current general principles of detection and techniques for the use of NATs are expected to remain valid.

HEPATITIS B VIRUS

HBV remains a very important infectious agent of human disease, with an estimated 350 million people chronically infected and over 2 billion people infected with the virus at some time in their lives. It is estimated that 1 to 2 million people die annually as a consequence of HBV infection, most of them in Asia.

Therapeutic and preventive options are available. An effective HBV vaccine is available and antiviral treatments using pegylated interferon-alpha (PEG-IFN) as well as nucleoside or nucleotide analogues are options for clinical use. Measurement of response to antiviral treatment is straightforward, with the appearance of anti-HBeAg (hepatitis B e antigen) antibodies as a primary endpoint for PEG-IFN in HBeAg-positive disease, and the control of HBV replication and consequent reduction of HBV DNA in plasma or serum as a primary endpoint for HBV-specific inhibitors. A consequence of using nucleoside and nucleotide analogues is, however, the appearance of mutations related to drug resistance.

Most clinical and virological data on nucleoside analogues is currently available on the drug lamivudine (also known as Epivir-HBV, GlaxoSmithKline, Research Triangle Park, North Carolina, USA) — the first registered nucleoside drug for the treatment of HBV. With the development of HBV DNA assays able to quantify the virus over a large dynamic range (from 50 to 10^9 copies per ml or currently expressed in World Health Organization [WHO] international units per milliliter [IU/ml]) and the development of sequence-based techniques to characterize even complete viral genomes, changes in viral replication due to antiviral effect or drug-resistant mutations are detected with relative ease. Specific mutations related to resistance are currently known for this drug, but one must be aware of the emergence of new and unknown changes in the HBV genome relating directly or indirectly to resistance (see Figure 11.1).

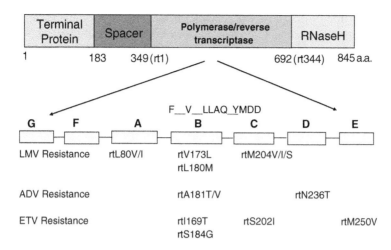

FIGURE 11.1 Mutations recognized and related to antiviral resistance to lamivudine (LMV), adefovir (ADV) and entecavir (ETV) are listed according to their locations in the domains (A through E) in the polymerase–reverse transcriptase domain of the HBV genome. The DNA polymerase gene can be divided into several segments, of which the polymerase–reverse transcriptase segment is important in replication. This segment can be divided into seven domains (A through G). Known mutations related to resistance are mentioned as described by Locarnini.[15]

ASSOCIATION OF GENOTYPE WITH NATURAL INFECTION AND RESPONSE TO THERAPY

Genotyping of HBV has attracted more interest since reports have indicated that different genotypes affect clinical outcome and response to antiviral treatment. Understanding the relationship between viral genotype and response to treatment could lead to improvements in patient management.

Patients with HBV genotypes C and D have lower response rates to interferon compared to those infected with genotypes A or B.[1,2] This may be related to the development of mutations in the basal core promoter region of genotypes C and D, compared to genotypes A and B. Studies in Taiwan, China, and Japan indicated that hepatocellular carcinoma (HCC) was more prevalent in patients infected with genotype C versus genotype B, and that patients with genotype B had better prognoses. In another study from India where HBV genotypes A and D are prevalent, patients infected with genotype D had more severe disease and developed more HCC. Although the evidence is not conclusive yet, indications of the clinical significance linking genotyping and disease outcome are growing.[3–5]

DEFINITION OF ANTIVIRAL EFFICACY AND MEASUREMENT OF DNA

Antiviral efficacy depends on the endpoint chosen. If efficacy is defined as seroconversion related to HBeAg or HBsAg (hepatitis B surface antigen), simple serological assays can be used. The most important endpoint for nucleoside or nucleotide analogues is a reduction of HBV DNA load in serum that implies a reduction or inhibition of viral replication and subsequently less progression of disease. Measurement of HBV DNA in serum has been a challenge for the last decade, mainly due to the large dynamic range of replication (up to $10^{10–11}$ copies per ml), the lack of standardization for all HBV genotypes but one (the WHO IU applies only to genotype A), and the inability to detect all genotypes with the same efficiency.

There are indications that a high HBV DNA viral load at baseline is associated with the emergence of YMDD variants related to lamivudine resistance, and that a low viral load at 3 months of therapy (defined as fewer than 1000 copies per ml) is a statistically significant predictor of patients not developing resistance during a 12- to 15-month follow-up period.[6,7] An antiviral drug that can inhibit replication over a million-fold (very efficient for an enzymatic process) is also able to reduce viral load from 9 $\log_{10}$ to 3 $\log_{10}$ copies per ml. This is desirable because the lower the viral load achieved during treatment, the less chance of appearance of resistance-related variants. In a small clinical study with a very sensitive HBV DNA assay, Ide et al.[8] observed no viral breakthrough in any of their eight patients with copy levels lower than 1.7 $\log_{10}$ copies per ml and a relation between increased viral copy number and earlier time point of appearance of viral breakthrough.

Although antiviral efficacy can be defined as a reduction of HBV DNA, the question remains as to what level of reduction should be achieved and measured. Patients with HBeAg seroconversion have HBV DNA levels below 10,000 copies per ml.[9] A sensitive HBV DNA assay should be able to detect down to 200 copies per ml, equivalent to approximately 50 IU. Due to the range of sensitivities required, real-time PCR (polymerase chain reaction) techniques meet the criteria for an accurate assessment of the effect of antiviral therapy.

DETECTION OF GENOTYPES AND RESISTANCE-RELATED SEQUENCE CHANGES

The ultimate characterization of the HBV genome is achieved by sequencing, and PCR-based methods are available to sequence the whole genome relatively easily. However, whole genome sequencing is not necessary to determine specific genotypes, since information from only a part of the gene encoding HBsAg is enough to determine the genotype accurately. A disadvantage of sequencing is that co-infection with multiple different genotypes usually cannot be detected.

Reports also indicate that in addition to mutations directly related to resistance located in the polymerase gene, changes in other parts of the HBV genome do occur. Lok and co-workers[10] have shown that a reversion of the pre-core stop codon and thus re-expression of HBeAg in lamivudine-resistant mutants is detectable. Torresi and co-workers showed that mutations in the overlapping HBsAg gene can even restore the replication phenotype of the virus.[11,12]

Although more nucleoside and nucleotide analogues have become available, most information is still related to lamivudine. The major genomic changes elicited by this drug are localized at positions rtL180 and rtM204, in the B and C domains of the polymerase (pol) gene, respectively.[13,14] The pattern of amino acid changes is well known due to prolonged usage of the drug,[15]although new changes such as the recently discovered rtM204S[16,17] are still being described. The rtM204S mutation does not seem to exert a major clinical impact, but it clearly demonstrates the continued need for detailed monitoring of genotypic changes as more drugs are used for longer periods. Six major patterns of lamivudine-resistance mutations have already been described, and it is clear that combinations of these patterns together with wild-type sequences can be present at any given time in a specific patient.

Resistance-related mutations have also been described for newly approved antivirals like adefovir dipivoxyl and entecavir, and it is likely that the number of mutations discovered will increase[18,19] (see Figure 11.1 for a summary). In addition, compensating mutations may arise during long-term treatment. These mutations are located in distant domains of the polymerase gene and are not directly involved in the binding of the nucleotides. However, they are able to change the three-dimensional structure of the pocket in which this binding occurs and therefore can enhance the effects of minor resistance-related changes. The effects of these mutations are not yet completely understood.[15]

VIRAL LOAD MEASUREMENTS AND DETECTION OF RESISTANCE-RELATED MUTATIONS

The general assumption is that antivirals should produce reductions of viral load, and thus treatment failure is usually defined as a reduction of viral load smaller than tenfold or a rebound of viral replication as determined by an increase in viral load measurements. To take into consideration variability in viral load measurements and viral replication, an increase of viral load is also typically defined as a tenfold increase compared to the lowest level measured earlier.

The nature of the assay used and the lower limit of detection achieved are two critical items. Furthermore, although it seems logical, an assay must be able to detect all HBV genotypes with equal sensitivity. Low detection limits enable earlier detection of viral breakthrough, and the achievement of low levels of HBV DNA in serum during treatment indicates that breakthrough is less likely or will be delayed.[8]

Current information clearly shows the clinical importance of an accurate and sensitive HBV DNA quantitative assay in combination with a sensitive assay to detect viral variants. However, one must not overlook the significance of mutations indirectly related to HBV. Although impractical in routine use, sequencing is necessary with the increased clinical use of newer nucleoside and nucleotide analogues. Assays based on restriction fragment length polymorphism (RFLP) and reverse hybridization have become more practical and routine, but both focus on specific changes only. Which assay eventually becomes more cost-effective and practical in clinical use depends on many factors, including changes in technology and changes in legislation. However, with the ability to detect variant populations in a more sensitive and quantitative manner, additional insights into viral population dynamics and the pathogenesis of HBV both during and after therapy are expected in the near future.

HEPATITIS C VIRUS

With the discovery of HCV in 1989, the mystery of infection with a non-A, non-B hepatitis agent could be resolved for a majority of patients. Infection with HCV is one of the leading causes of liver disease in the United States and is associated with cirrhosis and hepatocellular carcinoma. Furthermore, infection with HCV is currently the most common reason for liver transplantation. HCV is an RNA virus of the family Flaviviridae, and more than 6 genotypes and 50 subtypes have been discovered. For most patients (over 80%), infection with HCV results in a chronic infection that may not be noticed for many years.

GENOTYPES AND RESPONSE TO THERAPY

The classification of HCV genotypes was originally based on sequences in the 5' non-coding region of the virus, although accurate determination now uses sequences in the nonstructural protein NS5. HCV genotypes 1, 2, and 3 are distributed worldwide, and subtypes 1a, 1b, 2a, 2b, 2c, and 3a account for about 90% of the infections in North and South America, Europe and Japan. In the United States, genotype 1 accounts for up to 75% of all infections (for an overview see Simmonds[20]). HCV genotypes can be detected easily by sequence analysis or reverse hybridization techniques that also allow detection of multiple infections by different genotypes and/or subtypes.[21]

In general, a close association between HCV genotypes and geographical distribution exists. The importance of genotype was demonstrated in the 1990s when it was found that different genotypes responded differently to standard therapy (alpha-interferon at that time). The sustained response rate of standard interferon treatment (defined as non-detectable HCV RNA 6 months after end of treatment[22] was up to 9% for genotype 1 and 16 to 23% for genotypes 2 and 3. The treatment period was 24 to 48 weeks.

The addition of ribavirin to interferon treatment and an increase in the duration of treatment to 48 weeks increased the sustained response rates up to 31% for genotype 1 and around 65% for genotypes 2 and 3.[23,24] The availability of pegylated interferon (PEG-IFN, a drug that achieves higher blood levels and needs to be administered only once weekly) in 2001 increased the sustained response rate of combination therapy with ribavirin up to 80% for genotypes 2 and 3. Unfortunately, the sustained virological response rates remain under 50% for several groups of patients infected with genotype 1, for example, those who were non-responsive to prior treatment or have cirrhosis.

All current clinical studies show the genotype of the virus to be an independent predictor of treatment outcome, with genotype 1 always associated with a lower sustained response rate.

However, it is important to realize that most studies were performed in the United States and Western Europe, where genotypes 1 to 3 are over-represented. Data from large studies with patients infected with other HCV genotypes are still virtually unavailable. We have seen only limited indications that patients infected with genotype 4 respond to therapy in a similar manner to patients with genotype 1, while patients infected with genotype 5 respond to therapy in a similar manner to patients infected with genotypes 2 and 3.

METHODS AND APPLICATIONS OF RNA TESTS

Tests such as the enzyme immunoassay that detect antibodies against HCV are very reproducible and inexpensive. They are useful for screening at-risk populations and represent the methods of choice for initial screening of patients with liver disease. A negative test result is sufficient to exclude diagnosis of chronic HCV infection in an immunocompetent patient. A single positive qualitative test for HCV RNA confirms active viral replication.

Persistent HCV infection is diagnosed by detection of HCV RNA in the blood for at least 6 months. Several factors may be associated with spontaneous clearance of the virus, such as younger age, female gender, and certain histocompatibility complex genes. Repeat testing with a qualitative test (usually with a detection limit of 50 IU per ml) is not helpful in patient management, unless it can be used to determine whether an acute infection has resolved spontaneously.

Both signal and target amplification assays are available to measure HCV RNA levels. With the introduction of an international standard, normalization should be possible. However, significant variability exists among the available methods in parameters such as reported range, accuracy, and precision. Virological tests can be used to inform patient management decisions, determine optimal treatment schedules, and assess virological response rates to antiviral therapy.[25,26]

The 2002 consensus panel of the U.S. National Institutes of Health (NIH) recommended that for patients with HCV genotype 1, treatment should continue for 48 weeks using PEG-IFN in combination with ribavirin. Patients infected with genotype 2 or 3 should be treated for 24 weeks because therapy prolonged to 48 weeks shows no improvement in response rate. Genotyping is therefore very important, but one should keep in mind that the early determination of a likely sustained response is beneficial for a patient. Fortunately, initial response to treatment can be determined for patients infected by HCV genotypes 1 and 4 by measuring whether a drop in viral load can already be detected by week 12. This drop is expected to be at least 100-fold, and at this time point, a decision can be made whether continuation of treatment is an option (e.g., to reduce liver disease progression). A new decision point can also be introduced at week 24 of treatment for HCV genotypes 2 and 3. The RNAs of these HCVs are typically not detectable by qualitative assay in patients who will have sustained viral responses. Similarly, in patients infected with HCV genotypes 1 and 4, a qualitative assay can select those patients who had end-of-treatment virological responses and might have sustained virological responses if HCV RNA is undetectable at this time point. Of course, the most disappointing event for patients after termination of therapy is the relapse of viral replication. Because PCR-based assays have lower limits of detection around 50 IU per ml, some have speculated that using even more sensitive assays such as transcription-mediated amplification (TMA) with a lower limit of detection around 10 IU per ml may produce an improvement in the selection of patients who will or will not relapse after end of treatment.[27]

HUMAN IMMUNODEFICIENCY VIRUS

In 2004, 3.1 million people died of acquired immune deficiency syndrome (AIDS) and 39.4 million people are still living with human immunodeficiency virus (HIV). The vast majority live in countries where little or no treatment is available. Only 1.7 million people received anti-HIV treatment in developing countries out of 37.7 million living there (UNAIDS, 2004 AIDS epidemic update;

www.unaids.org) in January 2005. This overview will discuss the molecular methods used in developed countries for the management of HIV-infected individuals.

Antiretroviral Therapy

Five classes of antiretroviral drugs are currently available. Successful combination antiretroviral therapy (cART) seems to require three active drugs, preferably from two different drug classes. Triple nucleoside analogue therapy and Kaletra monotherapy (Abbott Laboratories, Abbott Park, Illinois, USA) are exceptions to this rule, although the efficacy of these combinations is still under investigation. Nucleoside analogue reverse transcriptase inhibitors (NRTIs) were the first antiretroviral drugs described[28] and they still serve as the backbones of most first-line therapies.

Protease inhibitors (PIs) and non-nucleoside analogue RT inhibitors (NNRTIs) were the next classes to be described. These drugs allowed full suppression of viral replication for many years along with recovery of the immune system.

Viral entry inhibitors represent the latest class of antiretroviral drugs. The two distinct types of entry inhibitors are (1) those inhibiting the fusion process (with T20 from Hoffmann-La Roche Ltd, Basel, Switzerland, the only current representative) and (2) those that inhibit binding to the second receptor. Of the latter group, several drugs that inhibit CXCR-4 and CCR-5, the two major second receptors used by HIV, are currently in Phase III clinical trials. It does not seem logical to assume that entry inhibitors will ever be used in first-line therapy since second receptor inhibitors inhibit either CXCR-4 or CCR-5 and up-front screening of second receptor usage for first-line regimens does not seem likely at present. T20 has the disadvantage that it cannot be administered orally. Recently an antiretroviral drug was described as belonging to a new class of nucleotide-competing reverse transcriptase inhibitors. Whether this drug will successfully complete clinical trials and achieve market distribution is unknown.

Quantification of HIV-1 RNA from the plasma of an infected individual is the most widely used molecular method for patient management. Developments in quantification techniques have run largely parallel to developments in the treatment of HIV-1. The first reports on studies of quantification of HIV-1 were published in the late 1980s and early 1990s — only a few years after the first reports on treatment of HIV-1 by monotherapies and dual therapies using NRTI.[28–30]

Multiple other markers and surrogate markers for the efficacy of cART including CD4 cell count, serum HIV-1 p24 antigen levels, and quantitative HIV-1 microcultures from peripheral blood mononuclear cells and plasma were investigated.[31] It was soon recognized that these markers do not correlate sufficiently with the efficacy of cART or were not sensitive enough. It should be mentioned, however, that the latest versions of p24 antigen detection assays are sufficiently sensitive to be used for patient management.

In-house and commercial assays for the quantitative detection of HIV-1 RNA were developed on different platforms (reverse transcriptase polymerase chain reaction [RT-PCR], branched DNA (bDNA), nucleic acid sequence-based amplification [NASBA]) and implemented in clinical diagnostic laboratories. With the introduction of protease inhibitors and non-nucleoside analogues for the treatment of HIV-1 in 1995, these highly sensitive assays became an essential part of the management of HIV-1-infected patients.[32] This was due to the fact that, in contrast to monotherapy and dual therapy, triple therapy resulted in reductions of viral loads to undetectable levels, defined initially as approximately 500 HIV-1 RNA copies per ml.

Protease inhibitors and viral load assays have also significantly improved our knowledge of the *in vivo* replication kinetics of HIV-1.[33] The seemingly enormous success of cART led to treatment strategies that were rather aggressive. HIV-1 viral load assays were improved and could eventually detect down to 50 HIV-1 RNA copies per ml. For research purposes, HIV-1 viral load assays that could detect down to 1 copy per ml were developed. Treatment guidelines in effect then suggested starting treatment very early based on low CD4 counts or regardless of CD4 counts in patients with high viral loads.[34]

Despite optimism that cART might ultimately cure HIV-1-infected patients, the level of plasma viremia determined the speed with which CD4 declined and therefore how fast an infected patient would progress to AIDS. This suggests that patients with high viral loads would need protection from fast progression to AIDS. The English treatment guidelines were the first to recognize that treatment would have to be life-long and not be initiated too early because of the long-term side effects of antiretroviral medication. Moreover, treatment should be targeted at preventing disease, and CD4 count is a better predictor than HIV-1 viral load of whether a patient will develop disease within the near future.

As a result, the current guidelines put less emphasis on viral load before start of treatment, and treatment is initiated generally only in patients with low CD4 counts or AIDS-defining opportunistic infections. After initiation of therapy, HIV-1 viral load becomes a more important parameter than CD4 count for two reasons. First, if the viral load is not under control, life-long restoration of the immune system is not likely to occur, and second, there are few options beyond treating HIV-1 to improve CD4 response.

RESISTANCE TO ANTIRETROVIRAL DRUGS

One of the major threats to successful treatment of HIV-1 is the development of antiretroviral drug resistance. If a patient is not sufficiently adherent, HIV-1 will start replicating in the presence of low drug concentrations and thus accumulate mutations that render the antiretroviral drugs ineffective. Resistance to one drug in a class results in both resistance to that drug and often also to other drugs in that class.[35] As a result, choosing a salvage regimen is often difficult and cannot be done based on drug history alone.

The mutations in the HIV-1 genome that confer resistance are known for most antiretroviral drugs. Sequencing regions of interest of the HIV-1 genome after therapy failure and analysis of the mutations found may allow composition of an effective new regimen. As an alternative genotyping method, the amino acid composition at only specific positions can be determined with either line probe assays or gene chips.[36] Studies of the effectiveness of genotyping versus standard of care are far from conclusive.[37,38] Multiple factors seem to influence the effect of genotyping on the outcome of the salvage regimen chosen, such as the availability of expert advice, cART experience of the patient, and types of drugs taken at the time of genotyping.

Several considerations are important for implementing genotyping systems in clinical diagnostic settings. First, several studies have shown that expert consultation is essential for composing a successfully active regimen. With the advent of easy-to-use commercial systems like the ViroSeq (Celera Diagnostics, Atlanta, Georgia, USA) and Trugene (Visible Genetics Inc., Toronto, Ontario, Canada) systems for mutation analysis and interpretation, more laboratories are able to offer sufficient analytical expertise along with services. Virological expertise in the clinical interpretation of genotyping results is, however, often lacking.

Second, cost-effective implementation of these assays is difficult and dependent on several issues. With current first-line regimens and the relatively high number of drugs with non-overlapping resistance profiles, resistance development can be predicted effectively and highly active second-line salvage therapies can be composed relatively easily on the basis of drug history.

For second- and third-line therapy failures, genotyping is often an essential part of patient management. If the new drugs currently undergoing development reach the market, effective third- and fourth-line salvage regimens can be composed without genotyping, using drugs with non-overlapping resistance profiles. It is therefore reasonable to assume that in the near future genotyping after therapy failure will become altogether cost-ineffective. With the accumulation of primary and secondary mutations in cART-experienced patients, predicting phenotypic resistance from sequence analysis has become more difficult. In composing so-called mega- and giga-HAART (highly active antiretroviral therapy) regimens involving more than four antiretroviral drugs, genotyping is often not better than standard of care and thus not cost-effective.

Third, virtually everything we know about genotyping and translating mutation analysis into activity of antiretroviral drugs *in vivo* is based on the most prevalent subtype (B) found in Europe and the United States. The number of non-subtype B infections in these countries, however, is growing. In other regions of the world where antiretroviral therapy (ART) is being introduced very rapidly, non-subtype B infections are already more prevalent. Since resistance development is different for different subtypes, predicting phenotype from genotype will become increasingly more difficult.[39,40] For example, after virological failure on nelfinavir (Pfizer, New York, USA), the D30N mutation will develop in subtype B strains but in general not in several other subtypes.

Fourth, when a patient is switched to another regimen, genotypic resistance will revert back to wild type if selective pressure is not exerted by the new regimen. Viruses containing the "old" resistance profiles will remain present at undetectable levels and rebound quickly once selective pressure is restored by subsequent regimens. Therefore it may be very helpful for management purposes to retain archived patient material for many years.

A growing matter of concern is the transmission of resistant strains that may seriously compromise ART in ART-naïve individuals. Prevalence of ART-resistant strains may vary from less than 5 to more than 25%. A recent study calculated that a prevalence above 5% is required to make genotyping cost-effective before start of therapy.[41] Such calculations can often not be extrapolated to other countries and depend on the way that physicians interpret genotyping. While some experts prefer to prescribe a regimen with a high barrier to resistance for patients with single resistance-associated mutations, that alone does not provide significant resistance; others will not change their management. The decision whether to perform genotyping before start of therapy is therefore dependent on local circumstances. Transmission of resistant strains has been shown to decline in some cohorts and rise in others.[42,43] Regular screening for resistant strains and regular evaluation of the choice for pre-therapy screening is thus sensible.

HIV-2 INCIDENCE

In contrast to the wealth of information on treatment and diagnostic tools for HIV-1, relatively little is known about HIV-2, the second virus causing AIDS. The reason is that HIV-2 is mainly present in West Africa and only limited numbers of HIV-2-infected individuals have been described in other parts of the world.[44] In Western European countries, the prevalence of HIV-2 ranges between 0.1 and 4% of the total number of HIV cases; Portugal has the highest incidence.

Little information is available about HIV-2 in the United States. At present, only about 12 laboratories worldwide offer HIV-2 viral load testing. Most use assays developed in-house and virtually nothing is known about how these tests compare to each other. Therefore, it is difficult to compare results of therapy.

Most antiretroviral drugs developed for the treatment of HIV-1 were tested *in vitro* against HIV-2. Unfortunately, only two classes of drugs seem to be active against HIV-2: NRTIs and PIs except amprenavir (Agenerase®, GlaxoSmithKline, Brentford, U.K.).[45] NNRTIs and T20 are not active against HIV-2, and the ability of second-receptor inhibitors to inhibit HIV-2 replication *in vivo* is not known. Due to the limited number of HIV-2-infected individuals, only two small prospectively followed cohorts of treated patients have been described.[46] We can conclude from the data obtained that a first-line regimen of two NRTIs plus a (boosted) PI can be safely used to treat HIV-2-infected individuals. First reports on nelfinavir-containing regimens, triple NRTIs, and Kaletra monotherapy seem to argue against using these regimens, although numbers are too small to allow definitive conclusions.

Several papers have been published on genotyping and phenotyping of HIV-2 after therapy failure.[46–48] The data point to significant differences between HIV-1 and HIV-2, especially with respect to resistance to NRTIs. It has, for instance, been shown that resistance in HIV-2 may develop against zidovudine (AZT) (Retrovir, GlaxoSmithKline) without mutations at any amino acid positions known to give resistance against AZT in HIV-1. This seriously compromises the ability to use HIV-2 genotyping in clinical practice.

Another serious problem is the rapid development of the multi-NRTI resistance mutation Q151M in a high percentage of HIV-2 patients failing NRTI-containing regimens. Such mutations are infrequent in HIV-1. Primary resistance mutations in HIV-2 protease seem to correspond with mutations in HIV-1 protease after failing PI-containing therapy, which suggests that genotyping HIV-2 protease by using knowledge obtained in HIV-1 infection can be used in the management of HIV-2 therapy failure.

EPSTEIN–BARR VIRUS

Epstein–Barr virus (EBV), an ubiquitous human gamma-herpes virus, infects more than 90% of humans and remains persistent for life in its host. Primary infection occurs predominantly in early childhood and is often asymptomatic or with non-specific symptoms. However, a primary EBV infection in a young adult often leads to infectious mononucleosis with pharyngitis. During this primary infection, EBV-infected B cells express a number of latent genes that drive lymphoproliferation.

One of these latent EBV genes is *EBNA1*, which binds to viral DNA and is responsible for the maintenance of EBV episomes in these replicating B cells. As a consequence of EBV-driven lymphoproliferation during the acute stage of infection, up to 1% of the total peripheral B cells are latently infected B cells, of which a small proportion will eventually undergo lytic replication, resulting in cell-free viremia in the peripheral blood. B cells, whether transformed by EBV or containing replicating virus, are highly immunogenic, resulting in an effective cytotoxic T cell and natural killer cell response that will eliminate most EBV-infected B cells during convalescence of the acute infection. Some EBV-infected B cells will persist, due to mechanisms allowing cells to avoid immune recognition, like down-regulation of EBV proteins. It is estimated that up to 50 EBV-infected cells per million peripheral B cells exist in persistent carriers.

BALANCE OF IMMUNOLOGICAL POWER

In immunocompetent persons, cytotoxic T cells predominantly control the latency of EBV. In immunosuppressed individuals, this immune-surveillance is unable to control the proliferation and outgrowth of EBV-infected B cells, resulting in an EBV lymphoproliferative disease (EBV-LPD). This is a serious complication of allogeneic stem-cell transplantation (allo-SCT) and solid organ transplantation.

Although the incidence of EBV-LPD is generally less than 2% after allo-SCT, the incidence may increase up to 20% in patients with established risk factors, such as unrelated donor SCT, the use of T cell-depleted (TCD) allografts, the use of anti-thymocyte globulin (ATG), and immunosuppression for the prevention or treatment of graft-versus-host-disease (GVHD). This EBV-related LPD is associated with a poor prognosis despite the use of several therapeutic options like anti-B cell monoclonal antibody therapy (rituximab, IDEC Pharmaceutical Corp., San Diego, California, USA),[49,50] donor lymphocyte infusion (DLI), and infusion of EBV-specific cytotoxic T cells.[51]

IDENTIFYING AND MONITORING AT-RISK PATIENTS

In an immunocompetent person, serology is still regarded as the method of choice for confirming acute versus persistent EBV infection. However, in transplant recipients, longitudinal monitoring of the nucleic acid of EBV in peripheral blood is increasingly recognized as the most valuable tool to determine a clinically relevant EBV infection and an important tool for the prediction of EBV-LPD and the therapeutic management of LPD.

Real-time-based PCR assays serve as the standard technology for determining the presence of EBV DNA in clinical materials. There is a clear correlation between an increase of viral load over time and an increased risk for the development of LPD in transplant patients, definitely in SCT. Without the introduction of more standardized assays and internationally accepted standards, the dynamic change (increase or decrease) in viral load in different compartments will most likely be used to correlate with clinical events instead of absolute EBV DNA levels. Increases above the level of 1000 copies EBV DNA per ml in cell-free compartments already serve as important triggers in patient management.

Another problem for the technological implementation of EBV viral load measurements is and has been the differences in clinical material used. Some measurements of changes in EBV DNA load were detected in peripheral blood mononuclear cells (PBMCs) and whole blood while other measurements were performed in serum or plasma.[52–57]

The goal of measuring EBV DNA is to determine thresholds for preemptive treatment or achieve highly sensitive assays predictive for the development of LPD — not to introduce the most sensitive assay available. Changes in viral load will probably be more predictive than absolute values alone. The variability of measurements among several studies may thus reflect differences in methods and materials used to quantify viral loads.

Several studies used PBMCs or whole blood as specimens; these allow highly sensitive detection of LPD but are not specific for this diagnosis. A significant number of transplantation patients showed similar or higher viral loads without the subsequent development of LPD.[57–59] Other studies cited plasma as an excellent sample for real-time PCR diagnosis of SCT-associated LPD.[56,57] Also, in solid-organ LPD, serum or plasma is suitable for EBV quantification. Wagner and co-authors clearly showed that real-time PCR measurements of viral load in plasma in their group of renal transplanted children and adolescents appeared to be more specific for the diagnosis of LPD than measurements in PBMCs.[56] Moreover, normalizing the detected EBV copy number to the amount of co-amplified genomic DNA could enhance the accuracy of measurements in PBMCs.

On the other hand, Stevens et al.[60] stated that in lung transplant recipients with late onset of LPD, increased EBV DNA loads in blood were restricted to the cellular compartment — a discrepancy warranting further research among the different lymphoproliferative diseases to determine whether plasma, serum, PBMCs, or whole blood would be best suited for EBV DNA measurement.

LYMPHOPROLIFERATIVE DISORDER AFTER BONE MARROW TRANSPLANTATION

Although stem cell transplantation is widely used and accepted as a therapy for hematologic malignancies, solid tumors, and non-malignant disorders, the neoplastic diseases are serious complications in long-term survivors. The risk of cancer in SCT recipients is four- to seven-fold higher than in the general population. Also, due to profound immunodeficiency in the first year after transplantation, LPD is the most common second disorder that frequently leads to malignancy, with the highest occurrence in the first half year.

Most of these early LPDs are EBV-associated, for which major risk factors in allogeneic SCT recipients are the use of human leukocyte antigen (HLA) mismatched related donors, T cell depletion, and the use of ATG as prophylaxis and treatment of acute graft-versus-host disease. Also, the number of cells used in the SCT is among the risk factors involved. Patients with several risk factors have the highest incidence of LPD.[61]

For a definitive diagnosis of LPD, a biopsy is required for morphological analysis, with detection of EBV antigens by immunohistochemistry and/or detection of EBV-encoded RNAs by *in situ* hybridization. The clonality of the B cell proliferation can be assessed for the immunoglobulin light-chain type. However, since the progression toward LPD can be very rapid, a diagnosis should be obtained as quickly as possible to allow prompt intervention.

Monitoring patients at risk for developing EBV-related LPD employs methods to measure EBV DNA loads. However, EBV reactivation is common in T cell-depleted allogeneic SCT. A study by van Esser et al.[62] showed that EBV reactivation was more common in the T cell-depleted group (54%) than the unmanipulated allogeneic SCT group (28%). However, the absolute EBV DNA levels at the peak of the first reactivation did not differ between the two groups. Also, recurrent reactivation was more common in the T cell-depleted group (16%) than in the unmanipulated allogeneic SCT group (3%). Several risk factors predicted first reactivation in univariate analysis, including T cell depletion and the use of ATG in the conditioning regimen, transplantation of an unrelated donor graft and a high CD34+ cell number of the graft. Following multivariate analysis, only the use of ATG and a high CD34+ cell count remained independently associated with EBV reactivation.

A change in EBV DNA load significantly predicts the development of EBV LPD in a quantitative manner. A stepwise tenfold increase of EBV DNA yielded a hazard ratio of 2.7 (95% confidence interval, 1.7–4.8) for patients receiving T cell-depleted grafts to develop EBV LPD.[50]

Absolute threshold values for developing EBV LPD cannot be given; however, a clear correlation exists between high viral load and the clinical diagnosis of LPD, both after SCT and solid organ transplantation. Lucas et al.[63] report similar findings using DNA extracted from PBMCs in a cohort of 195 patients receiving solid-organ transplantations. Wagner et al.[56] analyzed both plasma and PBMCs from renal transplant patients and observed greater specificity for the diagnosis of post-transplant lymphoproliferative disease in plasma, with both sensitivity and specificity of 100% for the EBV genome in levels above 10,000 copies per ml.

PREVENTION OF EBV-RELATED LYMPHOPROLIFERATIVE DISORDER

Employing methods to reduce the number of B cells as sources for EBV replication can reduce the risk of developing LPD in T cell-depleted grafts. B cell depletion of the donor graft is an option to reduce the incidence of EBV-related LPD. Preemptive infusion of EBV-specific cytotoxic T cells has been shown to reduce viral load and may prevent the development of LPD. This methodology is expensive and not yet easily implemented. One option is to use anti-B cell immunotherapy for patients with increased risks of developing EBV-LPD after SCT, using viral load measurements as guidance. However, this *in vivo* B cell depletion may impair immune status and should be used only for patients with the highest risk for LPD development.

The recent introduction of rituximab has provided a relatively simple and safe treatment option that has already been applied for the treatment of LPD. One option is to use rituximab for prevention in a selected patients based on more or less absolute quantitative EBV DNA levels measured in plasma based on established risk profiles. In a study performed by van Esser et al.,[50] a threshold value of 1000 copies per ml was used to initiate preemptive treatment. The study was intended to prevent EBV-related LPD, prevent mortality due to LPD, and ascertain whether viral reactivations in high-risk patients could be stopped.

Of the 49 patients monitored during the first 6 months after T cell-depleted allogeneic SCT, 17 showed reactivations of EBV reaching viral loads above the threshold of 1000 copies per ml. Two of those patients had already developed EBV-LPDs prior to the preemptive treatment. The 15 remaining patients were monitored intensively after the initiation of preemptive rituximab treatment (Figure 11.2). Fourteen showed complete remission characterized by the prevention of EBV LPD and the clearance of EBV DNA from plasma. This was reached after a median of 8 days (range of 1 to 46 days). However, it is also clear that a significant number of patients were most likely to clear the virus without preemptive treatment. One patient progressed to EBV-LPD but was able to clear the virus after a second infusion with rituximab and an infusion of donor lymphocytes.[50]

Two patients already clinically diagnosed as having EBV-related LPD were both able to clear the virus after two infusions with rituximab (Figure 11.3). Decisions could be guided by very intensive measurements of EBV viral load. Compared to an historic cohort, the incidence of EBV-related LPD could be reduced but not abolished by preemptive treatment guided by viral load measured in plasma. However, EBV-related mortality was significantly improved, based on a risk-

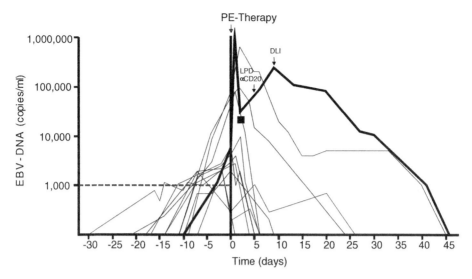

FIGURE 11.2 Preemptive anti-B cell immunotherapy for EBV. To monitor patients at risk for EBV LPD, EBV DNA load is measured regularly. At a level of 1000 copies per ml, the patients were recalled to the hospital and preemptive (PE) therapy with rituximab (anti-CD20 therapy) was initiated. Monitoring was performed almost daily (five times per week) until the EBV DNA level was measured twice below the limit of detection of the assay (50 copies per ml). If the signal did not decrease rapidly (or an increase was observed), a second infusion with rituximab or an infusion with donor lymphocytes (DLI) was given (thick line). The initiation of preemptive therapy is represented as day 0. Fine lines represent individual patients.

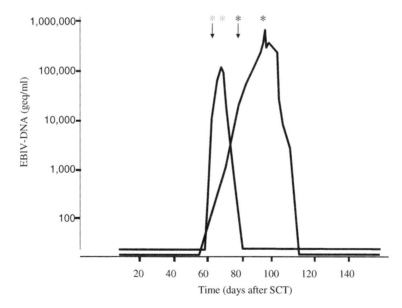

FIGURE 11.3 Viral load following therapy for established EBV LPD. Viral load dynamics in two patients with clinically diagnosed LPD after T cell-depleted SCT (arrows) prior to initiation of preemptive treatment with rituximab. Both patients reached complete remission combined with reductions of immune suppression after two infusions with rituximab (asterisks). Viral load was measured almost daily using established real-time detection methods in plasma.

adapted strategy in which the frequent monitoring of EBV reactivations and the early institution of a preemptive strategy played a central role. This information-based integrated approach between clinical and virological monitoring proved to be an optimal disease management concept.[50]

Now that effective therapies have become available for the treatment of LPD, the challenge will be to determine how to implement routine monitoring tests to diagnose complications as early as possible to identify patients who will require preemptive treatment. It is obvious that early identification and initiation of treatment will ultimately reduce morbidity and mortality. Defined thresholds are not yet completely established, but observed differences in threshold values are not large. Low threshold levels, whether in SCT or solid organ transplant patients, also indicate that detection of EBV DNA, most likely in plasma or mononuclear cells, must be sensitive and reproducible enough at levels as low as 100 to 500 copies per ml. Further standardization should facilitate the exchanges of data among laboratories.

CYTOMEGALOVIRUS

Cytomegalovirus (CMV) causes mostly mild or asymptomatic disease in immunocompetent persons. In contrast, in an immunocompromised host CMV may lead to such life-threatening organ diseases as CMV pneumonitis, gastrointestinal disease, and encephalitis. Primary infection with CMV leads to life-long virus latency in the host. Reactivation of the latent virus and re-infection with new CMV strains can occur, and both may be associated with severe clinical manifestations in immunocompromised persons.

Currently, both CMV infection and reactivation are of major clinical significance in transplant recipients. Disease caused by CMV in these patients can be due to viral infection, but CMV can also indirectly contribute to the loss of a transplanted organ and facilitate the establishment of other secondary infections. The rate of CMV-associated disease and death observed in these patients varies with different kinds of transplanted organs and is especially high in lung and bone marrow recipients (for a review see Patel and Paya[64]). Additional factors such as the combination of a CMV-seronegative recipient with a seropositive donor or the administration of ATG further increase the risk for severe CMV diseases.

For several years efficient antiviral drugs such as gancyclovir, foscarnet, and cidofovir have been available to inhibit CMV replication. Early application of these drugs may avoid the development of severe clinical disease, but will not replace early and rapid diagnosis of clinically relevant CMV infection. In patients undergoing primary CMV infection, the first detection of CMV in the host is of diagnostic significance. Sensitive PCR assays are the diagnostic tools of choice and allow the first detection of virus days earlier than other methods.

TIMING OF TREATMENT

CMV is latently present, however, in all persons who have been previously infected with CMV, and detection of CMV DNA in these seropositive patients is not necessarily identical with the development of a clinically significant CMV infection and thus does not always require therapeutic intervention. It is therefore a crucial point for control of CMV disease to define the appropriate time to start antiviral therapy.

Treatment at the onset of clinically evident CMV is usually too late to inhibit severe disease. In contrast, treatment in the absence of clinically significant CMV activity may impose senseless burdens on transplant patients, especially considering such drug side effects as bone marrow depletion and nephrotoxicity. Major efforts have been undertaken to define the right time point to start so-called preemptive therapy.

In the past, detection and quantification of CMV antigenemia was the basis for both diagnosis of increasing CMV replication and justification for preemptive therapy,[65] but numerous NAT assays now allow detection of CMV DNA in blood components. Because these tests are highly sensitive and because CMV DNA is also latently present in the blood of a substantial proportion of patients, only quantitation of CMV DNA has been found to be of clinical value.

Numerous studies have described various quantitative assays, both "home brewed" and commercial, and point to the possible clinical value of determining viral DNA loads in blood.[66–70] Detection and quantification of DNAemia was found to be even more advantageous than determination of the antigenemia level because DNAemia is a more direct marker for the presence of infectious virus in blood and it can be standardized and automated more easily than antigenemia.

The critical point in CMV DNAemia-guided therapy is defining the DNAemia level predictive of development of clinically significant CMV infection in individual patients that will require initiation of preemptive therapy. Several studies have addressed this aspect by comparing DNAemia levels with pp65 antigenemia quantities to determine a CMV DNA concentration indicative for preemptive therapy. Certain DNA cut-off levels have been defined by several investigators as indicative for preemptive therapy.[67,68,70,71]

In a more recent publication, Lilleri et al.[67] determined a cut-off of 100,000 to 300,000 copies of CMV DNA per ml of whole blood as indicative for preemptive therapy in solid organ transplant (SOT) recipients. For hematopoietic stem cell transplant recipients (HSCTRs) a level of 10,000 copies of CMV DNA per ml of whole blood was shown to have a high positive predictive value. The data obtained from different studies, however, vary due to different selections of samples, varying numbers of cells used for PCR, and lack of standardization of the pp65 antigenemia assay.

Several groups have addressed the question whether whole blood or plasma is the ideal specimen for detection of DNAemia. Both appear adequate. Determination of CMV DNA from whole blood has been shown to be more sensitive[69] and directly comparable to the pp65 assay, especially when using identical cell numbers for analysis.[68] When using plasma, one must consider the lower cut-off of clinically significant levels of DNAemia in this specimen type, shown recently to be at about 10,000 copies per ml for SOT patients.[67]

Plasma has also been found to be a useful specimen for follow-up and determination of preemptive therapy[66] and may be especially valuable for diagnosing patients with severe pancytopenia. Numerous factors must be considered to determine an exact and general DNAemia threshold value for preemptive therapy, including the patient group investigated, blood components or cell numbers used for testing, and differences in quantitative levels achieved by different assays.[67] Thus it seems questionable whether a general threshold can be defined for all transplantation centers. Some authors have suggested that each diagnostic center should determine its own clinically relevant cut-off levels.[68]

Detection and quantification of CMV mRNA transcripts has also been established. Immediately early and late viral transcripts have been quantified using nucleic acid sequence-based amplification (NASBA) technology.[72] Detection of late transcripts has been investigated as another candidate to indicate the start of preemptive therapy in SOT recipients, but to date it has not become a significant alternative to detection of DNAemia.

DISEASE IN IMMUNOCOMPROMISED PATIENTS

Emery et al. have shown the kinetics of CMV replication to be significantly associated with the development of CMV disease in transplant recipients.[73] Weekly quantitative investigation for CMV DNA in whole blood allows an assessment of replication curves. The rate of increase in CMV load between the last PCR-negative and the first PCR-positive sample was significantly faster in patients who developed CMV disease (0.33 $\log_{10}$ genomes per ml daily) than in those who remained free of symptoms (0.19 $\log_{10}$ genomes per ml daily).

Another problem in the follow-up of transplant recipients for CMV is that after cessation of therapy, recurrence of CMV replication can be observed in approximately 25 to 30% of cases. Also the kinetics of CMV viral load as determined by quantitative PCR may be useful in identifying patients who are more likely to have recurrences at a very early stage.[74] While the viral DNA load level was not found predictive for recurrence, the rate of increase in viral load was found to differ significantly. It was only 3.17 days in patients without recurrence in contrast to 8.8 days for those with CMV disease recurrence after cessation of antiviral therapy. Thus duration of therapy may be guided by such kinetic investigations.

While quantitative detection of CMV DNA in blood is an important tool for diagnosis and follow-up of CMV infections in transplant recipients, detection of viral DNA in other specimens such as urine and throat washings was not found to be as significantly predictive for the development of CMV disease. This is primarily due to the fact that fluids from specimens other than blood are not quantitatively defined. As an example, the significance of CMV DNA levels found in broncho-alveolar lavage (BAL) fluid is unclear, although presence of virus in the lung was supposed to be a more direct and earlier indication of developing lung complications especially in lung transplant patients.[75] Recently, it was shown that the clinical significance of the CMV DNA load in BAL could be clearly increased when the virus DNA concentration was recalculated to represent the CMV DNA concentration in the original lung epithelial lining fluid, which is diluted by the BAL procedure and present in BAL samples at varying concentrations.[76]

Severe and often lethal CMV end organ disease can also be found in patients with immuno-suppression due to HIV infection. Prior to the introduction of HAART, CMV-associated complications were seen frequently, including otherwise rare CMV-associated diseases such as CMV encephalitis or chorioretinitis. In CMV encephalitis, the detection of virus DNA in cerebrospinal fluid (CSF) by NAT is the method of choice for diagnosis leading to immediate initiation of antiviral treatment.[77,78] Since the introduction of HAART, the overall incidence of CMV end organ disease among HIV-infected subjects has dramatically declined. However, in untreated patients or patients with HAART failure, CMV disease may still occur. The usefulness of quantitative detection of CMV DNA for diagnosis of CMV disease has been shown.[79] CMV DNA quantification also serves as a means to identify CMV DNA levels in blood predictive for CMV end organ disease.[80,81]

Another group of patients in whom CMV infection may lead to life-threatening disease are newborns who have undergone intrauterine CMV infections. In these patients, NATs are significant for rapid and early identification of CMV infection. PCR assays allow the highly sensitive detection of viral DNA from the urine of the newborns[82] and have thus replaced other methods. In addition, NATs have been used for the prediction of the severity of children's CMV infections. Quantitative PCR showed that higher CMV DNA levels found in the amniotic fluids of pregnant women with primary CMV infections correlated with more severe fetal abnormalities.[81] In children with connatal CMV infections, the virus is sometimes detectable in the CSF, indicative of viral infection of the central nervous system (CNS). The virus levels in CSF were found to vary within a wide range (100 to 100 000 copies of CMV DNA per ml CSF), possibly reflecting the severity of the CNS disease.[77]

RESISTANCE TO ANTIVIRAL DRUGS

As noted earlier, CMV infection or reactivation can often be kept under control by the use of recently developed antiviral drugs (gancyclovir, foscarnet, and cidofovir). However, in the course of long-term treatment, drug-resistant strains may emerge (for an overview, see Emery[83]). Gancyclovir (GCV) is frequently used in first-line therapy, and two CMV gene loci are involved in the development of resistance against it: the UL97 phosphokinase involved in the phosphorylation of GCV and the UL54 DNA polymerase. Usually, the first evidence of resistance against GCV is the emergence of defined mutations in the *UL97* gene, resulting in a 5- to 12-fold increase of resistance.[84] Prolonged therapy results in further development of resistance mutations in the *UL54* gene.

Foscarnet and cidofovir resistance can also be observed, and resistance mutations are located in various parts of the CMV DNA polymerase gene (*UL54*). Development of CMV resistance against a given drug is indicated by an increase in viral load during therapy. Confirmation of resistance against gancyclovir is achieved mostly by genotyping. This analysis is preferred to phenotypic assays because it is more rapid and therefore of direct clinical value. Another advantage is that the emerging virus strains are analyzed directly from patient specimens, not after passage of the virus *in vitro* as necessary for phenotypic analyses that may lead to selective emergence of certain virus strains.

For detection of resistance against GCV, sequence analysis of the *UL97* gene with focus on amino acid changes is performed. Such changes were proven to be associated with a significant loss of sensitivity to gancyclovir. Sequencing, however, may not allow the detection of newly emerging resistant viruses present as minor strains in the host. To overcome this problem, point mutation assays[85] or analysis of restriction fragment length polymorphisms can be used for identification of known resistance mutations.

Genotypic detection of resistance mutations located in the *UL54* gene and associated with resistance against GCV, foscarnet, or cidofovir is still a challenge. The UL54 polymerase resistance mutations defined to date are located in various regions of the large *UL54* gene[86] and show in part cross-resistance against the different drugs. Genotypic resistance testing in the UL54 region usually requires direct sequencing. Considering the large size of the gene, the number of different mutations found, and the differentiation between resistance-associated gene variations and natural polymorphisms, *UL54* genotyping is not performed routinely by most centers and is still more suitable for research than for diagnostics.

ADENOVIRUSES

Adenoviruses (AdVs) are pathogens that may cause a variety of clinical diseases, including upper respiratory infections, enteritis, cystitis, and conjunctivitis or may lead only to asymptomatic infections. To date, 51 different human AdV serotypes have been discriminated. They are grouped into six major species designated A through F based on viral sequence and haemagglutinating properties.[87,88]

While AdVs usually cause benign infections in normal hosts, AdV infections in immunosuppressed patients may lead to severe life-threatening disease that can be caused by any of the AdV species.[88] Most frequently severe AdV disease is observed in bone marrow (BM) recipients.[89,90] AdV infections have been identified in up to 27% of patients receiving allogeneic BM transplants. The incidence of virus disease was found to be increasing within recent years and is especially high in children.[88,91] The clinical spectra of AdV infections in BM transplant patients can be similar to the spectra of immunocompetent persons, but AdV tends to disseminate and develop invasive disease in an immunocompromised host and is associated with very high mortality — up to 80%.[88,91,93,94] In this patient group, early and rapid diagnosis of AdV infection and identification of ongoing invasive disease are of particular importance.

The diagnosis of an AdV infection can be achieved by different methods such as virus culture and antigen testing or by the detection of viral nucleic acids by PCR assays from different specimens (stools, respiratory secretions, or sera). In immunocompromised patients, however, the highly sensitive NAT techniques have become the methods of choice especially for the detection and quantitation of virus from blood. In recent years, numerous groups have developed qualitative or quantitative PCR test systems for the detection of AdV DNA.[88,91,93–96] NAT diagnostics of AdV in immunocompromised hosts remain a challenge, complicated especially by two factors.

First, as mentioned above, 51 different AdV serotypes are known and each of them seems capable of causing invasive disease, although differences in frequency were noted in individual studies.[90–92,97] Thus all these strains should be detectable by a PCR assay used for diagnosis. This is, however, a significant problem because the sequence homologies among strains differ widely. While high homology is usual within species groups, sequence homology between species may be as low as 4%. Most PCR assays developed to date have concentrated on the hexon gene as the target of amplification. The hexon is the most highly conserved gene in the AdV genome, with at least a 50% homology among the strains.[92] Nevertheless, whether detection of all strains is possible solely by using one PCR assay is still questionable. Whether different strains may be detected at equal sensitivity is also questionable. For this reason, some studies concentrate on the detection of only a limited number of strains. Others have used panel of different PCR assays for diagnosis of all strains. A recent study has shown that the use of six different PCR assays is needed to equally detect all 51 AdV strains.[92] Whether such an approach is applicable to routine diagnosis is questionable.

The second major problem of AdV diagnostics in immunocompromised patients is relevant to other virus infections in this patient group, namely, what is the clinical significance of a positive AdV DNA PCR result? As mentioned above, different variations of AdV disease are observed in immunocompromised hosts. AdV frequently causes benign courses of infection, for example, self-limited enteritis in a substantial proportion of patients. However, the development of invasive disease also may be observed and is frequently associated with fatal outcomes.

Various approaches have been undertaken within the last few years to differentiate benign and invasive diseases and achieve early detection of change from one state to the other. As an important aspect aiding such differentiation, invasive disease has been associated with detection of virus at more than one patient site.[98] Other authors have shown that the presence of AdV DNA in peripheral blood (PB)[92] or in serum[93] is greatly associated with the development of invasive disease. Further analysis indicated that the amount of virus DNA in serum may also be a significant factor in predicting invasive AdV disease, and in particular the dynamics of viral replication in plasma or PB.[99] A tenfold increase of virus load in PB was found to be associated with development of disease. Although a general consensus strategy has not yet been established for early diagnosis of AdV disease in immunocompromised patients, it is obvious from the findings cited above that early diagnosis of invasive AdV disease will be based on quantitative PCR technology.

Early detection and prediction of invasive AdV disease hold great importance for the survival of patients. In contrast to other viral diseases, no satisfactory treatment schemes are available for AdV. However, a number of studies deal with various aspects of treatment of invasive AdV disease.[49,100] An increasing number of studies show that cidofovir treatment, although significantly nephrotoxic, may be a measure for dealing with AdV disease in immunocompromised patients.[49,101,102] A prospective study has shown that the effect of treatment can be controlled and predicted by monitoring viral load changes in blood plasma by real-time PCR after the first and the second infusions of cidofovir.[102]

In patients who have developed fatal invasive AdV disease in spite of treatment, decreases of viral DNA were found to be only minimal or not observed at all. In contrast, patients who recovered from AdV disease in follow-up showed a median decrease in AdV viral load of 1.4 $\log_{10}$ copies after the first infusion and 3 $\log_{10}$ copies after the second. The course of disease and the benefit of the treatment are somehow dependent on viral load at the onset of treatment.[102] Pre-existing high virus levels and a long delay between detection of infection and initiation of treatment were found to be predictors of treatment failure.[49] Thus, continuous follow-up screening for the presence and level of AdV DNA by quantitative PCR assays after transplantation is highly important for early diagnosis of invasive AdV disease, the initiation of therapeutic strategies, and prognosis of outcome.

PARVOVIRUS B19

Infection with parvovirus B19 is common and in the immunocompetent host it presents mostly as an asymptomatic infection or leads to the development of erythema infectiosum, sometimes complicated, in mostly female adults, by polyarthropathy (for overview, see Young and Brown[103]). Diagnosis of parvovirus B19 infection is commonly based on the detection of virus-specific antibodies or on the use of molecular techniques for the detection of viral DNA.[104]

During an acute B19 infection, viral titers often exceed 10^{10} genome copies per ml for a short time. High viremia levels then drop rapidly in the course of infection and persist at low levels for months or even years. In the acute phase, less sensitive hybridization assays can detect B19 infection; diagnosis of recent infection within weeks or months requires the use of more sensitive PCR assays. Several different primers and probes with high sensitivity capable of detecting fewer than 10 genome copies per reaction have been described.[105,106]

One significant complication observed with B19 infection is possible damage to fetuses when pregnant women become infected. Parvovirus B19 produces a lytic infection of human erythroid precursor cells, interrupting normal red blood cell production. Red blood cell volume of a fetus expands rapidly and cell survival is short; thus B19 infection may lead to severe anemia, generalized edema, and fetal death.

Timely intrauterine blood transfusions may increase the chances of fetal survival. For this reason, sensitive and early diagnosis of acute B19 infection in pregnant women is of great importance and PCR assays are frequently applied. Using a quantitative TaqMan-PCR assay, Knoll et al recently showed that at time of detection of a B19 infection virus, DNA levels between 7.2×10^2 and 2.6×10^7 copies per ml were identified in mothers' blood.[105] Viral level was not, however, found to be predictive for severity of B19 infection in the fetus, but this may have been due to the fact that the exact infection time point could not be assessed. B19 DNA was also detected in amniotic fluid and fetal materials. In most fetal samples, B19 DNA concentrations were several orders of magnitude higher compared to the corresponding B19 DNA load in amniotic fluid and maternal blood.

As a further complication of B19 infections, transient aplastic crisis was observed in individuals with increased red blood cell turn-over. In an immunocompromised host, infection may become persistent and lead to chronic anemia. In these cases, diagnosis is based on PCR assays that permit rapid detection of B19 infection.

In addition to the detection of B19 for diagnostic purposes, identification of the virus in blood products is highly important for avoiding transmission via transfusion. Plasma pools and plasma-derived medicinal products frequently contain B19 DNA detectable by PCR.[107] However, it has become obvious in recent years that the level of virus DNA is significant in determining whether patients receiving B19-containing blood products will become infected. It was shown in B19 seronegative plasma pool recipients that only plasma containing more than 10^7 genome copies per ml leads to infection or seroconversion in recipients.[104] Thus, the establishment of exactly standardized quantitative PCR assays is of utmost importance to screen blood and blood products and reliably show that they contain less than the maximum levels of B19 DNA recently defined as acceptable for transfusion.[104]

EPILOGUE

The detection of nucleic acids, whether by signal or target amplification technologies, has exerted an enormous positive influence on patient management by allowing the detection of the above-mentioned blood-borne viruses. The benefits for clinical virology include the capability of treating patients with more antiviral therapies, rapid monitoring of the effects of treatment, and the ability to determine which patients may derive benefits from a particular treatment by characterizing the virus or by determining viral load.

The problems surrounding standardization of assays and the nomenclature of the results are hurdles that must be surmounted. There have been discussions over the last decade about standardization, mostly focused on the detection of a few blood-borne viruses, in particular HBV, HCV, and HIV-1, because the first generation of commercially available amplification assays for the quantification of these viruses exhibited serious problems in detecting the different viral genotypes with equal sensitivity.[108 109]

The discussion about and implementation of the IU system to improve standardization certainly contributed to advancements in blood safety and screening of blood products.[110–112] In clinical virology, absolute quantitation levels and changes in viral load are highly relevant parameters, as numerous publications have shown in recent years. In blood screening, the focus centers more on standardizing lower limits of detection. Efforts therefore must be undertaken to introduce more standardization in more areas of clinical virology — beyond the three blood-borne viruses cited above. It is very important that quality control programs also focus their attention on all blood-borne viruses, similar to the programs of Quality Control of Molecular Diagnostics (QCMD, see www.qcmd.org), to enable accurate detection of all blood-borne viruses.

REFERENCES

1. Janssen HL, van ZM, Senturk H, Zeuzem S, Akarca US, Cakaloglu Y et al. Pegylated interferon alfa-2b alone or in combination with lamivudine for HBeAg-positive chronic hepatitis B: a randomised trial. *Lancet* 2005; 365 (9454): 123–129.
2. Kao JH, Chen PJ, Lai MY, Chen DS. Genotypes and clinical phenotypes of hepatitis B virus in patients with chronic hepatitis B virus infection. *J Clin Microbiol* 2002; 40 (4): 1207–1209.
3. Lindh M, Horal P, Dhillon AP, Norkrans G. Hepatitis B virus DNA levels, precore mutations, genotypes and histological activity in chronic hepatitis B. *J Viral Hepat* 2000; 7 (4): 258–267.
4. Rodriguez-Frias F, Buti M, Jardi R, Cotrina M, Viladomiu L, Esteban R et al. Hepatitis B virus infection: precore mutants and its relation to viral genotypes and core mutations. *Hepatology* 1995; 22 (6): 1641–1647.
5. Buti M, Cotrina M, Valdes A, Jardi R, Rodriguez-Frias F, Esteban R. Is hepatitis B virus subtype testing useful in predicting virological response and resistance to lamivudine? *J Hepatol* 2002; 36 (3): 445–446.
6. Lai CL, Dienstag J, Schiff E, Leung NW, Atkins M, Hunt C et al. Prevalence and clinical correlates of YMDD variants during lamivudine therapy for patients with chronic hepatitis B. *Clin Infect Dis* 2003; 36 (6): 687–696.
7. Puchhammer-Stöckl E, Mandl CW, Kletzmayr J, Holzmann H, Hofmann A, Aberle SW et al. Monitoring the virus load can predict the emergence of drug-resistant hepatitis B virus strains in renal transplantation patients during lamivudine therapy. *J Infect Dis* 2000; 181 (6): 2063–2066.
8. Ide T, Kumashiro R, Koga Y, Tanaka E, Hino T, Hisamochi A et al. A real-time quantitative polymerase chain reaction method for hepatitis B virus in patients with chronic hepatitis B treated with lamivudine. *Am J Gastroenterol* 2003; 98 (9): 2048–2051.
9. Kessler HH, Preininger S, Stelzl E, Daghofer E, Santner BI, Marth E et al. Identification of different states of hepatitis B virus infection with a quantitative PCR assay. *Clin Diagn Lab Immunol* 2000; 7 (2): 298–300.
10. Lok AS, Hussain M, Cursano C, Margotti M, Gramenzi A, Grazi GL et al. Evolution of hepatitis B virus polymerase gene mutations in hepatitis Be antigen-negative patients receiving lamivudine therapy. *Hepatology* 2000; 32 (5): 1145–1153.
11. Torresi J, Earnest-Silveira L, Civitico G, Walters T, Lewin S, Fyfe J et al. Restoration of replication phenotype of lamivudine-resistant hepatitis B virus mutants by compensatory changes in the "fingers" subdomain of the viral polymerase selected as a consequence of mutations in the overlapping S gene. *Virology* 2002; 299 (1): 88–99.
12. Torresi J. The virological and clinical significance of mutations in the overlapping envelope and polymerase genes of hepatitis B virus. *J Clin Virol* 2002; 25 (2): 97–106.
13. Allen MI, Deslauriers M, Andrews CW, Tipples GA, Walters KA, Tyrrell DL et al. Identification and characterization of mutations in hepatitis B virus resistant to lamivudine. *Hepatology* 1998; 27 (6): 1670–1677.
14. Niesters HG, Honkoop P, Haagsma EB, de Man RA, Schalm SW, Osterhaus AD. Identification of more than one mutation in the hepatitis B virus polymerase gene arising during prolonged lamivudine treatment. *J Infect Dis* 1998; 177 (5): 1382–1385.
15. Locarnini S. Hepatitis B viral resistance: mechanisms and diagnosis. *J Hepatol* 2003; 39 S124–S132.
16. Bozdayi AM, Uzunalimoglu O, Turkyilmaz AR, Aslan N, Sezgin O, Sahin T et al. YSDD: a novel mutation in HBV DNA polymerase confers clinical resistance to lamivudine. *J Viral Hepat* 2003; 10 (4): 256–265.

17. Niesters HG, de Man RA, Pas SD, Fries E, Osterhaus AD. Identification of a new variant in the YMDD motif of the hepatitis B virus polymerase gene selected during lamivudine therapy. *J Med Microbiol* 2002; 51 (8): 695–699.

18. Tenney DJ, Levine SM, Rose RE, Walsh AW, Weinheimer SP, Discotto L et al. Clinical emergence of entecavir-resistant hepatitis B virus requires additional substitutions in virus already resistant to lamivudine. *Antimicrob Agents Chemother* 2004; 48 (9): 3498–3507.

19. Angus P, Vaughan R, Xiong S, Yang H, Delaney W, Gibbs C et al. Resistance to adefovir dipivoxil therapy associated with the selection of a novel mutation in the HBV polymerase. *Gastroenterology* 2003; 125 (2): 292–297.

20. Simmonds P. Genetic diversity and evolution of hepatitis C virus — 15 years on. *J Gen Virol* 2004; 85 (Pt 11): 3173–3188.

21. Giannini C, Thiers V, Nousbaum JB, Stuyver L, Maertens G, Brechot C. Comparative analysis of two assays for genotyping hepatitis C virus based on genotype-specific primers or probes. *J Hepatol* 1995; 23 (3): 246–253.

22. Di Bisceglie AM, Martin P, Kassianides C, Lisker-Melman M, Murray L, Waggoner J et al. Recombinant interferon-alpha therapy for chronic hepatitis C: a randomized, double-blind, placebo-controlled trial. *New Engl J Med* 1989; 321 (22): 1506–1510.

23. McHutchison JG, Gordon SC, Schiff ER, Shiffman ML, Lee WM, Rustgi VK et al. Interferon alfa-2b alone or in combination with ribavirin as initial treatment for chronic hepatitis C. *New Engl J Med* 1998; 339 (21): 1485–1492.

24. Poynard T, Marcellin P, Lee SS, Niederau C, Minuk GS, Ideo G et al. Randomised trial of interferon α2b plus ribavirin for 48 weeks or for 24 weeks versus interferon α2b plus placebo for 48 weeks for treatment of chronic infection with hepatitis C virus. *Lancet* 1998; 352 (9138): 1426–1432.

25. National Institutes of Health. Consensus statement on management of hepatitis C. *NIH Consens State Sci* 2002; 19 (3): 1–46.

26. EASL International. Consensus statement from conference on hepatitis C, Paris, February 1999. *J Hepatol* 1999; 31 Suppl 1: 3–8.

27. Sarrazin C. Highly sensitive hepatitis C virus RNA detection methods: molecular backgrounds and clinical significance. *J Clin Virol* 2002; 25: S23–S29.

28. Yarchoan R, Klecker RW, Weinhold KJ, Markham PD, Lyerly HK, Durack DT et al. Administration of 3'-azido-3'-deoxythymidine, an inhibitor of HTLV-III/LAV replication, to patients with AIDS or AIDS-related complex. *Lancet* 1986; 1 (8481): 575–580.

29. Holodniy M, Katzenstein DA, Sengupta S, Wang AM, Casipit C, Schwartz DH et al. Detection and quantification of human immunodeficiency virus RNA in patient serum by use of the polymerase chain reaction. *J Infect Dis* 1991; 163 (4): 862–866.

30. Holodniy M, Katzenstein DA, Israelski DM, Merigan TC. Reduction in plasma human immunodeficiency virus ribonucleic acid after dideoxynucleoside therapy as determined by the polymerase chain reaction. *J Clin Invest* 1991; 88 (5): 1755–1759.

31. Machuca A, Gutierrez M, Mur A, Soriano V. Quantitative p24 antigenaemia for monitoring response to antiretroviral therapy in HIV-1 group O-infected patients. *Antivir Ther* 1998; 3 (3): 187–189.

32. Markowitz M, Saag M, Powderly WG, Hurley AM, Hsu A, Valdes JM et al. A preliminary study of ritonavir, an inhibitor of HIV-1 protease, to treat HIV-1 infection. *New Engl J Med* 1995; 333 (23): 1534–1539.

33. Ho DD, Neumann AU, Perelson AS, Chen W, Leonard JM, Markowitz M. Rapid turnover of plasma virions and CD4 lymphocytes in HIV-1 infection. *Nature* 1995; 373 (6510): 123–126.

34. Ioannidis JP, O'Brien TR, Goedert JJ. Evaluation of guidelines for initiation of highly active antiretroviral therapy in a longitudinal cohort of HIV-infected individuals. *AIDS* 1998; 12 (18): 2417–2423.

35. Miller V, Larder BA. Mutational patterns in the HIV genome and cross-resistance following nucleoside and nucleotide analogue drug exposure. *Antivir Ther* 2001; 6, Suppl 3: 25–44.

36. Wilson JW, Bean P, Robins T, Graziano F, Persing DH. Comparative evaluation of three human immunodeficiency virus genotyping systems: the HIV-GenotypR method, the HIV PRT GeneChip assay, and the HIV-1 RT line probe assay. *J Clin Microbiol* 2000; 38 (8): 3022–3028.

37. Tural C, Ruiz L, Holtzer C, Schapiro J, Viciana P, Gonzalez J et al. Clinical utility of HIV-1 genotyping and expert advice: the Havana trial. *AIDS* 2002; 16 (2): 209–218.

38. Panidou ET, Trikalinos TA, Ioannidis JP. Limited benefit of antiretroviral resistance testing in treatment-experienced patients: a meta-analysis. *AIDS* 2004; 18 (16): 2153–2161.

39. Grossman Z, Paxinos EE, Averbuch D, Maayan S, Parkin NT, Engelhard D et al. Mutation D30N is not preferentially selected by human immunodeficiency virus type 1 subtype C in the development of resistance to nelfinavir. *Antimicrob Agents Chemother* 2004; 48 (6): 2159–2165.

40. Ariyoshi K, Matsuda M, Miura H, Tateishi S, Yamada K, Sugiura W. Patterns of point mutations associated with antiretroviral drug treatment failure in CRF01_AE (subtype E) infection differ from subtype B infection. *J Acquir Immune Defic Syndr* 2003; 33 (3): 336–342.

41. Weinstein MC, Goldie SJ, Losina E, Cohen CJ, Baxter JD, Zhang H et al. Use of genotypic resistance testing to guide hiv therapy: clinical impact and cost-effectiveness. *Ann Intern Med* 2001; 134 (6): 440–450.

42. Bezemer D, Jurriaans S, Prins M, van Der HL, Prins JM, de WF et al. Declining trend in transmission of drug-resistant HIV-1 in Amsterdam. *AIDS* 2004; 18 (11): 1571–1577.

43. Little SJ, Holte S, Routy JP, Daar ES, Markowitz M, Collier AC et al. Antiretroviral-drug resistance among patients recently infected with HIV. *New Engl J Med* 2002; 347 (6): 385–394.

44. Schim van der Loeff MF, Aaby P. Towards a better understanding of the epidemiology of HIV-2. *AIDS* 1999; 13: S69–S84.

45. Witvrouw M, Pannecouque C, Van LK, Desmyter J, De CE, Vandamme AM. Activity of non-nucleoside reverse transcriptase inhibitors against HIV-2 and SIV. *AIDS* 1999;13 (12): 1477–1483.

46. van der Ende ME, Guillon C, Boers PH, Ly TD, Gruters RA, Osterhaus AD et al. Antiviral resistance of biologic HIV-2 clones obtained from individuals on nucleoside reverse transcriptase inhibitor therapy. *J Acquir Immune Defic Syndr* 2000; 25 (1): 11–18.

47. Damond F, Brun-Vezinet F, Matheron S, Peytavin G, Campa P, Pueyo S et al. Polymorphism of the human immunodeficiency virus type 2 (HIV-2) protease gene and selection of drug resistance mutations in HIV-2-infected patients treated with protease inhibitors. *J Clin Microbiol* 2005; 43 (1): 484–487.

48. Colson P, Henry M, Tivoli N, Gallais H, Gastaut JA, Moreau J et al. Polymorphism and drug-selected mutations in the reverse transcriptase gene of HIV-2 from patients living in southeastern France. *J Med Virol* 2005; 75 (3): 381–390.

49. Bordigoni P, Carret AS, Venard V, Witz F, Le FA. Treatment of adenovirus infections in patients undergoing allogeneic hematopoietic stem cell transplantation. *Clin Infect Dis* 2001; 32 (9): 1290–1297.

50. van Esser JW, Niesters HG, van der HB, Meijer E, Osterhaus AD, Gratama JW et al. Prevention of Epstein–Barr virus-lymphoproliferative disease by molecular monitoring and preemptive rituximab in high-risk patients after allogeneic stem cell transplantation. *Blood* 2002; 99 (12): 4364–4369.

51. Haque T, Wilkie GM, Taylor C, Amlot PL, Murad P, Iley A et al. Treatment of Epstein–Barr virus-positive post-transplantation lymphoproliferative disease with partly HLA-matched allogeneic cytotoxic T cells. *Lancet* 2002; 360 (9331): 436–442.

52. Kimura H, Morita M, Yabuta Y, Kuzushima K, Kato K, Kojima S et al. Quantitative analysis of Epstein–Barr virus load by using a real-time PCR assay. *J Clin Microbiol* 1999; 37 (1): 132–136.

53. Limaye AP, Huang ML, Atienza EE, Ferrenberg JM, Corey L. Detection of Epstein–Barr virus DNA in sera from transplant recipients with lymphoproliferative disorders. *J Clin Microbiol* 1999; 37 (4): 1113–1116.

54. Niesters HG, van Esser J, Fries E, Wolthers KC, Cornelissen J, Osterhaus AD. Development of a real-time quantitative assay for detection of Epstein–Barr virus. *J Clin Microbiol* 2000; 38 (2): 712–715.

55. Stevens SJ, Pronk I, Middeldorp JM. Toward standardization of Epstein–Barr virus DNA load monitoring: unfractionated whole blood as preferred clinical specimen. *J Clin Microbiol* 2001; 39 (4): 1211–1216.

56. Wagner HJ, Wessel M, Jabs W, Smets F, Fischer L, Offner G et al. Patients at risk for development of post-transplant lymphoproliferative disorder: plasma versus peripheral blood mononuclear cells as material for quantification of Epstein–Barr viral load by using real-time quantitative polymerase chain reaction. *Transplantation* 2001; 72 (6): 1012–1019

57. Wagner HJ, Fischer L, Jabs WJ, Holbe M, Pethig K, Bucsky P. Longitudinal analysis of Epstein–Barr viral load in plasma and peripheral blood mononuclear cells of transplanted patients by real-time polymerase chain reaction. *Transplantation* 2002; 74 (5): 656–664.

58. Scheenstra R, Verschuuren EA, de HA, Slooff MJ, The TH, Bijleveld CM et al. The value of prospective monitoring of Epstein–Barr virus DNA in blood samples of pediatric liver transplant recipients. *Transpl Infect Dis* 2004; 6 (1): 15–22.

59. Verschuuren EA, Stevens S, Hanekamp BB, de BC, Harmsen MC, Middeldorp JM et al. Patients at risk for post-transplant lymphoproliferative disease can be identified in the first months after lung transplantation by quantitative-competitive-EBV PCR. *J Heart Lung Transplant* 2001; 20 (2): 199.

60. Stevens SJ, Verschuuren EA, Pronk I, van der BW, Harmsen MC, The TH et al. Frequent monitoring of Epstein–Barr virus DNA load in unfractionated whole blood is essential for early detection of post-transplant lymphoproliferative disease in high-risk patients. *Blood* 2001; 97 (5): 1165–1171.

61. Wagner HJ, Rooney CM, Heslop HE. Diagnosis and treatment of post-transplantation lymphoproliferative disease after hematopoietic stem cell transplantation. *Biol Blood Marrow Transplant* 2002; 8 (1): 1–8.

62. van Esser JW, van der HB, Meijer E, Niesters HG, Trenschel R, Thijsen SF et al. Epstein–Barr virus (EBV) reactivation is a frequent event after allogeneic stem cell transplantation (SCT) and quantitatively predicts EBV-lymphoproliferative disease following T cell-depleted SCT. *Blood* 2001; 98 (4): 972–978.

63. Lucas KG, Filo F, Heilman DK, Lee CH, Emanuel DJ. Semiquantitative Epstein–Barr virus polymerase chain reaction analysis of peripheral blood from organ transplant patients and risk for the development of lymphoproliferative disease. *Blood* 1998; 92 (10): 3977–3978.

64. Patel R, Paya CV. Infections in solid-organ transplant recipients. *Clin Microbiol Rev* 1997; 10 (1): 86–124.

65. van der Bij W, Schirm J, Torensma R, van Son WJ, Tegzess AM, The TH. Comparison between viremia and antigenemia for detection of cytomegalovirus in blood. *J Clin Microbiol* 1988; 26 (12): 2531–2535.

66. Boeckh M, Huang M, Ferrenberg J, Stevens-Ayers T, Stensland L, Nichols WG et al. Optimization of quantitative detection of cytomegalovirus DNA in plasma by real-time PCR. *J Clin Microbiol* 2004; 42 (3): 1142–1148.

67. Lilleri D, Baldanti F, Gatti M, Rovida F, Dossena L, De GS et al. Clinically based determination of safe DNAemia cutoff levels for preemptive therapy or human cytomegalovirus infections in solid organ and hematopoietic stem cell transplant recipients. *J Med Virol* 2004; 73 (3): 412–418.

68. Meyer-Koenig U, Weidmann M, Kirste G, Hufert FT. Cytomegalovirus infection in organ-transplant recipients: diagnostic value of pp65 antigen test, qualitative polymerase chain reaction (PCR) and quantitative Taqman PCR. *Transplantation* 2004; 77 (11): 1692–1698.

69. Razonable RR, Brown RA, Wilson J, Groettum C, Kremers W, Espy M et al. The clinical use of various blood compartments for cytomegalovirus (CMV) DNA quantitation in transplant recipients with CMV disease. *Transplantation* 2002; 73 (6): 968–973.

70. Kalpoe JS, Kroes AC, de Jong MD, Schinkel J, de Brouwer CS, Beersma MF et al. Validation of clinical application of cytomegalovirus plasma DNA load measurement and definition of treatment criteria by analysis of correlation to antigen detection. *J Clin Microbiol* 2004; 42 (4): 1498–1504.

71. Li H, Dummer JS, Estes WR, Meng S, Wright PF, Tang YW. Measurement of human cytomegalovirus loads by quantitative real-time PCR for monitoring clinical intervention in transplant recipients. *J Clin Microbiol* 2003; 41 (1): 187–191.

72. Gerna G, Baldanti F, Middeldorp J, Lilleri D. Use of CMV transcripts for monitoring of CMV infections in transplant recipients. *Int J Antimicrob Agents* 2000; 16 (4): 455–460.

73. Emery VC, Sabin CA, Cope AV, Gor D, Hassan-Walker AF, Griffiths PD. Application of viral-load kinetics to identify patients who develop cytomegalovirus disease after transplantation. *Lancet* 2000; 355 (9220): 2032–2036.

74. Humar A, Kumar D, Boivin G, Caliendo AM. Cytomegalovirus (CMV) virus load kinetics to predict recurrent disease in solid-organ transplant patients with CMV disease. *J Infect Dis* 2002; 186 (6): 829–833.

75. Ibrahim A, Gautier E, Roittmann S, Bourhis JH, Fajac A, Charnoz I et al. Should cytomegalovirus be tested for in both blood and bronchoalveolar lavage fluid of patients at a high risk of CMV pneumonia after bone marrow transplantation? *Br J Haematol* 1997; 98 (1): 222–227.

76. Zedtwitz-Liebenstein K, Jaksch P, Bauer C, Popow T, Klepetko W, Hofmann H et al. Association of cytomegalovirus DNA concentration in epithelial lining fluid and symptomatic cytomegalovirus infection in lung transplant recipients. *Transplantation* 2004; 77 (12): 1897–1899.

77. Aberle SW, Puchhammer-Stockl E. Diagnosis of herpes virus infections of the central nervous system. *J Clin Virol* 2002; 25: S79–S85.

78. Cinque P, Vago L, Brytting M, Castagna A, Accordini A, Sundqvist VA et al. Cytomegalovirus infection of the central nervous system in patients with AIDS: diagnosis by DNA amplification from cerebrospinal fluid. *J Infect Dis* 1992; 166 (6): 1408–1411.

79. Yoshida A, Hitomi S, Fukui T, Endo H, Morisawa Y, Kazuyama Y et al. Diagnosis and monitoring of human cytomegalovirus diseases in patients with human immunodeficiency virus infection by use of a real-time PCR assay. *Clin Infect Dis* 2001; 33 (10): 1756–1761.

80. Erice A, Tierney C, Hirsch M, Caliendo AM, Weinberg A, Kendall MA et al. Cytomegalovirus (CMV) and human immunodeficiency virus (HIV) burden, CMV end-organ disease, and survival in subjects with advanced HIV infection (AIDS Clinical Trials Group Protocol 360). *Clin Infect Dis* 2003; 37 (4): 567–578.

81. Gouarin S, Gault E, Vabret A, Cointe D, Rozenberg F, Grangeot-Keros L et al. Real-time PCR quantification of human cytomegalovirus DNA in amniotic fluid samples from mothers with primary infection. *J Clin Microbiol* 2002; 40 (5): 1767–1772.

82. Demmler GJ, Buffone GJ, Schimbor CM, May RA. Detection of cytomegalovirus in urine from newborns by using polymerase chain reaction DNA amplification. *J Infect Dis* 1988; 158 (6): 1177–1184.

83. Emery VC. Progress in understanding cytomegalovirus drug resistance. *J Clin Virol* 2001; 21 (3): 223–228.

84. Chou S, Erice A, Jordan MC, Vercellotti GM, Michels KR, Talarico CL et al. Analysis of the UL97 phosphotransferase coding sequence in clinical cytomegalovirus isolates and identification of mutations conferring ganciclovir resistance. *J Infect Dis* 1995; 171 (3): 576–583.

85. Bowen EF, Johnson MA, Griffiths PD, Emery VC. Development of a point mutation assay for the detection of human cytomegalovirus UL97 mutations associated with ganciclovir resistance. *J Virol Methods* 1997; 68 (2): 225–234.

86. Emery VC, Griffiths PD. Prediction of cytomegalovirus load and resistance patterns after antiviral chemotherapy. *Proc Natl Acad Sci USA* 2000; 97 (14): 8039–8044.

87. Benko M, Harrach B. Molecular evolution of adenoviruses. *Curr Top Microbiol Immunol* 2003; 272: 3–35.

88. Hierholzer JC. Adenoviruses in the immunocompromised host. *Clin Microbiol Rev* 1992; 5 (3): 262–274.

89. Schilham MW, Claas EC, van ZW, Heemskerk B, Vossen JM, Lankester AC et al. High levels of adenovirus DNA in serum correlate with fatal outcome of adenovirus infection in children after allogeneic stem-cell transplantation. *Clin Infect Dis* 2002; 35 (5): 526–532.

90. Shields AF, Hackman RC, Fife KH, Corey L, Meyers JD. Adenovirus infections in patients undergoing bone marrow transplantation. *New Engl J Med* 1985; 312 (9): 529–533.

91. Flomenberg P, Babbitt J, Drobyski WR, Ash RC, Carrigan DR, Sedmak GV et al. Increasing incidence of adenovirus disease in bone marrow transplant recipients. *J Infect Dis* 1994; 169 (4): 775–781.

92. Lion T, Baumgartinger R, Watzinger F, Matthes-Martin S, Suda M, Preuner S et al. Molecular monitoring of adenovirus in peripheral blood after allogeneic bone marrow transplantation permits early diagnosis of disseminated disease. *Blood* 2003; 102 (3): 1114–1120.

93. Echavarria M, Forman M, van Tol MJ, Vossen JM, Charache P, Kroes AC. Prediction of severe disseminated adenovirus infection by serum PCR. *Lancet* 2001; 358 (9279): 384–385.

94. La Rosa AM, Champlin RE, Mirza N, Gajewski J, Giralt S, Rolston KV et al. Adenovirus infections in adult recipients of blood and marrow transplants. *Clin Infect Dis* 2001; 32 (6): 871–876.

95. Heim A, Schumann J. Development and evaluation of a nucleic acid sequence based amplification (NASBA) protocol for the detection of enterovirus RNA in cerebrospinal fluid samples. *J Virol Methods* 2002; 103 (1): 101–107.

96. Kidd AH, Jonsson M, Garwicz D, Kajon AE, Wermenbol AG, Verweij MW et al. Rapid subgenus identification of human adenovirus isolates by a general PCR. *J Clin Microbiol* 1996; 34 (3): 622–627.

97. Blanke C, Clark C, Broun ER, Tricot G, Cunningham I, Cornetta K et al. Evolving pathogens in allogeneic bone marrow transplantation: increased fatal adenoviral infections. *Am J Med* 1995; 99 (3): 326–328.

98. Baldwin A, Kingman H, Darville M, Foot AB, Grier D, Cornish JM et al. Outcome and clinical course of 100 patients with adenovirus infection following bone marrow transplantation. *Bone Marrow Transplant* 2000; 26 (12): 1333–1338.

99. Lankester AC, van Tol MJ, Claas EC, Vossen JM, Kroes AC. Quantification of adenovirus DNA in plasma for management of infection in stem cell graft recipients. *Clin Infect Dis* 2002; 34 (6): 864–867.

100. Bruno B, Gooley T, Hackman RC, Davis C, Corey L, Boeckh M. Adenovirus infection in hematopoietic stem cell transplantation: effect of ganciclovir and impact on survival. *Biol Blood Marrow Transplant* 2003; 9 (5): 341–352.

101. Legrand F, Berrebi D, Houhou N, Freymuth F, Faye A, Duval M et al. Early diagnosis of adenovirus infection and treatment with cidofovir after bone marrow transplantation in children. *Bone Marrow Transplant* 2001; 27 (6): 621–626.

102. Leruez-Ville M, Minard V, Lacaille F, Buzyn A, Abachin E, Blanche S et al. Real-time blood plasma polymerase chain reaction for management of disseminated adenovirus infection. *Clin Infect Dis* 2004; 38 (1): 45–52.

103. Young NS, Brown KE. Parvovirus B19. *New Engl J Med* 2004; 350 (6): 586–597.

104. Brown KE. Detection and quantitation of parvovirus B19. *J Clin Virol* 2004; 31 (1): 1–4.

105. Knoll A, Louwen F, Kochanowski B, Plentz A, Stussel J, Beckenlehner K et al. Parvovirus B19 infection in pregnancy: quantitative viral DNA analysis using a kinetic fluorescence detection system (TaqMan PCR). *J Med Virol* 2002; 67 (2): 259–266.

106. Manaresi E, Gallinella G, Zuffi E, Bonvicini F, Zerbini M, Musiani M. Diagnosis and quantitative evaluation of parvovirus B19 infections by real-time PCR in the clinical laboratory. *J Med Virol* 2002; 67 (2): 275–281.

107. Koppelman MH, Cuypers HT, Emrich T, Zaaijer HL. Quantitative real-time detection of parvovirus B19 DNA in plasma. *Transfusion* 2004; 44 (1): 97–103.

108. Damen M, Cuypers HT, Zaaijer HL, Reesink HW, Schaasberg WP, Gerlich WH et al. International collaborative study on the second EUROHEP HCV-RNA reference panel. *J Virol Methods* 1996; 58 (1–2): 175–185.

109. Zaaijer HL, ter Borg F, Cuypers HT, Hermus MC, Lelie PN. Comparison of methods for detection of hepatitis B virus DNA. *J Clin Microbiol* 1994; 32 (9): 2088–2091.

110. Holmes H, Davis C, Heath A, Hewlett I, Lelie N. An international collaborative study to establish the first international standard for HIV-1 RNA for use in nucleic acid-based techniques. *J Virol Methods* 2001; 92 (2): 141–150.

111. Saldanha J, Lelie N, Heath A. Establishment of the first international standard for nucleic acid amplification technology (NAT) assays for HCV RNA. *Vox Sang* 1999; 76 (3): 149–158.

112. Saldanha J, Gerlich W, Lelie N, Dawson P, Heermann K, Heath A. An international collaborative study to establish a World Health Organization international standard for hepatitis B virus DNA nucleic acid amplification techniques. *Vox Sang* 2001; 80 (1): 63–71.

12 Molecular Methods for the Diagnosis of Fungal Infections

Adam M. Bressler and Christine J. Morrison

CONTENTS

INTRODUCTION

Invasive mycoses are major causes of infectious morbidity and mortality, particularly in immuno-compromised or debilitated hosts. The rapid progression of disease in such susceptible populations makes an early and accurate diagnosis essential for optimum patient management and therapy. However, the signs and symptoms of fungal diseases are often nonspecific, making accurate clinical diagnosis difficult. Recent advances in computerized imaging technology have served to improve the diagnosis of many fungal infections but the utility of even such sophisticated methods still depends on the presence of often rare, pathognomonic disease characteristics.

A clinical diagnosis of most fungal infections must therefore rely on laboratory confirmation. Unfortunately, laboratory methods such as culture and direct microscopy that can confirm the diagnosis of many fungal diseases are insensitive. Positive culture and microscopy results can also be difficult to interpret in cases where the infecting microorganism is also a common human commensal (e.g., *Candida* species) or is ubiquitous in the environment (e.g., *Aspergillus* species). The detection of such organisms, particularly in specimens derived from nonsterile sites, may result from sample contamination rather than from infection. In contrast, the recovery from clinical materials of any of the endemic mycoses (*Histoplasma capsulatum, Blastomyces dermatitidis, Coccidioides immitis* and *posadasii, Paracoccidioides brasiliensis*, and *Penicillium marneffei*), that are not human commensals, is strong evidence for infection. Unfortunately, if microscopic examination of clinical materials is negative, and culture confirmation is attempted, a diagnosis may be delayed for several weeks.

Due to the limitations of culture and direct microscopy for the diagnosis of many fungal diseases, histopathologic methods have been employed as diagnostic adjuncts. Evidence of tissue invasion provides the most convincing evidence of infection. However, methods for obtaining biopsy tissues from immunosuppressed or severely debilitated patients are not without risk. In addition, obtaining biopsy tissue from immunocompetent patients, in whom infection with one of the endemic mycoses is suspected, may be excessively invasive because such infections may resolve spontaneously in this patient population. Also, histopathologic diagnosis is not completely straightforward because many fungi share similar tissue morphologies and few specific immunohistochemical stains are available to differentiate among fungal genera or species, including those for which certain anti-fungal drug therapies may be ineffective.

Other diagnostic methods, such as the detection of antibodies, may be of limited value in immunocompromised patients, who often display variable humoral responses. On the other hand, in immunocompetent individuals, antibodies against commensal or colonizing fungi may occur naturally and their presence may not indicate infectious processes. Therefore, tests to detect circulating antigens rather than antibodies have been developed as alternative means for diagnosing fungal infections.

Antigens, unlike antibodies, can be detected in the absence of a functioning immune system. Indeed, antigen detection tests may perform better in immunocompromised than immunocompetent hosts because the disease may be more severe and affect several organ systems, thereby producing higher concentrations of antigens in body fluids. Unfortunately, most antigen detection tests developed to date suffer from low sensitivity, attributed primarily to the rapid clearance of target antigens from the circulation. In attempts to produce a rapid, sensitive, and specific test for the diagnosis of fungal infections, nucleic acid-based detection methods have been developed in recent years.

Many nucleic acid-based detection tests have distinct advantages compared to conventional diagnostic methods. For example, newer molecular diagnostic procedures may be completed in a single day.[1,2] Nucleic acid-based tests are not dependent upon the recovery of viable microorganisms from clinical specimens or on a functioning immune system in a host. Only small quantities of DNA or RNA are required initially because target nucleic acids can be multiplied many-fold by current amplification methodologies. The ultimate advantage of molecular biology-based detection methods is that nucleic acid targets can be detected and identified directly from clinical materials, obviating the need to grow infecting organisms in cultures.

Most molecular diagnostic methods described to date employ polymerase chain reaction (PCR) amplification of single or multiple gene targets and most of these methods used target sequences within the ribosomal RNA (rRNA) gene complex. This region of the fungal genome includes both highly conserved and highly variable regions, allowing the selection of either pan-fungal or species-specific amplification targets, respectively. This chapter will review methods for the extraction, purification, amplification, and detection of fungal nucleic acids from clinical materials. Experimental tests developed for the diagnosis of fungal diseases using nucleic acid amplification and detection from body fluids, tissues, and blood culture bottles inoculated with blood from patients with fungemia will be included. Clinical studies of the past 10 to 15 years examining nucleic acid-based diagnosis of invasive aspergillosis (IA) and systemic candidiasis (IC) will be reviewed. These studies are outlined in Tables 12.1 and 12.2. A limited number of clinical studies have also been conducted to evaluate the use of molecular tests for the diagnosis of fungal diseases other than IA or IC, and these are outlined in Table 12.3.

Experimental nucleic acid-based diagnostic tests with the potential for rapid and highly specific identification of invasive fungal diseases have been described by research laboratories worldwide. Most studies to date have been limited to single research laboratories at individual institutions. Scarce data are available from multicenter studies aimed at determining overall test sensitivity and specificity, efficiency to predict clinical outcome, and value for monitoring antifungal drug therapy. In addition, commercial development of nucleic acid-based tests for the diagnosis of mycotic diseases has been limited to tests for culture confirmation[3-6] or the diagnosis of vulvovaginal candidiasis.[7]

Although a fluorescent *in situ* hybridization assay has been developed to microscopically identify a single species of *Candida* (*C. albicans*) in smears taken from positive blood culture bottles,[8,9] no system has yet been commercialized to detect fungal nucleic acids directly from tissues, blood, or other body fluids. One prototype system for the detection of *C. albicans* cells directly from whole blood is in commercial development. In this system, extraction of fungal DNA from whole blood is facilitated by a commercial kit (MagNA Pure LC DNA Purification Kit, Roche Diagnostics, Mannheim, Germany) used in conjunction with an automated extraction system (MagNA Pure LC System, Roche). Additional manual steps, including pretreatment with lytic enzymes, are required for maximum disruption of fungal cells before placement of samples into the automated extraction system.[10] Extracted samples are then transferred to an amplification and detection system (LightCycler, Roche).[2,10] Another kit, the LightCycler *Candida* (Roche), facilitates amplicon detection during real-time PCR amplification of fungal DNA. Commercialization of a completely closed system, from sample processing to amplification and detection, with no requirement for additional manual steps, would make nucleic acid-based diagnostic tests for fungal diseases more attractive for routine use in clinical laboratories.

Several obstacles must be overcome to optimize the extraction of fungal nucleic acids from clinical materials. While extraction of DNA or RNA from pure cultures of organisms is relatively straightforward, isolation and purification of nucleic acids from clinical materials is more problematic, in part because clinical materials often contain inhibitors of the PCR assay. Removal of host DNA and proteins can improve test sensitivity and, if whole blood is used, lysis of erythrocytes and leukocytes and removal of hemoglobin are essential.

Some researchers have used Percoll (Amersham Biosciences, Piscataway, New Jersey, USA) or Ficoll (Fluka Chemical Corp., Milwaukee, Wisconsin, USA) to first separate blood or body fluids into various fractions to facilitate recovery of fungal nucleic acids free from inhibitors.[11,12] Recovery of sufficient amounts of good quality fungal nucleic acid from clinical materials can also be challenging. Unlike pure cultures, clinical specimens often contain very low numbers of organisms.

For example, in one study, whole blood was collected into lysis centrifugation tubes and plated on a variety of growth media to determine the number of *Candida* cells initially present in each milliliter of whole blood. Of all positive blood cultures that grew *Candida* species, 92% contained ≤100 colony-forming units (CFU) per ml and 77% contained ≤10 CFU per ml[13]. An analytical sensitivity of 1 to 10 CFU per ml would therefore be required to maximize the usefulness of nucleic acid-based tests to detect *Candida* species directly from whole blood. On the other hand, the increased analytical sensitivity of PCR-based tests resulting from a million-fold or more amplification of the nucleic acid target makes such tests vulnerable to false positive results. Interpretation of PCR-based test results must therefore carefully consider the clinical presentation of the patient as well as the site from which the specimen was derived. Body fluids and tissues taken from sterile sites are preferred specimens for PCR testing so as to reduce the likelihood of false positive results.

EXTRACTION OF FUNGAL NUCLEIC ACID

There are perhaps as many different methods to extract fungal nucleic acids from clinical materials as there are laboratories conducting nucleic acid-based diagnostic testing (see Table 12.1 through Table 12.3). However, some basic commonalities exist and most are relevant for the isolation of both RNA and DNA from clinical materials. These include the necessity to first disrupt host cells and fungal cells to release intracellular nucleic acids, to extract and purify fungal nucleic acids free from inhibitors of the PCR, to amplify an appropriate nucleic acid target, and then to detect and/or quantify the amplicons produced.

Disruption of host cells can be accomplished primarily by hypotonic lysis with or without detergent. However, disruption of fungal cells is more complex. Fungal cell walls may contain chitin, mannan, and glucan — components that can make fungi resistant to breakage. In general, three methods for cell disruption from clinical materials have been described: enzymatic, mechanical, and chemical.

Enzymatic disruption generally uses zymolyase[11,14,15] or lyticase[12,16–19] to degrade chitin and glucan in fungal cell walls. The spheroplasts produced can then be osmotically, chemically, or mechanically disrupted. Because the cell walls of filamentous fungi are more resistant to enzymatic breakage compared to their yeast counterparts, most researchers use detergents or an alkalinization step to increase the DNA yield.[15,19] Various combinations of detergents, proteinase K, RNase (for DNA targets), DNase (for RNA targets), and boiling steps usually follow enzymatic digestion of the cell walls with zymolyase or lyticase.[11,14,15,17]

A simplified extraction method for the recovery of *Candida* species DNA from whole blood was described by Flauhaut et al.[20] who used a special lysis cocktail (1 N NaOH, 0.2 M sodium citrate, and 0.4 M *N*-acetylcysteine) to remove contaminating erythrocytes before applying the commercial QIAamp Tissue Kit (Qiagen AG, Basel, Switzerland; now the QIAamp DNA Mini Kit) to extract fungal DNA. Lysis buffer containing a chaotropic chemical lysis reagent and proteinase K were used to release fungal DNA. The DNA released was then purified using a silica-based spin column.[20] The QIAamp Tissue Kit and treatment of specimens with proteinase K alone performed equally as well as the kit combined with sequential treatment of samples with zymolyase and proteinase K.[20] This system, compared to other published nucleic acid extraction methods, had a number of advantages: it was faster, more reproducible, and easier to perform. The two major disadvantages were the need for a multistep manual procedure and the costly reagents employed.

Commercial development of DNA extraction kits has greatly simplified sample preparation methods from clinical materials. However, no kit has been accepted as the gold standard. Kits or reagents used for the recovery of fungal DNA include the High Pure PCR Template Preparation Kit (Roche), QIAamp DNA Mini Kit (Qiagen), the Masterpure DNA Purification Kit (Epicentre Technologies, Madison, Wisconsin, USA), the DNA-Pure Yeast Genomic Kit (CPG Inc., Lincoln Park, New Jersey, USA), the PUREGENE Yeast and Gram-Positive Bacteria DNA Isolation Kit (Gentra Systems, Inc., Minneapolis, Minnesota, USA), Dynabeads (Dynal Biotech ASA, Oslo, Norway), DNAzol (Molecular Research Center, Inc., Cincinnati, Ohio, USA), GeneReleaser (Bio-Ventures, Inc., Murfreesboro, Tennessee, USA), and the MagNA Pure LC DNA Purification Kit (Roche) to name only a few.

Despite the variety of extraction and purification reagents now available, efficiency and cost vary considerably. For example, Löffler et al.[21] compared five commercially available DNA extraction kits with an in-house method for efficiency to recover DNA from *C. albicans* and *Aspergillus niger* from whole blood. The QIAamp Tissue Kit, although more expensive than the other methods ($2.30 per sample compared to the least expensive commercial kit, $0.80 per sample, and to the in-house method, $0.10 per sample), facilitated a more analytically sensitive detection (1 to 10 CFU/ml versus 1000 CFU/ml) and could be completed in half the time required to perform the in-house method (4 hr versus 8 hr).

In another study, a noncommercial disruption and extraction method using lyticase, glass bead beating, and phenol was compared to four commercial kits for the recovery and amplification of *Aspergillus fumigatus* DNA from bronchoalveolar lavage (BAL) fluid and biopsied tissues.[22] The High Pure PCR Template Preparation Kit produced DNA that allowed a limit of sensitivity similar to the lyticase method (10 CFU/reaction), whereas the QIAamp DNA Mini Kit, Masterpure DNA Purification Kit, and DNA-Pure Yeast Genomic Kit were all one to two logs less sensitive than either of the former methods. Furthermore, the addition of a bead-beating step (beating samples with glass beads for 1 min before phenol extraction) improved the sensitivity of the lyticase method by an additional log but did not improve the sensitivity of the High Pure Kit method.

Other researchers employed chemical disruption of cell walls by boiling clinical specimens in the presence of guanidium thiocyanate, phenol, and Tris buffer.[23] This method was effective for the recovery of DNA from a variety of fungal genera (*A. fumigatus, C. albicans, C. immitis,* and *Cryptococcus neoformans*) when extracted from a number of different clinical materials (sputum, pleural and BAL fluids, cerebrospinal fluid, and tissues).[23] Finally, mechanical and chemical disruption methods that use combinations of glass beads and detergent, with or without boiling and with or without the addition of chaotropic agents, have been used to isolate fungal DNA.[24–27]

The isolation of DNA from filamentous molds has been successfully achieved using a commercial cell disruptor (FastPrep, Qbiogene, Carlsbad, California) and various chemical buffers. The FastPrep cell disruptor is essentially a bead-beater apparatus that contains holders for multiple tubes. Various combinations of beads of different shapes and compositions can be purchased (Qbiogene) and can be added to the test samples along with a specific lytic cocktail tailored to maximize the disruption of the fungus of interest.[28] The major advantage of this system is its capacity to mechanically disrupt several different samples simultaneously and to do so in a completely enclosed system so as to reduce the risks of aerosolization and environmental contamination.

PURIFICATION OF FUNGAL NUCLEIC ACIDS

Purification of fungal nucleic acids prior to PCR amplification has traditionally been accomplished using phenol–chloroform extraction and alcohol precipitation.[11,14,16,17,23,24] The development of commercial spin columns for DNA purification greatly simplified this process and eliminated the need for toxic chemicals.[2,20,21] An automated DNA extraction system developed by Roche (MagNA Pure LC)[10,29] has recently become commercially available. Two manual steps, including hypotonic lysis of erythrocytes and disruption of fungal cells by vortex mixing with glass beads, were first employed. These steps were then followed by application of the automated system incorporating a proprietary mixture of reagents and magnetic beads to extract and purify nucleic acids.[10,29]

The *in vitro* sensitivity of this method was 1 CFU/sample — a result comparable to results obtained when an in-house lyticase-plus-spin column purification method was employed.[10] Additional advantages of this system were increased rapidity, less hands-on time, greater reproducibility, and reduced cross-contamination compared to the in-house method.

Although traditional in-house methods have been shown to provide the greatest DNA yields at the lowest cost, these benefits must be balanced by practicality. Methods that may be acceptable for research purposes or for retrospective analyses may be difficult to incorporate into general clinical practice. Therefore, the application of complex enzymatic disruption methods, particularly in conjunction with phenol–chloroform extraction procedures, may not be feasible in a clinical laboratory setting.

AMPLIFICATION OF FUNGAL NUCLEIC ACIDS

Once nucleic acids have been extracted and purified, selection of an appropriate amplification target follows. The majority of publications to date have described methods to amplify and detect fungal DNA rather than RNA, although good success was reported for the detection of *Candida* and *Aspergillus* species RNA using a method termed nucleic acid sequence-based amplification (NASBA).[30–32] This method employs a series of isothermal amplifications of RNA using a T7 RNA polymerase. Amplification products are then detected by agarose gel electrophoresis or hybridization with specific probes.

The limit of test analytical sensitivity using hybridization probes directed to *18S* rRNA amplification products from *A. fumigatus* was reported to be 1 CFU (10-fold more sensitive than an established quantitative PCR method for DNA detection), although some cross-reactivity was reported with RNA from species of *Penicillium* and with *Alternaria alternata*[30]; the assay could be completed within 6 hr. Potential advantages to the use of RNA compared to DNA include isothermal amplification conditions, increased test sensitivity attributable to high transcript number, and the potential to determine cell viability.

In general, however, most nucleic acid-based diagnostic tests for mycotic diseases have been directed to the detection of DNA rather than RNA, perhaps to avoid the inherent problems associated with the generally more labile nature of RNA. In instances where DNA has been used as the amplification target, test sensitivity and specificity have been reported to vary according to the amplification target chosen. Several researchers used single or low copy number gene targets for PCR amplification that were very specific for a particular species or genus. Examples of such targets include genes for lanosterol 14α-demethylase,[13,33–35] actin,[12,24] heat-shock protein 90,[16] aspartic proteinase,[20] aspergillopepsin,[36] alkaline protease,[37,38] an 18-kDa immunoglobulin E-binding protein or ribotoxin,[39] and chitin synthetase.[17] Multicopy mitochodrial genes[40,41] were also used and found to be more sensitive targets than single copy genes.[42] However, mitochondrial genes may yield less reproducible results because they may vary in copy number.

A more widespread approach has been to use "universal" fungal or "pan-fungal" DNA primers and high copy number gene targets in an attempt to increase test sensitivity and broaden applicability.[15,18,25,26,43,44] For example, ribosomal RNA (rRNA) genes were reported to be present in 50 to 100 copies/fungal genome. Therefore, highly conserved regions of ribosomal DNA have been the most frequently employed amplification targets and included the *5.8S, 18S,* and *28S* rRNA genes as well as the more highly variable internal transcribed spacer (ITS) regions between these genes. Gene target regions that are conserved among all fungi but are not present in viruses, bacteria, or mammals are advantageous in that a PCR product can be obtained from all fungi using a single set of PCR primers and amplification conditions optimal for that single set of primers.

Although broad-range, pan-fungal primers offer the advantage of increased sensitivity to detect a wide variety of fungal targets, specificity may be reduced. For example, van Burik et al. developed a pan-fungal PCR assay using universal fungal primers and a long pan-fungal oligonucleotide probe.[19] The probe detected the majority of genera and species tested but could not differentiate DNA from species of *Candida* or *Aspergillus*. Because drug susceptibilities can differ among yeasts and molds and even among different species of *Candida*[45,46] and *Aspergillus*,[47,48] identification of the infecting species can be important for clinical management.

In addition, therapeutic treatment differs for aspergillosis, zygomycosis, scedosporiosis, and fusariosis, all of which may be clinically indistinguishable from one another and would not be differentiated by pan-fungal primers or probes. Thus, although amplicons produced using pan-fungal PCR primers can serve to screen for the presence of fungal DNA from many genera simultaneously in a given sample, probes specific for the most medically important fungal genera or species would need to be applied to correctly identify the infecting organism and guide clinical therapy.[49] In contrast, initial amplification with pan-fungal primers may be relevant in some cases, e.g., for the detection of rare or emerging fungal species for which no specific probe has yet been developed.

More commonly, however, researchers first used pan-fungal primers directed to the more conserved regions of the rRNA gene complex to amplify target DNA followed by specific detection of amplicons using genus- or species-specific probes directed to the more variable gene regions. For example, Einsele et al.[15] reported the use of pan-fungal primers and specific oligonucleotide probes to achieve a clinical sensitivity of 100% when two or more serial blood samples from patients with documented invasive fungal infections were tested.[15] Positive PCR results preceded radiological signs by a median of 4 days for 12 of 17 patients with either hepatosplenic candidiasis or pulmonary aspergillosis.

This method could also be used to monitor patient response to antifungal therapy as demonstrated by a decline in the number of PCR-positive samples obtained from patients responding to therapy.[15] Pan-fungal primers and species-specific probes were used by Shin et al.[25] to amplify *Candida* species DNA recovered directly from blood culture bottle fluid. A natural amplification of target DNA was allowed to occur as yeasts grew in blood culture bottles before aliquots were removed for DNA extraction. A very rapid mechanical disruption method that did not require expensive enzymes or phenol–chloroform extraction was then successfully employed.[25]

A simplified enzyme immunoassay format with *Candida* species-specific DNA probes directed to the internal transcribed spacer 2 (ITS2) region of the rRNA gene was used for the colorimetric detection of PCR amplicons. This method correctly identified *Candida* species from 150 blood culture bottles, including 73 positive blood culture bottle sets (aerobic and anaerobic) from 31 patients with candidemia. It also detected a case of mixed *C. albicans* and *C. glabrata* candidemia that was not detected by conventional culture methods because isolation plates were overgrown with coagulase-negative *Staphylococcus*. No false positive results were obtained using blood culture bottles from patients with bacteremias, from artificially produced non-*Candida* species fungemias, or from bottles without growth.

In a later study, Iwen et al.[50] used both the ITS1 and ITS2 internal transcribed spacer regions of the rRNA gene as PCR amplification targets and obtained similar results (100% sensitivity and 100% specificity compared to culture); *Candida* species were identified by DNA sequence analysis of the PCR products. These studies validate the usefulness of DNA recovered from blood culture bottles for the detection of candidemia caused by a number of different *Candida* species.[25,50]

Early studies described the application of a single set of primers and one round of PCR amplification to detect fungal DNA from clinical materials. In an attempt to increase sensitivity and/or specificity, nested PCR assays were later developed. These assays used an outer primer pair to produce the first set of amplicons and then a second set of primers internal to the first for a second round of PCR amplification. Two rounds of PCR amplification generally increased test sensitivity[1,33] by producing a larger number of amplification copies than would be produced using a single set of primers.

Ahmad et al.[51] used the same species-specific *Candida* probes developed earlier by Shin et al.[26] but used them to perform a semi-nested PCR assay. Universal fungal primers were used for the initial amplification round, and then the species-specific probe sequences of Shin et al. were used as internal primers for the nested PCR assay. Test clinical sensitivity was reported to be high (100%) using serum samples from 12 patients with positive *Candida* blood culture results; specificity was also high (100%) using serum from superficially colonized patients and from healthy subjects as controls. However, 9 of 16 (56%) patients with suspected disease and negative blood cultures for *Candida* species appeared positive in the nested PCR assay. These data indicate that either the nested PCR assay was more sensitive than culture for the detection of candidiasis or that the nested PCR assay was more prone to contamination and falsely positive results.

Increased susceptibility to contamination during sample manipulation and the added labor involved to perform nested PCR assays compared to standard PCR assays are disadvantages that limit the practical application of nested amplification assays in clinical laboratory settings. Studies from several laboratories have documented high levels of false positive results using nested PCR assays, including false positive results for negative control specimens obtained from patients with no evidence of fungal disease.[52–54]

Species-specific primers can be employed during the initial amplification step, obviating the need for a subsequent application of species-specific probes for identification. However, in such cases, a different primer pair is required for each fungal species to be identified. Therefore, the test sample must be divided into separate aliquots to be tested with each individual primer pair or, alternatively, a multiplex PCR system must be established. The disadvantage of a multiplex PCR system is that each primer pair must be optimized to function efficiently in the presence of the other primer pairs.

The use of multiple primers simultaneously is complex because optimal amplification conditions may differ among primers based on sequence variations. For example, melting temperatures and conformational structures can be strikingly different. In addition, in a multiplex system, primers must spatially compete for their specific target in a milieu of potentially interfering primers. Therefore, few multiplex PCR systems have been described to date for the detection of fungal DNA targets in clinical materials, and most have been limited to the identification of a small number of species.[55–58]

DETECTION OF AMPLIFIED FUNGAL DNA

Amplified DNA can be detected in a number of ways. The earliest method of detecting PCR amplicons used gel electrophoresis and ethidium bromide staining of the agarose gel, either with or without prior digestion of the amplicons with restriction enzymes to produce a recognizable digestion pattern.[14,16,34,44,52,59,60] Increased sensitivity was then achieved by Southern blotting and detection with radiolabeled probes,[16,33,43,61] and more recently with nonradioactive colorimetric probes.[15,19]

Identification of various species has also been achieved by comparing the different sizes of PCR amplicons produced (amplification product length polymorphisms, APLPs).[60] Others have used line-probe, reverse cross blot, and slot blot hybridization assays to detect PCR products.[23,35,59,62] The application of an enzyme immunoassay format using either colorimetric or fluorometric dyes to detect amplicons is one of the most easily adapted detection methods for use in a clinical laboratory.[18,20,25,63–65] This detection format is simple to perform, does not require any hazardous chemicals, has a sensitivity equal to that of Southern blotting,[66,67] and has a specificity of 100%.[25,67]

Other methods are referred to as "real-time PCR" because detection of the amplicon occurs as the PCR product is produced and quantitative results can be displayed graphically during amplicon generation. No post-amplification manipulations of product are required and results are obtained in real time. The absence of post-amplification manipulation helps prevent contamination and thereby protects specificity. Furthermore, since amplification and detection occur simultaneously, work flow is more efficient and the time required for the assay is reduced. Finally, the quantitative nature of real-time amplification and detection systems should theoretically make them useful tools for determining disease burden and patient response to therapy. Although studies have confirmed this expectation in an experimental animal model of aspergillosis,[68] few studies have attempted such analyses in humans.[2,15,52]

One such real-time, quantitative PCR method is the TaqMan system (Applied Biosystems, Inc., Foster City, California, USA). This test is a 5'-exonuclease assay that uses hybridization of a probe labeled at one end with a fluorescent reporter dye and at the other end with a quencher dye. No signal is generated while the quencher dye remains in close proximity to the reporter dye. As DNA amplification occurs, however, the 5' → 3' exonuclease activity of the *Taq* DNA polymerase separates the quencher dye from the reporter dye and allows a signal to be generated; the signal is then detected fluorometrically in the same system.

This test format was used by Shin et al.[26] to identify *Candida* species in 61 culture bottles inoculated with blood from candidemia patients. Universal fungal primers were used to amplify the ITS2 rDNA regions from all *Candida* species, and species-specific probes to this region were labeled with one of three fluorescent reporter dyes. Each dye emitted a characteristic wave length allowing up to three *Candida* species to be detected in a single reaction tube.

Probes correctly detected and identified 58 (95.1%) of 61 DNAs recovered from blood culture bottles, including specimens culture-positive for *C. albicans, C. parapsilosis, C. glabrata, C. krusei,* and *C. tropicalis.* No false positive results were obtained from bottles with no growth or from patients with bacteremias. This assay could detect and identify *Candida* to the species level in less than a day. Guiver et al.[69] later obtained similar results using the TaqMan system and probes to identify *C. albicans, C. kefyr, C. parapsilosis, C. krusei,* and *C. glabrata.*

The TaqMan system has also been used to identify *A. fumigatus* DNA in blood specimens. For example, Costa et al.[42] used a mitochondrial gene target and TaqMan technology to amplify and detect *A. fumigatus* DNA in whole blood. Whole blood fractions (serum, plasma, and white cell pellet) were compared as specimens for the optimum recovery of fungal DNA. Whereas serum and the white cell pellet were comparable for the recovery of *A. fumigatus* DNA, the yield from plasma was ten times lower. It was also determined that serum should be frozen as soon as possible after collection to prevent degradation of fungal DNA.

Another real-time PCR system employed to detect fungal DNA is Roche's LightCycler System, which allows the rapid amplification of fungal DNA in glass capillaries and the simultaneous detection of amplicons using fluorescence resonance energy transfer (FRET).[2,9,68] One DNA probe is labeled at the 5' end with the LightCycler Red fluorophore, and another probe that binds adjacent to the first is labeled at the 3' end with fluorescein. The fluorescein is excited by the light source within the LightCycler instrument and the energy emitted by the excited fluorescein is then transferred to the LightCycler Red fluorophore. The light emitted by the fluorophore is then measured and is proportional to the amount of DNA amplified.

The sensitivity of the LightCycler assay[2,10] to detect both *C. albicans* and *A. fumigatus* was comparable to the level previously published by the same authors (5 CFU/ml of whole blood) using a PCR-enzyme-linked immunosorbent assay (ELISA) for amplicon detection.[18] Addition of a commercial DNA extraction and purification system (MagNA Pure LC DNA Purification Kit, Roche) applied after lysis and removal of erythrocytes and disruption of *C. albicans* cells with glass beads allowed the detection of as little as 1 CFU/ml of whole blood.[10]

Some researchers used single probe hybridization reactions,[2,10] and others employed multiple probes, each with different melting curve characteristics, to detect multiple species simultaneously in a single reaction.[70] The latter method required extensive probe design and optimization of melting conditions for maximum performance. Others researchers used multiple primers simultaneously in conventional PCR reactions, followed by size differentiation of amplicons by agarose gel electrophoresis or microchip electrophoresis.[58]

Hendolin et al. used pan-fungal PCR primers and a multiplex liquid hybridization cocktail containing several probes in a single reaction mix to detect multiple species simultaneously.[71] The LightCycler format can also be used for species identification without species-specific probe hybridization. This is accomplished by employing species-specific primers and a SYBR Green DNA-binding dye to detect double-stranded amplification products in real time. A given species can then be identified by its characteristic melting profile. For example, Bu *et al.*[56] reported the detection and identification of *Candida* DNA in blood from 6 of 6 hematological malignancy patients with culture-proven *C. albicans* bloodstream infection using this method. No false positive results were observed using specimens from 8 of 8 blood culture-negative patients.

White et al.[72] used SYBR Green and melting curve analysis with a multispecies, Cy5-labeled *Candida* probe to detect but not differentiate seven species of *Candida* (*C. albicans, C. dubliniensis, C. glabrata, C. kefyr, C. krusei, C. parapsilosis,* and *C. tropicalis*). Proven cases of invasive fungal disease were defined by histopathological evidence of infection and/or positive culture from blood or other sterile body site, and probable cases were defined by host factors plus computed tomography and/or radiological evidence with supporting microbiological signs. Using these definitions, test sensitivity to detect a proven or probable case was reported as 95% using whole blood from 105 patients at high risk for invasive fungal infections (primarily patients with hematological malignancies, n = 77). The PCR test was found to be more sensitive than commercial antigen detection enzyme immunoassays (75%) and latex agglutination assays (25%).[72]

Bressler et al.[74] used a combination of TaqMan and LightCycler technologies by employing TaqMan 5'-exonuclease chemistry and detection of PCR amplicons in a LightCycler instrument. Universal, fungus-specific primers were used to amplify the entire ITS2 rRNA gene region from medically important *Aspergillus* species. An oligonucleotide probe targeting an ITS2 region conserved among all the medically important *Aspergillus* species was labeled at the 5' end with 6-carboxy-fluorescein and at the 3' end with a quencher dye; the LightCycler instrument was then used to detect PCR amplicons in real time.

Tissues for PCR analysis were processed by mechanical bead beating (Fast-Prep System, Qbiogene) and chemical lysis followed by automated DNA extraction (MagNA Pure DNA Purification Kit, Roche). Real-time PCR analysis resulted in the detection of fewer than 2 CFU/g tissue. DNA extraction, amplification, and detection could be accomplished in a single day.

The evidence regarding the sensitivity of real-time PCR assays compared to conventional or nested PCR assays is conflicting. Some studies demonstrated that the *in vitro* sensitivities of real-time assays equaled those of traditional PCR methods.[29,75]

Comparisons of studies are difficult, however, given the wide variety of test conditions employed among laboratories. Nonetheless, an examination of the studies outlined in Table 12.1 indicates a trend toward lower sensitivity for real-time assays. For example, the median sensitivity for real-time PCR assays was 73% (range, 46 to 85%) and the median sensitivity for standard PCR assays was 87% (range, 50 to 100%). Nested PCR assays demonstrated the greatest sensitivity (median, 100%; range, 64 to 100%; see Table 12.1). Confounding factors include testing serum or plasma instead of whole blood, using smaller volumes of test specimens, and employing less rigorous extraction methods; all may contribute to the reduced clinical sensitivity observed for real-time assays (Table 12.1).

A direct comparison of a real-time, dual hybridization probe assay with a nested PCR assay was described by Spiess et al.[73] Although the *in vitro* limits of detection were similar for both tests (13 fg versus 10 fg), the real-time assay failed to detect 43% of blood samples determined to be positive by the nested PCR assay. The same authors conducted a prospective study examining both methods using samples from consecutive patients.[54] Only those samples that gave positive results in the nested PCR assay were tested in the real-time PCR assay. Only 21.4% of the samples tested positive in the nested PCR assay were also positive by the real-time PCR format. However, direct comparisons of the methods used in this study were confounded because a mitochondrial cyto-chrome B gene was used as the amplification target for the real-time PCR assay, whereas an *18S* rRNA gene that should have been present in higher copy numbers was used as the target for the nested PCR assay. Also, the real-time PCR assay was designed to detect DNA from only *A. fumigatus* and *A. clavatus*, whereas the nested PCR assay was pan-generic; thus, the nested assay may have detected *Aspergillus* species that would not have been detected by the real-time PCR test.

A very high number of false positive results was reported for the nested PCR assay, including positive results for 37% of patients with no evidence of aspergillosis. It is therefore possible that samples considered positive in the nested PCR assay were, in fact, falsely positive. An examination of all studies cited in Table 12.1 reveals that the mean percentage of false positive results obtained for nested, standard, and real-time PCR assays were, respectively, 17, 11, and 5%; inclusion of possible data as falsely positive would significantly increase the percentage of falsely positive results regardless of the amplification method used (Table 12.1). Based on the significant advantages inherent in real-time PCR assays, including reduced susceptibility to contamination, increased rapidity, capacity to quantitate PCR amplicon production, and the development of compatible commercial extraction kits and automated detection systems, future development will no doubt include attempts to improve the sensitivity of real-time PCR assays for the diagnosis of fungal diseases.

In contrast to studies regarding the detection of *Aspergillus* species DNA, real-time PCR assays to detect *Candida* species DNA generally demonstrated increased sensitivity compared to either standard or nested PCR assays (Table 12.2). The mean sensitivity for real-time PCR assays was 96% (range, 88 to 100%), whereas the mean sensitivity for standard and nested PCR assays was 86% (range, 79 to 100%) and 79% (range, 44 to 100%), respectively. More studies were conducted on the recovery of *Candida* species DNA from whole blood than from serum, which may have impacted overall test sensitivity results.

Also, for studies examining the recovery of *Candida* DNA from blood and serum, nested PCR assays did not produce exceptionally high rates of false positive test results compared to standard or real-time assays. Indeed, standard PCR assays yielded higher percentages of false positive results than either of the other two methods (mean false positive results for standard, nested, and real-time assays were, respectively, 23, 5, and 3%). Very few studies placed patients into the "possible case" category, but, similar to studies regarding the molecular diagnosis of aspergillosis, inclusion of possible candidiasis cases as falsely positive decreased test specificity to a similar extent regardless of the amplification method used.

For the most significant and commonly encountered fungal genera and species, the application of pan-fungal PCR primers and microarray detection formats would seem a convenient method to

TABLE 12.1
Nucleic Acid-Based Methods for the Detection of *Aspergillus* Species DNA in Clinical Specimens

[References]	Study Design[a]	Specimen (ml)[b]	Extraction Method[c]	Species Detected, Gene Target[d]	Amplification Method[e]	Detection Method[f]	Patients with Positive PCR result/Number Tested (%) among IA[g] Groups		
							Proven or Probable	Possible	None or Controls
Einsele, 1997 [15]	R	WB (5–10)	CL, SPH, HA	Pan-fungal[h] *18S* rRNA	sPCR	SB	13/13 (100)	n/a[i]	1/100 (1)
Bretagne, 1998 [41]	R	Serum (0.2)	QA	*A. fumigatus/flavus*, mtDNA	cPCR	PCR-EIA	9/18 (50)	3/4 (75)	0/19 (0)
Skladny, 199 [59]	R	WB (5), BAL (10)	CL, PC	*Aspergillus* spp., *18S* rRNA	nPCR	AGE/EB	22/22 (100)	n/a	5/118 (4)
Williamson, 2000 [52]	R	Serum (0.6)	SPH, QT	*Aspergillus* spp., *28S* rRNA	nPCR	AGE/EB	13/13 (100)	5/5 (100)	4/19 (21)
Hebar, 2000 [82]	P	WB (5)	CL, SPH, HA	Pan-fungal, *18S* rRNA	sPCR	SB	8/8 (100)	n/a	17/76 (22)
Lass-Florl, 2001 [65]	P	WB (10)	SPH, QT	Pan-fungal, *18S* rRNA	sPCR	PCR-EIA	3/3 (100)	2/2 (100)	40/163 (25)
Buchheidt, 2001 [96]	P	WB (3–5), BAL (1.5)	CL, PC	*Aspergillus* spp., *18S* rRNA	nPCR	AGE/EB	33/36 (92); 13/13 (100)	n/a	36/242 (15); 9/87 (10)
Kami, 2001 [95]	R, P[j]	WB, Plasma (0.2)	QA	*Aspergillus* spp., *18S* rRNA	rtPCR	TM	26/33 (79)	n/a	7/89 (8)
Raad, 2002a [37]	P	WB (0.7)	CL, PC	AF, Afl, AN, AT[k] mtDNA;	sPCR	AGE/EB;	8/11 (73);	4/7 (57);	0/36 (0);
Raad, 2002b [38]	P	BAL (1)		*A. fumigatus/flavus*, alkaline protease		SB	22/32 (69)	6/18 (33)	14/199 (7)
Costa, 2002 [29]	R	Serum (0.5)	MP	*A. fumigatus/flavus*, mtDNA	rtPCR	LC FRET	14/20 (70)	n/a	0/30 (0)
Rantakokko-Jalava, 2003 [22]	R	BAL (1.5)	BB, SPH	*A. fumigatus*, mtDNA	rtPCR	LC FRET	8/11 (73)	4/5 (80)	5/83 (6)
Challier, 2004 [89]	R	Serum	QA	*A. fumigatus*, *28S* rRNA	rtPCR	TM	22/26 (85)	4/15 (27)	0/29 (0)

TABLE 12.1 (continued)
Nucleic Acid-Based Methods for the Detection of *Aspergillus* Species DNA in Clinical Specimens

[References]	Study Design[a]	Specimen (ml)[b]	Extraction Method[c]	Species Detected, Gene Target[d]	Amplification Method[e]	Detection Method[f]	Patients with Positive PCR result/Number Tested (%) among IA[g] Groups		
							Proven or Probable	Possible	None or Controls
Buchheid, 2004 [54]	P	WB (5), BAL (10)	CL, PC	*Aspergillus* spp., *18S* rRNA; *A. fumigatus/clavatus*, mtDNA	nPCR, rtPCR	AGE/EB, LC FRET	7/11 (64)	14/90 (16)	38/104 (37)
Kawazu, 2002 [94]	P	Plasma (0.2)	QA	*Aspergillus* spp., *18S* rRNA	rtPCR	TM	5/11 (46)	6/13 (46)	1/125 (1)

[a] P = prospective study; R = retrospective study
[b] WB = whole blood; BAL = bronchoalveolar lavage
[c] BB = bead beating; CL = cell lysis using detergent ± proteinase K; HA = heat plus alkaline conditions; MP = MagNA Pure Kit (Roche); PC = phenol and/or chloroform extraction; QA = QIAamp DNA Blood Kit (Qiagen); QT = QIAamp Tissue Kit (now available as QIAamp DNA Mini Kit, Qiagen); SPH = spheroplasting with lytic enzyme
[d] mtRNA = mitochondrial RNA gene; 18S and 28S = small and large subunit ribosomal RNA (rRNA) genes, respectively
[e] cPCR = competitive PCR; nPCR = nested PCR; rtPCR = real-time PCR; sPCR = standard PCR
[f] AGE = agarose gel electrophoresis; EB = ethidium bromide staining; PCR-EIA = PCR-enzyme immunoassay; SB = Southern blotting; TM, TaqMan 5′ exonuclease method; LC FRET = LightCycler fluorescence resonance energy transfer
[g] IA = invasive aspergillosis
[h] Pan-fungal primers followed by detection with generic- or species-specific probes
[i] n/a = no available data
[j] Retrospective evaluation followed by prospective validation
[k] Specific for AF, *A. fumigatus*; Afl, *A. flavus*; AN, *A. niger*; and AT, *A. terreus*

TABLE 12.2
Nucleic Acid-Based Methods for the Detection of *Candida* Species DNA or RNA in Blood or Serum

[References]	Sample[a] (ml)	Extraction Method[b]	Species Detected; Gene Target[c]	Amplification Method[d]	Detection Method[e]	Patients with Positive PCR Result/ Number Tested (%) among IC[f] Groups		
						Proven or Probable	**Possible**	**None or Controls**
Buchman, 1990 [14]	WB (0.1–5)	CL, SPH, boil, PC	*C. albicans*; *L1A1 (ERG11)*	sPCR	AGE/EB	2/2 (100)	1/1 (100)	2/14 (14)
Kan, 1993 [24]	Serum (0.1)	BB, CL, PC	CA, CG, CK, CP, CT; *Actin*	sPCR	PAGE (^{32}P)	11/14 (79)	n/a[b]	0/29 (0)
Jordan, 1994 [17]	Blood (0.2–5)	CL, SPH, PC	CA, CG, CP, CT; *CHS1*	sPCR	PCR-EIA	15/16 (93)	n/a	0/34 (0)
Rand, 1994 [44]	WB (0.5–3)	CL, SPH, PC	CA, CG, CP, CT, CV; *18S rRNA*	nPCR	AGE/EB, REA	11/25 (44)	n/a	2/34 (6)
Burgener-Kairuz, 1994 [33]	WB (0.15)	CL, SPH, PC	CA, CG; *L1A1 (ERG11)*	nPCR	SB	27/38 (71) CA, 7/7 (100) CG	n/a	2/39 (5) CA, 2/73 (3) CG
Burnie, 1997 [63]	Serum (0.2)	CL, boil, GT	*C. albicans*; *ITS rRNA*	sPCR	PCR-EIA	28/28 (100)	3/16 (19)	0/15 (0)
Einsele, 1997 [15]	WB (5–10)	CL, SPH, HA	CA, CT; *18S rRNA*	sPCR	SB	8/8 (100)	2/65 (3)	0/35 (0)
Flahaut, 1998 [20]	WB (0.9)	CL (NaOH), SPH, QT	*C. albicans*; *SAP*	sPCR	PCR-EIA	51/51 (100)	n/a	0/93 (0)
van Burik, 1998 [19]	WB (6.0)	CL, SPH, HA	CA, CG, CP, CT; *18S rRNA*	sPCR	SB	4/4 (100)	n/a	0/5 (0)
Morace, 1999 [34]	WB (3.0)	FC, CL, SPH, PC, G50	CA, CG, CT; *L1A1 (ERG11)*	sPCR	AGE/EB, REA	13/14 (93)	n/a	18/58 (31)
Posteraro, 2000 [35]	WB (3.0)	FC, CL, SPH, PC, G50	CA, CG; *ERG11 (L1A1)*	sPCR	Reverse cross-blot	6/6 (100)	n/a	2/8 (25)
Wahyuningsih, 2000 [64]	Serum (0.2)	QA	CA, CK; *ITS rRNA*	sPCR	PCR-EIA	7/8 (88)	3/9 (33)	0/27 (0)
Loeffler, 2000 [62]	WB (5.0)	SPH, QT	*C. albicans*; *18S rRNA*	rtPCR	LC FRET	2/2 (100)	n/a	0/50 (0)

TABLE 12.2 (continued)
Nucleic Acid-Based Methods for the Detection of *Candida* Species DNA or RNA in Blood or Serum

[References]	Sample (ml)[a]	Extraction Method[b]	Species Detected; Gene target[c]	Amplification Method[d]	Detection Method[e]	Patients with Positive PCR Result/Number Tested (%) among IC[f] Groups		
						Proven or Probable	Possible	None or Controls
Loeffler, 2003 [31]	WB (0.1)	CL, RN, QS	*C. glabrata*; *18S* rRNA	NASBA	ECL	12/12 (100)	9/16 (56)	0/22 (0)
Dendis, 2003 [60]	WB (0.5)	CL, SPH, GT, SiO_2 particles	CA, CG, CT; ITS rRNA	sPCR	AGE/EB, REA or APLP	5/5 (100)	n/a	0/6 (0)
						5/5 (100)	7/7 (100)	0/14 (0)
Maaroufi, 2004 [78]	Serum (0.4)	QA	CA, CP, CT; ITS rRNA	rtPCR	TM	7/8 (88)	1/1 (100)	0/15 (0)
Bu, 2005 [56]	WB (2.0)	CL, PG	*C. albicans*; ITS rRNA	rtPCR	LC	6/6 (100)	n/a	0/8 (0)
White, 2005 [72]	WB (5–10)	CL, SPH, HA	*Candida* spp; *18S* rRNA	rtPCR	LC	19/20 (95)	7/18 (39)	2/67 (3)

a WB = whole blood

b FC = Ficoll density separation of blood fractions; G50 = Sephadex G-50 column purification; GT = guanidium thiocyanate; PG = Puregene DNA Purification Kit (Gentra Systems); QS = QiaShredder spin purification column (Qiagen); RN = RNeasy Mini Kit (Qiagen); see Table 12.1, footnote c, for explanations of additional abbreviations

c *CHS1* = chitin synthase gene; ITS = internal transcribed spacer region of rRNA gene; *L1A1* (*ERG11*) = lanosterol demethylase gene; *SAP* = aspartyl proteinase gene; rRNA = ribosomal RNA gene

d NASBA = nucleic acid sequence based amplification (isothermal); nPCR = nested or hemi-nested PCR; rtPCR = real-time PCR; sPCR = standard PCR

e APLP = amplification product length polymorphism; ECL = electrochemiluminescence; PAGE = polyacrylamide gel electrophoresis; REA = restriction enzyme analysis; LC = LightCycler (Roche) melting curve analysis; see Table 12.1, footnote f, for explanations of additional abbreviations

f Patients with candidemia and/or histopathologic evidence of invasive, systemic candidiasis (IC) were considered proven cases

g Only *Candida* species detected in clinical specimens for a given study were listed; CA = *C. albicans*; CK = *C. krusei*; CG = *C. glabrata*; CP = *C. parapsilosis*; CT = *C. tropicalis*; CV = *C. viswanthii*

h n/a = no available data

detect and identify a large variety of fungal organisms simultaneously. Such assays require very small amounts of PCR product applied to the surface of a microarray platform. Hundreds of specific probes may be included in a single microarray and thus a large number of different fungi could be identified from a single pan-fungal amplification reaction. Data retrieval and analysis could be automated and the cost per sample minimized, particularly if microarrays were made in-house using oligonucleotide probes dotted onto inexpensive glass slides. Microarrays have been designed for identifying bacteria[76] and are under development for identifying and differentiating *Candida* species.[77] However, no commercial microarray system for the detection of fungal DNA extracted from clinical materials is currently available.

Fungal DNA extracted from clinical materials may also be subjected to comparative DNA sequence analysis after PCR amplification to identify infecting organisms. Limited human studies have described DNA sequence analysis for the identification of fungi from tissues and body fluids, and most represent sporadic case reports.

DNA sequence analysis has several limitations. For example, currently available public databases are not refereed and the quality of the sequences and the correctness of the identities of the organisms to which the sequences belong depend solely on the accuracy of the submitting author. Commercial databases are becoming available, although they are still very expensive and are not in complete agreement with phenotypic identification data.[3,4] The lack of concordance is often caused by incomplete databases or by differences between traditional phenotypic and newer molecular taxonomic identities of the organism under investigation. A uniform, standardized molecular taxonomic structure is still lacking for many fungal genera and species, and until such issues are resolved the results of direct DNA sequence analysis may be difficult to interpret.

SELECTION OF CLINICAL MATERIALS

PCR-based detection of fungal DNA has been applied to a wide variety of clinical materials. By far, whole blood is the most common specimen used for the detection of *Candida* species DNA[10,14,15,20,33,43,75] but serum or plasma,[16,24,51,63,64,78] sputum,[14,23] urine,[14,16] bile,[20,33] pus,[16,33] bronchoalveolar lavage,[20,23,33,40] peritoneal,[16,20,33] cerebrospinal,[33] pleural,[20] wound,[14] and intraocular fluids,[79,80] tissues,[20,23] and blood culture bottles[25,26,50] have all been used with varying degrees of success.

In general, the methods for the isolation and purification of DNA have remained essentially the same for any of these specimens; grinding in a micromortar or use of a bead-beating system was primarily used for tissue specimens and steps for those using whole blood included lysis of erythrocytes and leukocytes and repeated washing with detergent to remove inhibitors of the PCR assay. One advantage of the use of serum rather than whole blood is that no blood cell lysis steps are required. One author found no significant difference in test sensitivity upon use of serum instead of whole blood (93 versus 86% sensitivity) for the detection of *Candida* species DNA.[81] However, serum was more often PCR-positive than whole blood (27 versus 7% positive, P <0.05) in culture-negative blood samples.[81]

Specimens tested for the presence of *Aspergillus* species DNA included whole blood,[2,15,59,65,82,83] serum or plasma,[52,83] cerebrospinal fluid,[84,85] sputum,[23] tissues,[23,71,86,87] and BAL fluid.[23,27,40,59,86] The methodologies used for the isolation and purification of *Aspergillus* DNA from all samples were much the same except that for tissue examinations, some used bead-beating methods and commercial purification kits[67,71] and others used homogenization of tissue with grinders.[86,87] Homogenization with tissue grinders was commonly followed by beating with zirconium beads[86] or by suspension of homogenized samples in detergent followed by freezing.[87]

Whole blood was used as the test specimen for the recovery of *Aspergillus* species DNA by most researchers[2,15,37,59,65,82,83,88] and reported to be superior to plasma for optimum recovery in several cases. For example, Loeffler et al.[83] reported that whole blood and plasma allowed an equal recovery of DNA (limit of sensitivity = 10 CFU per ml). However, the sensitivity of PCR for *Aspergillus* DNA recovered from plasma was lower than that from whole blood. Nineteen specimens from three patients with proven aspergillosis were PCR-positive when either whole blood or plasma was examined; however, an additional 22 specimens from the same patients were PCR-positive only when whole blood was used. It was suggested that the presence in whole blood of white blood cells that may have contained phagocytosed fungal material and hence fungal DNA was responsible for the increased sensitivity.

Costa et al.[42] also reported that the white blood cell pellet from whole blood was superior to plasma for the recovery of *Aspergillus* species DNA (yield was 10-fold greater) but found serum which lacks white blood cells to be equivalent to whole blood. Therefore, whole blood, serum, and white blood cell fractions appear to be better specimens for the recovery of *Aspergillus* species DNA than plasma. However, in contrast to whole blood but similar to plasma, serum contains only free fungal DNA or perhaps circulating fungal fragments. How stable the free DNA or DNA in fungal fragments in the circulation may be remains uncertain. Extrapolations from animal and limited human studies suggest that the half-life of free DNA in circulation ranges from 5 to 120 minutes, after which it is degraded by endogenous DNases.

Not surprisingly, Costa et al.[42] reported that serum processed immediately gave better results than samples remaining at room temperature for 24 hr. However, results from other studies suggest that serum may be a good test specimen with equal sensitivity in PCR and provide superior ease of processing, transport, and storage compared to whole blood.[29,52,89] The collection of serum rather than other blood fractions provides another advantage in that serum can be used for other, nonmolecular tests run in parallel with the PCR test.[29] This may be helpful in making a final diagnosis; for example, serum is used for antigen detection assays to detect circulating galactomannan in the diagnosis of aspergillosis or mannan in the diagnosis of candidiasis.

Selection of appropriate clinical specimens for use in PCR assays is critical to the proper interpretation of the results obtained. Extreme care must be taken to avoid environmental contamination of samples. This is of particular concern when detecting DNA from *Aspergillus* and other mold species, as these organisms are ubiquitous in the environment. A positive PCR result using sputum or BAL, although suggestive of *Aspergillus* infection, may represent colonization rather than true infection. Up to 25% of BAL samples from healthy adults have been reported positive for *Aspergillus* species by PCR assay.[90] Blood can be obtained via less invasive procedures than BAL fluid and is a more practical clinical specimen for studies requiring repeated collection (such as those to evaluate the usefulness of PCR to follow disease progression or monitor drug therapy).

In contrast, repeated bronchial lavages may be hazardous for debilitated patients and are costly. In addition, PCR methods to detect aspergillosis using BAL specimens did not correlate well with the results of a commercially available galactomannan antigen detection test in one study[27] and the number of false positive results obtained using PCR was higher than those for the antigen detection test.[27,41]

PCR assay of BAL fluid to detect *Aspergillus* colonization, however, may be helpful for predicting patients who are likely to develop aspergillosis. For example, Einsele et al.[91] reported that five of seven patients whose BAL fluid was positive by PCR at time of transplant developed invasive aspergillosis a mean of 64 days after transplant. All five were granulocytopenic before transplant and had severe hematological malignancies. After transplant, high dose corticosteroids were administered to these patients for graft-versus-host disease. The two remaining patients who had PCR-positive BAL results at the time of transplant did not develop invasive aspergillosis. Unlike the others, however, these patients received only short-term or no immunosuppressive therapy. However, three patients who were PCR-negative at the time of transplant later developed invasive aspergillosis while on mechanical ventilation. The sensitivity and specificity of PCR-positive BAL samples for predicting the development of invasive aspergillosis were 63% and 98%, respectively.

In another study, Buchheidt et al.[54] described a PCR procedure using BAL fluid that yielded a test sensitivity of 100% and a test specificity of 93% among 67 patients with invasive pulmonary aspergillosis and controls. Concurrent PCR tests run on both the blood and BAL fluid from 45 patients, including 8 with proven IA, were 100% concordant.

Other studies documented the capacity of PCR assays to diagnose IA using tissue specimens.[67,68,74,92,93] Real-time, quantitative methods applied to tissue specimens may also provide insight into the burden of disease and patient response to therapy. O'Sullivan et al.[68] noted a direct correlation between tissue burden and PCR results. A statistically significant reduction in PCR-quantifiable *Aspergillus* DNA was reported in lung tissue after amphotericin B treatment.[68] At least one study successfully used a PCR method followed by comparative DNA sequence analysis to identify *Aspergillus* species DNA in human biopsy tissue.[93]

SENSITIVITY AND SPECIFICITY OF FUNGAL PCR ASSAYS

The literature for nucleic acid-based diagnosis of IA and candidiasis is very heterogeneous in terms of study design, case definitions, testing and treatment strategies, and data presentation. In an attempt to maximize the validity of data comparisons in Table 12.1, certain parameters were applied. Included studies were required to have followed consecutive patients at risk for IA with or without low risk controls or presented selected patients with appropriate controls. Only studies clearly defining criteria for positive cases were included, although not all used consensus guideline case definitions.

Since some of these studies were retrospective and no diagnostic gold standard existed at the time of the analysis, calculations of sensitivity and specificity were problematic. In determining test sensitivities, patients classified as *proven* or *probable* according to the study's definitions were considered to be true cases; those classified as *possible* or *suspected* were not. This tended to somewhat overestimate sensitivity and underestimate specificity. Also, the data represent the performance of a test among patients, not single samples. Most patients were sampled multiple times over an extended period.

The sensitivity of a PCR test to detect IA using the definition above varies considerably across studies (mean, 81%; range, 46 to 100%) and the specificities obtained were dependent upon how possible cases were viewed. By ignoring possible cases and examining only negative controls, overall test specificity was very high (mean, 90%; range, 63 to 100%). In fact, some possible cases may have represented true or early invasive fungal infections that resolved with treatment and/or neutrophil recovery. Thus, a positive test result for a possible case may, in fact, have represented a true case of IA.

However, not all possible cases yielded positive PCR test results and these data would negatively affect sensitivity. Alternatively, positive results for possible cases may represent falsely positive results. These data highlight the difficulties encountered when no gold standard diagnostic test is available for comparison purposes. In many instances, however, false positive results appeared in specimens from negative patients or controls with no evidence of infection; therefore, problems encountered with the classification of patients do not completely explain low test specificity results.

Another important factor in the interpretation of test results is the number of positive test results obtained per patient. If at least two consecutive positive PCR test results are required to define a true case, test specificity could be improved, sometimes at little cost to test sensitivity. Indeed, because multiple studies have demonstrated this finding, some researchers advocate a requirement for at least two positive PCR test results before a patient is considered to be a truly positive case.[52,65,94] However, data from one study demonstrated that requiring two positive PCR test results decreased sensitivity from 64 to 36% although specificity improved from 64 to 92%.[54]

Although it is difficult to compare results of studies because of multiple methodological variations, the overall trend was toward improved test sensitivity when whole blood rather than serum or plasma was used in a PCR test to detect IA (Table 12.1). This trend may be influenced, however, by the observation that studies examining whole blood were more likely to process larger sample volumes. Multiple studies used starting volumes of 5 to 10 ml of whole blood for DNA extraction (Table 12.1).

One report noted improved test sensitivity when more than 5 ml of blood instead of 200 μl was processed.[95] However, issues such as patient morbidity (restricting the volume of blood that may be safely collected) and the difficulty of processing large volumes of blood in a typical clinical laboratory setting may limit the sample volumes available in clinical practice, especially if blood is drawn frequently from the same high-risk patient to screen for the presence of infection. The application of newer, automated processing systems[29,75] that employ less complex DNA extraction protocols may result in less loss of DNA target during sample processing. Reduced loss of DNA target during processing may facilitate the use of smaller test volumes yet produce similar test sensitivities to those currently achieved only with larger sample volumes.

Test sensitivity for the detection of *Aspergillus* species DNA in various clinical specimens has been reported to range from 64 to 100% for whole blood,[15,19] 50 to 100% for serum,[29,41,52,89] 69% to 100% for BAL fluid,[15,38,59,61,96] and 100% for urine.[39] Specificity for detecting *Aspergillus* species DNA using these clinical specimens was reported to range from 63 to 100% for whole blood,[79] (if only a single positive specimen from a given patient was required for the patient to be considered positive) to 100% (if two specimens from a given patient were required for the patient to be considered positive) for serum, 63 to 100% for BAL fluid, and 88% for urine.

The limit of analytical sensitivity was reported as 1 to 10 CFU per ml whole blood[2,15] or 20 to 600 fg *A. fumigatus* DNA from whole blood,[15,39,54] 1 to 10 CFU per ml serum or plasma,[2,15,19] ≤10 fg *A. fumigatus* DNA from BAL fluid[54], and less than 2 CFU per g tissue.[74]

There appears to be a striking chronological trend in the literature toward a declining sensitivity to detect IA by PCR (Table 12.1). Many confounding factors may explain this phenomenon. As noted earlier, except for whole blood, newer studies were more likely to use smaller specimen volumes, less rigorous DNA extraction methods, and real-time PCR methods. However, even successive studies from the same group employing similar methods demonstrated progressive drops in test performance.[15,54,59,65]

Certain logical explanations may account for the observed reductions in sensitivity over time. For example, more recent studies were more likely to be prospective. Also, the establishment of consensus guidelines for fungal infections introduced more stringently defined proven, probable, and possible categories of patients. These changes more closely reflect the true performance of a diagnostic test rather than retrospective evaluations of selected cases and controls; prospective studies tend to include smaller numbers of proven and probable cases and greater numbers of possible cases. Clearly these differences affect positive and negative predictive values and also affect test sensitivity by decreasing the number of proven and probable cases.

Most studies included only small numbers of proven or probable cases (mean number of proven cases reported: 17.5 per study). In part, the more recent reduction in test sensitivity also reflects the evolution of the diagnosis, treatment and prevention of IA. Newer diagnostic tools such as high-resolution computed tomography scanning and serum galactomannan testing[97] are now more commonly used. Therefore, IA may be detected at earlier stages than previously possible and at a time when PCR test results may still be negative. Also, the increased use of more effective anti-fungal prophylaxis among high-risk patients may lead to reduced test sensitivity because of an earlier clearance of fungal DNA from the circulation.

Table 12.2 summarizes representative literature for the detection of nucleic acids from *Candida* species found in patient blood and blood fractions. Similar to studies examining IA, no standardized method or study design was followed by all researchers so that comparisons of studies were difficult. Nonetheless, the trend with time was to move from single copy amplification targets to more universal pan-fungal targets and develop real-time, quantitative methods for amplicon detection. The availability of manual commercial DNA extraction kits has made the isolation and purification of fungal nucleic acids more straightforward.

Test sensitivities have remained consistently high for studies conducted from 1997 forward but specificities varied considerably, depending upon how patients were classified into probable and possible disease categories. Only patients with confirmed candidemia or histopathological evidence of tissue invasion were considered proven cases of systemic candidiasis in Table 12.2.

Probable cases were also included in this category, but only if probable cases were defined by the study authors. However, if no probable or possible disease categories were designated, such patients were merged by default with patients with no evidence of disease (they were not culture- or histopathology-confirmed cases). Therefore, in such instances, test specificities were lower than those for studies where possible and probable cases were delineated.

The mean test sensitivity to detect *Candida* DNA extracted from whole blood was somewhat higher than for serum but the range for serum was narrower than for whole blood (mean for whole blood, 99%; range, 44 to 100% versus mean for serum, 89%; range, 79 to 100%). Surprisingly, the blood or serum volume used did not seem to directly impact test sensitivity results across studies (Table 12.2). This may indicate that the extraction and amplification methods used were more important factors than the volume of specimen tested for the detection of *Candida* DNA in contrast to the detection of *Aspergillus* DNA.

Unfortunately, most studies used very small numbers of clinical specimens for analysis, particularly for the number of proven or probable cases reported (mean number of proven cases reported: 14 per study). These data emphasize the need for multicenter studies to obtain sufficient sample size for critical interpretation and analysis of results.

Test sensitivity for the detection of *Candida* species DNA in various clinical specimens was reported to range from 44 to 100% for whole blood,[15,20,34,44] 79 to 100% for serum[24,63,64] (Table 12.2), 95 to 100% for blood culture bottles,[25,26,50] and 83 to 100% for all other tissues and fluids tested.[14,20,23,98] Specificity to detect *Candida* species DNA in these same clinical specimens ranged from 69 to 100% among all samples tested. The limit of test sensitivity ranged from 1 to 150 CFU per ml recovered from whole blood,[2,14,16,20,33,34,43,44] 50 to 100 CFU per ml for serum or plasma,[16,24,63] and 500 CFU per ml from blood culture bottles.[25]

Representative studies over the past decade using clinical materials for the nucleic acid-based diagnosis of non-IA, non-IC fungal infections are listed alphabetically by disease in Table 12.3. Most of these studies were conducted at single institutions, using limited numbers of clinical specimens. However, the number of studies[99–103] describing nucleic acid-based detection of DNA from *Pneumocytis jirovecii* (formerly *Pneumocystis carinii* var. *hominis*) for the diagnosis of pneumocystosis is greater than for the other diseases listed in Table 12.3. This may be because the infecting organism cannot be cultured *in vitro* and therefore the development of a nucleic acid-based diagnostic test to detect this organism was more critical than for agents of other fungal diseases.

However, microscopic examination of sputum and BAL fluid has been used successfully for the detection of *P. jirovecii* as a simple, nonmolecular alternative for the diagnosis of pneumocytosis. Unfortunately, *P. jirovecii* can also be a commensal colonizer of the human respiratory tract as well as an agent of pneumocystosis. Therefore, positive microscopic results may be difficult to interpret.

The general insensitivity of blood culture for the recovery of many fungi and the underlying respiratory nature of most of the diseases listed in Table 12.3 are perhaps the reasons that respiratory samples and tissues were the primary specimens examined. Tissue was obviously the specimen of choice for fungi associated with skin infections (*Malassezia, Sporothrix, Trichophyton,* and *Microsporum* species)[104–106] and a few studies described the extraction of fungal DNA from blood or serum specimens for the diagnosis of penicilliosis marneffei and trichosporonosis.[107,108] The sensitivity to detect DNA in blood and serum samples was no different from that for other specimens and varied widely (64 to 100%), as did the sensitivities for tissue diagnosis (62 to 100%).[109–111].

Specificity was lowest in studies examining lower respiratory tract and BAL specimens for the diagnosis of pneumocystosis (67 to 100%) and for skin samples examined for *Malassezia* species (22%); these data most probably reflect the colonizing nature of these organisms and the nonsterility of the sites from which the specimens were obtained. Nonetheless, specificity was reported to be high (100%) in other studies using BAL or sputum specimens from patients with other fungal infections, where colonization would not be expected, including those caused by *C. immitis*, *C. neoformans*, and *P. brasiliensis*[23,112,113] (Table 12.3). Oral wash fluid was shown to give a test sensitivity of 94% and a specificity of 90% in one study examining a real-time PCR assay for pneumocystosis; this compared to a sensitivity of 100% and a specificity of only 33% using lower respiratory tract fluid in the same study.[102]

Given the high sensitivity and specificity of the test using oral wash fluid and the less invasive nature of oral wash fluid collection compared to BAL fluid collection, it was suggested that oral wash fluid be used for the nucleic acid-based diagnosis of pneumocystosis. However, results from another study[100] did not find the use of oral wash fluid for the detection of *P. jirovecii* nucleic acid to be sufficiently sensitive (38% using oral wash fluid versus 86% for BAL fluid). Sensitivity to detect DNA from *C. neoformans* in cerebrospinal fluid was 100% in two separate studies[23,114] as was the sensitivity to detect DNA from a variety of fungal agents in ocular fluid or tissue.[80]

The extraction methods used on samples cited in Table 12.3 varied widely among laboratories and the greatest technical differences among fungal disruption methods were observed between those for *P. jirovecii* and those for other fungi. *P. jirovecii* disruption in clinical materials required only cell lysis and phenol–chloroform extraction steps compared to the more elaborate disruption and extraction methods required to obtain DNA from clinical materials containing other fungi. These data may reflect the structural differences in cell wall composition between *P. jirovecii* and the other fungi. By virtue of the chitin contained in their cell walls,[115] *P. jirovecii* cells are resistant to simple osmotic lysis. Nonetheless, *P. jirovecii* appears to be generally more susceptible to lysis than the other fungi.

PCR methods for the amplification of DNA from clinical materials containing the organisms listed in Table 12.3 were most likely to be nested or real-time PCR assays rather than conventional PCR assays; however, maximum test sensitivity was not dependent upon the use of a nested PCR assay. Only those studies conducted before the year 2000 used standard PCR assays, and the standard assays were reported to be 100% sensitive and 100% specific.[23] However, as in studies of IA and systemic candidiasis, the studies listed in Table 12.3 examined only small numbers of proven cases (average number of cases per study: 13.5) and were conducted at single institutions.

EFFECT OF ANTIFUNGAL THERAPY ON PCR RESULTS

By far, the greatest number of nucleic acid-based tests described in the literature related to the diagnosis of invasive aspergillosis and candidiasis. These diseases are the most prevalent invasive fungal diseases found clinically today,[116,117] they progress rapidly, and are often fatal in severely immunocompromised patients.[118,119] An early and specific diagnosis is therefore needed for the implementation of effective antifungal drug therapy to reduce morbidity and mortality. Although guidelines have recently been published,[49,120] no standardized treatment or prevention strategies for these diseases currently exist.

Many high risk patients receive fluconazole or voriconazole to prevent and/or treat yeast infections and itraconazole or voriconazole to prevent and/or treat mold infections. Several studies have reported that effective antifungal therapy reduces or eliminates fungal DNA from the circulation, which may lead to negative PCR results. Alternatively, specimens derived from patients with fungal infections that are aborted early in the course of infection by preemptive drug therapies may yield positive results in a PCR assay but would be classified as falsely positive.

The effect of early drug treatment or prophylaxis of high-risk patients on PCR test results was illustrated in a prospective study by Lass-Flörl et al.[121] Sixteen high-risk patients with radiographic findings suggestive of IA were treated with amphotericin B, liposomal amphotericin B, voriconazole, itraconazole, caspofungin, or a combination of these agents. Patients were classified as having proven, probable, or possible IA according to the European Organization for Research and Treatment of Cancer (EORTC) guidelines.[122]

TABLE 12.3
Nucleic Acid-Based Methods for the Detection of Non-*Candida*, Non-*Aspergillus* Species (Listed Alphabetically by Organism) Using Clinical Materials

[References]	Specimen (ml)[a]	Extraction Method[b]	Organism Detected; Gene Target[c]	Amplification Method[d]	Detection Method[e]	Proven[f] or Probable	None or Controls
						Patients with Positive PCR Result/ Number Tested (%) among Disease Groups	
Ferrer, 2001 [80]	Ocular (0.1–0.2)	FH, CL, PC	AA, AF,AN, CP, SP;[g] ITS rRNA	nPCR	DNA sequencing	6/6 (100)	1/4 (25)
Bialek , 2003 [109]	Tissue (5 μm)	X, CL, FH, QT	*Blastomyces dermatitidis*; *WI-1* gene	nPCR	DNA sequencing	8/13 (62)	0/45 (0)
Sandhu, 2005	Sputum (0.2)	GT, boil, PC	*Coccidioides immitis*;[h] 28S rRNA	sPCR	Slot blot	2/2 (100)	0/42 (0)
Bialek, 2004 [110]	Tissue (5 μm)	X, CL, FH, QT	*Coccidioides posadasii*; *Ag2/PRA* gene	nPCR	DNA sequencing	3/3 (100)	0/20 (0)
Sandhu, 1995	CSF (0.2); Tissue (5 μm)	GT, boil, PC	*Cryptococcus neoformans*; 28S rRNA	sPCR	Slot blot	2/2 (100)	0/42 (100)
Tanaka, 1996 [112]	BAL, sputum; tissue	CL, BB, PC; H, BB, PC	*Cryptococcus neoformans*; *URA5* gene	nPCR	SB	4/5 (80)	0/10 (0)
Rappelli, 1998 [114]	CSF (0.2)	GT, boil, PC	*Cryptococcus neoformans*; ITS rRNA	nPCR	AGE, EB	21/21 (100)	0/19 (0)
Bialek, 2002 [111]	Tissue (5 μm)	X, CL, FH, QT	*Histoplasma capsulatum*; 100-kDa-like gene	nPCR	DNA sequencing	20/20 (100)	0/50 (0)
Sugita, 2001 [104]	Skin	CL, boil, PC	*Malassezia* spp; ITS rRNA	nPCR	DNA sequencing	32/32 (100)	14/18 (78)
Gomes, 2000 [113]	Sputum (2.0)	H, CL, boil, PC	*Paracoccidioides brasiliensis*; *GP43* gene	nPCR	SB	11/11 (100)	0/1 (0)
Prariyachatigul, 2003 [108]	Blood (5.0)	QM	*Penicillium marneffei*; *18S* rRNA	nPCR	AGE, EB	2/2 (100)	0/17 (0)
Sandhu, 1999 [99]	BAL (1.0)	GT, boil, PC	*Pneumocystis carinii*;[h] 28S rRNA	sPCR	Slot blot	23/25 (92)	1/25 (4)
Huang, 1999 [100]	BAL (5–10), oral wash (10–15)	Boil DNA-ER	*Pneumocystis carinii*;[h] mtLSU rRNA gene	sPCR	SB	6/7 (86) [BAL]; 3/8 (38) [oral wash]	0/12 (0) [BAL]; 0/73 (0) [oral wash]

TABLE 12.3 (continued)

Nucleic Acid-Based Methods for the Detection of Non-*Candida*, Non-*Aspergillus* Species (Listed Alphabetically by Organism) Using Clinical Materials

[References]	Specimen (ml)[a]	Extraction Method[b]	Organism Detected; Gene Target[c]	Amplification Method[d]	Detection Method[e]	Patients with Positive PCR Result/ Number Tested (%) among Disease Groups	
						Proven[f] or Probable	None or Controls
Torres, 2000 [101]	BAL (1.0)	CL, PC	*Pneumocystis carinii*;[h] ITS rRNA	nPCR	Dot blot	25/29 (86)	4/29 (14)
Larsen, 2002 [102]	LRT (0.2); oral wash (0.2)	NucliSens	*Pneumocystis carinii*;[h] MSG gene	rtPCR	LC FRET	19/19 (100) [LRT]; 17/18 (94) [oral wash]	10/30 (33) [LRT]; 3/31 (10) [oral wash]
Flori, 2004 [103]	BAL (0.2)	QM	*Pneumocystis jirovecii*; MSG gene	rtPCR	LC FRET	11/11	21/139 (15)
Hu, 2003 [105]	Tissue (0.1 g)	Mince, CL, FH, PC	*Sporothrix schenkii*; 18S rRNA	nPCR	DNA sequencing	11/12 (92)	0/1 (0)
Turin, 2000 [106]	Tissue (2 mg)	H, SPH, PC	*Trichophyton, Microsporum* spp.; ITS rRNA	nPCR	AGE/EB	10/12 (83)	0/20 (0)
Nagai, 1999 [107]	Serum (0.1)	CL, heat	*Trichosporon* spp.; 28S rRNA	nPCR	SB	7/11 (64)	0/20 (0)

[a] BAL = bronchoalveolar lavage fluid; CSF = cerebral spinal fluid; LRT = lower respiratory tract specimens (BAL or induced sputum); ocular specimens included vitreous and aqueous fluids or corneal scrapings

[b] DNA-ER = DNA extraction reagent (Perkin-Elmer); FH = freezing followed by heating or boiling; H = homogenization by mincing, grinding, and/or shearing; NucliSens = NucliSens Kit (Organon Teknika); QM = QIAamp DNA Mini Kit (Qiagen); X = xylene deparaffinization; see Tables 12.1 and 12.2, footnotes c and b, respectively, for explanations of additional abbreviations

[c] *18S* and *28S* = small and large subunit ribosomal RNA (rRNA) genes, respectively; *Ag2/PRA* = antigen2/proline-rich antigen gene; *GP43* = 43,000-Da glycoprotein antigen gene; *ITS* = internal transcribed spacer region of rRNA gene; *MSG* = major surface glycoprotein gene; *mtLSU* = mitochondrial large subunit rRNA gene; *URA5* = orotate phosphoribosyltransferase gene; *WI-1* = Wisconsin-1 adhesion gene

[d] nPCR = nested or hemi-nested PCR; rtPCR = real-time PCR; sPCR = standard PCR

[e] See Table 12.1, footnote f, for explanations of abbreviations

[f] Confirmed by culture, histopathology, or direct microscopy

[g] AA = *Alternaria alternata*; AF = *Aspergillus fumigatus*; AN = *Aspergillus niger*; CP = *Candida parapsilosis*; SP = *Scedosporium apiospermum*

[h] Studies conducted before delineation of *C. posadasii* or renaming of *P. carinii* as *P. jirovecii*

Nine and seven patients with proven and probable IA were evaluated. Whole blood specimens collected before (n = 41) and during antifungal treatment (n = 67) were tested by a PCR-ELISA method. The sensitivity of the PCR assay using blood from proven cases of IA was 66% before antifungal therapy and 55% after therapy; blood from probable cases of IA showed test sensitivity of 57 and 42% before and after therapy, respectively.

It was concluded that the benefits of PCR diagnosis were more limited once antifungal drug treatment was initiated. It has also been suggested that persistent detection of fungal DNA in the blood may be a poor prognostic indicator.[52,88] However, many factors may influence the clearance of DNA from the circulation, including the burden of disease, degree of immunosuppression and/or mucosal disruption, and effectiveness of antifungal drug therapy.

Effect of PCR Results on Patient Management and Clinical Outcome

The true value of any diagnostic test is its ability to positively influence patient management or clinical outcome. The increased sensitivity of PCR tests compared to antigen or antibody detection tests holds promise for an earlier diagnosis of invasive fungal diseases, before the appearance of clinical or radiological signs.

A handful of reports suggest that positive PCR results can predict disease. For example, in a prospective study examining seven patients with *de novo* proven or probable IA, positive PCR results were obtained a median of 2 days (range, 1 to 23) and 9 days (range 2 to 34) prior to the appearance of clinical and radiological signs, respectively.[82] In a retrospective study, Williamson et al.[52] reported that a positive PCR test result preceded galactomannan antigen positivity in all 13 proven or probable cases of IA and preceded death by a median of 46 days.

In a study examining a nested PCR assay targeting the *18S* rRNA gene, PCR results were used to initiate or withhold empiric antifungal therapy in febrile neutropenic pediatric patients.[123] Thirty-one PCR-positive episodes were reported. In 19 of these episodes, blood cultures were positive for *Candida* species. The PCR result was available from 1 to 8 days earlier than positive culture results and led to the earlier initiation of antifungal drug therapy.

Unfortunately, real-time quantitative PCR assays have not improved the capacity of PCR tests to predict patient prognosis or treatment outcome. One reason for this may be that fungal DNA may circulate at very low concentrations, so that positive blood or serum levels are at or near the limit of detection.[29,89] However, even in cases where detectable DNA copy numbers were shown to be high, the PCR test still did not predict outcome.[73] However, this result may have been confounded by the fact that 14 of the 18 patients studied were extremely ill and died.

CONCLUDING REMARKS

Although strides have been made toward commercialization and standardization of nucleic acid-based diagnostic reagents and DNA extraction systems, it is clear that nucleic acid-based tests have not yet become the gold standard for the diagnosis of fungal infections. Currently, multiple positive PCR test results suggest that a patient has or is at risk for developing a fungal infection, but a negative PCR test result does not rule out a fungal infection.

The interpretation of a positive or negative test results remains problematic. Even if test specificity is good, low disease prevalence makes the positive predictive value low. Alternatively, even when a test has a high negative predictive value, low disease prevalence could result in a significant percentage of missed cases. This illustrates the difference between an episode-based and sample-based negative predictive value.

Because of the grave consequences of delayed treatment, a negative PCR test cannot be used to rule out invasive fungal infection. Rather, a positive or negative PCR result may represent a single, laboratory-based piece of evidence that may be used along with clinical, radiological, and other laboratory test results to guide the diagnosis and management of fungal infections. It is also not clear at present which nucleic acid-based test method is best. As this discussion has illustrated, many variables can affect test performance. Large, standardized comparative trials of nucleic acid-based test methods, preferably conducted as multicenter studies, are required to obtain sufficient sample sizes for robust statistical comparisons.

REFERENCES

1. White PL, Shetty A, Barnes RA. Detection of seven *Candida* species using the Light-Cycler system. *J Med Microbiol* 2003; 52 (3): 229–238.
2. Loeffler J, Henke N, Hebart H, Schmidt D, Hagmeyer L, Schumacher U, Einsele H. Quantification of fungal DNA by using fluorescence resonance energy transfer and the Light Cycler system. *J Clin Microbiol* 2000; 38 (2): 586–590.
3. Hall L, Wohlfiel S, Roberts GD. Experience with the MicroSeq D2 large-subunit ribosomal DNA sequencing kit for identification of filamentous fungi encountered in the clinical laboratory. *J Clin Microbiol* 2004; 42 (2): 622–626.
4. Hall L, Wohlfiel S, Roberts GD. Experience with the MicroSeq D2 large-subunit ribosomal DNA sequencing kit for identification of commonly encountered, clinically important yeast species. *J Clin Microbiol* 2003; 41 (11): 5099–5102.
5. Padhye AA, Smith G, Standard PG, McLaughlin D, Kaufman L. Comparative evaluation of chemiluminescent DNA probe assays and exoantigen tests for rapid identification of *Blastomyces dermatitidis* and *Coccidioides immitis*. *J Clin Microbiol* 1994; 32 (4): 867–870.
6. Stockman L, Clark KA, Hunt JM, Roberts GD. Evaluation of commercially available acridinium ester-labeled chemiluminescent DNA probes for culture identification of *Blastomyces dermatitidis, Coccidioides immitis, Cryptococcus neoformans*, and *Histoplasma capsulatum*. *J Clin Microbiol* 1993; 31 (4): 845–850.
7. Brown HL, Fuller DD, Jasper LT, Davis TE, Wright JD. Clinical evaluation of Affirm VPIII in the detection and identification of *Trichomonas vaginalis, Gardnerella vaginalis*, and *Candida* species in vaginitis/vaginosis. *Infect Dis Obstet Gynecol* 2004; 12 (1): 17–21.
8. Wilson DA, Joyce MJ, Hall LS, Reller LB, Roberts GD, Hall GS, Alexander BD, Procop GW. Multicenter evaluation of a *Candida albicans* peptide nucleic acid fluorescent *in situ* hybridization probe for characterization of yeast isolates from blood cultures. *J Clin Microbiol* 2005; 43 (6): 2909–2912.
9. Sogaard M, Stender H, Schonheyder HC. Direct identification of major blood culture pathogens, including *Pseudomonas aeruginosa* and *Escherichia coli*, by a panel of fluorescence *in situ* hybridization assays using peptide nucleic acid probes. *J Clin Microbiol* 2005; 43 (4): 1947–1949.
10. Schmidt K, Loeffler J, Hebart H, Schumacher U, Einsele H. Rapid extraction and detection of DNA from *Candida* species and research samples by MagNA Pure LC and the LightCycler system. *Biochemica* 2001; 2: 8–10.
11. Holmes AR, Lee YC, Cannon RD, Jenkinson HF, Shepherd MG. Yeast-specific DNA probes and their application for the detection of *Candida albicans*. *J Med Microbiol* 1992; 17 (5): 346–351.
12. Muncan P, Wise GJ. Early identification of candiduria by polymerase chain reaction in high risk patients. *J Urol* 1996; 156 (1): 154–156.
13. Kiehn TE. Bacteremia and fungemia in the immunocompromised patient. *Eur J Clin Microbiol Infect Dis* 1989; 8 (9): 832–837.
14. Buchman TG, Rossier M, Merz WG, Charache P. Detection of surgical pathogens by *in vitro* DNA amplification. Part I: rapid identification of *Candida albicans* by *in vitro* amplification of a fungus-specific gene. *Surgery* 1990; 108 (2): 338–347.
15. Einsele H, Hebart H, Roller G, Loffler J, Rothenhofer I, Muller CA, Bowden RA, van Burik J, Engelhard D, Kanz L, Schumacher U. Detection and identification of fungal pathogens in blood by using molecular probes. *J Clin Microbiol* 1997; 35 (6): 1353–1360.
16. Crampin AC, Matthews RC. Application of the polymerase chain reaction to the diagnosis of candidosis by amplification of an HSP 90 gene fragment. *J Med Microbiol* 1993; 39 (3): 233–238.
17. Jordan JA. PCR identification of four medically important *Candida* species by using a single primer pair. *J Clin Microbiol* 1994; 32 (12): 2962–2967.

18. Loffler J, Hebart H, Sepe S, Schumacher U, Klingebiel T, Einsele H. Detection of PCR-amplified fungal DNA by using a PCR-ELISA system. *Med Mycol* 1998; 36 (5): 275–279.
19. Van Burik JA, Myerson D, Schreckhise RW, Bowden RA. Panfungal PCR assay for detection of fungal infection in human blood specimens. *J Clin Microbiol* 1998; 36 (5): 1169–1175.
20. Flahaut M, Sanglard D, Monod M, Bille J, Rossier M. Rapid detection of *Candida albicans* in clinical samples by DNA amplification of common regions from *C. albicans*-secreted aspartic proteinase genes. *J Clin Microbiol* 1998; 36 (2): 395–401.
21. Loffler J, Hebart H, Schumacher U, Reitze H, Einsele H. Comparison of different methods for extraction of DNA of fungal pathogens from cultures and blood. *J Clin Microbiol* 1997; 35 (12): 3311–3312.
22. Rantakokko-Jalava K, Laaksonen S, Issakainen J, Vauras J, Nikoskelainen J, Viljanen MK, Salonen J. Semiquantitative detection by real-time PCR of *Aspergillus fumigatus* in bronchoalveolar lavage fluids and tissue biopsy specimens from patients with invasive aspergillosis. *J Clin Microbiol* 2003; 41 (9): 4304–4311.
23. Sandhu GS, Kline BC, Stockman L, Roberts GD. Molecular probes for diagnosis of fungal infections. *J Clin Microbiol* 1995; 33 (11): 2913–2919.
24. Kan VL. Polymerase chain reaction for the diagnosis of candidemia. *J Infect Dis* 1993; 168 (3): 779–783.
25. Shin JH, Nolte FS, Morrison CJ. Rapid identification of *Candida* species in blood cultures by a clinically useful PCR method. *J Clin Microbiol* 1997; 35 (6): 1454–1459.
26. Shin JH, Nolte FS, Holloway BP, Morrison CJ. Rapid identification of up to three *Candida* species in a single reaction tube by a 5′ exonuclease assay using fluorescent DNA probes. *J Clin Microbiol* 1999; 37 (1): 165–170.
27. Verweij PE, Meis JFGM, van den Hurk P, De Pauw BE, Hoogkamp-Korstanje JAA, Melchers WJG. Polymerase chain reaction as a diagnostic tool for invasive aspergillosis: evaluation in bronchoalveolar lavage fluid from low risk patients. *Serodiagn Immunother Infect Disease* 1994; 6: 203–208.
28. Muller FM, Werner KE, Kasai M, Francesconi A, Chanock SJ, Walsh TJ. Rapid extraction of genomic DNA from medically important yeasts and filamentous fungi by high-speed cell disruption. *J Clin Microbiol* 1998; 36 (6): 1625–1629.
29. Costa C, Costa JM, Desterke C, Botterel F, Cordonnier C, Bretagne S. Real-time PCR coupled with automated DNA extraction and detection of galactomannan antigen in serum by enzyme-linked immunosorbent assay for diagnosis of invasive aspergillosis. *J Clin Microbiol* 2002; 40 (6): 2224–2227.
30. Loeffler J, Hebart H, Cox P, Flues N, Schumacher U, Einsele H. Nucleic acid sequence-based amplification of *Aspergillus* RNA in blood samples. *J Clin Microbiol* 2001; 39 (4): 1626–1629.
31. Loeffler J, Dorn C, Hebart H, Cox P, Magga S, Einsele H. Development and evaluation of the Nuclisens Basic Kit NASBA for the detection of RNA from *Candida* species frequently resistant to antifungal drugs. *Diagn Microbiol Infect Dis* 2003; 45 (3): 217–220.
32. Widjojoatmodjo MN, Borst A, Schukkink RAF, Box ATA, Tacken NMM, van Gemen B, Verhoef J, Top B, Fluit AdC. Nucleic acid sequence-based amplification (NASBA) detection of medically important *Candida* species. *J Microbiol Methods* 1999; 38 (1–2): 81–90.
33. Burgener-Kairuz P, Zuber JP, Jaunin P, Buchman TG, Bille J, Rossier M. Rapid detection and identification of *Candida albicans* and *Torulopsis (Candida) glabrata* in clinical specimens by species-specific nested PCR amplification of a cytochrome P-450 lanosterol-α-demethylase (L1A1) gene fragment. *J Clin Microbiol* 1994; 32 (8): 1902–1907.
34. Morace G, Pagano L, Sanguinetti M, Posteraro B, Mele L, Equitani F, D'Amore G, Leone G, Fadda G. PCR-restriction enzyme analysis for detection of *Candida* DNA in blood from febrile patients with hematological malignancies. *J Clin Microbiol* 1999; 37 (6): 1871–1875.
35. Posteraro B, Sanguinetti M, Masucci L, Romano L, Morace G, Fadda G. Reverse cross blot hybridization assay for rapid detection of PCR-amplified DNA from *Candida* species, *Cryptococcus neoformans*, and *Saccharomyces cerevisiae* in clinical samples. *J Clin Microbiol* 2000; 38 (4): 1609–1614.
36. Kambouris ME, Reichard U, Legakis NJ, Velegraki A. Sequences from the aspergillopepsin *PEP* gene of *Aspergillus fumigatus*: evidence on their use in selective PCR idenfication of *Aspergillus* species in infected clinical samples. *FEMS Immunol Med Microbiol* 1999; 25 (3): 255–264.

37. Raad I, Hanna H, Sumoza D, Albitar M. Polymerase chain reaction on blood for the diagnosis of invasive pulmonary aspergillosis in cancer patients. *Cancer* 2002; 94 (4): 1032–1036.
38. Raad I, Hanna H, Huaringa A, Sumoza D, Hachem R, Albitar M. Diagnosis of invasive pulmonary aspergillosis using polymerase chain reaction-based detection of *Aspergillus* in BAL. *Chest* 2002; 121 (4): 1171–1176.
39. Reddy LV, Kumar A, Kurup VP. Specific amplification of *Aspergillus fumigatus* DNA by polymerase chain reaction. *Mol Cell Probes* 1993; 7 (2): 121–126.
40. Bretagne S, Costa JM, Marmorat-Khuong A, Poron F, Cordonnier C, Vidaud M, Fleury-Feith J. Detection of *Aspergillus* species DNA in bronchoalveolar lavage samples by competitive PCR. *J Clin Microbiol* 1995; 33 (5): 1164–1168.
41. Bretagne S, Costa J-M, Bart-Delabesse E, Dhedin N, Rieux C, Cordonnier C. Comparison of serum galactomannan antigen detection and competitive polymerase chain reaction for diagnosing invasive aspergillosis. *Clin Infect Dis* 1998; 26 (6): 1407–1412.
42. Costa C, Vidaud D, Olivi M, Bart-Delabesse E, Vidaud M, Bretagne S. Development of two real-time quantitative TaqMan PCR assays to detect circulating *Aspergillus fumigatus* DNA in serum. *J Microbiol Methods* 2001; 44 (3): 263–269.
43. Holmes AR, Cannon RD, Shepherd MG, Jenkinson HF. Detection of *Candida albicans* and other yeasts in blood by PCR. *J Clin Microbiol* 1994; 32 (1): 228–231.
44. Rand KH, Houck H, Wolff M. Detection of candidemia by polymerase chain reaction. *Mol Cell Probes* 1994; 8 (3): 215–222.
45. Wingard JR. Importance of *Candida* species other than *C. albicans* as pathogens in oncology patients. *Clin Infect Dis* 1995; 20 (1): 115–1225.
46. Nguyen MH, Peacock JE, Jr., Morris AJ, Tanner DC, Nguyen ML, Snydman DR, Wagener MM, Rinaldi MG, Yu VL. The changing face of candidemia: emergence of non-*Candida albicans* species and antifungal resistance. *Am J Med* 1996; 100 (6): 617–623.
47. Sutton DA, Sanche SE, Revankar SG, Fothergill AW, Rinaldi MG. *In vitro* amphotericin B resistance in clinical isolates of *Aspergillus terreus*, with a head-to-head comparison to voriconazole. *J Clin Microbiol* 1999; 37 (7): 2343–2345.
48. Lionakis MS, Lewis RE, Torres HA, Albert ND, Raad II, Kontoyiannis DP. Increased frequency of non-*fumigatus Aspergillus* species in amphotericin B- or triazole-pre-exposed cancer patients with positive cultures for aspergilli. *Diagn Microbiol Infect Dis* 2005; 52 (1): 15–20.
49. Pappas PG, Rex JH, Sobel JD, Filler SG, Dismukes WE, Walsh TJ, Edwards JE. Guidelines for treatment of candidiasis. *Clin Infect Dis* 2004; 38 (2): 161–189.
50. Iwen PC, Freifeld AG, Bruening TA, Hinrichs SH. Use of a panfungal PCR assay for detection of fungal pathogens in a commercial blood culture system. *J Clin Microbiol* 2004; 42 (5): 2292–2293.
51. Ahmad S, Mustafa AS, Khan Z, Al-Rifaiy AI, Khan ZU. PCR-enzyme immunoassay of rDNA in the diagnosis of candidemia and comparison with amplicon detection by agarose gel electrophoresis. *Int J Med Microbiol* 2004; 294 (1): 45–51.
52. Williamson ECM, Leeming JP, Palmer HM, Steward CG, Warnock D, Marks DI, Millar MR. Diagnosis of invasive aspergillosis in bone marrow transplant recipients by polymerase chain reaction. *Br J Haematol* 2000; 108 (1): 132–139.
53. Ferns RB, Fletcher H, Bradley S, Mackinnon S, Hunt C, Tedder RS. The prospective evaluation of a nested polymerase chain reaction assay for the early detection of *Aspergillus* infection in patients with leukaemia or undergoing allograft treatment. *Br J Haematol* 2002; 119 (3): 720–725.
54. Buchheidt D, Hummel M, Schleiermacher D, Spiess B, Schwerdtfeger R, Cornely OA, Wilhelm S, Reuter S, Kern W, Sudhoff T, Morz H, Hehlmann R. Prospective clinical evaluation of a LightCycler™-mediated polymerase chain reaction assay, a nested-PCR assay and a galactomannan enzyme-linked immunosorbent assay for detection of invasive aspergillosis in neutropenic cancer patients and haematological stem cell transplant recipients. *Br J Haematol* 2004; 125 (2): 196–202.
55. Chang HC, Leaw SN, Huang AH, Wu TL, Chang TC. Rapid identification of yeasts in positive blood cultures by a multiplex PCR method. *J Clin Microbiol* 2001; 39 (10): 3466–3471.
56. Bu R, Sathiapalan RK, Ibrahim MM, Al Mohsen I, Almodavar E, Gutierrez MI, Bhatia K. Monochrome LightCycler PCR assay for detection and quantification of five common species of *Candida* and *Aspergillus*. *J Med Microbiol* 2005; 54 (3): 243–248.

57. Li YL, Leaw SN, Chen J-H, Chang HC, Chang TC. Rapid identification of yeasts commonly found in positive blood cultures by amplification of the internal transcribed spacer regions 1 and 2. *Eur J Clin Microbiol Infect Dis* 2003; 22 (11): 693–696.

58. Fujita SI, Senda Y, Nakaguchi S, Hashimoto T. Multiplex PCR using internal transcribed spacer 1 and 2 regions for rapid detection and identification of yeast strains. *J Clin Microbiol* 2001; 39 (10): 3617–3622.

59. Skladny H, Buchheidt D, Baust C, Krieg-Schneider F, Seifarth W, Leib-Mosch C, Hehlmann R. Specific detection of *Aspergillus* species in blood and bronchoalveolar lavage samples of immuno-compromised patients by two-step PCR. *J Clin Microbiol* 1999; 37 (12): 3865–3871.

60. Dendis M, Horwath R, Michalek J, Ruzicka F, Grijalva M, Bartos M, Benedik J. PCR-RFLP detection and species identification of fungal pathogens in patients with febrile neutropenia. *Clin Microbiol Infect* 2003; 9 (12): 1191–1202.

61. Verweij PE, Latge JP, Rijs AJ, Melchers WJ, De Pauw BE, Hoogkamp-Korstanje JA, Meis JF. Comparison of antigen detection and PCR assay using bronchoalveolar lavage fluid for diagnosing invasive pulmonary aspergillosis in patients receiving treatment for hematological malignancies. *J Clin Microbiol* 1995; 33 (12): 3150–3153.

62. Loeffler J, Hebart H, Magga S, Schmidt D, Klingspor L, Tollemar J, Schumacher U, Einsele H. Identification of rare *Candida* species and other yeasts by polymerase chain reaction and slot blot hybridization. *Diagn Microbiol Infect Dis* 2000; 38 (4): 207–212.

63. Burnie JP, Golbang N, Matthews RC. Semiquantitative polymerase chain reaction enzyme immunoas-say for diagnosis of disseminated candidiasis. *Eur J Clin Microbiol Infect Dis* 1997; 16 (5): 346–350.

64. Wahyuningsih R, Freisleben HJ, Sonntag HG, Schnitzler P. Simple and rapid detection of *Candida albicans* DNA in serum by PCR for diagnosis of invasive candidiasis. *J Clin Microbiol* 2000; 38 (8): 3016–3021.

65. Lass-Florl C, Aigner J, Gunsilius E, Petzer A, Nachbaur D, Gastl G, Einsele H, Loffler J, Dierich MP, Wurzner R. Screening for *Aspergillus* spp. using polymerase chain reaction of whole blood samples from patients with haematological malignancies. *Br J Haematol* 2001; 113 (1): 180–184.

66. Fletcher HA, Barton RC, Verweij PE, Evans EG. Detection of *Aspergillus fumigatus* PCR products by a microtitre plate based DNA hybridisation assay. *J Clin Pathol* 1998; 51 (8): 617–620.

67. de Aguirre L, Hurst SF, Choi JS, Shin JH, Hinrikson HP, Morrison CJ. Rapid differentiation of *Aspergillus* species from other medically important opportunistic molds and yeasts by PCR-enzyme immunoassay. *J Clin Microbiol* 2004; 42 (8): 3495–3504.

68. O'Sullivan CE, Kasai M, Francesconi A, Petraitis V, Petraitiene R, Kelaher AM, Sarafandi AA, Walsh TJ. Development and validation of a quantitative real-time PCR assay using fluorescence resonance energy transfer technology for detection of *Aspergillus fumigatus* in experimental invasive pulmonary aspergillosis. *J Clin Microbiol* 2003; 41 (12): 5676–5682.

69. Guiver M, Levi K, Oppenheim BA. Rapid identification of candida species by TaqMan PCR. *J Clin Pathol* 2001; 54 (5): 362–366.

70. Selvarangan R, Bui U, Limaye AP, Cookson BT. Rapid identification of commonly encountered *Candida* species directly from blood culture bottles. *J Clin Microbiol* 2003; 41 (12): 5660–5664.

71. Hendolin PH, Paulin L, Koukila-Kahkola P, Anttila VJ, Malmberg H, Richardson M, Ylikoski J. Panfungal PCR and multiplex liquid hybridization for detection of fungi in tissue specimens. *J Clin Microbiol* 2000; 38 (11): 4186–4192.

72. White PL, Archer AE, Barnes RA. Comparison of non-culture-based methods for detection of systemic fungal infections, with an emphasis on invasive *Candida* infections. *J Clin Microbiol* 2005; 43 (5): 2181–2187.

73. Spiess B, Buchheidt D, Baust C, Skladny H, Seifarth W, Zeilfelder U, Leib-Mosch C, Morz H, Hehlmann R. Development of a LightCycler PCR assay for detection and quantification of *Aspergillus fumigatus* DNA in clinical samples from neutropenic patients. *J Clin Microbiol* 2003; 41 (5): 1811–1818.

74. Bressler A, McCaustland K, Holloway B, Morrison C. Real time PCR detection and quantification of *Aspergillus* species' DNA using a fluorescent pan-*Aspergillus* probe. *Progr Abstr* 2004; 65. Pre-sented at Advances against Aspergillosis, San Francisco, California, 9–11 Sept 2004, Abstract 25.

75. Loeffler J, Schmidt K, Hebart H, Schumacher U, Einsele H. Automated extraction of genomic DNA from medically important yeast species and filamentous fungi by using the MagNA Pure LC system. *J Clin Microbiol* 2002; 40 (6): 2240–2243.

76. Anthony RM, Brown TJ, French GL. Rapid diagnosis of bacteremia by universal amplification of 23S ribosomal DNA followed by hybridization to an oligonucleotide array. *J Clin Microbiol* 2000; 38 (2): 781–788.

77. Varnier OE, Bertolotti F, Ferrari D, Giacomazzi C, Martini L, McDermott JL, Soro O, Machetti M, Viscoli C, Thuroff E, Ulivi M, Landt O, Heiser V. Rapid identification of *Candida* species using a non-fluorescent DNA-based biochip. *Progr Abstr* 2004; 146. Presented at the 44th Annual Interscience Conference on Antimicrobial Agents and Chemotherapy, Washington, DC, 30 Oct-2 Nov 2004, Poster D-463.

78. Maaroufi Y, Ahariz N, Husson M, Crokaert F. Comparison of different methods of isolation of DNA of commonly encountered *Candida* species and its quantitation by using a real-time PCR-based assay. *J Clin Microbiol* 2004; 42 (7): 3159–3163.

79. Jaeger EEM, Carroll NM, Choudhury S, Dunlop AAS, Towler HMA, Matheson MM, Adamson P, Okhravi N, Lightman S. Rapid detection and identification of *Candida, Aspergillus*, and *Fusarium* species in ocular samples using nested PCR. *J Clin Microbiol* 2000; 38 (8): 2902–2908.

80. Ferrer C, Colom F, Frases S, Mulet E, Abad JL, Alio JL. Detection and identification of fungal pathogens by PCR and by ITS2 and 5.8S ribosomal DNA typing in ocular infections. *J Clin Microbiol* 2001; 39 (8): 2873–2879.

81. Bougnoux ME, Dupont C, Mateo J, Saulnier P, Faivre V, Payen D, Nicolas-Chanoine MH. Serum is more suitable than whole blood for diagnosis of systemic candidiasis by nested PCR. *J Clin Microbiol* 1999; 37 (4): 925–930.

82. Hebart H, Loffler J, Meisner C, Serey F, Schmidt D, Bohme A, Martin H, Engel A, Bunjes D, Kern WV, Schumacher U, Kanz L, Einsele H. Early detection of *Aspergillus* infection after allogenic stem cell transplantation by polymerase chain reaction screening. *J Infect Dis* 2000; 181 (5): 1713–1719.

83. Loeffler J, Hebart H, Brauchle U, Schumacher U, Einsele H. Comparison between plasma and whole blood specimens for detection of *Aspergillus* DNA by PCR. *J Clin Microbiol* 2000; 38 (10): 3830–3833.

84. Kami M, Shirouzu I, Mitani K, Ogawa S, Matsumura T, Kanda Y, Masumoto T, Saito T, Tanaka Y, Maki K, Honda H, Chiba S, Ohtomo K, Hirai H, Yazaki Y. Early diagnosis of central nervous system aspergillosis with combination use of cerebral diffusion-weighted echo-planar magnetic resonance image and polymerase chain reaction of cerebrospinal fluid. *Intern Med* 1999; 38 (1): 45–48.

85. Verweij PE, Brinkman K, Kremer HP, Kullberg BJ, Meis JF. *Aspergillus* meningitis: diagnosis by non-culture-based microbiological methods and management. *J Clin Microbiol* 1999; 37 (4): 1186–1189.

86. Melchers WJ, Verweij PE, van den Hurk P, van Belkum A, De Pauw BE, Korstanje JA, Meis JF. General primer-mediated PCR for detection of *Aspergillus* species. *J Clin Microbiol* 1994; 32 (7): 1710–1717.

87. Spreadbury C, Holden D, Aufauvre-Brown A, Bainbridge B, Cohen J. Detection of *Aspergillus fumigatus* by polymerase chain reaction. *J Clin Microbiol* 1993; 31 (3): 615–621.

88. Lass-Florl C, Gunsilius E, Gastl G, Bonatti H, Freund MC, Gschwendtner A, Kropshofer G, Dierich MP, Petzer A. Diagnosing invasive aspergillosis during antifungal therapy by PCR analysis of blood samples. *J Clin Microbiol* 2004; 42 (9): 4154–4157.

89. Challier S, Boyer S, Abachin E, Berche P. Development of a serum-based Taqman real-time PCR assay for dagnosis of invasive aspergillosis. *J Clin Microbiol* 2004; 42 (2): 844–846.

90. Bart-Delabesse E, Marmorat-Khuong A, Costa JM, Dubreuil-Lemaire ML, Bretagne S. Detection of Aspergillus DNA in bronchoalveolar lavage fluid of AIDS patients by the polymerase chain reaction. *Eur J Clin Microbiol Infect Dis* 1997; 16 (1): 24–25.

91. Einsele H, Quabeck K, Muller K-D, Hebart H, Rothenhofer I, Loffler J, Schaefer UW. Prediction of invasive pulmonary aspergillosis from colonisation of lower respiratory tract before marrow transplantation. *Lancet* 1998; 352 (9138): 1443.

92. Loeffler J, Kloepfer K, Hebart H, Najvar L, Graybill JR, Kirkpatrick WR, Patterson TF, Dietz K, Bialek R, Einsele H. Polymerase chain reaction of *Aspergillus* DNA in experimental animal models of invasive aspergillosis. *J Infect Dis* 2002; 185 (8): 1203–1206.

93. Imhof A, Schaer C, Schoedon G, Schaer DJ, Walter RB, Schaffner A, Schneemann M. Rapid detection of pathogenic fungi from clinical specimens using LightCycler real-time fluorescence PCR. *Eur J Clin Microbiol Infect Dis* 2003; 22 (9): 558–560.

94. Kawazu M, Kanda Y, Nannya Y, Aoki K, Kurokawa M, Chiba S, Motokura T, Hirai H, Ogawa S. Prospective comparison of the diagnostic potential of real-time PCR, double-sandwich enzyme-linked immunosorbent assay for galactomannan, and a (1-3)-β-D-glucan test in weekly screening for invasive aspergillosis in patients with hematological disorders. *J Clin Microbiol* 2004; 42 (6): 2733–2741.

95. Kami M, Fukui T, Ogawa S, Kazuyama Y, Machida U, Tanaka Y, Kanda Y, Kashima T, Yamazaki Y, Hamaki T, Mori S, Akiyama H, Mutou Y, Sakamaki H, Osumi K, Kimura S, Hirai H. Use of real-time PCR on blood samples for diagnosis of invasive aspergillosis. *Clin Infect Dis* 2001; 33 (9): 1504–1512.

96. Buchheidt D, Baust C, Skladny H, Ritter J, Suedhoff T, Baldus M, Seifarth W, Leib-Moesch C, Hehlmann R. Detection of *Aspergillus* species in blood and bronchoalveolar lavage samples from immunocompromised patients by means of a two-step polymerase chain reaction: clinical results. *Clin Infect Dis* 2001; 33 (9): 428–435.

97. Marr KA, Carter RA, Crippa F, Wald A, Corey L. Epidemiology and outcome of mould infections in hematopoietic stem cell transplant recipients. *Clin Infect Dis* 2001; 33 (9): 909–917.

98. Kirby A, Chapman C, Hassan C, Burnie J. The diagnosis of hepatosplenic candidiasis by DNA analysis of tissue biopsy and serum. *J Clin Pathol* 2004; 57 (7): 764–765.

99. Sandhu GS, Kline BC, Espy MJ, Stockman L, Smith TF, Limper AH. Laboratory diagnosis of *Pneumocystis carinii* infections by PCR directed to genes encoding for mitochondrial 5S and 28S ribosomal RNA. *Diagn Microbiol Infect Dis* 1999; 33 (3): 157–162.

100. Huang SN, Fischer SH, O'Shaughnessy E, Gill VJ, Masur H, Kovacs JA. Development of a PCR assay for diagnosis of *Pneumocystis carinii* pneumonia based on amplification of the multicopy major surface glycoprotein gene family. *Diagn Microbiol Infect Dis* 1999; 35 (1): 27–32.

101. Torres J, Goldman M, Wheat LJ, Tang X, Bartlett MS, Smith JW, Allen SD, Lee C-H. Diagnosis of *Pneumocystis carinii* pneumonia in human immunodeficiency virus-infected patients with polymerase chain reaction: a blinded comparison to standard methods. *Clin Infect Dis* 2000; 30 (1): 141–145.

102. Larsen HH, Masur H, Kovacs JA, Gill VJ, Silcott VA, Kogulan P, Maenza J, Smith M, Lucey DR, Fischer SH. Development and evaluation of a quantitative, touch-down, real-time PCR assay for diagnosing *Pneumocystis carinii* pneumonia. *J Clin Microbiol* 2002; 40 (2): 490–494.

103. Flori P, Bellete B, Durand F, Raberin H, Cazorla C, Hafid J, Lucht F, Sung RTM. Comparison between real-time PCR, conventional PCR and different staining techniques for diagnosing *Pneumocystis jiroveci* pneumonia from bronchoalveolar lavage specimens. *J Med Microbiol* 2004; 53 (7): 603–607.

104. Sugita T, Suto H, Unno T, Tsuboi R, Ogawa H, Shinoda T, Nishikawa A. Molecular analysis of *Malassezia* microflora on the skin of atopic dermatitis patients and healthy subjects. *J Clin Microbiol* 2001; 39 (10): 3486–3490.

105. Hu S, Chung WH, Hung SI, Ho HC, Wang ZW, Chen CH, Lu SC, Kuo Tt, Hong HS. Detection of *Sporothrix schenckii* in clinical samples by a nested PCR assay. *J Clin Microbiol* 2003; 41 (4): 1414–1418.

106. Turin L, Riva F, Galbiati G, Cainelli T. Fast, simple and highly sensitive double-rounded polymerase chain reaction assay to detect medically relevant fungi in dermatological specimens. *Eur J Clin Invest* 2000; 30 (6): 511–518.

107. Nagai H, Yamakami Y, Hashimoto A, Tokimatsu I, Nasu M. PCR detection of DNA specific for *Trichosporon* species in serum of patients with disseminated trichosporonosis. *J Clin Microbiol* 1999; 37 (3): 694–699.

108. Prariyachatigul C, Chaiprasert A, Geenkajorn K, Kappe R, Chuchottaworn C, Termsetjaroen S, Srimuang S. Development and evaluation of a one-tube seminested PCR assay for the detection and identification of *Penicillium marneffei*. *Mycoses* 2003; 46 (11–12): 447–454.

109. Bialek R, Cascante Cirera A, Herrmann T, Aepinus C, Shearn-Bochsler VI, Legendre AM. Nested PCR assays for detection of *Blastomyces dermatitidis* DNA in paraffin-embedded canine tissue. *J Clin Microbiol* 2003; 41 (1): 205–208.

110. Bialek R, Kern J, Herrmann T, Tijerina R, Cecenas L, Reischl U, Gonzalez GM. PCR assays for identification of *Coccidioides posadasii* based on the nucleotide sequence of the antigen 2/proline-rich antigen. *J Clin Microbiol* 2004; 42 (2): 778–783.

111. Bialek R, Feucht A, Aepinus C, Just-Nubling G, Robertson VJ, Knobloch J, Hohle R. Evaluation of two nested PCR assays for detection of *Histoplasma capsulatum* DNA in human tissue. *J Clin Microbiol* 2002; 40 (5): 1644–1647.

112. Tanaka K, Miyazaki T, Maesaki S, Mitsutake K, Kakeya H, Yamamoto Y, Yanagihara K, Hossain MA, Tashiro T, Kohno S. Detection of *Cryptococcus neoformans* gene in patients with pulmonary cryptococcosis. *J Clin Microbiol* 1996; 34 (11): 2826–2828.

113. Gomes GM, Cisalpino PS, Taborda CP, de Camargo ZP. PCR for diagnosis of paracoccidioidomycosis. *J Clin Microbiol* 2000; 38 (9): 3478–3480.

114. Rappelli P, Are R, Casu G, Fiori PL, Cappuccinelli P, Aceti A. Development of a nested PCR for detection of *Cryptococcus neoformans* in cerebrospinal fluid. *J Clin Microbiol* 1998; 36 (11): 3438–3440.

115. Baselski VS, Robison MK, Pifer LW, Woods DR. Rapid detection of *Pneumocystis carinii* in bronchalveolar lavage samples by using Cellufluor staining. *J Clin Microbiol* 1990; 28 (2): 393–394.

116. Rees JR, Pinner RW, Hajjeh RA, Brandt ME, Reingold AL. The epidemiological features of invasive mycotic infections in the San Francisco Bay area, 1992–1993: results of population-based laboratory active surveillance. *Clin Infect Dis* 1998; 27 (5): 1138–1147.

117. Trick WE, Fridkin SK, Edwards JR, Hajjeh RA, Gaynes RP. Secular trend of hospital-acquired candidemia among intensive care unit patients in the United States during 1989–1999. *Clin Infect Dis* 2002; 35 (5): 627–630.

118. McNeil MM, Nash SL, Hajjeh RA, Phelan MA, Conn LA, Plikaytis BD, Warnock DW. Trends in mortality due to invasive mycotic diseases in the United States, 1980–1997. *Clin Infect Dis* 2001; 33 (5): 641–647.

119. Wenzel RP. Nosocomial candidemia: risk factors and attributable mortality. *Clin Infect Dis* 1995; 20 (6): 1531–1534.

120. Stevens DA, Kan VL, Judson MA, Morrison VA, Dummer S, Denning DW, Bennett JE, Walsh TJ, Patterson TF, Pankey GA. Practice guidelines for diseases caused by *Aspergillus*. *Clin Infect Dis* 2000; 30 (4): 696–709.

121. Lass-Florl C, Gunsilius E, Gastl G, Freund M, Dierich MP, Petzer A. Clinical evaluation of *Aspergillus*-PCR for detection of invasive aspergillosis in immunosuppressed patients. *Mycoses* 2005; 48 (Suppl 1): 12–17.

122. Ascioglu S, Rex JH, de Pauw B, Bennett JE, Bille J, Crokaert F, Denning DW, Donnelly JP, Edwards JE, Erjavec Z, Fiere D, Lortholary O, Maertens J, Meis JF, Patterson TF, Ritter J, Selleslag D, Shah PM, Stevens DA, Walsh TJ, etc. Defining opportunistic invasive fungal infections in immunocompromised patients with cancer and hematopoietic stem cell transplants: an international consensus. *Clin Infect Dis* 2002; 34 (1): 7–14.

123. Lin M-T, Lu H-C, Chen W-L. Improving efficacy of antifungal therapy by polymerase chain reaction-based strategy among febrile patients with neutropenia and cancer. *Clin Infect Dis* 2001; 33 (10): 1621–1627.

13 Molecular Diagnostic Approaches in Infectious Disease

Leonard F. Peruski, Jr. and Anne Harwood Peruski

CONTENTS

CHALLENGES

Under normal circumstances, a clinical microbiology laboratory is faced with significant technical, clinical, and public health challenges. Not only must pathogens be identified, isolated, and characterized to permit the implementation of optimal therapies, but these tasks often must be accomplished under the multiple demands of limited time, organisms that are challenging to isolate and characterize, and the increasing need to define the functionality of a pathogen in terms of drug resistance, special virulence attributes, and epidemiological markers.

Compounding these pressures, a new challenge was thrust upon the clinical microbiology arena in October 2001: the threat of an intentional epidemic that clearly established that certain infectious disease (ID) agents would be used as weapons. Of these, agents such as *Bacillus anthracis* (anthrax) and variola major virus (smallpox) are considered to have the greatest potential for mass casualties and civil disruption. Also high on the prospective list are *Yersinia pestis, Francisella tularensis,* and the neurotoxins of *Clostridia botulinum.*[1–5] Lower ranking agents include *Burkholderia pseudomallei* and *mallei, Rickettsia* species, *Coxiella burnetti,* Venezuelan equine encephalitis virus, and the Marburg, Ebola, and influenza viruses.[1–5] Any of these agents will pose challenges to a microbiology laboratory tasked with their identification — usually under considerable pressure from both patient and public health perspectives to provide rapid answers.

In addition to these threats, several emerging infectious diseases also have the potential to produce significant public health consequences. They include Dengue fever, West Nile fever, and Rift Valley fever, as well as the recent reemergence of malaria in the eastern United States and avian influenza in Asia.[1-5] Compounding this already complex situation are the estimated 600 million international tourists annually, many with the potential to spread disease globally in a matter of hours.[3] The challenges facing modern clinical microbiologists and laboratorians appear overwhelming!

TECHNOLOGIES

Because of these combined threats posed by microbial pathogens, researchers face a need to rapidly identify such agents in clinical settings to improve both individual health and public health surveillance and epidemiology. A vast array of assay strategies based on the detection of nucleic acid signatures has emerged for surveillance and clinical diagnostic uses. Over the past 20 years, technologies have been developed or adapted to the challenges posed by these agents, permitting detection and identification within minutes to a few hours. In particular, the development of improved reagents and detection equipment has led to dramatic improvements in the sensitivity and specificity of nucleic acid-based systems, allowing an ever-increasing range of agents, virulence factors, and other genetic markers to be identified and quantitated.

Current assay systems have the necessary reliability, accuracy, and sensitivity to allow wider applications such as outpatient monitoring, large screening programs in developing countries, and point-of-care service. As a result of these continual improvements, nucleic acid-based assays are poised to serve as the new gold standard for the detection, identification, and characterization of microbial pathogens. Of these new technologies, the various incarnations of real-time PCR and related hybridization techniques are best suited to the clinical diagnostic laboratory because of their maturity as proven methods. Microarrays and variations of flow cytometry are emerging as the most promising high-throughput candidates for future adoption (Table 13.1).

Nucleic acid-based techniques are emerging as the new standards for the detection of microbial pathogens, in part because conventional gold standard techniques such as culture are either too slow or insufficiently sensitive, particularly when a patient has received antibiotics or when a clinician is working with fastidious organisms. Further, nucleic acid-based techniques can be extended far beyond mere detection and identification. Drug resistance, presence of virulence factors, and molecular typing can be accomplished with nucleic acid methodologies, albeit with varying degrees of success in light of the infancy of these applications. Because of this breadth of utility and ever-expanding promise of additional applications, nucleic acid-based techniques, whether used to interrogate suspect colonies isolated by conventional methods or used for direct detection in clinical specimens, are supplanting conventional identification strategies for pathogenic microorganisms in modern clinical microbiology laboratories.

These rapid diagnostic assays continue to evolve rapidly from the research bench to clinical pathogen detection, beginning first at reference laboratory level and then migrating to point-of-care-based systems. Throughput of the assays is also evolving in concert with the migration from research to the clinical bench, as single analyte detection gives way to multiple analyte detection and the focus shifts to methods using less invasive measures for obtaining specimens. Nucleic acid assays are typically more sensitive than the conventional assays that are being supplanted and often more fully characterize a pathogenic agent or host response to an agent. From a public health perspective, the advent of robust and reproducible nucleic acid assays has led to molecular approaches in epidemiology, with concordant improvements in pathogen tracking as well as better and faster assessments of epidemics and endemic diseases. From a clinical perspective, the migration of rapid nucleic acid technologies to points of care will enhance the detection, identification, and

TABLE 13.1
Comparison of Selected Assay Systems

Assay System	Sensitivity	Specificity	Throughput	Time	Relative Cost	Complexity	Comments
Culture and other conventional tests	High	High	Low–moderate	Days	Low–moderate	Low–moderate	Gold standard, largely limited to agents that can be propagated on culture media; may be labor- and time-intensive
Real-time PCR	High	High	Low–moderate	Minutes–hours	High	Moderate	Mature technology, increasingly vast array of assays and analytic systems
Microarrays	High	High	High	Hours	Very high	Moderate–high	At proof-of-concept stage; few commercial assays; extensive multiplexing capabilities and discriminatory power
Flow cytometry	High	High	High	Minutes	High–very high	Moderate–high	Based on existing technology; not as well developed as microarrays; lower ability to multiplex than microarrays; ability comparable to that of real-time PCR

characterization of microbial pathogens, with increasing beneficial impacts on medical care and patient outcomes.

This chapter will examine the role of nucleic acid detection assays in the clinical microbiology laboratory, looking at applications for the detection and identification of microbial pathogens (other than those most closely related to sexually transmitted diseases; see Chapter 9 and 10), strengths and limitations of the technologies, and examination of future directions and pending controversies relating to their use.

ISSUES IN ASSAY DEVELOPMENT AND VALIDATION

Development of nucleic acid-based assays is subject to two interrelated sets of metrics: quantitative and qualitative.[1, 4–9] Key among the former are sensitivity, specificity, and positive and negative predictive values. The more sensitive and specific an assay, the better it is. The higher the positive predictive and negative predictive values, the better the assay. Likewise, several qualitative metrics affect an assay including ease of operation, training required, sensitivity to contaminants and interferents, range of specimens that can be analyzed, and most importantly, evaluation against accepted standard methods. As part of the development process, assays must be optimized to achieve acceptable levels of performance based on these metrics. Once in use, finished assays must also be monitored continually for false positive and false negative results.

SPECIFICITY AND SENSITIVITY

As would be expected, all nucleic acid-based assays for the detection and identification of biological weapon and ID agents share one common trait: they require specific target sequences (Table 13.2). These target sequences in large part determine the specificities and sensitivities of the finished assays. As a result, sequences must be chosen carefully, selecting for those conserved among members of the pathogen of interest but also unique to that pathogen.

Criteria for selection include (but are not limited to) the ability to bind to a desired target with high affinity while at the same time retaining high specificity. Sensitivity is dependent on the ability to discriminate signal from background measurement at low target concentration and may depend on the assay format and instrumentation used for detection. Sensitivity is commonly determined using a statistically significant cut-off level above background.[5] Thus, both sensitivity and accuracy may be affected by the precision of signal and background measurements, and it is possible to increase the accuracy of an assay by increasing the statistically significant cut-off level. However, this effectively decreases its sensitivity.

Specificity of the target sequence is a principal determinant for assay quality. Qualitative factors such as the type of specimen or sample can exert great impact on both specificity and sensitivity, allowing the formation of spurious signals that mimic real targets or suppress their detection.

SIGNAL DETECTION

Assay sensitivity is also dependent on the signal used for measurement. Depending on the nature of the signal, the reactants may be detected visually, electronically, chemically, or physically, and a wide range of instruments can detect the presence of these labels with high degrees of sensitivity. Most assays use chromogenic, chemiluminescent, or fluorescent substrates that produce signals that are detectable visually or via sensor arrays. Fluorescent dyes and molecules capable of luminescent signals are the most commonly used labels. While usually more sensitive than other labeling chemistries, they may have high background levels because of their intrinsic fluorescent and luminescent properties.

TABLE 13.2
Nucleic Acid Targets for Selected Pathogens

Infectious Disease Agent	Nucleic Acid Targets
Bacillus anthracis	*pag, lef, cya, saspB, capA, capB,* Ba813
Brucella species	IS711, *ompAa,* 16S RNA
Burkholderia species	23S RNA
Clostridia botulinum	*boNTA, boNTB, boNTC, boNTD, boNTE, boNTF, boNTG*
Escherichia coli O157:H7	*stx*
Francisella tularensis	*tul4, fopA*
Helicobacter pylori	*cagA, vacA, iceA, babA, ureB*
Legionella pneumophila	*dotA,* 16S rRNA, *mip*
Listeria monocytogenes	*hylA, iap*
Mycobacterium tuberculosis	IS6110, *rpoB, katG, embB,* 16S rRNA
Shigella species	*ipaH, ial, stx*
Staphylococcus aureus	*sea, seb, sec, sed, see, seg, seh, sei,* TSST-1
Vibrio cholerae	*rtxA, epsM, mshA, tcpA, tox*
Yersinia pestis	*pst, caf1, yopM, pla,* f1 antigen
Leishmania species	kDNA, DNA polymerase gene
Pneumocystis jirovecii	beta-tubulin gene, mtLSUrRNA, *msg*
SARS coronavirus	N, M, S genes
Variola major (smallpox)	14-kDa fusion protein gene, hemagglutinin, DNA polymerase
Venezuelan equine encephalitis virus	E2, E3, ns1

OPTIMIZATION

Other key parameters affecting assay performance are annealing temperature, kinetics, reaction time, and reagent concentration. Optimization of each parameter is essential during the development process. Assay optimization usually involves an empiric approach in which one or at most two parameters are varied based on a matrix of potential reaction conditions. During this process, another key metric is determined: the limit of detection. Optimized assays are further tested empirically using a panel of target agents, molecular and environmental mimics, and potential interferents.[4] Such a specificity panel usually includes (or should include) genetic near neighbors along with materials or agents from environmental or biological sources likely to be components of samples. At first glance, optimization may seem tedious but on a single platform some of the basic parameters like time and temperature need only be optimized for one assay with subsequent assays for the same platform following the same general parameters.

MICROBIAL PATHOGENS AND REAL-TIME PCR

Since late 2001, the number of real-time PCR assays for infectious disease agents reported in the literature has skyrocketed as has the breadth of agents for which assays have been developed. Unfortunately, transitioning an assay from the laboratory bench to a clinical setting is not a trivial exercise and is subject to a number of practical constraints.[4–9]

As noted above, in addition to evaluation against the accepted standard method, an extensive range of metrics must be determined for each assay from analytical and clinical sensitivity and specificity to the effects of potential inhibitory substances and complex matrices to more complex statistical measures such as positive and negative predictive values. As a consequence, few of these reported assays have been commercialized (Table 13.3) and even fewer have undergone formal certification or licensure by an external independent agency for routine laboratory use. This, in turn, has limited to some extent the adoption of real-time PCR in clinical settings and remains the greatest challenge in clinicians' adoption of nucleic acid methods.

TABLE 13.3
Commercial Assays for Infectious Disease Agents

Agent	Type	Analysis System	Vendor	Approval	Comments
Bacteria					
B. anthracis	PCR	GeneXpert	Cepheid	RUO	Single-agent, two-target cartridge
	PCR	GeneXpert	Cepheid	RUO	Three-agent cartridge (*B. anthracis, F. tularensis, Y. pestis*)
	PCR	RAPIDS	Idaho Technologies	RUO	Three individual assays
	PCR	RealArt	artus	RUO	Available for range of real-time thermal cyclers
B. pertussis	PCR	SmartCycler	Cepheid	DIA	
Brucella species	PCR	RAPIDS	Idaho Technologies	RUO	Genus-specific only
Campylobacter species	PCR	RAPIDS	Idaho Technologies	RUO	Genus-specific only for *C. jejuni, coli, lari, upsaliensis*
	PCR	RealArt	artus	RUO	Available for range of real-time thermal cyclers
C. pneumoniae	PCR	SmartCycler	Cepheid	DIA	
C. botulinum	PCR	RAPIDS	Idaho Technologies	RUO	Type A neurotoxin isolates only
Enterococcus species	PCR	SmartCycler	Cepheid	DIA	
E. coli O157:H7	PCR	RAPIDS	Idaho Technologies	RUO	
F. tularensis	PCR	GeneXpert	Cepheid	RUO	Three-agent cartridge (*B. anthracis, F. tularensis, Y. pestis*)
	PCR	RAPIDS	Idaho Technologies	RUO	Two individual assays
L. monocytogenes	PCR	RAPIDS	Idaho Technologies	RUO	
	PCR	RealArt	artus	RUO	Available for range of real-time thermal cyclers
M. pneumoniae	PCR	SmartCycler	Cepheid	DIA	
M. avium	PCR	COBAS AmpliCor	Roche	RUO	
	HPA	AccuProbe	Gen-Probe		Culture identification
	HPA	AccuProbe	Gen-Probe		Culture identification
M. gordonae	HPA	AccuProbe	Gen-Probe		Culture identification
M. intracellulare	PCR	COBAS AmpliCor	Roche	RUO	Not available in U.S.
	HPA	AccuProbe	Gen-Probe		Culture identification
M. kansasii	HPA	AccuProbe	Gen-Probe		Culture identification
M. tuberculosis	PCR	COBAS AmpliCor	Roche	RUO	Not available in U.S.
	TMA/ HPA	Amplified MTD	Gen-Probe	DIA	FDA approved
	HPA	AccuProbe	Gen-Probe		Culture identification
	PCR	RealArt	artus	DIA	Available for range of real-time thermal cyclers
Salmonella species	PCR	RealArt	artus	RUO	Available for range of real-time thermal cyclers
S. aureus	PCR	SmartCycler	Cepheid	DIA	FDA approved for methicillin-resistant organisms (MRSA)

TABLE 13.3 (continued)
Commercial Assays for Infectious Disease Agents

Agent	Type	Analysis System	Vendor	Approval	Comments
Streptococcus, Group A	DNA probe	GASDirect	Gen-Probe		
Streptococcus, Group B	DNA probe	AccuProbe	Gen-Probe		Culture identification
	PCR	SmartCycler	Cepheid	DIA	FDA approved
	PCR	GeneXpert	Cepheid	DIA	FDA approved
Y. pestis	PCR	GeneXpert	Cepheid		
	PCR	RAPIDS	Idaho Technologies	RUO	Two individual assays
Virus					
Dengue fever virus	PCR	RealArt	artus	RUO	Available for range of real-time thermal cyclers
Enterovirus	PCR	SmartCycler	Cepheid	DIA	
	PCR	GeneXpert	Cepheid	DIA	FDA approved
	PCR	RealArt	artus	DIA	Available for range of real-time thermal cyclers
Influenza virus	PCR	SmartCycler	Cepheid	DIA	Detects both A and B strains
	PCR	RealArt	artus	RUO	Available for range of real-time thermal cyclers
Norovirus	PCR	SmartCycler	Cepheid	DIA	
Orthopox virus	PCR	RealArt	artus	RUO	Available for range of real-time thermal cyclers
Parvovirus B19	PCR	SmartCycler	Cepheid	DIA	
	PCR	RealArt	artus	DIA	Available for range of real-time thermal cyclers
Respiratory syncytial virus (RSV)	PCR	SmartCycler	Cepheid	DIA	
SARS	PCR	RealArt	artus	DIA	Available for range of real-time thermal cyclers
West Nile fever virus	TMA	Procleix	Gen-Probe	DIA	FDA IND
	PCR	RealArt	artus	DIA	Available for range of real-time thermal cyclers
Varicella-Zoster virus	PCR	RealArt	artus	DIA	Available for range of real-time thermal cyclers
Other					
Cryptosporidium	PCR	RAPIDS	Idaho Technologies	RUO	Two individual assays
E. histolytica	PCR	RealArt	artus	DIA	Available for range of real-time thermal cyclers
Plasmodium species	PCR	RealArt	artus	DIA	Available for range of real-time thermal cyclers

DIA = approved for diagnostic use in United States and/or European Union
FDA = U.S. Food and Drug Administration
HPA = hybridization protection assay
IND = investigational new drug
PCR = real-time polymerase chain reaction
RUO = research use only
TMA = transcription-mediated amplification

VIRUSES

Despite these hurdles, the wealth of real-time PCR assays and number of agents for which such assays have been developed, whether commercialized or from basic research reports, is amazing. Of the subspecialties that comprise the field of microbiology, real-time PCR has played the most significant role in virology, revealing relationships between agents and clinical disease. Real-time PCR has been used to ascertain multiple viral genotypes and to discriminate between agents in co-infections.[10–14]

Poorly understood relationships between specific viral agents and chronic disorders including malignancies and immunological syndromes have yielded considerable mechanistic and clinical detail to studies using real-time PCR as an investigative tool.[15–20] For example, Sidransky's group utilized real-time PCR to reveal that human papillomavirus DNA in sera of patients with head and neck squamous cell carcinomas may serve as a useful marker of early metastatic disease.[15] In another study using real-time PCR, MacKenzie and coworkers were able to rule out a role for herpes virus in common acute lymphoblastic leukemia.[20]

Other studies have determined viral load during different infective stages, establishing a role for real-time PCR in defining whether an infection is active, chronic, or in remission; the effectiveness of antiviral therapies; and host–pathogen interactions, all with practical outcomes on clinical treatment.[21–25]

Real-time PCR has also played a pivotal role in outbreak investigations and surveillance of newly emerged viruses, such as Hendra, West Nile fever, Ebola, and the lyssaviruses.[26–29] Epidemiological links between specific viral sequences and clinical disease states have been proven or supported by investigations drawing on the sensitivity and power of such assays.

An incredible range and depth of detection assays have been described for diverse viral agents from water- and food-borne pathogens like hepatitis A and E[30–35]; respiratory agents such as adenoviruses, rhinoviruses, respiratory syncytial virus, and the influenza viruses; and exotic agents like those producing smallpox and hemorrhagic fevers.[36–42]

BACTERIA

If nucleic acid-based detection has had the greatest impact to date in clinical virology, its impact on bacteriology is almost as remarkable. As with virology, real-time PCR and hybridization assays are the most widely used. The broad applications of this technology extend from pathogen identification to detection of specific virulence genes and other markers, and estimations of antibiotic resistance. Such applications have immediate benefits to patients because they can be used to determine infecting agents and also levels of infection and optimal therapeutic courses leading to reductions in hospitalization times, and the emergence of resistant strains into broader circulation.

Real-time PCR and hybridization assays have been applied widely to bacterial agents that are difficult to propagate. Not surprisingly, few pathogens have been the subjects of as many nucleic acid detection technologies as the mycobacteria. The range of assays, both commercialized and laboratory-reported, is incredibly extensive in depth and breadth. Commercial systems range from hybridization probe-based systems for the definitive identification of suspect colonies or blood cultures to real-time PCR assays that directly detect infections in pulmonary and nonpulmonary specimens.[43–51] Currently, probe-based systems are the most widely used within diagnostic and confirmatory settings, but real-time PCR is poised to supplant such systems due in part to its greater flexibility in analysis.

Legionella pneumophila is another respiratory pathogen of considerable clinical significance that is difficult to culture and isolate. Commercial systems to detect *Legionella* species based on nucleic acid technology entered the marketplace in the 1990s,[52–54] with the first real-time PCR assays for this pathogen reported around 2000.[55–57] More recently, promising multiplex real-time PCR assays have detected and differentiated several fastidious and clinically relevant respiratory pathogens.[58–61] These assays have been tested extensively on both environmental materials and clinical specimens, demonstrating excellent sensitivity and specificity in diverse samples such as sputum, aspirates, and biopsies.

Difficult to culture and identify even in reference laboratories, *Leptospira interrogans* is a bacterial pathogen of considerable significance in the developing world as a cause of febrile illness. It is usually diagnosed by serological assays in recovering patients. A newly reported real-time PCR assay detects members of this genus in clinical specimens, but at a significantly earlier time in the infection process, offering opportunities for earlier diagnosis and initiation of treatment.[62]

Borrelia species represent another important bacterial pathogen that is difficult to culture and identify by conventional means. Real-time PCR was exploited as early as 1999 to aid clinical diagnosis and detection.[63] Using environmental materials such as ticks led to considerable success in detecting *B. burgdorferi* and related bacteria, but success has not been as great with clinically relevant materials such as urine.[64–68]

More success has been seen with *Helicobacter pylori*, the causative agent of most peptic ulcers. Real-time PCR assays detected this pathogen in clinically relevant materials, including gastric biopsies, lavages, and fecal extracts,[69–72] and also detected potential resistance to antibiotics, permitting clinicians to modify treatment approaches to the needs of patients.[69,73–76] Detection of this pathogen by DNA-based approaches has been extensively covered in a recent review.[77]

Among the more common clinical pathogens, the detection and monitoring of antibiotic resistance in isolates of *Staphylococcus aureus*, *Staphylococcus epidermidis*, and *Enterococcus* species has also benefited from nucleic acid-based assays.[69,78] Treatment of fulminant and chronic syndromes ranging from meningitis and food poisoning to inflammatory bowel disease and sepsis has been significantly improved by these approaches through the rapid identification and characterization of the bacterial agents involved, enabling not only prompt individual treatment, but also tracing of the source of the outbreak.[78–80]

The proliferation of nucleic acid-based assays has been especially evident for microbial pathogens associated with biological warfare or terrorism, with real-time PCR the tool of choice. Since 2001 dozens of real-time PCR assays of varying utility have been reported for *B. anthracis*.[81–90] Such assays have provided rapid discrimination of fully virulent strains from harmless laboratory or vaccine strains. Additional assays have been reported for other potential threat agents such as *Coxiella burnetii*, *F. tularensis*, *V. cholerae*, and *Y. pestis*.

EUKARYOTES

In comparison with the wealth of available assays and those under development for bacterial and viral agents, a more limited number of nucleic acid-based assays have been described for eukaryotic pathogens. However, these assays, primarily real-time PCR techniques, have produced significant impacts on the study of fungal, parasitic and protozoan pathogens, defining relationships of microbial agent and pathology, drug resistance, and simple agent detection and identification.

Real-time PCR assays with clinical potential have been described for diverse pathogens such as *Pneumocystis jiroveci*,[91–100] *Leishmania* species,[101–110] and a wide range of water- and food-borne eukaryotic pathogens such as *Cryptosporidium parvum*,[111–119] *Cyclospora cayetanensis*,[120,121] *Giardia lamblia*,[122–126] and *Entamoeba histolytica*.[127,128] Recently, multiplexed real-time PCR assays have been reported for the combination of *E. histolytica* and *G. lamblia*.[122,129,130] Some have been sufficiently refined to be used to genotype or speciate some particularly challenging agents such as *C. parvum* and *Encephalitozoon* species.[119,131] Real-time PCR has also been used for the direct *in vivo* detection and quantitation of malarial parasites, as well as determining the maturation stage of *Plasmodium falciparum* by transcript profiling and ascertaining drug resistance.[132–134]

While generally not as important from a clinical or public health perspective, primarily because of the limited number of cases seen in most populations, the difficulty inherent in the culture and identification of fungal infections makes them especially attractive as targets for nucleic acid-based detection (reviewed by Hsu and coworkers[135]). Among the more common opportunistic and frank fungal pathogens, robust Q-PCR based assays have been described for *Candida albicans*,[135–142] *Coccidioides* species,[143,144] and *Cryptococcus neoformans*.[145]

EMERGING TECHNOLOGIES

Two emerging technologies of particular interest for nucleic acid-based detection of pathogens, gene microarrays and flow cytometry, are adaptations of existing hardware specifically for pathogen identification.[144, 146, 152] Both have already been demonstrated in concept but have not yet achieved practical use in clinical settings. A third technology known as host profile analysis is actually a new approach that draws on existing technologies such as real-time PCR to detect changes in a host that portend infection and clinical disease.[156, 157] It can also be adapted to work with microarray and flow cytometric methods. However, while promising in both theory and initial reports, host profiling has some significant (but not insurmountable) hurdles to overcome before it becomes a mainstream laboratory technology for infectious disease and pathogen identification.

MICROARRAYS

Nucleic acid arrays, commonly called microarrays, are poised to supplant real-time PCR in the infectious disease laboratory over the next decade. A key advantage of microarrays in infectious disease applications is the massively parallel nature of the technology, allowing nearly complete automation and consequent high sample throughput. Another key advantage is the volume of information that can be returned from the analysis of a single array.

Beyond simple organism identification, critical genetic characteristics of a pathogen, such as drug resistance, virulence, and epidemiological markers can in theory be probed simultaneously. Other significant advantages of array technology include reduced sample volumes and reagent needs. These advantages are offset to a considerable extent by the fragility and size of the current commercially available systems, along with initial cost and lack of commercially available arrays suitable for pathogen identification.

Probably the most exciting role for microarrays for infectious disease applications relates to the rapid identification and characterization of pathogens.[146,147] Because each microbial pathogen has a distinct genetic pedigree and a microarray interrogates this pedigree at the nucleotide level, this technology may offer the ultimate in multiplex analysis. The multi-pathogen identification microarray is a proof of this concept.[148] Sequences unique to 18 pathogenic microbes including prokaryotes, eukaryotes, and viruses were simultaneously screened and identified.

In other applications, microarrays have been used to discriminate different strains of *P. aeruginosa* and classify *M. tuberculosis* strains by drug sensitivity. In a dramatic display of the promise of this technology, microarrays were used to discriminate 56 genetically different strains of *B. anthracis*, quite likely the most monomorphic bacterial pathogen known.[149]

These specialized arrays, called resequencing microarrays, are of particular interest because of their ability to rapidly detect and characterize unknown variants of existing pathogens. In a proof of concept study reported at the 2004 Human Genome Meeting, scientists from the Health Protection Agency of the United Kingdom used a resequencing array to classify different *Neisseria meningitidis* strains. Besides correctly classifying 45 isolates identified by traditional methods, they also correctly classified 12 previously untypable isolates, demonstrating that not only can microarrays be used to identify known pathogens, but they can also recognize unusual or newly emerging strains as well.

FLOW CYTOMETRY

Flow cytometry is of particular interest to the clinical laboratory because the basic technology is common and well established for antibody–antigen interactions, and is adaptable to probing nucleic acid interactions.[150–154] Key advantages of this technology, in addition to familiarity and multiplex capabilities, are its high throughput and high degree of automation. With the advent of a new class of probes defined as microsphere-conjugated molecular beacons that can be used to detect nucleic acids in a multiplex manner, flow cytometry could be readily adapted to pathogen detection and identification.[150–154]

In a recent report using microspheres of different sizes and molecular beacons in two fluorophore colors, synthetic nucleic acid control sequences were specifically detected for three respiratory pathogens including the severe acute respiratory syndrome (SARS) coronavirus, demonstrating proof of concept for this hybrid technology using a known human pathogen.[155] An earlier study demonstrated the potential of this technology by correctly identifying 17 distinct bacterial species across the Gram-negative and Gram-positive spectra from a panel of laboratory isolates.[154]

As typical flow cytometers are able to detect up to four fluorescent channels, novel assays for the detection of infectious disease agents may allow multiplex detection of a nucleic acid panel in a single tube. Such panels could be used both to rapidly identify agents and ascertain their clinically relevant characteristics, for example, drug resistance, specific virulence properties, and relative pathogen load.

HOST PROFILE ANALYSIS

While not an assay system, host profile analysis represents a novel approach to the detection and identification of infectious disease agents through nucleic acid-based technology. Instead of directly identifying an offending pathogen by means of its genetic composition, host profile analysis determines the response of the patient to infection, identifying the agent by means of changes in the gene expression of the host that are unique in response to a given pathogen. For example, quantification of host cytokine profiles through analysis of mRNA is widely used to discriminate the early steps of immunological or pathophysiological responses. Increasingly it is being utilized for the evaluation of the disease status in select patients.

Recent work has exploited the utility of real-time PCR to ascertain levels of a broad range of interleukins, interferons, and other host factors and is now being extended to nucleic acid microarrays.[156] One particular technique, called CyProQuant-PCR (cytokine profiling quantitative PCR) can detect even a slight cytokine induction in a reproducible fashion.[157] While in its infancy as a technique, it is a promising tool for immunological monitoring of patients, especially because of its ability to be automated and maintain high throughput.

For example, when examining host response to infection with intracellular *C. pneumoniae*, *C. trachomatis*, and *S. typhimurium* pathogens, distinct expression profiles could be detected, demonstrating the ability to characterize pathogenic agents via host profile analysis.[158] *C. trachomatis* and *C. pneumoniae* infection induced CTGF, ETV4, NR4A2, DUSP4, DUSP5, GAS-1, EGR1 LIF, MIP-2, IER3, MCL-1, EPHA2, IL6, and IL8. *C. trachomatis* induced IL-11 Gro-alpha, GM-CSF, and fos-related antigen FRA- 1, while *C. pneumoniae* induced IL-8, ICAM-1, and prostaglandin endoperoxide-synthase 2 (Cox 2, PTGS2). Intracellular *Salmonella* infection caused major increases in IL-6 and IL-8 mRNA levels only.

Another study was able to discriminate infections caused by Gram-positive bacterial pathogens and those caused by Gram-negative agents through the analysis of the gene expression patterns of macrophages, providing a potential basis to rapidly discriminate these two bacterial populations in clinical infections.[159] In more recent reports, investigators exploited this approach to diagnose malaria and SARS.[157,160] While more limited in scope than the other reports mentioned, these studies suggest that cytokine profiling may be of particular interest when screening large populations for endemic and newly emergent diseases.

Should host profile analysis prove effective in the context of infectious disease studies, one public health issue that would benefit from such high throughput analysis is management of the clinical consequences of an intentional release of microbial pathogens, including those typically viewed as agents of biological warfare.[156] Profiling could be of critical importance to rapid assessment of the severity the of illness post-exposure. Since the anthrax outbreak of 2001 in the eastern United States, it is apparent that present approaches are highly specific but so labor- and time-intensive in large-scale use that they provide only the most general information on which to base clinical intervention.

PROBLEMS AND CONTROVERSIES

Despite considerable advances in nucleic acid detection for ID agents, significant issues remain that impede wider adoption of these technologies and their replacement of more conventional techniques as the new gold standards. Key among these are the following points.

First, the numbers of commercialized assays and cost-effective assays for some important pathogens is limited and should be expanded. While some rare ID agents such as *B. anthracis* are detectable by a large number of proven rapid assays, other more common and clinically relevant agents lack suitable assays.[161] Without greatly increased numbers of assays for more potential pathogens, nucleic acid-based testing is most likely to remain a "boutique" item, limited to a very few agents. In concert with assay development, well-defined mechanisms for the verification and validation of assays must be created. Specifically, uniform panels of pathogens, near-genetic neighbors, and potential interferents to be used in evaluating experimental assays are needed, along with standardized protocols for assay evaluation. Without this level of quality control, these assays will provide rapid but potentially inaccurate results, limiting their widespread adoption.

Second, more robust reagents, simpler protocols, and improved hardware are important for the adoption of these technologies by clinical laboratories. Many of the most common reagents remain designed to detect a single target and are based on wet chemistries, requiring specialized storage and potentially consuming considerable space. Some small steps have been taken to address this problem. For example, Idaho Technologies (Salt Lake City, Utah, USA, http://www.idahotech.com/) now offers lyophilized reagents off the shelf and will custom prepare such reagents, guaranteeing shelf life and performance. This nascent trend must continue and expand in order for these assays to gain broader acceptance beyond research and reference laboratories. In addition such assays must be multiplexed while retaining or even improving sensitivity and specificity. In like fashion, the hardware must become more robust, more automated, and more cost-effective. Thermocyclers should become even faster and less costly. Microarrays will need simpler operational parameters and will have to be much less expensive to foster adoption.

Finally, new guidelines for disease management and extensive quality control and standardization programs must be introduced in clinical laboratories to establish uniform standards of care and proficiency. These metrics must be created to enable laboratories to properly integrate these rapid assays into the clinical diagnostic process and provide clinicians with the tools to use the information from these rapid assays to make effective decisions for patient care.

CONCLUSIONS AND FUTURE PROSPECTS

Detection and identification of microbial pathogens in clinical settings is increasingly being accomplished by real-time PCR and hybridization assays. Advances in microarray technology and adaptation of flow cytometric methods portend the application of these methods to the direct detection of pathogens and/or patient profiling in the near future. Given the complexity of these approaches in terms of equipment requirements, reaction chemistries, and technical skills, there are three broad areas that are expected to be addressed in the near term for nucleic acid-based detection systems to become true point-of-care systems.

First, assays will increasingly be multiplexed, regardless of underlying technology, to provide differential diagnosis. Existing assays are largely designed to detect a single agent at a time, often rendering nucleic acid tests cumbersome when dealing with clinician needs for a rapid assessment of a differential diagnosis. Increasingly, solutions based on multiplexed chemistries and improved detection strategies are being developed.

One example of this approach was recently reported by Wellinghausen and co-workers at the University of Ulm, Germany.[162] They developed real-time PCR assays and algorithms for rapid detection of the most common clinically relevant bacteria from positive blood culture bottles, including *S. epidermidis, S. aureus, E. faecalis, E. faecium, Streptococcus* species, members of the Enterobacteriaceae, *Pseudomonas aeruginosa, Stenotrophomonas maltophilia, Acinetobacter* species, *Bacteroides* species, *H. influenzae*, and *N. meningitidis*. In blood, the PCR algorithm correctly detected over 98% of cultures with a single species of bacteria within a few hours. However, in mixed infections the sensitivity was lower, with 75% of cultures correctly detected; no false-positive results were detected. Detection failures appeared to result from concentrations of some of the pathogens present at concentrations below the limit of detection for the PCR assay but not for the gold standard culture method. This sort of approach, a multiplexed comprehensive assessment, must become increasingly common to speed the adoption of rapid nucleic acid technologies by clinical laboratories.

A similar approach has been described for the direct identification of enteric pathogens in stool specimens.[163] A real-time PCR procedure combined with a commercially available extraction kit allowed the discrimination of 17 species of food- or water-borne pathogens in less than 2 hours, including pathogenic variants of *E. coli, Salmonella* species, *Shigella* species, *Y. enterocolitica, Y. pseudotuberculosis, C. jejuni, V. cholerae, V. parahaemolyticus, V. vulnificus, Aeromonas* species, *S. aureus, C. perfringens*, and *B. cereus*. While the limits of detection were only 10^5 organisms per gram of stool, the rapidity of the procedure and breadth of agents detected were remarkable.

Second, detection of drug resistance will be accomplished by real-time methods. Clinical assessment of drug susceptibility typically takes several hours to days, with considerable potential consequences for a patient. More rapid methods are crucial for planning antimicrobial therapy and evaluating infection control procedures. Real-time PCR is now used in a limited fashion to detect clinically significant pathogens such as methicillin-resistant *S. aureus* (MRSA) and vancomycin-resistant enterococci (VRE), in addition to the identification and tracking of genetic elements involved in the assembly and spread of antimicrobial resistance.[164] While not without limitations, the speed (minutes to hours) and sensitivity (<100 gene copies) of this approach are outstanding relative to conventional testing methods. Of even greater promise for resistance screening and detection are microarrays. In promising pilot studies, microarrays provided more information in shorter periods and more cost effectively than traditional methods for the determination of drug resistance.[165,166]

Third, as the pace of genomics accelerates, so will the utility of nucleic acid detection technologies in clinical settings. As increasing numbers of genomes are solved to the nucleotide level and analyzed, more specific and general genetic markers will be identified for exploitation. General markers will allow related organisms including unknown ones to be detected, while specific markers will provide definitive identification and detect unusual attributes such as drug resistance or altered virulence. Features of a number of important but poorly explained human clinical syndromes such as Crohn's disease, cardiovascular disease, and reactive arthritis indicate potential microbial etiologies.[167–169] In these syndromes, the failure of cultivation-dependent microbial detection methods reveals our ignorance of microbial growth requirements. Sequence-based molecular methods, however, offer alternative approaches for microbial identification directly from host specimens found in settings of unexplained acute illnesses, chronic inflammatory diseases, and anatomic sites containing commensal microflora. The rapid expansion of genome sequence databases and advances in biotechnology present opportunities and challenges: identification of consensus sequences from which reliable, specific phylogenetic information can be inferred for all taxonomic groups of pathogens, broad-range pathogen identification on the basis of virulence-associated gene families, and use of host gene expression response profiles as specific signatures of microbial infection.[170]

Nucleic acid-based assays have proven indispensable for modern investigation of infectious diseases and their causative agents, speeding pathogen identification, elucidating pathologies, and identifying genetic markers of clinical importance. Most commercially available assays and those described in the scientific literature have enhanced the detection of microbial agents with respect to the more conventional approaches, increasing the attractiveness of this broad family of technologies in a clinical setting. As these technologies progress beyond infancy, they are poised to set new standards for reference assays in the clinical laboratory of the 21st century, promising not only to identify a specific agent that infects a patient, but revealing also its sensitivity to antimicrobial drugs, its range of virulence properties, and the prognosis of the patient. Armed with these technologies, the challenges facing modern clinical microbiologists and laboratorians may be less daunting than they seem.

REFERENCES

1. Andreotti PE, Ludwig GV, Peruski AH, Tuite JJ, Morse SS, Peruski LF, Jr. Immunoassay of infectious agents. *BioTechniques* 2003; 35 (10): 850–859.
2. Kortepeter MG, Parker GW. Potential biological weapons threats. *Emerging Infect Dis* 1999; 5 (4): 523–527.
3. Nichol ST, Arikawa J, Kawaoka Y. Emerging viral diseases. *Proc Natl Acad Sci USA* 2000; 97 (23): 12411–12412.
4. Peruski AH, Peruski LF, Jr. Immunological methods for detection and identification of infectious disease and biological warfare agents. *Clin Diagn Lab Immunol* 2003; 10 (4): 506–513.
5. Peruski LF, Jr., Peruski AH. Rapid diagnostic assays in the genomic biology era: detection and identification of infectious disease and biological weapon agents. *BioTechniques* 2003; 35 (4): 840–846.
6. Dumler JS, Valsamakis A. Molecular diagnostics for existing and emerging infections: complementary tools for a new era of clinical microbiology. *Am J Clin Pathol* 1999; 112 (Suppl 1): S33–S39.
7. Houpikian P, Raoult D. Traditional and molecular techniques for the study of emerging bacterial diseases: one laboratory's perspective. *Emerging Infect Dis* 2002; 8 (2): 122–131.
8. Ieven M. Molecular methods in the diagnostic microbiology laboratory: when to start and where to stop? *Verh K Acad Geneeskd Belg* 2000; 62 (1): 15–30.
9. Pfaller MA. Molecular approaches to diagnosing and managing infectious diseases: practicality and costs. *Emerging Infect Dis* 2001; 7 (2): 312–318.
10. Furuta Y, Ohtani F, Sawa H, Fukuda S, Inuyama Y. Quantitation of varicella zoster virus DNA in patients with Ramsay–Hunt syndrome and zoster sine herpete. *J Clin Microbiol* 2001; 39 (8): 2856–2859.
11. Jordens JZ, Lanham S, Pickett MA, Amarasekara S, Abeywickrema I, Watt PJ. Amplification with molecular beacon primers and reverse line blotting for the detection and typing of human papillomaviruses. *J Virol Methods* 2000; 89 (1–2): 29–37.
12. Kearns AM, Turner AJL, Taylor CE, George PW, Freeman R, Gennery AR. LightCycler-based quantitative PCR for rapid detection of human herpesvirus 6 DNA in clinical material. *J Clin Microbiol* 2001; 39 (8): 3020–3021.
13. Ohtani F, Furuta Y, Fukuda S, Inuyama Y. Herpes virus reactivation and serum tumor necrosis factor-alpha levels in patients with acute peripheral facial palsy. *Auris Nasus Larynx* 2001; 28 (Suppl): S145–S147.
14. Zerr DM, Huang ML, Corey L, Erickson M, Parker HL, Frenkel LM. Sensitive method for detection of human herpesviruses 6 and 7 in saliva collected in field studies. *J Clin Microbiol* 2000; 38 (5): 1981–1983.
15. Capone RB, Pai SI, Koch WM, Gillison ML, Danish HN, Westra WH, Daniel R, Shah KV, Sidransky D. Detection and quantitation of human papillomavirus (HPV) DNA in the sera of patients with HPV-associated head and neck squamous cell carcinoma. *Clin Cancer Res* 2000; 6 (11): 4171–4175.

16. Jabs WJ, Hennig H, Kittel M, Pethig K, Smets F, Bucsky P, Kirchner H, Wagner HJ. Normalized quantification by real-time PCR of Epstein–Barr virus load in patients at risk for post-transplant lymphoproliferative disorders. *J Clin Microbiol* 2001; 39 (2): 564–569.

17. Josefsson A, Livak K, Gyllensten U. Detection and quantitation of human papillomavirus by using the fluorescent 5' exonuclease assay. *J Clin Microbiol* 1999; 37 (3): 490–496.

18. Kennedy MM, Lucas SB, Russell-Jones R, Howells DD, Picton SJ, Bardon A, Comley IL, McGee JO, O'Leary JJ. HHV8 and female Kaposi's sarcoma. *J Pathol* 1997; 183 (4): 447–452.

19. Lo YMD, Chan LYS, Lo KW, Leung SF, Zhang J, Chan ATC, Lee JCK, Hjelm NM, Johnson PJ, Huang DP. Quantitative analysis of cell-free Epstein-Barr virus DNA in plasma of patients with nasopharyngeal carcinoma. *Cancer Res* 1999; 59 (6): 1188–1191.

20. MacKenzie J, Gallagher A, Clayton RA, Perry J, Eden OB, Ford AM, Greaves MF, Jarrett RF. Screening for herpesvirus genomes in common acute lymphoblastic leukemia, *Leukemia* 2001; 15 (3): 415–421.

21. Schutten M, Niesters HGM. Clinical utility of viral quantification as a tool for disease monitoring. *Expert Rev Mol Diagn* 2001; 1 (2): 153–162.

22. Clementi M. Quantitative molecular analysis of virus expression and replication. *J Clin Microbiol* 2000; 38 (6): 2030–2036.

23. Holodniy M, Katzenstein DA, Sengupta S, Wang AM, Casipit C, Schwartz DH, Konrad M, Groves E, Merigan TC. Detection and quantification of human immunodeficiency virus RNA in patient serum by use of the polymerase chain reaction. *J Infect Dis* 1991; 163 (4): 862–866.

24. Limaye AP, Jerome KR, Kuhr CS, Ferrenberg J, Huang M-L, Davis CL, Corey L, Marsh CL. Quantitation of BK virus load in serum for the diagnosis of BK virus-associated nephropathy in renal transplant recipients. *J Infect Dis* 2001; 183 (11): 1669–1672.

25. Nitsche A, Steuer N, Schmidt CA, Landt O, Siegert W. Different real-time PCR formats compared for the quantitative detection of human cytomegalovirus DNA. *Clin Chem* 1999; 45 (11): 1932–1937.

26. Gibb TR, Norwood DA, Jr., Woollen N, Henchal EA. Development and evaluation of a fluorogenic 5' nuclease assay to detect and differentiate between Ebola virus subtypes Zaire and Sudan. *J Clin Microbiol* 2001; 39 (11): 4125–4130.

27. Lanciotti RS, Kerst AJ. Nucleic acid sequence-based amplification assays for rapid detection of West Nile and St. Louis encephalitis viruses. *J Clin Microbiol* 2001; 39 (12): 4506–4513.

28. Smith IL, Halpin K, Warrilow D, Smith GA. Development of a fluorogenic RT-PCR assay (TaqMan) for the detection of Hendra virus. *J Virol Methods* 2001; 98 (1): 33–40.

29. Smith IL, Northill JA, Harrower BJ, Smith GA. Detection of Australian bat lyssavirus using a fluorogenic probe. *J Clin Virol* 2002; 25 (3): 285–291.

30. Abd El Galil KH, El Sokkary MA, Kheira SM, Salazar AM, Yates MV, Chen W, Mulchandani A. Combined immunomagnetic separation-molecular beacon-reverse transcription-PCR assay for detection of hepatitis A virus from environmental samples. *Appl Environ Microbiol* 2004; 70 (7): 4371–4374.

31. Hewitt J, Greening GE. Survival and persistence of norovirus, hepatitis A virus, and feline calicivirus in marinated mussels. *J Food Prot* 2004; 67 (8): 1743–1750.

32. Mansuy JM, Peron JM, Abravanel F, Poirson H, Dubois M, Miedouge M, Vischi F, Alric L, Vinel JP, Izopet J. Hepatitis E in the southwest of France in individuals who have never visited an endemic area. *J Med Virol* 2004; 74 (3): 419–424.

33. Orru G, Masia G, Orru G, Romano L, Piras V, Coppola RC. Detection and quantitation of hepatitis E virus in human faeces by real-time quantitative PCR. *J Virol Methods* 2004; 118 (2): 77–82.

34. Rezende G, Roque-Afonso AM, Samuel D, Gigou M, Nicand E, Ferre V, Dussaix E, Bismuth H, Feray C. Viral and clinical factors associated with the fulminant course of hepatitis A infection. *Hepatology* 2003; 38 (3): 613–618.

35. Weiss J, Wu H, Farrenkopf B, Schultz T, Song G, Shah S, Siegel J. Real time TaqMan PCR detection and quantitation of HBV genotypes A–G with the use of an internal quantitation standard. *J Clin Virol* 2004; 30 (1): 86–93.

36. Espy MJ, Cockerill FR, III, Meyer RF, Bowen MD, Poland GA, Hadfield TL, Smith TF. Detection of smallpox virus DNA by LightCycler PCR. *J Clin Microbiol* 2002; 40 (6): 1985–1988. Correction: *J Clin Microbiol* 2002; 40 (11): 4405.

37. Kulesh DA, Baker RO, Loveless BM, Norwood D, Zwiers SH, Mucker E, Hartmann C, Herrera R, Miller D, Christensen D, Wasieloski LP, Jr., Huggins J, Jahrling PB. Smallpox and pan-orthopox virus detection by real-time 3'-minor groove binder TaqMan assays on the Roche LightCycler and the Cepheid Smart Cycler platforms. *J Clin Microbiol* 2004; 42 (2): 601–609.
38. Kulesh DA, Loveless BM, Norwood D, Garrison J, Whitehouse CA, Hartmann C, Mucker E, Miller D, Wasieloski LP, Jr., Huggins J, Huhn G, Miser LL, Imig C, Martinez M, Larsen T, Rossi CA, Ludwig GV. Monkeypox virus detection in rodents using real-time 3' minor groove binder TaqMan® assays on the Roche LightCycler. *Lab Invest* 2004; 84 (9): 1200–1208.
39. Nitsche A, Ellerbrok H, Pauli G. Detection of orthopoxvirus DNA by real-time PCR and identification of variola virus DNA by melting analysis. *J Clin Microbiol* 2004; 42 (3): 1207–1213.
40. Olson VA, Laue T, Laker MT, Babkin IV, Drosten C, Shchelkunov SN, Niedrig M, Damon IK, Meyer H. Real-time PCR system for detection of orthopoxviruses and simultaneous identification of smallpox virus. *J Clin Microbiol* 2004; 42 (5): 1940–1946.
41. Panning M, Asper M, Kramme S, Schmitz H, Drosten C. Rapid detection and differentiation of human pathogenic orthopox viruses by a fluorescence resonance energy transfer real-time PCR assay. *Clin Chem* 2004; 50 (4): 702–708.
42. Sofi Ibrahim M, Kulesh DA, Saleh SS, Damon IK, Esposito JJ, Schmaljohn AL, Jahrling PB. Real-time PCR assay to detect smallpox virus. *J Clin Microbiol* 2003; 41 (8): 3835–3839.
43. Bruijnesteijn van Coppenraet ES, Lindeboom JA, Prins JM, Peeters MF, Claas ECJ, Kuijper EJ. Real-time PCR assay using fine-needle aspirates and tissue biopsy specimens for rapid diagnosis of mycobacterial lymphadenitis in children. *J Clin Microbiol* 2004; 42 (6): 2644–2650.
44. Cleary TJ, Roudel G, Casillas O, Miller N. Rapid and specific detection of *Mycobacterium tuberculosis* by using the Smart Cycler instrument and a specific fluorogenic probe. *J Clin Microbiol* 2003; 41 (10): 4783–4786.
45. Desjardin L, Chen Y, Perkins MD, Teixeira L, Cave MD, Eisenach KD. Comparison of the ABI 7700 System (TaqMan) and competitive PCR for quantification of IS*6110* DNA in sputum during treatment of tuberculosis. *J Clin Microbiol* 1998; 36 (7): 1964–1968.
46. Garcia de Viedma D, del Sol Diaz Infantes M, Lasala F, Chaves F, Alcala L, Bouza E. New real-time PCR able to detect in a single tube multiple rifampin resistance mutations and high-level isoniazid resistance mutations in *Mycobacterium tuberculosis*. *J Clin Microbiol* 2002; 40 (3): 988–995.
47. Lemaitre N, Armand S, Vachee A, Capilliez O, Dumoulin C, Courcol RJ. Comparison of the real-time PCR method and the Gen-Probe amplified *Mycobacterium tuberculosis* direct test for detection of *Mycobacterium tuberculosis* in pulmonary and nonpulmonary specimens. *J Clin Microbiol* 2004; 42 (9): 4307–4309.
48. Miller N, Cleary T, Kraus G, Young AK, Spruill G, Hnatyszyn HJ. Rapid and specific detection of *Mycobacterium tuberculosis* from acid-fast bacillus smear-positive respiratory specimens and BacT/ALERT MP culture bottles by using fluorogenic probes and real-time PCR. *J Clin Microbiol* 2002; 40 (11): 4143–4147.
49. Ruiz M, Torres MJ, Llanos AC, Arroyo A, Palomares JC, Aznar J. Direct detection of rifampin- and isoniazid-resistant *Mycobacterium tuberculosis* in auramine-rhodamine-positive sputum specimens by real-time PCR. *J Clin Microbiol* 2004; 42 (4): 1585–1589.
50. Sajduda A, Brzostek A, Poplawska M, Augustynowicz-Kopec E, Zwolska Z, Niemann S, Dziadek J, Hillemann D. Molecular characterization of rifampin- and isoniazid-resistant *Mycobacterium tuberculosis* strains isolated in Poland. *J Clin Microbiol* 2004; 42 (6): 2425–2431.
51. Shrestha NK, Tuohy MJ, Hall GS, Reischl U, Gordon SM, Procop GW. Detection and differentiation of *Mycobacterium tuberculosis* and nontuberculous mycobacterial isolates by real-time PCR. *J Clin Microbiol* 2003; 41 (11): 5121–5126.
52. Matsiota-Bernard P, Pitsouni E, Legakis N, Nauciel C. Evaluation of commercial amplification kit for detection of *Legionella pneumophila* in clinical specimens. *J Clin Microbiol* 1994; 32 (6): 1503–1505.
53. Saint CP. A colony based confirmation assay for *Legionella* and *Legionella pneumophila* employing the EnviroAmp Legionella system and seroagglutination. *Lett Appl Microbiol* 1998; 26 (5): 377–381.
54. Weir SC, Fischer SH, Stock F, Gill VJ. Detection of *Legionella* by PCR in respiratory specimens using a commercially available kit. *Am J Clin Pathol* 1998; 110 (3): 295–300.

55. Ballard AL, Fry NK, Chan L, Surman SB, Lee JV, Harrison TG, Towner KJ. Detection of *Legionella pneumophila* using a real-time PCR hybridization assay. *J Clin Microbiol* 2000; 38 (11): 4215–4218.

56. Hayden RT, Uhl JR, Qian X, Hopkins MK, Aubry MC, Limper AH, Lloyd RV, Cockerill FR. Direct detection of *Legionella* species from bronchoalveolar lavage and open lung biopsy specimens: comparison of LightCycler PCR, in situ hybridization, direct fluorescence antigen detection, and culture. *J Clin Microbiol* 2001; 39 (7): 2618–2626.

57. Wellinghausen N, Frost C, Marre R. Detection of legionellae in hospital water samples by quantitative real-time LightCycler PCR. *Appl Environ Microbiol* 2001; 67 (9): 3985–3993.

58. Khanna M, Fan J, Pehler-Harrington K, Waters C, Douglass P, Stallock J, Kehl S, Henrickson KJ. The pneumoplex assays, a multiplex PCR-enzyme hybridization assay that allows simultaneous detection of five organisms, *Mycoplasma pneumoniae, Chlamydia (Chlamydophila) pneumoniae, Legionella pneumophila, Legionella micdadei*, and *Bordetella pertussis*, and its real-time counterpart. *J Clin Microbiol* 2005; 43 (2): 565–571.

59. Maltezou HC, La-Scola B, Astra H, Constantopoulou I, Vlahou V, Kafetzis DA, Constantopoulos AG, Raoult D. *Mycoplasma pneumoniae* and *Legionella pneumophila* in community-acquired lower respiratory tract infections among hospitalized children: diagnosis by real time PCR. *Scand J Infect Dis* 2004; 36 (9): 639–642.

60. Templeton KE, Scheltinga SA, Sillekens P, Crielaard JW, van Dam AP, Goossens H, Claas ECJ. Development and clinical evaluation of an internally controlled, single-tube multiplex real-time PCR assay for detection of *Legionella pneumophila* and other *Legionella* species. *J Clin Microbiol* 2003; 41 (9): 4016–4021.

61. Welti M, Jaton K, Altwegg M, Sahli R, Wenger A, Bille J. Development of a multiplex real-time quantitative PCR assay to detect *Chlamydia pneumoniae, Legionella pneumophila* and *Mycoplasma pneumoniae* in respiratory tract secretions. *Diagn Microbiol Infect Dis* 2003; 45 (2): 85–95.

62. Palaniappan RU, Chang YF, Chang CF, Pan MJ, Yang CW, Harpending P, McDonough SP, Dubovi E, Divers T, Qu J, Roe B. Evaluation of lig-based conventional and real time PCR for the detection of pathogenic leptospires. *Mol Cell Probes* 2005; 19 (2): 111–117.

63. Pahl A, Kuhlbrandt U, Brune K, Rollinghoff M, Gessner A. Quantitative detection of *Borrelia burgdorferi* by real-time PCR. *J Clin Microbiol* 1999; 37 (6): 1958–1963.

64. Courtney JW, Kostelnik LM, Zeidner NS, Massung RF. Multiplex real-time PCR for detection of *Anaplasma phagocytophilum* and *Borrelia burgdorferi*. *J Clin Microbiol* 2004; 42 (7): 3164–3168.

65. Wagner EM, Schmidt BL, Bergmann AR, Derler AM, Aberer E. Inability of one-step real-time PCR to detect *Borrelia burgdorferi* DNA in urine. *J Clin Microbiol* 2004; 42 (2): 938.

66. Wang G, Liveris D, Brei B, Wu H, Falco RC, Fish D, Schwartz I. Real-time PCR for simultaneous detection and quantification of *Borrelia burgdorferi* in field-collected *Ixodes scapularis* ticks from the northeastern United States. *Appl Environ Microbiol* 2003; 69 (8): 4561–4565.

67. Piesman J, Schneider BS, Zeidner NS. Use of quantitative PCR to measure density of *Borrelia burgdorferi* in the midgut and salivary glands of feeding tick vectors. *J Clin Microbiol* 2001; 39 (11): 4145–4148.

68. Schwaiger M, Peter O, Cassinotti P. Routine diagnosis of *Borrelia burgdorferi* (sensu lato) infections using a real-time PCR assay. *Clin Microbiol Infect* 2001; 7 (9): 461–469.

69. Chisholm SA, Owen RJ, Teare EL, Saverymuttu S. PCR-based diagnosis of *Helicobacter pylori* Infection and real-time determination of clarithromycin resistance directly from human gastric biopsy samples. *J Clin Microbiol* 2001; 39 (4): 1217–1220.

70. He Q, Wang JP, Osato M, Lachman LB. Real-time quantitative PCR for detection of *Helicobacter pylori*. *J Clin Microbiol* 2002; 40 (10): 3720–3728.

71. Kobayashi D, Eishi Y, Ohkusa T, Ishige I, Suzuki T, Minami J, Yamada T, Takizawa T, Koike M. Gastric mucosal density of *Helicobacter pylori* estimated by real-time PCR compared with results of urea breath test and histological grading. *J Med Microbiol* 2002; 51 (4): 305–311.

72. Mikula, M., Dzwonek, A., Jagusztyn-Krynicka, K., and Ostrowski, J., Quantitative detection for low levels of *Helicobacter pylori* infection in experimentally infected mice by real-time PCR. *J Microbiol Methods* 2003; 55 (2): 351–359.

73. Glocker E, Kist M. Rapid detection of point mutations in the *gyrA* gene of *Helicobacter pylori* conferring resistance to ciprofloxacin by a fluorescence resonance energy transfer-based real-time PCR approach. *J Clin Microbiol* 2004; 42 (5): 2241–2246.

74. Lascols C, Lamarque D, Costa JM, Copie-Bergman C, Le Glaunec JM, Deforges L, Soussy CJ, Petit JC, Delchier JC, Tankovic J. Fast and accurate quantitative detection of *Helicobacter pylori* and identification of clarithromycin resistance mutations in *H. pylori* isolates from gastric biopsy specimens by real-time PCR. *J Clin Microbiol* 2003; 41 (10): 4573–4577.

75. Oleastro M, Menard A, Santos A, Lamouliatte H, Monteiro L, Barthelemy P, Megraud F. Real-time PCR assay for rapid and accurate detection of point mutations conferring resistance to clarithromycin in *Helicobacter pylori*. *J Clin Microbiol* 2003; 41 (1): 397–402.

76. Schabereiter-Gurtner C, Hirschl AM, Dragosics B, Hufnagl P, Puz S, Kovach Z, Rotter M, Makristathis A. Novel real-time PCR assay for detection of *Helicobacter pylori* infection and simultaneous clarithromycin susceptibility testing of stool and biopsy specimens. *J Clin Microbiol* 2004; 42 (10): 4512–4518.

77. Ruzsovics A, Molnar B, Tulassay Z. Review article: Deoxyribonucleic acid-based diagnostic techniques to detect *Helicobacter pylori*. *Aliment Pharmacol Ther* 2004; 19 (11): 1137–1146.

78. Probert WS, Bystrom SL, Khashe S, Schrader KN, Wong JD. 5′ exonuclease assay for detection of serogroup Y *Neisseria meningitidis*. *J Clin Microbiol* 2002; 40 (11): 4325–4328.

79. Bellin T, Pulz M, Matussek A, Hempen HG, Gunzer F. Rapid detection of enterohemorrhagic *Escherichia coli* by real-time PCR with fluorescent hybridization probes. *J Clin Microbiol* 2001; 39 (1): 370–374.

80. Ke D, Menard C, Picard FJ, Boissinot M, Ouellette M, Roy PH, Bergeron MG. Development of conventional and real-time PCR assays for the rapid detection of group B streptococci. *Clin Chem* 2000; 46 (3): 324–331.

81. Bell CA, Uhl JR, Hadfield TL, David JC, Meyer RF, Smith TF, Cockerill FR, III. Detection of *Bacillus anthracis* DNA by LightCycler PCR. *J Clin Microbiol* 2002; 40 (8): 2897–2902.

82. Bode E, Hurtle W, Norwood D. Real-time PCR assay for a unique chromosomal sequence of *Bacillus anthracis*. *J Clin Microbiol* 2004; 42 (12): 5825–5831.

83. Drago L, Lombardi A, Vecchi ED, Gismondo MR. Real-time PCR assay for rapid detection of *Bacillus anthracis* spores in clinical samples. *J Clin Microbiol* 2002; 40 (11): 4399.

84. Ellerbrok H, Nattermann H, Ozel M, Beutin L, Appel B, Pauli G. Rapid and sensitive identification of pathogenic and apathogenic *Bacillus anthracis* by real-time PCR. *FEMS Microbiol Lett* 2002; 214 (1): 51–59.

85. Espy MJ, Uhl JR, Sloan LM, Rosenblatt JE, Cockerill FR 3rd, Smith TF. Detection of vaccinia virus, herpes simplex virus, varicella-zoster virus, and *Bacillus anthracis* DNA by LightCycler polymerase chain reaction after autoclaving: implications for biosafety of bioterrorism agents. *Mayo Clin Proc* 2002; 77 (7): 624–628.

86. Higgins JA, Cooper M, Schroeder-Tucker L, Black S, Miller D, Karns JS, Manthey E, Breeze R, Perdue ML. A field investigation of *Bacillus anthracis* contamination of U.S. Department of Agriculture and other Washington, D.C. buildings during the anthrax attack of October 2001. *Appl Environ Microbiol* 2003; 69 (1): 593–599.

87. Hoffmaster AR, Meyer RF, Bowen MP, Marston CK, Weyant RS, Barnett GA, Sejvar JJ, Jernigan JA, Perkins BA, Popovic T. Evaluation and validation of a real-time polymerase chain reaction assay for rapid identification of *Bacillus anthracis*. *Emerging Infect Dis* 2002; 8 (10): 1178–1182.

88. Makino SI, Cheun HI, Watarai M, Uchida I, Takeshi K. Detection of anthrax spores from the air by real-time PCR. *Lett Appl Microbiol* 2001; 33 (3): 237–240.

89. Oggioni MR, Meacci F, Carattoli A, Ciervo A, Orru G, Cassone A, Pozzi G. Protocol for real-time PCR identification of anthrax spores from nasal swabs after broth enrichment. *J Clin Microbiol* 2002; 40 (11): 3956–3963.

90. Uhl JR, Bell CA, Sloan LM, Espy MJ, Smith TF, Rosenblatt JE, Cockerill FR 3rd. Application of rapid-cycle real-time polymerase chain reaction for the detection of microbial pathogens: the Mayo–Roche rapid anthrax test. *Mayo Clin Proc* 2002; 77 (7): 673–680.

91. Brancart F, Rodriguez-Villalobos H, Fonteyne PA, Peres-Bota D, Liesnard C. Quantitative TaqMan PCR for detection of *Pneumocystis jiroveci*. *J Microbiol Methods* 2005; 61 (3): 381–387.

92. Flori P, Bellete B, Durand F, Raberin H, Cazorla C, Hafid J, Lucht F, Sung RTM. Comparison between real-time PCR, conventional PCR and different staining techniques for diagnosing *Pneumocystis jiroveci* pneumonia from bronchoalveolar lavage specimens. *J Med Microbiol* 2004; 53 (7): 603–607.

93. Kaiser K, Rabodonirina M, Picot S. Real time quantitative PCR and RT-PCR for analysis of *Pneumocystis carinii hominis*. *J Microbiol Methods* 2001; 45 (2): 113–118.

94. Larsen HH, Kovacs JA, Stock F, Vestereng VH, Lundgren B, Fischer SH, Gill VJ. Development of a rapid real-time PCR assay for quantitation of *Pneumocystis carinii* f. sp. *carinii*. *J Clin Microbiol* 2002; 40 (8): 2989–2993.

95. Larsen HH, Masur H, Kovacs JA, Gill VJ, Silcott VA, Kogulan P, Maenza J, Smith M, Lucey DR, Fischer SH. Development and evaluation of a quantitative, touch-down, real-time PCR assay for diagnosing *Pneumocystis carinii* pneumonia. *J Clin Microbiol* 2002; 40 (2): 490–494.

96. Linke MJ, Rebholz S, Collins M, Tanaka R, Cushion MT. Noninvasive method for monitoring *Pneumocystis carinii* pneumonia. *Emerging Infect Dis* 2003; 9 (12): 1613–1616.

97. Meliani L, Develoux M, Marteau-Miltgen M, Magne D, Barbu V, Poirot JL, Roux P. Real time quantitative PCR assay for *Pneumocystis jirovecii* detection. *J Eukaryot Microbiol* 2003; 50 (Suppl): 651.

98. Ndam NG, Dumont B, Demanche C, Chapel A, Lacube P, Guillot J, Roux P. Development of a real-time PCR-based fluorescence assay for rapid detection of point mutations in *Pneumocystis jirovecii* dihydropteroate synthase gene. *J Eukaryot Microbiol* 2003; 50 (Suppl): 658–660.

99. Palladino S, Kay I, Fonte R, Flexman J. Use of real-time PCR and the LightCycler system for the rapid detection of *Pneumocystis carinii* in respiratory specimens. *Diagn Microbiol Infect Dis* 2001; 39 (4): 233–236.

100. Vestereng VH, Bishop LR, Hernandez B, Kutty G, Larson HH, Kovacs JA. Quantitative real-time polymerase chain-reaction assay allows characterization of pneumocystis infection in immunocompetent mice. *J Infect Dis* 2004; 189 (8): 1540–1544.

101. Bretagne S, Durand R, Olivi M, Garin JF, Sulahian A, Rivollet D, Vidaud M, Deniau M. Real-time PCR as a new tool for quantifying *Leishmania infantum* in liver in infected mice. *Clin Diagn Lab Immunol* 2001; 8 (4): 828–831.

102. Gomez-Saladin E, Doud CW, Maroli M. Short report: Surveillance of *Leishmania* sp. among sand flies in Sicily (Italy) using a fluorogenic real-time polymerase chain reaction. *Am J Trop Med Hyg* 2005; 72 (2): 138–141.

103. Mary C, Faraut F, Lascombe L, Dumon H. Quantification of *Leishmania infantum* DNA by a real-time PCR assay with high sensitivity. *J Clin Microbiol* 2004; 42 (11): 5249–5255.

104. Nicolas L, Milon G, Prina E. Rapid differentiation of Old World Leishmania species by LightCycler polymerase chain reaction and melting curve analysis. *J Microbiol Methods* 2002; 51 (3): 295–299.

105. Nicolas L, Prina E, Lang T, Milon G. Real-time PCR for detection and quantitation of *Leishmania* in mouse tissues. *J Clin Microbiol* 2002; 40 (5): 1666–1669.

106. Ogg MM, Carrion R, Jr., de Carvalho Botelho AC, Mayrink W, Correa-Oliveira R, Patterson JL. Short report: Quantification of Leishmaniavirus RNA in clinical samples and its possible role in pathogenesis. *Am J Trop Med Hyg* 2003; 69 (3): 309–313.

107. Rolao N, Cortes S, Rodrigues OR, Campino L. Quantification of *Leishmania infantum* parasites in tissue biopsies by real-time polymerase chain reaction and polymerase chain reaction-enzyme-linked immunosorbent assay. *J Parasitol* 2004; 90 (5): 1150–1154.

108. Schulz A, Mellenthin K, Schonian G, Fleischer B, Drosten C. Detection, differentiation, and quantitation of pathogenic *Leishmania* organisms by a fluorescence resonance energy transfer-based real-time PCR assay. *J Clin Microbiol* 2003; 41 (4): 1529–1535.

109. Svobodova M, Votypka J, Nicolas L, Volf P. *Leishmania tropica* in the black rat (*Rattus rattus*): persistence and transmission from asymptomatic host to sand fly vector *Phlebotomus sergenti*. *Microbes Infect* 2003; 5 (5): 361–364.

110. Vitale F, Reale S, Vitale M, Petrotta E, Torina A, Caracappa S. TaqMan-based detection of *Leishmania infantum* DNA using canine samples. *Ann NY Acad Sci* 2004; 1026: 139–143.

111. Amar CF, Dear PH, McLauchlin J. Detection and identification by real time PCR/RFLP analyses of *Cryptosporidium* species from human faeces. *Lett Appl Microbiol* 2004; 38 (3): 217–222.

112. Fontaine M, Guillot E. Development of a TaqMan quantitative PCR assay specific for *Cryptosporidium parvum*. *FEMS Microbiol Lett* 2002; 214 (1): 13–17.

113. Fontaine M, Guillot E. An immunomagnetic separation-real-time PCR method for quantification of *Cryptosporidium parvum* in water samples. *J Microbiol Methods* 2003; 54 (1): 29–36.

114. Fontaine M, Guillot E. Study of 18S rRNA and rDNA stability by real-time RT-PCR in heat-inactivated *Cryptosporidium parvum* oocysts. *FEMS Microbiol Lett* 2003; 226 (2): 237–243.

115. Higgins JA, Fayer R, Trout JM, Xiao L, Lal AA, Kerby S, Jenkins MC. Real-time PCR for the detection of *Cryptosporidium parvum*. *J Microbiol Methods* 2001; 47 (3): 323–337.

116. Limor JR, Lal AA, Xiao L. Detection and differentiation of *Cryptosporidium* parasites that are pathogenic for humans by real-time PCR. *J Clin Microbiol* 2002; 40 (7): 2335–2338.

117. MacDonald LM, Sargent K, Armson A, Thompson RC, Reynoldson JA. The development of a real-time quantitative-PCR method for characterisation of a *Cryptosporidium parvum in vitro* culturing system and assessment of drug efficacy. *Mol Biochem Parasitol* 2002; 121 (2): 279–282.

118. Tanriverdi S, Arslan MO, Akiyoshi DE, Tzipori S, Widmer G. Identification of genotypically mixed *Cryptosporidium parvum* populations in humans and calves. *Mol Biochem Parasitol* 2003; 130 (1): 13–22.

119. Tanriverdi S, Tanyeli A, Baslamisli F, Koksal F, Kilinc Y, Feng X, Batzer G, Tzipori S, Widmer G. Detection and genotyping of oocysts of *Cryptosporidium parvum* by real-time PCR and melting curve analysis. *J Clin Microbiol* 2002; 40 (9): 3237–3244.

120. Varma M, Hester JD, Schaefer FW 3rd, Ware MW, Lindquist HD. Detection of *Cyclospora cayetanensis* using a quantitative real-time PCR assay. *J Microbiol Methods* 2003; 53 (1): 27–36.

121. Verweij JJ, Laeijendecker D, Brienen EA, van Lieshout L, Polderman AM. Detection of *Cyclospora cayetanensis* in travellers returning from the tropics and subtropics using microscopy and real-time PCR. *Int J Med Microbiol* 2003; 293 (2–3): 199–202.

122. Amar CFL, Dear PH, McLauchlin J. Detection and genotyping by real-time PCR/RFLP analyses of *Giardia duodenalis* from human faeces. *J Med Microbiol* 2003; 52 (8): 681–683.

123. Bertrand I, Gantzer C, Chesnot T, Schwartzbrod J. Improved specificity for *Giardia lamblia* cyst quantification in wastewater by development of a real-time PCR method. *J Microbiol Methods* 2004; 57 (1): 41–53.

124. Guy RA, Xiao C, Horgen PA. Real-Time PCR assay for detection and genotype differentiation of *Giardia lamblia* in stool specimens. *J Clin Microbiol* 2004; 42 (7): 3317–3320.

125. Ng CT, Gilchrist CA, Lane A, Roy S, Haque R, Houpt ER. Multiplex real-time PCR assay using Scorpion probes and DNA capture for genotype-specific detection of *Giardia lamblia* on fecal samples. *J Clin Microbiol* 2005; 43 (3): 1256–1260.

126. Verweij JJ, Schinkel J, Laeijendecker D, van Rooyen MA, van Lieshout L, Polderman AM. Real-time PCR for the detection of *Giardia lamblia*. *Mol Cell Probes* 2003; 17 (5): 223–225.

127. Blessmann J, Buss H, Nu PAT, Dinh BT, Ngo QTV, Van AL, Alla MDA, Jackson TFHG, Ravdin JI, Tannich E. Real-Time PCR for detection and differentiation of *Entamoeba histolytica* and *Entamoeba dispar* in fecal samples. *J Clin Microbiol* 2002; 40 (12): 4413–4417.

128. Kebede A, Verweij JJ, Endeshaw T, Messele T, Tasew G, Petros B, Polderman AM. The use of real-time PCR to identify *Entamoeba histolytica* and *E. dispar* infections in prisoners and primary-school children in Ethiopia. *Ann Trop Med Parasitol* 2004; 98 (1): 43–48.

129. Guy RA, Payment P, Krull UJ, Horgen PA. Real-time PCR for quantification of *Giardia* and *Cryptosporidium* in environmental water samples and sewage. *Appl Environ Microbiol* 2003; 69 (9): 5178–5185.

130. Verweij JJ, Blange RA, Templeton K, Schinkel J, Brienen EAT, van Rooyen MAA, van Lieshout L, Polderman AM. Simultaneous detection of *Entamoeba histolytica, Giardia lamblia*, and *Cryptosporidium parvum* in fecal samples by using multiplex real-time PCR. *J Clin Microbiol* 2004; 42 (3): 1220–1223.

131. Wolk DM, Schneider SK, Wengenack NL, Sloan LM, Rosenblatt JE. Real-time PCR method for detection of *Encephalitozoon intestinalis* from stool specimens. *J Clin Microbiol* 2002; 40 (11): 3922–3928.

132. Blair PL, Witney A, Haynes JD, Moch JK, Carucci DJ, Adams JH. Transcripts of developmentally regulated *Plasmodium falciparum* genes quantified by real-time RT-PCR. *Nucleic Acids Res* 2002; 30 (10): 2224–2231.

133. Hermsen CC, Telgt DS, Linders EH, van de Locht LA, Eling WM, Mensink EJ, Sauerwein RW. Detection of *Plasmodium falciparum* malaria parasites *in vivo* by real-time quantitative PCR. *Mol Biochem Parasitol* 2001; 118 (2): 247–251.

134. Lee MA, Tan CH, Aw LT, Tang CS, Singh M, Lee SH, Chia HP, Yap EPH. Real-time fluorescence-based PCR for detection of malaria parasites. *J Clin Microbiol* 2002; 40 (11): 4343–4345.

135. Hsu MC, Chen KW, Lo HJ, Chen YC, Liao MH, Lin YH, Li SY. Species identification of medically important fungi by use of real-time LightCycler PCR. *J Med Microbiol* 2003; 52 (12): 1071–1076.

136. Bu R, Sathiapalan RK, Ibrahim MM, Al Mohsen I, Almodavar E, Gutierrez MI, Bhatia K. Monochrome LightCycler PCR assay for detection and quantification of five common species of *Candida* and *Aspergillus*. *J Med Microbiol* 2005; 54 (3): 243–248.

137. Chau AS, Mendrick CA, Sabatelli FJ, Loebenberg D, McNicholas PM. Application of real-time quantitative PCR to molecular analysis of *Candida albicans* strains exhibiting reduced susceptibility to azoles. *Antimicrob Agents Chemother* 2004; 48 (6): 2124–2131.

138. Frade JP, Warnock DW, Arthington-Skaggs BA. Rapid quantification of drug resistance gene expression in *Candida albicans* by reverse transcriptase LightCycler PCR and fluorescent probe hybridization. *J Clin Microbiol* 2004; 42 (5): 2085–2093.

139. Imhof A, Schaer C, Schoedon G, Schaer DJ, Walter RB, Schaffner A, Schneemann M. Rapid detection of pathogenic fungi from clinical specimens using LightCycler real-time fluorescence PCR. *Eur J Clin Microbiol Infect Dis* 2003; 22 (9): 558–560.

140. Maaroufi Y, De Bruyne JM, Duchateau V, Georgala A, Crokaert F. Early detection and identification of commonly encountered *Candida* species from simulated blood cultures by using a real-time PCR-based assay. *J Mol Diagn* 2004; 6 (2): 108–114.

141. Selvarangan R, Bui U, Limaye AP, Cookson BT. Rapid identification of commonly encountered *Candida* species directly from blood culture bottles. *J Clin Microbiol* 2003; 41 (12): 5660–5664.

142. White PL, Shetty A, Barnes RA. Detection of seven *Candida* species using the LightCycler system. *J Med Microbiol* 2003; 52 (3): 229–238.

143. Bialek R, Kern J, Herrmann T, Tijerina R, Cecenas L, Reischl U, Gonzalez GM. PCR assays for identification of *Coccidioides posadasii* based on the nucleotide sequence of the antigen 2/proline-rich antigen. *J Clin Microbiol* 2004; 42 (2): 778–783.

144. Delgado N, Hung CY, Tarcha E, Gardner MJ, Cole GT. Profiling gene expression in *Coccidioides posadasii*. *Med Mycol* 2004; 42 (1): 59–71.

145. Bialek R, Weiss M, Bekure-Nemariam K, Najvar LK, Alberdi MB, Graybill JR, Reischl U. Detection of *Cryptococcus neoformans* DNA in tissue samples by nested and real-time PCR assays. *Clin Diagn Lab Immunol* 2002; 9 (2): 461–469.

146. Call DR. Challenges and opportunities for pathogen detection using DNA microarrays. *Crit Rev Microbiol* 2005; 31 (2): 91–99.

147. Bryant PA, Venter D, Robins-Browne R, Curtis N. Chips with everything: DNA microarrays in infectious diseases. *Lancet Infect Dis* 2004; 4 (2): 100–111.

148. Wilson WJ, Strout CL, DeSantis TZ, Stilwell JL, Carrano AV, Andersen GL. Sequence-specific identification of 18 pathogenic microorganisms using microarray technology. *Mol Cell Probes* 2002; 16 (2): 119–127.

149. Zwick ME, Mcafee F, Cutler DJ, Read TD, Ravel J, Bowman GR, Galloway DR, Mateczun A. Microarray-based resequencing of multiple *Bacillus anthracis* isolates. *Genome Biol* 2005; 6 (1): R10.

150. Abravaya K, Huff J, Marshall R, Merchant B, Mullen C, Schneider G, Robinson J. Molecular beacons as diagnostic tools: technology and applications. *Clin Chem Lab Med* 2003; 41 (4): 468–474.

151. Antony T, Subramaniam V. Molecular beacons: nucleic acid hybridization and emerging applications. *J Biomol Struct Dyn* 2001; 19 (3): 497–504.

152. Chandler DP, Jarrell AE. Enhanced nucleic acid capture and flow cytometry detection with peptide nucleic acid probes and tunable-surface microparticles. *Anal Biochem* 2003; 312 (2): 182–190.

153. Kuhn H, Demidov VV, Doull JM, Fiandaca MJ, Gildea BD, Frank-Kamenetskii MD. Hybridization of DNA and PNA molecular beacons to single-stranded and double-stranded DNA targets. *J Am Chem Soc* 2002; 124 (6): 1097–1103.

154. Ye F, Li M-S, Taylor JD, Nguyen Q, Colton HM, Casey WM, Wagner M, Weiner MP, Chen J. Fluorescent microsphere-based readout technology for multiplexed human single nucleotide polymorphism analysis and bacterial identification. *Hum Mutat* 2001; 17 (4): 305–316.

155. Horejsh D, Martini F, Poccia F, Ippolito G, Di Caro A, Capobianchi MR. A molecular beacon, bead-based assay for the detection of nucleic acids by flow cytometry. *Nucleic Acids Res* 2005; 33 (2): E13.

156. Lin B, Vahey MT, Thach D, Stenger DA, Pancrazio JJ. Biological threat detection via host gene expression profiling. *Clin Chem* 2003; 49 (7): 1045–1049.

157. Boeuf P, Vigan-Womas I, Jublot D, Loizon S, Barale J-C, Akanmori BD, Mercereau-Puijalon O, Behr C. CyProQuant-PCR: a real time RT-PCR technique for profiling human cytokines, based on external RNA standards, readily automatable for clinical use. *BMC Immunol* 2005; 6 (1): 5.

158. Hess S, Peters J, Bartling G, Rheinheimer C, Hegde P, Magid-Slav M, Tal-Singer R, Klos A. More than just innate immunity: comparative analysis of *Chlamydophila pneumoniae* and *Chlamydia trachomatis* effects on host-cell gene regulation. *Cell Microbiol* 2003; 5 (11): 785–795.

159. Nau GJ, Schlesinger A, Richmond JFL, Young RA. Cumulative Toll-like receptor activation in human macrophages treated with whole bacteria. *J Immunol* 2003; 170 (10): 5203–5209.

160. Reghunathan R, Jayapal M, Hsu L-Y, Chng H-H, Tai D, Leung BP, Melendez AJ. Expression profile of immune response genes in patients with severe acute respiratory syndrome. *BMC Immunol* 2005; 6 (1): 2.

161. Shrestha NK, Shermock KM, Gordon SM, Tuohy MJ, Wilson DA, Cwynar RE, Banbury MK, Longworth DL, Isada CM, Mawhorter SD, Procop GW. Predictive value and cost-effectiveness analysis of a rapid polymerase chain reaction for preoperative detection of nasal carriage of *Staphylococcus aureus*. *Infect Control Hosp Epidemiol* 2003; 24 (5): 327–333.

162. Wellinghausen N, Wirths B, Franz AR, Karolyi L, Marre R, Reischl U. Algorithm for the identification of bacterial pathogens in positive blood cultures by real-time LightCycler polymerase chain reaction (PCR) with sequence-specific probes. *Diagn Microbiol Infect Dis* 2004; 48 (4): 229–241.

163. Fukushima H, Tsunomori Y, Seki R. Duplex real-time SYBR Green PCR assays for detection of 17 species of food- or waterborne pathogens in stools. *J Clin Microbiol* 2003; 41 (11): 5134–5146.

164. Sundsfjord A, Simonsen GS, Haldorsen BC, Haaheim H, Hjelmevoll SO, Littauer P, Dahl KH. Genetic methods for detection of antimicrobial resistance. *APMIS* 2004; 112 (11–12): 815–837.

165. Grimm V, Ezaki S, Susa M, Knabbe C, Schmid R, Bachmann TT. Use of DNA microarrays for rapid genotyping of TEM beta-lactamases that confer resistance. *J Clin Microbiol* 2004; 42 (8): 3766–3774. Erratum: *J Clin Microbiol* 2004; 42 (10): 4918.

166. Call DR, Bakko MK, Krug MJ, Roberts MC. Identifying antimicrobial resistance genes with DNA microarrays. *Antimicrob Agents Chemother* 2003; 47 (10): 3290–3295.

167. Shanahan F, O'Mahony J. The mycobacteria story in Crohn's disease. *Am J Gastroenterol* 2005; 100 (7): 1537–1538.

168. Alpert PT. New and emerging theories of cardiovascular disease: infection and elevated iron. *Biol Res Nurs* 2004; 6 (1): 3–10.

169. Toivanen A, Toivanen P. Reactive arthritis. *Best Pract Res Clin Rheumatol* 2004; 18 (5): 689–703.

170. Relman DA. Detection and identification of previously unrecognized microbial pathogens. *Emerg Infect Dis* 1998; 4 (3): 382–389.

14 Cancer Detection and Prognosis

Santiago Ropero and Manel Esteller

CONTENTS

INTRODUCTION

It is well known that tumors develop as a result of accumulated molecular-genetic or genomic alterations. In 1914, Bovery made the first suggestion that cancer has its origin in aberrations of the genome. This idea was supported by the evidence that cancer and cancer risk can be inherited, that mutagens can produce tumors, and that tumors are monoclonal in origin — the genetic characteristics of tumors are similar to those of the original transformed cells. Studies undertaken decades later revealed the involvement of many genes in cancer development.

Recent studies have revealed that alterations in human cancer arise from a wide variety of genomic changes, such as amplifications, translocations, deletions, and point mutations; any of these may be present in a given type of human tumor. Analysis of these genomic alterations led to the identification of oncogenes and tumor-suppressor genes involved in cancer development. At least 100 oncogenes have been identified whose malfunctions produce abnormal growth, and approximately 50 tumor-suppressor genes in the human genome are thought to predispose to familial cancer syndromes. These genes are involved in a range of cellular processes, such as transcriptional regulation, cell cycle control, programmed cell death, and genetic stability.

For example, characterization of amplified regions of the breast cancer genome has revealed the importance of certain oncogenes including HER2/neu and c-Myc. The studies that provide us with an understanding of the molecular basis of cancer development also provide new potential diagnostic and outcome markers and therapeutic targets for cancer patients.[1,2] Amplification of cyclin E and HER2/neu is associated with advanced stages of disease and poor clinical outcomes in ovarian cancer. The treatment of such patients with an antibody against HER2/neu (trastuzumab; Herceptin, Genentech, South San Francisco, California, USA) prolongs the disease-free intervals in patients with HER2/neu amplifications.[3] The molecular interactions between HER2 and estrogen receptor in breast cancer patients over-expressing HER2 and expressing estrogen receptor decrease the efficacy of a combined treatment directed against these two molecules.[4] Although numerous oncogenes and tumor-suppressor genes have already been identified, many more remain to be discovered and identified.

Cancer development is not restricted to the genetic changes described above; it can also be traced to epigenetic changes. Epigenetics deals with the inheritance of information based on gene-expression levels. The field of genetics concerns information transmitted on the basis of gene sequence. The main epigenetic modification in mammals, and particularly in humans, is methylation of cytosine nucleotides.

This chapter reviews some of the technologies used to detect the genetic and epigenetic changes that lead to alterations in the functions of the two main classes of genes recognized as playing roles in cancer (oncogenes and tumor-suppressor genes), and the use of these alterations as molecular markers for cancer detection and prognosis.

ONCOGENES

Proto-oncogenes control how often a cell divides and also its degree of differentiation. Certain mutations in a proto-oncogene produce an oncogene that is permanently activated, causing normal cells to grow out of control and become cancer cells. Proto-oncogenes can be activated by several mechanisms. First, activation can occur through gene amplification, causing more of the protein to be encoded and thereby enhancing its function. For example, amplification of HER2 is found in 20% of primary breast cancers.

Point mutations that produce increases in oncoprotein function represent another mechanism. They have been described for the Ras oncogene in colorectal, lung, and pancreatic cancers. Activating point mutations of ras codons 12, 13, and 61 prevents the interaction of Ras with GTPase-activating protein (GAP), maintaining ras protein in the activated GTP-bound state and enhancing downstream signaling events regulated by this oncoprotein, such as cell-cycle regulation.[5]

Chromosomal translocation is a third mechanism of oncogene activation. An example of this is seen in chronic myeloid leukemia (CML), in which chromosome 9 is translocated to chromosome 22. This translocation positions the ABL oncogene on chromosome 9 next to the breakpoint cluster region of the Philadelphia gene on chromosome 22. The resulting fusion gene produces a fusion protein with abnormal tyrosine kinase activity, which accounts for its transforming ability.

TABLE 14.1
Selected List of Most Representative Oncogenes Used for Clinical Diagnosis of Cancer

Oncogene	Function	Tumor Profile
	Receptors	
TB	Tyrosine kinase/EGF receptor	Glioblastoma and squamous cell cancer
HER2/neu	Tyrosine kinase	Breast and endometrial cancer
Ret	Tyrosine kinase	Thyroid cancer
	Signal transduction	
Abl	Tyrosine kinase	Chronic myelocytic leukemia
H-RAS	GTP binding protein	Bladder cancer
K-Ras	GTP binding protein	Pancreatic, lung, colon, and other cancers
	Transcription activators	
c-myc	DNA binding protein	Leukemias; lung, breast, and stomach cancer; neuroblastoma; glioma
L-myc	DNA binding protein	Lung cancer
N-myc	DNA binding protein	Neuroblastoma, retinoblastoma, and lung cancer
	Apoptosis regulators	
Bcl-2	Membrane protein/pro-apoptotic	B-cell lymphomas

Numerous oncogenes regulating a wide variety of cell functions have been characterized in various human cancers (Table 14.1),[6] allowing their use in cancer diagnosis and prognosis. However, the use of oncogenes with diagnostic values is not straightforward because there is no absolute association between any specific oncogenes and tumor types. Evidence indicates that oncogene activation is associated with poor prognosis. For example, HER2, which is over-expressed in 20% of breast cancer patients, has been correlated with poor prognosis and decreased patient survival.[7,8] Substantial evidence indicates that HER2 also affects the sensitivity of breast cancer cells to various anticancer treatments, such as chemotherapy, hormone therapy, radiation therapy, and cytokine therapy.[8,9]

TUMOR-SUPPRESSOR GENES

Tumor-suppressor genes are normal genes that act as negative regulators of cell growth or other functions that may affect invasive and metastatic potential, such as cell adhesion and regulation of protease activity. When tumor-suppressor genes do not work properly, cells can grow out of control, which may lead to cancer. Alfred Knudson's "two-hit" hypothesis argues that both alleles of the tumor-suppressor genes must be inactivated to bring about a complete loss of function.[10] The mechanisms for tumor-suppressor-gene inactivation include point mutation, deletion, insertion, translocation, and DNA methylation.

A cell's DNA methylation balance is dramatically altered in cancer by two major processes: normally heavily methylated regions of the genome become hypomethylated, while regions of the genome that are normally unmethylated become hypermethylated. CpG islands associated with tumor-suppressor genes are unmethylated in normal tissues, and often become hypermethylated during tumor formation, leading to gene silencing. The profile of promoter hypermethylation for cancer-associated genes differs for each primary human tumor[11–13] and cancer cell line.[14] A selection of hypermethyla-tion-silenced genes in human cancers is listed in Table 14.2.[13,15] CpG island hypermethylation of tumor-suppressor genes has three possible translational uses: as a marker for cancer cells, as a predictor of tumor behavior, and as a predictor of tumor response to therapeutic interventions.

TABLE 14.2
Selected List of Genes Silenced by CpG Island Hypermethylation in Human Cancer

Gene	Function	Tumor Profile
hMLH1	DNA mismatch repair	Colon, endometrium, stomach
BRCA1	DNA repair, transcription	Breast, ovary
p16^{INK4a}	Cyclin-dependent kinase inhibitor	Multiple types
p14^{ARF}	MDM2 inhibitor	Colon, stomach, kidney
p15^{INK4b}	Cyclin-dependent kinase inhibitor	Leukaemia
MGMT	DNA repair of O6-alkyl-guanine	Multiple types
GSTP1	Conjugation to glutathione	Prostate, breast, kidney
p73	p53 homologue	Lymphoma
LKB1/STK11	Serine/threonine kinase	Colon, breast, lung
ER	Estrogen receptor	Breast
PR	Progesterone receptor	Breast
AR	Androgen receptor	Prostate
PRLR	Prolactin receptor	Breast
RARβ2	Retinoic acid receptor β2	Colon, lung, head and neck
RASSF1A	Ras effector homologue	Multiple types
NORE1A	Ras effector homologue	Lung
VHL	Ubiquitin ligase component	Kidney, haemangioblastoma
Rb	Cell cycle inhibitor	Retinoblastoma
THBS-1	Thrombospondin-1, anti-angiogenic	Glioma
CDH1	E-cadherin, cell adhesion	Breast, stomach, leukaemia
CDH13	H-cadherin, cell adhesion	Breast, lung
FAT	Cadherin, tumor suppressor	Colon
HIC-1	Transcription factor	Multiple types
APC	Inhibitor of β-catenin	Aerodigestive tract
SFRP1	Secreted Frizzled-related protein 1	Colon
COX-2	Cyclo-ox genase-2	Colon, stomach
SOCS-1	Inhibitor of JAK/STAT pathway	Liver, myeloma
SOCS-3	Inhibitor of JAK/STAT pathway	Lung
GATA-4	Transcription factor	Colon, stomach
GATA-5	Transcription factor	Colon, stomach
SRBC	BRCA1-binding protein	Breast, lung
SYK	Tyrosine kinase	Breast
RIZ1	Histone/protein methyltransferase	Breast, liver
DAPK	Pro-apoptotic	Lymphoma, lung, colon
TMS1	Pro-apoptotic	Breast
TPEF/HPP1	Transmembrane protein	Colon, bladder
HOXA9	Homeobox protein	Neuroblastoma
IGFBP3	Growth factor-binding protein	Lung, skin
EXT1	Heparan sulfate synthesis	Leukemia, skin

IMPORTANCE OF GENETIC TESTING IN INHERITED TUMORS

The genetic alterations of oncogenes and tumor-suppressor genes that occur in cancer are classified into two main classes: germline and somatic alterations. Inherited abnormalities of tumor-suppressor genes have been found in several familial cancers. In many syndromes, only one gene is responsible for hereditary cancer. For example, APC is the gene responsible for familial adenomatous polyposis (FAP), whose penetration is nearly 100%.[16] Genetic testing is very useful for presymptomatic and final diagnoses of patients and for identifying the mutation carrier among family members. However, other familial cancer syndromes are associated with alterations in multiple gene function, as for example the mismatch repair genes hMSH2, hMLH1, hPMS1, and hPMS2 associated with hereditary non-polyposis colorectal cancer (HNPCC).[17–21]

The significance of genetic testing is different for all familial syndromes and must be tailored for every syndrome and for each case of every syndrome. Thus, it is very important to establish the value of genetic testing for the medical care of different hereditary cancer syndromes. Three groups can be distinguished. The first is comprised of the syndromes for which medical care may be adapted on the basis of the results of genetic testing. The second group includes the cancer syndromes in which medical benefit derived from the identification of genetic abnormalities is presumed but not yet proven. In the third group, germline genetic alterations have been identified in a small number of families that nevertheless derive no medical benefit from genetic testing. The American Society for Clinical Oncology (ASCO) recommends oncologists consider offering genetic testing for patients of the first two groups only.[22]

DNA ANALYSIS IN CANCER DETECTION

Over the last two decades, molecular techniques have been used to identify oncogenes and tumor-suppressor genes involved in cancer development. The primary aim of these studies is to use our knowledge of the expression and alterations of these genes to benefit patients, either as prognostic indicators for early diagnosis of tumors or eventually, to provide the means to develop targeted therapeutics that may allow the selective, efficient, and safe treatment of cancer.

Ideally, studies of alterations in genetic and biochemical processes at the molecular level help to establish definitive diagnoses and classification of tumors based on the recognition of complex profiles or unique molecular alterations that occur in specific tumor types. However, it is very difficult to achieve these objectives for several reasons: The crosstalk of different cancer-related pathways complicates the understanding of cancer biology; there is considerable heterogeneity in the targets and their functions among individuals with similar types of cancers; these targets are not absolutely specific to cancer cells; the effectiveness of the treatments is limited because the targets are affected by other factors and the functions of the targets may change over time.

The first genetic analyses were initially carried out in biopsy samples, but the discovery of free DNA in plasma opened new avenues for cancer detection and prognosis. Free DNA was identified 50 years ago in the sera of cancer patients and elevated levels of serum DNA were detected in patients with metastatic disease.[23,24] However, serum DNA is not specific to cancer, having been found at elevated levels in patients with severe infections and autoimmune diseases.

Plasma or serum from cancer patients is now analyzed to detect tumor markers such as oncogene mutations, microsatellite instability, hypermethylation of promoter regions, and viral DNA.[25] It is now possible to detect cancer cells using DNA from other body fluids such as saliva, urine, bronchoalveolar lavage, sputum, semen, and ductal lavage. The epithelial tumor grows and cancer cells are sloughed off the organ epithelium into body fluids. This makes it possible to detect molecular markers such as DNA mutation, methylation patterns, and microsatellite instability using these samples before the patients are symptomatic.

DNA can be analyzed for changes in gene copy number, chromosome translocations, deletions or losses of heterozygosity (LOH), telomere extension, microsatellite instability, promoter hypermethylation, and point mutations. For example, several studies show that the use of DNA hypermethylation to detect cancer cells in body fluids is a promising tool for early detection of tumor development, using methylation-specific PCR (MSP) as a quick, easy, non-radioactive, and sensitive method as we will discuss below.

DETECTION OF CHROMOSOMAL ABERRATIONS IN CANCER

Chromosome aberration is a distinctive feature of tumors that can be detected using cytogenetic and molecular methods. Among other effects, these genetic alterations provide one mechanism by which oncogenes and tumor-suppressor gene expression can be modified in cancer. The analysis of these recurrent abnormalities in a wide range of solid tumors revealed considerable variability in the degree to which tumor genomes are aberrant at the chromosomal level, from a few to several dozen.

The spectrum of genetic aberration differs in tumors that arise at different anatomical sites and in histologically distinct tumors that arise in the same anatomic location.[26,27] The tumor histology and aberration spectra also vary with the genetic make-up of an affected individual. For example, tumors in individuals with hereditary non-polyposis colorectal cancer are more likely to occur in the right colon and have diploid genomes, whereas cells of sporadic colorectal cancers are most often aneuploid.

Increasing numbers of molecular genetic technologies lend themselves to the study of chromosomal aberrations, such as chromosome banding, high-throughput analysis of LOH, comparative genome hybridization (CGH), fluorescence *in situ* hybridization (FISH), restriction landmark genomic scanning (RLGS), and representational difference analysis (RDA).

Banding Analysis

Banding analysis of metaphase chromosomes has been very useful for identifying causative chromosome rearrangements in leukemias and lymphomas.[28] However, it has been less successful with solid tumors because of the difficulty of obtaining representative high-quality metaphase chromosomes and because the extent of chromosome rearrangement complicates karyotype interpretation.

Fluorescence *In Situ* Hybridization

The use of the FISH technique using chromosome-specific probes[29,30] has significantly improved chromosome classification by increasing the specificity with which chromosomes or their subregions can be recognized. This method involves denaturing DNA in a tissue section followed by the application of a labeled fluorescent probe that is complementary to the sequence of interest. The different staining patterns permit the identification of a wide range of sequences, from those of the whole chromosome to specific loci, allowing the detection of DNA regions as small as 0.5 kb.

Combinatorial labeling strategies allow dozens of spatially separate loci or chromosomes to be analyzed simultaneously.[31,32] These techniques have dramatically increased the accuracy and sensitivity with which chromosome aberrations can be detected and classified, and clearly demonstrate the limitations of conventional banding analysis in the analysis of human malignancies.[33]

Disadvantages of FISH testing include the high cost, longer time required for slide scoring, requirement for a fluorescent microscope, inability to preserve slides for storage and review, and occasionally in identifying invasive tumor cells. FISH is very useful for detecting amplifications and deletions of a specific gene, changes in repetitive sequences located in centromeric and telomeric regions and chromosomal alterations as translocations, as well as marker chromosomes and ring chromosomes. For example, the HER2 amplification in breast cancer can be assessed in routine formalin-fixed paraffin-embedded clinical samples by FISH.[34] The assessment of amplification of HER2 gene by FISH is a tool to prescribe adjuvant treatment (trastuzumab) designed for the treatment of breast cancer over-expressing HER2.

Comparative Genome Hybridization

The application of FISH in solid tumors is still limited by the difficulty of interpreting complex karyotypes. This problem is solved, at least in part, by CGH, which maps changes in relative DNA sequence copy number onto normal metaphase chromosomes.[35] Total genomic DNAs from tumor and reference samples are labeled independently with different fluorochromes and cohybridized to normal sample preparations, along with excess unlabeled cot-1 DNAs that are repetitive sequences of the genome, to inhibit hybridization of labeled repeated sequences.

The ratio of the numbers of the two genomes that hybridize to each location on the target chromosomes is an indication of the relative copy numbers of the two DNA samples at that point in the genome. The principal advantages of this method are that it allows copy number to be determined throughout a complex genome with respect to a normal reference genome and it employs genomic DNA so that cell culture is not required.

The main limitations of CGH are that the resolution is limited to 10 to 20 Mb, it does not yield quantitative information about gene dosage, and it is unable to detect chromosome aberrations that do not produce changes in copy number. CGH analysis has been applied successfully in the identification of nonrandom loss or gain of specific chromosomal regions associated with pathologic classification, progression, prognosis, and clinical treatment of human neoplasia. For example, this technique has been used to identify recurrent gains on the chromosome arms 3q, 5p, 7p, 8q, 12p, and 20q in pancreatic carcinomas and losses on chromosomes 8p, 9p, 17p, 18q, and 19p.[36] Other studies using CGH found that the gain of chromosome 3q defines the transition from severe dysplasia to invasive carcinoma in cervical epithelium.[37]

Restriction Landmark Genomic Scanning

RLGS is another method that allows the identification of differences in copy number, and that is very useful for identifying mutations and polymorphisms. The genome is digested with a rare cutting enzyme and the resulting fragments are radioactively labeled at the ends of the fragments generated. The genome is digested with a second restriction enzyme and the fragments separated in a first-dimension gel. The DNA fragments are then digested in the gel with a third restriction enzyme and the resulting DNA fragments are electrophoretically separated in a second-dimension gel.

The sites recognized by the first enzyme are then detected using autoradiography or phosphor-imaging (Figure 14.1). The sequences corresponding to all restriction sites may be cloned from the second-dimension gel or identified using a NotI/EcoRV boundary library.[38,39] Applications of RLGS include high-speed construction of linkage maps, quantitative analysis of copy number at each landmark locus, and detection of mutations involving the restriction sites used to prepare the two-dimensional electrophorograms. The principal limitations are that many of the genetic alterations associated with cancer development are not associated with existing physical maps and that high-resolution analysis of regions of interest is not possible.

The application of these techniques has been extended beyond the mere diagnosis of chromosomal aberrations to functional and basic research directed to find new chromosomal aberrations in cancer. While FISH is probably the most used technique in clinical testing, the rest of the methods for detecting chromosomal aberrations can be used for this purpose but have been used preferentially for marker discovery.

DETECTION OF CANCER-ASSOCIATED MUTATIONS

Most mutations in cancer genes are not apparent at the cytogenetic level, so the use of methods to identify the genes and significant changes in their structures has become increasingly important. Somatic and inherited point mutations are salient features of molecular genetic changes in cancer because they can lead to the activation of oncogenes and the inactivation of tumor-suppressor genes. Traditionally, molecular detection methods entail the analysis of total cellular DNA using Southern blot or the polymerase chain reaction (PCR) in which regions of DNA are amplified and more easily identified.

Southern Blotting

Southern blot analysis has been very useful for detecting gene polymorphisms, deletions, insertions, and translocations that are recognized by a change in the length of the restriction fragment that is

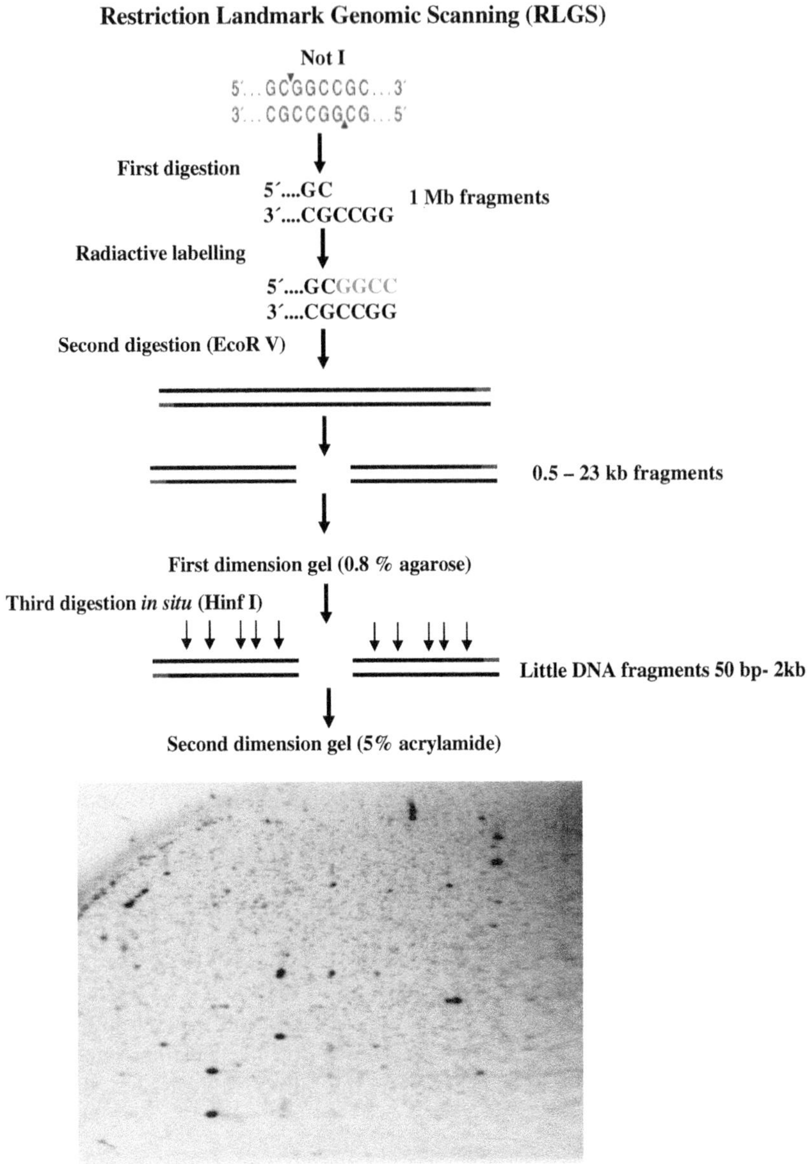

Restriction Landmark Genomic Scanning (RLGS)

Not I

5´...GC⌄GGCCGC...3´
3´...CGCCGG⌃CG...5´

First digestion

5´....GC   1 Mb fragments
3´....CGCCGG

Radiactive labelling

5´....GCGGCC
3´....CGCCGG

Second digestion (EcoR V)

0.5 – 23 kb fragments

First dimension gel (0.8 % agarose)

Third digestion *in situ* (Hinf I)

↓ ↓ ↓↓ ↓   ↓ ↓ ↓↓↓

Little DNA fragments 50 bp- 2kb

Second dimension gel (5% acrylamide)

FIGURE 14.1 Restriction landmark genomic scanning (RLGS) method. Genomic DNA is digested with a restriction enzyme (*Not*I) that recognizes typical sequences of CpG islands, allowing the identification of new methylated sequences. The fragments are radiolabeled and digested with a second restriction enzyme (*Eco*RV) that produces DNA fragments that are separated in a first-dimension gel. After electrophoresis, the DNA is digested *in situ* (in gel) with a third restriction enzyme and separated in a second-dimension gel. The restriction sites recognized by the first restriction enzyme are detected after x-ray exposure.

hybridized by a specific probe. Point mutations may cause the loss or gain of a restriction site recognition sequence, which also causes a change in the restriction fragment length.

The detection of allelic loss of DNA has been very useful in the detection of tumor-suppressor genes.[40] Southern blot analysis can also lead to the detection of amplified genes, in which the hybridization signal is intensified in the filter. The Southern blot has been used to detect amplification of MYCN and HER2 oncogenes in neuroblastoma and breast cancer patients, respectively.

However, these methods can be significantly hampered when tumor cell DNA extracted from a primary carcinoma sample is diluted by DNA from benign tissue and inflammatory cells.

PCR-Based Methods

Most of the diagnostic methods used to detect mutations and single-nucleotide polymorphisms (SNPs) require target-sequence amplification by PCR, followed by the determination of distinct sequence variants by means of a short hybridization probe or restriction endonucleases. The hybridization technique with an allelic point mutation-specific oligonucleotide probe (ASO),[41] the TaqMan PCR (Applied Biosystems, Foster City, California, USA) method,[42] and PCR-restriction fragment length polymorphism analysis (PCR-RFLP)[43] are included in this category. These methods are widely used for the molecular genetic diagnosis of familial tumors.

Although PCR can be used to detect single genetic mutations, most tumor types are known to contain several mutations in different oncogenes and tumor-suppressor genes. Thus, single PCR analysis is not effective for detecting multiple mutations associated with a given tumor type. Other complex technologies have been developed for the identification of multiple mutations. These are multiplex methods in which initial amplification of the target gene fragment from DNA by PCR is followed by mutation-specific ligation of small cDNA strands.

This approach has been used to detect common mutations in genes such as RAS, adenomatous polyposis coli (APC), and p53 in stool samples of patients with colorectal cancer.[44–46] For example, in the latter case, the sequence containing the p53 mutations was amplified by regular PCR and the mutations were detected with a mismatch ligation assay able to detect nine common mutations of the p53 gene.[46] These technologies are of great value for identifying known mutations or groups of mutations related to cancer development.

Other non-PCR-based methods such as mass spectroscopy are very useful for detecting groups of mutations. For example, mass spectroscopy has been used to identify a panel of RAS mutations in the sputum of patients with lung cancer and a specific p53 mutation at codon 259 in the sera of patients with hepatocellular cancer.[47,48] This methodology is more useful for identifying new mutations involved in cancer development than for its use as routine test for cancer diagnosis.

Molecular methods for detecting cancer-related mutations are highly valuable tools for detecting cancer at the early stages of development, micrometastasis in serum and early metastasis in lymph nodes, and for predicting responses to treatment and prognosis. For example, p53 mutations are good indicators for monitoring tumor spread into margins and draining lymph nodes in head and neck cancer patients. Such mutations are present in half of all head and neck tumors, and patients with this genetic disorder were found to face high risks of recurrence and poor overall survival.[49] Other studies carried out in patients with colorectal carcinoma have shown that the presence of mutations in RAS and p53 in lymph-node samples predicted poor outcomes.[50]

DETECTION OF INSERTIONS AND DELETIONS AND MICROSATELLITE INSTABILITY

Cancer cells are genetically unstable because of the inactivation of DNA repair genes. Some tumor types undergo inactivation of the DNA repair genes by promoter hypermethylation, thereby inducing the accumulation of mutations. Microsatellites are often expanded or deleted in these tumors. This characteristic, known as microsatellite instability (MSI), is another valuable clonal marker of cancer. Known DNA fragments containing insertions, deletions, or microsatellite loci can be amplified using specific primers, and the different fragment sizes can be analyzed by electrophoresis.

Capillary electrophoresis (CE) is effective for this purpose because it has several advantages over conventional electrophoresis. It is a high-throughput, high-resolution technique, with automatic operation, on-line detection of DNA, and associated automatic computer-based data acquisition. CE has been used to detect MSIs in cancer and, in particular, to analyze the polyadenine tracts of transforming growth factor-beta (TGF-β) type II receptors in colorectal cancer.[51,52]

MSI analysis has been used to detect head and neck squamous cell carcinoma (HNSCC) in exfoliated oral mucosal cells and thus provides a nonaggressive method for the diagnosis of this carcinoma.[53] Using saliva of HNSCC patients, MSI was detectable in 24 of 25 cases (96%), and loss of heterozygosity was identified in test samples in 19 of 31 cases (61%), indicating that MSI is a good marker for detecting tumor cells in a background of normal cells. It has also been used for small-cell lung carcinoma (SCLC) detection. MSI alterations have been detected in 76% of SCLC tumors and in 71% of plasma samples from patients with SCLC.[54] Finally, analysis of this genetic alteration has also been useful for monitoring responses to therapy. For example, MSI analysis of urine sediments serves as a good clinical tool for the detection of recurrent bladder cancer.[55]

DNA METHYLATION DETECTION IN CANCER

Molecular abnormalities associated with tumorigenesis include both genetic changes in DNA sequences and also epigenetic changes. Hypermethylation of CpG islands at the promoter region is the most common epigenetic phenomenon, causing transcriptional silencing of tumor-suppressor genes. In the past few years, we and other groups have extensively mapped the CpG islands that are aberrantly hypermethylated in cancer.

Such mapping of DNA methylation has highlighted the existence of a unique profile of hyper-methylated CpG islands that defines each tumor type.[11,13] Moreover, the promoter hypermethylation of CpG islands in tumor-suppressor genes occurs early in tumorigenesis. Thus, knowledge of CpG island hypermethylation of tumor-suppressor genes may be a valuable tool for early detection of a wide range of tumor types.

In fact, CpG island hypermethylation has been used as a tool for detecting cancer cells in several types of biological fluids and biopsy samples: bronchoalveolar lavage,[56] lymph nodes,[57] sputum,[58] urine,[59] semen,[60] ductal lavage,[61] and saliva. The application of DNA methylation as an epigenetic marker in clinical practice requires quick, non-radioactive, sensitive methods.

Two alternatives are currently used to study the distribution of 5-methylcytosine residues in particular DNA sequences: non-bisulfite and bisulfite methods. The first relies on the use of methylation-sensitive and -insensitive restriction endonucleases. One of the restriction enzymes of the isoschizomer pair is able to cut the DNA only when its target is unmethylated (MS-REs), whereas the other is not sensitive to methylated cytosines. Once DNA has been digested with MS-REs, the methylation status of a gene can be determined by Southern blot hybridization or PCR procedures.

The non-bisulfite methods are extremely specific but their limitation to specific restriction sites reduces their value. This problem can be avoided by a general process involving bisulfite modification of DNA, which comprises a wide range of techniques that allow the accurate quantitative determination of the methylation status of the allele. All bisulfite-associated methods require PCR amplification of the bisulfite-modified DNA and differences in methylcytosine patterns are displayed by methylation-dependent primer design (methylation-specific PCR, MSP), in conjunction with MS-REs (combined bisulfite restriction analyses [COBRA]), genomic sequencing, and other approaches.[62]

Bisulfite treatment converts all cytosine bases to uracil, except those that are methylated, are resistant to modification, and so remain as cytosines[63] (Figure 14.2A). This reaction is the basis for differentiating methylated and unmethylated DNA. Bisulfite modification of DNA requires prior DNA denaturation because only methylcytosines located in single strands are susceptible to conversion. Total conversion of cytosine to uracil is a critical step in this process because unconverted unmethylated cytosine could be a source of false-positive results for methylated alleles. One disadvantage of bisulfite conversion is that these manipulations can lead to DNA fragmentation. This problem can be minimized by increasing the concentration of sodium bisulfite and reducing

Different Steps in Bisulfite Sequencing and Methylation-specific PCR (MSP)

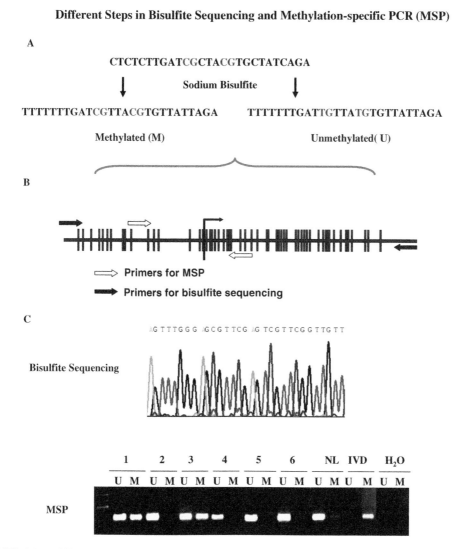

FIGURE 14.2 Different steps in bisulfite sequencing and methylation-specific PCR (MSP). A: Sodium bisulfite modification of DNA. This reaction converts all cytosines to uracils except those that are methylated. B: Primer design for bisulfite sequencing and MSP. CpG dinucleotides are represented as short vertical lines and the vertical arrow represents the transcription start site. Locations of bisulfite genomic sequencing PCR primers are indicated as black arrows and methylation-specific PCR primers as white arrows. C: Examples of DNA methylation analyses of tumor samples. Methylation map of a single DNA molecule obtained by direct sequencing of bisulfite-modified DNA. Analysis of the methylation status of a specific gene by methylation-specific PCR (MSP). U and M represent unmethylated and methylated DNA analyses, respectively. NL and IVD are DNAs from normal lymphocytes and *in vitro* methylated DNA, respectively. H₂0 represents a control reaction without DNA.

the time to complete the bisulfite reaction. Under these conditions, it is possible to obtain an optimal PCR product after only 4 hr of reaction.

Bisulfite Sequencing

Sequencing bisulfite-altered DNA is the most straightforward means of detecting cytosine methylation. In general, after denaturation and bisulfite modification, double-stranded DNA is obtained by primer extension and the fragment of interest is amplified by PCR.[64] The methylation map of the DNA fragment is then detected by standard DNA sequencing of the PCR products. One of the most important steps is the primer design. Bisulfite-converted DNA strands are no longer complementary, so primer design must be customized for each DNA chain. The primers should not contain CpG dinucleotides and should cover several cytosines that are not parts of the CpG dinucleotide in order to amplify both unmethylated and methylated alleles (Figure 14.2B).

Bisulfite sequencing provides whole methylation maps of a known DNA sequence (Figure 14.2C). However, the technical difficulty and labor-intensive nature of the method is a serious obstacle to screening large numbers of clinical samples.

Methylation-Specific PCR (MSP)

MSP is the most widely used technique for studying the methylation of CpG islands. The differences between methylated and unmethylated alleles that arise from sodium bisulfite treatment are the basis of MSP.[65] Methylation patterns must be determined in separate reactions. Thus, two sets of primers should be designed to anneal to the DNA. One primer set (U) will anneal to unmethylated DNA that has undergone chemical modification and a second primer set (M) will anneal with methylated DNA that has undergone chemical modification. As for bisulfite sequencing, primer design is a critical and complex component of the procedure.

As shown in Figure 14.2B, the primers for MSP should contain at least two or three CpGs and one should be at the 3′ end of the sense primer, so the unmethylated DNA is only amplified by the U primers in the unmethylated reaction and the methylated DNA is only amplified by the M primers in the methylated reaction. Moreover, the primers should also contain non-CpG cytosines in order to avoid false positive results produced by unmethylated but unconverted DNA (Figure 14.2B). The annealing temperature for both sets of primers should be the same and the PCR product should be between 80 and 175 bp. Both the unmethylated and methylated product reactions are visualized by ethidium bromide staining in a 6% polyacrylamide gel (Figure 14.2C).

The advantages of MSP are: (1) it avoids the use of restriction enzymes and resultant problems associated with incomplete enzymatic digestion; (2) it is very sensitive, thereby allowing the methylation statuses of small samples of DNA, even those from paraffin-embedded and microdissected tissues to be determined; (3) it is specific to relevant CpGs sites, not only those at a restriction site; and (4) it is not prone to false positive results. Due to its versatility, MSP has been widely proposed as a rapid and cost-effective clinical tool for studying CpG island hypermethylation in human cancer. For example, MSP has been successfully used to evaluate the responsiveness of human cancer patients to alkylating agents, showing that the methylation of the DNA repair gene MGMT was associated with regression of gliomas and lymphomas and prolonged overall and disease-free survival.[66,67]

In another study, MSP was used to detect promoter hypermethylation of both p16 and MGMT with a sensitivity up to 100% in sera from non-small cell lung cancer patients.[68] The authors showed that promoter hypermethylation can be detected up to 3 years prior to clinical diagnosis and may therefore significantly improve patient survival.[68] The main disadvantage of the MSP is its qualitative nature.

Fluorescence-Based Real-Time Quantitative PCR Analysis

Another MSP variation is MethyLight,[69] which uses fluorescence-based real-time PCR technology. Bisulfite modified DNA is amplified by fluorescence-based, real-time quantitative PCR using locus-specific PCR primers flanking an oligonucleotide probe with a 5′ fluorescence reporter dye and a 3′ quencher dye. Semiquantitative MethyLight data can be obtained by using a number of different probes for the sequence of interest. The main advantages of this method are its potential to allow the rapid screening of hundreds or thousands of samples and its sensitivity that allows it to detect

Combined Bisulfite Restriction Analysis (COBRA)

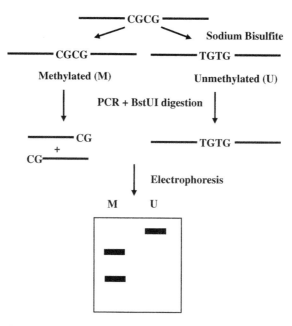

FIGURE 14.3 Combined bisulfite restriction analysis (COBRA). After sodium bisulfite modification, genomic DNA is amplified and digested with restriction enzymes that distinguish converted and unconverted CpG. The product of digestion is developed in an acrylamide gel and visualized by staining with ethidium bromide.

a single methylated allele in 10,000 unmethylated alleles. However, it requires expensive hybridization probes, and calibration curves must be derived for each set of conditions.

Combined Bisulfite Restriction Analysis (COBRA)

COBRA is a highly specific approach based on the creation or modification of a target for restriction endonuclease after bisulfite treatment. The method consists of a standard sodium bisulfite PCR treatment followed by restriction digestion that distinguishes methylated from unmethylated alleles and a quantitation step.[70] The primers used in the PCR do not contain CpG dinucleotides so the amplification step does not discriminate methylated and unmethylated alleles. One example of a restriction enzyme used in COBRA is BstUI. Its cleavage site (CGCG) is resistant to bisulfite modification when it is methylated but is transformed to TGTG when it is unmethylated (Figure 14.3).

Thus, DNA cleavage after bisulfite treatment if the restriction site is methylated and the cleavage products are proportional to the degree of methylation of the sequence analyzed. This is a specific and quantitative method, and thus more useful for analyzing the methylation degree of a clinical sample than for marker discovery. The main disadvantages are that COBRA analysis requires the complete chemical modification of DNA and it cannot identify all DNA sequences because it is confined to restriction targets.

Other Bisulfite-Based Techniques

Methylation-sensitive single nucleotide primer extension (Ms-SnuPE) employs bisulfite/PCR combined with single-nucleotide primer extension to analyze DNA methylation status quantitatively in a particular DNA region without using restriction enzymes.[71]

Methylation-sensitive single-strand conformation analysis (MS-SSCP) involves bisulfite modification of DNA that generates sequence disparities between methylated and unmethylated alleles, which can be resolved by SSCP.[72] In addition to classical non-denaturing slab gel electrophoresis, high-performance capillary electrophoresis (HPCE) can also be considered an alternative approach for resolving PCR nucleotide changes produced after bisulfite modification of DNA.[73]

Analysis of Global CpG Island Hypermethylation

Several molecular methods have been developed for detecting global changes in DNA methylation in CpG islands, including HPLC, methyl-acceptance assay, differential methylation hybridization, Ms-arbitrarily primed PCR, and RLGS. The use of RLGS for analyzing global CpG island hypermethylation requires methylation-sensitive restriction enzymes whose restriction sites are preferentially located in CpG islands. Two restriction enzymes are currently used: *Not*I and *Asc*I. The two-dimensional profiles of landmark sites of normal tissues and tumor samples are compared, allowing the identification of DNA sequences that are differentially methylated in tumor samples (Figure 14.1).

The CpG island hypermethylation analysis of tumor-suppressor genes may be a valuable tool in the essential transfer of research from laboratory to clinical practice. In contrast to genetic markers in which mutations occur at multiple sites and can be of very different types, promoter hypermethylation occurs only within CpG islands. The new technique that involves treating DNA with bisulfite, use of methylation-specific PCR primers to amplify the DNA, and genomic sequencing has made it possible for most laboratories and hospitals to study DNA methylation, even using archived pathological materials. Although these methods are probably the most commonly used for cancer detection in clinical studies, they can also be coupled with global genomic approaches such as CpG arrays or RLGS to establish molecular signatures of tumors based on DNA methylation markers.

RNA ANALYSIS IN CANCER DETECTION

The analysis of RNA allows the detection of changes in gene expression that represent one of the genetic alterations of cancer. It is now possible to isolate total RNA from tissue and body fluids, but this generally requires particular care to avoid the action of RNAse enzymes and deal with the low stability of RNA. More recent methods can extract degraded RNA from paraffin tissues.

The two basic methods for identifying and quantifying mRNA levels in cancer samples are Northern blot and reverse transcription-PCR (RT-PCR). Newer techniques such as microarray methods that allow measurement of differential expression of all genes including those of low abundance are more useful for identifying new cancer-related genes than for the diagnosis of cancer.

NORTHERN BLOTTING

Northern blot is essentially the same technique as the Southern blot. The difference is that intact mRNA is separated on the gel, blotted and detected using a probe of radiolabeled DNA. The method has been used to measure levels of mRNA expression and identify abnormalities in the size of the mRNA.

RT-PCR

RT-PCR does not involve the use of radioactivity and yields results more rapidly than Northern blotting. The detection of gene expression using RT-PCR in the plasma of cancer patients represents a noninvasive method for analyzing tumor expression profiles.

mRNA detection may be applicable to cancer detection. For example, the quantification of levels of tyrosinase mRNA can be used as a marker for the detection of melanoma cells.[74] Other studies have shown that the detection of telomerase mRNA, whose activity is commonly observed in a wide range of cancers, in the plasma and sera of patients with breast cancer, HCC, colorectal carcinoma, and follicular lymphoma may serve as a tool for cancer diagnosis.[75]

Altered levels of gene expression may occur in normal tissues and they can produce false positive results. For this reason, it is necessary to design quantitative approaches with cut-off values in order to differentiate expression levels in normal and tumor samples. Several examples illustrate the use of quantitative mRNA analysis for monitoring the responses of cancer patients to anticancer therapy. The increase in thymidylate synthase mRNA expression was related to multidrug resistance in human breast and colon cancer cell lines.[76] Low gene expression levels of dihydropyrimidine dehydrogenase, thymidilate synthase, and thymidine phosphorylase were associated with the responses of colorectal tumors to 5-fluororacil.[77] Taken together, these data indicate that the use of RT-PCR to analyze the expression of cancer-related genes in body fluids provides other clinical tools for cancer diagnosis and prognosis.

EXPRESSION MICROARRAYS

Microarrays constitute a group of miniaturized device technologies characterized by the common availability of hundreds or thousands of items including DNA sequences, RNA transcripts, and proteins that can be measured in a single experiment. The use of cDNA microarrays to analyze gene expression is the most common application of these technologies. Labeled mRNA samples are hybridized to arrays of cDNA clones or oligonucleotides derived from associated sequences, and the hybridization signals are detected using autoradiography, phosphorimaging or fluorescence imaging.

These technologies are high-throughput methods for identifying specific tumor biomarkers and molecular targets associated with biological and clinical phenotypes. This approach is very valuable for the molecular characterization of the distinct steps in cancer progression using culture cell lines and clinical materials. The use of well annotated tumor specimens has the potential for identifying target genes for novel diagnostic, prognostic, and therapeutic approaches.

This technology has been used to identify different patterns of gene expression in bladder tumors with different tumor stages and grades.[78] For example, the combination of expression microarrays of bladder tumors and bladder cancer cell lines allowed the identification of the tumor-suppressor role of KiSS-1 in bladder cancer progression. Lower transcript levels of KiSS-1 were observed in invasive bladder carcinomas as compared to superficial tumors and normal urothelium. Moreover, these ratios provided prognostic information. Lower expression of this gene was also observed in cells derived from advanced bladder tumors.[79]

CONCLUSIONS

Thousands of papers published every year reveal new genes implicated in cancer development, but few of these genes are really used as cancer markers. Thus, it is very important to establish specific molecular markers that allow the diagnosis of cancer. In this chapter, we have shown some of the most common methods for cancer detection and for discovering new molecular markers to be used in the future. Each method has its advantages and disadvantages, and a method should be selected according to the purpose of the analysis.

Incorporation of cancer marker analysis in clinical trials should be a crucial facet of marker development. Clinical trials should be designed to incorporate assays both to identify new markers and to correlate known markers with therapeutic responses and patient outcomes. New high-throughput techniques are necessary both for the development of new molecular markers for analyzing known markers that allow cancer detection in early stages of tumor development and for monitoring therapy response with high sensitivity and specificity.

REFERENCES

1. Kinzler KW, Vogelstein B. *Genetic Basis of Human Cancer*, 2nd, ed. Kinzler KW, Vogelstein B, Eds., Toronto: McGraw-Hill, 2002, 3.
2. Onyango P. Genomics and cancer. *Curr Opin Oncol* 2002; 14 (1): 79–85.
3. Zhang HT, Wang Q, Greene MI, Murali R. New perspectives on anti-*HER2/neu* therapeutics. *Drug News Perspect* 2000; 13 (6): 325–329.
4. Ropero S, Menendez JA, Vazquez-Martin A, Montero S, Cortes-Funes H, Colomer R. Trastuzumab plus tamoxifen: anti-proliferative and molecular interactions in breast carcinoma. *Breast Cancer Res Treat* 2004; 86 (2): 125–137.
5. Downward J. Targeting *RAS* signalling pathways in cancer therapy. *Nat Rev Cancer* 2003; 3 (1): 11–22.
6. Sanchez-Céspedes M, Sidransky D. DNA-based detection of neoplastic cells for clinical cancer management. *Rev Oncologia* 1999; 1: 284–290.
7. Slamon DJ, Clark GM, Wong SG, Levin WJ, Ullrich A, McGuire WL. Human breast cancer: correlation of relapse and survival with amplification of *HER2/neu* oncogene. *Science* 1987; 235 (4785): 177–182.
8. Colomer R, Montero S, Lluch A, Ojeda B, Barnadas A, Casado A, Massuti B, Cortes-Funes H, Lloveras B. Circulating *HER2* extracellular domain and resistance to chemotherapy in advanced breast cancer. *Clin Cancer Res* 2000; 6 (6): 2356–2362.
9. Ross JS, Fletcher JA. The *HER-2/neu* oncogene in breast cancer: prognostic factor, predictive factor, and target for therapy. *Stem Cells* 1998; 16 (6): 413–428.
10. Knudson AG Jr. Mutation and cancer: statistical study of retinoblastoma. *Proc Natl Acad Sci USA* 1971; 68 (4): 820–823.
11. Esteller M, Corn PG, Baylin SB, Herman JG. A gene hypermethylation profile of human cancer. *Cancer Res* 2001; 61 (8): 3225–3229.
12. Costello JF, Fruhwald MC, Smiraglia DJ, Rush LJ, Robertson GP, Gao X, Wright FA, Feramisco JD, Peltomaki P, Lang JC, Schuller DE, Yu L, Bloomfield CD, Caligiuri MA, Yates A, Nishikawa R, Su Huang H, Petrelli NJ, Zhang X, O'Dorisio MS, Held WA, Cavenee WK, Plass C Aberrant CpG island methylation has non-random and tumour-type-specific patterns. *Nat Genet* 2000; 24 (2): 132–138.
13. Esteller M. CpG island hypermethylation and tumor suppressor genes: a booming present, a brighter future. *Oncogene* 2002; 21 (35): 5427–5440.
14. Paz MF Fraga MF, Avila S, Guo M, Pollan M, Herman JG, Esteller M. A systematic profile of DNA methylation in human cancer cell lines. *Cancer Res* 2003; 63 (5): 1114–1121.
15. Esteller M. Dormant hypermethylated tumour suppressor genes: questions and answers. *J Pathol* 2005; 205 (2): 172–180.
16. Campbell WJ, Spence RA, Parks TG. Familial adenomatous polyposis. *Br J Surg* 1994; 81 (12): 1722–1733.
17. Leach FS, Nicolaides NC, Papadopoulos N,et al. Mutations of a mutS homologue in hereditary nonpolyposis colorectal cancer. *Cell* 1993; 75 (6): 1215–1225.
18. Fishel R, Lescoe MK, Rao MR, et al: The human mutator gene homolog MSH2 and its association with hereditary non polyposis colon cancer. *Cell* 1993; 75 (5): 1027–1038. Erratum: *Cell* 1994; 77 (1): 167.
19. Bronner CE, Baker SM, Morrison PT, et al: Mutation in the DNA mismatch repair gene homologue *hMLH1* is associated with hereditary non-polyposis colon cancer. *Nature* 1994; 368 (6468): 258–261.
20. Liu B, Parsons R, Papadopoulos N, et al: Analysis of mismatch repair genes in hereditary non-polyposis colorectal cancer patients. *Nat Med* 1996; 2 (2): 169–174.
21. Wijnen JT, Vasen HF, Khan PM, et al: Clinical findings with implications for genetic testing in families with clustering of colorectalcancer. *New Engl J Med* 1998; 339 (8): 511–518.
22. Statement of American Society of Clinical Oncology: genetic testing for cancer susceptibility. *J Clin Oncol* 1996; 14 (5): 1730–1736.

23. Shapiro B, Chakrabarty M, Cohn EM, Leon SA. Determination of circulating DNA levels in patients with benign or malignant gastrointestinal disease. *Cancer* 1983; 51 (11): 2116–2120.

24. Leon SA, Shapiro B, Sklaroff DM, Yaros MJ. Free DNA in the serum of cancer patients and the effect of therapy. *Cancer Res* 1977; 37 (3): 646–650.

25. Anker P, Lyautey J, Lederrey C, Stroun M. Circulating nucleic acids in plasma or serum. *Clin Chim Acta* 2001; 313 (1–2): 143–146.

26. Lassus H, Laitinen MP, Anttonen M, Heikinheimo M, Aaltonen LA, Ritvos O, Butzow R. Comparison of serous and mucinous ovarian carcinomas: distinct pattern of allelic loss at distal 8p and expression of transcription factor GATA-4. *Lab Invest* 2001; 81 (4): 517–526.

27. Smith JS, Alderete B, Minn Y, Borell TJ, Perry A, Mohapatra G, Hosek SM, Kimmel D, O'Fallon J, Yates A, Feuerstein BG, Burger PC, Scheithauer BW, Jenkins RB. Localization of common deletion regions on 1p and 19q in human gliomas and their association with histological subtype. *Oncogene* 1999; 18 (28): 4144–4152.

28. Rowley JD. Critical role of chromosome translocations in human leukemias. *Annu Rev Genet* 1998; 32: 495–519.

29. Pinkel D, Landegent J, Collins C, Fuscoe J, Segraves R, Lucas J, Gray J. Fluorescent *in situ* hybridization with human chromosome-specific libraries: detection of trisomy 21 and translocations on chromosome 4. *Proc Natl Acad Sci USA* 1988; 85 (23): 9138–9142.

30. Lichter P, Cremer T, Borden J, Manuelidis L, Ward DC. Delineation of individual human chromosomes in metaphase and interphase cells by *in situ* suppression hybridization using recombinant DNA libraries. *Hum Genet* 1988; 80 (3): 224–234.

31. Speicher MR, Gwyn Ballard S, Ward DC. Karyotyping human chromosomes by combinatorial multi-fluor FISH. *Nat Genet* 1996; 12 (4): 368–375.

32. Speicher MR, Ward DC. The coloring of cytogenetics. *Nat Med* 1996; 2 (9): 1046–1048.

33. Fleischman EW, Reshmi S, Sokova OI, Kirichenko OP, Konstantinova LN, Kulagina OE, Frenkel MA, Rowley JD. Increased karyotype precision using fluorescent *in situ* hybridization and spectral karyotyping precision in patients with myeloid malignancies. *Cancer Genet Cytogenet* 1999; 108 (2): 166–170.

34. Hicks DG, Tubbs RR. Assesment of the *HER2* status in breast cancer by fluorescence *in situ* hybridization: a technical review with interpretive guidelines. *Hum Pathol* 2005; 36 (3): 250–261.

35. Kallioniemi A, Kallioniemi OP, Sudar D, Rutovitz D, Gray JW, Waldman F, Pinkel D. Comparative genomic hybridization for molecular cytogenetic analysis of solid tumors. *Science* 1992; 258 (5083): 818–821.

36. Ghadimi BM, Schrock E, Walker RL, Wangsa D, Jauho A, Meltzer PS, Ried T. Specific chromosomal aberrations and amplification of the *AIB1* nuclear receptor coactivator gene in pancreatic carcinomas. *Am J Pathol* 1999; 154 (2): 525–536.

37. Heselmeyer K, Schrock E, du Manoir S, Blegen H, Shah K, Steinbeck R, Auer G, Ried T. Gain of chromosome 3q defines the transition from severe dysplasia to invasive carcinoma of the uterine cervix. *Proc Natl Acad Sci USA* 1996; 93 (1): 479–484.

38. Hirotsune S, Shibata H, Okazaki Y, Sugino H, Imoto H, Sasaki N, Hirose K, Okuizumi H, Muramatsu M, Plass C. Molecular cloning of polymorphic markers on RLGS gel using the spot target cloning method. *Biochem Biophys Res Commun* 1993; 194 (3): 1406–1412.

39. Smiraglia DJ, Fruhwald MC, Costello JF, McCormick SP, Dai Z, Peltomaki P, O'Dorisio MS, Cavenee WK, Plass C. A new tool for the rapid cloning of amplified and hypermethylated human DNA sequences from restriction landmark genome scanning gels. *Genomics* 1999; 58 (3): 254–262.

40. Hansen MF, Cavanee WK. Tumor suppressors: recessive mutations that lead to cancer. *Cell* 1988; 53 (2): 173–174.

41. Conner BJ, Reyes AA, Morin C, Itakura K, Teplitz RL, Wallace RB. Detection of sickle cell β^S-globin allele by hybridization with synthetic oligonucleotides. *Proc Natl Acad Sci USA* 1983; 80 (1): 278–282.

42. Livak KJ. Allelic discrimination using fluorogenic probes and the 5′ nuclease assay. *Genet Anal* 1999; 14 (5–6):143–149.

43. Ando M, Maruyama M, Oto M, Takemura K, Endo M, Yuasa Y. Higher frequency of point mutations in the c-K-ras 2 gene in human colorectal adenomas with severe atypia than in carcinomas. *Jpn J Cancer Res* 1991; 82 (3): 245–249.

44. Dong SM, Traverso G, Johnson C, Geng L, Favis R, Boynton K, Hibi K, Goodman SN, D'Allessio M, Paty P, Hamilton SR, Sidransky D, Barany F, Levin B, Shuber A, Kinzler KW, Vogelstein B, Jen J. Detecting colorectal cancer in stool with the use of multiple genetic targets. *J Natl Cancer Inst* 2001; 93 (11): 858–865.

45. Ahlquist DA, Skoletsky JE, Boynton KA, Harrington JJ, Mahoney DW, Pierceall WE, Thibodeau SN, Shuber AP. Colorectal cancer screening by detection of altered human DNA in stool: feasibility of a multitarget assay panel. *Gastroenterology* 2000; 119 (5): 1219–1227.

46. Traverso G, Shuber A, Levin B, Johnson C, Olsson L, Schoetz DJ Jr, Hamilton SR, Boynton K, Kinzler KW, Vogelstein B. Detection of *APC* mutations in fecal DNA from patients with colorectal tumors. *New Engl J Med* 2002; 346 (5): 311–320.

47. Jackson PE, Qian GS, Friesen MD, Zhu YR, Lu P, Wang JB, Wu Y, Kensler TW, Vogelstein B, Groopman JD. Specific p53 mutations detected in plasma and tumors of hepatocellular carcinoma patients by electrospray ionization mass spectrometry. *Cancer Res* 2001; 61 (1): 33–35.

48. Sun X, Hung K, Wu L, Sidransky D, Guo B. Detection of tumor mutations in the presence of excess amounts of normal DNA. *Nat Biotechnol* 2002; 20 (2): 186–189.

49. Brennan JA, Mao L, Hruban RH, Boyle JO, Eby YJ, Koch WM, Goodman SN, Sidransky D. Molecular assessment of histopathological staging in squamous-cell carcinoma of the head and neck. *New Engl J Med* 1995; 332 (7): 429–435.

50. Hayashi N, Arakawa H, Nagase H, Yanagisawa A, Kato Y, Ohta H, Takano S, Ogawa M, Nakamura Y. Genetic diagnosis identifies occult lymph node metastases undetectable by the histopathological method. *Cancer Res* 1994; 54 (14): 3853–3856.

51. Oto M, Suehiro T, Akiyama Y. Microsatellite instability in cancer identified by non-gel sieving capillary electrophoresis. *Clin Chem* 1995; 41 (3): 482–483.

52. Oto M, Koguchi K, Yuasa Y. Analysis of a polyadenine tract of the transforming growth factor-β type II receptor gene in colorectal cancers by non-gel-sieving capillary electrophoresis. *Clin Chem* 1997; 43 (5): 759–763.

53. Spafford MF, Koch WM, Reed AL, Califano JA, Xu LH, Eisenberger CF, Yip L, Leong PL, Wu L, Liu SX, Jeronimo C, Westra WH, Sidransky D. Detection of head and neck squamous cell carcinoma among exfoliated oral mucosal cells by microsatellite analysis. *Clin Cancer Res* 2001; 7 (3): 607–612.

54. Chen XQ, Stroun M, Magnenat JL, Nicod LP, Kurt AM, Lyautey J, Lederrey C, Anker P. Microsatellite alterations in plasma DNA of small cell lung cancer patients. *Nat Med* 1996; 2 (9): 1033–1035.

55. Steiner G, Schoenberg MP, Linn JF, Mao L, Sidransky D. Detection of bladder cancer recurrence by microsatellite analysis of urine. *Nat Med* 1997; 3 (6): 621–624.

56. Ahrendt SA, Chow JT, Xu LH, Yang SC, Eisenberger CF, Esteller M, Herman JG, Wu L, Decker PA, Jen J, Sidransky D. Molecular detection of tumor cells in bronchoalveolar lavage fluid from patients with early stage lung cancer. *J Natl Cancer Inst* 1999; 91 (4): 332–339.

57. Sanchez-Cespedes M, Esteller M, Hibi K, Cope FO, Westra WH, Piantadosi S, Herman JG, Jen J, Sidransky D. Molecular detection of neoplastic cells in lymph nodes of metastatic colorectal cancer patients predicts recurrence. *Clin Cancer Res* 1999; 5 (9): 2450–2454.

58. Palmisano WA, Divine KK, Saccomanno G, Gilliland FD, Baylin SB, Herman JG, Belinsky SA. Predicting lung cancer by detecting aberrant promoter methylation in sputum. *Cancer Res* 2000; 60 (21): 5954–5958.

59. Cairns P, Esteller M, Herman JG, Schoenberg M, Jeronimo C, Sanchez-Cespedes M, Chow NH, Grasso M, Wu L, Westra WB, Sidransky D. Molecular detection of prostate cancer in urine by GSTP1 hypermethylation. *Clin Cancer Res* 2001; 7 (9): 2727–2730.

60. Goessl C, Krause H, Muller M, Heicappell R, Schrader M, Sachsinger J, Miller K. Fluorescent methylation-specific polymerase chain reaction for DNA-based detection of prostate cancer in bodily fluids. *Cancer Res* 2000; 60 (21): 5941–5945.

61. Evron E, Dooley WC, Umbricht CB, Rosenthal D, Sacchi N, Gabrielson E, Soito AB, Hung DT, Ljung B, Davidson NE, Sukumar S. Detection of breast cancer cells in ductal lavage fluid by methylation-specific PCR. *Lancet* 2001; 357 (9265): 1335–1336.

62. Fraga MF, Esteller M. DNA methylation: a profile of methods and applications. *BioTechniques* 2002; 33 (3): 632–649.

63. Furuichi Y, Wataya Y, Hayatsu H, Ukita T. Chemical modification of tRNA-Tyr-yeast with bisulfite: a new method to modify isopentenyladenosine residue. *Biochem Biophys Res Commun* 1970; 41 (5): 1185–1191.

64. Clark SJ, Harrison J, Paul CL, Frommer M. High sensitivity mapping of methylated cytosines. *Nucleic Acids Res* 1994; 22 (15): 2990–2997.

65. Herman JG, Graff JR, Myohanen S, Nelkin BD, Baylin SB. Methylation-specific PCR: a novel PCR assay for methylation status of CpG islands. *Proc Natl Acad Sci* USA 1996; 93 (18): 9821–9826.

66. Esteller M, Garcia-Foncillas J, Andion E, Goodman SN, Hidalgo OF, Vanaclocha V, Baylin SB, Herman JG. Activity of the DNA repair gene *MGMT* and the clinical response of gliomas to alkylating agents. *New Engl J Med* 2000; 343 (10): 1350–1354.

67. Esteller M, Gaidano G, Goodman SN, Zagonel V, Capello D, Botto B, Rossi D, Gloghini A, Vitolo U, Carbone A, Baylin SB, Herman JG. Hypermethylation of the DNA repair gene O^6-methylguanine DNA methyltransferase and survival of patients with diffuse large B-cell lymphoma. *J Natl Cancer Inst* 2002; 94 (1): 26–32.

68. Esteller M, Sanchez-Cespedes M, Rosell R, Sidransky D, Baylin SB, Herman JG. Detection of aberrant promoter methylation of tumor suppressor genes in serum DNA from non-small cell lung cancer patients. *Cancer Res* 1999; 59 (1): 67–70.

69. Eads CA, Danenberg KD, Kawakami K, Saltz LB, Blake C, Shibata D, Danenberg PV, Laird PW. MethyLight: a high-throughput assay to measure DNA methylation. *Nucleic Acids Res* 2000; 28 (8): E32.

70. Xiong Z, Laird PW. COBRA: a sensitive and quantitative DNA methylation assay. *Nucleic Acids Res* 1997; 25 (12): 2532–2534.

71. Gonzalgo ML, Jones PA. Rapid quantitation of methylation differences at specific sites using methylation-sensitive single nucleotide primer extension (Ms-SNuPE). *Nucleic Acids Res* 1997; 25 (12): 2529–2531.

72. Bianco T, Hussey D, Dobrovic A. Methylation-sensitive, single-strand conformation analysis (MS-SSCA): A rapid method to screen for and analyze methylation. *Hum Mutat* 1999; 14 (4): 289–293.

73. Suzuki H, Itoh F, Toyota M, Kikuchi T, Kakiuchi H, Hinoda Y, Imai K. Quantitative DNA methylation analysis by fluorescent polymerase chain reaction single-strand conformation polymorphism using an automated DNA sequencer. *Electrophoresis* 2000; 21 (5): 904–908.

74. Foss AJ, Guille MJ, Occleston NL, Hykin PG, Hungerford JL, Lightman S. The detection of melanoma cells in peripheral blood by reverse transcription-polymerase chain reaction. *Br J Cancer* 1995; 72 (1): 155–159.

75. Chen XQ, Bonnefoi H, Pelte MF, Lyautey J, Lederrey C, Movarekhi S, Schaeffer P, Mulcahy HE, Meyer P, Stroun M, Anker P. Telomerase RNA as a detection marker in the serum of breast cancer patients. *Clin Cancer Res* 2000; 6 (10): 3823–3826.

76. Chu E, Drake JC, Koeller DM, Zinn S, Jamis-Dow CA, Yeh GC, Allegra CJ. Induction of thymidylate synthase associated with multidrug resistance in human breast and colon cancer cell lines. *Mol Pharmacol* 1991; 39 (2): 136–143.

77. Salonga D, Danenberg KD, Johnson M, Metzger R, Groshen S, Tsao-Wei DD, Lenz HJ, Leichman CG, Leichman L, Diasio RB, Danenberg PV. Colorectal tumors responding to 5-fluorouracil have low gene expression levels of dihydropyrimidine dehydrogenase, thymidylate synthase, and thymidine phosphorylase. *Clin Cancer Res* 2000; 6 (4): 1322–1327.

78. Thykjaer T, Workman C, Kruhoffer M et al. Identification of gene expression patterns in superficial and invasive human bladder cancer. *Cancer Res* 2001; 61: 2492–2499

79. Sanchez-Carbayo M, Capodieci P, Cordon-Cardo C. Tumor suppressor role of KiSS-1 in bladder cancer: loss of KiSS-1 expression is associated with bladder cancer progression and clinical outcome. *Am J Pathol* 2003; 162: 609–618.

15 Common Inherited Genetic Disorders

Madhuri Hegde and Sue Richards

CONTENTS

INTRODUCTION

Historically, molecular genetic testing developed in the mid-1980s in research laboratories where disease genes were being identified.[1-3] After identifying a disease gene, it was a small step to proceed to test patients for mutations that might cause the disease. Testing for common monogenic genetic disorders, such as cystic fibrosis,[4] the muscular dystrophies,[5] and fragile X syndrome[6] is now routinely performed in clinical testing laboratories in both academic and commercial settings. Such molecular genetic tests can be extremely specific (designed to detect a single mutation in a specific gene), or non-targeted (scanning whole genes to identify unknown mutations). The results can be qualitative (presence or absence of a mutation) or quantitative (number of repeat units or number of copies of a gene sequence present).

Testing laboratories use a wide variety of equipment and technologies, and new tests and technologies are continuously introduced. Genetic testing is now becoming integrated into all fields of medicine.[7]

This chapter considers several common inherited single gene disorders, molecular testing methods, and strategies. While many genetic tests exist, we excluded somatic mutations and infectious disease testing from our discussion even though they represent a very large segment of the testing industry. We have limited this discussion to only selected examples of genetic tests.

CHALLENGES OF GENETIC TESTING

Genetic testing has diverse applications. The same test used to diagnose a rare genetic condition also predicts its occurrence in asymptomatic family members, detects carriers, and identifies an affected fetus. Testing may be aimed at determining an appropriate therapeutic strategy or, in the absence of effective treatment, provide a prognosis or determine reproductive risk.[8-11] Molecular testing for single gene disorders has now moved into newborn screening programs in several states and the implementation has been a significant challenge. The purpose of screening is to identify newborns with the genetic disorder and rapidly initiate treatment to reduce symptoms and thus provide a better prognosis.

The American College of Medical Genetics (ACMG, http://www.acmg.net) recently recommended the addition of cystic fibrosis to the newborn screening panel. Some of the challenges have included programmatic issues of counseling and education for families, providing diagnostic confirmation services, developing and implementing testing in state laboratories, choosing appropriate technology, having access to trained genetics professionals, and performing such services on limited budgets.[1,2,12-15]

One of the biggest challenges in genetic testing is evaluating analytical and clinical validity for each genetic test before it enters the market. Assessing clinical utility is an even bigger challenge.[16] The Centers for Disease Control and Prevention (CDC) is currently addressing the evaluation of genetic testing for public health, particularly for common disorders. The assessment of clinical utility poses particular challenges for genetic tests that assess drug response or susceptibility to common diseases. In contrast to tests for single-gene disorders, risk association tests have limited predictive value.

For example, factor V Leiden, a gene variant of the factor V gene, confers an increased risk of venous thrombosis particularly in the homozygous state. However, a population-based study indicates that the cumulative risk of thrombosis in R506Q homozygotes by the age of 80 is only about 12%.[17-21] Other risk factors, including other gene variants and nongenetic factors such as cigarette smoking, immobility, pregnancy, surgery and oral contraception, influence whether venous thrombosis will occur even in the presence of factor V Leiden. Most importantly, most individuals with venous thrombosis do not have the factor V Leiden variant.

For such risk association tests, the gene test should be used only when gene–environment interactions are thoroughly understood and effective interventions are available to improve the health outcomes for people with the gene variant. Genetic testing, while often performed for purposes of risk association, requires a thorough understanding of the clinical consequences, and one challenge is to convey such information to a provider through a laboratory report. The good news is that technology is no longer a challenge because multiple platforms now support this simple SNP-based analysis and allow high-throughput testing.

A significant objective for genetic testing is improving quality assurance — a concern debated at length by the Secretary's Advisory Committee on Genetic Testing (SACGT, Office of Biotechnology Activities, U.S. National Institutes of Health, http://www4.od.nih.gov/oba/SACGHS.HTM). Many issues related to genetic testing are currently being addressed by a number of organizations, including the ACMG, CDC, CLSI (Clinical and Laboratory Standards Institute, formerly NCCLS), AMP (Association of Molecular Pathologists), CAP (College of American Pathologists), and multiple government agencies, academic institutions, and private foundations. Some of the immediate laboratory concerns include: (1) development of more comprehensive standards and guidelines for disease-specific genetic testing and professional policy statements for testing, (2) developing a network of available positive control materials used in quality control for genetic tests, (3) providing improved proficiency testing programs for laboratories, and (4) evaluation and standardization of interpretation and reporting of genetic tests.

The goal of quality assurance is to make a reproducibly perfect product that has clinical utility. A major concern in genetic testing is that the test validation process should be sufficiently thorough to prevent introduction of testing prematurely. Of equal concern is direct marketing to the consumer — a business strategy that has not gained the enthusiasm of the academic genetics community. As a result of these issues, the genetic testing industry is now in the process of defining a structure in which academia, industry, and government can work together to provide quality genetic testing.

Another challenge has been obtaining genetic testing for ultra-rare disorders in clinical laboratories. In the past, these tests were conducted in research laboratories and patient results were reported to clinicians who used the information in patient management and for family counseling. Such practice today is illegal. Fortunately, with the formation of the Rare Disorders Network, this practice is no longer a problem and all rare disorder testing is performed in CLIA-approved laboratories (Clinical Laboratory Improvement Amendments, http://www.fda.gov/cdrh/CLIA/). The challenge, however, continues as these rare disorders require sequence analysis and often involve very large genes. Recently, the ACMG has developed standards and guidelines for ultra-rare disorders to address these test validation issues. The next challenges will be implementation and expansion of these services.

Genetic testing faces many other challenges. Some immediate issues that directly affect laboratories surround intellectual property, in particular the restrictive licensing practices for gene patents practiced by numerous academic and commercial groups. Exclusive licenses, while common in the United States, are detrimental to the laboratory and clinical communities because they prevent multiple laboratories from offering particular genetic tests, thereby creating monopolies and quality assurance bottlenecks. Reimbursement from third party payers constitutes another challenge for clinical genetics laboratories. This issue is being addressed through ACMG's Economics Committee and by the development of new current procedure terminology (CPT) codes for genetic tests.

Laboratories performing genetic testing are challenged with multiple ethical, legal, and social issues such as determining if and when informed consent is necessary, duration of record retention, deciding whether residual specimens can be banked and utilized, respecting genetic privacy while performing and reporting appropriate family studies, and providing optimal educational materials for providers and patients. Because both federal and state regulations apply to testing laboratories, no single solution will answer all of these issues. The genetic testing industry is continually moving

TABLE 15.1
Fourteen Documented Trinucleotide Repeat Disorders

Disorder	Gene Locus	Gene Product	Trinucleotide Repeat	Normal Range	Pathological Range
HD	4p16.3	Huntingtin	CAG	7 to 26 (27 to 35 normal mutable)	36 to 39 reduced penetrance 40 and above affected
DRPLA	12p13.31	Atrophin-1	CAG	6 to 35	35 to 48 inconclusive 49 and above affected
Kennedy's disease	Xq11-q12	Androgen receptor	CAG	11 to 30	31 to 39 inconclusive 40 and above affected
Fragile X	Xq27.3	FMR-1	CGG	5 to 44	45 to 54 inconclusive 55 to 230 premutation 230 and above full mutation
FRDA	9q13	Frataxin	GAA	4 to 33	66 and above affected
Myotonic dystrophy	19q13.2-q13.3	Protein Kinase	CTG	5 to 35	36 to 49 inconclusive 50 and above affected
SCA1	6p23	Ataxin-1	CAG	6 to 44	39 to 83
SCA2	12q24.1	Ataxin-2	CAG	14 to 31	33 to 64
SCA3	14q32.1	Ataxin-3	CAG	12 to 40	54 to 86
SCA6	19p13	CACNA1A	CAG	4 to 20	20 to 31
SCA7	3p14-21.1	Ataxin-7	CAG	4 to 27	37 to >200
SCA 8	13q21	N/A	CTG	15 to 91	100 to 155
SCA 12	5q31-33	PPP2R2B	CAG	<29	66 to 93
SCA 17	6p27	TBP	CAG	25 to 42	45 to 63

forward, addressing these challenges, and developing new targets of interest, thus making genetic testing an exciting field of laboratory medicine.

SINGLE GENE DISORDERS

TRINUCLEOTIDE REPEAT DISORDERS

Trinucleotide repeat (TNR) disorders[22] constitute a class of diseases in which the mechanisms are excessive copies of triplet repeating units. Many TNR disorders show anticipation; successive generations are affected with earlier onset and more severe disease. Today, 14 TNR disorders have been documented (Table 15.1). Fragile X syndrome was the first TNR discovered and thus is considered the prototype for this class of disease.

FRAGILE X SYNDROME

Fragile X syndrome is a genetic condition[23] that causes a wide range of mental impairments, from mild learning disabilities to severe mental retardation and is the most common cause of inherited mental retardation (MR). A population prevalence figure of 1 in 4,000 or 2.4 in 10,000 has been reported on the basis of molecular genetic analysis.

In addition to mental impairment, fragile X syndrome is associated with a number of physical and behavioral characteristics. A carrier female typically has a 30 to 40% chance of giving birth to a mentally retarded male child and a 15 to 20% chance of having a mentally retarded female child.[24] Further, maternal family history frequently reveals a relative with mental retardation or developmental and learning disabilities. Most studies have dealt with recognition of this syndrome in older children and young adults, but many of the physical features, behavioral characteristics, and family history features are apparent earlier. The classic physical features of fragile X syndrome include a long face with a prominent jaw, large prominent ears, and post-pubertal macro-orchidism.[25-28]

In 1991, the fragile X gene (*FMR1*) was characterized and found to contain a tandemly repeated trinucleotide sequence (CGG) in the 5' untranslated region.[29] The number of CGG repeats in the *FMR1* genes in normal individuals varies from 5 to approximately 40. Premutations, alleles that are not disease-causing in an individual but increase in size in successive generations when transferred through the female germline, range from approximately 55 to 200 repeats. Full mutations are expansions of more than 200 repeats and are disease-causing. Alleles with approximately 41 to 54 copies of the repeat are intermediate size alleles that can be unstable and expand in successive generations or may be stably inherited.

A premutation is susceptible to expansion only after passage through a female meiosis, and larger premutations are more likely to expand to full mutations in successive generations. Males and females carrying premutations are unaffected. Male carriers are designated *normal transmitting* and pass the premutation to all their daughters. Their carrier daughters are unaffected, but are at risk of having affected children. Variable clinical severity is observed in both sexes. Males with full mutations are usually mentally retarded and show typical physical and behavioral features, while only one third of females with full mutations are severely affected.[30-35]

Some affected individuals, designated *mosaics*, exhibit both premutation and full mutation. Expansion of the trinucleotide repeat to more than 200 repeats is almost always associated with methylation of the promoter region of the gene and correlated gene inactivation. Although it is clear that methylation plays a role in determining phenotype, its effect on clinical severity is somewhat unpredictable, especially in females.

DNA studies have improved the accuracy of testing for fragile X syndrome.[36] The size of the trinucleotide repeat (detected by PCR) and the methylation status of the *FMR1* gene (detected by Southern analysis) are used for fragile X diagnosis as well as for carrier and prenatal diagnosis. PCR analysis utilizes flanking primers to amplify a fragment of DNA spanning the repeat region (Figure 15.1). The sizes of the PCR products reflect the approximate number of repeats in each

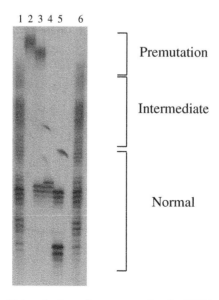

FIGURE 15.1 Fragile X PCR. Determination of repeat number in *FMR1* gene by PCR and probed with FMR1 probe. Lane 1, FRAX A PCR ladder; Lane 2, premutation male (86 repeats); Lane 3, premutation female (31, 82 repeats); Lane 4, normal male (32 repeats); Lane 5, normal female (20, 30 repeats); Lane 6, FRAX A PCR ladder.

A   B

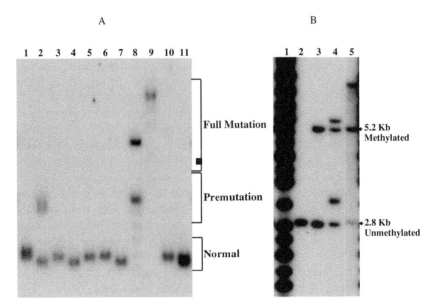

FIGURE 15.2 A: Fragile X Southern analysis. Genomic DNA digested with PstI and probed with FMR1 probe. Lane 1, normal female (32, 44 repeats); Lane 2, premutation female (20, 160 repeats); Lane 3, normal male (30 repeats); Lane 4, normal male (19 repeats); Lane 5, normal male (29 repeats); Lane 6, normal male (31 repeats); Lane 7, normal male (20 repeats); Lane 8, 100-bp ladder; Lane 9, full mutation male (>600 repeats); Lane 10, normal male (32 repeats); Lane 11, normal female (20, 30 repeats). B: Fragile X methylation. Genomic DNA digested with EcoRI and EagI to detect methylation status of *FMR1*. Lane 1, allelic marker; Lane 2, normal male (32 repeats); Lane 3, normal female (20, 30 repeats); Lane 4, premutation female (20, 140–165 repeats); Lane 5, expanded female (30, 46 repeats).

allele. The efficiency of the PCR reaction is inversely related to the number of CGG repeats, such that large expansions fail to yield a detectable product in the PCR assay. Thus, PCR is limited to detection of normal and premutation size alleles.

FMR1 analysis by Southern blotting (Figure 15.2A) allows simultaneous assays of both sizes of the larger repeats and methylation status (Figure 15.2B). A methylation-sensitive restriction enzyme that fails to cleave methylated sites is used to distinguish methylated and unmethylated alleles. A normal male has only one unmethylated allele, whereas a normal female has one methylated allele and one unmethylated allele due to random X inactivation. While Southern blot analysis is more labor-intensive than PCR and requires larger quantities of genomic DNA, it accurately detects alleles in all size ranges, even though precise sizing of the smaller alleles is not possible. Most clinical labs utilize both methods.

The ACMG has developed technical standards and guidelines for fragile X testing and laboratories should follow these guidelines. The following is a brief review and is not meant to be comprehensive.

Individuals for Whom Testing Should Be Considered

- Individuals of either sex with mental retardation, developmental delay, or autism, especially if they have (1) symptoms of fragile X syndrome, (2) a family history of fragile X syndrome, or (3) relatives with undiagnosed mental retardation
- Individuals seeking reproductive counseling who (1) have a family history of fragile X syndrome, (2) have a family history of undiagnosed mental retardation, (3) seek assisted reproductive options including egg donors
- Fetuses of known carrier mothers

Population carrier screening is not recommended at present except as part of a well defined clinical research protocol. The DNA test is accurate, but it is important to ensure that an effective program is in place to adequately inform tested populations of the meanings and implications of results. The nature of the *FMR1* mutation and its inheritance are complex issues, and testing necessitates appropriate follow-up counseling.

Approaches to Testing

- If an individual is being tested specifically for fragile X, DNA analysis for expanded CGG repeats is the method of choice.
- If the etiology of mental impairment is unknown, DNA analysis for fragile X syndrome should be performed as part of a comprehensive genetic evaluation that includes routine cytogenetic analysis. Cytogenetic studies are critical since constitutional chromosome abnormalities have been identified as frequently or more frequently than fragile X mutations in mentally retarded individuals.
- For an individual who is at risk due to a positive family history of fragile X syndrome, DNA testing alone is sufficient. If the diagnosis of the affected relative was based on previous cytogenetic testing for fragile X syndrome, then at least one affected relative should be included in DNA testing.
- Prenatal testing of a fetus is indicated following a positive carrier test in the mother. However, results from chorionic villus sampling (CVS) must be interpreted with caution because the methylation status of the *FMR1* gene is often not yet established in chorionic villi at the time of sampling. If a CVS is used, amniocentesis may be necessary to resolve an ambiguous result.

Case Presentation

A 6-year-old male presented to the pediatric genetics clinic with a personal and family history of mental retardation, frequent otitis media, and self-abusive behavior. A detailed pedigree (Figure 15.3) and family history showed X-linked inheritance in the family and genetic testing for fragile X syndrome was performed. Results of genetic testing indicated a full expansion of >300 CGG repeats, thus confirming the diagnosis of fragile X syndrome. As a result, testing was offered to other family members and the mother was advised that prenatal diagnosis was possible for her future pregnancies.

New Technology and Recent Advances

Advancement in the field of fluorescent technology and its application to molecular biology have proven useful in fragile X analysis. Labeling of PCR primers with fluorescent labels (TET, FAM, HEX) with product separation on automated capillary sequencers is now a reality and can be developed as "home brews" or with analyte-specific reagents (ASRs; Abbott Laboratories, Abbott Park, Illinois, USA, http://abbott.com).

Fluorescent technology has significant advantages over conventional radioactivity-based PCR amplification and gel-based electrophoresis, allowing detection of larger repeat sizes with higher accuracy. The platform offering high-throughput analysis with automated software calling of alleles is an attractive alternative. In addition, fluorescence-based testing is safer for technical staff and reduces labor and time for the assay. Methylation-specific PCR is also an attractive alternative to reduce turn around times and labor.[37–40]

HUNTINGTON'S DISEASE

Huntington's disease (HD) is a fatal autosomal-dominant genetic disease that destroys neurons in areas of the brain involved in emotions, intellect, and movement.[41] Its course is characterized by

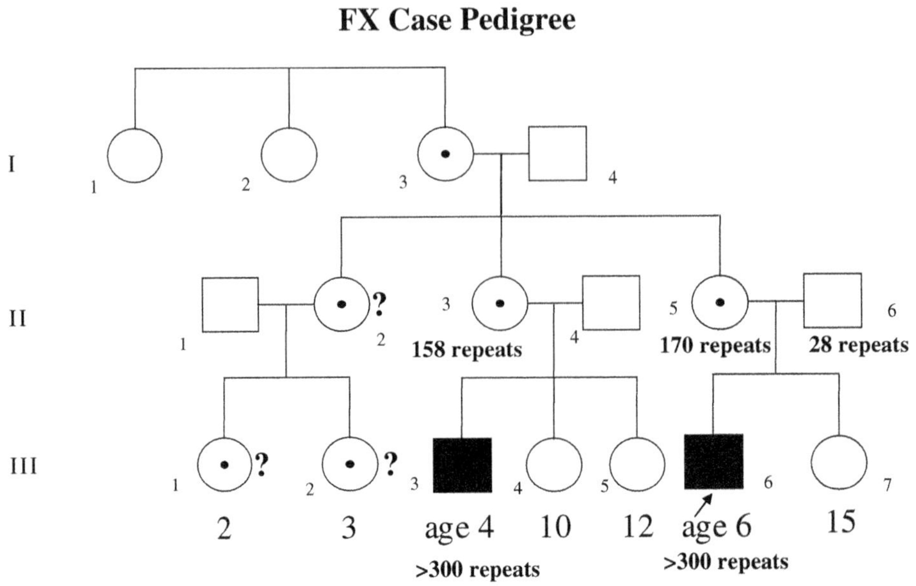

FX Case Pedigree

FIGURE 15.3 The proband III6 (age 6) was referred for genetic testing for fragile X syndrome and was found to have a repeat size of more than 300 repeats (full expansion). The mother of the proband II.5 was found to have repeat size of 170 in premutation range. Further genetic analysis indicated all females shown in the pedigree to carry premutations in the fragile X although the size of the premutation varied for each individual.

chorea (jerking uncontrollable movements of the limbs, trunk, and face), progressive loss of mental abilities, and the development of psychiatric disturbances. This central nervous system disorder presents with symptoms usually appearing in adults within the fourth or fifth decade of life, although age of onset varies. Within the same family, symptoms vary both in rate of progression and age of onset. Symptoms may include involuntary movements and loss of motor control.

Personality changes may occur, with loss of memory and decreased mental capacity. HD progresses without remission over 10 to 25 years and patients ultimately are unable to care for themselves. Juvenile onset HD occurs in approximately 16% of all cases. HD is panethnic and thus is not specific for any particular racial or ethnic group or sex.[42] HD is currently found in many different countries and ethnic groups around the world. The highest frequencies are found in Europe and countries of European origin such as the United States and Australia (40 to 100 cases per million people). The lowest documented frequencies are found in Africa, China, Japan, and Finland.

The genetic mutation, a CAG repeat expansion that occurs in the coding region of the *IT-15* gene located on chromosome 4, causes Huntington's disease.[43–45] Expansion of a polyglutamine tract in the huntingtin protein, called a TNR mutation, results in a gain of function role. Exactly how the TNR mutation alters the function of the protein is not well understood. Shortly after the discovery of the CAG repeat, genetic testing for HD was introduced.

Indications for Testing

HD testing is used for (1) an individual with symptoms of HD for confirmation of diagnosis; (2) a presymptomatic individual known to be at risk for HD when a first or second degree affected family member has a CAG repeat expansion; and (3) a fetus if one parent has demonstrated a CAG repeat expansion.

Test Interpretation

Normal individuals have 26 or fewer CAG repeats. Individuals with intermediate sized alleles in the range of 27 to 35 CAG repeats are not at risk for developing HD. However, intermediate alleles have been shown to expand upon passage in successive generations and are thus termed mutable alleles. Mutant alleles contain 36 or more repeats and are divided into two classes. Alleles containing 36 to 40 CAG repeats are characterized by reduced penetrance — only some individuals with alleles in this size range will develop clinical symptoms within their lifetimes.

Alleles of 41 or more repeats are associated with the development of HD. Although the number of repeats is loosely correlated with the age of onset, the association is not sufficiently strong to be predictive of age of onset and thus should not be used clinically for this purpose. Laboratories should follow the ACMG guidelines for appropriate size determination, interpretation, classification of alleles, and reporting of results.[46]

Presymptomatic Testing

Presymptomatic testing may involve patient enrollment in an approved HD testing center that provides a multidisciplinary approach to both pre- and post-test genetic and psychological counseling and neurological evaluation. Eligibility in such a program requires that the asymptomatic patient have a confirmed family history with a 50% risk for developing HD and be able to give informed consent for testing and evaluation. Confirmation by molecular analysis of the affected family member is strongly encouraged prior to presymptomatic testing of family members at risk.

Genetic counseling sessions help patients understand disease inheritance and the natural history of the disease along with the genetic testing options available and the risks and benefits of testing. The psychological counseling sessions help evaluate the readiness of a patient to undergo predictive testing since a positive diagnosis for this devastating disorder can be emotionally staggering. Following counseling and neurological evaluation, molecular testing is performed. While an expanded CAG repeat confirms the diagnosis of HD in a symptomatic patient, it predicts that an asymptomatic patient will develop HD sometime within his or her lifetime.[47,48] The size of the repeat, however, does not provide information about the age of onset of disease, although a rough correlation indicates that very large repeats result in earlier onset, such as those found in juvenile onset HD. However, it is important to understand that a negative test result for an asymptomatic at-risk individual is only meaningful if the affected family member has been shown to carry an expanded CAG repeat.[49]

Assay

The CAG repeats in the HD gene are found immediately 5' of a CCG repeat, which is also polymorphic in length.[50,51] The original primer sets that were used to size the CAG repeats included this CCG polymorphism and led to the misclassification of disease and normal alleles. As a result, a number of assays described amplify only the CAG repeats and are not confounded by CCG repeat size variation, and these are used in most clinical laboratories (Figure 15.4).

A very rare polymorphism has been described in the 3' primer used in some assays. This polymorphism can potentially disrupt primer binding to an HD chromosome and result in a false negative result. Apparent normal homozygous results should be evaluated for the presence of a polymorphism that results in failure to amplify one allele. This is a common pitfall for all PCR-based analyses. The use of alternative primers often allows resolution of two normal alleles in cases having identical numbers of CAG repeats but different numbers of CCG repeats.[52] Alternatively, a Southern blot using PstI-digested DNA hybridized to the 4G6P1.7 probe allows evaluation for large expansions and alleles that fail to amplify by PCR.

When testing for CAG repeat length, it is important to use a known repeat size standard that has been sequenced because repeat size can change during bacterial culture. In our experience,

Huntington Disease PCR

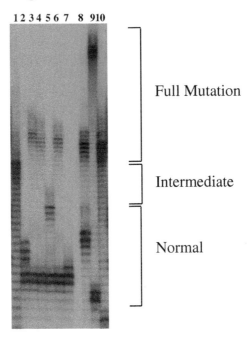

FIGURE 15.4. Determination of repeat number in HD gene by PCR. Lane 1, 35 repeat clone; Lane 2, normal male (17, 20 repeats); Lane 3, expanded female (17, 42 repeats); Lane 4, expanded male (17, 40 repeats); Lane 5, normal female (17, 27 repeats); Lane 6, expanded female (17, 41 repeats); Lane 7 normal female (17, 18 repeats); Lane 8, expanded female (23, 40 repeats); Lane 9, expanded female (15, 70 repeats); Lane 10, 40 repeat clone.

standards such as M13 ladders can be unreliable in size determination to the nearest repeat. If radioactively labeled primers are used, it is important to consider that end-labeled PCR products may migrate differently to products incorporating $P^{32}dCTP$ and adjust sizing accordingly. Recent advances by the National Institute for Standards and Technology (NIST) have been directed toward development and validation of size standards for fragile X testing, and it is likely that the next step will include other trinucleotide disorders such as HD.

Case Presentation

An asymptomatic 36-year-old male presented at clinic. He is an only child and has no children (Figure 15.5). Eight years ago, his mother was diagnosed with HD and is now deceased. Since no clinical testing was performed on this individual's mother, presymptomatic testing for the CAG repeat expansion was performed following appropriate evaluation and counseling through an HD predictive program. Testing indicated that he inherited an expanded CAG of 46 repeats and was counseled accordingly.

New Technology and Recent Advances

Fluorescent-based PCR assays for the CAG repeats and flanking CCG repeats are now performed in a number of U.S. clinical laboratories, although these tests are considered "home brews." The development of commercially available ASRs would be useful in the future for clinical laboratories offering these tests. Non-radioactive Southern blot assays are also widely used, which differentiate

HD Case Pedigree

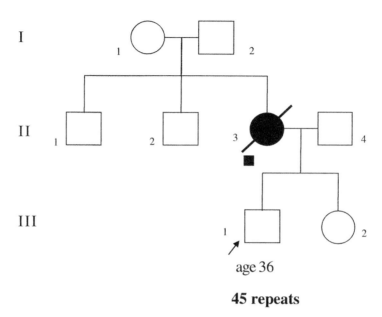

age 36

45 repeats

FIGURE 15.5 A 36-year old male III1 presented to the clinic with a family history of Huntington's disease. His deceased mother II3 had a clinical diagnosis of Huntington's disease but no genetic testing was done. After appropriate counseling in the predictive testing program, genetic testing was ordered and detected 46 repeats in this one copy of the HD gene of this individual, indicating full expansion in the HD gene.

true homozygosity from the failure of an allele to amplify and detect very large expansions such as those found in juvenile HD.

DUCHENNE MUSCULAR DYSTROPHY AND BECKER MUSCULAR DYSTROPHY

Duchenne muscular dystrophy (DMD) and Becker muscular dystrophy (BMD) are allelic X-linked recessive neuromuscular disorders that affect 1 in 3,000 and 1 in 10,000 live-born males, respectively.[53] DMD is characterized clinically by progressive loss of muscle strength; males become wheelchair-bound by the age of 12. Affected males manifest symptoms at an early age: difficulty in walking, weakness of the pelvic girdle muscles, and pseudohypertrophy of calf muscles. The heart muscle becomes enlarged, leading to dilated cardiomyopathy with death occurring by respiratory and cardiac failure in the early twenties.[54] BMD affects the same muscle groups as DMD but the age of onset is later, the symptoms are milder, and progression is slower.

Prior to the development of gene mutation-based testing, conventional screening for diagnosis of affected males was by clinical examination and included an assessment of creatine phosphokinase (CPK) level and muscle cell histology obtained only through an invasive biopsy.[53] Carrier screening for at-risk females was based on Bayesian analysis of CPK levels, family history including unaffected sons, and linkage analysis. Prenatal testing was either not available or inaccurate, and a female carrier faced the decision of terminating a pregnancy with a male fetus with a 50% chance of being affected or unaffected. Molecular analysis has revolutionized the approach to diagnosing these patients.[55–66]

The *dystrophin* gene which maps to Xp21 and results in DMD or BMD when defective is one of the largest genes described in humans — extending over 2.4 Mb in length and comprising 79 exons. The gene transcript is expressed in significant levels in muscle tissues and at lower levels in the brain. The gene product functions as a structural protein in muscle contraction, linking intracellular actin with an extracellular glycoprotein matrix complex. Loss or reduction of this product has serious biological and clinical consequences. Moreover, the genetics of DMD are complex.[67] Because this large gene has a high mutation rate, about one third of DMD and BMD cases represent *de novo* mutations. While most of the mutations are deletions (60%) or duplications (5 to 10%), about one third of patients have another type of mutation that escapes routine molecular analysis. Heterozygous carrier females generally do not exhibit disease phenotypes. However, some carrier females exhibit very mild phenotypes and very rarely can be severely affected. The manifestation of symptoms in these carrier females is usually due to non-random X chromosome inactivation or to an X:autosome translocation event. These features pose a diagnostic challenge for the analysis of DMD.

Diagnosis of DMD and BMD at the molecular level for deletions and duplications involves two methods. Currently, most laboratories offering DMD testing use multiplex PCR analysis as their first line of testing (Figure 15.6). This strategy is based upon the observation that the majority of deletions occur in two deletion-prone regions, located 500 and 1200 kb downstream of the 5′ end of the gene.[68,,69] Consequently, only a subset (about 20) of the 79 dystrophin exons are screened to detect the majority (approximately 98%) of deletions. Multiplex PCR is fast and easy and can provide a rapid result for prenatal diagnosis for a male fetus when the familial mutation is known.

Multiplex PCR amplification of genomic DNA is an efficient method of identifying deletion mutations in affected males and carrier females using ethidium bromide, SYBR Green dye staining, or fluorescence detection. In dosage analysis of carrier females, PCR amplification must remain in the logarithmic phase such that the amount of PCR product is directly proportional to the copy number of the target sequence. In case of a deletion carrier, the amount of product amplified from single (deleted) and from double copy (normal) should be in the ratio 1 to 2. In the case of a duplication carrier, this ratio will be 3 to 2.

Southern blot analysis is a second technique for screening for deletions and duplication mutations, and it is used to confirm mutations identified by multiplex PCR and to determine the extent of a deletion or duplication. Duplication mutations are more difficult to detect; results are inherently less accurate and require dosage analysis. Likewise, carrier testing for females is performed by dosage analysis following Southern blotting, requires quantitative analysis by densitometry to assess copy number, and is less accurate than testing in males.[70]

Carrier detection is even more technically challenging when an affected relative is not available for testing (Figure 15.7). Other methods utilized for carrier detection include pulsed field gel electrophoresis (PFGE), fluorescent *in situ* hybridization (FISH), and multiplex ligation-dependent probe amplification (MLPA).

DMD is a good example of a disorder in which genotype can predict phenotype most of the time. Deletion mutations in DMD patients are usually out of frame (frame shift mutations) and result in complete loss of the protein. In contrast, BMD patients generally have in-frame deletions that result in production of an internally deleted but partially functioning dystrophin protein. Therefore, it is important for laboratories to determine the nature and extent of these deletions, as they are often used in clinical management of patients and family counseling.

In the remaining 30% of DMD and BMD cases that exhibit no detectable deletion or duplication, the mutations are assumed to be either microdeletions or point mutations. Several methods have been suggested for the detection of point mutations in the *dystrophin* gene.[71,72] They include single strand conformational polymorphism (SSCP), heteroduplex analysis (HA), protein truncation testing (PTT), chemical mismatch cleavage analysis (CCM), denaturing gradient gel electrophoresis (DGGE), and denaturing high pressure liquid chromatography (DHPLC).

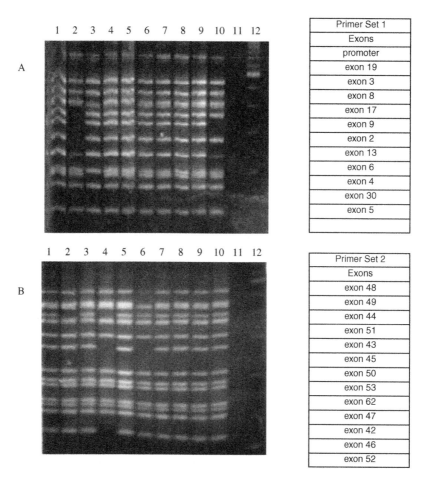

Primer Set 1
Exons
promoter
exon 19
exon 3
exon 8
exon 17
exon 9
exon 2
exon 13
exon 6
exon 4
exon 30
exon 5

Primer Set 2
Exons
exon 48
exon 49
exon 44
exon 51
exon 43
exon 45
exon 50
exon 53
exon 62
exon 47
exon 42
exon 46
exon 52

FIGURE 15.6 DMD *dystrophin* gene multiplex PCR. Lane 1 represents a normal control male. Lanes 2 through 10 represent amplification products of unrelated DMD males analyzed using multiplex I and multiplex II in panels A and B, respectively. Lane 2 in both panels A and B represents positive controls with deletion of exons 8 through 17 and exon 51, respectively. Lanes 3 and 10 in panel A represent DMD males with deletion of exon 4 and exons 13 through 17, respectively. Lanes 4 and 6 in panel B represent DMD males with deletions of exons 50 through 52 and 48 through 50, respectively. Lane 12 in both panels A and B represents 100-bp ladders. Lane 11 in both panels A and B represents no DNA. Panel A and panel B show amplification products using multiplex I and II, respectively.

An apparent variant identified by any these methods must be sequenced subsequently to confirm the presence of a mutation or polymorphism (Figure 15.8). Once a point mutation has been detected in the proband, carrier detection is possible by simply interrogating for the presence of the mutation event using direct sequencing, amplification refractory mutation system (ARMS), PTT analysis, or one of many single nucleotide polymorphism (SNP) detection methods (discussed in the section on cystic fibrosis later in this chapter). Approximately 70% of clinically diagnosed DMD and BMD patients who do not have deletion or duplication mutation in the *dystrophin* gene will be found to carry point mutations, generally nonsense mutations.[73,74] The size of a gene is a limiting factor to screening for point mutation within the gene but reduced sequencing costs have raised the possibility of full sequencing of the coding region.

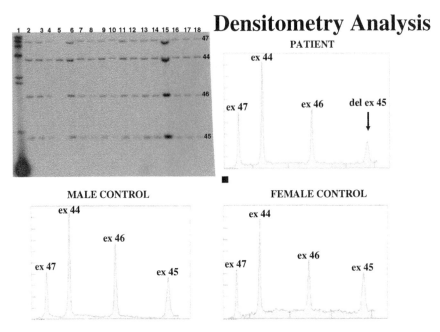

FIGURE 15.7 DMD *dystrophin* gene Southern blot analysis. Genomic DNAs were digested with HindIII and probed with cDNA47-4B. This probe covers exons 44 through 47 and each fragment represents an exon. Lane 1, 1-kb ladder; Lane 2, male control; Lanes 2 through 10, male patients; Lanes 11 through 17, female patients; Lane 18, female control. Lane 12 shows a female patient with a possible heterozygous deletion of exon 45 confirmed by densitometry analysis. Densitometric analysis was performed in comparison to male and female controls and a reduced dosage was found for exon 45 in the female patient, confirming a heterozygous deletion of exon 45.

Indications for Testing

- Confirmation of possible or definite clinical diagnosis of DMD or BMD in symptomatic individuals.
- Carrier testing for at-risk females when the mutation in the proband has been identified. All mothers of affected males should have carrier testing.
- Carrier testing for at-risk females having a DMD- or BMD-affected male relative who has not undergone molecular analysis.
- Prenatal testing for families in which the dystrophin mutation has been identified. Families with a definite diagnosis and no molecular lesion identified have the option of linkage analysis using markers within and surrounding the gene. Such analysis requires multiple family members and coordination between counselor and laboratory prior to testing to determine the best approach.

Approach to Clinical Testing

Molecular methods involve multiplex PCR amplification, quantitative PCR, Southern blotting, mutation scanning, and full sequence analysis of the *dystrophin* gene. In the absence of detectable deletion, immunohistochemical analysis of dystrophin protein in muscle cells may provide useful information (Figure 15.9). Dystrophin protein and DNA analyses coupled with clinical presentation are used to classify patients as DMD or BMD. Dystrophin protein is absent in DMD patients and positively correlates with out-of-frame deletions, while dystrophin is reduced in size or amount and deletions are in-frame in BMD patients.

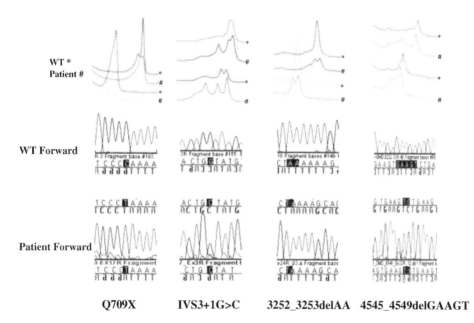

WT *
Patient #

WT Forward

Patient Forward

Q709X   IVS3+1G>C   3252_3253delAA   4545_4549delGAAGT

FIGURE 15.8 *(A color version follows page 270)* DMD *dystrophin* gene point mutation analysis. Wild-type DHPLC pattern (top) and patient DHPLC pattern (bottom) at two temperatures for each exon for different mutations. Sequence analysis showing confirmation of each mutation is shown below. DNA from patients having definite DMD or BMD but no deletion or duplication mutation detected by multiplex PCR or Southern blot was subjected to DHPLC to scan for other types of mutations. The DHPLC profile is shown for the wild-type compared to the patient. Following DHPLC, samples demonstrating a heteroduplex (sequence variant) were subjected to sequencing in both forward and reverse direction compared with the wild-type sequence. Four types of mutations are shown: a nonsense mutation Q709X resulting in the change of a glutamine to a stop codon at codon 709; a splicing mutation IVS3 + 1G > C that represents a change of a guanine to a cytosine at the first nucleotide position in intron 3; a deletion of two adenines at cDNA nucleotide position 3252; and a 5-bp deletion at cDNA position 4545.

These correlations can be made with >90% accuracy. While an ongoing debate concerns whether DNA or protein studies should serve as the gold standard and be the first test, it can be argued that DNA has the major advantage of being noninvasive. Protein analysis requires a muscle biopsy while DNA involves a simple blood draw. If an altered dystrophin protein is detected by immuno-histochemistry, transcripts may be useful in cases where the routine molecular analysis of deletion and duplication screening followed by point mutation analysis fails to identify a mutation.

Linkage analysis may be necessary to identify the at-risk X chromosome segregating in an affected family when no mutation has been identified in the gene even after exhaustive analysis. While linkage analysis is useful in the presence of a family history of DMD, its use is limited for sporadic cases that represent the most difficult counseling cases and thus are most challenging. An additional limitation of linkage analysis is that recombination between markers, even within the *dystrophin* gene, may lead to a wrong interpretation, and thus appropriate genetic counseling for these families is essential.

While molecular testing is relatively straightforward, interpretation of carrier testing is complicated and requires a good understanding of genetic mechanisms. Because of the high mutation rate with the *dystrophin* gene, mothers of affected males are not always carriers. A female having one affected son and no family history of muscular dystrophy is either a carrier, a somatic mosaic, a germline mosaic, or a noncarrier, with the son having a *de novo* mutation. A female who has a negative test result for a *dystrophin* gene mutation that has previously been identified in her affected son has about a 15% risk of having germline mosaicism and thus a chance of having another

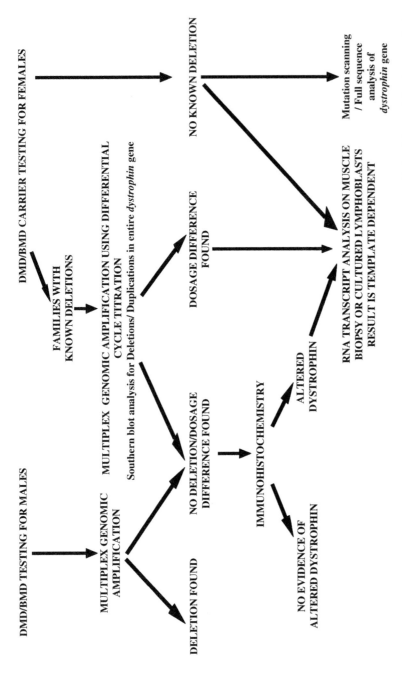

FIGURE 15.9 DMD flow chart for testing. Molecular testing for diagnosis of males includes multiplex PCR and Southern analysis for deletions and duplications. A positive test result confirms the diagnosis. Dystrophin protein analysis is indicated in the absence of a positive molecular test result. For an individual with a definite diagnosis and negative molecular test by deletion analysis, mutation scanning and sequencing is the next approach. Once a mutation is identified in the male proband, carrier detection in females is straightforward and can be done by dosage analysis for deletions or duplications or by sequence analysis for other types of mutations. If no mutation is identified in the male, then transcript analysis can be performed to identify deep intronic mutations that could affect splicing or create an alternative splice site.

affected son. DMD is one of the more complex genetic tests and should not be undertaken except by very specialized molecular genetics laboratories employing professionals who understand these complexities.

Case Presentation

A 5-year old male with progressive symmetrical muscle weakness, involvement of proximal limb muscles, and calf hypertrophy was referred to a genetics clinic. Muscle biopsy was performed and showed no detectable amount of dystrophin. Diagnostic testing for deletions and duplications in the *dystrophin* gene using multiplex PCR and Southern blotting were negative. The child received a clinical diagnosis of DMD even though no mutation was identified in the *dystrophin* gene. Using a mutation scanning test, a nonsense mutation was detected in exon 37 of the *dystrophin* gene, confirming the clinical diagnosis of DMD. No significant family history of DMD was noted and the mother tested negative for the mutation. She was given a 15% risk due to germline mosaicism and informed of the availability of prenatal testing for her future pregnancies.

New Technology and Recent Advances

Fluorescence technology offers a major advantage over conventional agarose gel staining methods, as the PCR products generated using fluorescently labeled primers can be accurately sized and quantified using an automated DNA sequencer.[75] Laboratories undertaking this analysis should be aware that the quality of primer synthesis is an important parameter in performing this assay because contaminating uncoupled dye has been a significant problem. The samples electrophoresed in the ABI model 3730XL DNA sequencer consistently size approximately 3 bp shorter compared to gel-based separation of products in an ABI model 3730XL platform. Thus, size interpretation should be modified accordingly.

Transcript (cDNA) analysis may prove to be a more appropriate resource for mutation screening than genomic DNA because it allows analysis of smaller numbers of fragments and thus is more efficient. Full sequence analysis of the transcript can also be performed to detect mutations. In addition, this technique can also identify large deletions or duplications in males and carriers with two different sized products. This method has limited use in clinical laboratories due to the requirement for RNA of good quality and sufficient quantity, which is oftentimes difficult to achieve due to collection and shipping issues. If this approach is used, laboratories may find it useful to use RNA extracted from muscle, which has been found to give the best results. While this approach has potential, it is not currently used in clinical laboratories in the U.S.

Diagnostic testing for promoter mutation and deep intronic mutations is likewise unavailable. Sequence analysis can identify variations in the promoter region, but the clinically functional analysis of the promoter variations and clinical correlation are significant challenges with no clear interpretations. It is likely that promoter mutations will aid some patients and thus future efforts should be directed toward complete gene analysis.

CYSTIC FIBROSIS

Cystic fibrosis (CF) is the most common inherited autosomal recessive disorder in the Caucasian population, with an incidence of 1 in 2500 and a carrier frequency of 1 in 25.[76] The disease is less common among African American and Hispanic American populations and rare in Asian Americans. CF is a life threatening disorder that causes severe lung damage and affects other organs, including the digestive, exocrine, and reproductive systems. The basic defect is found in cells that produce mucus, sweat, saliva, and digestive juices. Normally, these secretions are thin and slippery, but a defective gene in CF causes the secretions to become thick and sticky, trapping bacteria and leading to recurrent infections.

Respiratory failure is the most dangerous consequence of CF. Patients have a mean life expectancy of about 30 years with recurrent hospitalizations and a treatment regimens focused on management of the disease. Approximately 30,000 American adults and children are living with this disorder and, despite intensive research efforts, no significant improvements have been made in morbidity or mortality over the past decade.[77–80]

The specific signs and symptoms of CF can vary, depending on the severity of the disease. Approximately 85% of CF patients have pancreatic insufficiency and require continued treatment with digestive enzyme supplements. Males affected with CF are infertile due to congenital bilateral absence of the vas deferens (CBAVD). About 15% of CF newborns have meconium ileus, a mucous plug that blocks the intestines, and are first recognized by this feature.

Several states have now included CF testing in their newborn screening programs to identify infants and thus allow the early treatment associated with better disease prognosis.

The gold standard diagnostic for CF is a test that measures the amount of chloride in sweat. Consistently high chloride levels on multiple tests are indicative of CF. However, sweat testing is not always useful in newborns due to insufficient sweat production in early life. For this reason, molecular analysis is often the first diagnostic test performed.

In 1989, the *CFTR* gene on chromosome 7q31 was identified using a positional cloning approach and the major mutation in the Caucasian population designated DF508 was identified. The gene was shown to be a chloride transmembrane regulator channel with several important functional domains where mutations have deleterious consequences. Mutations were quickly identified and molecular analysis was rapidly introduced. More than 1,300 mutations have now been identified in the *CFTR* gene, including missense, nonsense, splicing, small deletions and insertions, and large gene deletions. The challenge has been to develop testing appropriate for screening of the most common mutations for carrier detection and comprehensive analysis for diagnostic purposes.

Recommendations for *CFTR* Testing

A National Institutes of Health (NIH) conference held in 1998 addressed recommendations for CF carrier testing and included individuals with family histories of CF, reproductive partners of CF patients, and individuals of the general population with no family histories of CF. The ACMG was challenged with developing a plan for implementation of widespread CF testing and as a result, now recommends a panel of 23 mutations for prenatal carrier screening (Table 15.2).

The recommendation targeted reproductive couples, including those planning pregnancies and those seeking prenatal care. As a result, the optimal setting for screening was viewed as the initial office visit with an obstetrician/gynecologist. The American College of Obstetricians and Gynecologists (ACOG) became actively involved with ACMG in the process of developing and implementing recommendations.

The recommendations addressed who should be tested and a guideline was developed to offer testing to individuals of Caucasian and Ashkenazic Jewish background and to make it available to African Americans, Hispanic Americans, and Asian Americans. The model of screening was considered and it was determined that either the sequential screening model (testing the female first, then testing her partner if she is positive) or the couple screening model (testing both partners simultaneously) was acceptable, and the choice was left to the clinician.

Guidelines for both testing and test interpretation, including recommendations, were developed and published. As a result, a large number of laboratories began CF testing and many that had been conducting such tests found their testing volume to increase substantially. The recommendation also spurred numerous technology developers to address high-throughput platforms and make ASRs available.

TABLE 15.2
Panel of 23 Mutations for Prenatal Carrier Screening Recommended by ACMG

Standard Mutation Panel			Reflex Tests
DF508	G542X	2789 + 5G → A	I506V,[a] I507V,[a] F508C[c]
R553X	R117H	W1282X	5T/7T/9T[b]
R1162X	R334W	A455E	
2184delA	3849 + 10 kbC → T	711 + 1G → T	
DI507	G551D	3659delC	
621 + 1G → T	1717 – 1G → A	N1303K	
G85E	R347P	R560T	
1898 + 1G → A	3120 + 1G → A		

[a] Test distinguishes a CF mutation from benign variants. I506V, I507V, and F508C are performed only as reflex tests for unexpected homozygosity for DF508 and/or DI507.
[b] 5T in cis can modify R117H phenotype or alone can contribute to congenital bilateral absence of vas deferens (CBAVD). 5T analysis is performed only as a reflex test for R117H positives.
(From *ACMG Guidelines*, www.acmg.net)

Indications for Testing

- Confirmation of clinical diagnosis of cystic fibrosis in newborns with meconium ileus, and in individuals with possible or definite diagnoses, including those with elevated sweat tests. A larger screening panel or sequence analysis may be necessary for diagnostic testing, particularly in non-Caucasian ethnic populations and in cases of mild variants. CBAVD patients (males with infertility) may have variant disorders and mutations in *CFTR*.
- Carrier testing for individuals with positive family histories of CF, partners of CF patients, couples planning pregnancies, and couples seeking prenatal care. Carrier testing is also recommended for gamete donors.
- Prenatal testing for the fetuses of carrier couples detected by screening as having a 1 in 4 risk of having an affected child and for fetuses of parents who already have CF children. In both cases, CF mutations in both parents must have been identified prior to prenatal testing.

Approach to Clinical Testing

The clinical spectrum ranges from a severe CF phenotype characterized by progressive lung disease, pancreatic insufficiency, male infertility, and elevated sweat chloride to a milder presentation that includes CBAVD in males. Genotype–phenotype correlations in *CFTR* are not sufficiently precise to be used clinically. Although laboratories vary in the number of mutations and methodologies used for CF testing, many offer a CF testing panel applied to both carrier screening and diagnostic cases. In some cases. they offer an expanded mutation panel for diagnostic testing.

For a review of the most common technologies used in *CFTR* testing and the technical recommendations, see the *ACMG Standards and Guidelines for Clinical Molecular Genetic Laboratories*.[81–84] These guidelines review and compare the technical aspects of the forward allele-specific oligonucleotide dot blot (home brew only), the reverse dot blot (commercially available from two vendors), the ARMS technology (available as an ASR), the oligonucleotide ligation assay (OLA; an ASR), and the liquid bead array (also available as an ASR).

These ASR technologies are widely used in laboratories across the U.S. and Europe. Technology continues to be a moving target with newer platforms rapidly advancing into the market place, including the NanoChip® (Nanogen, San Diego, California, http://www.nanogen.com/), Biochip (Clinical Micro Sensors, Pasadena, California, http://www.motorola.com/lifesciences/), and the Invader® (Third Wave, Madison, Wisconsin, http://www.twt.com/).

In certain diagnostic testing cases, for example, Hispanic American CF patients, full sequence analysis or a combination of scanning technologies and sequencing is utilized since many mutations found in this population are not included in the recommended panel for carrier testing. It is important that laboratories understand the difference in application of carrier versus diagnostic testing. The current recommendations only address prenatal carrier screening and clearly state that recommendations for diagnostic testing are different.

Assay

A number of technologies are available for mutation detection in the *CFTR* gene. Basically, they can be categorized into two areas. The first category detects known targeted mutations included in the recommended panel. These testing methods, all SNP analyses, include forward and reverse allele-specific oligonucleotide dot blots, amplification refractory mutation, liquid bead array technologies (all described in the *ACMG Standards and Guidelines* for *CFTR* testing), real-time polymerase chain reaction (PCR) coupled with melting curve analysis (allele-specific hybridization probes), oligonucleotide arrays, minisequencing with primer extension, the oligoligation assay (OLA; Applied Biosystems, Foster City, California, http://www.appliedbiosystems.com/), and MALDI-TOF (Sequenom, San Diego, California, http://www.sequenom.com/). Figure 15.10 depicts OLA results.

The second category detects unknown mutations using technologies that include direct sequencing and scanning such as SSCP, CSGE, and DHPLC. Both categories have advantages and limitations. For example, an assay based on allele-specific hybridization probes is specific but is limited to detecting only the specified target mutation panel. Methods for full gene analysis will detect a large number of mutations, but will also detect benign variants that do not cause disease, leading to difficulties in test interpretation.

While direct sequencing is the current gold standard for diagnostic testing, it is costly and labor-intensive for analyzing a large multiexon gene (27 exons) such as *CFTR*. For several testing methods for targeted analysis, a major advantage is that reagents are commercially available as ASRs and several may be headed for full-kit FDA approval in the near future. At the time of this writing, only one *CFTR* testing technology has obtained FDA approval: the Tag-It™ by TM Bioscience (Toronto, Ontario, Canada, http://www.tmbioscience.com/).

In most cases, testing is relatively straightforward. If a single mutation is identified in an asymptomatic individual, the individual is considered a carrier, while the presence of two mutations in a symptomatic patient confirms the diagnosis of CF. The exceptional mutation on the recommended panel is R117H, which can be either a benign polymorphism, an allele associated with CBAVD, or a CF disease-causing allele, depending upon whether it is located on the same or different chromosome as the 5T poly T variant.[85]

The *ACMG Standards and Guidelines* address the importance of reflex testing for 5T only in the presence of an R117H mutation. Testing laboratories that ignore this recommendation are doing a disservice to genetics by issuing confusing reports and failing to educate providers. Both failures have led clinicians to perform unnecessary prenatal testing resulting in terminations in some cases. Fortunately, such practices no longer occur in the U.S.

Full CFTR Test for 25 Mutations

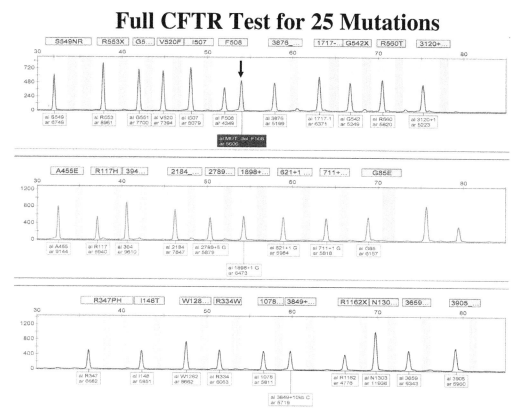

FIGURE 15.10 CF test using recommended panel of mutations. Results of CF25 mutation test performed using OLA and DF508 heterozygote are shown. OLA is a PCR-based assay using allele-specific probes of differing length to distinguish wild-type and mutant alleles. It has the advantage of being performed in a single multiplex, a high-throughput analysis using automated capillary electrophoresis with fluorescence detection, automated software interpretation, and commercial availability as an ASR.

Interpretation

It is important to understand that a negative test result does not completely rule out the possibility that an individual is affected with or is a carrier of CF.[81] Laboratories offering CF testing are required to provide residual carrier risk information on their laboratory reports. The risk differs, depending upon the racial or ethnic population or a family history of CF. It is therefore critical that laboratory directors performing CF testing have appropriate training with molecular genetics as their specialty. One of the major concerns is that non-genetics specialty laboratories are undertaking CF testing without employing trained genetics professionals experienced in test interpretation.

Case Presentation

A 25-year old pregnant Caucasian woman visits her obstetrician, is offered CF testing, and consents to testing. The laboratory performs testing for the panel of 23 recommended mutations using OLA and reports that the patient is a heterozygote carrier for the DF508 mutation. The report recommends testing her partner and genetic counseling. The partner, also Caucasian, tests negative and is given a risk of ~1 in 240 of being a CF carrier. During a genetic counseling session, the couple is informed that their risk of having a child affected with CF is approximately 1 in 1000 and that no further testing is indicated. This positive–negative couple scenario was in the past thought to be a counseling nightmare due to presumed increased anxiety of such couples. However, studies have shown that these concerns have not become issues in real world practice.

Technology Considerations

One challenge for *CFTR* mutation scanning using sequencing is test result interpretation when novel sequence variants are identified. Determining whether an alteration is a benign sequence variation or a disease-causing mutation can be a challenge, not only for CF diagnostic testing, but also with most other sequence-based tests. Interpretation can be even more difficult when a disease-causing mutation is associated with alterations at other sites that may be thousands of bases away. For these reasons, scanning the *CFTR* gene should be used cautiously in a clinical setting and only for diagnostic confirmation of CF. Its use for general population screening is limited because of the high cost and the uncertainty in interpreting alterations in carriers in the absence of data from pilot trials.

Additional technical support is needed from the commercial reagent manufacturers to help laboratories troubleshoot these ASR assays and provide better quality control. Manufacturers of ASRs have limited ability to assist laboratories with these assays, due in part to FDA regulation, while laboratories frequently seek guidance. The solution to this dilemma is for manufacturers to seek FDA approvals for their reagents.

Reflex carrier testing by a larger mutation panel than that currently recommended has been discouraged (Table 15.2). However, the panel was developed specifically to target a Caucasian population that represents the bulk of the CF population. Larger expanded panels that include mutations present in various ethnic/racial populations and thus raise the test sensitivity for these populations would be desirable.

The challenge to expanding the panel is not technical. It involves lack of sufficient clinical validation data for the less common mutations that have not undergone pilot trial testing. This pitfall led to the inclusion of the I148T allele in the original panel recommended for prenatal carrier screening. Soon after launching a nationwide screening program, it became apparent that the I148T allele was only a benign polymorphism and occurred 100-fold more often in the unaffected general population than predicted. This led to removal of the I148T allele upon recommendation review.[86,87] In the future it would be desirable to have data demonstrating a strong correlation between clinical phenotype and mutation prior to developing an extended panel, and this remains a challenge.

CANCER GENETICS

INHERITED COLON CANCER

Colorectal cancer (CRC) is the second leading cause of cancer-related death in the U.S. It accounts for over 57,000 deaths per year.[88–90] While the majority of colorectal cancer is not inherited, hereditary CRC accounts for less than 10% of total cases and primarily consists of familial adenomatous polyposis (FAP) and hereditary non-polyposis colon cancer (HNPCC).

FAMILIAL ADENOMATOUS POLYPOSIS

Familial adenomatous polyposis (FAP) is an autosomal-dominant hereditary cancer that accounts for approximately 1% of all CRC.[91–94] Classic FAP is characterized by hundreds to thousands of adenomatous polyps that usually appear in the distal colon during adolescence and progress to CRC. Extra-colonic manifestations including gastric polyps, osteomas, congenital hypertrophy of the retinal pigment epithelium (CHRPE), soft tissue tumors, thyroid carcinoma, and hepatoblastoma may also be present. Variants of FAP include attenuated FAP (AFAP) characterized by late onset and fewer polyps, Gardner syndrome associated with extra-colonic features, and Turcot syndrome associated with medulloblastoma.

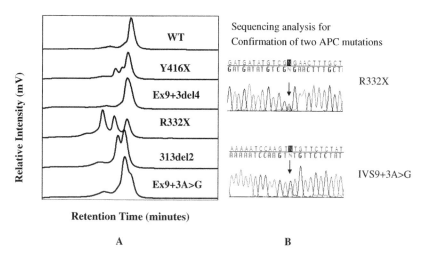

FIGURE 15.11 *(A color version follows page 270)* A: Scanning *APC* gene for mutations using DHPLC. DHPLC patterns of patients with various mutations in exon 9 of the *APC* gene identified by DHPLC. The top profile shows the wild-type pattern. All variants shown below have a pattern shifted to the left and include nonsense, deletions, and splicing mutations. The left and right panels represent the same samples analyzed under different denaturing conditions, showing differences in the heteroduplex patterns and pointing to the necessity for well defined conditions for analysis. B: Sequence analysis confirmation of two *APC* gene mutations in exon 9 shown in Figure A: R332X and IVS9+3A→G. Arrows point to sequence changes.

Mutations in the *APC* gene located on 5q cause FAP. The *APC* gene is relatively large, with 15 exons and approximately 8.5 Kb of coding sequence.[95] Exon 15 represents about two thirds of the gene, making molecular analysis challenging. The majority of germline mutations are truncating (~90%) and include small frame shift mutations and nonsense mutations, while the remainder consist of missense mutations (~4%) and gross alterations ~5%.[96] The position of the *APC* gene mutation may correlate with clinical phenotype; severe polyposis maps to the central region of *APC*, while attenuated FAP is usually due to mutations at the extreme 5′ or 3′ region or at the alternatively spliced exon 9.

The *APC* gene functions as a tumor suppressor with a gatekeeper function. While germline mutations in the *APC* gene lead to FAP, most CRC patients have sporadic mutations in the *APC* gene, suggesting a major role in tumor progression. One of the main functions of the APC protein is beta-catenin binding and down-regulation. In the absence of a functional APC protein, beta-catenin diffuses into the nucleus, interacts with transcription factors, and results in activation of growth factors, leading to unregulated cell growth and tumor formation.[97–99]

Truncating mutations in the *APC* gene can be detected using the protein truncation test, a cumbersome technique that requires radioactivity. A two-tiered comprehensive genetic testing strategy is recommended for FAP, which includes mutation scanning of the *APC* gene using DHPLC analysis followed by sequencing of targeted regions[100] (Figure 15.11A and Figure 15.11B). A similar condition overlapping with FAP results from recessive germline mutations in the *MYH* gene that is involved in base excision repair.[101] An algorithm for FAP–MYH testing is shown in Figure 15.12.

The second tier involves FISH (Figure 15.13), MLPA, or quantitative real time PCR analysis for detecting deletions and/or duplications in the *APC* gene. For patients whose clinical presentation and family history are consistent with *MYH*-associated polyposis, an assay for the two common mutations (Y165C and G382D) described as frequent in CRC patients in the European Caucasian population is recommended.[102] If results are negative, this assay can be followed up with sequencing for the entire coding region of the *MYH* gene.

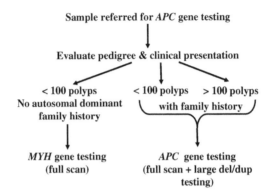

FIGURE 15.12 We recommend the FAP testing algorithm. If a patient has a strong family history of FAP, regardless of number of polyps, he or she should be tested for the *APC* gene. However, for patients with possible diagnosis of attenuated FAP who have fewer than 100 polyps and no family history of FAP, *MYH* gene testing is more likely to identify a mutation.

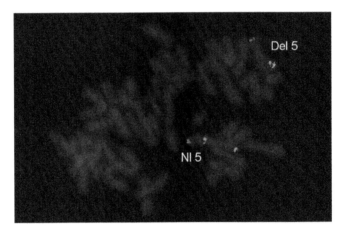

FIGURE 15.13 *(A color version follows page 270)* *APC* FISH. Confirmation of entire *APC* gene deletion detected by real-time PCR. FISH was performed using a biotin-labeled BAC RP11-107C15 as a probe. Arrow shows no hybridization of BAC RP11-107C15 probe, indicating entire *APC* gene deletion. The 5p telomere (green signals) and the 5q telomere (red signals) were used as control probes. Nl 5 indicates normal chromosome 5.

Case Presentations

A 29-year old male presented to an eye clinic for regular examination and was found to have CHRPE and an osteoma on his forehead, suggesting FAP (Figure 15.14). A detailed family history revealed that his two sisters died of colon cancer in their early twenties. Gastrointestinal (GI) examination revealed a carpet of adenomatous polyps. Genetic testing was performed for the *APC* gene mutations and a nonsense mutation was identified. Genetic counseling reviewed the clinical course of disease, monitoring and treatment options, and suggested testing for at-risk family members.

A 45-year old female presented at clinic with stomach complaints and rectal bleeding (Figure 15.15). A GI consultation and endoscopy diagnosed adenomatous polyps (colon cancer). The patient indicated no family history of FAP but testing for *APC* gene was performed since one third of FAP cases have *de novo* *APC* gene mutations. No mutation was detected in the *APC* gene and testing for the *MYH* gene was performed. Two heterozygous mutations (Y165C and G383D) were identified in this patient, confirming the diagnosis of *MYH*-associated polyposis. Since this is an autosomal-

FAP Case Pedigree

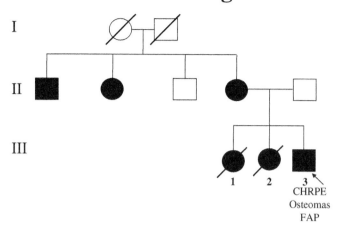

FIGURE 15.14 A 29-year-old male III3 presented to an eye clinic for regular examination that revealed CHRPE and an osteoma on his forehead, suggesting FAP. Family history was indicative of inherited colon cancer with two sisters (III1 and III2) who died of colon cancer in their early twenties. GI examination revealed a carpet of adenomatous polyps. Genetic testing was performed for the *APC* gene mutations and a nonsense mutation (Q1447X) was identified. This mutation is located in the β-catenin binding domain of the APC protein and was previously shown to be associated with FAP along with CHRPE and osteomas. These results confirmed the clinical diagnosis of FAP. Since a mutation in this affected family member has been documented, presymptomatic testing for at-risk individuals can be offered after appropriate genetic counseling.

MYH Case Pedigree

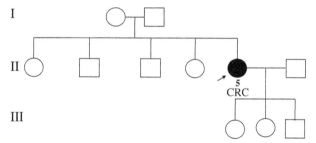

FIGURE 15.15 A 45-year-old female II5 presented with stomach complaints and rectal bleeding. GI consultation and endoscopy led to a diagnosis of adenomatous polyps (colon cancer). No mutation was detected in the *APC* gene. Since the patient had no previous family history of colon cancer, the possibility of autosomal-recessive inheritance was raised. Testing for *MYH* gene was performed. Two heterozygous mutations (Y165C and G383D) were identified, confirming the diagnosis of *MYH*-associated polyposis. Testing was recommended for the patient's reproductive partner, children, and other relatives.

recessive disease, testing was recommended for her reproductive partner and children. Testing was also offered to the patient's siblings although none had clinical presentations.

HEREDITARY NON-POLYPOSIS COLON CANCER

HNPCC is an autosomal-dominant disorder characterized by increased lifetime risk of early onset colorectal cancer as well as other cancers of the endometrium, stomach, small intestine, hepatobiliary system, kidneys, ureter, and ovaries. HNPCC, also known as Lynch syndrome, has been estimated to account for approximately 4% of all CRC cases.[103] Mutations in genes that predispose to HNPCC are highly penetrant, with mutation carriers estimated to have ~80% lifetime risk for developing colorectal cancer.

HNPCC is genetically heterogeneous due to mutations in the mismatch repair (MMR) gene family, primarily *MLH1, MSH2, MSH6* and less frequently in *PMS1* and *PMS2*.[104,105] The normal function of mismatch repair genes is to repair loop-out mismatches in DNA that escape the normal repair mechanism. This function is accomplished by two MutS complexes formed by these MMR proteins to both recognize a mismatch and determine which strand to copy. These mismatches frequently occur in microsatellites that have repeating units. While many microsatellites have no known function, some are located in or surround genes that play a role in the tumor development process. HNPCC patients carry germline mutations in one of the MMR genes, but a second mutational event that occurs somatically at the tumor site is necessary for the loss of mismatch repair function.

Two laboratory assays measure microsatellite instability (MSI) that occurs after Knudson's second-hit event. One assay involves the molecular approach of measuring microsatellite instability by function and compares the patient's normal and tumor cell DNA patterns at a number of microsatellites in or around genes involved in tumor development (Figure 15.16). The tumor DNA from an HNPCC patient will have an altered size allele with an increased or decreased number of bases. Approximately 95% of HNPCC patients will have MSI, while about 15% of sporadic CRC patients demonstrate MSI.[106]

Alternatively, an immunohistochemical assay that examines the tumor cells for the presence or absence of MMR proteins can be used to screen for HNPCC and direct testing for a specific gene. Both assays have been successfully utilized in screening large CRC populations to identify HNPCC patients. If immunohistochemistry shows absence of MLH1 protein and if the individual has no significant history of inherited CRC in the family, an assay for *MLH1* promoter methylation can be performed before mutation analysis of the *MLH1* gene (Figure 15.17). It is important to

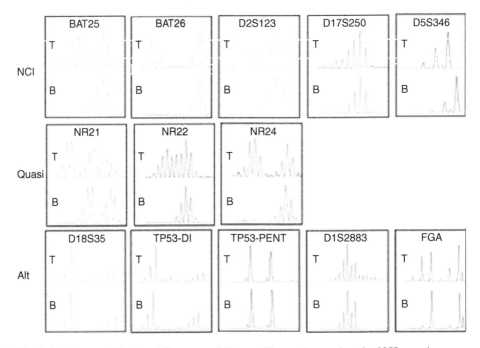

FIGURE 15.16 Microsatellite instability assay. Microsatellite markers using the NCI, quasimonomorphic, and alternate panel markers. Representative unstable and stable profiles are shown here. Each vertical panel represents a single marker, showing germ line and tumor DNA profiles. "T" represents profile of DNA extracted from tumor tissue and "B" represents profile of DNA extracted from normal tissue.

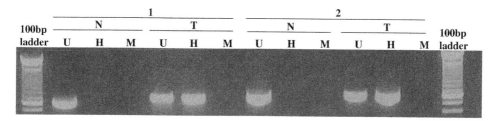

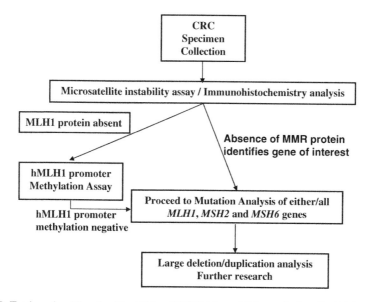

FIGURE 15.17 Methylation analysis of hMLH1 promoter. Amplification of the hMLH1 promoter region from the indicated normal (N) or adjacent tumor (T) DNAs before and after digestion with the indicated restriction endonucleases. The 1 and 2 indicate two cases with unmethylated germline and methylated hHMLH1 promoter from DNA extracted from normal tissue and tumor, respectively. Methylation-sensitive restriction enzymes were used to determine methylation status of the promoter. Such information is useful in separating sporadic from inherited CRC and thus identifying HNPCC patients (Figure 15.18 describes HNPCC testing). U = undigested. H = digested with HpaII. M = digested with MspI.

note that *MLH1* promoter methylation can be a secondary event to a germline mutation. Therefore family history is a very useful screening tool, as most HNPCC patients will have strong family histories and meet either Bethesda or Amsterdam criteria.[107–110]

Molecular diagnosis of HNPCC is complicated by the need to analyze multiple genes of large size. A comprehensive algorithm for HNPCC testing incorporates the following options: microsatellite instability analysis, immunohistochemical analysis for the MLH1, MSH2, and MSH6 proteins, DNA sequence analysis for germline mutations in the mismatch repair genes *MLH1*, *MSH2*, and *MSH6*, quantitative real time PCR analysis for gene deletions or duplications, and *MLH1* promoter methylation analysis (Figure 15.18). The possibility of finding a mutation in the MMR genes is

FIGURE 15.18 Testing algorithm for identifying HNPCC in a CRC population. Both blood and tumor are collected from the CRC patient and analyzed for microsatellite instability (MSI) by immunohistochemistry (IHC) for MLH1, MSH2, and MSH6, and for MLH1 promoter methylation status. Family history information is also reviewed. A positive MSI result is a candidate for further analysis. IHC is used to identify MMR protein absence in tumor tissue and thus indicate the gene for mutation analysis. A negative promoter methylation suggests that the tumor is not sporadic and also supports MMR gene analysis using DHPLC for identifying variants and sequencing to define precise changes.

higher in cases where testing is done systematically with comparison to clinical presentation and family history. This approach allows for greater sensitivity and a more directed molecular analysis for clinical testing in HNPCC.

Identification of HNPCC mutations is important for clinical surveillance and therapeutic intervention in carriers and genetic testing for at-risk relatives. Screening for this disorder in a CRC population is a strategy to identify carriers. Prognosis for HNPCC patients with MMR gene mutations is generally better than for sporadic CRC patients, although the patient may not benefit directly from genetic testing.

Assay

The large size of these genes (19 exons for *MLH1*, 16 exons for *MSH2,* and 10 exons for *MSH6*) encompassing 9241 nucleotides of coding sequence makes sequencing for germline mutations both labor-intensive and expensive. Prior to performing germline mutation analysis, DHPLC is an efficient method of screening for MMR gene mutations due to point mutations, small deletions, and small insertions. Oefner and others developed DHPLC as a rapid, semi-automated tool for detection of sequence variants.[111–113] The DHPLC platform meets the needs of clinical molecular laboratories because of its sensitivity, semi-automated operation, and cost effectiveness.[114] (Figure 15.19). Several approaches have previously been described to identify mutations in HNPCC.[114a]

Scanning methods including SSCP, CSGE, DGGE, and sequencing the entire coding region of the *MLH1* and *MSH2* genes are well established techniques for clinical diagnostic purposes. However, SSCP, CSGE, and DGGE lack sensitivity, and SSCP has inherent size limitations of 250 bp for fragment analysis with a sensitivity of 70 to 80%. While DGGE offers increased sensitivity, it is time-consuming and requires GC clamp primers; mutations in GC-rich regions may not be detected.

All scanning technologies fail to detect large gene deletions, and thus another technology is required. Large intragenic deletions can be detected using Southern blot analysis. More recently, techniques such as MLPA, real-time PCR, and long-range PCR have been used to detect large gene deletions on a clinical basis.[113a] It is important that sequencing laboratories include these technologies in order to provide more comprehensive testing for gene mutations.

Case Presentation

A 55-year old female presented to a GI clinic with a past history of polyps developed at age 48 (Figure 15.20). A detailed family history revealed colon and endometrial cancers in several family members. Mutation scanning for *MLH1*, *MSH2*, and *MSH6* genes was performed. Several variants were detected by DHPLC and further sequencing identified an insertion mutation in the *MSH6* gene that was interpreted as disease-causing. Genetic counseling for the patient covered the need for frequent monitoring and her risk for developing CRC and associated cancers. Genetic testing was recommended for other family members to identify those at risk who should be monitored more closely. HNPCC patients like this one usually present with strong family histories of CRC. The point of testing often is to identify the index case and perform pre-symptomatic testing for family members at risk.

New Technology and Recent Advances

With the completion of the human genome project, more genes have been found to be associated with colon cancer. These findings will translate into more genetic testing. Sequencing technology is now cost-effective to the point where many laboratories in the future may shift entirely to direct sequence analysis. Identification of large deletions in these genes using newer technologies including MLPA and real-time PCR has become routine practice.

Analysis of the clinical significance of sequence variations in general, and the regulatory domain in particular, remains a significant challenge. There is a need for the development of a universal

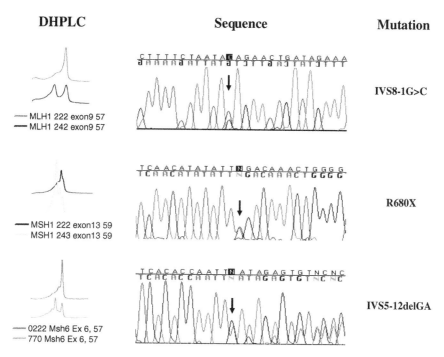

FIGURE 15.19 *(A color version follows page 270)* Identification of HNPCC by DHPLC and sequencing. Similar to the strategy for FAP testing, HNPCC analysis uses DHPLC to identify a variant and sequencing to define it. DHPLC patterns are shown to the left. The wild-type profile is shown directly above the patient profile. Also shown are a splicing mutation in *MLH1* (upper panel), a nonsense mutation in *MSH2* (middle panel), and a 2-base deletion creating a frameshift in *MSH6* (lower panel).

algorithm, including predictive software for splicing analysis and for evolutionary conservation analysis, protein modeling, and functional assays to address the clinical interpretation of sequence variants.

Inherited Breast Cancer

It has been estimated that approximately one in every 200 women in the general population carries a predisposing mutation in an autosomal-dominant susceptibility gene for breast cancer. One of these designated *BRCA*1 was the first major breast cancer gene to be isolated, followed by the *BRCA*2 gene.[115,116] Family history is an important risk factor for the development of breast cancer (Figure 15.21). Most breast cancer cases are diagnosed post-menopause and are sporadic. Only 5 to 10% of cases result from the autosomal-dominant inheritance of a mutated gene.

Selected breast and ovarian cancer families have been screened to determine the spectrum of germ-line mutations in the *BRCA*1 gene. This analysis, together with mutations reported in the *BRCA*1 gene mutation database (Breast Cancer Information Core [BIC, http://research.nhgri. nih.ov/bic/]) suggest mutation clustering at a limited number of sites within the *BRCA*1 and *BRCA*2 genes. These observations led Gayther et al. (1996)[117] to develop a multiplex heteroduplex assay to screen exons 2, 11A, 11B, and 20 for mutations simply and rapidly.

Further analysis of *BRCA*1 and *BRCA*2 genes for mutations led to the use of the PTT to screen for nonsense mutations in the *BRCA*1 and *BRCA*2 gene transcripts. The premises of this assay were the high proportion of truncating mutations in the *BRCA*1 and *BRCA*2 genes reported prior to 1998 and the distribution of these mutations throughout the gene, thereby making screening for specific mutations difficult.

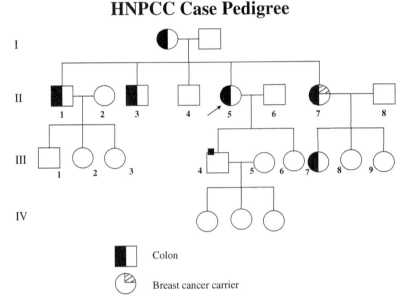

HNPCC Case Pedigree

FIGURE 15.20 A 55-year-old female II5 presented to the GI clinic with a history of polyps developed at age 48 and a family history of colon cancer. Several variants were detected by DHPLC in the *MLH1, MSH2,* and *MSH6* genes. Sequencing identified an insertion mutation in the *MSH6* gene 3336_3339insATGA. This insertion mutation was previously identified in other HNPCC patients and is expected to result in a truncated protein. The implications of identifying a MMR gene mutation in this individual and possible testing of her siblings and offspring and benefits of testing are important aspects of HNPCC testing.

Other mutation detection strategies such as allele-specific oligonucleotide hybridization, DGGE, and SSCP have also been used. However, while they appear to be efficient, they suffer from a lack of sensitivity. A mutation detection frequency of 10 to 14% for the mutation tested was reported for the *BRCA1* gene using a dual testing strategy of heteroduplex analysis and PTT (Second Australian Mutation Detection Workshop, 1999). Approximately 2 to 3% of *BRCA1* patients were found to carry duplications of exon 13; this 6-Kb duplication appears to be *Alu*–mediated. The high concentration of *Alu* sequences in the *BRCA1* gene suggests that duplications may be present in significant numbers of *BRCA1* families with breast cancer.

The hierarchical mutation detection strategy reported by Hegde et al.[118] has been designed for ease of use and sensitivity, but is also cost-effective, thus achieving a balance between comprehensiveness and labor and/or running costs. The contribution to current practice lies in streamlining a testing regime such that each tier of the process is manageable, while providing acceptable timelines for reporting. Figure 15.22 is a flow diagram of the mutation screening protocol for the *BRCA1* and *BRCA2* genes.

This strategy compares favorably with the heterogeneity of mutation detection strategies used by many testing laboratories to examine the *BRCA*1 and *BRCA*2 genes, with a reported detection frequency of 10 to 14%. The amplification and sequencing protocol reduces both handling errors and operator fatigue compared to other assays because all manipulations are carried out in a 96-well microtiter tray format using a multichannel pipette. The sample requirements for exon-by-exon sequencing are not complex and the assay has been established to amplify all the exons at one common annealing temperature.

Samples can be prepared for sequencing within one working day and the subsequent sequence compiled and compared within hours, although this is dependent upon the quality of the sequencing data. The entire analysis can be completed in a week from the time the sample arrives in the laboratory.

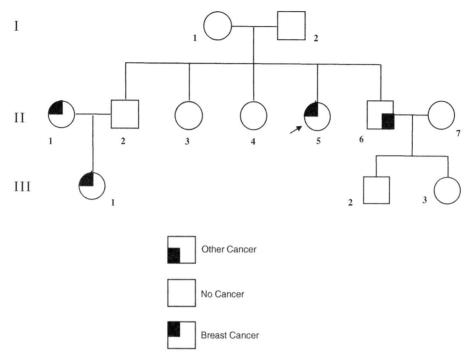

FIGURE 15.21 A 36-year-old female II5 was referred for genetic counseling. She had a strong family history of breast cancer: two one- or two-degree relatives on the same side of the family with breast or ovarian cancer including any of the following high risk features: bilaterality, diagnosis at age 40 or younger, breast and ovarian cancer in one individual. Genetic testing was recommended. A mutation was detected in the *BRCA1* gene 4184delTCAA mutation contained within codons 1355 and 1356 and resulting in a translational frame shift such that a TAG stop codon occurs at codon 1364. Implications of these results on the immediate family members (II1 and III1) are important considerations. Further presymptomatic testing can be offered to other at-risk members in the family.

One disadvantage of DNA-based sequencing of the *BRCA*1 gene exons is that the deletions of complete exons and errors in transcript processing that are unrelated to DNA sequence are not detected. The proportion of clinically important *BRCA*1 gene defects attributable to such abnormalities has been estimated between 5 and 15%. However, it is possible that deletions and duplications could be effectively screened by introducing a semi-quantitative assay of the exon amplification products prior to sequencing. This strategy could be based on multiplex exon amplification using fluorescently end-labeled primers or the incorporation of fluorescent dNTPs, and subsequent electrophoretic separation and quantitation or use of real-time technology.

FUTURE OF GENETIC TESTING

Over the past decade, the amount of scientific knowledge in the area of human genetics has expanded tremendously. Testing for genetic diseases with state-of-the-art technologies continues to evolve. More than 4000 disease genes in which specific variants have been associated with common disorders such as heart disease, diabetes, asthma, and cancer[119–121] have been identified. As more genes are known to be associated with these conditions and as genetic testing for them becomes possible, the number of available genetic tests is expected to increase. However, introduction of new tests should occur only after careful clinical validation studies; caution must be exercised in appropriate use and interpretation. It is important that laboratories offering these tests have a basic understanding of genetic principles and medicine.

Hierarchical Screening Strategy for Mutations in the *BRCA1* and *BRCA2* Genes

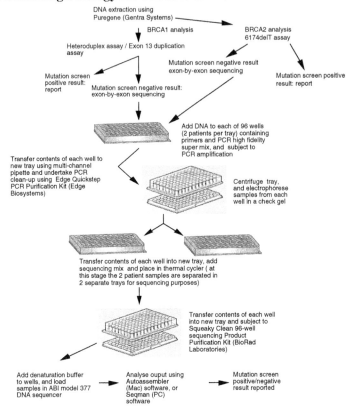

FIGURE 15.22 *BRCA* algorithm for mutation detection by sequencing. Molecular testing *BRCA 1* and *BRCA2* involves hierarchical testing strategy. The first tier of assays performed includes DHPLC or heteroduplex analysis of a few common mutations in the *BRCA1* and *BRCA2* genes and an assay for exon 13 duplication performed in parallel. If no mutation is detected, tier 2 testing involving full sequencing of the *BRCA1* and *BRCA2* genes can be performed. A positive test result confirms the diagnosis. Shown here are different stages in performing full sequencing of both genes. Because both genes are large, testing is performed in 96-well trays from PCR to sequencing and subsequent data analysis.

Technology drives the progress in this field. The Human Genome Project led to the development of powerful techniques and automation for mutation detection, and these technologies will continue to evolve in the future. The good news is that these powerful high-throughput methods are and will continue to be used in clinical laboratories. High density arrays of oligonucleotides and cDNAs are versatile tools that represent the future for molecular analysis of gene expression and mutation scanning, and these tests will be performed in CLIA-approved laboratories for clinical purposes. The potential applications for microarray technology in clinical laboratories in theory are unlimited and remain major areas of interest today. In the future, we expect these technologies to develop into FDA-approved reagents for disease-specific tests for use in clinical laboratories.

The applications for genetic testing will also continue to grow. Pharmacogenetics is rapidly moving toward genotyping for SNP associations that can be used to predict drug interactions and is expected to become a mainstream of genetic testing in the future, although its development currently remains in the research community. Predicting drug response based on cytochrome P450 genotype and drug therapy for lung cancer based on the presence of an epidermal growth factor receptor (EGFR) mutation are only a few examples of areas of rapid progress.[122]

While population-based testing for cystic fibrosis in a prenatal setting has become a reality in the past few years, such population-based testing will include more genetic tests in the future. The CDC has undertaken a review of genetic tests for relatively common disorders. The testing has the potential for improving public health by screening for conditions such as hereditary hemochromatosis, venous thrombosis associated with factor V Leiden, and prothrombin gene mutations. Other possibilities for both newborn and carrier screening include fragile X, particularly with the new finding of FXTAS (fragile X tremor and ataxia syndrome) in adult carriers of fragile X premutations.[123]

The availability of sequence information and the development of powerful technology for DNA-based testing will lead to identification of sequence variants with unknown significance in the disease process. Better algorithms for interpretation of unknown sequence variants with respect to clinical consequence are needed, including predictive software programs, functional assays, and better standards and guidelines for laboratories. Understanding the effects of these variants on protein function and expression will lead to better understanding of the clinical consequences of mutations. Development of more comprehensive testing regimes including mutation detection in a total approach to both diagnosis and patient management will continue to evolve and be guided through professional program practice guidelines.

The future is now. Ultra-rare disorder testing has become a reality in a handful of specialty clinical labs in the U.S. and is rapidly expanding. Higher-throughput clinical laboratories have developed robotically-driven sequencing units. No longer is clinical testing performed by research laboratories or sequencing facilities that are not CLIA-approved. Clinical laboratories now need more CAP proficiency testing made available for rare disorders, better access to positive control materials for test development, better partnering of academic and commercial laboratories, and better support from regulatory government agencies. Technology will continue to move forward. The only thing certain is change.

GENETIC TESTING FOR RARE DISEASES

Rare diseases occur infrequently in the general population. The medical community knows relatively little about them and, as a result, inadequate coverage is generally provided by the public health system. Lack of effective treatment is due both to a dearth of research and to the fact that developing drugs for a limited number of patients is not commercially profitable. Rare diseases affect a limited number of people in a population — defined as fewer than 1 in 2,000. While this number seems small, it translates to approximately 200,000 individuals. Most people represented by these statistics suffer from less frequently occurring diseases that affect fewer than 1 in 100,000 people. It is estimated that between 5,000 and 8,000 distinct rare diseases exist today and they affect 6 to 8% of the population in total. Due to their rarity, only particularly severe pathologies have been singled out as rare diseases. These diseases can almost always be characterized as:

- Serious, chronic, degenerative, and usually life threatening
- Disabling; quality of life is compromised due to a lack of autonomy
- Causing high levels of pain and suffering for the affected individual and his or her family
- Having no effective cure; symptoms can be treated to improve quality of life and life expectancy

Eighty percent of rare diseases have identified genetic origins and affect 3 to 4% of births. Other rare diseases are the results of bacterial or viral infections and allergies or have degenerative and proliferative causes. Symptoms of some rare diseases may appear at birth or in childhood, for example, infantile spinal muscular atrophy, neurofibromatosis, osteogenesis imperfecta, lysosomal storage disorders, chondrodysplasia, and Rett syndrome. Many others, such as Huntington's disease,

Crohn's disease, Charcot–Marie–Tooth disease, amyotrophic lateral sclerosis, Kaposi's sarcoma and thyroid cancer, only appear after adulthood is reached.

Rare diseases are characterized by a broad diversity of disorders and symptoms that vary not only from disease to disease but also from patient to patient suffering from the same disease.

The focus on rare diseases is a recent phenomenon. Until recently, public health authorities and policy makers largely ignored them. Some specific rare diseases are well known and screened for as a part of public policy when simple and effective preventive treatment is available. It is impossible to develop a public health policy specific to each rare disease. Medical and scientific knowledge about rare diseases is lacking. While the number of scientific publications about rare diseases, particularly those identifying new syndromes, continues to increase, fewer than 1,000 diseases benefit from any scientific knowledge. The acquisition and diffusion of scientific knowledge represent the vital bases for the identification of diseases and, most importantly, for research into new diagnostic and therapeutic procedures.

Almost everyone who has a rare disease encounters the same problems: delay in and failure of diagnosis, lack of information about the disease, failure to be referred to qualified professionals, lack of quality care and social benefits, poor coordination of in- and out-patient care, reduced autonomy, and difficulty reintegrating into working, social and family environments. Many rare diseases involve sensory, motor, mental and physical impairments. People affected by them are more vulnerable psychologically, socially, culturally, and economically. These difficulties can be reduced by the implementation of appropriate public policy.

Rare diseases frequently go undiagnosed due to a shortage of scientific and medical knowledge. At best, only some symptoms are recognized and treated. People can live for several years in precarious situations with undiagnosed diseases and lack competent medical attention because they remain excluded from the health care system. How well a rare disease is known determines how rapidly it is diagnosed and the quality of medical and social coverage. A patient's perception of the quality of his or her life is linked more to quality of care than to the gravity of the illness or the degree of the associated handicap.

For most rare diseases, no protocol exists for good clinical practice. Where one does exist, the knowledge remains isolated when it should be shared. Additionally, the segmentation of medical specialties is a barrier to the comprehensive care of a patient suffering from a rare disease. For some rare diseases, such as familial Mediterranean fever, fragile X syndrome, and cystic fibrosis, well targeted screening programs, treatment protocols, and defined medical, social, and educational programs exist in certain countries. New antenatal and asymptomatic phase screening methods for rare diseases allow effective treatment to be undertaken earlier, significantly improving quality and length of life.

Other screening programs should be introduced as soon as simple and reliable tests and effective treatments exist. Qualitative and quantitative progress in prognosis and clinical treatment is raising new public health questions about policies on generalized and targeted screening for certain diseases. Rare disease organizations have been created as a result of experiences gained by patients and their families who are so often excluded from health care systems and thus must take charge of their own diseases.

REFERENCES

1. Genetic testing: the complex issues involved in predicting potential future health problems. *Mayo Clin Womens Healthsource* 2005; 9 (3): 1–2.
2. Ensenauer RE, Michels VV, Reinke SS. Genetic testing: practical, ethical, and counseling considerations. *Mayo Clin Proc* 2005; 80 (1): 63–73.
3. Petersen GM. Genetic testing. *Hematol Oncol Clin North Am* 2000; 14 (4): 939–952.

4. Feldmann D, Guittard C, Georges MD, Houdayer C, Magnier C, Claustres M, Couderc R. Genetic testing for cystic fibrosis: evaluation of the Elucigene CF20 kit in blood and buccal cells. *Ann Biol Clin (Paris)* 2001; 59 (3): 277–283.

5. Worton RG, Thompson MW. Genetics of Duchenne muscular dystrophy. *Annu Rev Genet* 1988; 22: 601–629.

6. Pimentel MM. Fragile X syndrome (review). *Int J Mol Med* 1999; 3 (6): 639–645.

7. Guanti G. Genetic testing and surgeon decision. *Acta Chir Iugosl* 2004; 51 (2): 57–60.

8. Schwartz MK. Genetic testing and the clinical laboratory improvement amendments of 1988: present and future. *Clin Chem* 1999; 45 (5): 739–745.

9. Genetic testing kit on the horizon for heart disease. *Heart Advis* 2004; 7 (12): 2.

10. Fryer A. The genetic testing of children. *J Roy Soc Med* 1997; 90 (8): 419–421.

11. Elliott D. Genetic testing, organ transplantation, and an end to nondirective counseling. *Ann NY Acad Sci* 2000; 913: 240–247.

12. Kroese M, Zimmern RL, Sanderson S. Genetic tests and their evaluation: can we answer the key questions? *Genet Med* 2004; 6 (6): 475–480.

13. Wolpert CM, Schmidt MC. Genetic testing: understanding basics. *JAAPA* 2005; 18 (1): 48–52.

14. Breuning MH. Genetic Testing: From chromosomes to DNA, a revolution in prenatal diagnosis. *Eur J Hum Genet* 2005; 13 (5): 517–518.

15. Spigelman A, Burgess B, Groombridge C, Scott RJ. Genetic testing: a round table conversation. *Intern Med J* 2004; 34 (9–10): 587–589.

16. Taruscio D, Falbo V, Floridia G, Salvatore M, Pescucci C, Cantafora A, Marongiu C, Baroncini A, Calzolari E, Cao A et al. Quality assessment in cytogenetic and molecular genetic testing: the experience of the Italian Project on Standardisation and Quality Assurance. *Clin Chem Lab Med* 2004; 42 (8): 915–921.

17. Solymoss S. Factor V Leiden: who should be tested? *CMAJ* 1996; 155 (3): 296–298.

18. Creinin MD, Lisman R, Strickler RC. Screening for factor V Leiden mutation before prescribing combination oral contraceptives. *Fertil Steril* 1999; 72 (4): 646–651.

19. Press RD, Bauer KA, Kujovich JL, Heit JA. Clinical utility of factor V Leiden (R506Q) testing for the diagnosis and management of thromboembolic disorders. *Arch Pathol Lab Med* 2002; 126 (11): 1304–1318.

20. Sayinalp N, Haznedaroglu IC, Aksu S, Buyukasik Y, Goker H, Parlak H, Ozcebe OI, Kirazli S, Dundar SV, Gurgey A. The predictability of factor V Leiden [FV:Q(506)] gene mutation via clotting-based diagnosis of activated protein C resistance. *Clin Appl Thromb Hemost* 2004; 10 (3): 265–270.

21. Baris I, Koksal V, Etlik O. Multiplex PCR-RFLP assay for detection of factor V Leiden and prothrombin G20210A. *Genet Test* 2004; 8 (4): 381–383.

22. Sinden RR, Potaman VN, Oussatcheva EA, Pearson CE, Lyubchenko YL, Shlyakhtenko LS. Triplet repeat DNA structures and human genetic disease: dynamic mutations from dynamic DNA. *J Biosci* 2002; 27 (Suppl 1): 53–65.

23. Oostra BA, Willems PJ. A fragile gene. *Bioessays* 1995; 17 (11): 941–947.

24. Tassone F, Hagerman RJ, Loesch DZ, Lachiewicz A, Taylor AK, Hagerman PJ. Fragile X males with unmethylated, full mutation trinucleotide repeat expansions have elevated levels of FMR1 messenger RNA. *Am J Med Genet* 2000; 94 (3): 232–236.

25. Hoogeveen AT, Oostra BA. The fragile X syndrome. *J Inherit Metab Dis* 1997; 20 (2): 139–151.

26. Chakrabarti L, Davies KE. Fragile X syndrome. *Curr Opin Neurol* 1997; 10 (2): 142–147.

27. Garland EM, Vnencak-Jones CL, Biaggioni I, Davis TL, Montine TJ, Robertson D. Fragile X gene premutation in multiple system atrophy. *J Neurol Sci* 2004; 227 (1): 115–118.

28. Kooy RF, Willemsen R, Oostra BA. Fragile X syndrome at the turn of the century. *Mol Med Today* 2000; 6 (5): 193–198.

29. Verkerk AJ, Pieretti M, Sutcliffe JS, Fu YH, Kuhl DP, Pizzuti A, Reiner O, Richards S, Victoria MF, Zhang FP et al. Identification of a gene (FMR-1) containing a CGG repeat coincident with a breakpoint cluster region exhibiting length variation in fragile X syndrome. *Cell* 1991; 65 (5): 905–914.

30. Hammond LS, Macias MM, Tarleton JC, Shashidhar Pai G. Fragile X syndrome and deletions in FMR1: new case and review of the literature. *Am J Med Genet* 1997; 72 (4): 430–434.

31. Haataja R, Vaisanen ML, Li M, Ryynanen M, Leisti J. The fragile X syndrome in Finland: demonstration of a founder effect by analysis of microsatellite haplotypes. *Hum Genet* 1994; 94 (5): 479–483.

32. Rousseau F. The fragile X syndrome: implications of molecular genetics for the clinical syndrome. *Eur J Clin Invest* 1994; 24 (1): 1–10.

33. Nelson DL. The fragile X syndromes. *Semin Cell Biol* 1995; 6 (1): 5–11.

34. Toledano-Alhadef H, Basel-Vanagaite L, Magal N, Davidov B, Ehrlich S, Drasinover V, Taub E, Halpern GJ, Ginott N, Shohat M. Fragile-X carrier screening and the prevalence of premutation and full-mutation carriers in Israel. *Am J Hum Genet* 2001; 69 (2): 351–360.

35. Hagerman PJ, Hagerman RJ. The fragile-X premutation: a maturing perspective. *Am J Hum Genet* 2004; 74 (5): 805–816.

36. Biancalana V, Beldjord C, Taillandier A, Szpiro-Tapia S, Cusin V, Gerson F, Philippe C, Mandel JL. Five years of molecular diagnosis of Fragile X syndrome (1997–2001): a collaborative study reporting 95% of the activity in France. *Am J Med Genet A* 2004; 129 (3): 218–224.

37. Cukier P, Bachega TA, Mendonca BB, Billerbeck AE. Use of nonradioactive labeling to detect large gene rearrangements in 21-hydroxylase deficiency. *Rev Hosp Clin Fac Med Sao Paulo* 2004; 59 (6): 369–374.

38. Adhami F, Muller S, Hauser MT. Nonradioactive labeling of large DNA fragments for genome walking, RFLP and northern blot analysis. *Biotechniques* 1999; 27 (2): 314–320.

39. Nanba E, Kohno Y, Matsuda A, Yano M, Sato C, Hashimoto K, Koeda T, Yoshino K, Kimura M, Maeoka Y et al. Non-radioactive DNA diagnosis for the fragile X syndrome in mentally retarded Japanese males. *Brain Dev* 1995; 17 (5): 317–324.

40. Allefs JJ, Salentijn EM, Krens FA, Rouwendal GJ. Optimization of non-radioactive Southern blot hybridization: single copy detection and reuse of blots. *Nucleic Acids Res* 1990; 18 (10): 3099–3100.

41. Rasmussen A, Macias R, Yescas P, Ochoa A, Davila G, Alonso E. Huntington disease in children: genotype-phenotype correlation. *Neuropediatrics* 2000; 31 (4): 190–194.

42. Bates G. Huntingtin aggregation and toxicity in Huntington's disease. *Lancet* 2003; 361 (9369): 1642–1644.

43. Jones AL, Wood JD, Harper PS. Huntington disease: advances in molecular and cell biology. *J Inherit Metab Dis* 1997; 20 (2): 125–138.

44. Gusella JF, MacDonald ME. Huntington's disease. *Semin Cell Biol* 1995; 6 (1): 21–28.

45. Persichetti F, Srinidhi J, Kanaley L, Ge P, Myers RH, D'Arrigo K, Barnes GT, MacDonald ME, Vonsattel JP, Gusella JF et al. Huntington's disease CAG trinucleotide repeats in pathologically confirmed post-mortem brains. *Neurobiol Dis* 1994; 1 (3): 159–166.

46. Myers RH. Huntington's disease genetics. *NeuroRx* 2004; 1 (2): 255–262.

47. Purdon SE, Mohr E, Ilivitsky V, Jones BD. Huntington's disease: pathogenesis, diagnosis and treatment. *J Psychiatr Neurosci* 1994; 19 (5): 359–367.

48. Furtado S, Suchowersky O. Huntington's disease: recent advances in diagnosis and management. *Can J Neurol Sci* 1995; 22 (1): 5–12.

49. MacDonald ME, Gusella JF. Huntington's disease: translating a CAG repeat into a pathogenic mechanism. *Curr Opin Neurobiol* 1996; 6 (5): 638–643.

50. Benitez J, Fernandez E, Garcia Ruiz P, Robledo M, Ramos C, Yebenes J. Trinucleotide (CAG) repeat expansion in chromosomes of Spanish patients with Huntington's disease. *Hum Genet* 1994; 94 (5): 563–564.

51. Norremolle A, Riess O, Epplen JT, Fenger K, Hasholt L, Sorensen SA. Trinucleotide repeat elongation in the huntingtin gene in Huntington disease patients from 71 Danish families. *Hum Mol Genet* 1993; 2 (9): 1475–1476.

52. Lavedan CN, Garrett L, Nussbaum RL. Trinucleotide repeats (CGG)22TGG(CGG)43TGG(CGG)21 from the fragile X gene remain stable in transgenic mice. *Hum Genet* 1997; 100 (3–4): 407–414.

53. Love DR, Davies KE. Duchenne muscular dystrophy: the gene and the protein. *Mol Biol Med* 1989; 6 (1): 7–17.

54. Matsuo M. Duchenne muscular dystrophy. *SE Asian J Trop Med Public Health* 1995; 26 (Suppl 1): 166–171.

55. Gangopadhyay SB, Sherratt TG, Heckmatt JZ, Dubowitz V, Miller G, Shokeir M, Ray PN, Strong PN, Worton RG. Dystrophin in frameshift deletion patients with Becker muscular dystrophy. *Am J Hum Genet* 1992; 51 (3): 562–570.

56. Bassett DI, Bryson-Richardson RJ, Daggett DF, Gautier P, Keenan DG, Currie PD. Dystrophin is required for the formation of stable muscle attachments in the zebrafish embryo. *Development* 2003; 130 (23): 5851–5860.

57. Ginjaar IB, Kneppers AL, v d Meulen JD, Anderson LV, Bremmer-Bout M, van Deutekom JC, Weegenaar J, den Dunnen JT, Bakker E. Dystrophin nonsense mutation induces different levels of exon 29 skipping and leads to variable phenotypes within one BMD family. *Eur J Hum Genet* 2000; 8 (10): 793–796.

58. Nicholson LV, Johnson MA, Davison K, O'Donnell E, Falkous G, Barron M, Harris JB. Dystrophin or a "related protein" in Duchenne muscular dystrophy? *Acta Neurol Scand* 1992; 86 (1): 8–14.

59. Whittock NV, Roberts RG, Mathew CG, Abbs SJ. Dystrophin point mutation screening using a multiplexed protein truncation test. *Genet Test* 1997; 1 (2): 115–123.

60. Ehmsen J, Poon E, Davies K. The dystrophin-associated protein complex. *J Cell Sci* 2002; 115 (Pt 14): 2801–2803.

61. Samaha FJ, Quinlan JG. Dystrophinopathies: clarification and complication. *J Child Neurol* 1996; 11 (1): 13–20.

62. Beroud C, Carrie A, Beldjord C, Deburgrave N, Llense S, Carelle N, Peccate C, Cuisset JM, Pandit F, Carre-Pigeon F et al. Dystrophinopathy caused by mid-intronic substitutions activating cryptic exons in the DMD gene. *Neuromuscul Disord* 2004; 14 (1): 10–18.

63. von Moers A, van Landeghem FK, Cohn RD, Baumgarten E, Burger J, Stoltenburg-Didinger G. Dystrophinopathy in a boy with Chediak–Higashi syndrome. *Neuromuscul Disord* 1998; 8 (7): 489–494.

64. Fanin M, Danieli GA, Cadaldini M, Miorin M, Vitiello L, Angelini C. Dystrophin-positive fibers in Duchenne dystrophy: origin and correlation to clinical course. *Muscle Nerve* 1995; 18 (10): 1115–1120.

65. Fanin M, Hoffman EP, Saad FA, Martinuzzi A, Danieli GA, Angelini C. Dystrophin-positive myotubes in innervated muscle cultures from Duchenne and Becker muscular dystrophy patients. *Neuromuscul Disord* 1993; 3 (2): 119–127.

66. Witkowski JA. Dystrophin-related muscular dystrophies. *J Child Neurol* 1989; 4 (4): 251–271.

67. Sironi M, Cagliani R, Pozzoli U, Bardoni A, Comi GP, Giorda R, Bresolin N. The dystrophin gene is alternatively spliced throughout its coding sequence. *FEBS Lett* 2002; 517 (1–3): 163–166.

68. Beggs AH, Koenig M, Boyce FM, Kunkel LM. Detection of 98% of DMD/BMD gene deletions by polymerase chain reaction. *Hum Genet* 1990; 86 (1): 45–48.

69. Beggs AH, Kunkel LM. Improved diagnosis of Duchenne/Becker muscular dystrophy. *J Clin Invest* 1990; 85 (3): 613–619.

70. Medori R, Brooke MH, Waterston RH. Genetic abnormalities in Duchenne and Becker dystrophies: clinical correlations. *Neurology* 1989; 39 (4): 461–465.

71. Lenk U, Hanke R, Kraft U, Grade K, Grunewald I, Speer A. Non-isotopic analysis of single strand conformation polymorphism (SSCP) in the exon 13 region of the human dystrophin gene. *J Med Genet* 1993; 30 (11): 951–954.

72. Roberts RG, Barby TF, Manners E, Bobrow M, Bentley DR. Direct detection of dystrophin gene rearrangements by analysis of dystrophin mRNA in peripheral blood lymphocytes. *Am J Hum Genet* 1991; 49 (2): 298–310.

73. Roberts RG, Bobrow M, Bentley DR. Point mutations in the dystrophin gene. *Proc Natl Acad Sci USA* 1992; 89 (6): 2331–2335.

74. Roberts RG, Gardner RJ, Bobrow M. Searching for the 1 in 2,400,000: a review of dystrophin gene point mutations. *Hum Mutat* 1994; 4 (1): 1–11.

75. Schwartz LS, Tarleton J, Popovich B, Seltzer WK, Hoffman EP. Fluorescent multiplex linkage analysis and carrier detection for Duchenne/Becker muscular dystrophy. *Am J Hum Genet* 1992; 51 (4): 721–729.

76. Grody WW. Cystic fibrosis: molecular diagnosis, population screening, and public policy. *Arch Pathol Lab Med* 1999; 123 (11): 1041–1046.

77. Tummler B, Dork T, Kubesch P, Fislage R, Kalin N, Neumann T, Wulbrand U, Wulf B, Steinkamp G, von der Hardt H. Cystic fibrosis: the impact of analytical technology for genotype-phenotype studies. *Clin Chim Acta* 1993; 217 (1): 23–28.

78. Arquitt CK, Boyd C, Wright JT. Cystic fibrosis transmembrane regulator gene (CFTR) is associated with abnormal enamel formation. *J Dent Res* 2002; 81 (7): 492–496.

79. Irving RM, McMahon R, Clark R, Jones NS. Cystic fibrosis transmembrane conductance regulator gene mutations in severe nasal polyposis. *Clin Otolaryngol* 1997; 22 (6): 519–521.

80. Wu CC, Hsieh-Li HM, Lin YM, Chiang HS. Cystic fibrosis transmembrane conductance regulator gene screening and clinical correlation in Taiwanese males with congenital bilateral absence of the vas deferens. *Hum Reprod* 2004; 19 (2): 250–253.

81. Watson MS, Cutting GR, Desnick RJ, Driscoll DA, Klinger K, Mennuti M, Palomaki GE, Popovich BW, Pratt VM, Rohlfs EM et al. Cystic fibrosis population carrier screening: 2004 revision of American College of Medical Genetics mutation panel. *Genet Med* 2004; 6 (5): 387–391.

82. Richards CS, Haddow JE. Prenatal screening for cystic fibrosis. *Clin Lab Med* 2003; 23 (2): 503–530

83. Watson MS, Desnick RJ, Grody WW, Mennuti MT, Popovich BW, Richards CS. Cystic fibrosis carrier screening: issues in implementation. *Genet Med* 2002; 4 (6): 407–409.

84. Richards CS, Bradley LA, Amos J, Allitto B, Grody WW, Maddalena A, McGinnis MJ, Prior TW, Popovich BW, Watson MS et al. Standards and guidelines for CFTR mutation testing. *Genet Med* 2002; 4 (5): 379–391.

85. Quinzii C, Castellani C. The cystic fibrosis transmembrane regulator gene and male infertility. *J Endocrinol Invest* 2000; 23 (10): 684–689.

86. Buyse IM, McCarthy SE, Lurix P, Pace RP, Vo D, Bartlett GA, Schmitt ES, Ward PA, Oermann C, Eng CM et al. Use of MALDI-TOF mass spectrometry in a 51-mutation test for cystic fibrosis: evidence that 3199del6 is a disease-causing mutation. *Genet Med* 2004; 6 (5): 426–430.

87. Monaghan KG, Highsmith WE, Amos J, Pratt VM, Roa B, Friez M, Pike-Buchanan LL, Buyse IM, Redman JB, Strom CM et al. Genotype-phenotype correlation and frequency of the 3199del6 cystic fibrosis mutation among I148T carriers: results from a collaborative study. *Genet Med* 2004; 6 (5): 421–425.

88. Hamilton SR. Colon cancer testing and screening. *Arch Pathol Lab Med* 1999; 123 (11): 1027–1029.

89. Borum ML. Colorectal cancer screening. *Prim Care* 2001; 28 (3): 661–674.

90. Truszkowski JA, Summers RW. Colorectal neoplasms: screening can save lives. *Postgrad Med* 1995; 98 (5): 97–99, 103–106, 109–112.

91. de Silva DC, Fernando R. Familial adenomatous polyposis. *Ceylon Med J* 1998; 43 (2): 99–105.

92. Lal G, Gallinger S. Familial adenomatous polyposis. *Semin Surg Oncol* 2000; 18 (4): 314–323.

93. Macdonald F. Familial adenomatous polyposis. *Methods Mol Med* 2004; 92: 245–266.

94. Bodmer W. Familial adenomatous polyposis (FAP) and its gene, APC. *Cytogenet Cell Genet* 1999; 86 (2): 99–104.

95. Fodde R. The APC gene in colorectal cancer. *Eur J Cancer* 2002; 38 (7): 867–871.

96. Winesett DE. APC gene testing for familial adenomatous polyposis. *J Pediatr Gastroenterol Nutr* 1999; 28 (4): 452–454.

97. Bonneton C, Larue L, Thiery JP. The APC gene product and colorectal carcinogenesis. *C R Acad Sci III* 1996; 319 (10): 861–869.

98. Munemitsu S, Souza B, Muller O, Albert I, Rubinfeld B, Polakis P. The APC gene product associates with microtubules *in vivo* and promotes their assembly *in vitro*. *Cancer Res* 1994; 54 (14): 3676–3681.

99. Smith KJ, Johnson KA, Bryan TM, Hill DE, Markowitz S, Willson JK, Paraskeva C, Petersen GM, Hamilton SR, Vogelstein B et al. The *APC* gene product in normal and tumor cells. *Proc Natl Acad Sci USA* 1993; 90 (7): 2846–2850.

100. Wu G, Wu W, Hegde M, Fawkner M, Chong B, Love D, Su LK, Lynch P, Snow K, Richards CS. Detection of sequence variations in the adenomatous polyposis coli (APC) gene using denaturing high-performance liquid chromatography. *Genet Test* 2001; 5 (4): 281–290.

101. Cheadle JP, Sampson JR. Exposing the MYtH about base excision repair and human inherited disease. *Hum Mol Genet* 2003; 12 (2): R159–R165.

102. Gismondi V, Meta M, Bonelli L, Radice P, Sala P, Bertario L, Viel A, Fornasarig M, Arrigoni A, Gentile M et al. Prevalence of the Y165C, G382D and 1395delGGA germline mutations of the MYH gene in Italian patients with adenomatous polyposis coli and colorectal adenomas. *Int J Cancer* 2004; 109 (5): 680–684.

103. Olschwang S, Bonaiti C, Feingold J, Frebourg T, Grandjouan S, Lasset C, Laurent-Puig P, Lecuru F, Millat B, Sobol H et al. HNPCC syndrome (hereditary non polyposis colon cancer): identification and management. *Rev Med Interne* 2005; 26 (2): 109–118.

104. Scartozzi M, Bianchi F, Rosati S, Galizia E, Antolini A, Loretelli C, Piga A, Bearzi I, Cellerino R, Porfiri E. Mutations of hMLH1 and hMSH2 in patients with suspected hereditary nonpolyposis colorectal cancer: correlation with microsatellite instability and abnormalities of mismatch repair protein expression. *J Clin Oncol* 2002; 20 (5): 1203–1208.

105. Baba S. Molecular biological background of FAP and HNPCC, and treatment strategies of both diseases depend upon genetic information. *Nippon Geka Gakkai Zass* 1998; 99 (6): 336–344.

106. Ferrandez A, DiSario JA. Hereditary colorectal cancer: screening and management. *Curr Treat Opt Oncol* 2002; 3 (6): 459–474.

107. Wullenweber HP, Sutter C, Autschbach F, Willeke F, Kienle P, Benner A, Bahring J, Kadmon M, Herfarth C, von Knebel Doeberitz M et al. Evaluation of Bethesda guidelines in relation to microsatellite instability. *Dis Colon Rectum* 2001; 44 (9): 1281–1289.

108. Terdiman JP, Gum JR, Jr., Conrad PG, Miller GA, Weinberg V, Crawley SC, Levin TR, Reeves C, Schmitt A, Hepburn M et al. Efficient detection of hereditary nonpolyposis colorectal cancer gene carriers by screening for tumor microsatellite instability before germline genetic testing. *Gastroenterology* 2001; 120 (1): 21–30.

109. Kastl S, Gunther K, Merkel S, Hohenberger W, Ballhausen WG. Hereditary colonic carcinoma without polyposis (HNPCC) without satisfying the Amsterdam criteria. *Chirurgie* 2000; 71 (4): 444–447.

110. Syngal S, Fox EA, Li C, Dovidio M, Eng C, Kolodner RD, Garber JE. Interpretation of genetic test results for hereditary nonpolyposis colorectal cancer: implications for clinical predisposition testing. *JAMA* 1999; 282 (3): 247–253.

111. O'Donovan MC, Oefner PJ, Roberts SC, Austin J, Hoogendoorn B, Guy C, Speight G, Upadhyaya M, Sommer SS, McGuffin P. Blind analysis of denaturing high-performance liquid chromatography as a tool for mutation detection. *Genomics* 1998; 52 (1): 44–49.

112. Liu WO, Oefner PJ, Qian C, Odom RS, Francke U. Denaturing HPLC-identified novel FBN1 mutations, polymorphisms, and sequence variants in Marfan syndrome and related connective tissue disorders. *Genet Test* 1997; 1 (4): 237–242.

113. Jones AC, Austin J, Hansen N, Hoogendoorn B, Oefner PJ, Cheadle JP, O'Donovan MC. Optimal temperature selection for mutation detection by denaturing HPLC and comparison to single-stranded conformation polymorphism and heteroduplex analysis. *Clin Chem* 1999; 45 (8, Pt 1): 1133–1140.

114. Spiegelman JI, Mindrinos MN, Oefner PJ. High-accuracy DNA sequence variation screening by DHPLC. *Biotechniques* 2000; 29 (5): 1084–1092.

114a. Hegde M, Blazo M, Chong B, Prior T, Richards C. Assay validation for identification of hereditary nonpolyposis colon cancer-causing mutations in mismatch repair genes *MLH1*, *MSH2*, and *MSH6*. *J Mol Diagn* 2005; 7(4): 525–534.

115. Hill AD, Doyle JM, McDermott EW, O'Higgins NJ. Hereditary breast cancer. *Br J Surg* 1997; 84 (10): 1334–1339.

116. Sng JH. Hereditary breast cancer: a brief overview. *Ann Acad Med Singapore* 2000; 29 (3): 407–411.

117. Gayther SA, Pharoah PD, Ponder BA. The genetics of inherited breast cancer. *J Mammary Gland Biol Neoplasia* 1998; 3 (4): 365–376.

118. Hegde MR, Chong B, Fawkner MJ, Leary J, Shelling AN, Culling B, Winship I, Love DR. Hierarchical mutation screening protocol for the BRCA1 gene. *Hum Mutat* 2000; 16(5): 422–430.

119. Elias S. The human gynome. *Am J Obstet Gynecol* 2004; 190 (6): 1528–1533.

120. Makowski DR. The Human Genome Project and the clinician. *J Fla Med Assoc* 1996; 83 (5): 307–314.

121. Panchal J, Brandt EN, Jr. The Human Genome Research Project: implications for the healthcare industry. *J Okla State Med Assoc* 2001; 94 (5): 155–159.

122. Franklin WA, Veve R, Hirsch FR, Helfrich BA, Bunn PA, Jr. Epidermal growth factor receptor family in lung cancer and premalignancy. *Semin Oncol* 2002; 29 (1, Suppl 4): 3–14.

123. Greco CM, Berman RF, Martin RM, Tassone F, Schwartz PH, Chang A, Trapp BD, Iwahashi C, Brunberg J, Grigsby J, Hessl D, Becker EJ, Papazian J, Leehey MA, Hagerman RJ, Hagerman PJ. Neuropathology of fragile X-associated tremor/ataxia syndrme (FXTAS). *Brain* 2006; 129(1): 243–255.

16 Future Perspectives on Nucleic Acid Testing

Larry J. Kricka and Paolo Fortina

CONTENTS

INTRODUCTION

The molecular structure of DNA was established in 1953 and the genetic code was deciphered in 1966. There followed a rapid surge in both fundamental and applied nucleic acid research that laid the foundation for *in vitro* nucleic acid-based diagnostic tests and culminated in the draft sequence of the 3.2-gigabase human genome in 2000.[1,2] See also Smithsonian Institute Archives, DNA Sequencing and Video History Collection (http://siris-archives.si.edu).

Regulatory approval of *in vitro* nucleic acid tests has proceeded at a relatively cautious pace. In 1985, the U.S. Food and Drug Administration (FDA) granted the first clearance for a clinical diagnostic based on nucleic acid probe technology to Gen-Probe Inc. (San Diego, California, USA) for a culture confirmation test for Legionnaire's disease. In 1988, it approved a test for detecting infection by the human papillomavirus (ViraPap assay, Life Technologies Inc., Gaithersburg, Maryland, USA) as an indicator of risk for the development of cervical cancer. FDA approval of the first viral load test was granted in 1999 (Roche Amplicor HIV-1 Monitor Test [Roche Molecular Systems, Inc., Branchburg, New Jersey, USA). The first DNA-based laboratory test for an inherited disorder (a test for blood clotting abnormalities arising from Factor V Leiden and the Factor II genetic abnormalities) was approved by the FDA in 2003.

The issues confronting nucleic acid testing are diverse and incompletely resolved. For example, the medical significance of DNA sequence variation is still not fully understood. No current sequencing or testing method is either simple or direct, and this drives continuing innovation in sample preparation and testing technology. Finally, the advent of nucleic acid testing has generated legal and ethical issues related to access and dissemination of an individual's DNA sequence. This chapter reviews key issues in nucleic acid testing, surveys selected advances and new directions in nucleic acid assay technology since 2000, and explores some of the continuing challenges to the implementation of nucleic acid tests.

SCOPE OF NUCLEIC ACID TESTING

Nucleic acid testing has expanded beyond the detection of genetic mutations; it is now used in the screening and detection of cancer, cardiovascular diseases, neurological diseases, pediatric screening, and characterizing viruses and bacteria including biowarfare agents.[3] Even though nucleic acid manipulations and methodologies are common in the scientific community, society at large has yet to accept the full impacts of the various technological advancements and the compendia of information that these novel techniques offer. One branch of the medical community that is responding to this revolution is laboratory medicine where the power of molecular techniques in disease detection and prevention has already yielded significant clinical benefits. This is particularly true in the case of cancer.

In the 14th century, William of Ockham made a statement, "Pluralitas non est ponenda sine necessitate" (plurality should not be posited without necessity) that became known as Ockham's Razor. No statement could be more fitting to the process of diagnosing disease, but simplicity is elusive with diseases as complicated as cancer. However, as our understanding of how the genetic code interacts and reacts with our environment has increased, the focus of disease detection finally rested on DNA.

Studies of cancer have shown that key genetic changes occur in DNA before and during the course of the disease. These changes include the expression of tumor suppressor genes and various oncogenes. By studying the key changes occurring at various stages of the disease and in different kinds of cancers, it is hoped that it will be possible to track, detect, and diagnose various types of cancers.[4]

One approach incorporates the use of DNA microarrays to differentiate various hematologic malignancies such as leukemia and lymphomas. In a recent study, six known clinical subtypes of pediatric acute lymphoblastic leukemia were identified through gene expression signatures generated through DNA microarray studies.[5] Similar progress has been made in utilizing gene expression to analyze T cell leukemia and further characterize diffuse large B cell lymphoma.[6-8] Through the combined use of laser capture microdissection and high-throughput cDNA microarrays, it is possible to generate expression profiles for breast cancer cells.[9] The study revealed that 90 genes with significantly altered expression can be used as markers to diagnose and observe the progression of the disease. The multifactorial nature of cancer makes it one of the hardest diseases to track. The advancement of techniques such as gene expression profiling offers significant improvement in the ways we will diagnose different types of cancers in the future.[10]

Nucleic acid testing is also used to diagnose neurological conditions such as Huntington's disease,[11] Alzheimer's disease (AD; www.ahaf.org/alzdis/about/adabout.htm),[12] and Parkinson's disease (PD; www.ninds.nih.gov/parkinsonsweb).[13] The list includes testing for known mutations in the apolipoprotein E, presenilin 1 and 2, and *Parkin* genes found to be associated with the onset of the familial and juveniles form of AD and PD, respectively.[14,15] Recent findings also show that hereditary early-onset PD is caused by mutations in *PINK1*.[16] AD and PD are some of the most rampant geriatric diseases, with AD affecting about 100,000 people each year. The availability of reliable screening and diagnostic techniques will significantly affect the detection rate, thus providing the opportunity to delay the course of these diseases.

Genetic testing also is now more commonly used for prenatal diagnoses and newborn screening (www.acmg.net).[17] Most of the advancements relate to the fine tuning of various techniques such as multiplex polymerase chain reaction (PCR) and its derivatives and other techniques including microarray-based single nucleotide polymorphism (SNP) identification and gene expression profiling. These techniques are changing the way prenatal screening and prenatal genetic diagnoses are conducted world-wide.

One example of advancement in this field is the use of Papanicolaou (Pap) smear cell isolates to perform prenatal genetic screening of various genetic conditions ranging from Down's syndrome to cystic fibrosis. Serially enriched fetal cells were prepared from Pap smears and then analyzed by means of single-cell DNA fingerprinting. Multiplex fluorescent PCR was used to simultaneously perform DNA fingerprinting, sexing, and assaying for single gene defects and diagnosing chromosomal aneuploidy from a single cell within a single reaction.[18] The prenatal Pap smear is currently in a two- to three-year clinical test phase. Should it succeed, it will pave the way for a greater emphasis on nucleic acid testing in prenatal screening and diagnosis. This technique is less invasive than amniocentesis and chorionic villi sampling. It is also more affordable and will augment the effectiveness of the current biochemical prenatal tests.

Perhaps the major use of nucleic acid testing is in genotyping viral and bacterial agents. In 1999, the American Red Cross and 16 member laboratories of the America's Blood Centers started using nucleic acid testing to detect the presence of the human immunodeficiency virus (HIV) and hepatitis C virus through nucleic acid amplification (www.moffitt.usf.edu/pubs/ccj/v6n5/dept5.htm). Another advance has been the development of a multiple DNA line probe reverse hybridization assay (INNO-LiPA Mycobacteria v2, Innogenetics, Ghent, Belgium; www.innogenetics.com) that detects the genus Mycobacterium and differentiates 16 different mycobacterial strains present in an isolate.

According to the Constitution of the World Health Organization (WHO),[19] health is a state of complete physical, mental, and social wellbeing and not merely the absence of disease or infirmity. Historically, medicine was practiced to cure diseases. However, the realization that many diseases could be prevented paved way for two new roles for the medical community that it continues to try to perfect: disease diagnosis and prevention.

Nucleic acid testing is becoming one of the cornerstones in disease and pathogen detection. Extracting information from messages encoded in DNA is not an easy task, and this is one reason nucleic acid testing has been slow to dominate in the clinical laboratory. However, the development of new techniques and technologies ranging from new sample collection and processing methods to gene expression arrays underlines the undeniable potential of nucleic acid testing in the future of clinical laboratory diagnostic methods.

LEGAL AND ETHICAL ISSUES

The rapid, sustained growth and commercialization of nucleic acid testing led to concerns about the consequences of patenting nucleic acid sequences and the ethical issues associated with the availability of genetic information.[20–24] In particular, testing for mutations in the *BRCA1* and *BRCA2* genes (linked with increased risk of familial breast and ovarian cancers) fueled a debate about patenting of genes and genetic tests and access of patients and clinical laboratories to tests and testing that continues to this day.

Opponents of gene patenting advocate that no disease gene should be patented as this inhibits free access to genes and gene sequences. They have concerns that gene patenting could lead to single providers for tests and a consequent control of access to testing and retesting. Proponents of gene patenting counter these objections with arguments based on economics, namely, that investments in developing nucleic acid-based tests and costly regulatory submissions are unlikely without the guarantee of a limited term monopoly provided by a patent.

The defining nature of DNA sequence information in terms of an individual's likelihood of developing disease or passing on disease traits has led to ethical concerns about confidentiality and misuse of the results of genetic tests. Foremost has been the concern about the impact of the results of genetic testing on job security and insurability. Despite current safeguards for the confidentiality of patient medical records, the specter of misuse will continue to ensure a conservative approach to many aspects of genetic testing for the foreseeable future.

NUCLEIC ACID DETECTION TECHNOLOGY

The growth phase of nucleic acid testing has spawned many different detection technologies and assay strategies but no consensus yet indicates which is best. The ideal assay would not require label or amplification steps and would have single molecule sensitivity, along with the desirable but difficult-to-realize advantage of rapid analysis at low cost. Today many different labels and detection reactions are employed in both qualitative and quantitative assays, often in combination with an amplification step, such as PCR. We have not yet reached the goal of an ideal assay, and development of new labels and detection methods, binding agents, and nonamplification strategies is a continuing effort.

Prominent among the emerging technologies are those based on microminiaturized devices (microchips, microarrays) and nanoscale devices (nanoparticles, carbon nanotubes, nanopores). These are providing new ways to detect nucleic acids and simplify the overall analytical process and may provide technologies for performing nucleic acid tests, not only in clinical laboratories, but also in the more demanding point-of-care environment.

NONTRADITIONAL SPECIMENS

Traditionally, the specimen used for nucleic acid analysis was a cell extract, but more conveniently accessed extracellular sources of DNA continue to attract attention. The finding that DNA and RNA are present in blood plasma fueled extensive investigations of the utility of analyzing cell-free nucleic acids.[25–30] For example, total cell-free fetal DNA in second trimester serum revealed 1.7 times more Down's syndrome cases than controls, and when added to quadruple marker screening (serum inhibin A, human chorionic gonadotropin, unconjugated estriol, alpha-fetoprotein), it improved the detection rate (86% versus 81% at a 5% false-positive rate).[31]

Pleural fluid DNA has also been shown to be useful for classifying pleural effusions. Cell-free DNA (measured as beta-globin sequence) is significantly higher in exudates than transudates.[32] Urine can also be a source of DNA for analysis. In renal transplant patients, cell-free DNA from a donor kidney can be detected in urine and provide a new marker for monitoring transplant engraftment. Increased levels of urine DNA were associated with graft rejection.[33]

LABEL-FREE NUCLEIC ACID ANALYSIS

A long-cherished but unrealized ambition has been a direct assay for a nucleic acid that combines simplicity and sensitivity. Based on current knowledge, a limited number of intrinsic properties of DNA could form the basis of a direct assay.

The mass of a nucleic acid is one attribute that has been used in a label-free assay in conjunction with a quartz crystal microbalance, but has not reached a level of analytical performance that would render it suitable for routine application.[34,35] The electrochemical properties of nucleic acids, in particular the guanine bases, also provide a route to label-free assays.[36] One example of this strategy is employed in an assay based on a phosphonic acid-functionalized polypyrrole layer coated onto the surface of a platinum electrode. DNA capture probes are immobilized via magnesium ions that bind to both the phosphonic acid on the polypyrrole layer and the phosphonic acid residues on the DNA probe. The presence of the DNA on the electrode surface hinders chloride ion exchange at the electrode and changes the shape of the cyclic voltammogram, thus allowing detection of a specific hybridization event (e.g., complementary and noncomplementary 27-mer).[37]

Another route to label-free DNA analysis measures the change in the surface charge at a microfabricated silicon field-effect sensor when target DNA hybridizes to a DNA probe immobilized on the sensor via a poly-L-lysine layer. This assay can detect nanomolar DNA concentrations.[38]

STAINS AND LABELS FOR NUCLEIC ACID ASSAYS

The simplest form of DNA testing is direct staining of total DNA or DNA subjected to some type of separation technique. Several reagents (e.g., cyanine dyes) suitable for selectively detecting single-stranded or double-stranded DNA have been developed.[39] These reagents rely on simple binding to the nucleic acid (staining) and offer incremental advances in terms of sensitivity and also compatibility with PCR.[40]

Most nucleic acid testing is based on a sandwich-type assay or a variant of this procedure that involves the binding of a labeled probe to a target captured by a nucleic acid capture probe immobilized onto a solid support. Some key factors governing choice of a label in this type of testing include stability, sensitivity, rapidity and ease of detection, overall cost for the label, and associated detection reagents and measuring system. Also, the possibility of modulating a property of a label when a labeled probe binds to a binding molecule is highly desirable, as it can form the basis of a more convenient nonseparation assay. Some recently evaluated labels include redox active tags based on Ru and Os bipyridyl complexes[41] and ferrocene complexes.[42]

Another strategy has been to develop releasable labels, for example a TAMRA (5-carboxytetramethylrhodamine) dye label can be released from guanine nucleobases modified with TAMRA by oxidation under mild conditions after hybridization.[43] Photocleavable biotin labels have also been developed to isolate DNA products of primer extension reactions on streptavidin-coated surfaces for matrix-assisted laser desorption/ionization time-of-flight (MALDI-TOF) mass spectrometric analysis.[44] The biotin is attached to a nucleotide via a nitrobenzyl linker that can be cleaved by near-ultraviolet light under mild conditions.

CAPTURE AND DETECTION PROBES

The mainstay of a nucleic acid assay has been complementary DNA or RNA as a binding agent. Other molecules with specific recognition properties for a nucleic acid sequence include histones, peptide nucleic acids (PNAs), and single-stranded binding protein (SSB).

Histones are involved in the packing of DNA in cells and have binding properties for DNA. Histone 1 conjugated to dextran-coated superparamagnetic nanoparticles (10-nm diameter magnetite core) has been used to bind target DNA adsorbed to silica particles. Binding was detected by measuring magnetic permeability of the magnetic tracer, and a detection limit for plasmid DNA of 52 µg/mL was achieved.[45]

PNAs bind to DNA with greater specificity and strength than complementary DNA and have thus attracted interest as capture and detection probes in nucleic acid analysis. For example PNA probes have been employed for analysis of mtDNA to detect mutations associated with mitochondrial encephalomyopathy, lactic acidosis, stroke-like episodes (MELAS),[46] and in SNP assays with MALDI-TOF MS as the detection system.[47]

Escherichia coli single-stranded binding protein binds to single-stranded DNA, and selectivity of binding forms the basis of an assay strategy that discriminates single stranded capture probe from streptavidin-labeled capture probe–target hybrids that accumulate on a biotin-modified carbon paste electrode surface. The SSB is labeled with a gold nanoparticle (5-nm diameter). This probe will bind only to single-stranded, streptavidin-labeled capture probe on the electrode, and hence produce an oxidation signal. No binding and consequently no signal is produced from double-stranded capture probe–target hybrids. This assay had a detection limit of 2.17 pM for a DNA 21-mer.[48]

Nucleic Acid Immobilization Techniques

Immobilization of a nucleic acid target or capture probe is a key requirement of many types of nucleic acid assays. The explosive growth of microarray assays has focused attention on this aspect of assay development. Polymers are a less expensive alternatives to glass or gold as substrates for DNA immobilization to form microarrays. The utility of polymer surface for DNA immobilization has been studied using copolymers of polystyrene and maleic acid or maleic anhydride. A 26-mer was immobilized electrostatically to polylysine or covalently bound to the different polymer surfaces and surface-bound DNA characterized using a number of different techniques such as atomic force microscopy (AFM). Immobilization yields on the carboxy-rich surfaces on a glass slide were comparable to those on unmodified glass surfaces, and the polymer surfaces have been recommended for microarray applications.[49] Another immobilization strategy exploits a 3-D-like linker system based on amino-reactive acrylic and epoxy functional groups on a glass surface for covalent immobilization of DNA via nucleobases.[50]

More sophisticated solid phases provide a dual function–immobilization of reagents together with enhancement of signal generation. For example, functionalized magnetic particles provide a means of immobilizing DNA and, under influence of rotating magnetic fields, act as microelectrodes for electrochemiluminescence.[51]

Nonseparation Assays

A nonseparation assay that has no requirement to separate labeled probe hybridized to target and nonhybridized labeled probe is both analytically simple and convenient. This type of mix-and-measure assay is rapid and amenable to automation. The luminescent oxygen channeling immuno assay (LOCI), molecular beacon, and TaqMan type assays have become popular.[52] For example, TaqMan allele discrimination utilizes PCR amplification of the region of DNA surrounding the SNP and concomitant hybridization of two allele-specific probes, each labeled with a quencher dye and distinct reporter dyes.[53] The DNA polymerase used in the assay has 5′ to 3′ nuclease activity that cleaves the probe if hybridized, thus releasing the reporter dye from the quencher, allowing signal to be detected in real time or at assay end point. Applied Biosystems (Foster City, California, USA) recently chose 150,000 SNPs for genotyping utilizing this technology, and >100,000 have been validated for research use only and are currently available (Assays on Demand). SNPs that are not included in Assays on Demand can be genotyped using TaqMan technology with custom-designed primers and probes chosen using ABI's Assays by Design service.

A recent type of nonseparation assay uses a cationic water-soluble conjugated polymer containing a 2,1,3-benzothiadiazole chromophore that electrostatically complexes with a PNA–Cy5 capture probe. Binding of target DNA to the complexed probe changes the conformation of the polymer complex. This alters the fluorescence resonance energy transfer (FRET) properties and provides the basis for detection of hybridization events. If no DNA is present, the solution containing the polymer–PNA–Cy5 dye is blue when irradiated with excitation light, and this shifts to red in the presence of complementary single-stranded DNA.[54]

A molecular beacon is a probe functionalized at one end with a fluorophore and at the other with a quencher. Molecular beacons have served successfully as DNA detection reagents. A different way of exploiting this analytical strategy is using a gold surface assembled on a quartz substrate to quench a fluorescent rhodamine reporter; 5'-thiolated beacons (3'-rhodamine label) specific for *Staphylococcus aureus Fem*A and *mec*R methicillin-resistance genes were immobilized onto gold. Increases in signal of up to 26-fold were observed when the beacons bound to specific target and disrupted the energy transfer to the gold.[55]

An important application for nonseparation assays is real-time monitoring of PCR reactions. One recent strategy utilized a quenched single-stranded primer labeled with a dye and a quencher separated by a section of DNA encoding a restriction enzyme recognition sequence (FRED, or fluorescent restriction enzyme detection). As soon as the primer is incorporated into a double-stranded DNA amplicon during PCR, it is cleaved by a thermal-stable restriction enzyme (*Bst*N1) present in the reaction mixture, and cleavage causes the dye to regain its fluorescence properties.[56]

Another elegant nonseparation assay strategy employs an intrasterically inactivated inhibitor DNA enzyme construct comprising a DNA capture probe linked on one end to cereus neutral protease and at the other to a phosphoramidite inhibitor of the enzyme. The enzyme activity of this conjugate is inhibited due to binding of the inhibitor to the active site of the enzyme. Binding of target DNA to the probe triggers activation of an enzyme by removing the inhibition and the resulting enzyme activity was detected using an EDANS (5-[2-aminoethyl] amino-1-naphthalene-sulfonic acid)-peptide-DABCYL (4-dimethylaminophenylazobenzoic acid) substrate. A detection limit of 10 fmol (100 pM) of a 24-mer was achieved in a PCR-independent assay.[57]

AMPLIFICATION AND NONAMPLIFICATION METHODS

PCR is the dominant target amplification technique and has a solid patent position. Both commercial and scientific motives fuel the search for alternative target or probe amplification methods that might be patently distinct, and would be faster and less prone to cross-contamination. A recent contender is linked linear amplification[58] using multiple cycles of primer extension reactions and a series of up to 18 nested primers and a thermostable DNA polymerase. The protocol was tested in assays for beta-globin gene and Factor V Leiden mutations, and was found to yield amplification comparable to PCR and was more resistant to carry-over contamination.

Another nucleic acid amplification method, loop-mediated isothermal amplification (LAMP), amplifies DNA under isothermal conditions.[59] It uses a DNA polymerase and a set of four primers designed to recognize six sequences on the target DNA. The LAMP product has a stem–loop structure and approximately 10^9 copies can be synthesized in <1 hr. LAMP has been applied to the detection of a range of infectious agents such as human herpes virus 6 (HHV-6) and could detect 25 copies per tube.[60] The LAMP procedure produces large amounts of pyrophosphate, and the turbidity produced when pyrophosphate reacts with magnesium ions provides a simple way to monitor the reaction in real time.[61]

MULTIPLEXED ASSAYS

A multiple simultaneous analysis capability has become important in nucleic acid testing. In routine molecular diagnosis, it simplifies testing for multiple mutations, e.g., the 25 most prevalent mutations and 6 polymorphisms in the cystic fibrosis transmembrane regulatory gene.[62] In a research setting, it allows massive parallel testing for thousands of genes, e.g., RNA expression assays.[63–65]

Accurate monitoring of the individual assays that comprise a multiplex can be achieved using sample or reagents immobilized in a two-dimensional array on a flat surface (see also the microarray section) or on a set of particles, each of which has a unique signature.

Hybridization to synthetic two-dimensional oligonucleotide arrays provides highly parallel analysis. For example, high-density oligonucleotide arrays synthesized by Affymetrix (Santa Clara, California, USA) have been used to assay large numbers of alleles simultaneously.[66,67] These arrays are synthesized with eight immobilized probes per SNP (four per strand). Each probe spans the SNP, with the central base at the SNP location. The central base of each of the four probes at the SNP position contains A, C, G, and T, respectively.

Upon hybridization to the sample, the probe containing the SNP will strongly hybridize to the perfect match probe and the other probes less strongly, for homozygous loci. Heterozygous loci will have two probes with strong hybridization. The ability to automatically call bases from array data is performed using the ABACUS (adaptive background genotyping scheme) algorithm that has been shown to provide high base-call rates with very high accuracy when combined with Affymetrix arrays and applied to both homozygous and heterozygous alleles.[68]

One advantage of this system is that the probes are produced in parallel, allowing for tens of thousands of SNPs to be determined in parallel. Hybrids of probe and biotinylated targets are stained using streptavidin–phycoerythrin conjugate, and fluorescence associated with specific probe features is detected using a confocal scanner. Comparative fluorescence signals among sets of perfectly matched and mismatched oligonucleotides are used to detect polymorphisms. A disadvantage is the inflexible nature of the array synthesis technology. For each new set of SNPs to be interrogated, a new set of masks must be manufactured.

Another SNP screening method that utilizes a two-dimensional array is the ParAllele technique (ParAllele Bioscience, South San Francisco, California, USA). This method uses molecular inversion probes (~100 to 120 bp) that are complementary on either end to the flanking sequence of a SNP. The probe hybridizes to the sample forming a "padlock" structure that is filled in and then intramolecularly ligated to form a single-stranded circle. Following extensive nuclease digestion, the circular structure is linearized with uracylglycosylase and the single-stranded DNA is amplified across the SNP using universal primers.[69] Products are sorted by tag arrays (e.g., Affymetrix GeneChip®) specific to unique tag sequences on each probe on a photolithographically generated array.

An elegant application for a two-dimensional array is to screen for a range of mutations using DNA sequencing by hybridization (SBH). This was initially proposed in 1987 and has proven to be a useful, inexpensive alternative to traditional direct DNA sequence analyses, including the Sanger dideoxy chain-termination method. SBH is an accurate tool for detection of base substitutions, insertions, and deletions in p53.[70]

This approach is an indirect sequencing method where probes designed to read all possible sequences in any DNA sample are hybridized to DNA template molecules. The probes are overlapping in sequence to allow the complete resequencing of any template of interest. Computer algorithms have been developed to analyze the hybridization patterns of the probes.

A further advance on SBH is combinatorial SBH (cSBH) utilizing a DNA ligation step. Two sets of universal short probes are used, one attached to a solid support (glass slide) and the other free in solution and labeled with a fluorophore. Unlabeled target DNA template is mixed with DNA ligase and the solution-phase labeled probe set, and hybridized to the support-bound probes. When both array-bound and solution-phase labeled probe hybridize to the target DNA at contiguous complementary positions, they are covalently linked by DNA ligase, creating one long labeled probe attached to the array surface. The combinatorial process generates all possible probes that are complementary to the target. A standard array reader scores fluorescent signals at each array position and cSBH software deconvolutes the scanned image to generate a complete sequence readout of the template PCR product.[71]

Various types of particles have been developed for multiplexed assays. An innovative multiplex strategy uses microminiature glass rods that are doped with rare earth ions to produce fluorescent micro-barcodes.[72] An aluminosilicate glass is doped with a particular combination of rare earth ions (e.g., cerium, dysprosium) such that it has a characteristic fluorescence. Individual 3.5 mm × 3.5 mm squares of the differently doped glass are then stacked in a predetermined order to define a bar code. The stack of glass is fused and then drawn into a 20-μm thick, 100-μm wide ribbon fiber that is then scribed every 20 μm using a laser to produce the micro-barcodes.

Advantages of these multiplexed labels are that they have narrow emission bands (10 to 20 nm) and high quantum efficiencies and are relatively inert to most solvents. Also, more than a million combinations are possible, giving this technology an enormous multiplexing capacity. The micro-bar codes are activated for reaction with DNA probes by coating the surfaces with gamma-aminopropylsilane. The positively charged amine groups interact with the negatively charged DNA backbone and immobilize the DNA by multiple electrostatic interactions. The micro-bar codes are decoded by imaging using a spectral imager on a fluorescence microscope.

Multimetal microrods constructed from different-thickness (nm to μm) layers of metals (gold, silver, platinum, palladium, nickel, cobalt, copper) provide the basis for yet another microparticle-based strategy.[73] The striping pattern can be determined by optical reflectivity, and the utility of these microrods has been demonstrated in a sandwich DNA probe assay for a 24-mer target using a 12-mer capture probe immobilized on the microrod and a tetramethylrhodamine-labeled 12-mer detection probe.

An interesting way of using microminiature particles is embodied by the randomly ordered BeadArray™ technology commercialized by Illumina (San Diego, California, USA).[74,75] This consists of ~50,000 individual fibers fused together into a hexagonally packed matrix that can hold up to ~50,000 beads, each ~3 μm in diameter and spaced ~5 μm apart. To assemble a randomly ordered fiber optic array, a collection of bead types, each with a distinct oligonucleotide capture probe, is pooled. An etched fiberoptic bundle then is dipped into the bead pool, allowing individual beads to assemble into the microwells at the bundle's end. Since the assembly of beads into wells is a random process, the location and identity of beads in the array must be determined post-assembly by decoding, a proprietary Illumina-based technology.

CAPILLARY ELECTROPHORETIC METHODS

Capillary electrophoresis (CE) is an effective and efficient method for sizing and quantitating nucleic acids, especially in a microchip format.[76] Three companies now offer microchip CE systems (Agilent Technologies, Inc., Palo Alto, California, USA; Hitachi, Ltd., Tokyo, Japan; Shimadzu Corporation, Kyoto, Japan). Early work confirmed the benefits of this technology compared with conventional gel-based methods for analysis of DNA and RNA mixtures (e.g., PCR reaction products).[77–82]

MASS SPECTROMETRIC METHODS

Mass spectrometry in its different forms has emerged as an important analytical technique in clinical diagnostics. It has also been applied to nucleic acid analysis. MALDI-TOF has found a number of applications,[83] for example, it has been used for multiplexed detection (12 polymorphisms in 8 genes involved in folate or homocysteine metabolism and 100 cystic fibrosis gene mutations),[84,85] multiplex genotyping (30-fold multiplexing for genotyping the p53 gene),[86] frameshift mutations,[78] allele discrimination, expression profiling, and DNA methylation analysis.[88] Electrospray ionization tandem MS has been combined with high-performance liquid chromatography (HPLC) for analysis of short oligonucleotides produced by PCR and restriction digestion in order to detect point mutations of c-K-Ras gene in colon cancer.[89]

TABLE 16.1
Applications of Microtechnology and Nanotechnology in Nucleic Acid Testing

Technology Type	Application
Microtechnology	
Microfluidic chip	Cell isolation and lysis, DNA sizing (capillary electrophoresis), PCR, LCR, nucleic acid purification
Lab-on-a-chip	Integrated DNA analysis (sample preparation and analysis)
Microarray	Sequencing, mutation and SNP analysis, gene expression analysis
Nanotechnology	
Nanoparticle	Labeling
Nanotube	Labeling, AFM probes
Nanopore	Sequencing

MICROTECHNOLOGY AND NUCLEIC ACID TESTING

Microminiaturization and microchip devices are areas of intensive research and commercial development. Microchip devices (microfluidic chips, bioelectronic chips, microarrays) have found a number of advantageous applications in nucleic acid analysis. Applications include DNA enrichment,[90] DNA extraction from cells,[91] PCR,[92] capillary electrophoretic-based mutation detection (e.g., *BRCA1* and *BRCA2* in an allele-specific PCR reaction in glass microchips),[77] continuous-flow PCR and reverse transcriptase PCR,[93] and SNP analysis (Table 16.1).[94]

LAB-ON-A-CHIP

Significant integration of the analytical steps in DNA analysis has been achieved in a microchip format (lab-on-a-chip).[95] For example, parallel processing of different samples for nucleic acid purification has been accomplished on microfluidic chips that perform cell isolation, cell lysis, DNA or mRNA purification, and recovery in nanoliter volumes. Such a microchip device was used to extract mRNA from a single cell.[96,97] Other chips combine PCR with capillary electrophoresis,[98] DNA sample preparation and capillary electrophoresis,[99] and simultaneous DNA amplification and electrochemical detection[100] or electrophoretic analysis.[101]

MICROARRAYS

DNA or oligonucleotide microarrays have been highly successful and widely adopted microanalytical devices.[102,103] In research, the role of microarrays is firmly established, but their role in routine genetic testing is less clear. Quality control of massively parallel hybridization reactions on a single analytical test device is a significant challenge that has not yet been adequately addressed in the context of routine clinical laboratory testing.[104] Work continues on the refinement of the technical aspects of the basic microarray methodology. For example, probe design is an important aspect of microarray performance, and this has been the subject of a detailed study that investigated the effect of probe length (30- to 100-mer) and the numbers and positions of mismatches on hybridization.[105]

Microarrays are usually constructed on glass or silicon surfaces. Some alternatives include polypyrrole and gold surfaces. A polypyrrole surface was used to construct a DNA chip that accomplished the rapid and reliable detection of K-Ras mutation.[106] A gold surface was employed in a chemo-responsive porous diffraction grating in conjunction with nanoparticle-labeled probes for real-time and sensitive (40 to 900 fM) DNA detection. Specific binding of the target to a complementary 3' probe immobilized on an array of 5 μm × 5 μm gold squares that constituted a

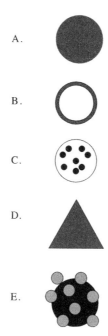

FIGURE 16.1 Classification of nanoparticle labels. A: Solid nanoparticle. B: Core-shell nanoparticle. C: Loaded or doped nanoparticle. D: Shaped nanoparticle. E: Decorated nanoparticle.

grating was detected with a 5′ probe on 13-nm diameter gold nanoparticles. A contributing factor to the sensitivity of this assay was the modulation of the local refractive index by the nanoparticles as a result of plasmon absorption.

The change in diffraction efficiency was measured using three lasers (two helium–neon and a diode-pumped Nd/YAG) and silicon photodiode detectors. This set-up was used to detect anthrax lethal factor sequence with a detection limit of 40 fM in a direct nonamplification assay. One possible use of this method is target multiplexing by means of nanoparticle labels of different size, shape, or composition.[107]

NANOTECHNOLOGY AND NUCLEIC ACID ANALYSIS

Nanotechnology is the branch of technology dealing with dimensions and tolerances of 0.1 nm to 100 nm or with the manipulation of individual atoms or molecules. It offers the next source of analytical tools for DNA analysis, although implementation and commercialization may be lengthy processes.[108–110] Concerns are already emerging about possible adverse impacts of nanomaterials.[111] Nevertheless, the pace of research and development and the scope of achievements indicate that nanotechnology will become central to many aspects of everyday life, and healthcare in particular. In the context of nucleic acid analysis, the most important areas of nanotechnology are nanoparticles, nanotubes, nanopores, and scanning microscopy.

NANOPARTICLES

Research and development activities related to the use of nanoparticles as labels (Figure 16.1) are extensive and ongoing. A nanoparticle may be composed of a single material such as gold[112] or contain several materials in a core shell design. Nanoparticles may be impregnated (doped) or produced in different shapes (nanorods, nanoprisms, nanodisks).[113,114] Nanoparticle composition, size, and shape can influence optical signature of a particle and this provides a route to developing multicolor labels.

A core shell nanoparticle used as a label provides a number of advantages. It avoids aggregation problems encountered with some types such as silver nanoparticles. Particle aggregation can be reduced for silver nanoparticles (e.g., 12 nm diameter) by growing a thin monolayer of gold onto the surfaces of the particles.[115] Core shell nanoparticles also combine the beneficial properties of both materials used for their construction. For example, gold-coated copper particles can be effectively detected by anodic stripping voltametry at a conducting polypyrrole surface. This particle combines the good voltametric properties of the copper core and the surface modification properties of the gold shell and was used in an assay that detected 5 pmol/L of a 24-mer target.[116]

A variant of the core shell is a decorated nanoparticle. The increase in mass when a particle is decorated with another material is useful in certain types of assays. For example, a 10-nm nanoparticle can be used as a label in a microgravimetric quartz-crystal microbalance-based nucleic acid assay The signal from the 10-nm gold nanoparticle label can be amplified by increasing the mass of the particle through deposition of gold on its surface via a gold-catalyzed reduction of gold chloride in presence of hydroxylamine The assay sensed $<1 \times 10^{-15}$ M of a 27-mer DNA target.[117]

Impregnating or doping a nanoparticle is a way to impart a defined detection property. For example, electroactive tris(2,2′-bipyridyl)cobalt(III) can be doped into silica nanoparticles (average diameter of 70 nm) and the doped particle used as a label. A detection limit of 2×10^{-10} mol/L was attained for a 24-mer using this strategy.[118]

Nanoparticles can also be doped with fluorophores. The advantage of this type of particle as a label is that the silicon shields the fluorophores and increases photostability. For example, TAMRA-dextran was encapsulated into 60-nm diameter silica particles and the particles employed in a sandwich assay for a 27-mer. The assay detected 0.8 fM and this represented a 10,000-fold signal enhancement compared to using a single TAMRA label.[119]

The shape of a nanoparticle can affect its properties. Nanoprisms (100-nm edges) have been fabricated in silver.[114] Nanoparticles of this shape interact with light differently from spherical particles and as a consequence have a different color. This is an intriguing difference and provides a route to multiplexed assays in which the nanoparticle labels are all made from exactly the same material but rely on differences in shape to achieve unique optical signals, as opposed to using nanoparticles made from different materials such as ZnS, CdS, and PbS.

An alternative multiplexing strategy involves the combination of gold nanoparticles and Raman-active dyes.[120] For example, 13-nm diameter gold particles coated with Cy3-labeled oligonucleotide probe bind to target captured by probes immobilized on a microarray chip. A silver deposition step coats a layer of silver onto the captured nanoparticles and this acts as a promoter of Raman scattering. Locations on the microarray chip appear gray where particles have been captured and the chip can be scanned via a flatbed scanner. This surface-enhanced Raman scattering (SERS) method can be adapted for multiplexing by using different dyes (e.g., Cy3, Cy5, Cy3.5, Texas Red) that produce distinct SERS signals. A detection limit of 20 fM has been achieved and further optimization promises an even lower limit.

CARBON NANOTUBES

Nanometer diameter single wall and multiwall carbon nanotubes (CNTs) and nanotubes fabricated from other materials (BN, GaN, BC, organic polymers) represent a fast-growing area of research and development. Nanotubes have already found a number of useful applications, serving as tips for atomic force microscopes, electrodes in batteries and capacitors, interconnects in electronic chips, and in field-emission flat-panel displays (http://nanotechweb.org).[121]

CNTs can be considered nano-sized containers or carriers and they can be filled or coated with biological molecules. This property has been exploited to provide amplification in recognition and transduction events in a nucleic acid assay. An alkaline phosphatase enzyme label was immobilized on a CNT attached to a DNA probe to achieve a loading of approximately 9600 enzyme molecules per CNT. This was then used in a sandwich assay in combination with a capture probe immobilized on a magnetic bead. A second application for CNTs in this assay was to use a CNT layer on an electrode to concentrate the alpha-naphthol electroactive product released by the alkaline phosphatase from the alpha-naphthyl phosphate substrate for the enzyme label. The assay detected 820 copies (1.3 zmol) of an oligonucleotide target.[122]

Gold nanotube membranes have been recently developed for specific DNA molecular recognition. The nanotubes are formed in the pores of a 6-μm thick filter by deposition of gold. A 30-mer oligonucleotide hairpin molecule is attached to the inside of the 12-nm diameter gold nanotube and this provides the molecular recognition for a complementary DNA molecule and transports the DNA through the membrane. Single base transport selectivity was obtained, and this strategy could form the basis of an assay technique in the future.[123]

NANOPORES

A nanopore provides the basis of a novel way of sequencing DNA.[124] Alpha-hemolysin self-assembles in a lipid bilayer to form a nanopore with an internal diameter of 1.5 nm at its narrowest point. An electrical field can be used to draw a single strand of DNA through the pore, and as it traverses the pore, it modulates the ionic current flowing through the pore. The amplitude and duration of the blockade are dependent on the base composition of the DNA. It has been possible to obtain signatures characteristic of a nucleotide sequence. The next step for this technology is to refine the technique to distinguish single nucleotides.

Estimates for the sequencing rate range from 1,000 to 10,000 bases/sec, and this far exceeds the current rate of approximately 30,000 bases/day using conventional sequencers. Moves are now underway to replace the biological nanopore with a solid state nanopore (3 to 10 nm in diameter) fabricated in a silicon nitride membrane[125] or with a nanochannel that would stretch the DNA molecule and simplify the sequencing.[126]

SCANNING MICROSCOPY

The different types of microscopes such as the atomic force microscope (AFM) and scanning tunneling microscope (STM) have revolutionized our ability to detect and observe individual atoms and molecules. A single nucleic acid molecule and hybridization events on that molecule can be directly monitored by microscopy. Probes labeled with different-sized labels are used in hybridization experiments. Scanning of a target revealed the locations of hybridized probes and the amplitude of deflection of the CNT attached to the AFM tip signified the size of the label attached to the probe.[127] This approach offers a direct method for analysis of nucleic acids and may vie with nanopore-based techniques in the future.

A carbon nanotube attached to an AFM tip also provided a useful tool for chromosomal analysis.[128] The nanotube probe can be used in the dynamic force mode to dissect tiny chromosomal fragments, and then after dissection, topographic profiles of the fragments can be obtained with the carbon nanotube probe in ambient condition.

EXTERNAL QUALITY ASSESSMENT AND QUALITY CONTROL

External Quality Assessment (EQA) or proficiency testing is an important component of assuring the quality of laboratory analyses and various national and international schemes have been available for chemistry tests (e.g., glucose, sodium, cholesterol, urea) for several decades. EQA programs have been developed for various molecular tests and began as pilot schemes in the early 1990s.[129–133] One goal of an EQA scheme is to improve the overall analytical performances of laboratories. This effect is evident, for example, in results from the European scheme for cystic fibrosis that has shown a steady decrease in the number of laboratories making genotype errors from approximately 35% in 1996 to 10% in 2000.

A recent Italian program evaluated DNA extraction, amplification, and analysis of PCR products.[134] Not surprisingly, it found that results varied in all stages of the analytical procedure. A scoring system was devised to provide a general estimation of quality, and on this basis 7 of 31 participating laboratories produced poor or unacceptable results. Quality concerns are also evident from the results of a European proficiency testing program (51 laboratories in 19 countries) for the molecular detection and quantitation of hepatitis B virus DNA. Only half of the laboratories enrolled achieved performance scores in the good and adequate categories for either of two panels of specimens circulated.[135]

Nucleic acid testing is a relatively young discipline and quality assurance (QA) and quality control (QC) are still in active phases of development. Validation of equipment is an important component of quality control and assurance. A thermocycler for PCR is an essential instrument for most nucleic acid tests, and the performance of thermocyclers has a direct bearing on analytical quality. The impact of thermocycler operation on PCR-based test results was studied in an inter-laboratories survey involving two standardized assays in 18 laboratories using 33 cyclers of various kinds.[136] This study revealed significant variation in reaction yield, inter-block variation in results, and highlighted the importance of temperature calibration of thermocyclers for improving repeatability of data. QA and QC must be continuing priorities to allow further improvements in the quality of molecular diagnostic testing and attain the accuracy and precision levels currently achieved by less complicated serum chemistry tests.

POINT-OF-CARE TESTING

Point-of-Care Testing (POCT) has been very successful for assessing blood glucose and fertility hormones, most notably detecting human chorionic gonadotropin in home pregnancy testing. A continuing theme has been the development of devices for nucleic acid tests at points of care. No FDA-approved point-of-care nucleic acid-based tests presently exist, and it is likely to be a long time before such tests make the difficult transition to a place on the waived list (http://www.access-data.fda.gov/scripts/cdrh/cfdocs/cfClia/analyteswaived.cfm). A key question centers on what tests are suitable or needed in a point-of-care setting.

The recent concern about releases of biological warfare agents into the environment in the course of a terrorist attack has provided a logical and justifiable reason for point-of-care tests to detect such agents in the environment and screen exposed individuals. A likely consideration for such tests is their reliability and how quickly they can produce results. Expansion of the test menu to other nucleic acid targets would also raise ethical considerations over the misuse of this genetic information. Nevertheless, technology is propelling nucleic acid testing toward a full point-of-care capability. For example, a lateral flow device of the type successfully employed in point-of-care immunoassays has been applied to nucleic acid testing. The detection phase of a PCR-based screening test for human papillomavirus 16 (HPV16) has been adapted to a lateral flow assay format using 400-nm diameter up-converting phosphor particles as labels. This type of label is excited by IR light, emits in the visible range, and is background-free (no natural occurring up-converting substances). This assay was 100-fold more sensitive than a comparable assay that used colloidal gold as the label.[137,138]

CONCLUSIONS

Nucleic acid manipulation and testing are commonplace in the scientific community and central to all future health care, although the medical significance of DNA sequence variation is not yet fully understood. Many nucleic acid testing methods in use currently are neither simple nor direct. There is still a need for further improvements of such assays and this unmet need is driving continuing innovation in sample preparation methods and detection technology. One promising approach involves advances in microtechnology and nanotechnology that are providing new directions for simplifying assay procedures (e.g., lab-on-a-chip analyzers) and enhancing detection sensitivity. Finally, society at large has not fully accepted nucleic acid technology, and concerns about the dissemination of an individual's DNA sequence and the impact of this information on that individual in society remain.

REFERENCES

1. International Human Genome Sequencing Consortium. Initial sequencing and analysis of the human genome. *Nature* 2001; 409 (6822): 860–921. Errata: *Nature* 2001; 411 (6838): 720 and *Nature* 2001; 412 (6846): 565–566.
2. Venter JC, Adams MD, Myers EW, Li PW, Mural RJ et al. The sequence of the human genome. *Science* 2001; 291 (5507): 1304–1351. Errata: *Science* 2001; 292 (5523): 1838 and *Science* 2002; 295 (5559): 1466.
3. Fortina P, Surrey S, Kricka LJ. Molecular diagnostics: hurdles for clinical implementation. *Trends Mol Med* 2002; 8 (6): 264–266.
4. Golub TR. Toward a functional taxonomy of cancer. *Cancer Cell* 2004; 6 (2): 107–108.
5. Ebert BL, Golub TR. Genomic approaches to hematologic malignancies. *Blood* 2004; 104 (4): 923–932.
6. Rosenwald A, Wright G, Chan WC, Connors JM, Campo E, Fisher RI, Gascoyne RD et al. The use of molecular profiling to predict survival after chemotherapy for diffuse large B-cell lymphoma. *New Engl J Med* 2002; 346 (25): 1937–1947.
7. Bullinger L, Dohner K, Bair E, Frohling S, Schlenk RF, Tibshirani R, Dohner H, Pollack JR. Use of gene-expression profiling to identify prognostic subclasses in adult acute myeloid leukemia. *New Engl J Med* 2004; 350 (16): 1605–1616.
8. Valk PJM, Verhaak RGW, Beijen MA, Erpelinck CAJ, van Doorn-Khosrovani SBvW, Boer JM, Beverloo HB, Moorhouse MJ, van der Spek PJ, Lowenberg B, Delwel R. Prognostically useful gene-expression profiles in acute myeloid leukemia. *New Engl J Med* 2004; 350 (16): 1617–1628.
9. Sgroi DC, Teng S, Robinson G, LeVangie R, Hudson JR, Jr., Elkahloun AG. *In vivo* gene expression profile analysis of human breast cancer progression. *Cancer Res* 1999; 59 (22): 5656–5661.
10. Irish JM, Hovland R, Krutzik PO, Perez OD, Bruserud O, Gjertsen BT, Nolan GP. Single cell profiling of potentiated phospho-protein networks in cancer cells. *Cell* 2004; 118 (2): 217–228.
11. Potter NT, Spector EB, Prior TW. Technical standards and guidelines for Huntington disease testing. *Genet Med* 2004; 6 (1): 61–65.
12. Post SG, Whitehouse PJ, Binstock RH, Bird TD, Eckert SK, Farrer LA, Fleck LM, Gaines AD, Juengst ET, Karlinsky H, Miles S, Murray TH, Quaid KA, Relkin NR, Roses AD, St. George-Hyslop PH, Sachs GA, Steinbock B, Truschke EF, Zinn AB. The clinical introduction of genetic testing for Alzheimer's disease: an ethical perspective. *JAMA* 1997; 277 (10): 832–836.
13. Hughes AJ, Daniel SE, Ben Shlomo Y, Lees AJ. The accuracy of diagnosis of Parkinsonian syndromes in a specialist movement disorder service. *Brain* 2002; 125 (4): 861–870.
14. Rogaeva EA, Fafel KC, Song YQ, Medeiros H, Sato C, Liang Y, Richard E, Rogaev EI, Frommelt P, Sadovnick AD, Meschino W, Rockwood K, Boss MA, Mayeux R, St. George-Hyslop P. Screening for PS1 mutations in a referral-based series of AD cases: 21 novel mutations. *Neurology* 2001; 57 (4): 621–625.

15. Kitada T, Asakawa S, Hattori N, Matsumine H, Yamamura Y, Minoshima S, Yokochi M, Mizuno Y, Shimizu N. Mutations in the *parkin* gene cause autosomal recessive juvenile parkinsonism. *Nature* 1998; 392 (6676): 605–608.

16. Valente EM, Abou-Sleiman PM, Caputo V, Muqit MMK, Harvey K, Gispert S, Ali Z, Del Turco D, Bentivoglio AR, Healy DG, Albanese A, Nussbaum R, Gonzalez-Maldonado R, Deller T, Salvi S, Cortelli P, Gilks WP, Latchman DS, Harvey RJ, Dallapiccola B, Auburger G, Wood NW. Hereditary early-onset Parkinson's disease caused by mutations in *PINK1*. *Science* 2004; 304 (5674): 1158–1160.

17. Wheeler PG, Smith R, Dorkin H, Parad RB, Comeau AM, Bianchi DW. Genetic counseling after implementation of statewide cystic fibrosis newborn screening: two years' experience in one medical center. *Genet Med* 2001; 3 (6): 411–415.

18. Irwin DL, Bryan JL, Chan FY, Matthews PL, Healey SC, Peters M, Findlay I. Prenatal diagnosis of tetrasomy 18p using multiplex fluorescent PCR and comparison with a variety of techniques. *Genet Test* 2003; 7 (1): 1–6.

19. International Health Conference. Constitution of the World Health Organization. *Bull World Health Organ* 2002; 80 (12): 983–984.

20. Leonard DGB. Improved method for diagnosis of Charcot–Marie–Tooth Type 1A: patent pending? *Clin Chem* 2001; 47 (5): 807–808.

21. Merz JF, Kriss AG, Leonard DGB, Cho MK. Diagnostic testing fails the test. *Nature* 2002; 415 (6872): 577–579.

22. Cho MK, Illangasekare S, Weaver MA, Leonard DGB, Merz JF. Effects of patents and licenses on the provision of clinical genetic testing services. *J Mol Diagn* 2003; 5 (1): 3–8.

23. Meurer MJ. Pharmacogenomics, genetic tests, and patent-based incentives. *Adv Genet* 2003; 50: 399–426.

24. Carroll AM, Coleman CH. Closing the gaps in genetics legislation and policy: a report by the New York state task force on life and the law. *Genet Testing* 2001; 5 (4): 275–280.

25. Rijnders RJ, Christiaens GC, Bossers B, van der Smagt JJ, van der Schoot CE, de Haas M. Clinical applications of cell-free fetal DNA from maternal plasma. *Obstet Gynecol* 2004; 103 (1): 157–164.

26. Chiu RWK, Chan LYS, Lam NYL, Tsui NBY, Ng EKO, Rainer TH, Lo YMD. Quantitative analysis of circulating mitochondrial DNA in plasma. *Clin Chem* 2003; 49 (5): 719–726.

27. Lo YM. Circulating nucleic acids in plasma and serum: an overview. *Ann NY Acad Sci* 2001; 945: 1–7.

28. Pertl B, Bianchi DW. Fetal DNA in maternal plasma: emerging clinical applications. *Obstet Gynecol* 2001; 98 (3): 483–490.

29. Wong BC, Lo YM. Cell-free DNA and RNA in plasma as new tools for molecular diagnostics. *Expert Rev Mol Diagn* 2003; 3 (6): 785–797.

30. Chen Xq, Bonnefoi H, Pelte MF, Lyautey J, Lederrey C, Movarekhi S, Schaeffer P, Mulcahy HE, Meyer P, Stroun M, Anker P. Telomerase RNA as a detection marker in the serum of breast cancer patients. *Clin Cancer Res* 2000; 6 (10): 3823–3826.

31. Farina A, LeShane ES, Lambert-Messerlian GM, Canick JA, Lee T, Neveux LM, Palomaki GE, Bianchi DW. Evaluation of cell-free fetal DNA as a second-trimester maternal serum marker of Down syndrome pregnancy. *Clin Chem* 2003; 49 (2): 239–242.

32. Chan MHM, Chow KM, Chan ATC, Leung CB, Chan LYS, Chow KCK, Lam CW, Lo YMD. Quantitative analysis of pleural fluid cell-free DNA as a tool for the classification of pleural effusions. *Clin Chem* 2003; 49 (5): 740–745.

33. Li Y, Hahn D, Zhong XY, Thomson PD, Holzgreve W, Hahn S. Detection of donor-specific DNA polymorphisms in the urine of renal transplant recipients. *Clin Chem* 2003; 49 (4): 655–658.

34. Su X. Covalent DNA immobilization on polymer-shielded silver-coated quartz crystal microbalance using photobiotin-based UV irradiation. *Biochem Biophys Res Commun* 2002; 290 (3): 962–966.

35. Zhou XC, Huang LQ, Li SF. Microgravimetric DNA sensor based on quartz crystal microbalance: comparison of oligonucleotide immobilization methods and the application in genetic diagnosis. *Biosens Bioelectron* 2001; 16 (1–2): 85–95.

36. Drummond TG, Hill MG, Barton JK. Electrochemical DNA sensors. *Nat Biotechnol* 2003; 21 (10): 1192–1199.

37. Thompson LA, Kowalik J, Josowicz M, Janata J. Label-free DNA hybridization probe based on a conducting polymer. *J Am Chem Soc* 2003; 125 (2): 324–325.

38. Fritz J, Cooper EB, Gaudet S, Sorger PK, Manalis SR. Electronic detection of DNA by its intrinsic molecular charge. *Proc Natl Acad Sci USA* 2002; 99 (22): 14142–14146.

39. Kricka LJ. Stains, labels and detection strategies for nucleic acids assays. *Ann Clin Biochem* 2002; 39 (Pt 2): 114–129.

40. Bengtsson M, Karlsson HJ, Westman G, Kubista M. A new minor groove binding asymmetric cyanine reporter dye for real-time PCR. *Nucleic Acids Res* 2003; 31 (8): e45.

41. Weizman H, Tor Y. Redox-active metal-containing nucleotides: synthesis, tunability, and enzymatic incorporation into DNA. *J Am Chem Soc* 2002; 124 (8): 1568–1569.

42. Yu CJ, Wan Y, Yowanto H, Li J, Tao C, James MD, Tan CL, Blackburn GF, Meade TJ. Electronic detection of single-base mismatches in DNA with ferrocene-modified probes. *J Am Chem Soc* 2001; 123 (45): 11155–11161.

43. Okamoto A, Tanaka K, Saito I. A nucleobase that releases reporter tags upon DNA oxidation. *J Am Chem Soc* 2004; 126 (2): 416–417.

44. Bai X, Kim S, Li Z, Turro NJ, Ju J. Design and synthesis of a photocleavable biotinylated nucleotide for DNA analysis by mass spectrometry. *Nucleic Acids Res* 2004; 32 (2): 535–541.

45. Abrahamsson D, Kriz K, Lu M, Kriz D. A preliminary study on DNA detection based on relative magnetic permeability measurements and histone H1 conjugated superparamagnetic nanoparticles as magnetic tracers. *Biosens Bioelectron* 2004; 19 (11): 1549–1557.

46. Hancock DK, Schwarz FP, Song F, Wong LJ, Levin BC. Design and use of a peptide nucleic acid for detection of the heteroplasmic low-frequency mitochondrial encephalomyopathy, lactic acidosis, and stroke-like episodes (MELAS) mutation in human mitochondrial DNA. *Clin Chem* 2002; 48 (12): 2155–2163.

47. Ren B, Zhou JM, Komiyama M. Straightforward detection of SNPs in double-stranded DNA by using exonuclease III/nuclease S1/PNA system. *Nucleic Acids Res* 2004; 32 (4): e42.

48. Kerman K, Saito M, Morita Y, Takamura Y, Ozsoz M, Tamiya E. Electrochemical coding of single-nucleotide polymorphisms by monobase-modified gold nanoparticles. *Anal Chem* 2004; 76 (7): 1877–1884.

49. Ivanova EP, Pham DK, Brack N, Pigram P, Nicolau DV. Poly(L-lysine)-mediated immobilisation of oligonucleotides on carboxy-rich polymer surfaces. *Biosens Bioelectron* 2004; 19 (11): 1363–1370.

50. Hackler L Jr, Dorman G, Kele Z, Urge L, Darvas F, Puskas LG. Development of chemically modified glass surfaces for nucleic acid, protein and small molecule microarrays. *Mol Divers* 2003; 7 (1): 25–36.

51. Weizmann Y, Patolsky F, Katz E, Willner I. Amplified DNA sensing and immunosensing by the rotation of functional magnetic particles. *J Am Chem Soc* 2003; 125 (12): 3452–3454.

52. Foy CA, Parkes HC. Emerging homogeneous DNA-based technologies in the clinical laboratory. *Clin Chem* 2001; 47 (6): 990–1000.

53. Livak KJ. SNP genotyping by the 5′-nuclease reaction. *Methods Mol Biol* 2003; 212: 129–147.

54. Liu B, Bazan BC.. Interpolyelectrolyte complexes of conjugated copolymers and dna: platforms for multicolor biosensors. *J Am Chem Soc* 2004; 126 (7): 1942–1943.

55. Du H, Disney MD, Miller BL, Krauss TD. Hybridization-based unquenching of DNA hairpins on au surfaces: prototypical "molecular beacon" biosensors. *J Am Chem Soc* 2003; 125 (14): 4012–4013.

56. Cairns MJ, Turner R, Sun L-Q. Homogeneous real-time detection and quantification of nucleic acid amplification using restriction enzyme digestion. *Biochem Biophys Res Commun* 2004; 318 (3): 684–690.

57. Saghatelian A, Guckian KM, Thayer DA, Ghadiri MR. DNA detection and signal amplification via an engineered allosteric enzyme. *J Am Chem Soc* 2003; 125 (2): 344–345.

58. Reyes AA, Ugozzoli LA, Lowery JD, Breneman JW, III, Hixson CS, Press RD, Wallace RB. Linked linear amplification: a new method for the amplification of DNA. *Clin Chem* 2001; 47 (1): 31–40.

59. Notomi T, Okayama H, Masubuchi H, Yonekawa T, Watanabe K, Amino N, Hase T. Loop-mediated isothermal amplification of DNA. *Nucleic Acids Res* 2000; 28 (12): e63i–e63vii.

60. Ihira M, Yoshikawa T, Enomoto Y, Akimoto S, Ohashi M, Suga S, Nishimura N, Ozaki T, Nishiyama Y, Notomi T, Ohta Y, Asano Y. Rapid diagnosis of human herpesvirus 6 infection by a novel DNA amplification method, loop-mediated isothermal amplification. *J Clin Microbiol* 2004; 42 (1): 140–145.

61. Mori Y, Nagamine K, Tomita N, Notomi T. Detection of loop-mediated isothermal amplification reaction by turbidity derived from magnesium pyrophosphate formation. *Biochem Biophys Res Commun* 2001; 289 (1): 150–154.

62. Strom CM, Clark DD, Hantash FM, Rea L, Anderson B, Maul D, Huang D, Traul D, Tubman CC, Garcia R, Hess PP, Wang H, Crossley B, Woodruff E, Chen R, Killeen M, Sun W, Beer J, Avens H, Polisky B, Jenison RD. Direct visualization of cystic fibrosis transmembrane regulator mutations in the clinical laboratory setting. *Clin Chem* 2004; 50 (5): 836–845.

63. Cheval L, Virlon B, Billon E, Aude JC, Elalouf JM, Doucet A. Large-scale analysis of gene expression: methods and application to the kidney. *J Nephrol* 2002; 15 (Suppl 5): S170–S183.

64. Armstrong SA, Golub TR, Korsmeyer SJ. MLL-rearranged leukemias: insights from gene expression profiling. *Sem Hematol* 2003; 40 (4): 268–273.

65. Feroze-Merzoug F, Schober MS, Chen YQ. Molecular profiling in prostate cancer. *Cancer Metastasis Rev* 2001; 20 (3–4): 165–171.

66. Kennedy GC, Matsuzaki H, Dong S, Liu Wm, Huang J, Liu G, Su X, Cao M, Chen W, Zhang J, Liu W, Yang G, Di X, Ryder T, He Z, Surti U, Phillips MS, Boyce-Jacino MT, Fodor SP, Jones KW. Large-scale genotyping of complex DNA. *Nat Biotechnol* 2003; 21 (10): 1233–1237.

67. Matsuzaki H, Dong S, Loi H, Di X, Liu G, Hubbell E, Law J, Berntsen T, Chadha M, Hui H, Yang G, Kennedy GC, Webster TA, Cawley S, Walsh PS, Jones KW, Fodor SPA, Mei M. Genotyping over 100,000 SNPs on a pair of oligonucleotide arrays. *Nat Methods* 2004; 1 (2): 109–111.

68. Cutler DJ, Zwick ME, Carrasquillo MM, Yohn CT, Tobin KP, Kashuk C, Mathews DJ, Shah NA, Eichler EE, Warrington JA, Chakravarti A. High-throughput variation detection and genotyping using microarrays. *Genome Res* 2001; 11 (11): 1913–1925.

69. Hardenbol P, Baner J, Jain M, Nilsson M, Namsaraev EA, Karlin-Neumann GA, Fakhrai-Rad H, Ronaghi M, Willis TD, Landegren U, Davis RW. Multiplexed genotyping with sequence-tagged molecular inversion probes. *Nat Biotechnol* 2003; 21 (6): 673–678.

70. Drmanac R, Drmanac S. Sequencing by hybridization arrays. *Methods Mol Biol* 2001; 170: 39–51.

71. Cowie S, Drmanac S, Swanson D, Delgrosso K, Huang S, du Sart D, Drmanac R, Surrey S, Fortina P. Identification of APC gene mutations in colorectal cancer using universal microarray-based combinatorial sequencing-by-hybridization. *Hum Mutat* 2004; 24 (3): 261–271.

72. Dejneka MJ, Streltsov A, Pal S, Frutos AG, Powell CL, Yost K, Yuen PK, Muller U, Lahiri J. Rare earth-doped glass microbarcodes. *Proc Natl Acad Sci USA* 2003; 100 (2): 389–393.

73. Nicewarner-Pena SR, Freeman RG, Reiss BD, He L, Pena DJ, Walton ID, Cromer R, Keating CD, Natan MJ. Submicrometer metallic barcodes. *Science* 2001; 294 (5540): 137–141.

74. Oliphant A, Barker DL, Stuelpnagel JR, Chee MS. BeadArray™ technology: enabling an accurate, cost-effective approach to high-throughput genotyping. *BioTechniques* 2002; 32 (Suppl 2): S56–S61.

75. Gunderson KL, Kruglyak S, Graige MS, Garcia F, Kermani BG, Zhao C, Che D, Dickinson T, Wickham E, Bierle J, Doucet D, Milewski M, Yang R, Siegmund C, Haas J, Zhou L, Oliphant A, Fan JB, Barnard S, Chee MS. Decoding randomly ordered DNA arrays. *Genome Res* 2004; 14 (5): 870–877.

76. Mitchelson KR. The use of capillary electrophoresis for DNA polymorphism analysis. *Mol Biotechnol* 2003; 24 (1): 41–68.

77. Tian H, Brody LC, Fan S, Huang Z, Landers JP. Capillary and microchip electrophoresis for rapid detection of known mutations by combining allele-specific DNA amplification with heteroduplex analysis. *Clin Chem* 2001; 47 (2): 173–185.

78. Nachamkin I, Panaro NJ, Li M, Ung H, Yuen PK, Kricka LJ, Wilding P. Agilent 2100 bioanalyzer for restriction fragment length polymorphism analysis of the *Campylobacter jejuni* flagellin gene. *J Clin Microbiol* 2001; 39 (2): 754–757.

79. Chen L, Ren J. High-throughput DNA analysis by microchip electrophoresis. *Comb Chem High Throughput Screen* 2004; 7 (1): 29–43.

80. Kataoka M, Inoue S, Kajimoto K, Sinohara Y, Baba Y. Usefulness of microchip electrophoresis for reliable analyses of nonstandard DNA samples and subsequent on-chip enzymatic digestion. *Eur J Biochem* 2004; 271 (11): 2241–2247.

81. Garcia L, Kindt A, Bermudez H, Llanos-Cuentas A, De Doncker S, Arevalo J, Quispe Tintaya KW, Dujardin JC. Culture-independent species typing of neotropical *Leishmania* for clinical validation of a PCR-based assay targeting heat shock protein 70 genes. *J Clin Microbiol* 2004; 42 (5): 2294–2297.

82. Liu Y, Ganser D, Schneider A, Liu R, Grodzinski P, Kroutchinina N. Microfabricated polycarbonate CE devices for DNA analysis. *Anal Chem* 2001; 73 (17): 4196–4201.
83. Jurinke C, Oeth P, van den Boom D. MALDI-TOF mass spectrometry: a versatile tool for high-performance DNA analysis. *Mol Biotechnol* 2004; 26 (2): 147–164.
84. Meyer K, Fredriksen A, Ueland PM. High-level multiplex genotyping of polymorphisms involved in folate or homocysteine metabolism by matrix-assisted laser desorption/ionization mass spectrometry. *Clin Chem* 2004; 50 (2): 391–402.
85. Wang Z, Milunsky J, Yamin M, Maher T, Oates R, Milunsky A. Analysis by mass spectrometry of 100 cystic fibrosis gene mutations in 92 patients with congenital bilateral absence of the vas deferens. *Hum Reprod* 2002; 17 (8): 2066–2072.
86. Kim S, Ulz ME, Nguyen T, Li CM, Sato T, Tycko B, Ju J. Thirty-fold multiplex genotyping of the p53 gene using solid phase capturable dideoxynucleotides and mass spectrometry. *Genomics* 2004; 83 (5): 924–931.
87. Ruparel H, Ulz ME, Kim S, Ju J. Digital detection of genetic mutations using SPC-sequencing. *Genome Res* 2004; 14 (2): 296–300.
88. Gut IG. DNA analysis by MALDI-TOF mass spectrometry. *Hum Mutat* 2004; 23 (5): 437–441.
89. Lleonart ME, Ramon y Cajal S, Groopman JD, Friesen MD. Sensitive and specific detection of K-ras mutations in colon tumors by short oligonucleotide mass analysis. *Nucleic Acids Res* 2004; 32 (5): e53.
90. Dai J, Ito T, Sun L, Crooks RM. Electrokinetic trapping and concentration enrichment of DNA in a microfluidic channel. *J Am Chem Soc* 2003; 125 (43): 13026–13027.
91. Chung YC, Jan MS, Lin YC, Lin JH, Cheng WC, Fan CY. Microfluidic chip for high efficiency DNA extraction. *Lab Chip* 2004; 4 (2): 141–147.
92. Kricka LJ, Wilding P. Microchip PCR. *Anal Bioanal Chem* 2003; 377 (5): 820–825.
93. Obeid PJ, Christopoulos TK, Crabtree HJ, Backhouse CJ. Microfabricated device for DNA and RNA amplification by continuous-flow polymerase chain reaction and reverse transcription-polymerase chain reaction with cycle number selection. *Anal Chem* 2003; 75 (2): 288–295.
94. Schmalzing D, Belenky A, Novotny MA, Koutny L, Salas-Solano O, El-Difrawy S, Adourian A, Matsudaira P, Ehrlich D. Microchip electrophoresis: a method for high-speed SNP detection. *Nucleic Acids Res* 2000; 28 (9): e43i–e43vi.
95. Anderson RC, Su X, Bogdan GJ, Fenton J. A miniature integrated device for automated multistep genetic assays. *Nucleic Acids Res* 2000; 28 (12): e60.
96. Hong JW, Quake SR. Integrated nanoliter systems. *Nat Biotechnol* 2003; 21 (10): 1179–1183.
97. Hong JW, Studer V, Hang G, Anderson WF, Quake SR. A nanoliter-scale nucleic acid processor with parallel architecture. *Nat Biotechnol* 2004; 22 (4): 435–439.
98. Lagally ET, Emrich CA, Mathies RA. Fully integrated PCR-capillary electrophoresis microsystem for DNA analysis. *Lab Chip* 2001; 1 (2): 102–107.
99. Paegel BM, Yeung SH, Mathies RA. Microchip bioprocessor for integrated nanovolume sample purification and DNA sequencing. *Anal Chem* 2002; 74 (19): 5092–5098.
100. Lee TM, Carles MC, Hsing IM. Microfabricated PCR-electrochemical device for simultaneous DNA amplification and detection. *Lab Chip* 2003; 3 (2): 100–105.
101. Rodriguez I, Lesaicherre M, Tie Y, Zou Q, Yu C, Singh J, Meng LT, Uppili S, Li SF, Gopalakrishnakone P, Selvanayagam ZE. Practical integration of polymerase chain reaction amplification and electrophoretic analysis in microfluidic devices for genetic analysis. *Electrophoresis* 2003; 24 (1–2): 172–178.
102. Schena M, Ed. DNA *Microarray Biochip Technology*. Eaton Publishing: Natick, MA, 2000; 297 pp.
103. Chittur SV. DNA microarrays: tools for the 21st century. *Comb. Chem. High Throughput Screen* 2004; 7 (6): 531–577.
104. Kricka LJ, Master SJ. Validation and quality control of protein microarray-based analytical methods, in *Microarrays in Clinical Diagnostics*, Joos TO, Fortina P, Eds. Human Press, Inc.: Totowa, NJ; ch 14.
105. Letowski J, Brousseau R, Masson L. Designing better probes: effect of probe size, mismatch position and number on hybridization in DNA oligonucleotide microarrays. *J Microbiol Methods* 2004; 57 (2): 269–278.
106. Lopez-Crapez E, Livache T, Marchand J, Grenier J. K-*ras* mutation detection by hybridization to a polypyrrole DNA chip. *Clin Chem* 2001; 47 (2): 186–194.

107. Bailey RC, Nam JM, Mirkin CA, Hupp JT. Real-time multicolor DNA detection with chemoresponsive diffraction gratings and nanoparticle probes. *J Am Chem Soc* 2003; 125 (44): 13541–13547.

108. Mazzola L. Commercializing nanotechnology. *Nat Biotechnol* 2003; 21 (10): 1137–1143.

109. Why small matters. *Nat Biotechnol* 2003; 21 (10): 1113.

110. Jain RK. Molecular regulation of vessel maturation. *Nat Med* 2003; 9 (6): 685–693.

111. Colvin VL. The potential environmental impact of engineered nanomaterials. *Nat Biotechnol* 2003; 21 (10): 1166–1170.

112. Csaki A, Moller R, Fritzsche W. Gold nanoparticles as novel label for DNA diagnostics. *Expert Rev Mol Diagn* 2002; 2 (2): 187–193.

113. Chen S, Webster S, Czerw R, Xu J, Carroll DL. Morphology effects on the optical properties of silver nanoparticles. *J Nanosci Nanotechnol* 2004; 4 (3): 254–259.

114. Jin R, Cao Y, Mirkin CA, Kelly KL, Schatz GC, Zheng JG. Photoinduced conversion of silver nanospheres to nanoprisms. *Science* 2001; 294 (5548): 1901–1903.

115. Cao Y, Jin R, Mirkin CA. DNA-modified core-shell Ag/Au nanoparticles. *J Am Chem Soc* 2001; 123 (32): 7961–7962.

116. Cai H, Zhu N, Jiang Y, He P, Fang Y. Cu–Au alloy nanoparticles as oligonucleotide labels for electrochemical stripping detection of DNA hybridization. *Biosens Bioelectron* 2003; 18 (11): 1311–1319.

117. Weizmann Y, Patolsky F, Willner I. Amplified detection of DNA and analysis of single-base mismatches by the catalyzed deposition of gold on Au nanoparticles. *Analyst* 2001; 126 (9): 1502–1504.

118. Zhu N, Cai H, He P, Fang Y. Tris(2,2′-bipyridyl)cobalt(III)-doped silica nanoparticle DNA probe for the electrochemical detection of DNA hybridization. *Anal Chim Acta* 2003; 481: 181–189.

119. Zhao X, Tapec-Dytioco R, Tan W. Ultrasensitive DNA detection using highly fluorescent bioconjugated nanoparticles. *J Am Chem Soc* 2003; 125 (38): 11474–11475.

120. Cao YC, Jin R, Mirkin CA. Nanoparticles with Raman spectroscopic fingerprints for DNA and RNA detection. *Science* 2002; 297 (5586): 1536–1540.

121. Dresselhaus MS, Dresselhaus G, Avouris P. Carbon Nanotubes: Synthesis, Structure, Properties and Applications. Springer-Verlag: Berlin. 2001; 477 pp.

122. Wang J, Liu G, Jan MR. Ultrasensitive electrical biosensing of proteins and DNA: carbon-nanotube derived amplification of the recognition and transduction events. *J Am Chem Soc* 2004; 126 (10): 3010–3011.

123. Kohli P, Harrell CC, Cao Z, Gasparac R, Tan W, Martin CR. DNA-functionalized nanotube membranes with single-base mismatch selectivity. *Science* 2004; 305 (5686): 984–986.

124. Deamer DW, Akeson M. Nanopores and nucleic acids: prospects for ultrarapid sequencing. *Trends Biotechnol* 2000; 18 (14): 147–151.

125. Li J, Gershow M, Stein D, Brandin E, Golovchenko JA. DNA molecules and configurations in a solid-state nanopore microscope. *Nat Mater* 2003; 2 (9): 611–615.

126. Austin R. Nanopores: the art of sucking spaghetti. *Nat Mater* 2003; 2 (9): 567–568.

127. Woolley AT, Guillemette C, Cheung CL, Housman DE, Lieber CM. Direct haplotyping of kilobase-size DNA using carbon nanotube probes. *Nat Biotechnol* 2000; 18 (7): 760–763.

128. Iwabuchii S, Mori T, Ogawa K, Sato K, Saito M, Morita Y, Ushiki T, Tamiya E. Atomic force microscope-based dissection of human metaphase chromosomes and high resolutional imaging by carbon nanotube tip. *Arch Histol Cytol* 2002; 65 (5): 473–479.

129. Dequeker E, Cuppens H, Dodge J, Estivill X, Goossens M, Pignatti PF et al. Recommendations for quality improvement in genetic testing for cystic fibrosis: European Concerted Action on Cystic Fibrosis. *Eur J Hum Genet* 2000; 8 (Suppl 2): S2–S24.

130. Dequeker E, Cassiman J-J. Genetic testing and quality control in diagnostic laboratories. *Nat Genet* 2000; 25 (3): 259–260.

131. Ferns GA, O'Dowd D, Wark G, Collins N. Molecular diagnostics in routine practice: quality issues and application to complex disease. *Ann Clin Biochem* 2003; 40 (Pt 4): 309–312.

132. Muller CR. Quality control in mutation analysis: European Molecular Genetics Quality Network (EMQN). *Eur J Pediatr* 2001; 160 (8): 464–467.

133. DNA Advisory Board. Quality assurance standards for forensic DNA testing laboratories. *Forensic Sci Commun* 2000; 2: 1–14.

134. Raggi CC, Pinzani P, Paradiso A, Pazzagli M, Orlando C. External quality assurance program for PCR amplification of genomic DNA: an Italian experience. *Clin Chem* 2003; 49 (5): 782–791.

135. Valentine-Thon E, van Loon AM, Schirm J, Reid J, Klapper PE, Cleator GM. European proficiency testing program for molecular detection and quantitation of hepatitis B virus DNA. *J Clin Microbiol* 2001; 39 (12): 4407–4412.

136. Saunders GC, Dukes J, Parkes HC, Cornett JH. Interlaboratory study on thermal cycler performance in controlled PCR and random amplified polymorphic DNA analyses. *Clin Chem* 2001; 47 (1): 47–55.

137. Corstjens PL, Zuiderwijk M, Nilsson M, Feindt H, Niedbala RS, Tanke H.J. Lateral-flow and up-converting phosphor reporters to detect single-stranded nucleic acids in a sandwich-hybridization assay. *Anal Biochem* 2003; 312 (2): 91–100.

138. Corstjens P, Zuiderwijk M, Brink A, Li S, Feindt H, Niedbala RS, Tanke H. Use of up-converting phosphor reporters in lateral-flow assays to detect specific nucleic acid sequences: a rapid, sensitive DNA test to identify human papillomavirus type 16 infection. *Clin Chem* 2001; 47 (10): 1885–1893.

.

17 Bridging the Gap between Analytical and Clinical Validation

Mel Krajden

CONTENTS

INTRODUCTION TO LABORATORY VALIDATION

This chapter will articulate the interplay between analytical and clinical validation from an infectious disease laboratory diagnostic perspective. The primary emphasis will be on validation components relating to qualitative or quantitative nucleic acid testing, although many of the concepts apply to serological or other analytical methods. Diagnostic laboratories play increasingly important roles in supporting modern health care delivery by establishing the presence of pathogens or illness in an individual, either through the direct detection of an agent or through the detection of a specific immune response.

To support individual diagnosis and care, laboratories typically confirm the presence or absence of a microorganism or monitor how the agent or pathogen responds to a therapeutic intervention. Through the analysis of population-based test results, laboratories may also play an important role in providing information for policy planning and public health response. More recently, the application of molecular tools to fingerprint microorganisms can be used to provide information on the similarity or heterogeneity of organisms and strains and thereby support epidemiological tracking or detection of outbreaks.[1-4] In order to assure users that test results are both accurate and relevant, laboratories must be able to provide comprehensive validation of the tests they perform.

In order to provide comprehensive test validation, laboratories must use processes to ensure that procedural elements have been followed and that test performance has been verified. Laboratory validation has been addressed by a number of regulatory bodies. As defined by the International Standards Organization (ISO), laboratory validation is the confirmation, through the provision of objective evidence, that requirements for a specific intended use or application have been fulfilled.[5] According to the World Health Organization (WHO), validation is the action or process of proving that a procedure, process, system, equipment, or method used works as expected and achieves the intended result.[6]

The Clinical and Laboratory Standards Institute (formerly the National Committee for Clinical Laboratory Standards) lists the components of the validation process to include quality control, proficiency testing, employee competency, instrument calibration, and correlation of laboratory test findings with clinical status.[7] The various components of laboratory and clinical validation include a complex and dynamic interdisciplinary process of learning how to best detect an analyte or target of interest and provide an estimate of the clinical utility of its detection.

Analytical and clinical validation is about process control and involves the identification of contributing factors that affect test accuracy, reproducibility, and interpretation. Laboratorians must recognize that the amount of process control required depends on the nature of the agent, the degree to which test reagents have been validated, the technological platform used for testing, the assurance that specific regulatory approvals have been addressed, and that evolving knowledge regarding test utility is incorporated into practice. For analytical or clinical conclusions to be considered reliable requires statistically verifiable and reproducible results and a quality assurance process capable of aligning the various process elements with emerging knowledge. This chapter will illustrate the complex interrelationship of laboratory and clinical validation by describing how the components of pre-analytical, analytical and post-analytical test validation help to ensure that laboratories provide accurate and effective diagnostic, monitoring, and epidemiological support.

Pre-analytical components typically represent specimen procurement aspects, including collection of an optimum specimen in a biologically and clinically relevant time frame and ensuring that the handling procedures preserve specimen integrity. The analytical components involve the processes of detecting and measuring the analyte or target of interest and include ancillary quality control procedures for assuring the results are valid and accurate.

Post-analytical components are concerned with the way the test information impacts on the clinical management of the individual or how the information affects broader policy decisions. Post-analytical components involve the timeliness and mechanisms of reporting as well as how test results are interpreted and acted upon. The more information conveyed to the user regarding the analytical and clinical performance of a test, the more likely that the appropriate clinical or policy decisions will be made.

PRE-ANALYTICAL COMPONENTS

As outlined above, the pre-analytical components must ensure that the appropriate type of specimen is collected in a biologically relevant time frame and the handling procedures preserve specimen integrity. Understanding the biological behavior of an agent in the host is important because this information can help define the best type of specimen and the time frame after infection to optimize detection of the pathogen. For example, testing for hepatitis C virus RNA within the first few days after infection may lead to a false negative test result, as typically 1 to 2 weeks are required before viral RNA is detectable.[8-10] In contrast, serological tests require 5 or more weeks until an immunocompetent host is able to produce an antibody response.[9-11]

The optimum specimen for detection is also affected by the biological behavior of the agent. For example, hepatitis C is a hepatotrophic virus that replicates predominantly in the liver, and the amount of viral nucleic acid detected in hepatic tissue is substantially higher than that found in blood.[12–15] Nonetheless, measurement of hepatitis C RNA viral load in plasma or serum has been shown to be a suitable surrogate to confirm active infection and to provide information on the relative amount of viral replication in the host. Successful pharmacological treatment of hepatitis C typically leads to clearance of detectable hepatitis C virus RNA from blood and from hepatic tissue.[14,16] For clinical purposes, testing for hepatitis C virus RNA in blood samples is easier and safer than obtaining liver biopsy tissue and therefore the most practical specimen to confirm active infection. For other pathogenic agents such as influenza virus, respiratory fluids are required to optimize detection of infection.[17–20]

Once a sample is obtained, it is important to ensure that the analyte or target of interest is stable and that specimen integrity is maintained during handling and transport. Numerous publications have highlighted the fact that anticoagulants, storage times, and transport times affect specimen integrity.[21–27] Inappropriately handled samples may produce inaccurate analyte or target quantification or yield false negative test results and therefore lead to inappropriate care management.[28–31]

Pre-analytical validation components also include the ability to verify the time of collection, handling conditions, storage temperatures, and time of receipt of the specimen in the laboratory. Similarly, the handling and storage conditions of extraction and other test reagents must be tracked and monitored to ensure that their composition, concentration, and function are maintained. Other important validation components include documentation that personnel have been adequately trained and that automated equipment is correctly calibrated and functioning appropriately. For more information on pre-analytical components of test validation, the reader is referred to relevant ISO, WHO and Clinical and Laboratory Standards Institute documents.[5–7]

ANALYTICAL COMPONENTS

In the most simplistic sense, accurate detection or quantification of an analyte or target requires an assay capable of generating a reproducible relationship between the amount of input target in the sample and the output signal. Design and development of reproducible assays depend on the use of standards, reference materials, and calibrators that contain known amounts of the analytes or targets. Increasingly these assays are performed and the output signals are assessed by highly automated and technologically complex equipment.

This section will demonstrate how standards, reference materials, and calibrators are required to ensure inter- and intra-laboratory analytical test accuracy and reproducibility. This will be followed by an outline of critical assay parameters and definitions that will provide insights into how the various components of assay design may affect assay performance.

Analytical Standards

An analytical standard is a measured amount of analyte or modified analyte used to ensure that the constituents of the analytical process function as expected. Since standards are complex to create or define, availability is typically driven by demonstrated commercial or clinical needs.[32,33] For many analytes or pathogens, reliable standards may not be available and laboratories must try to develop their own analytical standards by addressing the challenges discussed below.

Naturally derived analytical standards involve the use of known quantities of viral particles, bacteria, and other materials that contain genomic RNA and/or DNA or plasmids and have undergone extensive physical or biochemical quantification. Synthetic standards typically involve the use of oligonucleotides that have been synthesized *in vitro*, the use of recombinant phage to generate single-stranded DNA, cloning of the target into plasmids or unrelated organisms, or the generation of recombinant RNA transcripts. The derived nucleic acid is then quantified by physical or bio-chemical methods. It must be recognized that synthetic standards may not necessarily perform analytically as standards derived directly from a pathogen.[34] Further studies are typically required to assure that the analytical standard performs as expected. Ideally a synthetic or naturally derived analytical standard should be diluted in a matrix that mimics as closely as possible the physical and chemical behavior of the host's body fluid (e.g., serum, plasma, cerebrospinal fluid).

The literature frequently refers to a gold standard in the context of an assay or methodology (e.g., viral culture) that serves as the accepted standard for identifying the presence or absence of an agent or target. A gold standard should not be confused with a laboratory assay analytical standard, which is a known quantity of an analyte or target that has been verified through multiple physical or biochemical methods. As analytical standards contain precisely measured amounts of an analyte or target, their availability and consistent use by laboratories facilitates intra- and inter-assay performance monitoring and is very useful when systematic methodological assessments must be performed.

Reference Materials

As noted above, analytical standards are critical for systematic inter- and intra-assay performance assessment. Through the use of these standards, different laboratories can "tune" their assays to generate comparable results when analyzing the same specimen and ensure consistent lot-to-lot assay performance. While manufacturers may have resources to produce their own characterized standards, these standards typically have not been shared with other manufacturers or non-commercial laboratories.[35] This practice has limited the abilities of different manufacturers and non-commercial laboratories to standardize assay performance across different technical platforms. In response, WHO has played a key role in defining internationally agreed-upon reference materials[33,36-38] and the ready availability of these reference materials supports the development and standardization of analytical standards.[33,35,36,38]

The selection of reference materials requires consideration for known genetic variants, mutants, genotypes, and quasi-species in order to minimize the risk of developing a reference material that is not representative of the range of circulating agents found in the population. Poor selection of a reference material can adversely impact its suitability to standardize assays necessary for accurate detection or quantification. Reference materials are typically quantified in terms of international units (IU) per unit volume and are available to manufacturers and noncommercial laboratories.[36] It must be recognized that once these international reference materials are developed, additional studies are still required to ensure that the materials are sufficiently representative of circulating variants, mutants involving genetically similar genotypes, subtypes, etc.[34,39-43]

Calibrators

A calibrator involves a measured amount of analyte that can be co-amplified with the target sequence (i.e., an internal calibrator) or can be run in parallel as an external calibrator to generate an external calibration curve.[40,44-48] In both cases, the signal generated from the specimen is compared to one or more of the calibrators. For target amplification tests, an internal calibrator can be constructed of nucleic acid that uses the same binding primer site as the target of interest but has a deletion, insertion, or heterogeneous sequence that is internal to the primer binding site.[44,46-48] This allows the calibrator to be distinguished from the target of interest when amplified. Through the use of a known input concentration of the calibrator and measuring the output signals obtained from both the calibrator and the analyte or target of interest after amplification, one can accurately calculate the amount of input analyte or target present in the specimen prior to amplification.

The advantage of using an internal calibrator results from its addition at the time of specimen extraction so that it undergoes the same extraction, amplification, and detection processes as the specimen itself and therefore can help gauge whether the extraction, amplification, and detection processes were adequate or whether the sample may have contained inhibitory substances. It is important to limit the amount of internal calibrator used in a reaction well because amplification of the internal calibrator may compete for reagents in the sample well. This might affect the amplification efficiency of the analyte or target and thereby reduce overall test sensitivity. The degree to which amplification of the internal calibrator affects test sensitivity must be assessed by the laboratory or the manufacturer.

An external calibrator contains a measured amount of target or modified target; it is tested in the same run as the specimen but not in the same tube or well. Different amounts of target can be run to generate an external calibration curve. The signal results obtained from the specimens can then be compared to those generated by the external calibrators.[40,45] Since external calibrators are not present in the sample reaction well, they do not compete with the target for assay substrates and therefore do not modify the detection of the analyte or target. However, interfering substances or other factors, such as a mispipetting of reagents in a reaction well, that may affect sample well reaction efficiency can alter the detection or quantification of the analyte or target despite the fact that the external calibrator performs or is measured within range. Thus, one limitation of external calibrators is that the actual amplification performance in the individual sample well is not directly measured.

Another approach to better assess amplification or detection performance within a given specimen involves the detection of single copy host cellular genes such as beta-globin or actin.[44,49] Detection of host cellular genes can be used to verify a number of analytical components of an assay such as the sufficiency of cellular material in the sample (a measure of sample adequacy) and appropriate functioning of the amplification or detection reagents. Detection of a housekeeping gene does not verify whether an analyte or target of interest was present in the specimen because the primers used to amplify the target are different. Nor does detection of a housekeeping gene verify that an assay quantified the analyte accurately. As discussed above, the use of an internal control or calibrator would typically be required to ensure accurate quantification of an analyte.

Another important strategy involves the quantification of a host gene or multiple host genes that can be used as a reference to normalize the expression of the host gene or genes relative to other analytes or targets in the same sample.[49] This permits a laboratory to assess the amount of expression of the analyte or target relative, for example, to host cellular gene activation or suppression. Co-amplification of host genes, especially in the presence of a strong inflammatory response, for example, with an excess of white blood cells, may swamp or affect the quantitative detection of the analyte or target of interest because host gene co-amplification, like co-amplification of an internal calibrator, leads to competition for reagents in the sample well. Because plasma and serum typically do not contain sufficient quantities of residual cellular material, detection of host gene expression is limited to samples that contain sufficient cellular material.

DEFINITIONS AND CONCEPTS RELATING TO ANALYTICAL ASSAY COMPONENTS

THEORETICAL ASSAY CONCEPTS

As discussed earlier, accurate detection or quantification of an analyte or target requires an assay capable of generating a reproducible relationship between the amount of input target in the sample and the output signal. On a theoretical basis, as the concentration of the analyte or target increases, the signal produced by the assay will typically follow a sigmoid-shaped curve (Figure 17.1). Samples containing concentrations below the limit of detection of the assay will yield signals similar to the background noise of the assay. As the amount of input target in the sample is increased, a linear or near-linear signal response occurs over the assay's dynamic range or range of quantification.

Theoretical Detection Assay

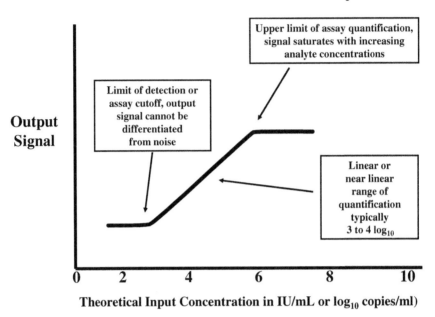

FIGURE 17.1 Relationship between increasing input concentrations of an analyte or target and the output signal in a theoretical detection assay.

The analytical lower limit of quantification is defined by the concentration at which the analyte or target can be detected with acceptable precision and accuracy. When the upper limit of quantification is reached, the signal saturates. Most conventional assays have a dynamic range of about 3 to 4 $\log_{10}$.[40,45,50] Modern real-time nucleic acid amplification assays can have dynamic ranges between 5 to 7 $\log_{10}$ or more.[39,51–53] Samples containing concentrations beyond the upper limit of the quantification range should be diluted and retested for accurate quantification.[50,54]

Many analytical factors are important for relating the input concentration to the output signal of a given assay. Table 17.1 provides a simplistic summary of the major assay parameters required to achieve analytically correct measurements of an analyte or target. The analytical accuracy of an assay assesses the degree of agreement between the measured analyte or target and the true value of that analyte.[7,55] Analytical accuracy is typically assessed through comparison of a new method against an established test or a different type of test (see section titled "Analytical Sensitivity and Specificity versus Clinical or Diagnostic Sensitivity and Specificity").

Precision refers to the agreement of independent test results under defined conditions. Intra-assay precision refers to test reproducibility in the same analytical run. Inter-assay precision concerns test reproducibility among different runs. Precision among days, sites, lots, and batches, can also be assessed. Table 17.1 also includes definitions relating to analytical sensitivity[7,55] including analytical limit of detection, analytical limit of quantification, and clinical reportable range. Definitions of analytical specificity, clinical or diagnostic sensitivity, and clinical or diagnostic specificity are included in Table 17.1 and will also be addressed in the "Analytical Sensitivity and Specificity versus Clinical or Diagnostic Sensitivity and Specificity" section.

To ensure that an assay generates consistent and accurate results within runs, between runs, and in different testing laboratories, it is necessary to verify the relationship of the input concentration and the output signal through the use of internal or external calibrators as outlined previously. As noted earlier, the availability of standardized reference materials has dramatically improved the ability of different laboratories and manufacturers to develop and calibrate their assays.[33,36] Only

TABLE 17.1
Analytical and Diagnostic Assay Definitions

Term	Definition
Accuracy	How closely test method result compares with true (accepted reference) value of substance or analyte subjected to measurement; analytical accuracy may not be identical to clinical accuracy (see clinical or diagnostic sensitivity and positive predictive values below)
Precision	Reproducibility or repeatability of result produced by test method; precision is determined under stipulated conditions, e.g, within-run, -day, -site, -batch, -lot
Analytical sensitivity	Smallest amount of analyte that can be reproducibly detected by a test
Analytical limit of detection	Lowest concentration of analyte that can be consistently detected (typically, in 95% of samples tested under routine clinical laboratory conditions)
Analytical limit of quantification or analytic measurement range	Lowest and highest concentrations of analyte that can be quantified with acceptable precision and accuracy under routine clinical laboratory conditions, i.e., without dilution, concentration, or other pretreatment
Clinical reportable range	Range of analyte values that a test method can measure, allowing for specimen dilution, concentration, or other pretreatment to extend analytical limit of quantification or measurement range
Analytical specificity	Ability of a test or procedure to correctly identify or quantify the analyte or substance of interest, i.e., to yield either a negative result or a result below a given threshold when the analyte or substance is absent
Clinical or diagnostic sensitivity	Proportion of subjects with a specified clinical disorder whose test results are positive or above a defined threshold when the clinical disorder or disease is present
Clinical or diagnostic specificity	Proportion of subjects who do not have a specific clinical disorder whose tests results are negative or below a defined threshold
Positive predictive value	Proportion of true positives divided by the sum of true positives and false positives identified when a test is applied to a given population
Negative predictive value	Proportion of true negatives divided by the sum of true negatives and false negatives identified when a test is applied to a given population

through the shared use of reference materials can manufacturers and independent laboratories develop assays capable of generating reproducible and accurate test results that are relatively independent of technology platforms.

To illustrate the importance of understanding the laboratory and clinical implications of an assay's lower limit of detection, dynamic range or range of quantification, and upper limit of quantification, the detection and quantification of hepatitis C virus RNA will be used as an example. Figure 17.2 demonstrates the range of concentrations of hepatitis C RNA in $\log_{10}$ IU/ml found in a population of 369 individuals undergoing baseline quantitative hepatitis C RNA testing prior to receiving pegylated interferon and ribavirin treatment. Individuals in the population had hepatitis C virus RNA loads ranging from 2 to 7 $\log_{10}$ IU/mL in their plasma. However, the quantitative COBAS AMPLICOR HCV MONITOR v2.0 assay (Roche Diagnostics, http://www.roche-diagnostics.com) has a dynamic range varying from 600 to 850,000 IU/mL, 600 to 500,000 IU/mL, and more recently 600 to 700,000 IU/mL (Figure 17.3a).[11,50] To detect an individual whose hepatitis C virus RNA viral load lies below the lower limit of quantification or the quantification range of this assay (i.e., below 600 IU/mL) requires the use of a more sensitive test such as the qualitative COBAS AMPLICOR HCV Test v2.0 that has a lower limit of detection (around 50 IU/mL) or the

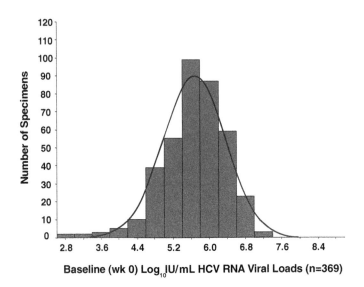

Baseline (wk 0) Log$_{10}$ IU/mL HCV RNA Viral Loads (n=369)

FIGURE 17.2 Frequency distribution of baseline Log$_{10}$ IU/mL of hepatitis C virus RNA viral loads in patients undergoing combination pegylated interferon and ribavirin therapy.

Versant HCV RNA qualitative assay (Bayer Healthcare, http://www.bayerhealthcare.com) with a lower limit of detection of approximately 10 IU/mL.[56–58]

From a clinical perspective, knowing the information highlighted above, it is clear that for diagnostic purposes the more sensitive commercial qualitative hepatitis C virus assays should be used to confirm active infection. For example, in chronic hepatitis C virus-infected treatment, naïve individuals, typically 3% or fewer, have low hepatitis C RNA viral loads.[59] If testing was performed using the currently less sensitive commercial quantitative assays, individuals with low viral loads would be missed, yielding false negative test results.

An analytically incorrect result can also occur when testing individuals whose hepatitis C virus RNA concentrations lie beyond the upper limit of the dynamic range or range of quantification of the assay. Specimens that contain hepatitis C virus RNA concentrations beyond the range of quantification of the assay saturate the signal output and require dilution and retesting for accurate quantification to take place.

Based on the range of viral load distributions outlined in Figure 17.2, approximately 45 to 55% of samples would need to be diluted and retested when tested with the current COBAS AMPLICOR HCV MONITOR v2.0. If dilution and retesting are not done, samples with high concentrations of hepatitis C virus RNA will not be correctly quantified. As illustrated in Figure 17.3b, a greater number of higher viral load samples would fall within the quantification range of the Versant HCV RNA 3.0 assay (bDNA; 615 to 7,700,000 IU/ml), which would reduce the number of samples requiring dilution and retesting.[41] However, qualitative nucleic acid testing would still be needed to detect specimens with low viral loads.[59]

With the availability of real-time amplification technologies, the dynamic range or range of quantification of assays is now approaching 4 to 7 log$_{10}$ or more.[39,50–53,60] The use of real-time PCR assays such as TaqMan (Applied Biosystems, Foster City, California, USA, http://www2.applied-biosystems.com) will begin to allow reliable quantification almost over the complete range of concentrations detected in individuals infected with hepatitis C virus and other pathogens (Figure 17.3c). These points illustrate how understanding the principles of lower limit of detection, dynamic range or range of quantification, and upper limit quantification of assay are necessary to accurately detect or quantify the amount of nucleic acid in a specimen. This information is also a critical prerequisite for generating effective laboratory and clinical correlations as outlined below.

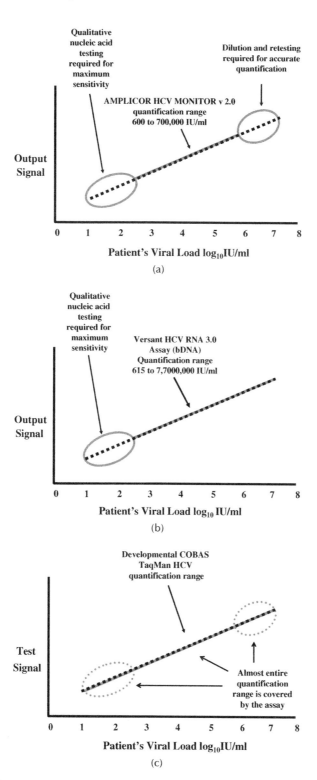

FIGURE 17.3 Relationship between range of a patient's theoretical hepatitis C virus RNA viral load and the abilities of different assays to detect or quantify hepatitis C RNA.

Why Accurate Quantification of Analytes or Targets Is Important

This section will use the quantification and detection of hepatitis C virus RNA to illustrate how critical accurate laboratory detection and quantification are for proper patient management. At present, approximately 55% of hepatitis C virus-infected individuals can be virologically cured of their infection by combination pegylated interferon and ribavirin treatment.[14,61,62] For individuals infected with hepatitis C virus genotype 1, cure rates of approximately 45% are achieved with 48 weeks of therapy. A virological cure is unlikely to occur unless the HCV RNA viral load decreases by at least 2 $\log_{10}$ or 100-fold after 12 weeks of treatment.[61,63,64] As a result, monitoring for a treatment-induced reduction in HCV RNA viral load by 12 weeks is now used to identify the 20 to 25% of individuals who are early virological nonresponders.

Early identification of nonresponders has critical clinical importance because the pegylated interferon and ribavirin combination is expensive and poorly tolerated. It has been shown that only >3% of individuals who have not responded by 12 weeks will clear the infection with additional therapy. By individualizing the treatment course and eliminating ineffective therapy, unnecessary side effects are prevented and costs are substantially reduced. In contrast, about 75 to 80% of individuals infected with hepatitis C virus genotypes 2 or 3 will be virologically cured with 24 weeks of treatment. Most individuals, approximately 97%, will demonstrate early virological responses. Thus performing viral load testing for individuals infected with genotypes 2 and 3 is not as relevant for patient management. More recently, however, it has been shown that early virological response can be used to guide shorter courses of treatment to further reduce unnecessary costs and side effects.[65–68]

The previous section outlines how laboratory testing can directly support the clinical management of hepatitis C virus infection. Consistent test accuracy and reproducibility are clearly important elements of a laboratory's role, but as the costs of specific tests and technologies continue to be scrutinized, clinical utility must be effectively documented.[69–71] In order to assure that the correct test is performed at the correct time, a comprehensive total quality management system must be in place to constantly refine the diagnostic and monitoring algorithms as new knowledge is generated. Total quality management will be addressed later in the chapter. As outlined below, other factors may influence test accuracy and reproducibility and have an important effect on the generation of valid test results.

Interfering Substances

Another component of analytical validation is assessing whether *endogenous* or *exogenous* substances may interfere with the detection or quantification of a given analyte or target. A number of approaches may be used to assess the impacts of potential interfering substances on assay performance. The first approach involves the addition of a known amount of analyte or target to a sample that may contain interfering substances, e.g., heparinized blood.[25] The addition of a known amount of the analyte allows one to determine whether the detection of the analyte or internal calibrator is affected by the presence of the *exogenous* potential interfering substance.

A second approach involves obtaining specimens from one or more persons with a particular disease or disorder, such as hyperlipidemia or end stage renal failure, that may contain one or more *endogenous* interfering substances. The laboratorian then spikes the samples with known amounts of the analyte or target and determines whether an *endogenous* substance that may accompany a particular clinical illness interferes with analyte detection or quantification. One disadvantage of using clinical samples is that a sample might contain more than one interfering substance that could confound the analysis and assessment.

To address this possibility, a third approach involves the addition of known quantities of purified substances that might potentially interfere with the detection of the analyte or target. For example, known quantities of bilirubin, triglycerides, hemoglobin, therapeutic drugs, and anticoagulants such as heparin or EDTA can be added to a clinical sample or to an artificial matrix and the impact of the addition on the detection and quantification of the analyte or target of interest can be assessed.[7,41]

By performing these types of analyses, a laboratory can help assure that the detection or quantification of an analyte is not affected by common and potentially interfering exogenous or endogenous substances. It must also be recognized that it is impossible to test every specific drug or compound for potential assay interference. Typically, the commonly encountered substances are evaluated. It should be noted that other sources of inhibitory substances can include residual detergents, extraction reagents, and inhibitory materials found in the plastics used for specimen procurement or analysis.[44]

SAMPLE CONTAMINATION OR CARRY-OVER

Many pathogens produce high concentrations of nucleic acid that are frequently detected in blood during the course of acute or chronic infection of a host.[15,40,72,73] This, coupled with the fact that modern nucleic acid amplification assays (e.g., PCR or signal amplification) are capable of amplifying analytes, targets, and/or signals by 10^6- fold or greater, has important repercussions for the risk of cross-contamination of specimens before or during testing.

Laboratories performing nucleic acid detection and quantification must pay particular attention to ensure that they use procedures to minimize the risk of specimen cross-contamination. An example is droplet aerosolization from an adjacent positive specimen cross-contaminating a negative specimen. Alternatively, amplicon products from a reactive specimen can also contaminate a negative specimen and yield a false positive test result.

Errors can occur during the pre-analytical phase at the time of sample aliquoting or during the analytical phase if contaminated reagents, pipette tips, or other laboratory equipment are used. Avoidance of contamination can be achieved through rigorous techniques that can in many cases be automated. Amplicon contamination of PCR products can also be minimized by the incorporation of uracil-containing amplification products, followed by pre-digestion of reaction mixes with uracil-N-glycosylase.[44] The use of closed system, real-time nucleic acid amplification assays such as the TaqMan can minimize the risk of amplicon contamination if the sample reaction well remains closed after amplification. As expected, specimen carry-over or cross-contamination can have an important impact on assay specificity.

It is also important to be able to distinguish poor specificity due to cross-contamination or carry-over from an intrinsic assay design flaw. For example, if an assay used a non-specific probe, it could falsely identify the presence of an analyte or target in a sample because the probe may have cross-reacted with another substance in the specimen. This represents a problem with analytical specificity which is different from a technical cause of reduced specificity that relates to specimen carry-over or splashes between reaction wells. Improvements in design methodology, such as the use of closed systems, can reduce the risk of technical errors that can cause false positive test results. They will not address flaws in probe design or poorly designed detection instruments. It is important for laboratorians to understand these two different sources of errors so that they can use a systematic approach to process redesign to improve test specificity if possible.

ERRORS DUE TO SPECIMEN MISLABELING

Specimen mislabeling and the processing and analysis of an incorrect sample are important issues for clinical diagnostic laboratories. The exact frequency of these types of errors is difficult to quantify, but the more automated the procurement and analytical processes are, the less likely are these types of errors to occur. In some cases, labeling errors may be difficult to distinguish from errors due to specimen cross-contamination. For example, rates of cross-contamination for *Mycobacterium tuberculosis* have been shown to vary from 0.5 to 3%.[74-77] My experience at a large testing facility suggests that mislabeling errors can occur approximately 0.1 to 1% of the time.

Detection of mislabeling and sample handling errors is difficult to identify unless a previously negative result becomes positive, a previously positive result becomes negative, or a major discrepancy between clinical diagnosis and laboratory test result becomes apparent. The prevalence rate of a pathogen in a population affects the probability of identifying a mislabeling or sample testing error. For example, when testing a population that is 99% anti-HIV negative, an error in an anti-HIV negative result will not be detected until a false positive test result is produced and recognized. This is statistically infrequent in low prevalence populations because an error in a negative result is likely to remain undetected in a situation where most test results are negative. Similarly, because 90 to 95% of women undergoing prenatal rubella antibody tests are seropositive, a labeling error might not be detected until a discordant test result is noted. In this regard, the availability of longitudinal data on an individual can help laboratories and clinicians detect aberrant test results that may arise from testing the wrong specimen.

MOLECULAR FINGERPRINTING OF MICROORGANISMS

An evolving role for some laboratories includes the application of molecular tools to assist in epidemiological tracking of outbreaks by molecular "fingerprinting" of microorganisms. A variety of molecular tools have been adapted to assess the similarities and heterogeneities of organisms and strains.[1-4] While molecular typing includes distinguishing different *genera,* the same approaches can be used to fingerprint epidemiologically related organisms that share detectable characteristics that are different from control organisms. The ability of a typing method to distinguish epidemiologically related from unrelated isolates is referred to as its discriminatory power.

As outlined by Harrington and Bishai,[4] many different molecular tools have been used for typing and subtyping microorganisms. However, the widespread availability of PCR and automated sequencing equipment has dramatically increased the ability to identify and fingerprint microorganisms in a timely fashion using sequence-based procedures. The power of sequence-based molecular fingerprinting comes from its applicability to all organisms and strains; its ability to accurately and statistically quantify sequence variations or differences; the fact that closely related organisms and strains can be distinguished and distinct organisms and strains grouped; the sequences can be stored in local databases and readily compared to the results obtained by other laboratories worldwide; appropriate software is readily available; the cost is relatively affordable; and the turn-around time is fast.

Figure 17.4 demonstrates the sequence similarities of a number of hepatitis C virus strains based on sequencing the hepatitis C virus E1–E2 hypervariable region. The figure illustrates how related strains can be epidemiologically linked on the basis of their hypervariable sequence similarity. For example, a set of specimens labeled CBS (Canadian Blood Services) ST6, ST7, ST9, and ST10 represent duplicate sequence results of a blood donor and the probable source of the infection. These epidemiologically linked specimens demonstrated a 99% nucleotide sequence homology to each other and the sequences were substantially different from other CBS control specimens from the same geographic region that are also included in the figure. This sequence-based information was used to molecularly confirm the transmission to the blood donor from the suspected source — which was consistent with the epidemiological history. These results were reported elsewhere using a different primer set.[78]

Figure 17.4 also illustrates specimens 1880-1a-B.M. (13 years) and 1881–1a-J.K (17 years), both of whom were shown to have been infected at birth by the same mother. Despite their different ages, the sequence homology in the HCV hypervariable region was highly conserved. This figure also displays almost identical sequence homology between a mother and child (MD 01016 and MD 01017).

This section is not intended to focus on the epidemiology of the hepatitis C virus. It is intended to show that effective linkage between clinical and epidemiological information is an absolute requirement for correctly interpreting the implications of a microorganism's fingerprint pattern.

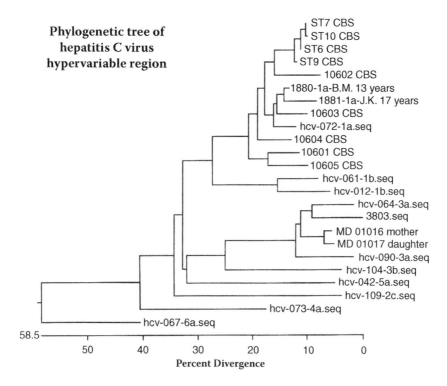

Phylogenetic tree of hepatitis C virus hypervariable region

ST7 CBS
ST10 CBS
ST6 CBS
ST9 CBS
10602 CBS
1880-1a-B.M. 13 years
1881-1a-J.K. 17 years
10603 CBS
hcv-072-1a.seq
10604 CBS
10601 CBS
10605 CBS
hcv-061-1b.seq
hcv-012-1b.seq
hcv-064-3a.seq
3803.seq
MD 01016 mother
MD 01017 daughter
hcv-090-3a.seq
hcv-104-3b.seq
hcv-042-5a.seq
hcv-109-2c.seq
hcv-073-4a.seq
hcv-067-6a.seq

58.5

50   40   30   20   10   0

Percent Divergence

FIGURE 17.4 Phylogenetic tree of hepatitis C virus hypervariable region sequences demonstrating percent divergence among different hepatitis C virus isolates.

Cases cannot be linked effectively without controls and effective molecular fingerprinting requires an in-depth understanding of the technology used. For example, direct PCR product sequencing typically identifies the predominant sequence in a specimen and may obscure the detection of low frequency mixed infections that may require cloning and subsequent sequencing.[1–4]

Perhaps the most important element of this section is emphasizing the importance of effective laboratory validation, without which the accuracy and reproducibility of a molecular fingerprint can be questioned, and a test's discriminatory power potentially diminished. A laboratory must ensure that cross-contamination of specimens or amplicon contamination has not led to a false positive test result. In addition, epidemiologically unrelated controls must be available and accurately assessed to allow for accurate data analysis.

An important question for the laboratorian is how much validation is necessary. For example, how many specimens must be tested to validate an assay, how many controls are needed, and how much validation should occur on an ongoing basis to ensure that test results continue to be valid? Unfortunately, no simple answers exist because the number of tests required to provide a statistically valid result depends on the technology used, the nature of the analyte or target in question, the assay design, how critical the result is, and the resources available to the laboratory. When performing forensic testing, an error can have catastrophic consequences and hence every effort must be made to assure that all the component processes are valid and that expected variations that can occur in control specimens are both recognized and properly interpreted.

POST-ANALYTICAL COMPONENTS

Post-analytical test components cover a broad array of issues relating to how test information impacts on the clinical management of an affected individual and how knowledge of test results from a population influences public health policy. It can also include the timeliness and mechanisms of reporting as well as how the test results are interpreted or acted upon. A challenging question is to define where the analytical components intersect with the post-analytical issues. The following section will illustrate how difficult it can be to separate the analytical components from the clinical and diagnostic components of a test result.

ANALYTICAL SENSITIVITY AND SPECIFICITY VERSUS CLINICAL OR DIAGNOSTIC SENSITIVITY AND SPECIFICITY

Table 17.2 outlines how individuals might be classified diagnostically by a gold standard test.[79,80] Test effectiveness is typically reported in terms of sensitivity, specificity, and positive and negative predictive values:

$$\text{Sensitivity} = \text{True Positive}/(\text{True Positive} + \text{False Negative})$$

$$\text{Specificity} = \text{True Negative}/(\text{True Negative} + \text{False Positive})$$

$$\text{Positive Predictive Value} = \text{True Positive}/(\text{True Positive} + \text{False Positive})$$

$$\text{Negative Predictive Value} = \text{True Negative}/(\text{True Negative} + \text{False Negative})$$

When compared to a gold standard, a perfect test would demonstrate 100% sensitivity, specificity, and predictive values. However, in the "real world," gold standard tests and diagnoses are moving targets.

With the adoption of increasingly more sensitive tests involving nucleic acid amplification and other analytically sensitive technologies, articulating the true meaning of a test's sensitivity, specificity, and predictive values in the context of real world scenarios is very complex. For example, while the sensitivity and specificity of a test are regarded as independent of the prevalence of a disease, the positive or negative predictive values vary substantially with disease prevalence. When testing a low prevalence population, the impact of using a test with limited specificity magnifies the false identification of individuals as infected with an agent. Furthermore, in a statistician's desire to succinctly define the ability of a test to identify an illness or specified clinical disorder, the test sensitivity and specificity terms are often misused.

Both laboratorians and clinicians must understand the significant difference between the analytical sensitivity and the clinical or diagnostic sensitivity of a test. The analytical sensitivity is the smallest amount of analyte that can be reproducibly detected by a test. This is distinctly different from clinical or diagnostic sensitivity that is generally considered to reflect the ability of a test to correctly identify individuals who have an illness or specified clinical disorder.[81]

TABLE 17.2
Parameters Used to Assess Assay Sensitivity and Specificity

		Gold Standard Diagnosis		
		Positive	Negative	Total
Test	Positive	True positive	False positive	True positive + false positive
Result	Negative	False negative	True negative	False negative + true negative
	Total	True positive + false negative	False positive + true negative	N = true positive + false positive + false negative + true negative

Analytical specificity is the ability of a test to accurately distinguish the analyte of interest from other substances in the sample. The clinical or diagnostic specificity is the ability of a test to identify people who do not have an illness or specified clinical disorder (who are correctly identified as negative by testing). Unfortunately, many publications do not clearly describe or articulate the difference between analytical versus clinical sensitivity and specificity. As a result, the implications of a positive or negative test for diagnosis and management are not clearly conveyed.[69,79,80]

The importance of distinguishing analytical and clinical sensitivity and specificity can be illustrated by two viral illnesses. The first example is a short-lived infection associated with influenza virus that is not known to result in a chronic infectious state. Individuals typically manifest symptoms only during the active phase of infection and the virus is typically cleared by the host response or in rare circumstances causes overwhelming infection resulting in death.[17–20]

From a clinical perspective, influenza-like symptoms are not easily distinguishable from other respiratory viral infections and therefore specific tests are required for accurate clinical diagnosis. In the case of influenza virus infection, a test with maximum analytical sensitivity and specificity would serve to enhance the ability to detect active infection during the early incubation period and better define the phase of nasopharyngeal shedding.[17–20] Using a test with enhanced sensitivity can change our knowledge of the natural history of an infection by identifying infected individuals earlier during their incubation phase or by providing more information on the duration of viral shedding. Therefore, for an agent such as influenza virus, the congruence is clear: using a test with greater analytical sensitivity also serves to enhance the clinical sensitivity.[17–20]

In contrast, the biological behaviors of other pathogens may not yield congruence between enhanced analytical sensitivity and clinical or diagnostic sensitivity. Examples include infection with cytomegalovirus, Epstein–Barr virus, and human papillomavirus. Each of these viruses may cause a spectrum of asymptomatic to symptomatic illness, but once the infection is controlled by the immune system, the virus remains in a dormant or latent state or perhaps replicates at subclinical levels under the control of the host's immune response.[72,82,83]

High levels of virus replication are more likely to be associated with acute infection or with symptoms, but as noted above, all infected individuals are expected to have latent or subclinical virus present.[72,82,83] Therefore, the clinical manifestations and the amount of viral replication are dependent in part on when the person was infected, his or her age, and underlying immunological status.

The biological behaviors of agents such as cytomegalovirus, Epstein–Barr virus, and human papillomavirus pose important challenges for highly sensitive analytical assays, because using a test with maximum analytical sensitivity may confound the clinical or diagnostic sensitivity if the test is so sensitive that latent or subclinical infections are identified. In this case, detection of latent infection provides an analytically correct result, but the test has limited or poor clinical specificity because the person may not manifest clinical illness. A test with 100% analytical sensitivity actually provides poor clinical specificity by effectively producing a clinical false positive test result. Such tests may, however, be useful for identifying early acute infections where maximum sensitivity may be important.

One approach to reconciling the relationship between the analytical sensitivity of a test and its diagnostic or clinical specificity involves the use of receiver operating characteristic curve analysis.[80] This approach has been especially useful for defining clinically relevant thresholds or cut-offs that balance analytical sensitivity with the clinical or diagnostic specificity of a test. While receiver operating characteristic curve analysis is relevant for all diagnostic tests, such analysis is particularly relevant for quantitative analytical tests involving a need to correlate how increases in microbial load due, for example, to cytomegalovirus viremia, can be correlated with the probability of clinical cytomegalovirus disease.[84–86]

Figure 17.5 outlines a theoretical receiver operating characteristic curve. On the y axis is plotted the *analytical* sensitivity in percent. The rate of false positivity or the percent reduction in clinical or *diagnostic* specificity (1 – clinical specificity) is plotted on the x axis. A perfect test would yield

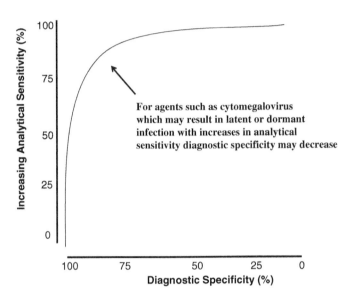

FIGURE 17.5 Theoretical receiver operator characteristic curve outlining how increasing analytical sensitivity may reduce diagnostic specificity.

100% sensitivity and specificity from an analytical perspective, but as the analytical sensitivity increases, the clinical or diagnostic specificity decreases, especially for illnesses in which latent infection can be present without clinical illness. This is clearly highlighted by the work of Humar et al.[85] and Pellegrin et al.[86] who illustrate how using tests with increasing cytomegalovirus DNA test sensitivity actually decreases the diagnostic specificity, especially when the agent causes chronic infection.

These examples illustrate how difficult it is to use contemporary statistical terminology to effectively outline the analytical and clinical performance measures of tests. It is absolutely crucial for a laboratorian to both understand and be able to effectively convey to a clinician the complex interrelationship of a test result and its clinical interpretation. The ability to convey this type of information is necessary to bridge a very important gap in the knowledge of clinicians who are faced with an ever-increasing array of technically complex tests. The complex interrelationship between tests and clinical interpretations highlights the need for better total quality management of laboratory and clinical test validation.

CONTINUOUS QUALITY IMPROVEMENT OR TOTAL QUALITY MANAGEMENT

A comprehensive total quality management program integrates quality control, quality assurance, and proficiency testing. Laboratories typically participate in regional, national, and/or international proficiency programs to evaluate test performance and interpretive accuracy.[7,87] Such programs are critical to assist in evaluating inter-laboratory reliability and reproducibility. Pre-analytical quality control and assurance includes the ability to document appropriate collection and handling of specimens and the use of methods for detecting and acting to correct process and procedural errors.

Analytical quality control and assurance involves confirming that test reagents function according to manufacturers' specifications, abnormal results are recognized, and control measures are implemented to ensure that the analytical testing process is both accurate and reproducible. This chapter has highlighted the complexity of analytical procedures and the challenges in ensuring run-to-run and lot-to-lot test kit reproducibility.

The extent of quality control and assurance activities required typically depends on whether a test has achieved regulatory approval, has been cleared for research use only or investigational use, or involves a manufacturer's test that has been adapted or modified for off-labeled use. When analyte-specific reagents (ASRs) are used, the manufacturer typically optimizes the reagents to ensure that under optimal conditions accuracy and reproducibility are achieved. This is not the case when a laboratory uses home-brewed or in-house assays. Therefore, the nature of the test as well as the volume and frequency of testing will impact on the amount of quality control and assurance necessary to ensure valid test results.

The role of the laboratorian in overseeing the process control elements of analytical and clinical validation involves the ability to identify the factors that contribute to test accuracy, reproducibility, and interpretation. The amount of process control required depends on the complexity of the technology used and how rapidly new knowledge alters the utility of detecting a specific analyte or target of interest.

While the generally required validation elements are articulated by various regulatory bodies,[5–7,71] these elements must be adapted to the nature of the disease tested, the degree to which test reagents have been validated within the laboratory or by the manufacturer, and the technological platform used for testing. They must ensure that intellectual property restrictions or specific regulatory approvals have been addressed. If the currently accepted standards of laboratory validation are not met or are insufficiently documented by the facility or the manufacturer, then a laboratory's leaders face the risk of litigation. As this chapter has outlined, in order for analytical or clinical conclusions to be considered robust, results must be statistically verifiable and reproducible and be constantly re-evaluated through a quality assurance process that incorporates emerging knowledge. If emerging knowledge identifies improved test utility or test deficiencies, a laboratory has an implied duty to inform.

Quality assurance of post-analytical components can be particularly challenging for certain illnesses. For some infections, multiple analytical tests are used to assess the clinical status of the infection. Clinicians may not be aware of the appropriate tests to request or be knowledgeable about the precise interpretation of the results. Clinical information is often not conveyed to the testing laboratory, making it difficult to select the appropriate tests.

As illustrated in the theoretical assay concepts section, qualitative hepatitis RNA testing is presently the appropriate test to diagnose active hepatitis C virus infection. However, monitoring the response to treatment requires baseline and follow-up viral load testing. The management of hepatitis C virus infections is illustrative of the kinds of pre-analytical, analytical, and post-analytical components of laboratory testing that must be effectively linked to produce clinically valid test results. Although published guidelines can help increase the appropriateness of test orders, ensuring that diagnostic guidelines address the broad clinical presentations of different pathogens and that the information is current remains an important challenge. This may require consensus processes involving professional or community recommendations.

Another challenge involves transitioning new tests from research into clinical service. This may involve specialized test technology that may be out of the reach of smaller volume laboratories and a process for proper utilization management of a test after it has been transitioned. In some cases, small volume or orphaned tests may be limited to research purposes until they attain sufficient commercial support for broader application.

Another reality is that clinicians often deal with multiple laboratories, each of which may be involved in testing different markers. As a result, no comprehensive summary of test results may be available. Individuals may also be seen by multiple clinicians who may not have the results of previous tests. This can lead to needless and costly repeat testing on the part of a clinician. Similarly, a laboratory may not have access to a previous result that might help it ensure that the appropriate testing algorithm is performed. In the long term, and especially when test complexity for monitoring and diagnosis is high, it is likely that an electronic longitudinal patient record of sequential test results would be an effective way of ensuring that test results are available to the clinician to support best clinical practices.

Because of the rapid evolution of both diagnostic techniques and therapeutic interventions, a need clearly exists for greater cooperation between clinicians, laboratories, researchers, and regulatory authorities to better define the analytical and clinical performance characteristics of tests. Ideally, this should involve a total quality management framework that supports a comprehensive approach across pre-analytic, analytic, and post-analytic laboratory components. This quality management framework should focus on providing accurate diagnostic testing and cost-effective therapeutic monitoring; should enhance our ability to collect and analyze aggregate test results for policy planning; and make more effective use of modern molecular tools to help identify and fingerprint microorganisms when appropriate.

SUMMARY

This chapter has attempted to bridge the gap between laboratory validation that fundamentally outlines the processes to consistently generate accurate and reproducible test results and the recognition that the biological behavior of an agent in a host and test results will impact on care management or policy decisions. While the chapter has not gone into exhaustive detail about a variety of different pathogenic agents and test methodologies, it has illustrated key areas that are often omitted in publications. In particular, while pre-analytic, analytic, and post-analytic components are often conceptualized as separate entities, some of which are under the direct control of a laboratory, it is clear that laboratory and clinical validation cannot be separated.

Highlighted in the chapter was the critical need for optimizing analytical accuracy and reproducibility as a key element in properly assessing analytical and diagnostic sensitivity and specificity. An improved understanding of the relationship between analytical sensitivity and specificity and clinical utility is an absolute requirement to help laboratories ensure that their tests are accurate, valid, and clinically useful and therefore worth performing. Demonstrating test utility will be an increasing challenge for laboratories as test complexity increases and health care cost pressures force greater scrutiny of laboratory reimbursement.

ACKNOWLEDGMENT

I thank Maureen Rowlands for her superb editorial assistance.

REFERENCES

1. Ou CY, Ciesielski CA, Myers G, Bandea CI, Luo CC, Korber BT, et al. Molecular epidemiology of HIV transmission in a dental practice. *Science* 256 (5060), 1165–1171, 1992.
2. Esteban JI, Gómez J, Martell M, Cabot B, Quer J, Camps J, González A, Otero T, Moya A, Esteban R, Guardia J. Transmission of hepatitis C virus by a cardiac surgeon. *New Engl J Med* 334 (9), 555–560, 1996.
3. Harpaz R, VonSeidlein L, Averhoff FM, Tormey MP, Sinha SD, Kotsopoulou K, Lambert SB, Robertson BH, Cherry JD, Shapiro CN. Transmission of hepatitis B virus to multiple patients from a surgeon without evidence of inadequate infection control. *New Engl J Med* 334 (9), 549–554, 1996.
4. Harrington SM, Bishai WR. Molecular epidemiology and infectious diseases, in *Infectious Disease Epidemiology: Theory and Practice*, Nelson KE, Master-Williams C, Graham NMH, Eds. Jones & Bartlett, Mississauga, Ontario, 2004.
5. International Organization for Standardization. *Quality Management Systems: Fundamentals and Vocabulary*. ISO 9000, Geneva, 2000.

6. World Health Organization. *Glossary of Terms for Biological Substances Used for Texts of the Requirements.* Expert Committee on Biological Standardization, Geneva, 1995.
7. NCCLS [now Clinical and Laboratory Standards Institute]. Qualitative molecular methods for infectious diseases: approved guideline, MM6-A. Wayne, Pennsylvania, 2003.
8. Farci P, Alter HJ, Wong D, Miller RH, Shih JW, Jett B, Purcell RH. A long-term study of hepatitis C virus replication in non-A, non-B hepatitis. *New Engl J Med* 325, 98-104, 1991.
9. Krajden M. Hepatitis C virus diagnosis and testing. *Can J Publ Health* 91, Suppl 1, S34–S39, 2000.
10. Jackson BR, Busch MP, Stramer SL, Aubuchon JP. The cost-effectiveness of NAT for HIV, HCV, and HBV in whole-blood donations. *Transfusion* 43 (6), 721–729, 2003.
11. Pawlotsky JM. Use and interpretation of virological tests for hepatitis C. *Hepatology* 36 (5, Suppl 1), S65–S73, 2002.
12. Pessoa MG, Terrault NA, Detmer J, Kolberg J, Collins M, Hassoba HM, Wright, TL. Quantitation of hepatitis G and C viruses in the liver: evidence that hepatitis G virus is not hepatotropic. *Hepatology* 27 (3), 877–880, 1998.
13. Wong DK, Yuen MF, Tse E, Yuan H, Sum SS, Hui CK, Lai CL. Detection of intrahepatic hepatitis B virus DNA and correlation with hepatic necroinflammation and fibrosis. *J Clin Microbiol* 42 (9), 3920–3924, 2004.
14. McHutchison JG, Poynard T, Esteban-Mur R, Davis GL, Goodman ZD, Harvey J, Ling MH, Garaud JJ, Albrecht JK, Patel K, Dienstag JL. Hepatic HCV RNA before and after treatment with interferon alone or combined with ribavirin. *Hepatology* 35 (3), 688–693, 2002.
15. Neumann AU, Lam NP, Dahari H, Gretch DR, Wiley TE, Layden TJ, Perelson AS. Hepatitis C viral dynamics *in vivo* and the antiviral efficacy of interferon-alpha therapy. *Science* 282 (5386), 103–107, 1998.
16. Tsuda N, Yuki N, Mochizuki K, Nagaoka T, Yamashiro M, Omura M, Hikiji K, Kato M. Long-term clinical and virological outcomes of chronic hepatitis C after successful interferon therapy. *J Med Virol* 74 (3), 406–413, 2004.
17. Boivin G, Cote S, Dery P, De Serres G, Bergeron MG. Multiplex real-time PCR assay for detection of influenza and human respiratory syncytial viruses. *J Clin Microbiol* 42 (1), 45–51, 2004.
18. Erdman DD, Weinberg GA, Edwards KM, Walker FJ, Anderson BC, Winter J, Gonzalez M, Anderson LJ. GeneScan reverse transcription-PCR assay for detection of six common respiratory viruses in young children hospitalized with acute respiratory illness. *J Clin Microbiol* 41 (9), 4298–4303, 2003.
19. Hayden FG, Atmar RL, Schilling M, Johnson C, Poretz D, Paar D, Huson L, Ward P, Mills RG. Use of the selective oral neuraminidase inhibitor oseltamivir to prevent influenza. *New Engl J Med* 341 (18), 1336–1343, 1999.
20. Templeton KE, Scheltinga SA, Beersma MF, Kroes AC, Claas EC. Rapid and sensitive method using multiplex real-time PCR for diagnosis of infections by influenza a and influenza B viruses, respiratory syncytial virus, and parainfluenza viruses 1, 2, 3, and 4. *J Clin Microbiol* 42 (4), 1564–1569, 2004.
21. Grant PR, Kitchen A, Barbara JA, Hewitt P, Sims CM, Garson JA, Tedder RS. Effects of handling and storage of blood on the stability of hepatitis C virus RNA: implications for NAT testing in transfusion practice. *Vox Sang* 78 (3), 137–142, 2000.
22. Ginocchio CC, Wang XP, Kaplan MH, Mulligan G, Witt D, Romano JW, Cronin M, Carroll R. Effects of specimen collection, processing, and storage conditions on stability of human immunodeficiency virus type 1 RNA levels in plasma. *J Clin Microbiol* 35 (11), 2886–2893, 1997.
23. Krajden M, Comanor L, Rifkin O, Grigoriew A, Minor JM, Kapke GF. Assessment of hepatitis B virus DNA stability in serum by the Chiron quantiplex branched-DNA assay. *J Clin Microbiol* 36, 382–386, 1998.
24 Krajden M, Minor JM, Rifkin O, Comanor L. Effect of multiple freeze-thaw cycles on hepatitis B virus DNA and hepatitis C virus RNA quantification as measured with branched-DNA technology. *J Clin Microbiol* 37 (6), 1683–1686, 1999.
25. Holodniy M, Kim S, Katzenstein D, Konrad M, Groves E, Merigan TC. Inhibition of human immunodeficiency virus gene amplification by heparin. *J Clin Microbiol* 29, 676–679, 1991.
26. Holodniy M, Mole L, Yen-Lieberman B, Margolis D, Starkey C, Carroll R, Spahlinger T, Todd J, Jackson JB. Comparative stabilities of quantitative human immunodeficiency virus RNA in plasma from samples collected in VACUTAINER CPT, VACUTAINER PPT, and standard VACUTAINER tubes. *J Clin Microbiol* 33, 1562–1566, 1995.

27. Lee DH, Li L, Andrus L, Prince AM. Stabilized viral nucleic acids in plasma as an alternative shipping method for NAT. *Transfusion* 42 (4), 409–413, 2002.

28. Boeckh M, Huang M, Ferrenberg J, Stevens-Ayers T, Stensland L, Nichols WG, Corey L. Optimization of quantitative detection of cytomegalovirus DNA in plasma by real-time PCR. *J Clin Microbiol* 42 (3), 1142–1148, 2004.

29. Boeckh M, Woogerd PM, Stevens-Ayers T, Ray CG, Bowden RA. Factors influencing detection of quantitative cytomegalovirus antigenemia. *J Clin Microbiol* 32 (3), 832–834, 1994.

30. Nesbitt SE, Cook L, Jerome KR. Cytomegalovirus quantitation by real-time PCR is unaffected by delayed separation of plasma from whole blood. *J Clin Microbiol* 42 (3), 1296–1297, 2004.

31. Schafer P, Tenschert W, Schroter M, Gutensohn K, Laufs R. False-positive results of plasma PCR for cytomegalovirus DNA due to delayed sample preparation. *J Clin Microbiol* 38 (9), 3249–3253, 2000.

32. Saldanha J, Lelie N, Heath A. Establishment of the first international standard for nucleic acid amplification technology (NAT) assays for HCV RNA, *Vox Sang* 76 (3), 149–158, 1999.

33. Valentine-Thon E. Quality control in nucleic acid testing: where do we stand? *J Clin Virol* 25, Suppl 3, S13–S21, 2002.

34. Heermann KH, Gerlich WH, Chudy M, Schaefer S, Thomssen R. Quantitative detection of hepatitis B virus DNA in two international reference plasma preparations. *J Clin Microbiol* 37 (1), 68–73, 1999.

35. Zaaijer HL, ter Borg F, Cuypers HTM, Hermus MC, Lelie PN. Comparison of methods for detection of hepatitis B virus DNA. *J Clin Microbiol* 32 (9), 2088–2091, 1994.

36. Saldanha J. Validation and standardisation of nucleic acid amplification technology (NAT) assays for the detection of viral contamination of blood and blood products. *J Clin Virol* 20 (1-2), 7–13, 2001.

37. Saldanha J, Gerlich W, Lelie N, Dawson P, Heermann K, Heath A. An international collaborative study to establish a World Health Organization international standard for hepatitis B virus DNA nucleic acid amplification techniques, *Vox Sang* 80 (1), 63–71, 2001.

38. Collins ML, Zayati C, Detmer JJ, Daly B, Kolberg JA, Cha TA, Irvine BD, Tucker J, Urdea MS. Preparation and characterization of RNA standards for use in quantitative branched DNA hybridization assays. *Anal Biochem* 226, 120–129, 1995.

39. Cook L, Ng KW, Bagabag A, Corey L, Jerome KR. Use of MagNA pure LC automated nucleic acid extraction system followed by real-time reverse transcription PCR for ultrasensitive quantitation of hepatitis C virus RNA. *J Clin Microbiol* 42 (9), 4130–4136, 2004.

40. Niesters HG, Krajden M, Cork L, de Medina M, Hill M, Fries E, Osterhaus AD. A multicenter study of the Digene Hybrid Capture II signal amplification technique for detection of hepatitis B virus DNA in serum samples and testing of EUROHEP standards. *J Clin Microbiol* 38 (6), 2150–2155, 2000.

41. Elbeik T, Surtihadi J, Destree M, Gorlin J, Holodniy M, Jortani SA, Kuramoto K, Ng V, Valdes R Jr, Valsamakis A, Terrault NA. Multicenter evaluation of the performance characteristics of the Bayer Versant HCV RNA 3.0 assay (bDNA). *J Clin Microbiol* 42 (2), 563–569, 2004.

42. Mellor J, Hawkins A, Simmonds P. Genotype dependence of hepatitis C virus load measurement in commercially available quantitative assays. *J Clin Microbiol* 37 (8), 2525–2532, 1999.

43. Triques K, Coste J, Perret JL, Segarra C, Mpoudi E, Reynes J, Delaporte E, Butcher A, Dreyer K, Herman S, Spadoro J, Peeters M. Efficiencies of four versions of the AMPLICOR HIV-1 MONITOR test for quantification of different subtypes of human immunodeficiency virus type 1. *J Clin Microbiol* 37 (1), 110–116, 1999.

44. Neumaier M, Braun A, Wagener C. Fundamentals of quality assessment of molecular amplification methods in clinical diagnostics. *Clin Chem* 44 (1), 12–26, 1998.

45. Hendricks DA, Stowe BJ, Hoo BS, Kolberg J, Irvine BD, Neuwald PD, Urdea MS, Perrillo RP. Quantitation of HBV DNA in human serum using a branched DNA (bDNA) signal amplification assay. *Am J Clin Pathol* 104, 537–546, 1995.

46. Mayerat C, Burgisser P, Lavanchy D, Mantegani A, Frei PC. Comparison of a competitive combined reverse transcription-PCR assay with a branched-DNA assay for hepatitis C virus RNA quantitation. *J Clin Microbiol* 34 (11), 2702–2706, 1996.

47. Nolte FS, Fried MW, Shiffman ML, Ferreira-Gonzalez A, Garrett CT, Schiff ER, Polyak SJ, Gretch DR. Prospective multicenter clinical evaluation of AMPLICOR and COBAS AMPLICOR hepatitis C virus tests, *J Clin Microbiol* 39 (11), 4005–4012, 2001.

48. Rodriguez-Lazaro D, D'Agostino M, Pla M, Cook N. Construction strategy for an internal amplification control for real-time diagnostic assays using nucleic Acid sequence-based amplification: development and clinical application. *J Clin Microbiol* 42 (12), 5832–5836, 2004.

49. Paik S, Shak S, Tang G, Kim C, Baker J, Cronin M, Baehner FL, Walker MG, Watson D, Park T, Hiller W, Fisher ER, Wickerham DL, Bryant J, Wolmark N. A multigene assay to predict recurrence of tamoxifen-treated, node-negative breast cancer. *New Engl J Med* 351 (27), 2817–2826, 2004.

50. Pawlotsky JM. Diagnostic testing in hepatitis C virus infection: viral kinetics and genomics. *Semin Liver Dis* 23, Suppl 1, 3–11, 2003.

51. Watzinger F, Suda M, Preuner S, Baumgartinger R, Ebner K, Baskova L, Niesters HG, Lawitschka, A, Lion T. Real-time quantitative PCR assays for detection and monitoring of pathogenic human viruses in immunosuppressed pediatric patients. *J Clin Microbiol* 42 (11), 5189–5198, 2004.

52. Kleiber J, Walter T, Haberhausen G, Tsang S, Babiel R, Rosenstraus M. Performance characteristics of a quantitative, homogeneous TaqMan RT-PCR test for HCV RNA. *J Mol Diagn* 2 (3), 158–166, 2000.

53. Germer JJ, Harmsen WS, Mandrekar JN, Mitchell PS, Yao JD. Evaluation of the Cobas TaqMan HCV test with automated sample processing using the MagNA pure LC instrument. *J Clin Microbiol* 43 (1), 293–298, 2005.

54. Krajden M, Minor J, Cork L, Comanor L. Multi-measurement method comparison of three commercial hepatitis B virus DNA quantification assays. *J Viral Hepat* 5 (6), 415–422, 1998.

55. Seabrook JM, Hubbard RA. Achieving quality reproducible results and maintaining compliance in molecular diagnostic testing of human papillomavirus. *Arch Pathol Lab Med* 127 (8), 978–983, 2003.

56. Krajden M, Ziermann R, Khan A, Mak A, Leung K, Hendricks D, Comanor L. Qualitative detection of hepatitis C virus RNA: comparison of analytical sensitivity, clinical performance, and workflow of the Cobas Amplicor HCV test version 2.0 and the HCV RNA transcription-mediated amplification qualitative assay. *J Clin Microbiol* 40 (8), 2903–2907, 2002.

57. Lee SC, Antony A, Lee N, Leibow J, Yang JQ, Soviero S, Gutekunst K, Rosenstraus M. Improved version 2.0 qualitative and quantitative AMPLICOR reverse transcription-PCR tests for hepatitis C virus RNA: calibration to international units, enhanced genotype reactivity, and performance characteristics. *J Clin Microbiol* 38 (11), 4171–4179, 2000.

58. Hendricks DA, Friesenhahn M, Tanimoto L, Goergen B, Dodge D, Comanor L. Multicenter evaluation of the Versant HCV RNA qualitative assay for detection of hepatitis C virus RNA. *J Clin Microbiol* 41 (2), 651–656, 2003.

59. Krajden M, Shivji , R, Gunadasa , K, Mak , A, McNabb , G, Friesenhahn , M, Hendricks , D, Comanor L. Evaluation of the core antigen assay as a second-line supplemental test for diagnosis of active hepatitis C virus infection. *J Clin Microbiol* 42 (9), 4054–4059, 2004.

60. Barbeau JM, Goforth J, Caliendo AM, Nolte FS. Performance characteristics of a quantitative TaqMan hepatitis C virus RNA analyte-specific reagent. *J Clin Microbiol* 42 (8), 3739–3746, 2004.

61. Fried MW, Shiffman ML, Reddy KR, Smith C, Marinos G, Goncales FL Jr, Haussinger D, Diago M, Carosi G, Dhumeaux D, Craxi A, Lin A, Hoffman J, Yu J. Peginterferon alfa-2a plus ribavirin for chronic hepatitis C virus infection, *New Engl J Med* 347 (13), 975–982, 2002.

62. Hadziyannis SJ, Sette H Jr, Morgan TR, Balan V, Diago M, Marcellin P, Ramadori G, Bodenheimer H Jr, Bernstein D, Rizzetto M, Zeuzem S, Pockros PJ, Lin A, Ackrill AM. Peginterferon-alpha-2a and ribavirin combination therapy in chronic hepatitis C: a randomized study of treatment duration and ribavirin dose. *Ann Intern Med* 140 (5), 346–355, 2004.

63. Torriani FJ, Rodriguez-Torres M, Rockstroh JK, Lissen E, Gonzalez-Garcia J, Lazzarin A, Carosi G, Sasadeusz J, Katlama C, Montaner J, Sette H Jr, Passe S, De Pamphilis J, Duff F, Schrenk UM, Dieterich DT. Peginterferon Alfa-2a plus ribavirin for chronic hepatitis C virus infection in HIV-infected patients. *New Engl J Med* 351 (5), 438–450, 2004.

64. Davis GL. Monitoring of viral levels during therapy of hepatitis C. *Hepatology* 36 (5, Suppl 1), S145–S151, 2002.

65. Sato Y, Tokuue H, Kawamura N, Nezu-Yajima S, Nakajima H, Ishida H, Takahashi S. Short-term interferon therapy for chronic hepatitis C patients with low viral load. *Hepatogastroenterology* 51 (58), 968–972, 2004.

66. Fujiyama S, Chikazawa H, Honda Y, Tomita K. Effective interferon therapy for chronic hepatitis C patients with low viral loads. *Hepatogastroenterology* 50 (51), 817–820, 2003.

67. Tabaru A, Narita R, Hiura M, Abe S, Otsuki M. Efficacy of short-term interferon therapy for patients infected with hepatitis C virus genotype 2a. *Am J Gastroenterol* 100 (4), 862–867, 2005.

68. Mangia A, Santoro R, Minerva N, Ricci G. L, Carretta V, Persico M, Vinelli F, Scotto G, Bacca D, Annese M, Romano M, Zechini F, Sogari F, Spirito F, Andriulli A. Peginterferon alfa-2b and ribavirin for 12 vs. 24 weeks in HCV genotype 2 or 3. *New Engl J Med* 352 (25), 2609–2617, 2005.

69. Troyer DA, Mubiru J, Leach RJ, Naylor SL. Promise and challenge: markers of prostate cancer detection, diagnosis and prognosis. *Dis Markers* 20 (2), 117–128, 2004.

70. Kattan MW. Evaluating a new marker's predictive contribution. *Clin Cancer Res* 10 (3), 822–824, 2004.

71. Amos J, Grody W. Development and integration of molecular genetic tests into clinical practice: the U.S. experience. *Expert Rev Mol Diagn* 4 (4), 465–477, 2004.

72. Zanghellini F, Boppana SB, Emery VC, Griffiths PD, Pass RF. Asymptomatic primary cytomegalovirus infection: virologic and immunologic features. *J Infect Dis* 180 (3), 702–707, 1999.

73. Nowak MA, Bonhoeffer S, Hill AM, Boehme R, Thomas HC, McDade H. Viral dynamics in hepatitis B virus infection. *Proc Natl Acad Sci USA* 93, 4398–4402, 1996.

74. Ruddy M, McHugh TD, Dale JW, Banerjee D, Maguire H, Wilson P, Drobniewski F, Butcher P, Gillespie SH. Estimation of the rate of unrecognized cross-contamination with mycobacterium tuberculosis in London microbiology laboratories. *J Clin Microbiol* 40 (11), 4100–4104, 2002.

75. Richardson M, Carroll NM, Engelke E, Van Der Spuy GD, Salker F, Munch Z, Gie RP, Warren RM, Beyers N, Van Helden PD. Multiple *Mycobacterium tuberculosis* strains in early cultures from patients in a high-incidence community setting. *J Clin Microbiol* 40 (8), 2750–2754, 2002.

76. Fitzpatrick L, Braden C, Cronin W, English J, Campbell E, Valway S, Onorato I. Investigation of laboratory cross-contamination of *Mycobacterium tuberculosis* cultures. *J Clin Infect Dis* 38 (6), e52–e54, 2004.

77. Nguyen LN, Gilbert GL, Marks GB. Molecular epidemiology of tuberculosis and recent developments in understanding the epidemiology of tuberculosis. *Respirology* 9 (3), 313–319, 2004.

78. Larke B, Hu YW, Krajden M, Scalia V, Byrne SK, Boychuk LR, Klein J. Acute nosocomial HCV infection detected by NAT of a regular blood donor. *Transfusion* 42 (6), 759-65, 2002.

79. Einstein AJ, Bodian CA, Gil J. Relationships among performance measures in the selection of diagnostic tests. *Arch Pathol Lab Med* 121 (2), 110–117, 1997.

80. Henderson AR. Assessing test accuracy and its clinical consequences: a primer for receiver operating characteristic curve analysis. *Ann Clin Biochem* 30 (Pt 6), 521–539, 1993.

81. Saah AJ, Hoover DR. "Sensitivity" and "specificity" reconsidered: the meaning of these terms in analytical and diagnostic settings. *Ann Int Med* 126, 91–94, 1997.

82. Gandhi MK, Khanna R. Human cytomegalovirus: clinical aspects, immune regulation, and emerging treatments, *Lancet Infect Dis* 4 (12), 725–738, 2004.

83. Thorley-Lawson DA, Gross A. Persistence of the Epstein–Barr virus and the origins of associated lymphomas. *New Engl J Med* 350 (13), 1328–1337, 2004.

84. Emery VC, Cope AV, Bowen EF, Gor D, Griffiths PD. The dynamics of human cytomegalovirus replication *in vivo*. *J Exp Med* 190 (2), 177–182, 1999.

85. Humar A, Gregson D, Caliendo AM, McGeer A, Malkan G, Krajden M, Corey P, Greig P, Walmsley S, Levy G, Mazzulli T. Clinical utility of quantitative cytomegalovirus viral load determination for predicting cytomegalovirus disease in liver transplant recipients. *Transplantation* 68 (9), 1305–1311, 1999.

86. Pellegrin I, Garrigue I, Binquet C, Chene G, Neau D, Bonot P, Bonnet F, Fleury H, Pellegrin JL. Evaluation of new quantitative assays for diagnosis and monitoring of cytomegalovirus disease in human immunodeficiency virus-positive patients. *J Clin Microbiol* 37 (10), 3124–3132, 1999.

87. Lefrere JJ, Roudot-Thoraval F, Lunel F, Alain S, Chaix ML, et al. Expertise of French laboratories in detection, genotyping, and quantification of hepatitis C virus RNA in serum. *J Clin Microbiol* 42 (5), 2027–2030, 2004.

Index

A

CPT. *See* Current procedure terminology
CRC. *See* Colorectal cancer
Creatine phosphokinase, 383
cRNA. *See* Complementary RNA
Cryptosporidium, 337
cSBH. *See* Combinatorial sequencing by hybridization
CSF. *See* Cerebrospinal fluid
Current procedure terminology, 375
%CV. *See* Percent coefficient of variation
CVS. *See* Chorionic villus sampling
Cyclic amplification methods, 4–6
    amplification of RNA, 6
    ligase chain reaction, 5–6
    polymerase chain reaction, 4–5
Cycling Probe (ID Biomedical), 21–24, 35, 37–38,
        313–314, 321–322
CYP450 test, 80–81
Cystic fibrosis, 21, 374, 385, 389–394, 405–406, 415, 419,
        421, 426
    assay, 392–393
    case presentation, 393–394
    CFTR testing, 390–391
    clinical testing, 391–392
    indications for testing, 391
    interpretation, 393
    technology considerations, 394
Cytochrome P450 pharmacogenetic test, 80–81
Cytogenetics, 127–128
Cytomegalovirus, 21, 25, 30, 33–34, 37–38, 40–42, 52,
        124, 126, 197, 201, 261, 276, 288–291,
        447
    immunocompromised patients, disease in,
        289–290
    resistance to antiviral drugs, 290–291
    treatment timing, 288–289
Cytopathic effect, 231

D

DA. *See* Discrepant analysis
DAB. *See* Diaminobenzidine
DABCYL. *See* 4-dimethylaminophenylazobenzoic acid
DASH. *See* Dynamic allele-specific hybridization
Data from different sources, combining, in RNA expression
        profiling, 183
ddNTP. *See* Dideoxynucleoside triphosphate
Denaturation, 117–118
Denaturing gradient gel electrophoresis, 160, 384, 400,
        402
Denaturing high performance liquid chromatography, 160
Dengue fever virus, 337
2'-deoxyguanosine, 22, 123
Deoxyribonucleic acid
Deoxyuridine 5-triphosphate, 15
Detection of amplification products, 9–11
    endpoint detection, 9–11
        direct sequence analysis, 11
        forward hybridization, 9
        restriction-fragment length polymorphism analysis,
            11

reverse hybridization, 9–11
    Southern blotting, 9
Detection probes, nucleic acid testing, 417–418
DFA. *See* Direct fluorescent antibody
DGGE. *See* Denaturing gradient gel electrophoresis
dHPLC. *See* Denaturing high performance liquid
        chromatography
DIA. *See* Approved for diagnostic use
Diaminobenzidine, 28, 119–120
Dideoxynucleoside triphosphate, 155
Differentially expressed genes, RNA expression profiling,
        identifying, 179
DIG. *See* Digoxigenin
Digoxigenin, 115–117, 119–120, 123, 127, 225
4-dimethylaminophenylazobenzoic acid, 12, 419
Direct fluorescent antibody, 212–214, 216, 221, 230,
        255
Direct sequence analysis, 11
Discrepant analysis, 213–215
Disease applications, 209–456
    analytical validation, 435–456
    bacterial sexually transmitted diseases, 211–241
    blood-borne viruses, 275–299
    cancer detection, prognosis, 353–371
    clinical validation, 435–456
    fungal infections, 301–330
    future perspectives, 413–433
    herpes simplex, 243–273
    human papillomaviruses, 243–273
    inherited genetic disorders, 373–411
    molecular diagnostic approaches, 331–352
DLI. *See* Donor lymphocyte
DMD. *See* Duchenne muscular dystrophy
DNA analysis, 76–78
    bladder cancer, 77
    breast cancer, 76
    cancer, 78, 357–366
    lung cancer, 76
    mouth cancer, 76
DNA methylation detection, cancer detection, prognosis,
        362–366
    bisulfite sequencing, 363–364
    bisulfite-based techniques, 365–366
    combined bisulfite restriction analysis, 365
    fluorescence-based real-time quantitative polymerase
        chain reaction analysis, 364–365
    global CpG island hypermethylation analysis, 366
    methylation-specific polymerase chain reaction, 364
DNA sequence analysis, 80–81, 201–202
DOE. *See* U.S. Department of Energy
Donor lymphocyte, 284, 287
Double chromogenic *in situ* hybridization, 119–120
Double-stranded DNA probes, 115
Duchenne muscular dystrophy, 383–389
    case presentation, 389
    clinical testing, 386–389
    indications for testing, 386
    new technology and recent advances, 389
dUTP. *See* Deoxyuridine 5-triphosphate
Dynamic allele-specific hybridization, 153–154
Dynamic range, 32–34

N

Printed and bound by CPI Group (UK) Ltd, Croydon, CR0 4YY

23/10/2024

01778251-0008